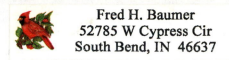

PHYSICS
A World View

third edition

About the Authors

Larry D. Kirkpatrick

Author

Larry D. Kirkpatrick *is professor of physics at Montana State University and adjunct curator of astronomy at the Museum of the Rockies. He served eight years as academic director and coach of the U.S. Physics Team that competes in the International Physics Olympiad each summer. Kirkpatrick is currently serving as vice president of the American Association of Physics Teachers and will serve as president-elect, president, and past president in 1998, 1999, and 2000, respectively. Kirkpatrick received a B.S. in physics from Washington State University in 1963 and a Ph.D. in experimental high-energy physics from MIT in 1968. He is an avid sports fan and loves country western dancing.*

Gerald F. Wheeler

Author

Gerald F. Wheeler *is executive director of the National Science Teachers Association. Prior to that he was professor of physics at Montana State University, director of MSU's Science/Math Resource Center, program director (Public Understanding of Science and Technology) at the American Association for the Advancement of Science, and professor of physics at Temple University. Wheeler received his B.S. in 1963 from Boston University with a major in science education and his Ph.D. at SUNY–Stony Brook in 1972 in experimental nuclear physics. Between undergraduate and graduate school, he taught high school physics. He enjoys cooking, photography, and old cars.*

PHYSICS
A World View

third edition

Larry D. Kirkpatrick

Montana State University

Gerald F. Wheeler

National Science Teachers Association

Saunders Golden Sunburst Series

Saunders College Publishing

Harcourt Brace College Publishers

Fort Worth   Philadelphia   San Diego   New York   Orlando   Austin
San Antonio   Toronto   Montreal   London   Sydney   Tokyo

Publisher: John Vondeling
Associate Editor: Marc Sherman
Project Editor: Anne Gibby
Production Manager: Charlene Catlett Squibb
Art Director: Lisa Caro
Text Designer: Chazz Bjanes
Cover Designer: Kathleen Flanagan
Product Managers: Nick Agnew, Angus McDonald

Cover credit: The physicist's view of cycling motion complements the artist's view. Photographic Images by Alex Pietersen.

Printed in the United States of America

PHYSICS: A WORLD VIEW, third edition

ISBN 0-03-020052-0

Library of Congress Catalog Card Number: 97–66893

7 8 9 0 1 2 3 4 5 6   048   10 9 8 7 6 5 4 3 2 1

We dedicate this book to our five children—

Jennifer, Monica, Solveiga, Soren, and Peter—
who in their very different ways have taught us how
to view the world again.

Preface

This textbook is intended for a conceptual course in introductory physics for students majoring in fields other than science, mathematics, or engineering.

Writing this book has been an exercise in translation. We have attempted to take the logic, vocabulary, and values of physics and communicate them in an entirely different language. In some areas the physics is so abstract that it took creative bridges to span the gulf between the languages. A good job of translating requires careful attention to both languages, that of the physicist and that of the student. We are indebted to the many students who shared their confusions with us and wrestled with the clarity of our translations. We are equally indebted to the many physicists who shared our search for the proper word or metaphor that comes closest to capturing the abstract, elusive idea.

Mathematics is the structural foundation for all of the physics world view. As stated above, this textbook translates most of the ideas into longer, less tightly structured sentences. Still, the mathematics holds much of the beauty and power of physics, and we want to offer a glimpse of this for students whose mathematical background is adequate. For that reason, we have written a mathematical supplement that delves deeper into the mathematical structure of the physics world view. A full description of the mathematical supplement, *Physics: A Numerical World View,* is given in the ancillary section of this preface.

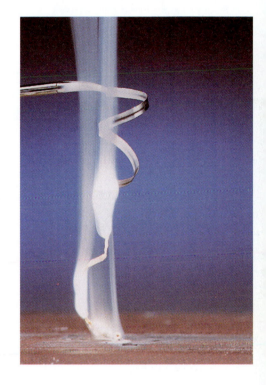

Objectives

The main objective of this physics textbook is to provide non-science-oriented students with a clear and logical presentation of some of the basic concepts and principles of physics in an appropriate language. Our overriding concern has been to choose topics and ideas for students who will be taking only this course in physics. We continually reminded ourselves that this may be our one chance to describe the way physicists look at the world and test their ideas. We chose topics that convey the essence of the physics world view. As an example of this concern, we have placed more modern physics—specifically the theories of relativity—in the first half of the book rather than toward the end, as in most traditional textbooks. We also describe the historical development of quantum physics carefully in order to show why various atomic models—models that make common sense—fail to explain the experimental evidence. At the same time, we have attempted to motivate students through practical examples that demonstrate the role of physics in other disciplines and in their everyday lives.

Coverage

The topics covered in this book are the fundamental topics in classical and modern physics. The book is divided into nine parts. The first Interlude precedes Chapter 5. Each Interlude sets the theme for that section. Part I (Chapters 1–4) deals with the fundamentals of motion, ending with a careful look at gravity, our most familiar force. Part II (Chapters 5–6) re-examines motion through an investigation of two fundamental conservation laws: momentum and energy. In the beginning of Part III (Chapters 7–10), we set the stage for expanding our understanding of energy by investigating the structure of matter, first macroscopically, then microscopically. This part ends with heat and thermodynamics. Part IV (Chapters 11–13) explores the concepts involved in classical, special, and general relativity. Part V (Chapters 14–15) develops the basic properties of wave phenomena and applies them to a study of music. It also gives the reader a background that will be helpful in understanding much of quantum physics. Part VI (Chapters 16–18) covers the study of light and optics, starting with the general question of the basic nature of light, covering interesting applications, and ending with consequences of the wave nature of light. Part VII (Chapters 19–21) covers the basic concepts in electricity and magnetism. In Part VIII (Chapters 22–23) we develop the story of the quantum, starting with the discovery of the electron and ending with quantum physics. The final section of the textbook, Part IX (Chapters 24–26), takes the student deeper into the study of the structure of matter by looking at the nucleus and eventually the fundamental particles.

New to the Third Edition

After soliciting comments from physics teachers and students, we carefully considered each suggestion and used many of them in reworking the entire text. We simplified explanations of some phenomena, updated developing areas, such as elementary particles and lasers, and added new explanatory material on the global positioning system, lasers, and reactions caused by collisions. There is also increased coverage of the physics of biology and medicine.

In order to make the text friendlier to those who prefer to skip over the more mathematical treatments, we've isolated this material into 25 Math Boxes. Care has been taken to allow readers to skip over these boxes without loss of continuity.

We have also reexamined each drawing for clarity and to see if it accomplished its intended functions. Twenty-one figures were modified or entirely redrawn to remove ambiguities or to make the meaning more transparent. Many of the photographs have been changed to improve their functionality. The legends for numerous illustrations and photographs have been rewritten or improved. Legends have been added to all chapter opening photographs.

To emphasize that physics is an evolving science, we have replaced the 16 Physics Updates with 31 new ones. These are released by the American Institute of Physics to science writers nationwide and highlight some of the advances in physics and related sciences that have taken place during the past three years.

The number of conceptual questions has been increased by 62 for a total of 1514. An additional 119 questions were replaced and 57 revised. The 504 exercises include 7 new, 21 replaced, and 245 revised.

Thematic Paths

A big part of the flow of any course rests on decisions made by you. For this reason, the textbook contains more material than can be covered in an introductory course in one term. Even we teach the course differently. One of us likes to stress thermodynamics, while the other spends more time on optics. It is possible to take many routes through the material, depending on your interests and the interests of your students. To illustrate some possibilities, we have compiled seven different paths that can be used for semester-long courses. Each thematic path uses about one-half of the material presented in the text. These seven different emphases (or thematic paths) might be called Physical Science, Electricity and Magnetism, Optics, Energy, Vibrations and Waves, Relativity, and Elementary Particles.

Physical Science emphasizes those topics that are basic to both physics and chemistry. After studying motion and the concepts of momentum and energy, this emphasis delves into the structure and states of matter, heat and thermodynamics, the basic properties of waves, and ends up with atomic physics.

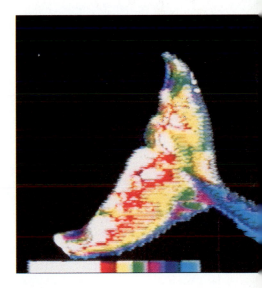

In the *Electricity and Magnetism* course, we begin with motion and the concepts of momentum and energy, skip to the chapter on waves, and then to the three chapters on electricity, magnetism, and electromagnetism. We conclude with the two chapters on atomic physics.

The *Optics* emphasis also begins with motion and waves. It then covers most of the three chapters on light and finishes with atomic physics.

The *Energy* course of study begins with motion, momentum, and mechanical energy. It then covers the two chapters on thermal energy and thermodynamics. After the chapter on waves, the course skips to the three chapters on electricity and magnetism, including electromagnetic waves. The course ends with a study of the nucleus and nuclear energy.

Vibrations and Waves covers motion and energy, and then concentrates on the wave properties of music, light, electromagnetism, and the quantum-mechanical atom.

The *Relativity* option yields a very different course. After a study of the basics of motion, momentum, and energy, the course includes the three chapters on classical, special, and general relativity. There are many ways to complete this course; we favor finishing with some of the properties of light.

For those interested in the search for the ultimate building blocks of the Universe, we suggest the *Elementary Particles* emphasis. The study of motion is followed by the chapter on the structure of matter. The amount of classical and special relativity will depend on the time you have. We then study waves and the wave aspect of light before moving on to selected topics in electricity. The main part of this course is the chapters on atomic and nuclear physics with elementary particles as the capstone.

We have detailed these seven theme-based courses in tabular form in the *Instructor's Resource Manual.*

Features

Most instructors would agree that the textbook selected for a course should be the student's major "guide" for understanding and learning the subject matter. Furthermore, a textbook should be written and presented to make the material accessible and easier to teach, not harder. With these points in mind, we have included many pedagogical features to enhance the usefulness of our text for both you and your students.

Organization: The book contains a story line about the development of the current physics world view. It is divided into nine parts: the fundamentals of motion and gravity; the conservation laws of momentum and energy; the structure of matter, including heat and thermodynamics; the theories of relativity; wave phenomena and sound; light and optics; electricity and magnetism; the story of the quantum; and the nucleus and fundamental particles. Each part includes an overview of the subject matter to be covered in that part and some historical perspectives.

Style: We have written the book in a style that is clear, logical, and succinct, in order to facilitate students' comprehension. The writing style is somewhat informal and relaxed, which we hope students will find appealing and enjoyable to read. New terms are carefully defined, and we have tried to avoid jargon.

Mathematical Level: The mathematical level in the text has been kept to a minimum, with some limited use of algebra and geometry. Equations are presented in words as well as in symbols. For those desiring a higher mathematical presentation, we have written a mathematical supplement, *Physics: A Numerical World View.* The text is available shrink-wrapped with this ancillary.

We have not shied away from using numbers where they assist in developing a more complete understanding of a concept. On the other hand, we have rounded off the values of physical constants to help simplify the discussion. For example, we use 10 (meters per second) per second for the acceleration due to gravity, except when discussing the law of universal gravitation, where the additional accuracy is needed to understand the development.

Math Boxes: More mathematical material has been placed into Math Boxes. These 25 boxed features allow those students following a more mathematical track to find them easily, while allowing others to skip over this mathematical material without loss of continuity. See, for example, "Computing Acceleration" on page 14, or "Computing Conservation of Kinetic Energy," on page 133.

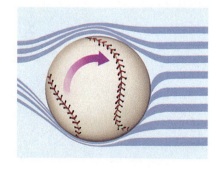

Illustrations: The readability and effectiveness of the text material are enhanced by the large number of figures, diagrams, photographs, and tables. Full color is used to add clarity to the artwork and to make it realistic. For example, vectors are color-coded for each physical quantity. Three-dimensional effects are produced with the use of color and airbrushed areas, where appropriate. Many of the illustrations show the development of a phenomenon over time as a series of "snapshots." To illustrate the flow of time, we have added a *clock icon* in these drawings. The color photographs have been carefully selected, and their accompanying captions serve as an added instructional tool. A complete description of the pedagogical use of color appears at the front of the book.

Chapter Opening Questions and Answers: Each chapter begins with an inquiry that is answered at the end of the chapter. These focus the student's attention on an important aspect of each chapter.

In-text Questions: Questions to stimulate thinking are given at key spots throughout each chapter and are set off by a colored screen. These 187 questions allow students to immediately test their comprehension of the concepts discussed. The answers to these questions are in footnotes on the same page. Most questions could also serve as a basis for initiating classroom discussions.

Conceptual Questions and Exercises: An extensive set of questions and exercises is included at the end of each chapter, with a total of 1514 questions and 504 exercises. The questions and exercises are presented in pairs, meaning that each even-numbered question or exercise has a similar odd-numbered one immediately preceding it. This arrangement allows the student to have one question or exercise with an answer in the back of the book and a very similar one without an answer. The pairing also allows you to discuss one exercise in class and assign its "partner" for homework. We have also included a small number of challenging questions and exercises, which are indicated by an asterisk printed next to the number. Answers to the odd-numbered questions and exercises are given at the end of the book. Answers to all questions and solutions to all exercises are included in the *Instructor's Resource Manual.*

Physics on Your Own: Each chapter contains several projects and simple experiments for students to do on their own. The 91 projects have been designed to require a minimum of apparatus and to illustrate the concepts presented in the text. These illustrate the experimental aspects of physics and the application of physics to our everyday lives.

Special Topics: All chapters include optional special topic boxes to expose students to various practical and interesting applications of physical principles. Some of the special topics include mirages, liquid crystal displays, fluorescent colors, holograms, gravity waves, radon, and superconductivity.

Biographical Sketches: In addition to the historical perspectives provided in the eight Interludes between the major parts of the text, we have added short biographies of important scientists throughout the text to give a greater historical emphasis without interrupting the development of the physics concepts.

Physics Updates: We have included 31 news releases known as *Physics Updates* to emphasize that physics is an evolving science. *Physics Updates* are released by the American Institute of Physics to science writers nationwide and highlight some of the advances in physics and related sciences that have taken place during the past three years. These news releases can serve as the starting points for independent studies by students.

Important Concepts: Important statements and equations are highlighted in several ways for easy reference and review.

- Important principles and equations are boxed for easy reference.
- Marginal notes are used to highlight important statements, equations, and concepts in the text.
- Each chapter ends with a summary reviewing the important concepts of the chapter and the key terms. If terms introduced in previous chapters are reused, they are relisted for convenient reference.

Units: The international system of units (SI) is used throughout the text. The British engineering system of units (conventional system) is used to a limited extent in the chapters on mechanics, heat, and thermodynamics to help the student develop a better feeling for the sizes of the SI units.

Appendices: Three appendices are provided at the end of the text, one on the metric system, one on powers-of-ten notation, and one on the Nobel laureates in physics.

Endpapers: Tables of physical data and other useful information, including fundamental constants and physical data, and standard abbreviations of units appear on the endpapers. In addition, the front endpaper includes the color code for all figures and diagrams.

Ancillaries

The following ancillaries are available to accompany this text:

Physics: A Numerical World View: This mathematical supplement, keyed to the text, looks at the development of the physics world view with more emphasis on mathematics. Sections that have an extended, parallel presentation in the mathematical supplement have the Σ symbol to the right of the section title. The supplement is available shrink-wrapped with the text at no additional cost.

Instructor's Resource Manual: The *Instructor's Resource Manual* contains answers to all questions and solutions to all exercises in the text and all problems in the math supplement. Each chapter in this manual has teaching tips on each text section and additional information about integrating the Physics Demonstration Videotape or Videodisc in your course.

Test Bank: To help you prepare quizzes and exams, the Test Bank provides over 1000 multiple-choice, short answer, numerical, and graphical questions covering every chapter in the text. Answers to all questions are included in the Test Bank.

ExaMaster+™ Computerized Test Banks: The Test Bank is available in computerized form for IBM, Macintosh, and Windows users. The Exa-Master+™ Computerized Test Banks allow you to preview, select, edit, and add items to tailor tests to your course or select items at random, add or edit graphics, and print up to 99 different versions of the same test and answer sheet. ExamRecord™ gradebook software is available with the IBM® PC version.

Overhead Transparency Acetates: The conceptual physics student often has difficulty gleaning all the information presented in physics diagrams. Therefore, a collection of 100 full-color transparency acetates of important figures from the text is available to adopters. These transparencies feature large print for easy viewing in the classroom.

A World View of Environmental Issues: A popular ancillary distributed free to students when purchasing the text is an environmental supplement based on an NSF grant to the Temple University physics department. It explores the following topics from a basic physics perspective: (1) the scientific method, (2) how to research environmental issues using databases and the Internet, (3) nuclear power plants and nuclear waste disposal, (4) indoor air pollution including radon and sick building syndrome, (5) stratospheric ozone depletion and the greenhouse effect, (6) electromagnetic waves and effects on humans, and (7) the mass media and its portrayal of environmental issues.

Physics Demonstration Videotape/Videodisc: J.C. Sprott of the University of Wisconsin, Madison, has prepared 1 3/4 hours of video demonstrating a wide variety of simple physics experiments. The 74 demonstrations are divided into 12 subject areas, covering the major topics in the introductory physics course. The *Instructor's Resource Manual for the Physics Demonstration Videotape* gives helpful hints about integrating the videotape into your lecture and replicating demonstrations. More information about correlating the video to the text is also available in the *Instructor's Resource Manual* to this text.

Saunders College Publishing may provide complimentary instructional aids and supplements or supplement packages to those adopters qualified under our adoption policy. Please contact your sales representative for more information. If as an adopter or potential user you receive supplements you do not need, please return them to your sales representative or send them to:

Attn: Returns Department
Troy Warehouse
465 South Lincoln Drive
Troy, MO 63379

Acknowledgments

Physicists and physics teachers who gave freely of their time to explore the many options of explaining the physics world view with a minimum of mathematics include our colleagues John Carlsten, William Hiscock, Robert Swenson, and George Tuthill from Montana State University, as well as Arnold Arons (University of Washington), Larry Gould (University of Hartford), and Bob Weinberg (Temple University). We appreciate the special efforts of Montana State University photography graduate David Rogers for many of the photographs used in the text.

The following reviewers of the second edition provided valuable input for the current revision:

Philip Baringer, *University of Kansas*
Louis Cadwell, *Providence College*
Jorge Cossio, *North Miami-Dade Community College*
Gary DeLeo, *Lehigh University*
Mark Miksic, *Queens College*
Thomas Smith, *Mount San Antonio College*
George Smoot, *University of California at Berkeley*
Jan Yarrison-Rice, *Miami University of Ohio*

Reviewers who played an important role in the development of the second edition of the text include:

Jeffrey Collier, *Bismarck State Community College*
Leroy Dubeck, *Temple University*
John R. Dunning, Jr., *Sonoma State University*
Joseph Hamilton, *Vanderbilt University*
Roger Herman, *Pennsylvania State University*
Robert Lieberman, *Cornell University*
Richard Lindsay, *Western Washington University*
Robert A. Luke, *Boise State University*
Allen Miller, *Syracuse University*
John Mudie, *Modesto Junior College*
Steven Robinson, *Colorado State University*
Carl Rosenzweig, *Syracuse University*
James Watson, *Ball State University*

Reviewers of the first edition, who helped shape the content and structure of the current edition, include:

John C. Abele, *Lewis & Clark College*
David Buckley, *East Stroudsburg University*
Robert Carr, *California State University, Los Angeles*
Art Champagne, *Princeton University*
David E. Clark, *University of Maine*
Robert Cole, *University of Southern California*
John R. Dunning, Jr., *Sonoma State University*
Abbas Faridi, *Orange Coast College*
Simon George, *California State University, Long Beach*
Patrick C. Gibbons, *Washington University*

Robert E. Gibbs, *Eastern Washington University*

Tom J. Gray, *Kansas State University*

Roger Hanson, *University of Northern Iowa*

Steven Hoffmaster, *Gonzaga University*

Michael Hones, *Villanova University*

Sardari L. Khanna, *York College*

David A. Krueger, *Colorado State University*

Jean P. Krish, *University of Michigan*

Leon R. Leonardo, *El Camino College*

Robert A. Luke, *Boise State University*

Peter McIntyre, *Texas A&M University*

Paul Nachman, *New Mexico State University*

Van E. Neie, *Purdue University*

Barton Palatnick, *California State Polytechnic University, Pomona*

Cecil G. Shugart, *Memphis State University*

Leonard Storm, *Eastern Illinois University*

Paul Varlashkin, *East Carolina University*

Leonard Wall, *California State Polytechnic University, San Luis Obispo*

James Watson, *Ball State University*

Robert Wilson, *San Bernardino Valley College*

John M. Yelton, *University of Florida*

After giving serious consideration to each of the reviewers' suggestions, we made the final decisions and, therefore, accept the responsibility for any errors, omissions, and confusions that might remain in the text. We would, of course, appreciate receiving any comments that you might have. Send comments and suggestions to Larry D. Kirkpatrick, Physics, Montana State University, Bozeman MT 59717-3840 or via e-mail at kirkpatrick@physics.montana.edu.

Finally, we would like to thank the staff at Saunders, who brought a level of professionalism and enthusiasm for the challenge that we have never experienced before. Special thanks go to John Vondeling, Publisher; Marc Sherman, Senior Associate Editor; Anne Gibby, Senior Project Editor; Lisa Caro, Art Director; Dena Digilio-Betz, Photo Researcher; and George Kelvin for his beautiful illustrations.

Larry D. Kirkpatrick
Gerald F. Wheeler
Montana State University
August 1997

The publisher and authors have gone to great measures to ensure as error-free a text as is humanly possible. We are so confident of the success of this effort that we are offering $5.00 for any first time technical error you may find. Note that we will pay only for each error the first time it is brought to our attention. Also note that this offer is valid only for technical errors in physics—not questions of grammar or style, nor for typographer's errors. Please write to John Vondeling, Publisher, and include your social security number.

To the Student

This course could be one of the most challenging experiences that you will ever have—except for first grade. But then you were too young to notice.

What happened in first grade? Well, you learned to read and that was really hard. First you had to learn the names of all those weird little squiggles. You had to learn to tell a *b* from a *d* from a *p*. Even though they looked so much alike, you did it, and it even seemed like fun. Then you learned the sounds each letter represented, and that was not easy because the capitals looked different but made the same sound, and some letters could have more than one sound.

Then one day your teacher put some letters on the board. First, the letter *C*, and you all knew it could make a Kuh or Suh. Then your teacher wrote the letter *A*. That had lots of possibilities. Finally, a *T*, which luckily had only one sound. You tried out several combinations including Kau-AAH-Tuh. Then suddenly someone shouted out in triumph, "That isn't kuh-aah-tuh! It's a small furry animal with a long skinny tail that says 'meow'." And your world was never the same again. When your car paused at an eight-sided red sign, you sounded out *stop* and understood how the drivers knew what to do. You saw the words *ice cream* on the front of a store and knew you wanted to go in.

If this book works, you will become aware of a whole world you never noticed before. You will never walk down a street, ride in a car, or look in a mirror without involuntarily seeing an extra dimension. There are times when you will have to memorize what symbols mean—just like in first grade. There will be times when you will confuse things that seem as much alike as *b*, *d*, and *p* once did, until you suddenly see how different they are. And there will be times when you will look at a combination of events and equations helplessly reciting Kuh-AAH-Tuh in total frustration. This has happened to all of us. But then the moment of insight will come, and you will see whole new images fitting together. You will see the *C-A-T* and will experience fully, and consciously, the exhilaration you felt in first grade.

So, welcome to one of the most challenging (and rewarding) courses you have ever taken in your life. If you work at it and let it happen, this experience will change your world view forever.

Larry D. Kirkpatrick
Gerald F. Wheeler
Montana State University
Bozeman, MT 59717

Adapted from an article by Barbara Wolff, "An Introduction to Physics—Find the CAT." *The Physics Teacher* **27,** 427.

Contents Overview

Contents

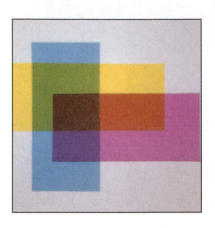

PHYSICS
A World View

third edition

PROLOGUE

A scientist's view of the human cannonball differs from that of a poet.

On Building a World View

Physics is the study of the material world. It is a search for patterns, or rules, for the behavior of objects in the Universe. This search covers the entire range of material objects, from the smallest known particles—so small that it is meaningless to discuss what they look like—to the astronomical objects—millions of times bigger than our Sun or so dense that a thimbleful of material would weigh a billion tons. The search also covers the entire span of time, from the primordial fireball to the ultimate fate of the Universe. Within this vast realm of space and time, the searchers have one goal: to comprehend the course of events in the whole world—to create a world view.

The phrase "world view" has a fairly elastic meaning. When we think about world views, the interpretations can stretch from the philosophic to the poetic. In physics, the world view is a shared set of ideas that represents the current explanations of how the material world operates. These include some rather common constructs, like gravity and mass, as well as strange sounding ones, like quarks and black holes.

The physics world view is a dynamic one. Ideas are constantly being proposed, debated, and tested against the material world. Some survive the scrutiny of the community of physicists; some don't. The inclusion of new ideas often forces the rejection of previously accepted ones. Some firmly accepted ideas in the world view are very difficult to discard, but in the long run experimentation wins out over personal biases. The model of the atom as a miniature Solar System was reluctantly given up because the experimental facts just didn't support it; it was replaced by a mathematical model that's difficult to visualize.

A few years before his death Albert Einstein described the process of science to a lifelong friend, Maurice Solovine. Solovine was not a scientist but apparently enjoyed discussing science with his famous friend. In one of their last exchanges, Solovine wrote that he had trouble understanding a certain passage in one of Einstein's essays. The next week Einstein wrote back, carefully explaining his view of the process of science. Figure P-1(a) is a reproduction of a diagram from this letter. Figure P-1(b) shows our interpretation of the diagram.

In his text Einstein explains the diagram. The lower horizontal line represents the "real world." The curved line on the left signifies the creative leap a scientist makes in attempting to explain some phenomena. The leap is intuitive, and although it may be very insightful, it is not scientific. The scientific process begins, Einstein explains, when the scientist takes the idea, or axiom, and develops consequences based on it. These consequences are illustrated by a number of smaller circles connected to the axiom with lines. A very powerful axiom has a large number of consequences. The final task in the process of science is to test these consequences against the material world. In Einstein's drawing these tests are vertical arrows returning to the world. If there is no match between the predicted consequences and the real world, the idea is (scientifically) worthless. If there is a match, there is hope that the idea has merit. With the publication of the idea, the community of scientists is brought into the process, and the original work is often modified.

Ideas for which there are simple physical models are easier to understand and accept. There is comfort in "picturing" electrons and protons as tiny balls, but those physicists who yearned for the electron to be a miniature billiard ball never got their wish. Cosmologist Sir Hermann Bondi, commenting on doing physics in realms beyond the range of direct human experience, said, "We should be

> I do not know what I may appear to the world, but to myself I seem to have been only like a boy playing at the sea-shore, and diverting myself in now and then finding a smoother pebble or a prettier shell than ordinary, whilst the great ocean of truth lay all undiscovered before me.
>
> —Isaac Newton

(a)

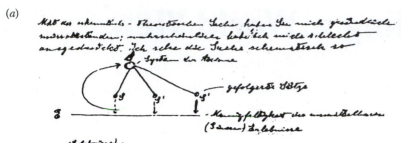

(b)

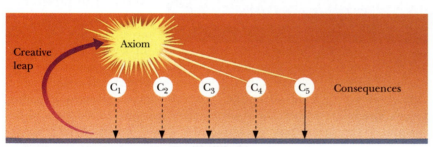

(a) Reproduction of a diagram from one of Albert Einstein's letters. (b) Our interpretation of this diagram.

surprised that the gas molecules behave so much like billiard balls and not surprised that electrons don't." This advice, though correct, may leave one feeling like Alice in *Through the Looking Glass* listening to the White Queen say that one should be able to "believe six impossible things before breakfast."

Although some of the ideas in physics might appear to be contrary to common sense, they do in fact make sense. Normally, things that make sense—that don't violate your intuition, or common sense—are those that fit into your past experience. Common sense is a personal world view. Like the physics world view, your common sense is built on a large experimental base. The difference between what makes sense to you and what makes sense to a physicist is, in part, due to different ranges of experience. Where our observations are limited by the range of human sensations, the physicist has instruments that bring

ultrahuman sensations into consideration. Our twinkling star is the physicist's window into the Universe.

So, without common sense to guide them about whether to accept a new idea, how do physicists decide which ones to adopt? Acceptance is based on whether the idea works, how well it fits into the world view, and if it is better than the old explanations. Although the most basic criteria for accepting an idea are that it agrees with the results of past experiments and successfully predicts the outcome of future experiments, acceptance is a human activity. Because it is a human activity, it has subjective aspects. The phrases "how well it fits" and "if it is better" imply opinions. Ideas have appeal, some more than others.

If an idea is very general, having many consequences in the Einsteinian sense, it can replace many separate ideas. It is regarded as more funda-

Telescopes extend the range of human senses. (Scott Goldsmith/Tony Stone Images)

mental and thus more appealing. It is possible to construct a different explanation for each observation. For example, a scheme could be created to explain the disappearance of water from an open container, and another, unrelated idea could be employed to account for the fluidity of water, and so on. An idea about the structure of liquids (not just water) that could be used to explain these phenomena and many others would be a highly valued replacement for the collection of separate ideas.

The simplicity of an idea also influences opinions about its worth. If more than one construct is proposed to explain the same phenomena and if they all predict the experimental results equally well, the most appealing idea is the simplest one. Although elaborate (Rube Goldberg) constructions are cute in cartoons, they hold very little value in the building of a physics world view.

Incredible as an idea may be initially, physicists seem to become more and more comfortable with it the longer it remains in their world view. Most physicists are comfortable with the relativistic notions of slowed-down time and warped space; when the ideas were first introduced, however, they caused quite a stir. As more and more experimental results support an idea, it gains stature and becomes a more established part of our beliefs. But, even if an idea be-

comes very familiar and comfortable, it is still tentative. Experimental results can never prove an idea; they can only disprove it. If the predictions are borne out, the best that can be claimed is, "So far, so good."

Our goal in writing this book is to help you view the world differently. We describe the building of a physics world view and share with you some of the results. We will start in areas that are familiar, where your common sense serves you well, and chart a course through less-traveled areas. Although there is no end to this journey, the book must stop. We hope it will stop at a new place, a place you have never been before.

We also hope that you will find joy in the process similar to that expressed by Isaac Newton.

> I do not know what I may appear to the world, but to myself I seem to have been only like a boy playing on the sea-shore, and diverting myself in now and then finding a smoother pebble or a prettier shell than ordinary, whilst the great ocean of truth lay all undiscovered before me.

"IN LAYMAN'S TERMS? I'M AFRAID I DON'T KNOW ANY LAYMAN'S TERMS."

1

The blurred image of the car clearly shows that it's moving. To appear blurred in the photograph, the car had to be in different places during the time the shutter was open. If we know how fast the car is moving, can we determine how long it will take to reach its destination? (See page 21 for the answer to this question.)

Describing Motion

The blurring of the car's image shows that it is moving.

A property common to everything in the Universe is change. Some things are big, some are small; some are red, some have no color at all; some are rigid, some are fluid; but they all are changing. In fact, change is so important that the fundamental concept of time would be meaningless without it.

Change even occurs where seemingly there is none. Water evaporating, colors fading, flowers growing, and stars evolving are all examples of changes that are beyond our casual observations. Also beyond our sensations is the fact that these changes are a result of the motion of material, often at the submicroscopic level. Because change—and thus motion—is so pervasive, we begin our exploration of the ideas of physics with a study of motion.

Within our commonsense world view we generally group all motions together, simply observing that an object is moving or that it is not moving. Actually, there is an extraordinary diversity of motion, ranging from the very simple to the extremely complicated. Fortunately, the complex motions—ones more common in our everyday experiences—can be understood as combinations of simpler ones. For example, the Earth's motion is a combination of a daily rotation about its axis and an annual revolution around the Sun. Or, closer to home, the motion of a football can be treated as a combination of a vertical rise and fall, a horizontal movement, and a spinning about an axis.

We therefore begin our discussion of motion by trying to describe and understand the simplest kinds of motion. This will yield a conceptual framework within our world view from which even the most complicated motions, such as those associated with a hurricane or with a turbulent waterfall, can be understood.

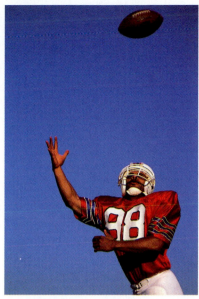

The motion of a football is a combination of three simpler motions.

Average Speed

Imagine driving home from school. For simplicity, assume that you can drive home in a straight line. Normally, you might describe this trip in terms of the time it takes. If pressed for a more detailed account, you would probably give the distance or the actual route taken, adding points of interest along the way. For our purposes we need to develop a more precise description of motion.

First, we note that your position continually changes as you drive. Second, we observe that it takes time to make the trip. These two fundamental notions—space and time—are at the core of our concept of motion. Furthermore, different positions along the trip can be matched with different times.

One relationship between space and time can be illustrated by answering the question, "How fast were you going?" Actually, there are two ways to answer this question; one way looks at the total trip, whereas the other considers the moment-by-moment details of the trip. For the total-trip description we use the concept of an **average speed,** which is defined as the total distance traveled divided by the time it took to cover this distance.

$$\text{average speed} = \frac{\text{distance traveled}}{\text{time taken}}$$

We can write this relationship more efficiently by using symbols as abbreviations

$$\bar{s} = \frac{d}{t}$$

where $\bar{s}$ is the average speed, d is the distance traveled, and t is the time taken for the trip. A bar is often used over a symbol to indicate its average value.

This ratio of distance over time gives the average rate at which the car's position changes. Speed is a quantitative measure of how rapidly the change takes place. The definition of average speed states a particular relationship between the concepts of space and time. If any two of the three quantities are known, the third is determined.

There is a humorous story about a small-time country farmer that illustrates this relationship. The farmer was visited by his big-time cousin. Anxious to make a good impression, the host spent the morning showing his cousin around his small farm. At lunch the cousin could not resist the urge to brag that on his ranch he could get into his car in the early morning and drive until the sun set and he would still be on his property. The country farmer thought for a moment and said, "I had a car like that once."

Mile markers and your wristwatch can be used to calculate your average speed.

kilo = 1000
centi = 1/100

Measuring Space and Time

To measure speed, we need a device for measuring distance, such as a ruler, and one for measuring time, such as a clock. Most highways have "mile markers" along the side of the road so maintenance and law enforcement officials can accurately find certain locations. These mile markers and your wristwatch give you all the information you need to determine average speeds.

To be useful, measuring devices must be calibrated; that is, we must agree on the units to be used. Most countries use the metric system. The United States is one of the last to use the English system, which is based on the foot, pound, and second.

The United States is slowly adopting the metric system, the primary one used in science, so we should gain some familiarity with it. On the other hand, our principal goal is to establish connections between your commonsense world view and the physics world view. Learning the metric system at the same time as beginning your study of physics is complicated because you need to develop a feeling for the new units as well as the scientific ideas. As a compromise, we will use the metric system predominantly, but will give the approximate English equivalents in parentheses when it is useful.

The unit of time, the second, is the same in both systems. Distances in the metric system are measured in meters (abbreviated m) or units which are multiples of ten longer or shorter than the meter. One meter is slightly longer than 1 yard, a **kilo**meter (= 1000 meters) is a little more than ½ mile, and a **centi**meter (= 1/100 meter) is a little less than ½ inch. More on the metric system can be found in Appendix A.

Assuming that we begin "thinking metric," speeds have units such as meters per second (m/s) or kilometers per hour (km/h). A person walks about 1½ meters per second, and a car traveling at 55 miles per hour is going approximately 90 kilometers per hour.

Physics on Your Own   Estimate the average speed of an everyday object such as a falling leaf, a falling snowflake, or a wave of water traveling from one end of your bathtub to the other.

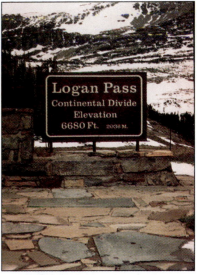

Some highway signs give the elevation in feet and meters.

Images of Speed Σ

From the earliest cave drawings to modern time-lapse photography, it has been a part of human nature to try to represent our experiences. Artists, as well as scientists, have devised many ways of illustrating motion. A blurred painting or photograph such as that in Figure 1-1 is one way to "see" motion. One difference between the artist and the scientist is that the scientist uses the representations to analyze the motion.

A clever image of motion that also provides a way of measuring the speed of an object is the multiple-exposure photograph in Figure 1-2. Photographs such as these are made in a totally dark room with a stroboscope (normally just called a "strobe") and a camera with an open shutter. A strobe is a light source that flashes at a constant, controllable rate. The duration of each flash is very short (about 10-millionths of a second), producing a still image of the moving object.

If the strobe flashes ten times per second, the resulting photograph will show the position of the object at time intervals of ¹⁄₁₀ of a second.

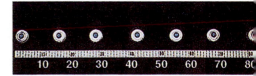

Figure 1-2   A strobe photograph of a moving puck shows its position at different times.

Figure 1-1   The blurring of the background tells us that the race car is moving.

Table 1-1 Calculations of the average speed of the puck between flashes in Figure 1-2		
Position at each flash (cm)	**Distance traveled between flashes (cm)**	**Average speed between flashes (cm/s)**
2		
	13	130
15		
	13	130
28		
	14	140
42		
	13	130
55		
	13	130
68		
	14	140
82		

Thus, we can "freeze" the motion of the object into a sequence of individual events and use this representation to measure its average speed within each interval.

As an example of measuring average speed, let's determine the average speed of the puck in Figure 1-2. During the time required to make this photograph, the puck travels from a position near the 2-centimeter mark to one near the 82-centimeter mark, a total distance of 80 centimeters. Since there are seven images, there are six intervals and the total time taken is six times the time between flashes; that is, 0.60 second. Therefore, the average speed is

$$\bar{s} = \frac{d}{t} = \frac{80 \text{ cm}}{0.60 \text{ s}} = 130 \text{ cm/s}$$

We can also determine the average speed of the puck between each pair of adjacent flashes. This is done in Table 1-1. Allowing for the uncertainties in reading the values of the positions of the puck, the average speeds in the last column can be considered to be the same. Therefore, the puck was probably traveling at a constant speed.

Suppose that you live 40 miles from school and it took you 2 hours to drive home. Your average speed during that trip was

$$\bar{s} = \frac{d}{t} = \frac{40 \text{ miles}}{2 \text{ hours}} = 20 \frac{\text{miles}}{\text{hour}}$$

This means that, on the average, you traveled a distance of 20 miles during each hour of travel. This answer is read "20 miles per hour" and is often written as 20 miles/hour, or abbreviated as 20 mph. It is important to include the units with your answer. A speed of "20" does not make any sense. It could be 20 miles per hour or 20 inches per year, very different average speeds.

Actually, you probably weren't moving at 20 miles per hour during much of your trip. At times you may have been stopped at traffic lights; at other times you traveled at 50 miles per hour. The use of average speed disregards the details of the trip. In spite of this, the concept of average speed is a useful notion.

QUESTION   What is the average speed of an airplane that flies 3000 miles in 6 hours?

COMPUTING AVERAGE SPEED Σ

If you know the average speed, you can determine other information about the motion. For instance, you can obtain the time needed for a trip. Suppose you plan to drive a distance of 60 miles, and past experience tells you that you'll probably be able to maintain an average speed of 50 miles per hour. How long will the trip take? Without consciously doing any calculation, you can probably guess that the answer is a little over 1 hour. How do you get a more precise answer? You divide the distance traveled by the average speed.

$$t = \frac{d}{\bar{s}}$$

$$\text{time taken} = \frac{\text{distance traveled}}{\text{average speed}}$$

For our example we obtain

$$t = \frac{d}{\bar{s}} = \frac{60 \text{ miles}}{50 \text{ miles/hour}} = 1.2 \text{ hours}$$

You can also calculate how far you could drive if you traveled with a specified average speed for a specified time.

$$d = \bar{s}t$$

$$\text{distance traveled} = \text{average speed} \times \text{time taken}$$

Suppose, for example, you plan to maintain an average speed of 50 miles per hour on an upcoming trip. How far can you travel if you drive an 8-hour day?

$$d = \bar{s}t = \left(50 \, \frac{\text{miles}}{\text{hour}} \right)(8 \text{ hours}) = 400 \text{ miles}$$

Therefore, you would expect to drive 400 miles each day.

ANSWER   Using our definition for average speed, we have $\bar{s} = d/t =$ (3000 miles)/(6 hours) = 500 mph.

Fastest and Slowest

The fastest speed in the Universe is the speed of light, 300 million meters each second (186,000 miles per second), and the slowest speed is, of course, zero. Between those extremes is a vast range of speeds. With the exception of a few, very small subatomic particles that have been catapulted through huge electrical voltages or energized by nuclear reactions, most things move at speeds close to the slow end of the range.

The fastest large objects are planets moving at speeds up to 107,000 mph. Our own Earth orbits the Sun at 67,000 mph. (Even this admittedly large speed is 10,000 times slower than the speed of light.) Closer to home, but still in space, the Apollo spacecraft returned to Earth traveling at 25,000 mph and the Space Shuttles orbit the Earth at 17,500 mph.

Our people-carrying machines have a wide range of speeds, from supersonic airplanes with a record speed of 2193 mph (the Lockheed SR-71A) to moving stairways that approximate fast walking speeds of 4 mph. In between we have record speeds set by passenger planes (the French Concorde) at 1450 mph, jet-powered race cars at 633 mph, road cars at 217 mph, road motorcycles at 186 mph, human-powered vehicles at 65.5 mph, and elevators at 22.7 mph.

When we give up our machines, we slow down considerably. The fastest recorded human speed is Donovan Bailey's 100-meter dash in 9.84 seconds, or about 22.7 mph. The fastest female sprinter is Florence Griffith Joyner at 10.49 seconds (21.3 mph). (In the time it took either sprinter to run the race, a beam of light could go to the Moon, bounce off a mirror, and return to Earth with 7 seconds to spare!) As the distances get longer, human speeds slow: The 1-mile record is held by Nourredine Morceli in a time of 3 minutes and 44.39 seconds, which corresponds to a little more than 16 mph. The record pace for a marathon is a little over 12 mph.

Other animals range from slow (three-toed sloths that creep at 0.07 mph, giant turtles that lumber along

Gail Devers wins the gold medal in the 100-meter dash at the 1996 Olympic Games.

at 0.23 mph, and sea otters that swim at 6 mph) to the very fast (killer whales and sailfish, which swim at 35 and 68 mph, respectively, and cheetahs, reported to run up to 63 mph). In 1973, Secretariat set the record for the Kentucky Derby by running 1-1/4 miles in 1 minute 59.2 seconds for an average speed of almost 38 mph. The streamlined Peregrine falcon diving for its prey has been clocked at 217 mph.

Nobody knows the slowest moving object. A good candidate for a natural motion is a continent drifting at 1 centimeter per year, or 0.7 billionths of a mile per hour. Recently a machine was built for testing stress corrosion that moves at a million millionth of a millimeter per minute, or 37 billion billionths of a mile per hour. At this rate it would take about 2 billion years to move 1 meter!

Source: Guinness Book of Records, Bantam Books, 1996.

Instantaneous Speed Σ

The notion of average speed is limited in most cases. Even something as simple as your trip home from school is a much richer motion than our concept of average speed indicates. For example, it doesn't distinguish the parts of your trip when you were stopped, waiting for a traffic light to change, from those parts when you were exceeding the speed limit. The simple question, "How fast were you going as you passed Third and Vine?" is not answered by knowing the average speed.

To answer the question, "How fast were you going at a specific point?" we need to consider a new concept known as the **instantaneous speed.** This more complete description of motion tells us how fast you were traveling at any instant during your trip. Since this is the function of your car's speedometer (Fig. 1-3), the idea is not new to you, although its precise definition might be.

Actually, the definitions of average and instantaneous speeds are quite similar. They differ only in the size of the time interval involved. If we want to know how fast you are going at a given instant, we must study the motion during a very small time interval. The instantaneous speed is equal to the average speed over a time interval that is infinitesimally small.

As a first approximation in measuring the instantaneous speed, we could measure how far your car traveled during $\frac{1}{10}$ of a second and calculate the average speed for this time interval. But $\frac{1}{10}$ of a second is not infinitesimally small. With precise equipment we could measure time intervals of $\frac{1}{100}$ of a second, $\frac{1}{1000}$ of a second, or an even smaller interval. How small an interval do we need? Actually, what we need is an "instant" of time.

Recall that average speed is the ratio of distance traveled divided by the time interval. But what happens to this ratio as the interval becomes smaller and smaller? As we use smaller time intervals, the distance traveled also gets smaller. Although each element of the ratio becomes vanishingly small, the ratio tends toward a single value. Instantaneous speed is simply the limiting value of the average speed as the time interval gets exceedingly small—in principle, infinitesimally small. It is the instantaneous speed rather than the average speed that plays an important role in the analysis of nearly all realistic motions.

Figure 1-3   A speedometer tells you the car's instantaneous speed, in this case 60 mph.

instantaneous speed is the limiting value of the average speed

Speed with Direction   Σ

We have made a lot of progress in attempting to accurately represent motion. However, as we develop the rules for explaining (and thus predicting) the behavior of objects in the next chapter, we will need to go further. Objects do more than speed up and slow down. They can also change direction, sometimes keeping the same speed, but at other times changing both their speed and their direction. Either the average speed or the instantaneous speed tells us how fast an object is moving, but neither tells us the direction of motion. If we are discussing a vacation trip, direction doesn't seem important; you obviously know in which direction you're going. However, we are trying to develop rules of motion for all situations, and the direction is as important as the speed. You can get a sense for this by remembering situations where there is an abrupt change in direction, maybe a car you were riding in swerved sharply. The squeal of the tires and your own body's reaction are clues that there are new factors involved when an object changes direction.

In the physics world view we combine speed and direction into a single concept called **velocity.** The combination is actually quite simple;

velocity equals speed with a direction

when we talk of an object's velocity, we give the speed (for example, 15 miles per hour), and just add the direction (north, to the left, or 30 degrees above the horizontal). The speed is known as the **magnitude** of the velocity; it gives its size. We use the symbol v to represent the magnitude of the instantaneous velocity.

Quantities that have both a size and a direction are called **vectors.** Vectors do not obey the normal rules of arithmetic. We will study the rules for combining vector quantities in Chapter 2. For now, it is only important to realize that the direction of the motion can be as important as the speed.

For the rest of this chapter we will only deal with objects traveling in one dimension. They might be going left and right, east and west, or up and down, but not turning corners. Notice that by doing this we eliminate motions as simple as a home run. The payoff is that we learn to manipulate the new concepts before tackling the many realistic, but more difficult, situations.

Acceleration

Since the velocities of many things are not constant, we need a way of describing how velocity changes. We now define a new concept, called **acceleration,** that describes the rate at which velocity changes. The magnitude of the average acceleration $\bar{a}$ of an object is the change in its velocity v divided by the time Δt it takes to make that change:

$$\frac{\text{average}}{\text{acceleration}} = \frac{\text{change in velocity}}{\text{time taken}}$$

$$\bar{a} = \frac{\Delta v}{\Delta t}$$

The symbol Δv is called "delta vee." The delta symbol Δ is used to represent a change in a quantity. Thus, Δv represents the change in velocity and must not be thought of as the product of Δ and v. To calculate the change in velocity, we subtract the velocity at the beginning of the time interval from the velocity at the end. As we did with speed, we can speak either of the average acceleration or the instantaneous acceleration, depending on the size of the time interval.

The units of acceleration are a bit more complicated than those of speed and velocity. Remember that the units of velocity are distance divided by time: for example, miles per hour or meters per second. Since

Physics Update

A hypervelocity launcher has accelerated a quarter-inch disk of metal to a speed of 15.8 kilometers/second, or about 36,000 miles per hour, a record for a macroscopic object. For comparison, the Space Shuttle's orbital speed is 17,500 mph, while the speed for total escape from the Earth is 25,000 mph. The tremendous acceleration ensues from the following sequence: A gun fires a piston, which compresses a column of hydrogen gas, which moves a specially sculpted impactor down a barrel, where it strikes the projectile. The launcher, developed at Sandia National Laboratory, is currently used for studying the effects of space debris colliding with the proposed orbiting space station.

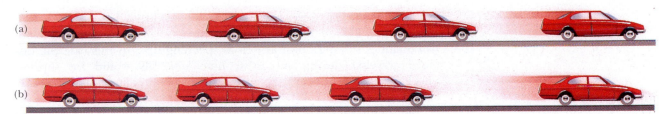

Figure 1-4   Strobe drawings of two cars. Which car is accelerating and which is traveling at a constant speed?

acceleration is the change in velocity divided by the time interval, its units are (distance per time) per time: for example, (kilometers per hour) per second or (meters per second) per second.

The concept of acceleration is probably familiar to you. We talk about one car having "better acceleration" than another. This usually means that it can obtain a high speed in a shorter time. For instance, a Dodge Grand Caravan can accelerate from 0 to 60 miles per hour in 10.7 seconds; a Ford Taurus requires 8.9 seconds and a Chevrolet Corvette requires only 5.2 seconds.

> **QUESTION**   Which car has the largest average acceleration?

Another way of becoming more familiar with acceleration is by experiencing it. For example, when an elevator begins to move up (or down) rapidly, the sensation you get in your stomach is due to the elevator (and you) quickly changing speed. Astronauts feel this when the Space Shuttle blasts off from its launchpad. Exciting examples of the same effect can be achieved on a roller coaster. In fact, amusement parks can be thought of as places where people pay money to experience the effects of acceleration.

In contrast, you don't feel motion when you're traveling in a straight line at a constant speed; that is, motion with zero acceleration. The motion you do feel when riding in a car on a straight highway is due to small vibrations of the car. (These vibrations are tiny changes in direction, small accelerations of the car caused by bumps in the road.)

If you are not accelerating but rather are viewing a moving object, how can you tell whether or not it is accelerating? One way is to take a strobe photograph of its motion. Figure 1-4 is a drawing of two such pictures. Which of the two corresponds to the car accelerating?

If you answered car (b), you have a qualitative understanding of acceleration. Car (a) travels the same distance during each time interval and therefore is traveling at a constant speed. Car (b) travels farther during each successive time interval; it is accelerating.

> **ANSWER**   The Corvette has the largest average acceleration since it reaches 60 miles per hour in the shortest time interval.

Amusement parks sell the thrills of acceleration.

Even if the car were slowing down, it would be accelerating. (We don't usually use the word *deceleration* in physics because the word *acceleration* includes slowing down as well as speeding up.) In this case the distance traveled during successive time intervals would be shorter. Acceleration refers to any change in speed or direction; that is, to any change in velocity.

Acceleration is a vector quantity. When the acceleration is in the same direction as the velocity, the speed of the object is increasing. When the acceleration and the velocity point in opposite directions, the object is slowing down. The idea of an acceleration having a direction might seem a little abstract and, perhaps, unnecessary. A car's velocity obviously has a direction and probably seems easier to comprehend. However, as we continue our study of acceleration we will see many examples in which the direction of the acceleration has physical consequences. The discussion of accelerations due only to a change in the direction of the velocity appears in Chapter 3.

COMPUTING   ACCELERATION Σ

The ideas of acceleration and velocity can be further illustrated with a numerical example. Consider a car moving along a straight highway at 40 mph that speeds up to 60 mph during an interval of 20 seconds. What is the car's average acceleration? Using the symbols v_i and v_f to represent the initial and final velocities, we have

$$\bar{a} = \frac{\Delta v}{t} = \frac{v_f - v_i}{t} = \frac{60\text{ mph} - 40\text{ mph}}{20\text{ s}} = \frac{20\text{ mph}}{20\text{ s}} = 1\text{ mph/s}$$

The car accelerates at 1 (mile per hour) per second. That is, during each second, its speed increases by 1 mph. If, on the other hand, the car made this change in velocity in 10 seconds, our new calculation would yield an average acceleration of 2 mph per second. These calculations illustrate that acceleration is more than just a change in velocity; it is a measure of the rate at which the velocity changes.

A First Look at Falling Objects

With these few ideas we can now look at a common motion, a ball falling near the Earth's surface. How does the ball fall? Does it fall faster and faster until it hits the ground? Or does it reach a certain speed and then remain at that speed for the duration of its fall? Does the rate at which it falls depend on its weight? For example, would a cannonball fall faster than a feather?

Questions about this rather simple motion have fascinated scientists since at least the time of Aristotle (4th century B.C.). They turned out to be quite difficult to answer. In fact, modern answers to these questions were not given until early in the 17th century.

Until that time the accepted answers were those attributed to Aristotle. He thought that the weight of an object and the medium through which it was falling determined its downward speed. Obviously, a cannonball falls faster than a feather if they both fall in the air. Also, a cannonball falls more slowly in molasses than it does in air. Aristotle proposed that the speed of a falling object is equal to the weight divided by the resistance of the medium.

At first sight this may seem perfectly plausible. If we apply his theory, however, we find that it contradicts common experience. Namely, it follows from Aristotle's theory that a 10-pound rock would fall ten times as fast as a 1-pound rock. Presumably, Aristotle never checked his prediction by actually dropping two such rocks and seeing what happened. If he had, he would have discovered that his theory was wrong.

You can perform an equivalent experiment to test the theory. Hold a heavy book (a physics text is quite appropriate) and a piece of paper at equal heights above the floor and drop them simultaneously. Which falls faster? Now repeat the experiment, but this time wad the paper into a tight ball. How do the results differ now?

Immoderate Genius

Galileo Galilei.

On February 15, 1564, Galileo Galilei was born in Pisa, Italy, into the family of a Florentine cloth merchant. As a boy he was schooled in Latin, Greek, and the humanities at the local monastery until his family relocated to Florence where his father assumed the boy's education in mathematics and music. The young Galileo mastered the lute while learning advanced mathematics, physics, and astronomy.

At the age of 17, Galileo returned to the town of his birth to study medicine. Before completing his training, however, Galileo withdrew from medical school due to conflicts with his mentors caused by his curiosities in the sciences. At age 25, Galileo enlisted the aid of his father's friends in academia and received an appointment as professor of mathematics at the University of Pisa. Other prestigious appointments followed, as Galileo engaged in a variety of scientific pursuits.

Galileo studied time, motion, floating bodies, the nature of heat, and the construction of telescopes and microscopes. In 1610 he achieved fame for improving previous telescope designs to allow magnification of the heavens up to 32 times. Galileo observed features of the Moon, sunspots, and various planets, including Venus in its various phases. He discovered four of Jupiter's moons and showed that the Milky Way consists of an enormous number of stars.

Galileo's many observations supported the Copernican Sun-centered (heliocentric) theory of the Solar System. In this belief, Galileo directly contradicted the Roman Catholic Church dogma of the day, which held that the Earth was the stationary center of the Universe; such a model as proposed by Aristotle was supported by

Legend has it that Galileo dropped balls from this leaning tower in Pisa while developing his ideas about free fall.

biblical references and was held sacred. To deny this view was considered heresy, a crime that carried the highest penalty of being burned at the stake. In 1615, the Church began its investigation of Galileo's beliefs, and in 1633, after publishing *Dialogue Concerning the Two World Systems,* he was taken to Rome, charged with heresy, and sentenced to life imprisonment. Later, the sentence was commuted to house arrest at his villa at Arcetri.

In his last years, Galileo reflected upon his life in a manuscript entitled *Discourses and Mathematical Discoveries Concerning Two New Sciences,* which was smuggled out of Italy and published in Holland in 1638. After completion of this work, Galileo became blind and died while still under house arrest in 1642.

References: Broderick, S. J. *Galileo: The Man, His Work, His Misfortunes.* New York: Harper & Row, 1964. AIP Niels Bohr Library.

In the first case the book fell much faster than the paper. This is in qualitative agreement with Aristotle's claim. But the second case certainly disagrees with Aristotle. Using the Aristotelian rule to predict the fall of the paper and book conflicts with reality. If the book is 500 times heavier than the paper, according to Aristotle, the book should drop at a speed

500 times faster than the paper. This means that the book would hit the floor before the paper dropped 1/100 of a foot! Clearly, Aristotle was wrong.

Our experiment with the book and paper might lead you to believe that in the absence of any resistance, as, for example, in a vacuum, two objects would fall side by side, independent of their weights. This means that a cannonball and a feather would fall together in a vacuum. This was the opinion of Galileo Galilei, an Italian physicist of the 17th century.

Galileo is often called the founder of modern science, due as much to his style of building a physics world view as to his particular contributions. His style was characterized by a strong desire to verify his theories with measurements. That is, he performed experiments to check his ideas. His goal was simple: to find rules of nature—often in the form of equations—that expressed the results of his investigations.

His work led to some new ideas about motion. He developed the concepts and mathematical language necessary to describe motion. For example, he invented the concept of acceleration.

Although Galileo was unable to directly test his ideas about free fall, he did suggest the *thought* experiment illustrated in Figure 1-5. Imagine dropping three identical objects simultaneously from the same height. The Aristotelians would agree that the three would fall side by side. Now imagine repeating the experiment but with two of the objects close to each other. Nothing significant has changed, so there will again be a three-way tie. Finally, consider the situation where the two are touching. Since there was a tie before, there will be no dragging of one on the other, and again there will be a tie. But if the two are touching, they can be considered to be a single object that is twice as big. Consequently, big and small objects fall at the same rate!

A common present-day test of this idea is to drop two objects, such as a coin and a feather (a cannonball is somewhat impractical), inside a plastic tube from which the air has been removed. In a vacuum, the race always ends in a tie. An ultramodern demonstration of this was conducted by astronauts on the Moon. Since there is no atmosphere on the Moon, a hammer and a feather fell at the same rate.

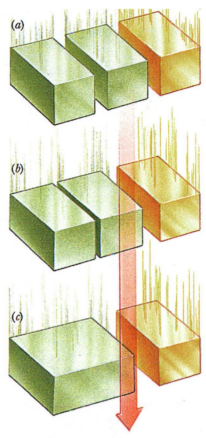

Figure 1-5 Galileo's thought experiment. All objects fall at the same rate in a vacuum.

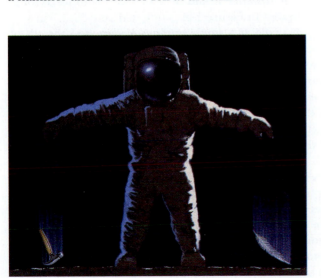

A hammer and feather dropped on the Moon hit the ground at the same time because there is no air.

Physics on Your Own   Check Galileo's statement that two objects of different weights fall at the same rate. Simultaneously drop balls out of a third- or fourth-floor window. Does the air cause the balls to fall at different rates? Make an estimate of the effects of air resistance by dropping two balls of approximately the same weight but different sizes, such as a solid rubber ball and a tennis ball.

Free Fall: Making a Rule of Nature

If we can find a rule for the motion of an object in free fall, it will be equally valid for all objects, heavy or light. Although we will obtain a rule that strictly speaking is only valid in a vacuum, it will be useful in many other situations—whenever the effects of air resistance can be ignored.

The quantitative measurement of a freely falling object was impossible with the technology of Galileo's time. Things simply happened too fast. To get around this, he studied a motion very similar to free fall, that of a ball starting from rest and rolling down a ramp. As the ramp is inclined more and more, the velocity of the ball increases. When the ramp is vertical, the ball falls freely. Galileo hoped (correctly) that he would be able to ignore the complications of rolling and be able to deduce the rule for free fall before the ramp became too steep and the ball's motion too fast to measure.

Galileo chose to measure the time the ball took to travel the length of the ramp. He then measured the time for different fractions of the ramp's length. Experimenting at this single ramp angle, he discovered a relationship between the distance the ball traveled and the time it took. When the ball started from rest, the distance traveled was proportional to the square of the time taken. This means that if the time interval was twice as long, the distance would be $2 \times 2 = 2^2 = 4$ times as long; if the time was three times as long, the distance would be $3 \times 3 = 3^2 = 9$ times as long. This relationship is illustrated in Figure 1-6.

After establishing this rule at a single angle, he increased the tilt of the ramp and started over again. At steeper angles he found what you might expect intuitively; the ball traveled farther in the same time. The exciting discovery, however, was that at each new ramp angle he discovered the *same* rule between distance and time squared. This relationship held *independently* of the ramp angle, a fact that was crucial to the success of his proposal to extrapolate these results to the free-fall situation. He correctly concluded that the ball would obey the same relationship if the ramp was vertical.

Using modern techniques we are able to take a strobe photograph (Fig. 1-7) of a falling ball and "see" its motion. The ball is clearly not moving at a constant speed. We can tell this by noting that the distances between the images are continually increasing. In fact, Galileo showed that the ball falls with a constant acceleration.

Constant acceleration means that the speed changes by the same amount during each second. If, for example, we find that the ball's speed changed by a certain amount during 1 second at the beginning of the

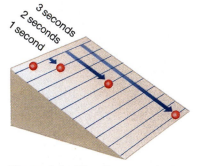

Figure 1-6   The distance a ball travels from rest down a ramp is proportional to the square of the time taken.

flight, then it would change by the same amount during any other 1 second of its flight.

We now know that for the case of the vertical ramp (free fall) the acceleration is about 9.8 (meters per second) per second [32 (feet per second) per second]. This value is known as the acceleration due to gravity and varies slightly from place to place on the surface of the Earth. Students at Montana State University in Bozeman have determined that the value of the acceleration due to gravity in the basement of the physics building is 9.80097 (meters per second) per second. For convenience in calculations we will usually round this value off to 10 (meters per second) per second.

At any time during the fall, if we know its speed and its acceleration, we can calculate how fast the ball will be moving 1 second later. Assume that the ball is traveling 40 meters per second with an acceleration of 10 (meters per second) per second. This acceleration means that the speed will change by 10 meters per second during each second. Since the ball is speeding up, 1 second later the ball will be traveling with a speed of 50 meters per second. One second after that the speed will be 60 meters per second, and so on.

COMPUTING   FREE FALL Σ

Using the definitions for average speed and acceleration, Galileo was able to show that the distance d a ball traveled beginning at rest was given by

$$d = \frac{1}{2}at^2 \qquad t = \sqrt{\frac{2d}{a}} \quad (= g)$$

where t is the time elapsed since the ball was released and a is the constant acceleration of the ball.

> **QUESTION**   How deep would the well be if you hear the splash after 3 seconds?

Let's use this relationship to solve a problem. Pretend that while walking in the woods, you discover an abandoned well. You decide to find out how deep it is by dropping a rock into it. Assume that you hear the splash 2 seconds later and that it takes a negligible time for the sound to travel up the hole. You can determine the depth of the well by multiplying the time by itself (2×2, in this case) and then multiplying this by one-half the acceleration due to gravity

$$d = \frac{1}{2}at^2 = \frac{1}{2}\left(10\ \frac{\text{m}}{\text{s}^2}\right)(2\ \text{s})^2 = 20\ \text{m}$$

Therefore, the well is 20 meters (66 feet) deep. Note that in calculations we write (m/s)/s as m/s². These can be seen to be equivalent because $\frac{\text{m}}{\text{s}} \cdot \frac{1}{\text{s}} = \frac{\text{m}}{\text{s}^2}$.

ANSWER   The rock falls a distance of 45 meters in 3 seconds.

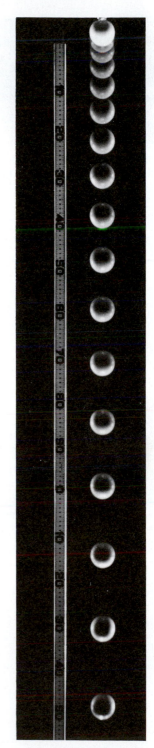

Figure 1-7   This strobe photograph of a falling ball shows that the ball has a constant acceleration.

Air resistance prevents the skydiver from having a constant acceleration.

As a final example of free fall, consider a thrill-seeking skydiver who jumps from a plane and decides not to pull the parachute cord until 30 seconds have elapsed. Our diver accelerates at a rate of 10 meters per second each second. Thus, at the end of ½ minute (assuming our skydiver can resist pulling the cord for that long) the speed will be about 300 meters per second. That is about 675 mph! Actually, as we will see in Chapter 2, this description of free fall is quite inaccurate when the effect of air resistance becomes important.

Physics on Your Own   Hold a dollar bill so that it is vertical. Have your friend hold his thumb and index finger on each side of the middle of the bill. Tell him that he can keep the dollar if he can catch it when you let go.

To catch the bill, he must be able to react within 0.13 second. Very few people can do this. However, if he is lucky and anticipates your move, you lose. If you let him hold his fingers near the bottom of the bill, he has 0.18 second to react. Many people can do this.

Starting with an Initial Velocity

What happens if the object is already in motion when we start our observations? Suppose, for example, a ball is thrown vertically upward. Our experience tells us that it will slow down, stop, and then fall. Examining strobe photographs of objects thrown vertically upward shows that the behavior of the rising object is symmetrical to that of the same object falling. The velocity changes by 10 meters per second during each second. In fact, the strobe photograph in Figure 1-7 could just as well have been a photograph of a ball rising. (Of course, what goes up must come down; in taking the photograph we would have to close the camera's shutter before the ball started back down.)

QUESTION   If you could throw the ball with a vertical speed of 40 meters per second, how long will it take to reach its maximum height? How high will it go?

We can use the symmetry between motion vertically upward and downward in answering the following question: If you throw a ball vertically upward with an initial speed of 20 meters per second, how long will it take to reach its maximum height? Ignoring air resistance, we know that the ball slows down by 10 meters per second during each second. Therefore, at the end of 1 second, it will be going 10 meters per second. At the end of 2 seconds, it will have an instantaneous speed of zero. It takes the ball 2 seconds to reach the top of its path.

ANSWER   It will take 4 seconds to reach a maximum height of 80 meters.

A Subtle Point

Let's pause at the end of this chapter to emphasize the fact that Galileo used experiment and reasoning to discover a pattern in nature. He was able to discern the motion of objects subject only to the pull of Earth's gravity. Using simple mathematics and the rule that he formulated, you can calculate the outcome of future experiments. In a limited, but very real, way you can predict the future. Predictions based on this rule are not the crystal-ball type popularized in science fiction, but they represent a very real accomplishment. The discovery of patterns and the creation of rules of nature are central in physicists' attempts to build a world view.

SUMMARY

We began building a physics world view with the study of motion because motion is a dominant characteristic of the Universe. We can obtain data about the motion of objects from strobe photographs. The average speed $\bar{s}$ of an object is the distance d it travels divided by the time t it takes to travel this distance, $\bar{s} = d/t$. The units for speed are distance divided by time, such as meters per second or kilometers per hour.

Instantaneous speed is equal to the average speed taken over a time interval that is infinitesimally small. Speed in a given direction is known as velocity, a vector quantity.

Acceleration is the change in velocity divided by the time it takes to make the change, $a = \Delta v/\Delta t$. The units for acceleration are equal to those of speed divided by time such as (meters per second) per second or (kilometers per hour) per second.

Acceleration is also a vector. When the acceleration is in the same direction as the velocity, the object's speed increases; when they point in opposite directions, the speed decreases.

Galileo reasoned that all objects fall at the same rate in the absence of any air resistance. Furthermore, he discovered that these freely falling objects fall with a constant acceleration of about 10 (meters per second) per second.

> **CHAPTER**
>
> **1**
>
> **REVISITED**
>
> We can determine how long it will take a car to reach its destination if we know its average speed and how far it has to go. The travel time is equal to the distance divided by the average speed.

KEY TERMS

acceleration: The change in velocity divided by the time it takes to make the change. Measured in units such as (meters per second) per second. An acceleration can result from a change in speed, a change in direction, or both.

average speed: The distance traveled divided by the time taken. Measured in units such as meters per second or miles per hour.

centi: A prefix meaning $\frac{1}{100}$. A centimeter is $\frac{1}{100}$ meter.

instantaneous speed: The limiting value of the average speed as the time interval becomes infinitesimally small. The magnitude of the velocity.

kilo: A prefix meaning 1000. A kilometer is 1000 meters.

magnitude: The size of a vector quantity. For example, speed is the magnitude of the velocity.

vector: A quantity with a magnitude and a direction. Examples are velocity and acceleration.

velocity: A vector quantity that includes the speed and direction of an object.

CONCEPTUAL QUESTIONS*

1. Describe the motion of the pucks in the strobe photographs. Assume the pucks move from left to right and do not retrace their paths.

(a)

(b)

2. Describe the motion depicted in the following strobe drawing.

3. Where is the speed the fastest in the following strobe drawing?

4. Where does the ball shown in the strobe drawing have the slowest speed?

5. Sketch a strobe drawing for the following description of a caterpillar moving along a straight branch. The caterpillar begins from rest and slowly accelerates to a constant speed. It then speeds up to a higher constant speed. Finally it gets tired and stops for a rest.

6. A car is driving along a straight highway at a constant speed when it hits a mud puddle, slowing it down. After the puddle the driver resumes the original speed. Make a strobe drawing for this motion.

7. Imagine a ball rolling in a straight line. Sketch strobe drawings for each of the following situations:

   a. The ball rolls from left to right at a constant speed.

   b. The ball rolls from left to right at a constant speed that is greater than in (a).

   c. What would the last drawing look like if the time between flashes were half as long?

8. Sketch strobe drawings for each of the following motions:

   a. A ball rolls from left to right and continually speeds up.

   b. A ball rolls from right to left and continually slows down. (How do these two drawings differ?)

   c. A ball rolls from left to right. Initially it speeds up, but when it reaches the middle of its path, it starts slowing down.

9. Approximately how many meters long is a midsize car?

10. About how long is a football field in meters?

11. Which (if either) has the greater average speed, a car that travels from milepost 78 to milepost 83 in 5 minutes, or one that travels from milepost 178 to milepost 183 in 5 minutes?

12. Which (if either) has the greater average speed, a truck that travels from milepost 111 to milepost 120 in 10 minutes, or one that travels from milepost 212 to milepost 220 in 10 minutes?

13. Approximately how many meters per second can a person run?

14. What is the approximate cruising speed of a Boeing 747 in miles per hour and in kilometers per hour?

15. How might you estimate your speed if the speedometer in your car is broken?

16. How might you estimate the speed of a jogger?

17. Why is it not correct to say that time is more important than distance in determining speed?

18. How does instantaneous speed differ from average speed?

19. A truck driver averages 92 kilometers per hour between 2 P.M. and 6 P.M. Can you determine the speed of the bus at 4 P.M.?

20. An ancient marathoner covered the first 20 miles of the race in 4 hours. Can you determine how fast he was running when he passed the 10-mile marker?

21. What are the units of the physical properties measured by a stopwatch, an odometer, and a speedometer?

22. Which of the following can be used to measure an average speed: stopwatch, odometer, or speedometer? An instantaneous speed?

*Most of the questions (and exercises) have been paired; most odd-numbered questions are followed by an even-numbered question that is very similar. Short answers to most odd-numbered questions (and exercises) are provided at the back of the book. The more challenging questions are indicated by an asterisk.

23. What is the essential difference between speed and velocity?

24. If you are told that a car is traveling 65 mph east, are you being given the car's speed or its velocity?

25. Which way do you accelerate when you apply the brakes on your bicycle?

26. What is the direction of the acceleration of a car that is traveling east and increasing its speed?

27. In what sense can the brakes on your bicycle be considered an "accelerator"?

28. Which of the following (if any) could not be considered an "accelerator" in an automobile—gas pedal, brake pedal, or steering wheel?

29. In which of the two strobe photographs is the acceleration of the puck constant? Explain your reasoning. Assume that the puck moves from left to right without retracing its path.

(a)

(b)

30. Describe the motion of the ball shown in each of the strobe drawings. (Assume the ball moves from left to right.)

31. Assume that an airplane accelerates from 550 mph to 555 mph, a car accelerates from 60 mph to 65 mph, and a bicycle accelerates from 0 to 5 mph. If all three vehicles accomplish these changes in the same length of time, which one (if any) has the largest acceleration?

32. If a runner accelerates from 0 to 10 mph in 2 seconds and a cheetah accelerates from 0 to 45 mph also in 2 seconds, which one has the larger acceleration?

33. A race car accelerates from 120 mph to 128 mph in 2 seconds. A runner accelerates from 0 to 10 mph in 2 seconds. Which has the larger acceleration?

34. Which is larger, an acceleration from 55 mph to 60 mph or one from 60 mph to 64 mph if both take place over the same period of time?

35. When we say that light objects and heavy objects fall at the same rate, what assumption(s) are we making?

36. Free fall near the surface of the Moon can be described as motion with a constant _____.

37. What happens to the acceleration of a ball in free fall if its mass is doubled?

38. Two balls have the same size but are made from different materials: one from rubber and the other from lead. How do their accelerations compare if they are dropped?

39. How did the ideas of Galileo and Aristotle concerning the motion of a freely falling object differ?

40. A sheet of paper and a book fall at different rates unless the paper is wadded up into a ball, as shown in the figure. How would Galileo and Aristotle account for this?

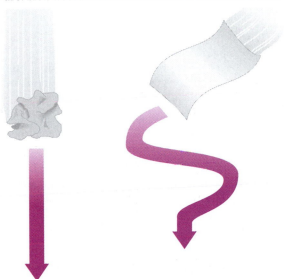

41. A student decides to test Aristotle's and Galileo's ideas about free fall by simultaneously dropping a 20-pound ball and a 1-pound ball from the top of a grain elevator. The two balls have the same size and shape. What actually happens?

42. A ping-pong ball and a golf ball have approximately the same size but very different masses. Which hits the ground first if you drop them simultaneously from a tall building and do not neglect air resistance?

43. The Moon is a nice place to study free fall because it has no atmosphere. If an astronaut on the Moon simultaneously drops a hammer and a feather from the same height, which one hits the ground first?

44. A penny and a feather are placed inside a long cylinder and the air is pumped out. The cylinder is inverted. Which hits the bottom first—the penny or the feather?

45. How would you describe the motion of a cylinder rolling down a straight ramp?

46. How would you describe the motion of a ball rolling up a straight ramp?

*47. What assumption(s) did Galileo make in the experiments that led him to conclude that the distance fallen from rest is proportional to the square of the time?

*48. How did Galileo conclude that the acceleration during free fall is constant?

49. How (if at all) does the acceleration of a ball thrown upward differ from one that is dropped?

50. How (if at all) does the acceleration of a cylinder rolling up a ramp differ from one that is rolling down the ramp?

*51. If we ignore air resistance, acceleration during free fall is constant. How do you suppose the acceleration would change if we do not ignore air resistance? Explain your reasoning.

52. If we neglect air resistance, during which of the first 5 seconds of free fall does a ball's speed change the most?

EXERCISES

1. How long is a 100-m dash in yards if 1 m = 1.094 yd?

2. If 1 in. = 2.54 cm, how tall (in centimeters) is a 6-ft model?

3. To be eligible to enter the Boston Marathon, a race that covers a distance of 26.2 miles, a runner must be able to finish in less than 3 h. What minimum average speed must be maintained to accomplish this?

4. A trucker drives a distance of 400 km (250 miles) between noon and 5 P.M. What was the average speed of the truck during this time? Can you figure out the speed of the truck at 2 P.M.?

5. In 1989, Ann Transom broke the U.S. women's record for a 24-h run by covering a distance of 143 miles. What was her average speed?

6. The 10,000-m run world record is 26 min 43.63 s. What was the runner's average speed in m/s?

7. How far can a van travel in 12 h at an average speed of 70 mph?

8. If an airplane averages 550 mph, how far can it fly in 6 h?

9. How far would the Space Shuttle travel during a 12-day mission at an average speed of 17,500 mph?

10. If you can walk at an average speed of 1.4 m/s, how far can you walk in an hour?

11. If a cheetah runs at 25 m/s, how long will it take a cheetah to run a 100-m dash? How does this compare with human times?

12. How many hours would be required to make a 5000-km trip across the United States if you average 100 km/h?

13. At an average speed of 150 mph, how long would it take a race car to complete a 500-mile race?

14. If a runner can average 4 mph, can he complete a 100-mile super marathon in less than 24 h?

15. A Volkswagen is stopped at a traffic light. When the light turns green, the driver accelerates so that the car is going 10 m/s after 5 s. What is the car's average acceleration?

16. What average acceleration does it take to stop a car traveling at 60 mph in 8 s?

17. If a Cessna 172 requires 20 s to reach its liftoff speed of 120 km/h, what is its average acceleration?

18. A car speeds up from 50 mph to 70 mph to pass a truck. If this requires 6 s, what is the average acceleration of the car?

19. A rock climber accidentally drops a piton. If it is traveling 7 m/s as it passes you, how fast will it be going 1 s later? Two seconds later?

20. A window washer drops a bucket of water. If it hits the ground traveling at 45 m/s, how fast was it moving 1 s earlier? Two seconds earlier?

21. You accidentally drop a watch from the roof of a 14-story building. Later while sweeping up the pieces on the ground, you establish that the watch stopped working 3 s after you dropped it. How tall is the building?

22. A rock is dropped into an abandoned mine and a splash is heard 4 s later. Assuming that it takes a negligible time for the sound to travel up the mine shaft, determine the depth of the shaft and how fast the rock was falling when it hit the water.

23. A ball is dropped from a height of 80 m. Construct a table showing the height of the ball and its speed at the end of each second until the ball hits the ground.

24. A ball is fired vertically upward at a speed of 30 m/s. Construct a table showing the height of the ball and its velocity at the end of each second until the ball hits the ground.

*25. A dummy is fired vertically upward from a cannon with a speed of 40 m/s. How long is the dummy in the air? What is its maximum height?

*26. You decide to launch a ball vertically so a friend located 45 m above you can catch it. What is the minimum launch speed you can use? How long after the ball is launched will your friend catch it?

2

An object moving through a medium must contend with resistance to its motion. For instance, bike racers streamline their bikes and clothing to minimize the air resistance. The effects of air resistance are very noticeable when skydivers open their parachutes. The acceleration is not constant; in fact, at some point the acceleration becomes zero. What happens to the speed of the falling object at this time? (See p. 47 for the answer to this question.)

Explaining Motion

The descent of the U.S. Army Parachute Team, the "Golden Knights," is slowed by air resistance.

What is meant by explaining motion? Don't motions just happen? These questions can probably be debated by philosophers for hours. In the physics world view, to explain something means to create a scheme, or model, that can predict the outcome of experiments. These experiments don't have to be elaborate; they can be as simple as throwing a baseball or looking at a rainbow.

Physicists try to create a set of ideas that explains how the world might work. Notice the word *might;* we have no proof that the ideas are correct or unique. There may be equally good (or better) schemes yet to be discovered.

Is it reasonable to expect that rules of nature exist? Motions appear to be reproducible. That is, if we start out with the same conditions and do the same thing to an object, we get the same resulting motion. The same motion occurs regardless of when the experiment is done—the results on Monday match those on Tuesday—and regardless of whether the experiment is done in San Francisco or in Philadelphia. (Imagine the consequences if this were not true and the "experiment" were throwing a baseball. Baseball teams would need a different pitcher for every day of the week and for every ballpark!) This reproducibility is a necessary condition for even attempting to search for a set of rules that nature obeys. Einstein reflected this idea when he commented that he believed God was cunning but not malicious. His point was that although the rules of nature might be difficult to find, as we search for them we realize that they do not change.

An Early Explanation

Aristotle developed an explanation of motion that lasted for nearly 2000 years. Many of Aristotle's ideas make common sense—they were based on our most common experiences.

Aristotle believed that the world was composed of four elements: earth, water, air, and fire. These four were the building blocks of the material world. Each substance was a particular combination of these four elements. If this seems naive to you, consider our modern world view. We take chemical elements, each with its own special attributes, and combine them to form compounds that have quite different attributes. For example, we take hydrogen, a very explosive gas, and oxygen, the element required for combustion, and combine them to form water that we use to fight fires!

Each Aristotelian element had its own natural place in the hierarchy of the Universe. Earth, the heaviest, belonged to the lowest position. Water was next, then air and fire. Aristotle reasoned that if any of these were out of its hierarchical position, its natural motion would be to return. These natural motions occurred in straight lines, toward or away from the center of the Earth.

It is interesting to note that if you try to test this part of Aristotle's world view, it works! Put some water and earth (dirt) in a glass and wait. Watch as each element settles into its natural place (Fig. 2-1).

For other than natural motions, Aristotle would probably challenge you to think of your own experiences. To move something you have to

(a)

(b)

Figure 2-1   In the Aristotelian world view water rises while earth falls.

make an effort. Even though we have developed machines to make the effort for us, we still agree that an effort must be made and that after the effort is stopped, the object comes to rest.

Although this seems reasonable, there are problems with the Aristotelian explanation. For example, objects don't stop immediately. An arrow continues to fly even after is loses contact with the bow string. The Aristotelian explanation invoked an interaction between the arrow and the air. As the arrow moves through the air it creates a partial vacuum behind it. The air, rushing in behind the arrow to fill the void, pushes on the arrow and causes the continued motion. This explanation, however, predicts that motion without an effort is impossible in a vacuum, while seeming to imply that in air it is perpetual.

The Beginnings of Our Modern Explanation

What is the motion of an object when there is nothing external trying to change its motion? One might guess, in agreement with Aristotle, that the only natural motion of an object is to return to the Earth; otherwise it has no motion—it remains at rest. Medieval thinkers agreed that objects have this tendency not to move.

Let's do a simple experiment. Give this book a brief push across a table or desk. Although the book starts in a straight line at some particular speed, it quickly slows down and stops. It certainly seems clear that it is natural for an object to come to rest and remain at rest.

Remaining at rest is a natural state. However, there is another state of motion which is just as natural, but not nearly as obvious. Suppose you were to repeat this book-pushing experiment on a surface covered with ice. The book would travel a much greater distance before coming to rest. Our explanation of the difference in these two results is that the ice is slicker than the desktop. Different surfaces interact with the book with different strengths. The book's interaction with the ice is less than that with the wood. What would you predict would happen to the book if the surface were perfectly slick? The book would not slow down at all; it would continue in a straight line at a constant speed forever. Stated differently, when the interaction is reduced to zero, the book's motion is constant.

Thus, it seems that a natural motion is one in which the speed and direction are constant. Note that this statement covers the object both "at rest" and "moving." An interaction with an external agent is required to cause an object to change its velocity. If left alone, it would naturally continue in its initial direction with its initial speed.

Galileo reached this same conclusion by deducing the outcome of a thought experiment. He thought about the motion of a perfectly round ball placed on a tilted surface free of "all external and accidental obstacles." He noted—presumably from the same experiences we have all had—that a ball rolling down a slope speeds up [Fig. 2-2(a)]. Conversely, if the ball were rolling up the slope, it would naturally slow down. The ball experiences an interaction on the falling slope that speeds it up and an interaction on the rising slope that slows it down.

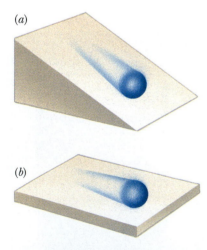

(*a*)

(*b*)

Figure 2-2 Galileo's thought experiment. (a) On a tilted surface a ball's speed changes. (b) On a level surface a ball's speed and direction are constant.

Now, Galileo asked himself what would happen to the same ball if it were placed on a level surface. Nothing. Since the surface does not slope, the ball would neither speed up nor slow down [Fig. 2-2(b)]. Galileo's genius was in realizing that the ball's lack of acceleration would be true regardless of whether the ball was at rest or moving. The ball would continue its motion forever.

It is important to remember that this is another of Galileo's thought experiments and not an account of an actual experiment. He assumed that there were no resistive interactions between the ball and the surface. There was no friction. By doing this, he was able to strip motion of its earthly aspects and focus on its essential features.

Galileo was the first to suggest that constant-speed, straight-line motion was just as natural as at-rest motion. This property of remaining at rest or continuing to move in a straight line at a constant speed is known as **inertia.**

The common use of the word *inertia* usually refers to an emotional state, one of feeling sluggish. Often when people say something has a lot of inertia, they are referring to their difficulty in getting it moving. (Sometimes they are talking about themselves.) As you build your physics world view, it is important to distinguish between the everyday uses of words and the usage within physics.

> **QUESTION**   What is the main difference between the everyday usage of the word *inertia* and its use in physics?

You already have a vast set of experiences that are directly related to inertia as it is used here. If something is at rest, it takes an interaction of some kind to get it moving. A magician uses the inertial property of cups and saucers when abruptly pulling the tablecloth from beneath them. If the tablecloth is smooth enough, the interaction with the cloth will be small, and the dishes will remain (nearly) at rest. Figure 2-3 shows the before and after photographs of this trick.

> **Physics on Your Own**   Impress your friends by pulling a tablecloth from under some dishes. To increase your chances of avoiding a disaster, you should do the following: Use a smooth, hemless cloth about the size of a pillowcase. Don't be timid; pull the cloth quickly in a downward direction across the straight edge of the table. Choose some dishes that are stable (and cheap!).

Another example of how inertia can be fascinating is the circus strongman who boasts of his strength by asking someone to hit him on the head with a sledgehammer. The strongman always does this demonstration holding a big, heavy block on his head. The inertia of the block is

> **ANSWER**   The physics usage also includes the idea that objects tend to keep moving.

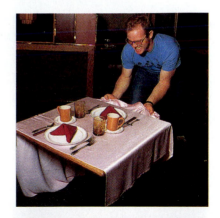

Figure 2-3   The tablecloth can be pulled from under the dishes because of their inertia.

large enough that the blow from the sledgehammer does little to move it. The happy strongman only has to be strong enough to hold up the block!

> **Physics on Your Own**   Suspend an object by a thin thread, as shown in Figure 2-4. Exert a steady pull on the string tied to the bottom of this object until one string breaks. (Watch that the falling object doesn't hit your hands or feet.) Which string breaks? Now exert a very abrupt pull on the lower string. Which string breaks? How do you explain your observations?

Figure 2-4   The puzzle of identical strings: Which string breaks?

But there is more to inertia than getting things moving. If something is already moving, it is difficult to slow it down or speed it up. For example, remember drying your wet hands by shaking them. When you stop your hands abruptly, the water continues to move and leaves your hands. In a similar way, seat belts counteract your body's inertial tendency to continue forward at a constant speed when the car suddenly stops.

However, all objects do not have the same inertia. For example, imagine trying to stop a baseball and a cannonball, each of which is moving at 150 kilometers per hour (about the speed a major league pitcher throws a baseball). The cannonball has more inertia and, as you can guess, requires a much larger effort to stop it. Conversely, if you are the pitcher trying to throw them, you would find it much harder to get the cannonball moving.

Although Galileo did not fully explain motion, he did take the first important step and, by doing so, radically changed the way we view the motion of objects. His work profoundly influenced Isaac Newton, the originator of our present-day rules of motion.

Newton's First Law of Motion

Isaac Newton, an Englishman, was born a few months after Galileo's death. Although he is probably best known for his work on gravitation, his most profound contribution to our modern world view is his three laws of motion. Like Galileo, Newton was interested in the interactions that occur while an object is in motion rather than in its final destination. He formulated Galileo's observations into what is now called **Newton's first law of motion.** It is also referred to as the **law of inertia.**

Newton's first law

> Every object remains at rest or in motion in a straight line at constant speed unless acted on by an unbalanced force.

This law incorporates Galileo's idea of inertia and introduces a new concept, **force.** In the Newtonian world view, the book sliding across the table slows down and stops because there is a force (called friction) that opposes the motion. Similarly, a falling rock speeds up because there is a force (called gravity) continually changing its speed. In short, there is no acceleration unless there is a net, or unbalanced, force.

Diversified Brilliance

Isaac Newton

On December 25, 1642, Isaac Newton was born in Woolsthorpe in Lincolnshire, England. Newton's father—a country gentleman in Lincolnshire—had died three months before he was born, leaving him to be reared by his mother. When he was three, his mother married a local pastor and moved a few miles away, leaving young Newton in the care of the housekeeper.

At the age of 14, Newton returned home from school at the request of his mother to help work on the family farm. He proved to be not much of a farmer, however, and spent most of his time reading. Times when he could be alone, Newton would amuse himself by building model windmills powered by mice, water-clocks, sundials, and kites carrying fiery lanterns, which frightened the country folk. A local schoolmaster recognized Newton's abilities and helped him enter Trinity College at Cambridge at the age of 18, where he would receive his bachelor of arts degree four years later, in 1665.

Later that same year, the bubonic plague raged through the English countryside and, consequently, the university was closed. Newton returned to Woolsthorpe, and the next 18 months proved to be his most productive. It was during this interlude that Newton developed his theories and ideas about optics, the laws of motion, celestial mechanics, calculus, and his famous law of gravity. After the Great Plague, Newton returned to Cambridge, where he was appointed professor of mathematics at the age of 26. From here, Newton went on to develop a reflecting telescope, one that utilized a mirror to collect light instead of a lens, which was used in earlier models. Newton also published his most notable book—with the help of Edmund Halley—*Principia Mathematica Philosophiae Naturalis* (*Mathematical Principles of Natural Philosophy*) in 1687.

In 1701 Newton was appointed master of the mint, and in 1703 he was elected president of the Royal Society, a position he retained until his death. Isaac Newton's honors did not end there, however; in 1705 Queen Anne knighted him in recognition of his many accomplishments, forever changing his name to Sir Isaac Newton. This was the first knighthood given for scientific achievement.

Sir Isaac Newton's life was not all discoveries and honors. In fact, he spent a great deal of his later life quarreling with fellow scientists. Robert Hooke accused Newton of stealing some of his ideas about gravity and light. Newton also fought bitterly with Gottfried Leibniz, a German mathematician who claimed to have developed calculus first, and Christian Huygens, who worked independently on the wave theory of light. In 1727 Sir Isaac Newton became seriously ill, and on the 20th of March one of the greatest physicists of all time died.

Adapted from Serway R.A. and Faughn J.S. *College Physics.* Philadelphia: Saunders College Publishing, 1992.

Additional references: AIP Niels Bohr Library; Manuel F.E. *A Portrait of Isaac Newton.* Cambridge: Harvard University Press, 1968.

All of us have an intuitive understanding of forces; casually speaking, a force is a push or a pull. But it should be noted that the concept of force is a human construct. Because we have grown up with forces as a part of our personal world view, most of us feel quite comfortable with them. But we don't actually see forces. We see objects behave in a certain way, and we infer that a force is present. In fact, alternative world views have been developed that do not include the concept of force. This concept, however, has greatly aided the process of building a physics world view.

Although the concept of force includes much more than our intuitive ideas of push and pull, we use these thoughts as a beginning. A force can be defined in terms of the observed behavior of objects. For example, a

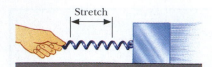

Figure 2-5 The stretch of the spring is a measure of the applied force.

"force-measurer" (Fig. 2-5) constructed with rubber bands or springs would allow us to quantify our observations of force by measuring the amount of stretch using some arbitrary scale. We will have more to say about these devices after we learn about Newton's second law.

Another important characteristic of forces is that they are directional, meaning that the direction of the force is as important as its size. Different results are produced by forces of the same size when applied in different directions. Imagine a skater coasting across the ice. A force in the direction of the original motion increases the skater's speed. A force of the same size applied in the opposite direction slows the skater down. So we need to incorporate this difference into our understanding of motion. As you might guess from the discussion in Chapter 1, we will do this by treating forces as vectors.

Remember that Newton's first law refers to the *unbalanced* force. In many situations there is more than one force on an object. There is only an unbalanced force if the vector sum of the forces is not zero. When two forces of equal size act along a straight line but in opposite directions, they cancel each other. In this case, the forces tend to stretch or compress the object, but the unbalanced, or *net*, force is zero. The "helpers" in Figure 2-6 could each be exerting a very large force, but if the forces are equal in size and opposite in direction, there is no unbalanced force on the cart.

The converse of this also holds. When we observe an object with no acceleration, we infer that there is no unbalanced force on that object. If you see a car moving at a constant speed on a level, straight highway, you infer that the frictional forces balance the driving forces. This is not to say that there are no forces acting on the car, because there are many. The crucial point is that the vector sum of all of these forces is zero; there is no unbalanced, or net, force acting on the car.

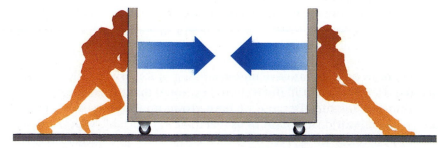

Figure 2-6 Equal forces acting in opposite directions cancel and the cart does not accelerate.

ANSWER Since the speed and direction are constant, there is no acceleration and the net force must be zero.

The Net Force

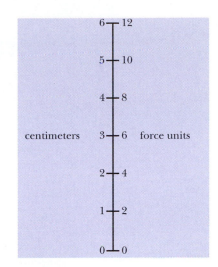

Figure 2-7   A scale for force vectors. One centimeter represents two units of force.

When we have more than one force acting on an object, we need to determine their combined, or net, effect. To find out if there is a net force, we must be able to combine, or add, forces. But adding forces is a little tricky because of their directional property. If we combine a force of 3 units with a force of 4 units and they act along a line, the net force is either 7 (if they're both pointing in the same direction) or 1 (if they're pointing in opposite directions). If these forces are not parallel to each other, but at some angle, the net force can have any value from 1 to 7.

This property may not be a conscious part of your world view, but you have certainly had experiences with the directional aspects of force. If somebody pushes you in one direction while somebody else pushes you in another direction, you intuitively know in which direction you are going to fall. You don't need an elaborate scheme to determine this. In physics, however, we must go beyond these qualitative experiences and develop rules for combining forces.

Mathematical rules have been developed for combining quantities that have directional properties. Quantities that obey these rules are called vectors and can be used to represent forces. We can represent any vector by an arrow; its length represents the size of the quantity and its direction represents the direction of the quantity. To complete this representation, we assign a convenient scale to our drawing. For instance, we let 1 centimeter (cm) in the drawing in Figure 2-7 represent 2 units of force. Then an arrow that is 4 centimeters long represents a force of 8 units acting in the direction the arrow points.

> **QUESTION**   How many units of force would a 10-centimeter arrow represent?

In texts, vector quantities are represented by boldface symbols **F** and in handwritten materials with an arrow over the symbol $\vec{F}$. The size, or magnitude, of the vector quantity is represented by an italic symbol. Therefore, the force is written as **F**, and the magnitude of the force is written as *F*. In all of our drawings, force vectors will be shown in blue; other vectors will be shown in other colors.

We can combine vectors using a graphic method and the scale shown in Figure 2-7. Consider the two forces illustrated in Figure 2-8(a). What is the net force produced by the two forces? The answer is obtained by placing the tail of one of the vectors on the head of the other, being careful not to change the direction or size of either vector. The sum of these two vectors is the vector that reaches from the tail of the first to the head of the second, as illustrated in Figure 2-8(b). This is the net force. To obtain

> **ANSWER**   Since each centimeter represents 2 units of force, a 10-centimeter arrow would represent 20 units of force.

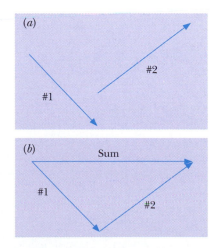

Figure 2-8   To add the two vectors in (a), place the tail of one on the head of the other. (b) The result is the vector from the tail of the first to the head of the second.

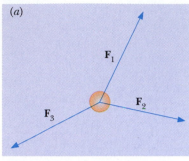

(a)

$\mathbf{F}_1$

$\mathbf{F}_2$

$\mathbf{F}_3$

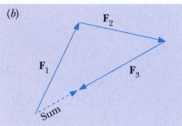

(b)

$\mathbf{F}_2$

$\mathbf{F}_1$

$\mathbf{F}_3$

Sum

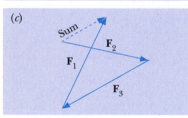

(c)

Sum

$\mathbf{F}_2$

$\mathbf{F}_1$

$\mathbf{F}_3$

Figure 2-9   The three forces acting on the ball (a) can be added to find the net force (b). The order in which the forces are added (c) does not matter.

the size of this net force, we measure the length of the arrow with a ruler and use the scale of our drawing to convert this length to force units. In the particular arrangement illustrated, the arrow is 4 centimeters long and therefore represents 8 units of force.

Consider the three forces acting on the ball shown in Figure 2-9(a). Each force is 4.6 units long and is represented by an arrow 2.3 centimeters in length. To add these forces we move $\mathbf{F}_2$ (without changing its orientation) and place its tail on the head of $\mathbf{F}_1$, as shown in Figure 2-9(b). We then place the tail of $\mathbf{F}_3$ on the head of $\mathbf{F}_2$. The sum of these three forces is the arrow from the tail of $\mathbf{F}_1$ to the head of $\mathbf{F}_3$. This arrow measures 1.25 centimeters long and, therefore, represents a force of 2.5 units acting toward the upper right.

Whenever a new method is adopted, it is a good idea to find a situation in which either the new or old one can be used to obtain an answer. Although such a check doesn't guarantee the validity of the new method, it is a necessary condition. If the vectors are parallel, the new, graphic method must be equivalent to our normal rules of arithmetic.

> **QUESTION**   What is the net force produced by a force of 8 units acting to the left and 3 units acting to the right?

Let's check this by adding two parallel forces, where one is acting to the right with a force of 4 units and the other is acting to the left with a force of 3 units, as shown in Figure 2-10(a). Using our graphic method, we again combine the two by taking the second vector and placing its tail on the head of the first. The sum is a vector that points to the right and is 1 unit long, as in Figure 2-10(b). Combining them with our more familiar rules, we assign a positive value to one direction (say, the right) and a negative value to the opposite direction. Adding the signed numbers, +4 and −3, yields a force of +1; that is, one unit of force acting to the right, in agreement with our graphic method.

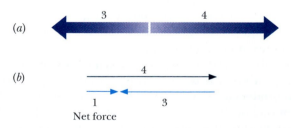

(a)

3   4

(b)

4

1   3

Net force

Figure 2-10   (a) The two opposing forces add to give a net force of one unit to the right (b).

> **ANSWER**   If we assign negative values to the left, we have −8 + 3 = −5, or the net force is 5 units acting to the left.

Physics Update

Atomic force microscopes (AFMs) can directly measure forces at the submicroscopic level. For instance, scientists at the Naval Research Lab (NRL) have measured the force between two complementary strands of DNA. The NRL researchers now hope to use a device based on AFM technology to detect biomolecules. They have developed a "force amplified biological sensor," which will soon be capable of detecting minute amounts of various biological species, such as cells, proteins, viruses, and bacteria. Employed in a working device, an array of such biosensors would be able to perform immunoassays (the process by which the presence of antigens is detected) in about 10 minutes, much faster than other methods at these small concentrations.

The Second Law

Newton's first law tells what happens when there is no net force acting on an object; the speed and direction don't change. If there is a net force, the object accelerates and thus changes its motion. Newton's second law describes the relationship between a net force and the resulting acceleration.

Our development of the second law will discuss some simple experiments that illustrate this relationship before we state it formally. Assume that we have a collection of identical springs (the force measurers mentioned earlier), a collection of objects, and all the necessary equipment to measure accelerations. Further assume that if we stretch a spring by a fixed amount and maintain this stretch, the spring exerts a constant force.

Since the second law describes the net force, we need a situation in which the frictional forces are so small that they can be disregarded and any other forces are balanced so that the forces that we apply are the only ones affecting the acceleration. A horizontal air-hockey table is a good experimental surface. The hockey puck rides on a cushion of air, so it experiences very small frictional forces.

If we pull a hockey puck with a spring stretched by a certain amount and maintain the direction and amount of force even while the puck moves, we find that the puck experiences a constant acceleration in the direction of the force. After doing this many times with differing amounts of stretch, we conclude that a constant net force produces a constant acceleration. Furthermore, the direction of the acceleration is always in the same direction as the force.

Let's now compare the results we obtained when pulling the puck with one spring with what happens when we pull the puck with two springs. When two springs are pulling "side by side," as shown in Figure 2-11, the force is twice as large as that of the single spring. If we stretch each of the two springs by the same amount as before, we find that the two springs produce twice the acceleration. If we use three springs, they produce three times the acceleration, and so on. In general, we find that

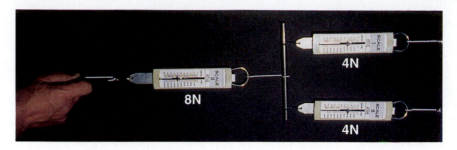

Figure 2-11 Two springs pulling "side by side" exert twice the force of one spring, as indicated by the scale readings.

the acceleration of an object is proportional to the net force acting on it. This relationship will be part of the second law.

But this is not the entire story. Imagine pushing on a cannonball with the same force that is used on the hockey puck. Intuition tells you that the acceleration of the cannonball will be smaller. If pressed for a reason, you might respond, "Because there is more 'stuff' in a cannonball." Or, you might say, "It weighs more." We will soon see that although the term *weight* is not technically correct, these intuitive feelings lead in the correct direction.

We build on these feelings by investigating how inertial behavior depends on the amount of matter in the object. The **mass** of an object is a measure of the amount of matter in the object. We assume that masses combine in the simplest possible way: The masses add. Therefore, the combined mass of two identical objects is twice the mass of one of them.

Again, we use a spring to pull on one of the hockey pucks and record its acceleration. We then look at the acceleration of two pucks (somehow tied together) pulled by a single spring. If the spring is stretched by the same amount as before, the acceleration is one-half the original. Likewise, one spring pulling on three pucks yields one-third the acceleration, and so on. This relationship is called an **inverse proportionality,** where the word *inverse* indicates that the changes in the two values are opposite each other. If the mass is *increased* by a certain multiple, the acceleration produced by the force is *reduced* by the same multiple.

Notice that the more mass an object has, the more force it takes to produce a given acceleration. This means that the object has more inertia. We therefore take an object's mass to be the measure of the amount of inertia the object possesses.

Newton put the two preceding ideas together into one of the most important physical rules of nature ever proposed. This rule states that the acceleration of an object is equal to the net force on the object divided by its inertial mass and can be written symbolically as

$$\text{acceleration} = \frac{\text{net force}}{\text{mass}}$$

$$\mathbf{a} = \frac{\mathbf{F}_{net}}{m}$$

where we have written the acceleration and the net force as vectors to emphasize that they always point in the same direction.

Any such mathematical equation can be rearranged using algebra. **Newton's second law of motion** is more commonly written as

$$\mathbf{F}_{net} = m\mathbf{a}$$

net force = mass × acceleration

> The net force on an object is equal to its mass times its acceleration and points in the direction of the acceleration.

Newton's second law

The second law describes a specific relationship between three quantities: net force, mass, and acceleration. Although we have a prescription for determining the numerical value of an acceleration, we have not yet done this for the other two quantities. We have a choice to make. We can choose a standard spring stretched by a specific amount as our definition of 1 unit of force, or we can take a certain amount of matter and define it as 1 unit of mass, or we could even choose the two units independently.

Historically, a certain amount of matter was chosen as a mass standard. It was assigned the value of 1 **kilogram** (abbreviated kg). It is roughly the mass of this book. The value of the unit force is then defined in terms of the observed acceleration of this standard mass. The force needed to accelerate a 1-kilogram mass at 1 (meter per second) per second is called 1 **newton** (N), in honor of Isaac Newton. The gravitational force on a very small apple is about 1 newton.

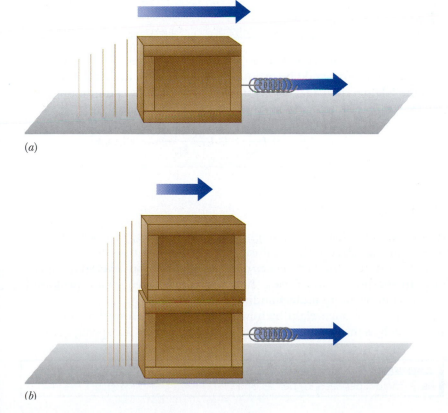

(a)

(b)

The acceleration of two masses pulled by identical springs is one-half as large as for a single mass.

In the United States, a commonly used unit of force is the *pound* (lb). The unit of mass, a *slug*, is used so seldom that you may never have heard of it. One pound is the force required to accelerate a mass of 1 slug at 1 (foot per second) per second.

Once again we have found a pattern in nature. We can use Newton's second law to predict the motion of objects before we actually do the experiment. For example, suppose that we ask what net force is needed to accelerate a 5-kilogram object at 3 (meters per second) per second. Applying the second law, we have

$$F_{\text{net}} = ma = (5 \text{ kg})(3 \text{ m/s}^2) = 15 \text{ kg·m/s}^2 = 15 \text{ N}$$

In using any rule of nature, we must use a consistent set of units. The units are an integral part of the rules of nature. In the preceding case, when accelerations are measured in (meters per second) per second, the masses must be in kilograms, and the forces in newtons. The combination kg · m/s² is equal to a newton.

> **QUESTION**   Suppose that in this situation you discovered that there was a 5-newton force of friction opposing the motion. What acceleration would you get if you applied the same 15-newton force?

We can use the second law to ask other questions. For example, what acceleration would be produced by a 2-newton net force acting on the 5-kilogram object? Rearranging the second law and putting in the values of mass and force, we get

$$a = \frac{F_{\text{net}}}{m} = \frac{2 \text{ N}}{5 \text{ kg}} = 0.4 \, \frac{\text{N}}{\text{kg}} = 0.4 \, \frac{\text{kg·m/s}^2}{\text{kg}} = 0.4 \text{ m/s}^2$$

Mass, Force, and Weight

Mass is often confused with weight, which in turn is confused with the force of gravity. Part of the confusion lies in the fact that they are all related. In addition, the differences don't come up in our everyday experience. In the physics world view, however, the differences are profound and thus important to understanding motion.

We measure our **weight** by how much we can compress a calibrated spring, such as that in a bathroom scale. We compress the spring because

> **ANSWER**   The net force would be 10 newtons, so the acceleration would be 2 (meters per second) per second.

the Earth is attracting us; we are being pulled downward. Our weight depends on the strength of this gravitational attraction. If we were on the Moon, our weight would be less because the Moon's gravitational force on us would be less.

Our mass, however, is *not* dependent on our location in the Universe. It is a constant property that depends only on how much there is of us. If we were far, far away from any planet or other celestial body, we would be weightless but not massless. And, because we are not massless, the force required to accelerate us is still given by Newton's second law.

The idea of weightlessness fascinates science fiction writers. Some of them, however, confuse the concepts of mass and weight. Contrary to some fictional accounts of weightlessness, the laws of motion still hold in these situations. For example, suppose that while far out in space, where all gravitational forces are negligible, you float across your spacecraft and collide with a wall. You won't just bounce off, feeling no pain. The wall provides a force to slow and reverse your motion. Newton's second law tells us that this force depends only on your mass and the acceleration you experience, both of which are the same as here on Earth. If the force can break bones on Earth, it can do the same in the spacecraft. Being weightless does not mean that you are massless. Similarly, imagine a huge trailer truck in outer space "hanging" from a spring scale. Although the scale would read zero, if you tried to kick the truck, you would find that it resisted moving.

The marketplace is another place where mass and weight are often confused. We talk about buying "a pound of butter." A pound is a unit of weight and is determined by how much the butter stretches a spring in the scales. It is a measure of the gravitational attraction and varies slightly from place to place. The shopper doesn't really care about the weight of the butter but is interested in purchasing a certain amount of butter; the important thing is its mass. Stores use spring scales by calibrating them with standard masses to compensate for the value of the local gravitational attraction.

This confusion between mass and weight will not be resolved by switching over to the metric system. We will probably still refer to the standard masses as "weights" and the process of determining the amount of butter as "weighing." What is really meant by saying "the butter weighs 1 kilogram" is that the amount of butter has a weight that is equal to the weight of 1 kilogram. Because that is quite a mouthful, people will probably refer to the weight of the butter as being 1 kilogram.

A 1-kilogram mass near the surface of the Earth has a weight of 9.8 newtons, or about 2.2 pounds. Therefore, a "pound" of butter has a weight a little less than 5 newtons and a mass a little less than ½ kilogram. When the United States is fully converted to the metric system, butter will most likely be purchased by the half kilogram, as is done presently in most of the world.

Although the distinctions between mass, weight, and the force of gravity are not important in the marketplace, the spacecraft example demonstrates that we have to be careful when discussing these concepts in physics.

Although floating astronauts are weightless, they still have masses and obey Newton's second law.

Shoppers use supermarket scales to determine the masses of the produce.

Weight

The force causing an object to accelerate toward the Earth's surface is just the force of gravity. We usually neglect the rotation of the Earth and commonly equate the weight **W** of an object with the force of gravity acting on it. In the idealized situation of no air resistance described in Chapter 1, we concluded that all objects near the Earth's surface fall at a constant acceleration. Let's represent the acceleration due to gravity by the symbol **g.** If we replace **F** by **W** and **a** by **g** in Newton's second law, we obtain

$$\mathbf{W} = m\mathbf{g}$$

weight = mass × acceleration due to gravity

This is just a mathematical way of saying that the weight of an object is proportional to its mass and directed downward.

COMPUTING WEIGHT Σ

As a numerical example, let's calculate the weight of a child with a mass of 25 kilograms.

$$W = mg = (25 \text{ kg})(10 \text{ m/s}^2) = 250 \text{ N}$$

Therefore, the child has a weight of 250 newtons (about 55 pounds).

> **QUESTION**   What is the weight of a wrestler who has a mass of 120 kilograms?

This process can be reversed to obtain the mass of a dog that has a weight of 150 newtons.

$$m = \frac{W}{g} = \frac{150 \text{ N}}{10 \text{ m/s}^2} = 15 \text{ kg}$$

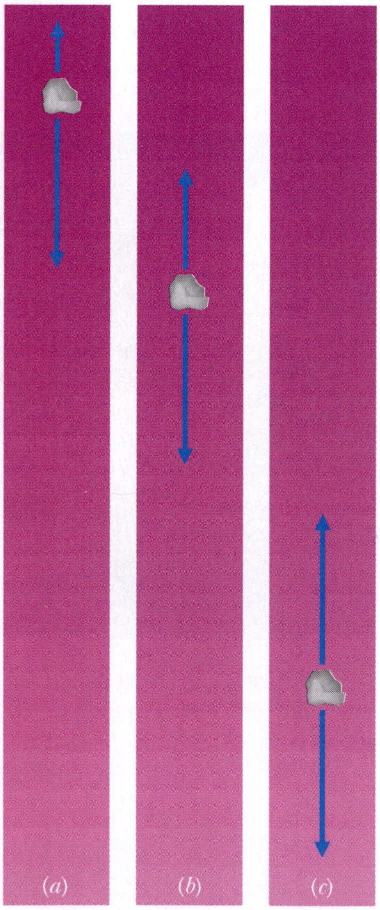

Free Fall Revisited

Objects falling on Earth don't fall in a vacuum but through air, which offers a resistive force to the motion. Thus, in realistic situations, a falling object has two forces acting on it simultaneously: the weight acting downward and the air resistance acting upward. Among other factors, the force due to air resistance depends on the speed of the object. The greater the speed, the greater the air resistance. You can experience this by sticking your hand out a car window as the car's speed increases.

With these facts in mind, consider the downward motion of a falling rock. Initially, it falls at a low speed and the air resistance is small. There is a net force downward which is equal to the weight of the rock minus the

Figure 2-12 As the speed of a falling rock increases, the force of the air resistance increases until it equals (c) the weight of the rock.

ANSWER   1200 newtons.

force of the air resistance [Fig. 2-12(a)]. Because there is a net force, the rock accelerates, thus increasing its speed. As the rock speeds up, however, its weight remains constant while the air resistance increases [Fig. 2-12(b)]. Thus, the net force and the acceleration decrease. The rock continues to speed up but at a decreasing rate. Eventually the rock reaches a speed for which the air resistance equals the weight [Fig. 2-12(c)]. There is no longer a net force acting on the rock and it stops accelerating—its speed remains constant. This maximum speed is called the **terminal speed** of the object.

The terminal speeds of different objects are not necessarily equal. Even if the shape and size of two objects are identical—thus having identical frictional forces—the objects have different terminal speeds if they have different masses. In all cases, though, the object continues to accelerate until the frictional force is the same size as its weight. The value of the terminal speed is determined by a combination of many factors: the size, shape, and weight of the object as well as the properties of the medium. A BB has a much larger terminal speed than a feather, primarily

Terminal Speeds

How fast something can move depends on the forces that retard its motion as well as those that propel it forward. The primary retarding forces are frictional forces, often due to the medium through which the object moves. A 100-meter race in 3 feet of water would produce times far slower than the current record of 9.84 seconds in air. But, even in air, there is resistance to motion. A clean, waxed car has a measurable increase in gas mileage over a dirty car.

When the retarding forces equal the propelling forces, there is no net force on the object and the object stops accelerating; it reaches a constant speed known as the terminal speed. Minimizing the retarding forces increases the terminal speed.

Streamlining an object minimizes its air resistance. In 1980, Steve McKinney set the world unpowered land-speed record by paying a great deal of attention to minimizing air resistance and friction. He skied down a 40° slope at slightly over 200 kilometers per hour (125 mph!). The current downhill skiing record is held by Jeffrey Hamilton at 241 kilometers per hour (150 mph). The woman's record is held by Karine Dubouchet at 225 kilometers per hour (104 mph). The Peregrine falcon, already streamlined, dives for prey at speeds up to 350 kilometers per hour.

The effect of minimizing air resistance was convincingly demonstrated by U.S. Air Force Captain

Downhill skiers gain speed by reducing their air resistance and the resistance of the skis with the snow.

Joseph Kittinger when he jumped from a balloon at 31,000 meters and attained a speed of more than 1000 kilometers per hour (625 mph) after falling approximately 4000 meters. At this speed he nearly broke the sound barrier! He was then slowed down as the atmosphere became more dense.

Source: Guinness Book of Records, Bantam Books, 1996.

because the feather's shape creates a resistive force that quickly becomes comparable to its weight.

Let's look again at the skydivers discussed in Chapter 1. Assuming no air resistance, we calculated that the skydivers would be falling at a speed of 1080 kilometers per hour (675 mph) after only 30 seconds. As skydivers know, the maximum speed they can obtain near sea level is a little over 300 kilometers per hour (190 mph). Obviously, air resistance is responsible for this difference. Skydivers also know that their terminal speed can be altered by changing their shape by falling "feet-first" or "spread-eagle"—the larger the surface of the object facing into the wind, the greater the air resistance.

Galileo versus Aristotle

Recall that in Chapter 1 we discussed the views of Aristotle and Galileo on the subject of falling objects, and clearly came down on the side of Galileo. Now it seems that we are agreeing with Aristotle! If a falling object reaches a terminal speed whose value is determined by the object's weight and its interaction with the medium, wasn't Aristotle correct? It might seem so.

The motion of a bowling ball dropped from a great height shows that each man correctly described a part of the ball's motion. Initially, before the air resistance becomes significant, the ball exhibits constant acceleration, as hypothesized by Galileo. As the air resistance grows, the acceleration is not constant, but decreases to zero. From that point on, the object travels at a constant speed, as described by Aristotle.

Each person described different extremes of the motion: Galileo, the extreme of negligible air resistance; Aristotle, the extreme of maximum air resistance. One might, naively, suggest that we determine how much of the motion is accelerated and how much is at a constant speed. We could then award a physics prize to the person whose explanation holds for the longest time. Aristotle would win; we only need to drop the object from higher and higher positions, making the constant-speed portion of the fall as large as desired.

But building a physics world view doesn't always progress by choosing on such a basis. Galileo has fared better in the eyes of science historians because his idealization stripped away the nonessentials of falling motion and thus uncovered the more fundamental behavior of motion. Galileo therefore paved the way for Newton's work, which explains the entire motion of a freely falling object (including Aristotle's observations). As long as all the forces acting on an object are known, the resulting acceleration can be calculated.

Friction

Newton's insight can be turned around; rather than predicting the motion from the forces, we can use an object's motion to tell us something about the forces acting on it. Imagine pushing horizontally on a large wooden crate [Fig. 2-13(a)]. Suppose that at first you don't push hard enough to move the crate. If it doesn't move, there is no acceleration and, according to Newton's second law, there can be no net force on the crate. This means that there must be at least one other force canceling out the push. This other force is the force of friction exerted on the box by the floor. As long as the crate does not move, the frictional force must be equal in size and opposite in direction to your applied force. This force is called **static friction** to distinguish it from the frictional force that occurs when the crate moves.

This static frictional force seems a bit mysterious. Because it is equal to the force you exert, the frictional force is small if you push with a small force. But if you push with a large force, the frictional force is large [Fig. 2-13(b)]. It is a force that opposes the applied force and ceases to exist when the applied force is removed. The static frictional force can have any value from zero up to a maximum value determined by the surfaces and the weight of the crate.

If your applied force exceeds the maximum static frictional force, the crate accelerates in the direction of your applied force. Although the crate is now sliding, there is still a frictional force [Fig. 2-13(c)]. The value of this **sliding friction** is less than the maximum value of the frictional force when the crate is at rest. Unlike air resistance, *sliding friction has a constant value, independent of the speed of the object.*

Physics on Your Own   With a simple wooden block and a long rubber band, you can verify the behavior of static and sliding friction. Connect the rubber band to the block with a thumb tack, and slowly pull on the block. The stretch of the band provides a visual indication of the force you are applying. If the block does not move, the static force is equal but opposite in direction to the force of the rubber band. Continue to increase your pull. What happens?

Repeat the experiment, with the block sliding across the table at a constant speed. How does the stretch of the rubber band now compare with its stretch in the static situation?

It is important to understand the difference between static and sliding friction when making an emergency stop in an automobile. Since you want to stop the car as quickly as possible, you want to have the maximum frictional force with the road. This occurs when the tires are rolling because the surface of the tire is not sliding along the surface of the road and it is the larger static friction that is important. Therefore, you should not brake so hard as to make the tires skid. The same thing occurs when a car takes a corner too fast. Once the tires start to skid, the frictional force is reduced, making it difficult to recover from the skid. If you've ever been in one of these unfortunate situations, you may recall how fast the car slides once it starts to skid.

A significant recent advance in the automotive industry is based on the fact that static friction is greater than sliding friction. Antilock brakes are a computer-controlled braking system that keeps the wheels from skidding, and thus maximizes the frictional forces. Sensors monitor how fast the wheels are rotating and continuously feed the data to an onboard

(*a*)

(*b*)

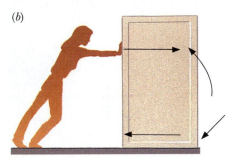

(*c*)

Figure 2-13   The static frictional force is equal and opposite to the applied force if the box does not accelerate. The applied force can be small (a) or large (b) as long as it doesn't cause the box to move. (c) The sliding frictional force has a constant value independent of the speed.

computer. The computer controls the braking by repeatedly applying and releasing pressure to the brake pads. Without antilock braking, drivers who jammed on the brakes hoping to avoid danger caused the wheels to lock, which often results in a loss of control as well as an increase in the stopping distance.

We Need a Third Law

There is still one more Newtonian law to consider. Imagine that you are playing tennis and have just hit a ball. The racket exerts a force on the ball that causes it to move. The high-speed photograph in Figure 2-14 shows that the strings of the tennis racket are pushed back at the same time the ball is flattened. The ball is squashed by the force of the racket *on the ball* and, at the same time, the racket strings are stretched by the force of the ball *on the racket*. At the same time the racket is exerting a force on the ball, the ball is exerting an opposite force on the racket.

If you wish to pursue this point further, find a friend who will help with a simple experiment. Give your friend a shove. At the same time you are pushing, you will feel a force being exerted on you, regardless of whether your friend pushes back. If you can, try this wearing ice skates or roller blades; it will be even more dramatic.

Let's carry this one step further. Lean on a wall. Notice the force of the wall pushing back on you. If this force did not exist, you would fall forward.

It could have been these kinds of experiences that led Newton to his third law of motion. He realized that there is no way to push something without being pushed yourself. For every force there is always an equal and opposite force. The two forces act on different objects, are the same size, and act in opposite directions. Formally, we state **Newton's third law** as follows:

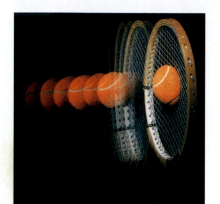

Figure 2-14   A high-speed photograph illustrating Newton's third law. The ball exerts a force on the strings and the strings exert an equal and opposite force on the ball.

Newton's third law

> If an object exerts a force on a second object, the second object exerts an equal force back on the first object.

Because Newton referred to these forces as action and reaction, they are often known as an action–reaction pair. However, because the two forces are equivalent, it doesn't matter which one is called the action and which the reaction. Another statement of the third law might be: For every action there is an equal and opposite reaction.

Forces always occur in pairs. In Newton's words, "If you press a stone with your finger, the finger is also pressed by the stone." These forces *never act on the same body*. When you press the stone with your finger, you exert a force *on the stone*. The reaction force is *acting on you*.

Consider a ball with a weight of 10 newtons falling freely toward the Earth's surface. Ignoring air resistance, there is only one force acting on the ball; the Earth's gravity is pulling it downward with a force of 10 newtons. What is the second force in the action–reaction pair? Since the first force can be labeled *Earth on ball* $\mathbf{W}_{eb}$, we just reverse the order of the objects to get the second label; it must be *ball on Earth* $\mathbf{W}_{be}$ (Fig. 2-15).

Therefore, Newton's third law tells us that the ball must be exerting an upward force on the Earth of 10 newtons. Although your common sense may tell you that the Earth must exert a larger force because it is so much larger, this is not true. No matter what the origin of the forces, Newton's third law tells us that the forces must be equal in size and opposite in direction.

But if this is true, why doesn't the Earth accelerate toward the ball? It does, but if we put the values into Newton's second law, we find that the Earth's mass is so large that its acceleration is miniscule. The Earth does accelerate, but we don't notice it.

A very important point concerning third-law forces is that one of the forces acts on the ball, while the other acts on the Earth; one causes the ball to accelerate, the other causes the Earth to accelerate. Because the two forces act on different objects, the two forces cannot cancel; if these are the only two forces acting, both objects accelerate.

Without the third law there would be paradoxical events in the Newtonian world view. Consider, for example, why a person doesn't fall through the floor. There is a gravitational force pulling the person down. If this were the only force acting on the person, according to Newton's second law, the person would accelerate downward through the floor. Since the person is not accelerating, the net force on the person must be zero. Thus, the question arises, What is the force that balances the downward gravitational force?

Newton's third law provides the answer. The Earth attracts the person with a force that we can label W_{ep}, as shown in Figure 2-16. The person pushes down on the floor with a force that we label F_{pf}. According to the third law, the floor pushes upward on the person with a force F_{fp} that is equal in size and opposite in direction to F_{pf}. Therefore, there are two forces acting on the person: the person's weight W_{ep} and the upward force of the floor F_{fp}. Although these two forces are equal in size and act in opposite directions, they are not third-law forces; they both act on the person. They are equal and opposite because the person is not accelerating and, therefore, by Newton's second law the net force on the person must be zero. Note that the third-law force associated with the person's weight W_{ep} is the gravitational force of the person acting on the Earth W_{pe}.

> **QUESTION**   A branch exerts an upward force on an apple in a tree. What is the third-law companion to this force?

When you fire a rifle, it recoils. Why? As explained by the third law, when the rifle exerts a forward force on the bullet (by virtue of an explosion), the bullet simultaneously exerts an equal force on the rifle but in the backward direction. But why doesn't the rifle accelerate as much as

> **ANSWER**   The downward force of the apple on the branch. Note that it is not the downward force of gravity on the apple. Although these forces are equal and opposite, both act on the apple and cannot be action–reaction forces.

Figure 2-15   The Earth exerts a force W_{eb} on the ball. According to Newton's third law the ball exerts an equal and opposite force W_{be} on the Earth.

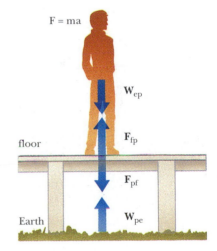

Figure 2-16   The Earth exerts a force W_{ep} on the person which causes the person to exert a force F_{pf} on the floor. By Newton's third law the floor exerts an equal and opposite force F_{fp} on the person. Although W_{ep} and F_{fp} are equal and opposite, they are not an action–reaction pair; they both act on the person.

Figure 2-17 As the man walks to the left, he exerts a force on the boat that causes the boat to move to the right.

the bullet? The force of the rifle on the bullet produces a large acceleration because the mass of the bullet is small. The force of the bullet on the rifle is the same size, but produces a small acceleration because the mass of the rifle is large. Incidentally, if you want a rifle to recoil less, you could make it more massive. On the other hand, if we could make a superstrong but very lightweight rifle with a mass less than the bullet, it would be more dangerous to shoot the rifle than to stand at the target!

Even the common act of walking is possible only because of reaction forces. In walking you must have a force exerted on you in the direction of your motion. And yet the force you produce is clearly in the opposite direction. The solution to this apparent paradox lies in the third law. As you walk, you exert a force against the floor (down and backward); the floor therefore exerts a force back, causing you to go forward (and up a little). If there is any sand or loose earth where you walk, you can see that it has been pushed *back*. If you want a clearer demonstration of the fact that you push backward against the floor, try walking on a skateboard or in a rowboat (Fig. 2-17). But be careful!

Physics on Your Own   Determine your weight on a bathroom scale placed in an elevator when the elevator is stopped, as it accelerates upward, as it travels between floors at a constant speed, and as it stops. What do you expect to happen? Explain your reasoning in each case.

SUMMARY

According to Newton's first law of motion, an object at rest remains at rest and an object in motion remains in motion with a constant velocity unless acted on by a net outside force. The net force is determined by adding all the forces acting on an object according to the rules for combining vector quantities. The converse of Newton's first law is also true: If an object has a constant velocity (including the case of zero velocity), the unbalanced, or net, force acting on the object must be zero.

If there is a net force, the object accelerates with a value given by Newton's second law of motion, $\mathbf{a} = \mathbf{F}_{net}/m$. The direction of the acceleration is always the same as the net force. In the metric system the unit of mass is the kilogram, and the unit of force is the newton.

Newton's second law of motion can be used to study frictional forces. Static friction can range from zero to a maximum value that depends on the force pushing the surfaces together and the nature of these surfaces. Sliding friction has a constant value less than the maximum static value. A special kind of friction is air resistance, which varies with the speed of the falling object. As the air resistance acting on a falling object becomes equal to the object's weight, the acceleration goes to zero and the object falls at its terminal speed.

If we assume that we can ignore the effects of the rotating Earth, the weight of an object close to the Earth's surface is given by $\mathbf{W} = m\mathbf{g}$, where $\mathbf{g}$ is the acceleration due to gravity, about 10 (meters per second) per second downward. Weight, a force, should not be confused with mass.

There is no such thing as an isolated force. All forces occur in pairs that are equal in size, opposite in direction, and act on the two different objects that are interacting. As you stand on the floor, you exert a downward force *on the floor*. According to Newton's third law, the floor must exert an upward force *on you* of the same size. These two forces do not cancel as they act on different objects—one on you and one on the floor. The other force acting on you is the force of gravity.

CHAPTER

2

REVISITED

The acceleration of an object in free fall gets smaller and smaller because the force due to the air resistance increases as the speed of the object increases. Therefore, the net force continually decreases. When the net force equals zero, the acceleration also becomes zero and the speed assumes a constant value. This speed is known as the terminal speed.

KEY TERMS

acceleration: The change in velocity divided by the time it takes to make the change. Measured in units such as (meters per second) per second. An acceleration can result from a change in speed, a change in direction, or both.

force: A push or a pull. Measured by the acceleration it produces on a standard, isolated object, $\mathbf{F} = m\mathbf{a}$. Measured in newtons.

inertia: An object's resistance to a change in its velocity.

inverse proportionality: A relationship in which a quantity is related to the reciprocal of a second quantity.

kilogram: The standard international (SI) unit of mass, the approximate mass of 1 liter of water. A kilogram of material weighs about 2.2 pounds on Earth.

law of inertia: Newton's first law of motion.

mass: A measure of the quantity of matter in an ob-

ject. The mass determines an object's inertia. Measured in kilograms.

newton: The SI unit of force. A net force of 1 newton accelerates a mass of 1 kilogram at a rate of 1 (meter per second) per second.

Newton's first law of motion: Every object remains at rest or in motion in a straight line at constant speed unless acted on by an unbalanced force.

Newton's second law of motion: $\mathbf{F}_{net} = m\mathbf{a}$ The net force on an object is equal to its mass times its acceleration. The net force and the acceleration are vectors that always point in the same direction.

Newton's third law of motion: If an object exerts a force on a second object, the second object exerts an equal force back on the first object.

sliding friction: The frictional force between two surfaces in relative motion. This force does not depend very much on the relative speed.

static friction: The frictional force between two surfaces at rest relative to each other. This force is equal and opposite to the net applied force if the force is not large enough to make the object accelerate.

terminal speed: The speed obtained in free fall when the upward force of air resistance is equal to the downward force of gravity.

velocity: A vector quantity that includes the speed and direction of an object.

weight: $\mathbf{W} = m\mathbf{g}$ The force of gravitational attraction of the Earth for an object. This definition is modified in Chapter 11 for accelerating systems such as elevators and spacecraft.

CONCEPTUAL QUESTIONS

1. Was it necessary for Galileo to be able to measure time in order to develop his law of inertia?

2. Did Galileo need to be able to measure mass in order to develop his law of inertia?

3. If you give this book a shove so that it moves across a tabletop, it slows down and comes to a stop. How can you reconcile this observation with Newton's first law?

4. Assume that you are pushing a car across a level parking lot. When you stop pushing, the car comes to a stop. Does this violate Newton's first law?

5. What can you say about the forces acting on an automobile that is traveling down a straight stretch of highway at a constant speed?

6. How does the net force on the first subway car compare with that on the last one if the subway train has a constant velocity?

7. Describe the thought experiment that led Galileo to conclude that straight-line motion with a constant speed was a natural motion.

8. What idealization(s) did Galileo make in his thought experiment that led to the law of inertia?

9. Why does a tassel hanging from the rearview mirror appear to swing backward as the car accelerates in the forward direction?

10. When dogs finish swimming, they often shake themselves to dry off. What is the physics behind this?

11. Assume that you're not wearing your seat belt and the car stops suddenly. Why would your head hit the windshield?

12. Modern automobiles are required to have headrests to protect your neck during collisions. For what type of collision are these headrests the most effective?

13. Why does a blacksmith use an anvil when hammering a horseshoe?

Question 13.

14. How can you use only one hand to tear a single sheet of toilet paper from a roll?

15. Give an example where the use of the word *inertia* to describe something in everyday life is closely related to its use in physics.

16. How would you determine if two objects have the same inertia?

17. How are force and inertia related?

18. Is mass or weight more closely related to inertia?

19. What are the largest and smallest forces you can get by adding a 5-newton force to an 8-newton force?

20. What are the largest and smallest forces you can obtain by adding a 4-newton force and a 5-newton force?

21. What kind of motion does a constant, nonzero net force produce?

22. How would you describe the motion of a ball rolling down a straight ramp in terms of its velocity and its acceleration? What can you deduce about the forces acting on the ball?

23. If the net force on a boat is directed due east, what is the direction of the acceleration of the boat? Would your answer change if the boat had a velocity due north but the net force still acted to the east?

24. If the net force on a hot-air balloon is directed vertically upward, what is the direction of the acceleration of the balloon? What would be the direction of the acceleration if the balloon were being blown westward (with the net force still acting vertically upward)?

25. Two people are pushing in opposite directions on a very large ball. How can you tell if one person is exerting more force on the ball than the other?

26. How does the net force on the first subway car compare with that on the last one if the subway train has a constant acceleration?

27. If you triple the net horizontal force applied to a cart, what happens to the acceleration?

28. What happens to the acceleration of a rocket if the net force on it is doubled?

29. If a car can accelerate at 4 (meters per second) per second, what will its acceleration be when towing another car like itself?

30. What happens to the acceleration of an object when the mass is cut in half but the force remains the same?

31. How do the concepts of mass and weight differ? In what ways are they related?

32. When an astronaut works on the Moon, is either her mass or her weight the same as on Earth? Explain.

33. What are the units of the force of gravity, weight, and mass?

34. If you buy a sack of potatoes labeled 5 kilograms, are you buying the potatoes by mass or by weight?

35. What happens to the weight of an object if you double its mass?

36. What happens to the weight of an object if you take it from Earth to the Moon, where the acceleration due to gravity is only one-sixth as large?

37. If a golf ball and a ping-pong ball are simultaneously dropped from the same height, they do not reach the ground at the same time. How would Aristotle explain this? How would Galileo?

38. Under what conditions will a golf ball and a ping-pong ball that are dropped simultaneously from the same height reach the ground at the same time?

39. What is the acceleration of a parachutist who has reached terminal speed?

40. A marble dropped into a bottle of liquid soap quickly reaches a terminal speed. What is the acceleration of the marble just before it hits the bottom of the bottle?

41. Will a parachutist have a larger terminal speed before or after the parachute is deployed?

42. Will a marble have a larger terminal speed in syrup or water? Why?

43. A friend falsely claims, "Newton's first law doesn't work if there is any friction." How would you correct this claim?

44. One of your classmates falsely asserts, "Newton's second law only works when there are no frictional forces." How would you correct this assertion?

45. How would you explain the fact that you can push on a railroad boxcar and yet it doesn't move? What does Newton's second law say about this situation?

46. You push on a desk, but it doesn't move. How can you reconcile this with Newton's first law?

47. A force of 300 newtons is required to push a box across the floor at a constant speed. What is the net force acting on the box? Explain your reasoning.

48. What force is required to pull a dog in a wagon along a level sidewalk, as in the figure, at a constant speed if the frictional force is 400 newtons?

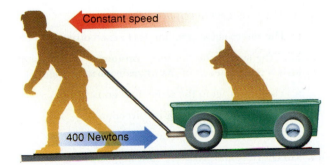

Question 48.

49. State Newton's three laws of motion in your own words.

50. Make up situations that illustrate each of Newton's three laws.

51. The Earth exerts a force of 500,000 newtons on one of the communications satellites. What force does the satellite exert on the Earth?

52. Is the force that the Sun exerts on the Earth bigger, smaller, or the same size as the force that the Earth exerts on the Sun? Explain your reasoning.

53. What is the net force on an apple that weighs 6 newtons when you hold it at rest?

54. Suppose you are holding an apple that weighs 6 newtons. What is the net force on the apple just after you drop it?

55. Why do the cannons aboard pirates' ships roll backward when they are fired?

56. Why does a tennis racket slow down when it hits a ball?

57. Describe the force(s) that allow you to walk across a room.

*58. We often say that the engine supplies the forces that propel a car. This is an oversimplification. What are the forces that actually move the car?

59. The figure shows a ball hanging by a string from the ceiling. Identify the action–reaction pairs in this drawing. Argue that the sizes of all six forces are the same.

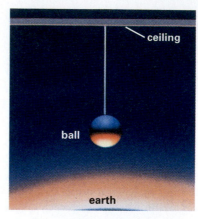

Question 59.

60. A ball with a weight of 40 newtons is falling freely toward the surface of the Moon. What force does this ball exert on the Moon?

*61. If the force exerted by a horse on a cart is equal and opposite to the force exerted by the cart on the horse, as required by Newton's third law, how does the horse manage to move the cart?

*62. Since the forces demanded by Newton's third law of motion are equal in magnitude and opposite in direction, how can anything ever be accelerated?

EXERCISES

1. Find the size of the net force produced by a 5-N and a 9-N force in each of the following arrangements:

   a. The forces act in the same direction.

   b. The forces act in opposite directions.

   c. The forces act at right angles to each other.

2. Find the size of the net force produced by a 4-N and an 8-N force in each of the following arrangements:

   a. The forces act in the same direction.

   b. The forces act in opposite directions.

   c. The forces act at right angles to each other.

3. Two horizontal forces act on a wagon, 500 N forward and 350 N backward. What force is needed to produce a net force of zero?

4. A skydiver is being pulled down by the Earth with a force of 970 N. If there is also an air resistance of 400 N, what additional force would be needed to maintain a constant speed?

5. What is the acceleration of a 1000-kg car if the net force on the car is 3000 N?

6. What is the acceleration of a 500-kg buffalo if the net force on the buffalo is 2000 N?

7. The net horizontal force on a 50,000-kg railroad boxcar is 4000 N. What is the acceleration of the boxcar?

8. A 30-06 bullet has a mass of 0.010 kg. If the average force on the bullet is 9000 N, what is the bullet's average acceleration?

9. What net force is needed to accelerate 50 kg at 2 m/s^2?

10. If a sled with a mass of 40 kg is to accelerate at 3 m/s^2, what net force is needed?

11. A salesperson claims a 1200-kg car has an average acceleration of 3 m/s^2 from a standing start to 100 km/h. What average net force is required to do this?

12. If a 30-kg instrument has a weight of 50 N on the Moon, what is its acceleration when it is dropped?

13. If a skydiver has a net force of 400 N and an acceleration of 5 m/s^2, what is the mass of the skydiver?

14. A child on roller skates undergoes an acceleration of 0.6 m/s^2 due to a horizontal net force of 18 N. What is the mass of the child?

15. A 2-kg ball is falling in a vacuum near the surface of the Earth. What are the direction and size of each force acting on the ball?

16. A 3-kg ball has been thrown vertically upward. If we ignore the air resistance, what are the direction and size of each force acting on the ball while it is traveling upward?

17. If a space explorer with a mass of 90 kg has a weight of 1080 N on a newly discovered planet, what is the acceleration due to gravity on this planet?

18. A fully equipped astronaut has a mass of 150 kg. If the astronaut has a weight of 555 N standing on the surface of Mars, what is the acceleration due to gravity on Mars?

19. What applied force is needed to accelerate 50 kg at 2 m/s^2 if the frictional force is 200 N?

20. A crate has a mass of 24 kg. What applied force is required to produce an acceleration of 3 m/s^2 if the frictional force is known to be 90 N?

21. If a pull of 220 N accelerates a 40-kg child on ice skates at a rate of 5 m/s^2, what is the frictional force of the skates with the ice?

22. A skydiver with a mass of 80 kg is accelerating at 4 m/s^2. What is the force of the air resistance acting on the skydiver?

Newton's first law states that an object naturally travels in a straight line. Yet in nearly every motion we observe, the objects execute much more complicated motions. Gymnasts twist and turn, and cars maneuver tight corners. What causes these motions? And what happens to you when you're part of the motion—say, as a passenger in a car rounding a corner? (See p. 71 for the answer to this question.)

Motions in Space

6 Atlanta 1996

The motion of the gymnast is a combination of a rotation about her center of mass and a translation through space.

S o far, we have restricted our discussion to straight-line, or one-dimensional, motion. Most motions, however, take place in more than one dimension; most commonly in three-dimensional space.

Going from one dimension to two or three dimensions is less difficult than you might anticipate because all the motions can be divided into separate motions in each of the three dimensions. A knuckleball's motion can be thought of as the sum of three separate motions: up/down, left/right, and near/far (along the line between the pitcher and catcher). This separation means that we can apply the laws that were developed for one dimension to many common motions in space.

Our first consideration will be to look at two-dimensional motion; that is, motion confined to a flat surface. Adding this second dimension allows us to study a variety of motions, from the path of a thrown ball to the Earth's annual journey around the Sun.

Circular Motion

The motion of the Earth around the Sun, the Moon around the Earth, a race car around a circular track, and a ball swinging on the end of a string are examples of circular (or nearly circular) motion. Each of these occurs in a two-dimensional plane.

As we look at these motions, we start with the simplest situation, motion where the speed of the object remains constant. In the one-dimensional cases we studied in the previous chapter, constant speed implied the absence of a net force. This is not the case in two dimensions. An object moving in a circular path at a constant speed must have a force acting on it.

To see this, imagine whirling a ball on the end of a string in a circle above your head, as in Figure 3-1. Because you have to pull on the string to do this, the string must be exerting a force on the ball. And because a string can only exert a force along its length, this force must act toward the center of the circle. It has the special name **centripetal force,** which means *center-seeking* force.

It is important to distinguish between the adjectives centri*petal* (center-seeking) and centri*fugal* (center-fleeing). The word *centripetal* is rarely used in our everyday language. The word *centrifugal* is much more common and often mistakenly substituted for *centripetal*. The force we are discussing, the centripetal force, is directed toward the center of the circle. We will discuss centrifugal effects further in Chapter 11.

If you cut the string, the ball no longer moves in a circle but flies off in a straight line (if we ignore the force of gravity). It would continue to travel in the direction of its velocity at the time the string was cut due to its inertia. The ball's path (Fig. 3-2) is perpendicular to the string and tangent to the circular path. Without applied forces, the ball exhibits straight-line motion, as described by Newton's first law.

A car racing on an unbanked, circular track (again, at a constant speed) doesn't have a string anchoring it to the center of the track. The centripetal force, in this case, is provided by the frictional force of the track acting on the car's tires *toward the center* of the track. An oil slick on

Figure 3-1   The force acting inward along the string keeps the ball in a circular path.

a centripetal force is a center-seeking force

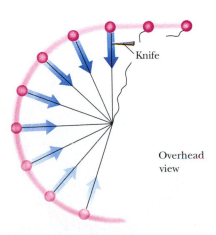

Figure 3-2   When the string is cut, the ball travels in a straight line tangent to the circular path.

the track would greatly reduce the frictional force, resulting in a potentially disastrous situation.

> **QUESTION**   Assuming the frictional force goes to zero when the car hits the oil slick, describe its motion.

Let's reexamine Newton's first law of motion. It says that an object will go in a *straight line* unless acted on by a net force. Therefore, any time an object changes direction, there must be a net force acting on it. Newton's first law—the law of inertia—applies to direction as well as speed. Conversely, if the object is traveling in a straight line, there can be no net force acting to either side.

This discussion can be generalized to an even larger class of motions. Any slight change in direction can be thought of as momentarily traveling along an arc of a circle, as shown in Figure 3-3. This requires that a force act toward the inside of the turn. For this reason, it is dangerous to drive your car on icy, curved roads—you need friction (traction) to turn the corners.

As mentioned in Chapter 2, the law of inertia applies to all objects everywhere. If a spacecraft was far from all massive bodies (where gravitational forces can be ignored), it would coast in a straight line. The only time NASA would need to turn on its engines would be to make "midcourse corrections" to change the spacecraft's direction or its speed.

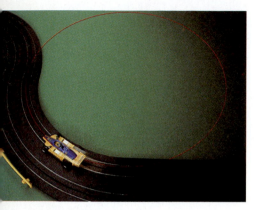

Figure 3-3   As the race car turns the corner, it momentarily moves along a circular path.

Acceleration Revisited

We have established that an object moving in a circle with constant speed must have a net force acting on it. As a consequence, according to Newton's second law, the object must be accelerating. This acceleration is equal to the change in velocity divided by the time it takes to make the change. In this case the velocity changes direction without changing size. If this seems a little bizarre—that an object whose speed remains constant is accelerating—remember that a vector such as velocity can change in two ways.

Like the force vector, the velocity vector can be represented by an arrow, as shown in Figure 3-4. The length of the arrow in this case tells us the speed (for example, 1 centimeter equals 5 meters per second), and the direction of the arrow is the direction of motion. The velocity vectors corresponding to circular motion at a constant speed are tangent to the circular path and have equal lengths. Although they have equal lengths, they are different vectors because their directions are different.

The change in velocity is determined by subtracting the initial velocity vector from the final one. Suppose that the initial velocity of an object

Figure 3-4   The velocity vector for circular motion with constant speed changes direction, but not size.

> **ANSWER**   The car would slide in a straight line in the direction it was traveling at the time it hit the oil slick.

is represented by the arrow $\mathbf{v}_i$ and its final velocity by the arrow $\mathbf{v}_f$ [Fig. 3-5(a)]. Then, we need to calculate $\mathbf{v}_f - \mathbf{v}_i$. We can apply our rule for vector addition from Chapter 2 to evaluate this expression by rewriting it as the sum of the final velocity and the negative of the initial velocity, $\mathbf{v}_f - \mathbf{v}_i = \mathbf{v}_f + (-\mathbf{v}_i)$, where the negative of a vector means that it points in the opposite direction. To subtract the velocities, we turn $\mathbf{v}_i$ around and place its tail on the head of $\mathbf{v}_f$ (making certain not to change their original orientations or lengths). The change in velocity $\Delta\mathbf{v}$ is also a vector and is represented by an arrow drawn from the tail of $\mathbf{v}_f$ to the head of $-\mathbf{v}_i$, as shown in Figure 3-5(b).

You can verify that this procedure for subtracting vectors is correct. Since $\Delta\mathbf{v}$ is the change in the velocity, adding it to the initial velocity $\mathbf{v}_i$ (using the rules of vector addition developed earlier) yields the final velocity $\mathbf{v}_f$. Figure 3-5(c) illustrates this.

The fact that the acceleration is defined as a vector ($\Delta\mathbf{v}$) divided by a number (the time elapsed) means that acceleration is also a vector. Its direction is the same as that of $\Delta\mathbf{v}$ and its size is obtained by dividing the size of $\Delta\mathbf{v}$ by the elapsed time.

$$\mathbf{a} = \frac{\Delta\mathbf{v}}{\Delta t}$$

Unlike the velocity vector, which points in the direction of motion, the direction of the acceleration vector is not intuitive. The safest procedure is to not make an intuitive guess, but to formally subtract the velocity vectors to obtain the direction of the acceleration vector.

Let's check this definition for acceleration against the familiar case of straight-line motion, where the object changes speed but not direction. If the object just speeds up, the final velocity vector will be longer than the initial vector but will still point in the same direction. This is shown in Figure 3-6(a). To determine $\Delta\mathbf{v}$, we again add the negative of the initial velocity to the final velocity and draw $\Delta\mathbf{v}$ from the tail of $\mathbf{v}_f$ to the head of $-\mathbf{v}_i$, as shown in Figure 3-6(b). The change in velocity is just the numerical difference in the lengths of the velocity vectors (that is, the difference in the initial and final speeds); it points in the same direction as $\mathbf{v}_f$. Therefore, the acceleration is in the same direction as the velocity, and the speed increases.

QUESTION   Which way would the acceleration point if the object were slowing down?

This more complete definition of acceleration is often greeted with disbelief and bewilderment by the new physics student. "How can you claim an acceleration when the speed stays constant?" is a typical reaction.

ANSWER   In this case the final velocity vector is the shorter one. Adding the negative of the longer initial velocity to it yields a change in velocity that points backward.

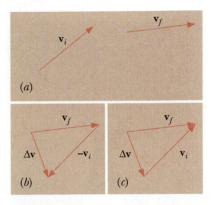

Figure 3-5   (a) Arrows $\mathbf{v}_i$ and $\mathbf{v}_f$ represent initial and final velocities with the same speeds. (b) $\Delta\mathbf{v} = \mathbf{v}_f - \mathbf{v}_i$ is the change in the velocity. (c) The initial velocity plus the change in velocity equals the final velocity.

$$\text{acceleration} = \frac{\text{change in velocity}}{\text{time taken}}$$

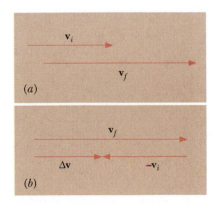

Figure 3-6   (a) Two parallel velocity vectors. (b) Subtracting the two velocities by adding the negative of the initial velocity $\mathbf{v_i}$ to the final velocity $\mathbf{v_f}$ gives the change in velocity $\Delta\mathbf{v}$.

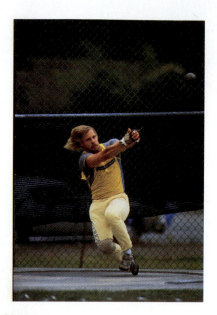

The hammer thrower provides the centripetal force required to make the hammer go in a circle.

centripetal acceleration
$$= \frac{\text{speed squared}}{\text{radius}}$$

The reason for the confusion is that the concept of acceleration has a more complex meaning in physics than in everyday language.

Although this difference in meaning may seem strange and perhaps even unnecessary, remember our goal is to describe motion completely. We must include the observed fact that a change in *just* the direction (without a change in speed) requires a net force and, therefore, according to Newton's second law, there must be an acceleration.

Acceleration in Circular Motion

In the case of circular motion at a constant speed, we know that the force is inward along the circle's radius. We find the **centripetal acceleration** by subtracting two velocity vectors separated by a short time interval. This gives a change in velocity that points close to the center of the circle. As

---

COMPUTING   CENTRIPETAL ACCELERATION Σ

A more detailed analysis of circular motion with a constant speed tells us that the centripetal acceleration a is equal to the object's speed v squared divided by the radius r of the circle.

$$a = \frac{v^2}{r}$$

As an example, we can calculate the centripetal acceleration of a 0.2-kilogram ball traveling in a circle with a radius of 1 meter every second. Since this circle has a circumference of $2\pi r = 6.3$ meters, the ball has a speed of 6.3 meters per second.

$$a = \frac{v^2}{r} = \frac{(6.3 \text{ m/s})^2}{1 \text{ m}} = 40 \text{ m/s}^2$$

Since $F_{\text{net}} = ma$ is always valid, we can use this result to find the centripetal force required to produce this circular motion.

$$F_{\text{net}} = ma = (0.2 \text{ kg})(40 \text{ m/s}^2) = 8.0 \text{ N}$$

QUESTION   If you double the speed of the ball traveling in a circle, what happens to the centripetal force?

ANSWER   The force quadruples.

we shorten the time interval, the change in velocity points closer to the center. As the time interval approaches zero, the instantaneous change in velocity points at the center. This agrees with Newton's second law since the instantaneous acceleration must always point in the same direction as the net force.

You can get a feeling for how the centripetal force depends on the parameters of the motion by twirling a ball on a string above your head and noticing how hard you have to pull on the string. If you increase the speed of the ball while keeping the radius the same, you have to pull harder. This tells you that the force increases as the speed increases. You can also increase the radius (lengthen the string) while keeping the speed along the circle the same. Increasing the radius requires less pull on the string. Therefore, the force varies inversely with the radius of the circular path.

Projectile Motion

When something is thrown or launched near the surface of the Earth, it experiences a constant vertical gravitational force. Motion under these conditions is called **projectile motion.** It occurs whenever an object is given some initial velocity and thereafter travels in a trajectory subject only to the force of gravity. Examples of projectile motion range from human cannonballs to the forward pass in football if we ignore the effects of air resistance (which we will do throughout the remainder of this chapter).

The study of projectile motion is simplified because the motion can be treated as two mutually independent, perpendicular motions, one horizontal and the other vertical. This reduces a complicated situation to two independent, one-dimensional motions that we already know how to handle.

Even though this division works, its consequences are often difficult to accept. Suppose a bullet is fired horizontally from a pistol and simultaneously another bullet is dropped from the same height. Which bullet hits the ground first?

Many students incorrectly answer, "The dropped bullet hits first." The time it takes to reach the ground is determined by the vertical motion. Although the two bullets have different horizontal speeds, their vertical motions are identical *if we ignore the air resistance.* Therefore, they take the same time to reach the ground. Note that we do not claim that they travel the same distance or with the same velocity; clearly the fired bullet has a much greater velocity and consequently travels much farther by the time it hits the ground.

This result is convincingly demonstrated in the strobe photograph in Figure 3-7. The ball on the right was fired in a horizontal direction at the same time the ball on the left was dropped. The horizontal lines are included as a visual aid to help you see that the vertical race ends in a tie.

Which bullet will hit the ground first if they are simultaneously released from the same height?

Figure 3-7 Strobe photograph of two balls. The ball on the right was fired horizontally; the other was dropped.

Banking Corners

It takes a centripetal force to make a car go around a corner. On a flat parking lot or an unbanked curve, this force comes from the frictional interaction between the tires and the road. The maximum size of this frictional force varies a little bit with the type of tire and the type of surface, but depends mostly on the weather conditions, whether the surface is dry, wet, or icy. The maximum frictional force does not depend very much on the speed. However, as the speed of the car increases for a given turning radius, the centripetal force provided by friction must also increase. At some speed, the car is moving too fast and the frictional forces are simply not large enough for the car to make it around the curve. Disaster may result!

There is another way to get a centripetal force. By tilting, or banking, the road, the demands on the frictional forces can be reduced. Let's first consider the case when there is no frictional force. As shown in the figure, the car's weight acts vertically downward, pushing the car against the road. Since there is no frictional force, the road can only exert a force on the car that is perpendicular to the road. (Since the car does not accelerate in the vertical direction, we know that the vertical part of the force due to the road must just cancel out the car's weight.) As shown in the figure, the sum of these two forces on the car—the force due to gravity and the force due to the road—is a horizontal force acting toward the inside of the curve. It is this horizontal force that provides the centripetal acceleration that makes the car go around the curve.

For a given bank to the curve, there is a well-determined centripetal force and a corresponding centripetal acceleration. It therefore requires a specific speed to execute circular motion with a radius that matches the curve. This speed is the design speed of

the curve and is the speed that does not require any frictional forces between the tires and the road. This is the speed that you want to use if the road is icy. If your speed is higher, the circle your car follows will be larger than that of the roadway and you will skid off the outside of the curve. If your speed is lower than the design speed, the centripetal force will cause your car's path to have too small a radius and the car will slide off the road to the inside!

The relationship between the design speed and the banking angle is precise and can be calculated. To get a feeling for this, consider a curve with a radius of 100 meters. For a bank angle of 13°, the design speed is found to be 15 meters per second (34 mph). Doubling the design speed to 30 meters per second requires a bank angle of 43°, and tripling the design speed to 45 meters per second (100 mph) requires a bank angle of 65°.

In actuality, the centripetal force is provided by the banking of the curve and the force of friction. If the car is traveling faster than the design speed of the curve, the frictional force acts toward the inside of the curve to provide the extra centripetal force needed. However, if the car is traveling slower than the design speed, the frictional force acts outward, reducing the net centripetal force. Thus, friction means that you can safely execute corners at speeds above and below their design speeds. For instance, the corner with the 13° bank described in the previous paragraph could be taken at speeds up to 30 meters per second (67 mph).

The centripetal force experienced by a car going around a curve increases as the weight of the car increases. This is fortunate because the larger mass of the car requires a greater centripetal force to make the car go around the corner. Therefore, the design speed of the curve is the same for all cars. Can you imagine what the road signs would be like if this were not true?

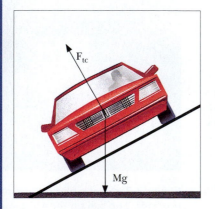
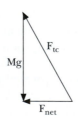

The forces acting on a car going around a banked curve with no friction.

Banking corners allows cars to take curves at high speeds.

Figure 3-8   Which washer will hit the floor first?

Let's look at the motion of the ball thrown in Figure 3-9. Assuming that gravity is the only force acting on the ball, the acceleration of the ball must be in the vertical direction. Therefore, only the vertical speed changes; the horizontal speed remains constant throughout the flight.

How do we describe the vertical motion? The ball takes off with a certain initial vertical speed and has a constant downward acceleration throughout the flight. On the upward part of its flight, the downward acceleration slows the ball's vertical speed by 10 meters per second during each second. At some point the vertical speed becomes zero (point A in Fig. 3-9). Then the ball starts its vertical descent, speeding up at 10 (meters per second) each second.

Although the vertical speed is zero at point A, the ball is still moving horizontally. The horizontal spacings between the images are equal, indicating that the horizontal speed is constant along the entire path. This curve also explains why basketball players such as Michael Jordon appear to "hang in the air" when they drive to the basket for a slam dunk. Note that near the top of the curve, the ball travels very little in the vertical direction while it covers a much larger distance horizontally.

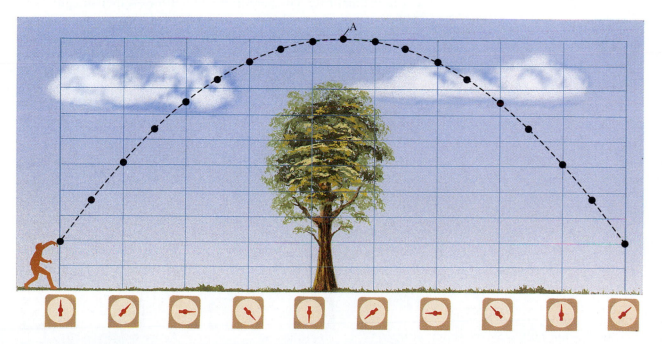

Figure 3-9   Strobe drawing of a thrown ball's motion. Note that the horizontal motion has a constant speed.

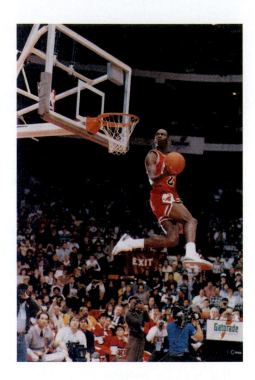

Michael Jordan appears to defy
gravity as he slam dunks the ball.

To further our understanding, we contrast the motion of a projectile
on the Earth with its corresponding motion on a mythical planet called
Narang, a planet with **No A**ir **R**esistance **A**nd **N**o **G**ravity. We fire a bullet
from a horizontal rifle. On Narang the bullet travels in a straight, horizontal line at a constant velocity, as it experiences no force after it leaves

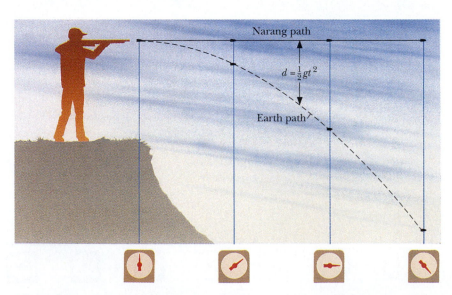

Figure 3-10   Paths of bullets fired horizontally on Narang and Earth. The distance between the two paths is due to free fall.

the rifle barrel. On Earth the bullet's path differs because the Earth's gravity affects its vertical motion. The horizontal motion of the bullet, however, is exactly the same on each planet, as shown in Figure 3-10.

Imagine now that we fire the rifle at an upward angle. On Narang the bullet again follows a straight line, but this time its path is inclined upward. On Earth the bullet continually falls downward from the straight-line path. At any time during the flight, the distance between the two paths is equal to the distance an object would free-fall during the same elapsed time (Fig. 3-11).

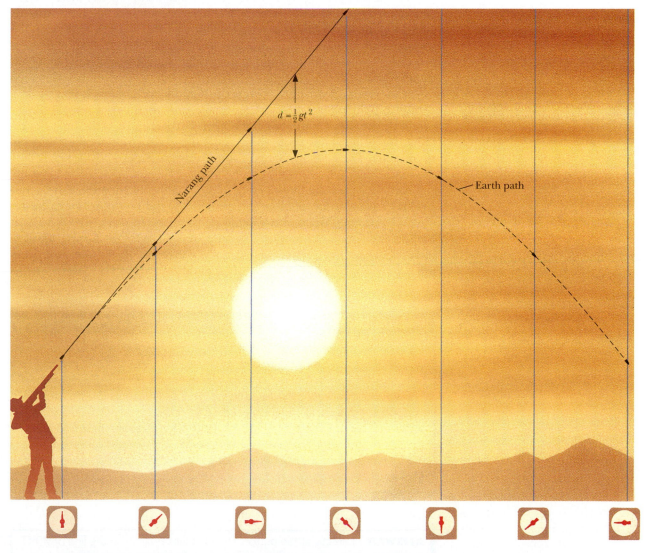

Figure 3-11 Paths of bullets fired at an upward angle on Narang and Earth. The distance between the two paths is due to free fall.

Figure 3-12 Tranquilizing a gorilla in a tree. Where should you aim?

A Thrown Wrench

When something is thrown, it usually undergoes another motion. As it moves though space, it often spins or tumbles. We need to ask if it is legitimate to treat these motions as independent of each other.

The strobe photograph in Figure 3-13 show the top view of a wrench sliding on a nearly frictionless surface. The wrench is doing two things: It is moving along a path and it is rotating. Motion along a path is called **translational motion.** Notice that the spot marked with the taped black **x** is moving in a straight line and at a constant speed; it is moving at a constant velocity.

If we mentally shrink any object so that its entire mass is located at a certain point, the translational motion of this new, very compact object would be the same as the original object. Furthermore, if the object is rotating freely, it rotates about this same point. This point is called the **center of mass.** The center of mass of the wrench in Figure 3-13 is marked by the black **x.**

Although realistic motions are a combination of rotational and translational motions, the photograph shows that the complexity can be reduced by treating them as two separate motions. In other words, we can look at the rotational motion as if the object were not moving along a path and look at the translational motion as if the object were not rotating. We will take advantage of this separation in the remainder of this chapter.

Figure 3-13 Strobe photograph of a wrench sliding across a horizontal surface. Notice that the black **x** moves with a constant velocity.

Physics on Your Own Make a small hole in a ping-pong ball so that you can load up one side with glue. After the glue has dried, throw the ball to a friend. Try putting some spin on the ball. What happens and why?

Rotations

A net force on an ordinary, extended object—not a compact point mass—can sometimes cause rotation as well as translation. In situations where this happens and the object is free, it will rotate about its center of mass, as with the thrown wrench in Figure 3-13. If it is fixed at some point, or about some axis, the rotation will be about that point or axis. A door pivoting on its hinges is an example of this second situation.

The rules for rotational motion have many analogies with translational motion. The distance traveled is no longer measured in ordinary distance units, but in an angular measure such as degrees or revolutions. **Rotational speeds** are measured in terms of the angular rotation divided by the time. The units used to express this measurement could be degrees per second or revolutions per minute (rpm). Modern compact discs spin at variable rates between 200 and 500 revolutions per minute. **Rotational acceleration,** like its translational counterpart, is a measure of the rate at which the rotational speed changes.

In addition, Newton's laws have the same form for rotational motion as they do for translational motion. The first law says that the natural motion is one in which the rotational speed is constant. If the object is not rotating, it remains so. If it is rotating, it continues to rotate with the same rotational speed. This can be seen with the thrown wrench in Figure 3-13. If you imagine riding along with the wrench, you see that the wrench rotates about the **x** by the same amount between successive flashes. That is, its rotational speed is constant.

Torque

A change in the rotational speed can only occur when there is a net external interaction on the object. This interaction involves forces, but unlike translational motion, the locations at which the forces act are as

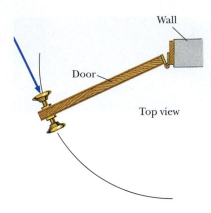

Figure 3-14 The torque exerted on the door is equal to the product of the distance from the hinge and the force applied to the doorknob.

important as their sizes and directions. The same force can produce different effects depending on where and in which direction it is applied.

You can experiment with these ideas by exerting different forces on a door. If you push directly toward the hinges or pull directly away from them, the door does not rotate. Rotations only occur when a horizontal force is applied in any other direction; the largest occurs when the force is perpendicular to the face of the door.

Even when you apply the force in the perpendicular direction, you get different results depending on where you push. Try opening the door by pushing at different distances from the hinges. The largest effect occurs when the force is applied farthest from the hinges. That's why doorknobs are put there!

Physics on Your Own   A young child can easily beat you in a game of door "push"-of-war if you carefully choose the points where each of you pushes. Discover where you should choose the points.

The rotational analog of force in translational motion is called **torque** and combines the effects of the force on the door and the distance from the hinge. If you push on the doorknob, the door will move along a circular path with a radius equal to the distance from the hinge to the doorknob, as shown in Figure 3-14. If we restrict ourselves to the case when the force is perpendicular to the radius, the torque τ is equal to the radius r multiplied by the force F.

torque = radius × force

$$\tau = rF$$

Since torque is equal to a product, we can see why the same applied force can produce different torques on an object. The torque is increased if the force is applied farther from the center of rotation. This fact is useful if you have to loosen a stubborn nut. The biggest torque occurs when you push or pull the wrench at the spot farthest from the nut. We can make the distance longer (for the really stubborn nuts) by slipping a pipe over the wrench.

Imagine that you have a flat tire and one of the nuts is stuck tight. Suppose further that your wrench is 0.30 meter long. How much torque could you apply by stepping on the end of the wrench?

If you weigh 500 N, the maximum torque you could apply would be

$$\tau = rF = (0.30 \text{ m})(500 \text{ N}) = 150 \text{ N} \cdot \text{m}$$

QUESTION   Suppose that this is not enough but you found a pipe in your car that could be slipped over the wrench, tripling its effective length. What torque could you now apply?

ANSWER   Since the torque is a product of the force and the distance, you would get a torque that is three times as large as the original, or 450 N·m.

When there is more than one applied force, situations can arise when the net force is zero but the net torque is not zero. In other words, a pair of forces that produces no translational acceleration can still produce rotational acceleration. The two forces on the stick in Figure 3-15 are equal in size and opposite in direction, but they do not act along the same line. The stick accelerates clockwise because the torques are nonzero and act in the same rotational direction.

Figure 3-16 shows two girls on a seesaw. Each girl's weight multiplied by her distance from the pivot point gives the torque that she applies to the board. If the torques are equal and opposite, there will be no rotational acceleration. Of course, if this was all that happened, playing on a seesaw would be dull. The seesaw's motion alternates between two rotations—first in one direction, then in the other. The momentary torque that makes the transition from one direction to the other is provided when a child pushes off the ground with her feet.

Two people with quite different weights are still able to balance the seesaw. The lighter person sits farther from the pivot point, as shown in Figure 3-17. This equalizes the torques. The extra distance compensates for the reduced force due to the smaller weight.

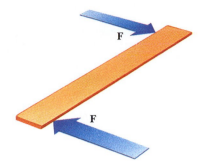

Figure 3-15   Two equal but opposite forces can produce a rotational acceleration if they do not act along the same line.

Figure 3-16   If a seesaw is balanced, the torques exerted by the two children are equal.

Figure 3-17   The heavier person sits closer to the pivot point to equalize the torques.

(a)   *(b)*

Figure 3-18   (a) Two masses taped near the ends of a meterstick have a large rotational intertia. (b) When the masses are moved closer to the center, the rotational inertia is considerably less.

Extended arms help children maintain their balance on an abandoned railroad track.

Rotational Inertia   Σ

The rotational acceleration depends on the object as well as the net torque. If you push on the door to a bank vault and a door in your house, you get different rotational accelerations. In the translational case, the same net force produces different accelerations for different inertial masses. In the rotational case, the same net torque produces different rotational accelerations, but now the acceleration depends on more than the object's mass; the distribution of the mass is also important.

Consider a "dumbbell" arrangement of a meterstick with a ½-kilogram mass taped on each end. Holding the dumbbell at its center and rotating it back and forth demonstrates convincingly that a large torque is required to give it a substantial rotational acceleration [Fig. 3-18(a)]. This is completely analogous to the translational inertial properties we have encountered. The rotational analog of inertia is **rotational inertia.**

Changing the arrangement gives different results. If the two masses are moved closer to the center of the meterstick, it is much easier to start and stop the rotation [Fig. 3-18(b)]. The dumbbell has less rotational inertia even though no mass was removed. Simply changing the distribution of the mass changed the rotational inertia; it is larger the farther the mass is located from the point of rotation.

Losing one's balance on the high wire amounts to gaining a rotation off the wire. Tightrope walkers increase their rotational inertia by carrying long poles. Their increased rotational inertia helps them maintain their balance by allowing them more time to react. We naturally do something like this when we try to keep our balance. Picture yourself walking on a railroad track. Where are your arms?

Physics on Your Own Balance a broom on the palm of your hand so that the handle is vertical. Is it easier when the bristles are up or down? Why?

Center of Mass

The center-of-mass concept is useful for examining the effect of gravity on objects. Rather than dealing with an incredibly large number of forces acting on the object, we treat the object as if the total force (that is, its weight) acts at the center of mass. By doing this, we can account for the translational and rotational motions of the object.

Now we need a way of finding the center of mass. This could be determined by a mathematical averaging procedure that considers the distribution of the object's mass. A certain amount of mass on one side of the object is balanced, or averaged, with some mass on the other side. But there are easier ways.

Finding the center of mass for a regularly shaped object is fairly simple. The symmetry of the object tells us that the center of mass must be at the geometric center of the object. It is interesting to note that there does not have to be any mass at that spot; a hollow tennis ball's center of mass is still at its geometric center.

QUESTION Where would you expect the center of mass of a donut to be?

Locating the center of mass of an irregularly shaped object is a little more difficult. However, since the weight can be considered to act at the center of mass, we can locate it with a simple experiment. Hang the object from some point along its surface so it is free to swing, as in Figure 3-19(a). The object will come to rest in a position where there is no net torque on it. At this position the weight acts along a vertical line through the support point. Therefore, the center of mass is located someplace on this vertical line. Now suspend the object from another point, establishing a second line. Since the center of mass must lie on both lines, it must be at the intersection of the two lines [Fig. 3-19(b)].

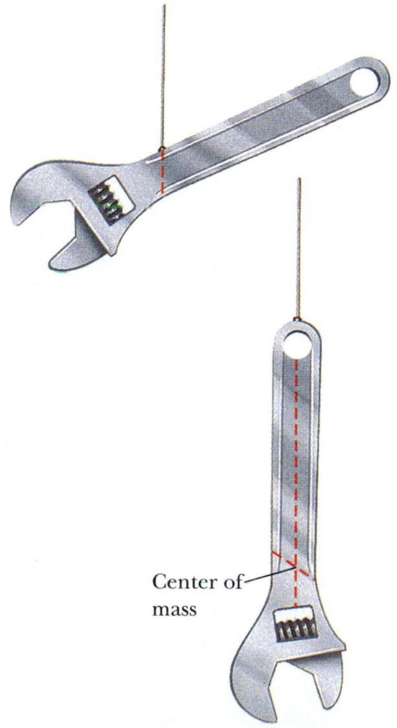

Figure 3-19 The center of mass of the wrench is located at the intersection of the vertical lines obtained by hanging the wrench from two or more places.

Physics on Your Own Guess the location of the center of your home state. You can check your guess by taping a map of your state onto a piece of cardboard. After cutting around the edges of the state, suspend it from several points, as shown in Figure 3-20, to locate its center.

ANSWER Because the donut is approximately symmetric, its center of mass is near the center of the hole.

Floating in Defiance of Gravity

Many performers—long jumpers, basketball players, and ballet dancers, to name a few—would like to defy gravity and stay in the air longer. Regardless of their desires, however, the force of gravity is ever present and the center of mass of the projectile—be it a pebble or a performing artist—follows the parabolic path discussed in this chapter. The only control over the path of the center of mass is in the initial conditions at the moment of launch. A different angle or different launch velocity alters the projectile's path. However, once the object is launched the center of mass follows the predetermined path.

The grand jeté in ballet seems to be a contradiction. In this popular ballet move, dancers execute a running leap across the stage, creating a seemingly floating motion that suspends them in the air longer than gravity should allow.

There are two parts to this illusion. First, all objects spend most of the time-of-flight near the peak of the motion. By counting the images of the ball in Figure 3-9, it is easy to verify that the ball spends half of the time in the top one-quarter of the vertical space. During this time, the object is moving mostly horizontally.

The second part of the illusion depends on the dancer's skill. Although the dancer's center of mass follows a parabolic path, a skillful dancer can change the position of the center of mass within her body during the flight. This allows the head and torso to stay at a nearly fixed height for a longer time. As illustrated in Figure B, the movements of the arms and legs raise and lower the location of the center of mass within the body. So, during the beginning and end of the jump, the dancer's arms and legs are pointing downward, keeping the center of mass low. During the middle of the flight, the dancer's arms are up and the legs are outstretched, raising the center of mass relative to the rest of the dancer's body. The total effect is a flattening of the path of the head and torso.

Reference: Laws, Kenneth. *The Physics of Dance*. New York: Schirmer, 1984.

Figure A.

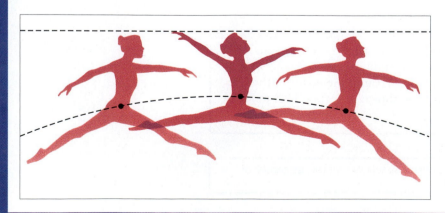

Figure B. The dancer raises and lowers the center of mass within the body during the grand jeté.

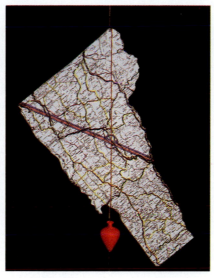

Figure 3-20   The center of the state of Vermont is located by hanging it from two points.

Stability

We can extend our ideas about rotating objects to see why some things tip over easily and others are quite stable. Picture a child making a tall tower out of toy blocks. Much to the child's delight, the tower always tips over. But why does this happen? Clearly there are taller structures in the world than this child's tower.

We answer this by looking at the stability of a one-block tower. In Figure 3-21(a) the left side of the block is slightly above the table. If let go in this position the block's weight (acting at the center of mass) provides a counterclockwise torque about the right edge. The force of the table on the block acts through this edge but produces no torque since it acts at the pivot point. Thus, the net torque is counterclockwise and the block falls back to its original position.

Now in Figure 3-21(b) the block is tilted far enough so that the weight acts to the right of the pivot point. The weight produces a clockwise torque and the block falls over. The block tips over whenever its center of mass is beyond the edge of the base.

As the child's toy tower gets taller and its center of mass gets higher, the amount the tower has to sway before the center of mass passes beyond the base gets smaller. We can make the tower more stable by keeping it short and/or widening its base.

If you get bumped while standing with your feet close together, you begin to fall over. To stop this, you quickly spread your legs and increase your support base. Car manufacturers promote superwide wheel bases because this innovation makes the car more stable.

Figure 3-21   (a) When the center of mass is above the base, the block returns to its upright position. (b) When the center of mass is beyond the base, the block topples over.

Physics on Your Own   Stand with your back to a wall with your heels placed against the wall. Carefully try to bend over and touch your toes. Why can't you do this?

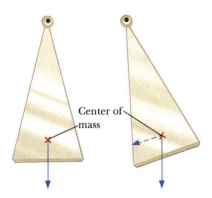

Figure 3-22   Stable equilibrium occurs when the center of mass is located below the point of suspension.

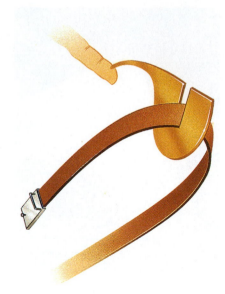

Figure 3-23   The center of mass of the sky hook and belt is located directly below the pivot point.

Tightrope walking is difficult because the support base (the wire's thickness) is so small. A slight lean to the left or right puts the center of mass past the support point and creates a torque. The torque produces a rotation in the same direction as the initial lean, making the situation worse. Such a situation is known as **unstable equilibrium.**

The most stable arrangement occurs when the center of mass is below the support point, as in Figure 3-22. As the center of mass sways left or right, the torque that is created rotates the object back to the original orientation. This situation is known as **stable equilibrium.**

Physics on Your Own   Cut a piece of plywood into the shape shown in Figure 3-23. Try hooking the "sky hook" over your fingers as indicated. Try this again after you have inserted a stiff belt into the slot as shown. Where is the center of mass of the sky hook and belt?

SUMMARY

All multidimensional motion can be divided into separate motions—translation in each of the three dimensions (up/down, left/right, and near/far) and rotation about an axis. The laws for one-dimensional motion apply to each dimension separately.

An object moving in a circular path at a constant speed must have a force acting on it. This centripetal force causes the circular motion; without it the object would fly off in a straight line, moving in the direction of its velocity at the time of release. Whenever there is any change in the velocity of an object, it experiences an acceleration and therefore must experience a net force. A net force with a constant magnitude acting perpendicular to the velocity produces circular motion with constant speed.

Projectile motion results from the constant, downward force of gravity. Again, these problems are simplified by the fact that the horizontal and vertical motions are independent. Assuming that air resistance is negligible, the only acceleration is in the vertical direction, the direction of the force of gravity. Therefore, the vertical motion is just that of free fall. The horizontal speed remains constant throughout the flight.

Objects can also rotate or revolve around some axis, and this can happen whether they are fixed or moving through space. These motions—rotational and translational—are also independent of each other. The rotation of a free body takes place about the center of mass. For translational motion, all of an object's mass can be considered to be concentrated at its center of mass.

The rules for rotational motion are similar to the rules for translational motion. The distance "traveled" is an angular measure, rotational speeds are angles divided by time, and rotational accelerations are changes in rotational speeds divided by time. Newton's laws for rotational motion have the same form as the laws for translational motion. A change in rotational speed only occurs when there is a net torque on the object. The torque τ is equal to the radius r multiplied by the perpendicular force F, or $\tau = rF$, and has units of newton-meters. Rotational inertia depends on the distance of the mass from the axis of rotation as well as on the mass itself.

The stability of an object depends on the torques produced by its weight (acting at the center of mass) and on the supporting force(s).

CHAPTER

3

REVISITED

An object changes direction whenever a net force acts on it and changes its rotation whenever there's a net torque. The car turns the corner because the pavement exerts a force on its tires that provides the centripetal force toward the center of the curve. If you are a passenger in the car, your inertia keeps you moving forward in a straight line. The frictional forces between you and the car's seat are usually large enough to provide the force that makes you follow the same curve. You may get any additional needed force from the door. You feel that you've been pushed outward against the door, when, in fact, the door came to you and is pushing you into the curved path.

KEY TERMS

center of mass: The balance point of an object. This location has the same translational motion as the object would if it were shrunk to a point.

centripetal acceleration: The acceleration of an object moving along a curved path. For uniform circular motion, the acceleration points toward the center of the circle and has a magnitude given by v^2/r.

centripetal force: The force causing an object to change directions. For uniform circular motion, the force is directed toward the center of the circle and has a magnitude given by mv^2/r.

projectile motion: A type of motion that occurs near the surface of the Earth when the only force acting on the object is that of gravity.

rotational acceleration: The change in rotational speed divided by the time it takes to make the change.

rotational inertia: The property of an object that measures its resistance to a change in its rotational speed.

rotational speed: The angle of rotation or revolution divided by the time taken. Measured in units such as degrees per second or revolutions per minute.

stable equilibrium: An equilibrium position or orientation to which an object returns after being slightly displaced.

torque: The rotational analog of force. It is equal to the radius multiplied by the force perpendicular to the radius.

translational motion: Motion along a path.

unstable equilibrium: An equilibrium position or orientation from which an object leaves after being slightly displaced.

CONCEPTUAL QUESTIONS

Important: Ignore the effects of air resistance in the following questions and exercises.

1. The figure shows a race track with identical cars at points a, b, and c. The cars are moving clockwise at constant speeds. Draw arrows indicating the direction of the net force on each car and the instantaneous velocity of each car. In what direction would car a travel if there were an oil slick at point a? Why?

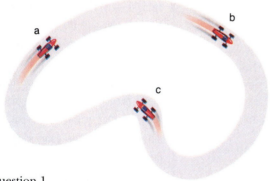

Question 1.

2. A motorcycle drives through a vertical loop-the-loop at constant speed, as shown in the figure. Draw arrows to show the directions of the instantaneous velocity and the net force on the motorcycle at points a, b, and c.

3. What is the force that causes the Space Shuttle to orbit the Earth?

4. What is the force that allows a person on roller blades to turn a corner? What happens if this force is not strong enough?

5. A child rides on a carousel. In which direction does each of the following vectors point: (a) velocity, (b) change in velocity, (c) acceleration, and (d) net force?

6. If you are riding on a merry-go-round, in which directions do your velocity, change in velocity, acceleration, and net force point?

7. How do the velocity vectors differ on opposite sides of the path for uniform circular motion? How are they the same?

8. How are the acceleration vectors on opposite sides of the path for uniform circular motion the same? How do they differ?

9. An object executes circular motion with a constant speed whenever a net force of a constant magnitude acts perpendicular to the velocity. What happens to the speed if the force is not perpendicular to the velocity?

*10. An object is acted on by a force that always acts perpendicular to the velocity. If the force continually in-

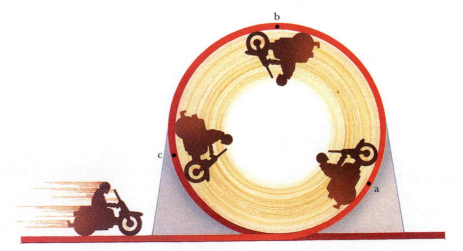

Question 2.

creases in magnitude, does the speed change? Draw a sketch of the path of the object.

*11. A vine is just strong enough to support Tarzan when he is hanging straight down. However, when he tries to swing from tree to tree, the same vine breaks at the bottom of the swing. How could this happen?

*12. Imagine that you swing a bucket in a vertical circle at constant speed. Will you need to exert more force when the bucket is at the top of the circle or at the bottom? Explain.

13. What force(s) allow(s) you to turn a bicycle while riding on a flat parking lot?

14. Most footraces take place on unbanked tracks. How do the racers turn the corners?

15. The Earth executes a nearly circular orbit around the Sun. What does this tell you about the speed of the Earth along this orbit?

*16. According to Newton's third law, the Earth exerts a force on the Sun. Does the Sun move in a circular path as the Earth goes around it each year? How can this idea be used to determine if nearby stars have planets?

17. An ashtray slides along a frictionless table at a constant velocity and then sails off the edge. Draw all the forces acting on the ashtray while it is on the table and while it is in the air.

18. A playful astronaut decides to throw rocks on the Moon. What forces act on the rock while it is in the "vacuum"? (We can't say "air"!)

19. A left fielder throws a baseball toward home plate. At the instant the ball reaches its highest point, what are the directions of the ball's velocity, the net force on the ball, and the ball's acceleration?

20. A quarterback throws a long pass toward the end zone. At the instant the ball reaches its highest point, what are the directions of the ball's velocity, the net force on the ball, and the ball's acceleration?

21. A rock dropped from 1.8 meters above the surface of Mars requires 1.0 second to reach the ground. Would it require a shorter, a longer, or the same time if the rock were thrown horizontally from this height with a speed of 10 meters per second?

22. A hammer dropped on the surface of Mars falls with an acceleration of 3.7 (meters per second) per second. Would its acceleration be smaller, larger, or the same if it were thrown horizontally at 6 meters per second?

23. A 2-kilogram ball is thrown horizontally at a speed of 12 meters per second. At the same time a 1-kilogram ball is dropped from the same height. Which ball hits level ground first?

24. A 1-kilogram ball is thrown horizontally with a speed of 20 meters per second. At the same time a 2-kilogram ball is thrown horizontally at the same height, but with half the speed. Which ball hits level ground first?

25. A fearless bicycle rider announces that he will jump the Beaver River Canyon. If he does not use a ramp, but simply launches himself horizontally, is there any way that he can succeed? Why?

Question 25.

26. A physics student reports that upon arrival on planet X, she promptly sets up the "gorilla-shoot" demonstration. She does not realize that the gravity on planet X is stronger than it is on Earth. Should she modify her procedure? If so, how? If not, why?

27. In football and soccer, it is often desirable to give up some of the distance a kick travels to gain hang time, the time the ball remains in the air. How does the kicker do this?

28. The irons in golf have faces that make different angles with the shaft of the club. How does this affect the distance traveled and maximum height of the golf ball?

Question 28.

29. A football is poorly thrown so that it tumbles end over end. What point of the football follows the trajectory expected for projectile motion?

30. Describe the motion of a spinning ball that has some extra mass added to one side.

31. How do the units for rotational speed and rotational acceleration differ?

32. Which of the following are not proper units for rotational acceleration: revolutions per minute, (revolutions per minute) per second, or (degrees per second) per second?

33. If the object shown in the figure is fixed, but free to rotate about point A, which force will produce the larger torque? Why?

Questions 33 & 34.

34. If the object shown in the figure is not fixed and point A is the object's center of mass, which force will produce translational motion without rotation?

35. Use the concept of torque to explain how a crowbar or a claw hammer works.

36. Apply the concept of torque to explain how wheelbarrows allow you to transport a heavy load with a lifting force less than the weight of the load.

37. The torque that you can apply to the rear wheel of a 10-speed bicycle depends on which gear you are using. Why?

38. Why is it harder to ride a tricycle than a bicycle up a hill?

39. What is needed to change the rotational velocity of an object?

40. What do we call an object's resistance to a change in its rotational velocity?

41. An object's rotational inertia _____ with an increase in mass and _____ as the mass is moved closer to the center.

42. Which would be harder to rotate about its center, a 12-foot-long 2" × 4" or a 6-foot-long 4" × 4"?

43. A flywheel with a large rotational inertia is often attached to the drive shaft of automobile engines. What purpose does the flywheel serve?

44. Would you have a larger rotational inertia in the tuck, pike, or layout positions? (See Fig. 5-5 if you do not know what these diving positions are.)

45. If a solid disc and a hoop have the same mass and radius, which would have the larger rotational inertia about its center of mass? Why?

46. A solid sphere and a solid cylinder are made of the same material. If they have the same mass and radius, which one has the larger rotational inertia about its center? Why?

47. Make up an example that clearly illustrates the meaning of Newton's first law for rotation.

48. Make up an example that clearly illustrates the meaning of Newton's second law for rotation.

49. In the text we found the center of mass at the intersection of two lines. If you suspend the same object used in that experiment from a third point, this line would pass through the intersection of the first two. Why?

*50. How would you determine the center of mass of an automobile?

51. Where is the center of mass of the figurine resting on the pedestal in the figure?

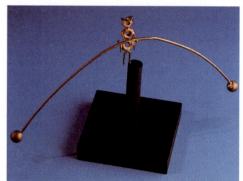

Question 51.

52. A spoon and a fork can be suspended beyond the edge of a glass by using a flat toothpick, as shown in the fig-

Question 52.

ure. Where is the center of mass of the spoon–fork combination?

53. Using diagrams, show why an ice cream cone is more stable when it is placed upside down rather than on its tip.

54. Why do you feel more secure standing on the bottom step of a step ladder rather than the top step?

55. How does "weighting" a pair of dice change the probabilities of throwing certain numbers?

56. Is it possible to "fix" a roulette wheel so that you will win? Can you do it without the tampering being obvious?

57. If you stand facing a wall with your toes touching the wall, you cannot stand up on your toes. Why?

58. It is possible (and quite likely) for a high jumper's center of mass to pass *under* the bar while the jumper passes *over* the bar as in the figure. Explain how this is possible.

Question 58.

*59. Assume that you make a balance by drilling a hole through the 50-centimeter mark of a meterstick and suspending it with a string. If you drill the hole near one edge of the meterstick, should you hang it so that the hole is near the top or the bottom edge? Why?

*60. Why is a rowboat more stable than a canoe?

EXERCISES

1. Find the size and direction of the change in velocity for each of the following initial and final velocities.

   a. 3 m/s west to 6 m/s west.

   b. 6 m/s west to 3 m/s west.

   c. 3 m/s west to 6 m/s east.

2. What is the change in velocity for each of the following initial and final velocities?

   a. 90 km/h right to 120 km/h right.

   b. 90 km/h right to 120 km/h left.

3. What is the size and direction of the change in velocity if the initial velocity is 3 m/s south and the final velocity is 4 m/s west?

4. What is the change in velocity of a car that is initially traveling west at 60 km/h and then drives 100 km/h toward the north?

5. A cyclist turns a corner with a radius of 40 m at a speed of 20 m/s.

   a. What is the cyclist's acceleration?

   b. If the cyclist and cycle have a combined mass of 90 kg, what is the force causing them to turn?

6. A 50-kg person on a merry-go-round is traveling in a circle with a radius of 2 m at a speed of 2 m/s.

   a. What acceleration does the person experience?

   b. What is the centripetal force? How does it compare with the person's weight?

7. Given that the distance from the Earth to the Moon is 3.8×10^8 m, that the Moon takes 27 days to orbit the Earth, and that the mass of the Moon is 7.4×10^{22} kg, what is the acceleration of the Moon and the size of the attractive force between the Earth and the Moon?

8. Given that the radius of the Earth's orbit is 1.5×10^{11} m and its mass is 6×10^{24} kg, what is the acceleration of the Earth and the size of the attractive force between the Sun and the Earth?

9. A baseball is hit with a horizontal speed of 14 m/s and a vertical speed of 14 m/s upward. What are these speeds 1 s later?

10. What are the horizontal and vertical speeds of the baseball in the previous question 2 s after it is hit?

11. An ashtray slides across a table with a speed of 0.6 m/s and falls off the edge. If it takes 0.4 s to reach the floor, how far from the edge of the table does it land?

12. A car drives off a vertical cliff at a speed of 29 m/s (60 mph). If it takes 4 s for it to hit the ground, how far from the base of the cliff does it land?

13. A tennis ball is hit with a vertical speed of 10 m/s and a horizontal speed of 20 m/s. How long will the ball remain in the air? How far will it travel horizontally during this time?

14. If a baseball is hit with a vertical speed of 20 m/s and a horizontal speed of 12 m/s, how long will the ball remain in the air? How far will it go?

15. If a CD makes 900 revolutions in 3 min, what is the CD's average rotational speed?

16. What is the rotational speed of the hand on a clock that indicates the seconds?

17. If the rotational speed of a CD changes from 500 rpm to 200 rpm as it plays 30 min of music, what is the CD's average rotational acceleration?

18. If it takes a CD 2 s to reach its initial rotational speed of 500 rpm when it is turned on, what is its average acceleration?

19. What torque does a 140-N salmon exert on a 1.5-m-long fishing pole if the pole is horizontal and the salmon is out of the water?

20. A pirate with a mass of 90 kg stands on the end of a plank that extends 2 m beyond the gunwale. What torque is needed to keep him from falling into the water?

21. A child with a mass of 20 kg sits at a distance of 3 m from the pivot point of a seesaw. Where should a 15-kg child sit to balance the seesaw?

22. Two children with masses of 24 and 36 kg are sitting on a balanced seesaw. If the lighter child is sitting 3 m from the center, where is the heavier child sitting?

23. What mass would you hang on the right side of the system in the figure to balance it—that is, to make the clockwise and counterclockwise torques equal?

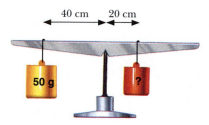

Exercise 23.

24. Is the system shown in the figure balanced? If not, which end will fall? Explain your reasoning.

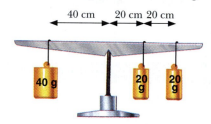

Exercise 24.

Gravity

CHAPTER

4

A popular legend has it that Newton had his most creative thought while watching an apple fall to the ground. He made a huge conceptual leap by equating the motion of the apple to that of the Moon and developing the concept of gravity. How far does gravity reach? Do we see any evidence of gravity elsewhere in the Universe? (See p. 97 for the answer to this question.)

Stars are born when molecules and dust in these clouds coalesce due to their mutual gravitational attraction. This photograph of the "Eagle Nebula" was taken with the Hubble Space Telescope Wide Field and Planetary Camera.

In the physicist's view of the world, there are four fundamental forces: the gravitational, the electromagnetic, the weak, and the strong. The most familiar force in our everyday lives is gravity. Every school-age child "knows" that objects fall because of gravity. But what is gravity? Saying that it is what makes things fall doesn't tell us much.

Is gravity some material like a fluid or a fog? Or, is it something more ethereal? No one knows. Because we have given something a name doesn't mean we understand it completely. We do understand gravity in the sense that we can precisely describe how it affects the motion of objects. For instance, we have already seen how to use the concept of gravity to describe the motion of falling objects. We can do more. By looking carefully at the motions of certain objects we can develop an equation that describes this attractive force between material objects and explore some of its consequences.

On the other hand, we cannot answer questions like "What is gravity?" or "Why does gravity exist?"

The Concept of Gravity

The concept of gravity hasn't always existed. It was conceived when changes in our world view required a new explanation of why things fell to the Earth. When the Earth was believed to be flat, gravity wasn't needed. Objects fell because they were seeking their natural places. A stone on the end of a string hung down because of its tendency to return to its natural place. "Up" and "down" were absolute directions.

The realization that the Earth was spherical required a change in perspective. What happens to those unfortunate people on the other side of the Earth who are upside-down? But the change in thinking was made without gravity. The center of the Earth was at the center of the Universe and things naturally moved toward this point. Up and down became relative but the location of the center of the Universe became absolute.

Gravity was also not needed to understand the motion of the heavenly bodies. The earliest successful scheme viewed the Earth as the center of the Universe with each of the celestial bodies going around the Earth in circular orbits. Perpetual, circular orbits were considered quite natural for celestial motions; little attention was given to the causes of these motions. Aristotle did not recognize any connections between what he saw as perfect, heavenly motion and imperfect, earthly motion. He stated that circular motion with constant speed was the most perfect of all motions and thus the natural heavenly motion needed no further explanation.

This changed slightly with a new view of heavenly motion by Nicholas Copernicus, a 16th-century Polish scientist and clergyman. He proposed that the planets (including the Earth) go around the Sun in circular orbits, and the Moon orbits the Earth. This is essentially the scheme still taught in schools today. A hint of a concept of gravity appears in Copernicus' work. He felt that the Sun and Moon would attract objects near their surfaces—each would have a local gravity—but he had no concept of that attractive influence spreading throughout space.

A hundred years later, Johannes Kepler, a German mathematician and astronomer, suggested that the planets move because of an interac-

Celestial Mathematics

Johannes Kepler

Johannes Kepler was born on December 27, 1571, in Weil der Stadt, near what is now Stuttgart, Germany. He was a sickly child whose ill health plagued him throughout his life. Early on, smallpox left him prone to illness, and in later life he was cursed with eczema and abscesses. His poor health destroyed any chance of his becoming a farmer or obtaining any other laborious type of occupation; as a result, his parents placed him in a Lutheran seminary. Although Kepler was originally enrolled as a student of religion, his mathematical ability flourished and in 1594 he accepted a position as a mathematics professor in Graz.

Part of Kepler's responsibility as a mathematics professor was to prepare an astrological calendar. Being an exceedingly learned man with astute political observations, Kepler included in his calendar predictions of unrest in Austria. His predictions came to pass and his popularity grew among the peasants as well as the royalty. Kepler's interest in the planets and his ability with calculations later led him to produce a book entitled *Mysterium Cosmographicum.*

Kepler's book led to a lifelong friendship with Galileo and an offer of collaboration from the great astronomer Tycho Brahe. Kepler was eager to work with Tycho's data, though he had trouble prying it out of him. Tycho recognized Kepler's promise and wanted his assistance. Although their relationship was often stormy, their common need overcame their mutual friction.

At Tycho's death, Kepler succeeded him in official positions. More important, however, was that Kepler took over Tycho's data. With these voluminous observations, Kepler was able to calculate the orbits of planets with precision and develop his famous laws of planetary motion. He became an astute observer of the heavens. In 1604, Kepler observed a supernova. It was the second supernova to appear in 32 years and was also seen by Galileo. Kepler's study of it has led us to call it "Kepler's supernova."

In 1611 Kepler's life took a turn for the worse: He was deeply affected by the death of his wife, and the local ministry decided to deny him communion because of his liberal religious beliefs. His remarriage meant happier times, but even these were shattered when his mother was accused of being a witch. The ensuing legal battle left Kepler drained.

Despite his troubles, in 1618, Kepler published what he considered to be his greatest work, *Harmonices Mundi.* This book went to great lengths to show that one could derive musical scales from the velocities of the planets. After completing this work, Kepler felt that he had brought harmony to the heavens. In 1630, Johannes Kepler succumbed to illness and died, leaving his large family destitute.

Adapted from Pasachoff, J.M. *Astronomy: From the Earth to the Universe,* 4th ed. Philadelphia: Saunders College Publishing, 1991.

---

tion between them and the Sun. Kepler also moved us away from the assumption that the planets traveled in circular paths. After many years of trial and error, Kepler correctly deduced that the orbits of the planets were *ellipses.* Furthermore, the planets do not have constant speeds in their journeys around their elliptical paths but speed up as they approach the Sun and slow down as they move farther away.

Influenced by early, important work on magnetism, Kepler postulated that the planets were "magnetically" driven along their paths by the Sun. Since this work occurred shortly before the acceptance of the idea of inertia, Kepler did not realize that a force was not needed to drive the planets along their orbits but a force was needed to cause the orbits to be curved. Kepler postulated an interaction reaching from the Sun to the various planets and driving the planets, but he didn't consider the possibility of any interaction between planets themselves; the Sun reigned supreme, a metaphor for his God, from which everything else gained strength.

Newton developed our present, common view of gravity. He started by saying there was nothing special about the rules of nature that he had developed for use on Earth. They should also apply to heavenly motions. As we learned in the previous chapter, anything traveling in a circle must be accelerating. The acceleration must be toward the center of the circle and the object must, therefore, have a net force acting on it. Newton went searching for this force.

Launching an Apple

Creativity often involves bringing together ideas or things from seemingly unrelated areas. After an artist or scholar has done it, the connection often seems obvious to others. Newton made such a synthesis between motions on the Earth and motions in the heavens.

Legend has it that he made his intellectual leap while contemplating such matters and seeing an apple fall. As a tribute to this legend and as an aid to our understanding, we discuss the hypothetical problem of launching an apple into orbit around the Earth. The transition of the Earth-bound apple into heavenly orbits provides an analogy of Newton's intellectual leap. For the launch, we will only use the rules of motion that

Figure 4-1 An unsuccessful launch results in projectile motion.

Legend has it that Newton conceived his law of universal gravitation while observing an apple fall from a tree in his yard.

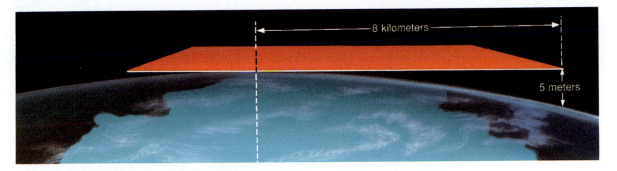

Figure 4-2   A plane touching a spherical Earth is five meters above the Earth's surface at a distance of 8 kilometers from the point of contact.

we have developed for use on Earth. (Incidentally, our discussion parallels the discussion in Newton's famous book *Principia;* the apple part is ours.)

To accomplish this space-age thought experiment, we need a very high place to stand and a very, very strong arm. For each launch we will throw the apple horizontally. On our first attempt we throw the apple with an ordinary speed and find that the apple follows a projectile path (Fig. 4-1) like the ones we discussed in Chapter 3.

On our next attempt, imagine that we throw the apple much faster. The apple still falls to the ground but the path is no longer like the first one. If the apple travels very far, the Earth's curvature becomes important. The force of gravity points in slightly different directions at the beginning and end of the path.

Normally, we are not aware of the curvature of the Earth's surface because the Earth is so huge. Over large distances, however, this cannot be ignored. If we imagine that the Earth is perfectly smooth—without hills and valleys—and construct a large horizontal plane at our location as shown in Figure 4-2, the Earth's surface will be 5 meters below the plane at a distance of 8 kilometers away. (This is about 16 feet at a distance of 5 miles.)

Imagine, then, that on our next attempt to launch the apple, we throw it with a speed of 8 kilometers per second. (This may seem slow, but it is 18,000 mph!) During the first second, the apple drops 5 meters. But so does the surface of the Earth. Thus, the apple is still moving horizontally at the end of the first second. The motion during the next second is a repeat of that during the first. And so on. The apple is in orbit.

If air resistance is negligible, the speed remains constant. The only force is the force of gravity. It is a centripetal force acting perpendicular to the instantaneous velocity. By throwing the apple hard enough, we have changed the motion from projectile to circular. Unlike projectile motion, our apple will not come down even though it is continually falling. The illustration used by Newton in his discussion of this is reproduced in Figure 4-3.

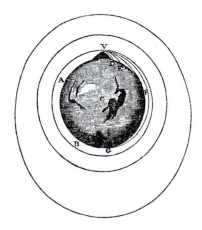

Figure 4-3   This illustration was used in Newton's *Principia* in the discussion of launching an object into orbit around the Earth.

Newton's Gravity

Newton felt that the laws of motion that worked on the surface of the Earth should also apply to motion in the heavens. Because the Moon orbits the Earth in a nearly circular orbit, it must be accelerating toward the Earth. According to the second law, any acceleration requires a force. He believed that if this force could be shut off, the Moon would no longer continue to move along its circular path, but would fly off in a straight line like a stone from a sling.

The genius of Newton was in relating the cause of this heavenly motion to earthly events. Newton felt that the Moon's acceleration was due to the force of gravity—the same gravity that caused the apple to fall from the tree. How could he demonstrate this? First, he calculated the acceleration of the Moon. Because the distance to the Moon and the time it took the Moon to make one revolution were already known, he was able to calculate that the Moon accelerated 0.00272 (meters per second) per second. This is a very small acceleration. In one second the Moon moves about 1 kilometer along its orbit, but only falls 1.4 millimeters (about ½₀ inch in 0.6 mile).

In contrast to the Moon's acceleration, the apple has an acceleration of 9.80 (meters per second) per second and falls about 5 meters in its first second of flight. (In earlier discussions we rounded the acceleration off to 10 [meters per second] per second for ease of computation. The small difference is important here.) We can compare these two accelerations by dividing one by the other:

$$\frac{\text{acceleration of Moon}}{\text{acceleration of apple}}$$

$$\frac{0.00272 \text{ m/s}^2}{9.80 \text{ m/s}^2} = \frac{1}{3600}$$

Why are these two accelerations so different? The mass of the Moon is certainly much larger than that of the apple. But that doesn't matter. As we saw in Chapter 1, freely falling objects all have the same acceleration independent of their masses.

However, the accelerations of the apple and the Moon weren't equal. Could Newton's idea that both motions were governed by the same gravity be wrong? Or could the rules of motion that he developed on Earth not apply to heavenly motion? Neither. Newton reasoned that the Moon's acceleration is smaller because the gravitational attraction of the Earth is smaller at larger distances; it is "diluted" by distance.

How did the force decrease with increasing distance? Retracing Newton's reasoning is impossible since he didn't write about how he arrived at his conclusions, but he may have used the following kind of reasoning.

Many things get less intense the farther you are from their source. Imagine a paint gun that can spray paint uniformly in all directions. Suppose the gun is in the center of a sphere of radius 1 meter and at the end of a minute of spraying, the paint on the inside wall of the sphere is 1 millimeter thick. If we repeat the experiment with the same gun but with a sphere that is 2 meters in radius, the paint will be only ¼ millimeter thick because a sphere with twice the radius has a surface area that is four times the original (Fig. 4-4). If the sphere has three times the radius, the surface is nine times bigger and the paint is $(\frac{1}{3})^2 = \frac{1}{9}$ as thick. The thickness

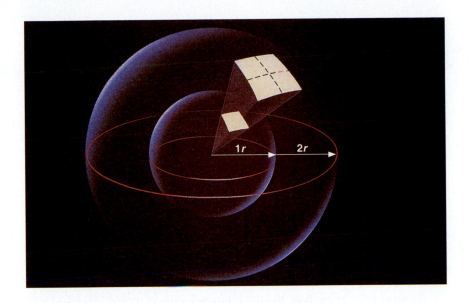

Figure 4-4   If the sphere's radius is doubled, the sphere's surface increases by a factor of four.

of the paint decreases as the square of the radius of the sphere increases. This is known as an **inverse-square** relationship. A force reaching into space could be diluted in a similar manner.

> **QUESTION**   If the sphere were 4 meters in radius, how thick would the paint be?

Newton may also have received encouragement for this explanation by working backward from observational results on the motion of the planets that had been developed by Kepler. Kepler found a relationship that connected the orbital periods of the planets with their average distances from the Sun. Using Kepler's results and an expression for the acceleration of an object in circular motion, we can show that the force decreases with the square of the distance. This means that if the distance between the objects is doubled, the force is only one-fourth as strong. If the distance is tripled, the force is one-ninth as strong. And so on. This relationship is shown in Figure 4-5.

Figure 4-5   The force on a 0.1 kilogram mass at various distances from the Earth. Notice that the force decreases as the square of the distance.

> **ANSWER**   It would be $(\frac{1}{4})^2 \times 1$ millimeter = $\frac{1}{16}$ millimeter thick.

> **QUESTION**   What happens to the force of gravity if the distance the two objects is cut in half?

The obvious test of the notion of gravity, of course, was to see if the relationship between distance and force gave the correct relative accelerations for the apple on Earth and the orbiting Moon. Newton could use this rule and make the comparison. The distance from the center of the Earth to the center of the Moon is about 60 times the radius of the Earth. The Moon is, therefore, 60 times farther away from the center of the Earth than the apple. Thus, the force at the Moon's location—and its acceleration—should be 60^2, or 3600, times smaller. This is in agreement with the previous calculation. The data available in Newton's time were not as good as those we have used here, but were good enough to convince him of the validity of his reasoning. Modern measurements yield more precise values and agree that *the gravitational force is inversely proportional to the square of the distance.*

Newton now knew how gravity changed with distance: The force of Earth's gravity exists beyond the Earth and gets weaker the farther away you go. But there are other factors. He already knew that the force of gravity depended on the object's mass. His third law of motion said that the force exerted on the Moon by the Earth was equal in strength to that exerted on the Earth by the Moon; they attracted each other. This symmetry indicated that both masses should be included in the same way. *The gravitational force is proportional to each mass.*

The Law of Universal Gravitation

Having made the connection between celestial motion and motion near the Earth's surface with a force that reaches across empty space and pulls objects to the Earth, Newton took another, even bolder, step. He stated that the force of gravity existed between *all* objects, that it was truly a *universal* law of gravitation.

The boldness of this assertion becomes apparent when one realizes that the force between two ordinary-sized objects is extremely small. Clearly, as you walk past a friend, you don't feel a gravitational attraction pulling you together. But that is exactly what Newton was claiming. Any two objects have a force of attraction between them; his rule for gravity is a **law of universal gravitation.**

Putting everything together, we arrive at an equation for the gravitational force:

law of universal gravitation

$$F = G\frac{M_1 M_2}{r^2}$$

where M_1 and M_2 are the masses of the two objects, r is the distance between their centers, and G is a constant that contains information about the strength of the force.

> **ANSWER**   The force becomes four times as strong.

Although Newton arrived at this conclusion when he was 24 years old, he didn't publish his results for over 20 years. This was partly due to one unsettling aspect of his work. The distance that appears in the relationship is the distance from the center of the Earth. This means that the mass of the Earth is assumed to be concentrated at a point at its center. This might seem like a reasonable assumption when considering the force of gravity on the Moon; the Earth's size is irrelevant when dealing with these huge distances. But what about the apple on the Earth's surface? In this case, the apple is attracted by mass that is only a few meters away, mass that is 13,000 kilometers away, as well as all the mass in between (Fig. 4-6). It seems less intuitive that all this would somehow act like a very compact mass located at the center of the Earth. But that is just what happens. Newton was eventually able to show mathematically that the sum of the forces due to each cubic meter of the Earth is the same as if all of them were concentrated at its center.

This result holds if the Earth is spherically symmetric. It doesn't have to have a uniform composition, it need only be composed of a series of spherical shells, each of which has a uniform composition. In fact, 1 cubic meter of material near the center of the Earth has almost four times the mass of a typical cubic meter of surface material.

Newton applied the laws of motion and the law of universal gravitation extensively to explain the motions of the heavenly bodies. He was able to show that three observational rules developed by Kepler to describe planetary motion were a mathematical consequence of his work. Kepler's rules were the results of years of work reducing observational data to a set of simple patterns. Newton showed theoretically that his laws yielded the same patterns.

By the 18th century, scientists were so confident of Newton's work that when a newly discovered planet failed to behave "properly," they assumed that there must be other, yet to be discovered, masses causing the

Figure 4-6 The sum of all the forces on the apple exerted by all portions of the Earth acts as if all the mass were located at the Earth's center.

deviations. When Uranus was discovered in 1781, a great effort was made to collect additional data on its orbit. By going back to old records, additional times and locations of its orbit were determined. Although the main contribution to Uranus' orbit is the force of the Sun, the other planets also have their effects on Uranus. In this case, however, the calculations still differed from the actual path by a small amount. The deviations were explained in terms of the influence of an unknown planet. This led to the discovery of Neptune in 1846.

This still didn't completely account for the orbits of Uranus and Neptune; a search began for yet another planet. The discovery of Pluto in 1930 still left some discrepancies. Although the search for new planets continues, analysis of the paths of the known planets indicates that any additional planets must be very small and/or very far away.

Physics Update

The closest extrasolar planet yet discovered (June 1996) orbits the star Lalande 21185, only 8.1 light-years from Earth (a light-year is a distance of 6 trillion miles). George Gatewood of the University of Pittsburgh observed a telltale wobble in the light coming from the star, indicating the presence of a Jupiter-sized planet circling the star in a Saturn-sized orbit. (The planet cannot be observed directly, so astronomers look for motion of stars due to the gravitational attraction of the planets.) Gatewood's data even hinted at the possibility of other planets in the same star system. Also, another planet has been found by Geoff Marcy of San Francisco State and Paul Butler of Berkeley, who announced two new planets in January 1996. Their new find is a Jupiter-sized planet orbiting the star Rho Cancri (40 light-years from Earth) at a distance of only 10% of Earth's distance from the Sun. It completes a "year" in only about two Earth weeks.

The Value of G

Even though Newton had an equation for the gravitational force, he couldn't use it to actually calculate the force between two objects; he needed to know the value of the constant G. The way to get this is to measure the force between two known masses separated by a known distance. However, the force between two objects on Earth is so tiny that it couldn't be detected in Newton's time.

It was over 100 years after the publication of Newton's results before Henry Cavendish, a British scientist, developed a technique that was sensitive enough to measure the force between two masses. Modern measurements yield the value

gravitational constant

$$G = 0.000000000067 \ \text{N} \cdot \text{m}^2/\text{kg}^2 = 6.7 \times 10^{-11} \ \text{N} \cdot \text{m}^2/\text{kg}^2$$

(See Appendix B for an explanation of this notation.) Putting this value into the equation for the gravitational force tells us that the force between two 1-kilogram masses separated by 1 meter is only 0.000000000067 newton. This is minuscule compared with a weight of 10 newtons for each

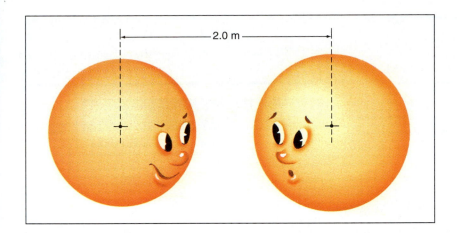

|← 2.0 m →|

The attraction between these spheroidal friends depends on the distance between their centers.

mass. The small value of *G* explains why two friends don't feel their mutual gravitational attraction when standing near each other.

Cavendish referred to his experiment as one that "weighs" the Earth although it would have been more accurate to claim that it "massed" the Earth. His point, though, was important. By measuring the value of *G*, Cavendish made it possible to accurately determine the mass of the Earth for the first time. The acceleration of a mass near the surface of the Earth depends on the value of *G*, the mass of the Earth, and the radius of the Earth. Since he now knew the values of all but the mass of the Earth, he could calculate it. The mass of the Earth is 5.98×10^{24} kilograms. That's about a million, million, million, million times larger than your mass.

Newton had worked on the problem of determining the value of *G*. Knowing that one could calculate the value if only the mass of the Earth were known, he made an estimate of the mass of the Earth. He used the known value of the radius of the Earth and estimated the average mass of each cubic meter of the Earth from an examination of materials near the Earth's surface. His value was in pretty good agreement with modern values but he had no way of verifying it.

Once the mass of the Earth is known, we can use the law of universal gravitation to calculate the acceleration due to gravity near the surface of the Earth:

$$g = \frac{F}{M} = \frac{GM_E}{R_E^2}$$

where M_E is the mass of the Earth and $R_E = 6370$ kilometers is the radius of the Earth. Plugging in the numerical values yields $g = 9.8$ (meters per second) per second, as expected.

Because the Earth orbits the Sun, the Sun's mass can also be calculated with the Cavendish results. In making these computations it is assumed that the value of *G* measured on Earth is valid throughout the Solar System. This cannot be proven. On the other hand, there is no evidence to the contrary, and this assumption gives consistent results. Newton made this same claim more than 100 years earlier.

How Much Do You Weigh?

In the 1600s people didn't imagine that anybody would one day travel to distant planets. Nevertheless, Newton's law of universal gravitation allowed them to predict what they would weigh if they ever found themselves on another planet.

According to Newton's law of universal gravitation, a person's weight on a planet depends on the mass and radius of the planet as well as the mass of the person. Your weight on Jupiter compared with that on Earth depends on these factors. Assuming that Jupiter is the same size as Earth (it isn't) but that it has 318 times as much mass as Earth (it does) means that you would weigh 318 times as much on this fictitious Jupiter as on Earth. Actually, Jupiter's diameter is 11.2 times that of Earth. Since the law of universal gravitation contains the radius squared in the denominator, your weight is actually reduced by a factor of 11.2 squared, or 125. Combining these two factors means that you would tip a Jovian bathroom scale at $^{318}/_{125} = 2\frac{1}{2}$ times your weight on Earth. You would be lightest on Pluto, weighing only 8 percent of your Earth weight because Pluto's small mass decreases your weight more than its small radius increases it. Your weight on each of the planets is given in the table.

On the Moon the force of gravity is only $\frac{1}{6}$ that on the Earth. Thus, an astronaut's weight is only $\frac{1}{6}$ of what it would be on Earth. This means that astronauts can jump higher and will fall more slowly, as we have seen from the television images transmitted to Earth during the lunar explorations. Vehicles designed for lunar travel would collapse under their own weight on Earth.

Weights on Each of the Planets		
Planet	**Relative**	**150-lb Person**
Mercury	0.38	57 lb
Venus	0.91	136
Earth	1.00	150
Mars	0.38	57
Jupiter	2.53	380
Saturn	1.07	160
Uranus	0.92	138
Neptune	1.18	177
Pluto	0.08	12

The lunar rover would collapse under its own weight if used on Earth.

COMPUTING GRAVITY   Σ

Let's calculate the gravitational force between two friends. To make the calculation of this force easier, we make one unrealistic assumption; we assume that the friends are spheres! This allows us to use the distance between their centers as their separation and still get a reasonable answer. Assuming that the friends have masses of 70 and 85 kilograms (about 154 and 187 pounds, respectively) and are standing 2.0 meters apart, we have

$$F = \frac{(6.7 \times 10^{-11} \text{ N} \cdot \text{m}^2/\text{kg}^2)(70 \text{ kg})(85 \text{ kg})}{(2.0 \text{ m})^2} = 1.0 \times 10^{-7} \text{N}$$

This very tiny force is about one 10-billionth (10^{-10}) of either friend's weight.

Gravity Near the Earth's Surface   Σ

In earlier chapters we assumed that the gravitational force on an object was constant near the surface of the Earth. We were able to do this because the force changes so little over the distances in question. In fact, to assume otherwise would have unnecessarily complicated matters.

Near the surface of the Earth, the gravitational force decreases by one part in a million for every 3 meters (about 10 feet) gain in elevation. Therefore, an object that weighs 1 newton at the surface of the Earth would weigh 0.999999 newton at an elevation of 3 meters. An individual with a mass of 50 kilograms has a weight of 500 newtons (110 pounds) in New York City; this person would weigh about 0.25 newton (one ounce) less in mile-high Denver.

The variations in the gravitational force result in changes in the acceleration due to gravity. The value of the acceleration—normally symbolized as g—is nearly constant near the Earth's surface. As long as one stays near the surface, the distance between the object and the Earth's center changes very slightly. If an object is raised one kilometer (about ½ mile) the distance changes from 6378 kilometers to 6379 kilometers and g only changes from 9.800 (meters per second) per second to 9.797 (meters per second) per second.

However, even without a change in elevation, g is not strictly constant from place to place. The Earth would need to be composed of spherical shells, with each shell being uniform. This is not the case. Underground salt deposits have less mass per cubic meter and give smaller values of g than average, whereas metal deposits produce larger g values. Therefore, measurements of g can be used to locate large-scale underground ore deposits. By noting variations, geologists can map regions for further exploration.

The passengers in this airplane weigh less because of their altitude.

Physics Update

New measurements of the Newtonian gravitational constant G, the number that determines the strength of gravity, depart significantly from the accepted value established in the 1980s. G is the least well known of all the fundamental constants; the accepted value of 6.6726×10^{-11} m³/kg·s² is known with a relatively high uncertainty of 0.01%. G is arguably the most difficult constant to measure because, among other reasons, gravity is the weakest of all forces and it is impossible to shield delicate measurements from the gravitational influences of buildings and other nearby objects. Underscoring this difficulty, in 1995 three scientists from international labs reported new measurements of G which disagreed widely with one another and with the standard value. The University of Wuppertal (Germany) value was 0.7% below the accepted value, the New Zealand Measurement Standards Laboratory measurements were 0.07 to 0.08% below, and the German Bureau of Standards value was 0.6% above. Although the techniques differed, the groups all essentially determined G by measuring the gravitational effects of cylindrical masses acting on objects suspended above the ground. Researchers at Los Alamos, the lab which helped set the 1980s standard, are undertaking a new measurement of G which may be five times as precise as current measurements and may shed light on these puzzling results.

Satellite dishes are aimed at communications satellites in geosynchronous orbits above the Earth's Equator.

GEOS weather satellites orbit the Earth once each day maintaining fixed locations above the Equator.

QUESTION   Given the fact that water has less mass per cubic meter than soil and rock, would you expect the value of *g* to be smaller or larger than average over a lake?

Satellites Σ

Newton's theory also predicts the orbits of artificial satellites orbiting the Earth. By knowing how the force changes with distance from the Earth, we know what accelerations—and, consequently, other orbital characteristics—to expect at different altitudes. For instance, a satellite at a height of 200 kilometers should orbit the Earth in 88.5 minutes. This is close to the orbit of the satellite *Vostok 6* that carried the first woman, Valentina Tereskova, into Earth orbit in June 1963. Its orbit varied in height from 170 to 210 kilometers and had a period of a little over 88 minutes.

Physics on Your Own   Many satellites have north–south orbits with periods of approximately 90 minutes. Search the night sky near the North Star until you locate one of these satellites moving southward. Estimate the time this satellite spends above the horizon. Why is this time much, much shorter than 45 minutes?

ANSWER   A cubic meter of water would provide less attraction than a cubic meter of soil and rock. Therefore, the value of *g* would be smaller.

The higher a satellite's orbit, the longer it takes to complete one orbit. The Moon takes 27.3 days; *Vostok 6* took 88 minutes. It is possible to calculate the height that a satellite would need to have a period of 1 day. With this orbit, if the satellite were positioned above the Equator, it would appear to remain fixed directly above one spot on the Earth—an orbit called geo (Earth) synchronous. Such geosynchronous satellites have an altitude of 36,000 kilometers, or about 5½ Earth radii, and are useful in establishing worldwide communications networks. Backyard satellite dishes that are used to pick up television signals point to geosynchronous satellites. The first successful geosynchronous satellite was *Syncom II,* launched in July 1963. Some geosynchronous satellites are used to monitor the weather on Earth.

> **QUESTION**   If you could spot a geosynchronous satellite in the sky, how could you distinguish it from a star?

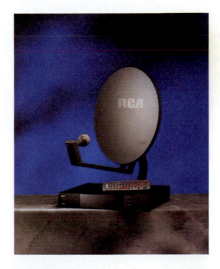

The 18" TV dishes receive digital signals from geosynchronous satellites.

> **Physics on Your Own**   Estimate the locations of the weather satellites from the views of the United States they present during the evening weather forecast and your knowledge of the possible orbits of geosynchronous satellites.

Any space probe requires the same computations as those done for satellites; NASA's computers calculate the trajectories for space flights using Newton's laws of motion and the law of gravitation. The forces on the spacecraft at any time depend on the positions of the other bodies in the Solar System. These can be calculated with the gravitation equation by

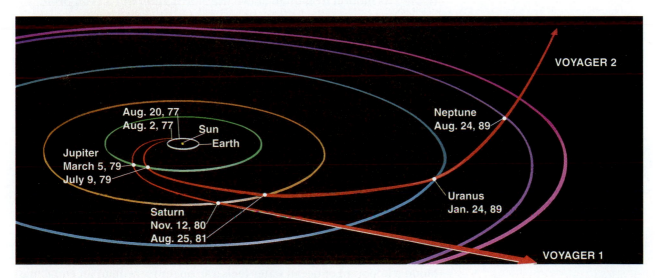

Diagram of the paths of *Voyagers 1* and *2* through the Solar System.

> **ANSWER**   The satellite remains in the same location in the sky, whereas the stars drift westward as the Earth rotates under them.

inserting the distance to and the mass of each body. The net force produces an acceleration of the spacecraft that changes its velocity. From this, the computer calculates a new position for the spacecraft. It also calculates new positions for the other celestial bodies, and the process starts over. In this manner the computer plots the path of the spacecraft through the Solar System.

Tides

No one was able to explain why we have tides before Newton's work with gravity. Some things were known. The tides are due to bulges in the surface of the Earth's oceans. There are two bulges, one on each side of the Earth as shown in Figure 4-7. The occurrence of tides at a given location is due to the rotation of the Earth. Imagine for simplicity that the bulges are stationary—pointing in some direction in space—and that the Earth is rotating. Each point on the Earth passes through both bulges in 24 hours and we have high tides at these times. Low tides occur halfway between the bulges. So we have two low and two high tides each day.

What wasn't known was why the Earth had these bulges. Newton claimed they were due to the force of gravity on the Earth by the Moon. The Earth exerts a gravitational force on the Moon which causes the Moon to orbit it. But the Moon exerts an equal and opposite force on the Earth that causes the Earth to orbit the Moon. Actually, both the Earth and the Moon orbit a common point located between them. This point is the center of mass of the Earth–Moon system. Because the Earth is so much more massive than the Moon, the center of mass is much closer to the Earth. In fact, its location is inside the Earth as shown in Figure 4-8. The Earth's orbital motion would look more like a wobble to somebody viewing its motion from high above the North Pole. But it is an orbit.

Figure 4-7   Exaggerated ocean bulges. As the Earth rotates, the bulges appear to move around the Earth's surface.

> **QUESTION**   Are the forces between the Earth and Moon a Newtonian third law pair?

Because the Earth has an orbital motion around the Moon, we can use the conclusions developed for the Moon's motion to help us understand Earth's tides. Namely, since we concluded that the Moon is continually falling toward the Earth, the Earth then is continually falling toward the Moon. This centripetal acceleration toward the Moon is the key to understanding tidal bulges.

Forget momentarily that the Earth is moving along its orbit and just consider the Earth falling toward the Moon as in Figure 4-9. This acceleration is the major contributor to the tides. Because the strength of the Moon's gravity gets weaker with increasing distance, the force on different parts of the Earth is different. For example, on the side nearest the Moon a kilogram of ocean water feels a stronger force than an equal mass

> **ANSWER**   Yes, one of the forces is the force of the *Earth on Moon,* and the other is the force of the *Moon on Earth.*

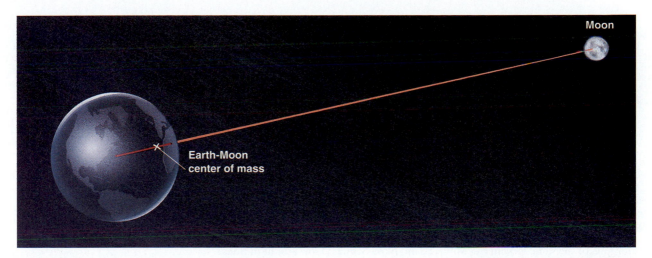

Figure 4-8 The center of mass of the Earth–Moon system is located inside the Earth.

of rock at the Earth's center. Similarly, a kilogram of material on the far side of the Earth feels a smaller force than both a kilogram on the near side and the one at the center.

If there are different-sized forces at different spots on the Earth, there are different accelerations for different parts. Parts of the Earth outrace other parts in their fall toward the Moon. Material on the side of the Earth facing the Moon tries to get ahead, while the material on the other side lags behind. Of course, the Earth has internal forces keeping it together that eventually balance these inequalities. But we do end up with a stretched-out Earth.

Although this reasoning accounts for the occurrence of the two high tides each day, it is too simple to get the details right. We observe that high tides do not occur at the same time each day. This happens because the Moon orbits the rotating Earth once a month. The normal time interval between successive high tides is 12 hours and 25 minutes. High tides do not occur when the Moon is overhead, but later—as much as 6 hours later. This is due to such factors as the frictional and inertial effects of the water and the variable depth of the ocean.

Although the height difference between low and high tides in the middle of the ocean is only about 1 meter, the shape of the shoreline can greatly amplify the tides. The greatest tides occur in the Bay of Fundy on the eastern seaboard between Canada and the United States; there the maximum range from low to high tide is 16 meters (54 feet)!

We would also expect to observe solar tides since the Sun also exerts a gravitational pull on the Earth and the Earth is "falling" toward it. These do occur, but their heights are a little less than one-half of those due to the Moon. This value may seem too small, taking into consideration that the Sun's gravitational force on the Earth is about 180 times greater than the Moon's. The solar effect is so small because it is the difference in the force from one side of the Earth to the other that matters and not the absolute size. The tides due to the planets are even smaller, that of Jupiter being less than one ten-millionth that due to the Sun.

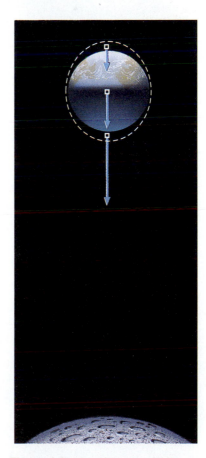

Figure 4-9 Equal masses of the Earth experience different gravitational forces due to their different distances from the Moon. The effect is exaggerated in the diagram.

Photographs of the same view in the Minas Basin, Nova Scotia, at low and high tides.

The continents are much more rigid than the oceans. Even so, the land experiences measurable tidal effects. Land areas may rise and fall as much as 23 centimeters (9 inches). Because the entire area moves up and down together, we don't notice this effect.

> **QUESTION**   Is the height of the high tide related to the phase of the Moon; that is, is it higher when the Sun and Moon are on the same side of the Earth (new moon), when they are on opposite sides (full moon), or when they are at right angles to each other (first- or third-quarter moon)?

How Far Does Gravity Reach?

The law of gravitation has been thoroughly tested within the Solar System. It accounts for the planets' motions, including their irregularities due to the mutual attraction of all the other planets.

What about tests outside the Solar System? We haven't sent probes out there. We are fortunate, however, as nature has provided us with ready-made probes. Astronomers observe that many stars in our galaxy revolve around a companion star. These binary star systems are the rule rather than the exception. These pairs revolve around each other in exactly the way predicted by Newton's laws.

Occasionally, a star is spotted that appears to be alone yet is moving in a circular path. Our faith in Newton's laws is so great that we assume a companion star is there; it is just not visible. Some of these invisible stars have later been detected because of signals they emit other than visible light.

> **ANSWER**   The highest high tides and the lowest low tides occur near new and full moons when the Earth, Moon, and Sun are in a line.

Photographs of star clusters show that the gravitational interaction occurs between stars. In fact, measurements show that all the stars in the Milky Way Galaxy are rotating about a common point under the influence of gravity. This has been used to estimate the total mass of the Galaxy and the number of stars in it. The Milky Way Galaxy is very similar in size and shape to our neighboring galaxy, the Andromeda Galaxy.

Such successes are a remarkable witness to Newton's genius. For over two centuries, scientists applied his laws of motion and the law of gravitation without discovering any discrepancies. However, some exceptions to the Newtonian world view were eventually discovered. It should not take away from his fame to admit these exceptions. They only occur when we venture very far from the realm of our ordinary senses. In the world of very high velocities and extremely large masses, we must replace Newton's ideas with the theories of special and general relativity (Chapters 12 and 13). In the world of the extremely small, we must use the theories of quantum mechanics (Chapter 23). It should be noted, however, that when these newer theories are applied in the realm where Newton's laws work, the new theories give the same results.

We also do not know if the value of G varies with time. No such variation has been detected, but a small variation with time could exist. Since the measurements of G are still limited in accuracy, it has been suggested that NASA orbit two satellites about each other. Accurate knowledge of the satellites' masses as well as their orbital data would give a more accurate value for G.

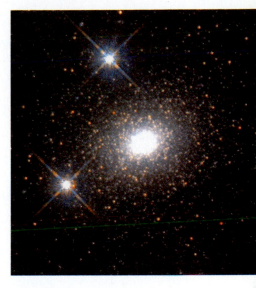

Star clusters provide evidence of gravity at work between stars.

The Andromeda Galaxy is similar to our own Milky Way Galaxy.

The Field Concept   Σ

Implicitly, we have assumed the force between two masses to be the result of some kind of direct interaction—sort of an action-at-a-distance interaction. This type of interaction is a little unsettling since there is no direct pushing or pulling mechanism in the intervening space. Gravitational effects are evident even in situations where there is a vacuum between the masses.

Another way to imagine the interaction is to consider one mass as somehow modifying the space surrounding it and then the other mass reacting to this modified space. This is the **field** concept. The field concept divides the task of determining the forces on a mass into two distinct parts: determining the field from the first mass and then calculating the force that this field exerts on the second mass.

If this were the only purpose of the field idea, it would play a minor role in our physics world view. In fact, it probably seems like we are trading one unsettling idea for another. However, as we continue our studies we will find that the field takes on an identity of its own and is a valuable aid in understanding these and many other phenomena.

gravitational field

By convention, the value of the **gravitational field** at any point in space is equal to the force experienced by one unit of mass if it was placed at that point. Because force is a vector quantity, the gravitational field is a

"THAT WRAPS IT UP—THE MASS OF THE UNIVERSE."

vector field; it has a size and a direction at each point in space. It is often convenient to talk about the gravitational field rather than the gravitational force. The strength of the gravitational force depends on the object being considered, whereas the strength of the gravitational field is independent of the object. The gravitational field at the surface of the Earth is 9.8 newtons per kilogram as expected.

SUMMARY

Although no one knows what gravity is or why it exists, we can accurately describe how gravity affects the motions of objects. The same laws of motion work on Earth and in the heavens. Newton's universal law of gravitation states that there is a gravitational attraction between every pair of objects given by

$$F = G\,\frac{M_1 M_2}{r^2}$$

where M_1 and M_2 are the masses of the two objects, r is the distance between their centers, and G is the gravitational constant. The value of G was first determined by Cavendish and is believed to be constant with time and space.

Ignoring air resistance, a projectile launched with a horizontal speed of 8 kilometers per second will go into orbit around the Earth. The projectile's speed and altitude remain constant and it does not return to Earth even though it is continually falling. The higher a satellite's orbit, the longer it takes to complete one orbit. A satellite with a period of 1 day and positioned above the Equator would appear to remain fixed in the sky. The Moon, a natural satellite, takes 27.3 days to complete one orbit around the Earth.

The force of gravity can be considered constant when the motion occurs over short distances near the Earth's surface. However, there are small variations in the acceleration due to gravity with latitude, elevation, and the types of surface material. At larger distances, the force decreases as the square of the distance. Stars in binary systems revolving around each other and the motion of stars within galaxies support this idea.

The value of the gravitational field at any point in space is equal to the force experienced by one unit of mass placed at that point.

C H A P T E R
4
R E V I S I T E D

Studying the motions of distant celestial objects as well as the distributions of mass within multiparticle systems such as globular clusters and galaxies provides direct evidence that gravity extends throughout the entire visible Universe.

KEY TERMS

field: A region of space that has a number or a vector assigned to every point.

force: A push or a pull. Measured by the acceleration it produces on a standard, isolated object, $F = ma$. Measured in newtons. (See the law of universal gravitation.)

gravitational field: The gravitational force experienced by one unit of mass placed at a point in space.

inverse-square: A relationship in which a quantity is related to the reciprocal of the square of a second quantity. An example is the law of universal gravitation; the force is inversely proportional to the square of the distance. If the distance is doubled, the force decreases by a factor of four.

law of universal gravitation: All masses exert forces on all other masses. The force F between any two objects is given by $F = GM_1 M_2/r^2$, where G is a universal constant, M_1 and M_2 are the masses of the two objects, and r is the distance between their centers.

mass: A measure of the quantity of matter in an object. The mass determines an object's inertia and the strength of its gravitational interaction. Measured in kilograms.

CONCEPTUAL QUESTIONS

1. What force (if any) drives the planets along their orbits?

2. What force (if any) causes the planets to execute (nearly) circular orbits?

3. If we are able to launch an apple horizontally with a speed of 40 meters per second, what path will it follow?

*4. Approximately what trajectory would you expect if we launched an apple with a speed of 10 kilometers per second?

5. The Earth exerts a gravitational force of 50,000 newtons on a satellite. What force does the satellite exert on the Earth?

6. Is the size of the gravitational force the Earth exerts on the Moon smaller than, larger than, or the same size as the force the Moon exerts on the Earth?

7. If an apple were placed in orbit at the same distance from Earth as the Moon, what acceleration would the apple have?

8. How does the average acceleration of the Moon about the Sun compare with that of the Earth about the Sun?

9. What happens to the surface area of a cube when the length of each side is doubled?

10. What happens to the volume of a cube if the length of each side is doubled? This is known as a cubic relationship.

11. Which of the following statements about the Moon is correct?

   a. The Moon has a constant velocity.

   b. There is no net force acting on the Moon.

   c. The Earth exerts a stronger force on the Moon than the Moon exerts on the Earth.

   d. The Moon experiences a centripetal acceleration toward the Earth.

12. Which of the following statements about the Moon is *not* correct?

   a. The acceleration due to gravity on the Moon is weaker than on Earth.

   b. The Earth's gravitational pull on the Moon equals the Moon's gravitational pull on Earth.

   c. There is a net force acting on the Moon.

   d. The Moon is not accelerating.

13. According to the law of universal gravitation, the force increases if the masses _____ and/or the distance between the objects _____.

14. Why didn't Newton have to know the mass of the Moon to obtain the law of universal gravitation?

15. According to the law of universal gravitation, the gravitational influence of the Earth gets weaker with distance. How far does it extend?

Question 6.

16. Please comment on the following statement made by a TV newscaster during an Apollo flight to the Moon: "The spacecraft has now left the gravitational force of the Earth."

17. How was the numerical value of *G*, the gravitational constant, determined?

*18. Why don't we define the value of *G* to be numerically equal to one, matching what we did with the proportionality constant in Newton's second law?

19. If a satellite in a circular orbit above the Earth is continually "falling," why doesn't it quickly return to Earth?

20. *Skylab* caused quite a commotion when it returned to Earth in July 1979. Why would it suddenly return to Earth after it had been in orbit for many years?

Question 20.

21. A person falling off a cliff feels weightless (and helpless!). Is the gravitational force on the person zero?

22. An astronaut in the Space Shuttle feels weightless. Is the gravitational force on the astronaut zero?

*23. If the Earth were hollow, but still had the same mass and radius, would your weight be different?

*24. If the Earth had half its present mass and half of its present radius, would your weight be different?

25. Why don't two people feel the gravitational attraction between them?

26. Why don't two pucks on an air-hockey table accelerate toward each other due to their mutual gravitational attraction?

27. What does it mean to "weigh" the Earth?

*28. How might you "weigh" the Moon?

29. The Sun exerts a gravitational force on the Moon. Does the Moon orbit the Sun?

30. Draw a picture of the Moon's orbit as viewed from a point far above the north pole of the Sun.

*31. Why do we use the form $W = mg$ for the gravitational force on a object near the Earth, but the form $F = GM_1M_2/r^2$ when the object is far from Earth?

32. Would you expect the value of *g* to be larger or smaller than normal over a large deposit of uranium ore?

33. How could we determine the mass of a planet such as Venus, which has no moon?

34. Why can we not determine the mass of the Moon by noting that it orbits the Earth in a nearly circular orbit? What can we do to determine the Moon's mass?

35. Would you weigh more in New York City or in Denver? Why?

36. Would you weigh more in a boat or on shore? Why?

37. What do you think would happen to the Moon's orbit if the gravitational attraction between the Moon and the Earth were slowly growing weaker?

38. What changes would occur in the Solar System if the gravitational constant *G* were slowly getting smaller?

39. Some of NASA's Earth satellites remain above a single location on Earth. Why don't these geosynchronous satellites fall to Earth under the influence of gravity?

40. Is it possible for an Earth satellite to remain "stationary" over Seattle? Why, or why not?

41. How would you tell the difference between a geosynchronous satellite and a star without looking through a telescope?

*42. Assume that NASA fails in its attempt to put a communications satellite into geosynchronous orbit. If the orbit is too small, what apparent motion will the satellite have as seen from the rotating Earth?

43. Newton's third law says that the gravitational force exerted on the Earth by the Moon is equal to that exerted on the Moon by the Earth. Why is it, then, that the Earth doesn't appear to orbit the Moon?

*44. Would you expect the Sun to be at rest in the center of the Solar System?

45. Would you expect to observe tides on the Great Lakes?

*46. Why would the inertia and friction of water cause the tides to occur after the Moon passes overhead?

47. Jupiter rotates once in 9 hours 50 minutes. How long is it between high tides in its atmosphere?

*48. The Moon is observed to keep the same side facing the Earth at all times. If the Moon had oceans, how much time would elapse between its high tides?

49. At which positions of the Moon in the figure do the highest tides occur?

50. At which positions of the Moon in the figure do the lowest tides occur?

51. Which positions of the Moon in the figure correspond to the largest difference between high tide and low tide?

52. Which positions of the Moon in the figure correspond to the smallest difference between high tide and low tide?

*53. In *The Jupiter Effect,* authors John Gribbin and Stephen Plagemann claim that the extra tidal force produced when all the planets lie along one line might be enough to trigger an earthquake along the San Andreas Fault in California. What do you think about the possibility?

54. How does the strength of the Earth's gravitational field fall off with increasing distance?

Questions 49–52.

EXERCISES

1. Given that the radius of the Earth is 6400 km, calculate the acceleration of an apple in orbit just above the surface of the Earth.

2. The speed of the Moon in its orbit is approximately 1 km/s and the Earth–Moon distance is 380,000 km. Show that these numbers yield an acceleration for the Moon that is very close to that given in the text.

3. What is the acceleration due to gravity at a distance of one Earth radius above the Earth's surface?

4. If you were located halfway between the Earth and the Moon, what acceleration would you have toward Earth? (Neglect the gravitational force of the Moon as it is much less than that of the Earth.)

5. The gravitational force between two metal spheres in outer space is 3600 N. How large would the force be if the two spheres were three times farther apart?

6. Two spacecraft in outer space attract each other with a force of 2 million N. What would the attractive force be if they were one-half as far apart?

7. The gravitational force between two metal spheres in outer space is 2000 N. How large would this force be if one of the spheres had twice the mass?

8. How would the Sun's gravitational force on the Earth change if the Earth had twice its present mass? Would the Earth's acceleration change?

9. How does the Earth's gravitational force on you compare when standing on the Earth and when riding in a satellite 6400 km above the Earth's surface? The Earth's radius is 6400 km.

10. By what factor is the gravitational force on an astronaut reduced when the spacecraft is 4 Earth radii away from the center of the Earth?

11. If the Earth shrank until its diameter were only one-half its present size without changing its mass, what would a 1-kg mass weigh at its surface?

12. If the Earth expanded to twice its diameter without changing its mass, what would happen to the value of g?

13. What is the gravitational force between two 5-kg iron balls separated by a distance of 0.1 m? How does this compare with the weight of either ball?

14. The masses of the Moon and Earth are 7.4×10^{22} kg and 6×10^{24} kg, respectively. The Earth–Moon distance is 3.8×10^8 m. What is the size of the gravitational force between the Earth and the Moon? Does the acceleration of the Moon produced by this force agree with the value given in the text?

15. If an astronaut in full gear has a weight of 1200 N on Earth, how much will the astronaut weigh on the Moon?

16. The acceleration due to gravity on Mars is about one-third that on Earth. What would a 40-kg scientific instrument weigh on Mars?

17. By what factor is the Earth's gravitational field reduced at a distance of 5 Earth radii away from the center of Earth?

18. The radius of Venus' orbit is approximately 0.7 times that of the radius of Earth's orbit. How much stronger is the Sun's gravitational field at Venus than at the Earth?

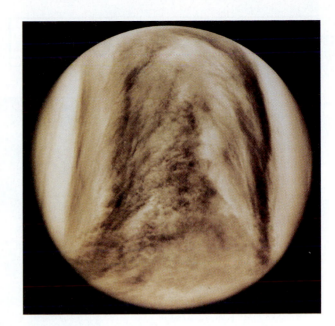

Exercise 18.

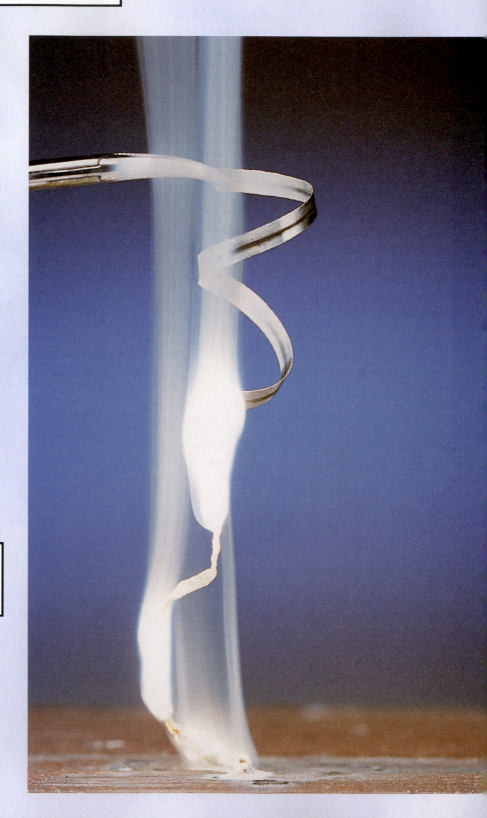

A piece of magnesium ribbon burns in air to form a white powder.

The Discovery of Invariants

One theme of the physics world view has been that change is an essential part of the Universe. In fact, change has been so pervasive in our journey that you might be surprised to learn that there are some things that don't change. There are a number of quantities that are, in fact, constant, or invariant. For something to be invariant, the numerical value associated with it must be constant, meaning that we obtain the same value at all times. We say that the quantity is conserved. In our everyday language, we often use the word *conserve* to mean that something is saved, or at least used sparingly. In physics, when something is conserved, it remains constant; its value does not change.

According to Swiss child psychologist Jean Piaget, an essential part of an individual's development is the establishment of invariants, or conservation rules. That is, as we process sensory data, we build constructs that are relatively permanent components of our personal world views. Most of this building occurs instinctively at an early age. We certainly don't go around consciously searching for invariants.

> **For something to be invariant, the numerical value associated with it must be constant, meaning that we obtain the same value at all times.**

A striking example of such a learned invariant that occurs at a very early age is the permanence of the physical size of objects. Although the size of the image on your retina gets smaller as an object moves away from you, common sense tells you that the object remains the same size. How do you know this? Piaget would claim that you have seen objects return to their original sizes many times and, therefore, you formed this invariant.

The concept that objects have a permanent size is so much a part of our common sense that our brain overrides the sensory information it receives. When you perceive objects as being far away through depth perception clues, you automatically compensate for the smaller retinal images. Artists can create nonsensical situations by providing our brain with conflicting information, as in the optical illusion shown in the photographs.

One of the first invariants discovered was the quantity of matter. When a piece of wood burns, it loses most of its mass. This loss is quite obvious if it is burned on an equal-arm balance. As the wood burns, the balance rises continually, indicating a

Which figure is taller?

decrease in mass. Where does the lost mass go? Does it actually disappear, or does it simply escape into the air?

Burning wood in a closed container gives different results. Provided there is enough air to allow the wood to burn and the products of the burning—the ashes and smoke—don't escape, the equal-arm balance remains level. Thus, the mass of the closed system does not change.

Near the end of the 18th century, the chemist Antoine Lavoisier did a large number of experiments showing that mass does not change when chemical reactions take place in closed flasks. These investigations led to the generalization that the mass of any closed system is invariant. Lavoisier's generalization is known as a law of nature; the law of **conservation of mass.** (When we study relativity and nuclear physics, we will discover that this law has to be modified. However, for ordinary physical and chemical processes this law is obeyed to a high degree of accuracy and is very useful in analyzing many processes.)

Other invariants have been discovered. Imagine a moving billiard ball hitting a stationary billiard ball head-on. After the collision, the first ball stops, and the second has a velocity that is equal to the original velocity of the first ball. It appears that something is transferred from the first ball to the second. Christian Huygens, a contemporary of Galileo and Newton, suggested that a "quantity of motion" is invariant in this collision. This quantity of motion is transferred from one ball to the other.

Discovering invariants is difficult. A lot of effort has been spent and will continue to be spent searching for them because the rewards are worth it. Their discovery yields powerful generalizations in the physics world view.

In principle, we can use Newton's second law to predict the motion of any object, whether it be a

As the wood burns, the left-hand side rises.

leaf, an airplane, a billiard ball, or even a planet. All we have to do is determine the net force acting on the object and calculate the resulting acceleration. Solving these equations, however, often turns out to be quite difficult. To use the second law we need to know all the forces acting on the object at all times. This is complicated for most motions, as the forces are often not constant in direction or size.

The use of invariants often allows us to bypass the details of the individual interactions. In some cases, such as those involving nuclear forces, it is not a case of trying to avoid these troubling complications, but rather taking the only possible path; we don't know the forces well enough to apply Newton's laws.

In the next two chapters we will look at two important invariants—momentum and energy. These two "quantities of motion" allow us to look at the behaviors of objects and systems of objects from an entirely new point of view.

Momentum

The collisions of billiard balls illustrate the law of conservation of momentum.

CHAPTER

5

Collisions send objects scattering in seemingly random directions. Yet nature has a scheme—a rule—that governs the path of each individual particle in collisions and explosions, whether it is a group of billiard balls or colliding galaxies. If the rule predicts the future, can it also determine the past? (See p. 122 for the answer to this question.)

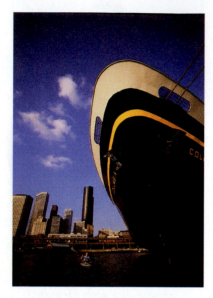

The kayak and the ship have different momenta even when they are moving at the same speed because of their different masses.

linear momentum = mass × velocity

An object is initially at rest. If you push on it for a certain time, it acquires a certain motion. We could describe the motion in terms of its velocity. But we could also use your sensations as a measure of the amount of motion. Suppose, for example, you try to stop the object. There is a definite difference in the sensation of catching a ball that is moving fast from that of catching one that is moving more slowly. The faster ball requires a larger effort to stop and therefore has more of this quantity of motion.

But there is more to this quantity of motion. All objects with the same velocity do not require the same effort to stop them. Consider a kayak and a battleship, each moving at 5 kilometers per hour. Which of these would you prefer to stop? The mass of the object is also an important part of this quantity of motion.

Linear Momentum

The examples discussed above are included in a new concept that quantifies motion. The **linear momentum** of an object is defined as the product of its inertial mass and its velocity. Momentum is a vector quantity which has the same direction as the velocity. Using the symbol **p** for momentum, we write the relationship as

$$\mathbf{p} = m\mathbf{v}$$

The adjective *linear* distinguishes this from another kind of momentum that we will discuss later in this chapter. Unless there is a possibility of confusion, this adjective is usually omitted.

There is no special unit for momentum as there is for force; the momentum unit is simply that of mass times velocity (or speed), that is, kilogram-meter per second (kg · m/s).

An object may have a large momentum due to a large mass and/or a large velocity. A slow battleship and a fast bullet have large momenta.

> **QUESTION**   Which has the greater momentum, an 18-wheeler parked at the curb or a Volkswagen rolling down a hill?

The word *momentum* is often used in our everyday language in a much looser sense but it is still roughly consistent with its meaning in the physics world view; that is, something with a lot of momentum is hard to stop. You have probably heard someone say, "We don't want to lose our momentum!" Coaches are particularly fond of this word.

> **ANSWER**   Since the truck has zero velocity, its momentum is also zero. Therefore, the VW has the larger momentum as long as it is moving.

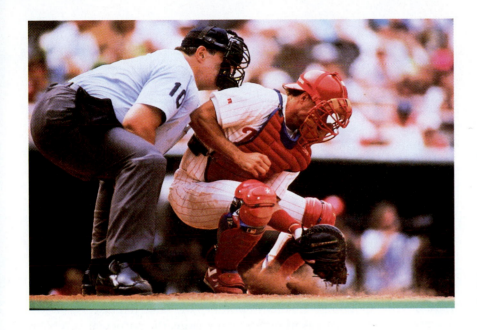

The catcher is protected from the baseball's momentum.

Changing an Object's Momentum

The momentum of an object changes if its velocity and/or its mass changes. We can obtain an expression for the amount of change by rewriting Newton's second law ($\mathbf{F} = m\mathbf{a}$) in a more general form. Actually, Newton's original formulation is closer to the new form. Newton realized that mass as well as velocity could change. His form of the second law says that the net force is equal to the change in the momentum divided by the time required to make this change.

$$\mathbf{F}_{net} = \frac{\Delta(m\mathbf{v})}{\Delta t}$$

net force = $\dfrac{\text{change in momentum}}{\text{time taken}}$

If we now multiply both sides of this equation by the time Δt, we get an equation that tells us how to produce a change in momentum.

$$\mathbf{F}_{net}\Delta t = \Delta(m\mathbf{v})$$

impulse = net force × time
        = change in momentum

This relationship tells us that this change is produced by applying a net force to the object for a certain time. The interaction that changes an object's momentum—a force acting for a time interval—is called **impulse.** Impulse is a vector quantity that has the same direction as the net force.

Since impulse is a product of two things, there are many ways to produce a particular change in momentum. For example, two ways of changing an object's momentum by 10 kilogram-meters per second are to exert a net force of 5 newtons on the object for 2 seconds, or to exert 100 newtons for 0.1 second. They each produce an impulse of 10 newton-seconds and therefore a momentum change of 10 kilogram-meters per second. The units of impulse (newton-seconds) are equivalent to those of momentum (kilogram-meters per second).

QUESTION   Which of the following will cause the larger change in the momentum of an object—a force of 2 newtons acting for 10 seconds or a force of 3 newtons acting for 6 seconds?

Modern cars employ air bags to increase stopping times to protect passengers.

Although the momentum change may be the same, certain effects depend on the particular combination of force and time. Suppose you had to jump from a second-story window. Would you prefer to jump onto a wooden or a concrete surface? Intuitively, you would choose the wooden one. Our commonsense world view tells us that jumping onto a surface that "gives" is better. But why is this so?

You undergo the same change in momentum with either surface; your momentum changes from a high value just before you hit to zero afterwards. The difference is in the time needed for the collision to occur. When a surface gives, the collision time is longer. Therefore, the net force must be correspondingly smaller to produce the same impulse.

Since our bones break when forces are large, the particular combination of force and time is important. For a given momentum change, a short collision time could cause large enough forces to break bones. You may break a leg landing on the concrete. On the other hand, the collision time with wood might be large enough to keep the forces in a huge momentum change from doing any damage.

This idea has many applications. Dashboards in cars are covered with foam rubber to increase the collision time during an accident. New cars are built with shock-absorbing bumpers to minimize damage to cars, and with air bags to minimize injuries to passengers. The barrels of water or sand in front of median dividers serve the same purpose. Stunt people are able to leap from amazing heights by falling onto large air bags that increase their collision times on landing. Volleyball players wear knee pads. Small pieces of styrofoam are used as packing material in shipping boxes to smooth out the bumpy rides.

Even without a soft surface, we have learned how to increase the collision time when jumping. Instead of landing stiff-kneed, we bend our knees immediately upon colliding with the ground. We are then brought to rest gradually rather than abruptly.

Physics on Your Own   Play catch with a raw egg. After each successful toss take one step away from your partner. How do you catch the egg to keep from breaking it?

A pole vaulter lands on thick pads to increase the collision time and thus reduce the force.

ANSWER   The larger impulse causes the larger change in the momentum. The first force yields (2 N)(10 s) = 20 N · s; the second (3 N)(6 s) = 18 N · s. Therefore, the first impulse produces the larger change.

Landing the Hard Way: No Parachute!

An extreme example of minimizing the effects of momentum change occurred during World War II. A Royal Air Force rear gunner jumped (without a parachute!) from a flaming Lancaster bomber flying at 5500 meters (18,000 feet). He attained a terminal speed (no pun intended) of over 54 meters per second (120 mph), but survived because his momentum change occurred in a series of small impulses with some branches of a pine tree and a final impulse from 46 centimeters (18 inches) of snow. Because this took a longer time than hitting the ground directly, the forces were reduced. Miraculously, he only suffered scratches and bruises.

The record for surviving a fall without a parachute is held by Vesna Vulovic. She was serving as a hostess on a Yugoslavian DC-9 that blew up at 33,330 feet (10,160 meters) in 1972. She suffered many broken bones and was hospitalized for 18 months after being in a coma for 27 days.

Source: *Guinness Book of Records*, Bantam Books, 1996.

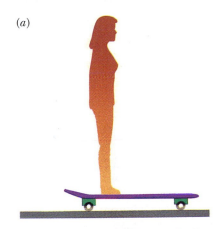

Conservation of Linear Momentum

Imagine standing on a giant skateboard that is at rest [Fig. 5-1(a)]. What is the total momentum of you and the skateboard? It must be zero, since everything is at rest. Now suppose that you walk on the skateboard. What happens to the skateboard? When you walk in one direction, the skateboard moves in the other direction as shown in Figure 5-1(b). An analogous thing happens when you fire a rifle: The bullet goes in one direction and the rifle recoils in the opposite direction.

These situations can be understood even though we don't know the values of the forces—and thus the impulses—involved. We start by assuming that there are no net forces external to the objects. In particular, we assume that the frictional forces are negligible and that any other external force—such as gravity—is balanced by other forces.

When you walk on the skateboard, there is an interaction. The force you exert on the skateboard is, by Newton's third law, equal and opposite to the force the skateboard exerts on you. The times during which these forces act on you and the skateboard must be the same since there is no way that one can touch the other without also being touched. As you and the skateboard each experience the same force for the same time, you must each experience the same-sized impulse and, therefore, the same-sized change in momentum.

But impulse and momentum are vectors, so their directions are important. Because the impulses are in opposite directions, the changes in the momenta are also in opposite directions. Thus, your momentum and that of the skateboard still add to zero. In other words, even though you and the skateboard are moving and, individually, have nonzero momenta, the total momentum remains zero. Notice that we arrived at this conclusion without considering the details of the forces involved. It is true for all forces between you and the skateboard.

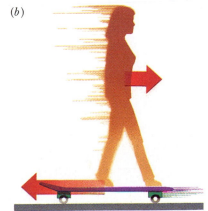

Figure 5-1   (a) Person and skateboard at rest have zero momentum. (b) When the person walks to the right, the board moves to the left, keeping the total momentum zero.

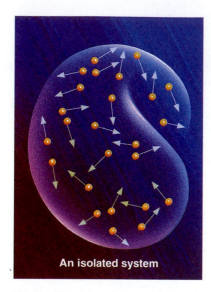

An isolated system

The total linear momentum of an isolated system is conserved. A system is isolated if there's no external force acting on it.

conservation of linear momentum

QUESTION   Suppose that the skateboard has half your mass and that you walk at a velocity of 1 meter per second to the left. Describe the motion of the skateboard.

Because the changes in the momenta of the two objects are equal in size and opposite in direction, the value of the total momentum does not change. We say that the total momentum is **conserved.**

We can generalize these findings. Whenever any object is acted on by a force, there must be at least one other object involved. This other object might be in actual contact with the first, or it might be interacting at a distance of 150 million kilometers, but it is there. If we widen our consideration to include all of the interacting objects, we gain a new insight.

Consider the objects as a system. Whenever there is no net force acting on the system from the outside (that is, the system is isolated, or closed), the forces that are involved act only between the objects within the system. As a consequence of Newton's third law, the total momentum of the system remains constant. This generalization is known as the law of **conservation of momentum.**

The total linear momentum of a system does not change if there are no net external forces.

This means that if you add up all of the momenta now and leave for a while, when you return and add them again, you will get the same number even if the objects were bumping and crashing into each other while you were gone. In practice, we apply the conservation of momentum to systems where the net external force is zero or the effects of the forces can be neglected.

You experience conservation of momentum first hand when you try to step from a small boat onto a dock. As you step toward the dock, the boat moves away from the dock and you may fall into the water. Although the same effect occurs when we disembark from an ocean liner, the large mass of the ocean liner reduces the velocity given it by our stepping off. A large mass requires a small change in velocity to undergo the same change in momentum.

Although we don't notice it, the same effect occurs whenever we walk. Our momentum changes; the momentum of something else must therefore change in the opposite direction. The something else is the Earth. Because of the enormous mass of the Earth, its speed need only change by an infinitesimal amount to acquire the necessary momentum change.

ANSWER   The skateboard must have the same momentum but in the opposite direction. Since it has half the mass, its speed must be twice as much. Therefore, its velocity must be 2 meters per second to the right.

COMPUTING Momentum $\boxed{\Sigma}$

Let's calculate the recoil of a rifle. A 150-grain bullet for a 30-06 rifle has a mass m of 0.01 kilograms and a muzzle velocity v of 900 meters per second (2000 mph). Therefore, the momentum p of the bullet is

$$p = mv = (0.01 \text{ kg})(900 \text{ m/s}) = 9 \text{ kg} \cdot \text{m/s}$$

Because the total momentum of the bullet and rifle was initially zero, conservation of momentum requires that the rifle recoil with an equal momentum in the opposite direction. If the mass M of the rifle is 4.5 kilograms, the speed V of its recoil is given by

$$V = \frac{p}{M} = \frac{9 \text{ kg} \cdot \text{m/s}}{4.5 \text{ kg}} = 2 \text{m/s}$$

If you do not hold the rifle snugly against your shoulder, the rifle will hit your shoulder at this speed (4.5 mph!) and hurt you.

QUESTION   Why does holding the rifle snugly reduce the recoil effects?

Collisions

Interacting objects don't need to be initially at rest for conservation of momentum to be valid. Suppose a ball moving to the left with a certain momentum crashes head-on with an identical ball moving to the right with the same-sized momentum. Before the collision, the two momenta are equal in size but opposite in direction, and since they are vectors, they add to zero.

After the collision the balls move apart with equal momenta in opposite directions. Because the masses of the balls are the same, the speeds after the collision are also the same. These speeds depend on the type of ball. The speeds may be almost as large as the original speeds in the case of billiard balls, quite a bit smaller in the case of lead balls, or even zero if the balls are made of a soft putty and stick together. In all cases the two momenta are the same size and in opposite directions. The total momentum remains zero.

QUESTION   A firecracker at rest in deep space explodes. What happens to its linear momentum?

ANSWER   Holding the rifle snugly increases the recoiling mass (your mass is now added to that of the rifle) and therefore reduces the recoil speed.

ANSWER   Because the forces are internal to the firecracker, its total linear momentum remains zero. If you add up all of the momenta of the fragments taking into account their directions, they will sum to zero.

We can use the conservation of momentum to measure the speed of fast-moving objects. For example, consider determining the speed of an arrow shot from a bow. We first choose a movable, massive target—a wooden block suspended by strings. Before the arrow hits the block [Fig. 5-2(a)], the total momentum of the system is equal to that of the arrow (the block is at rest). After the arrow is embedded in the block [Fig. 5-2(b)], the two move with a smaller, more measurable speed. The final momentum of the block and arrow just after the collision is equal to the initial momentum of the arrow. Knowing the masses, the arrow's initial speed can be determined.

Another example of a small, fast-moving object colliding with a much more massive object is graphically illustrated by one of your brave(?) authors, who lies down on a bed of nails as shown in Figure 5-3. (This in itself may seem like a remarkable feat. However, your author does not have

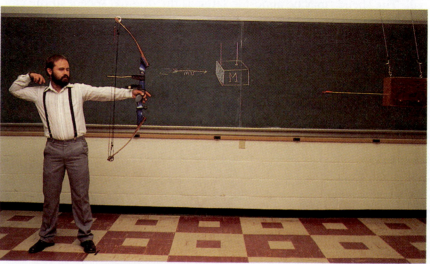

Figure 5-2 Determining the velocity of an arrow using conservation of momentum. The momentum of the block and arrow after the collision is equal to the momentum of the arrow before the collision.

(a)

(b)

Figure 5-3   Your author demonstrates some physics by letting a student break a concrete block on his chest while lying on a bed of nails.

to be a fakir with mystic powers, because he knows that the weight of his upper body is supported by 500 nails so that each nail only has to support 0.4 pound. It takes approximately 1 pound of force for the nail to break the skin.) Once on the bed of nails, he places a board on his chest and tops that off with a concrete block as shown in Figure 5-3(a). He then invites a student to break the concrete block with a sledgehammer [Fig. 5-3(b)].

This dramatic demonstration illustrates several ideas. The board on the chest spreads out the blow so that the force on any one part of the chest is small. Because it takes time for the hammer to break through the concrete block, the collision time is increased and, therefore, the force is decreased even further. Finally, momentum is conserved in the collision, but the much larger mass of your author ensures that the velocity imparted to his body is much less than the velocity of the hammer. This means that his body is only slowly pushed down onto the nails and the additional force that each nail must exert to stop his body is small. Therefore, your author's back is not perforated and he wins the admiration of his students without sacrificing his body.

Investigating Accidents

Accident investigators use conservation of momentum to reconstruct automobile accidents. Newton's laws can't be used to analyze the collision itself because we do not know the detailed forces involved. Conservation of linear momentum, however, tells us that regardless of the details of the crash, the total momentum of the two cars remains the same. The total momentum immediately before the crash must be equal to that immediately after the crash. Because the impact takes place over a very short time, we normally ignore frictional effects with the pavement and treat the collision as if there were no net external forces.

As an example, consider a rear-end collision. Assume that the front car was stopped and the two cars locked bumpers on impact. From an

The initial speeds of these cars can be determined by analyzing the collision.

analysis of the length of the skid marks made *after* the collision and the type of surface, the total momentum of the two cars just after the collision can be calculated. (We will see how to do this in Chapter 6; for now, assume we know their total momentum.) Because one car was stationary, the total momentum before the crash must have been due to the moving car. Knowing that the momentum is the product of mass and velocity (*mv*), we can compute the speed of the car just before the collision. We can thus determine whether the driver was speeding.

COMPUTING A Collision

Let's use conservation of momentum to analyze this collision. For simplicity, assume that each car has a mass of 1000 kilograms (a metric ton) and that the cars traveled along a straight line. Further assume that we have determined that the speed of the two cars locked together was 10 meters per second (about 22 mph) after the crash. The total momentum after the crash was equal to the total mass of the two cars multiplied by their combined speed.

$$p = (m_1 + m_2)\,v = (1000 \text{ kg} + 1000 \text{ kg})\,(10 \text{ m/s}) = 20{,}000 \text{ kg} \cdot \text{m/s}$$

But because momentum is conserved, this was also the value before the crash. Before the crash, however, only one car was moving. So if we divide this total momentum by the mass of the moving car, we obtain its speed.

$$v = \frac{p}{m} = \frac{20{,}000 \text{ kg} \cdot \text{m/s}}{1000 \text{ kg}} = 20 \text{ m/s}$$

The car was, therefore, traveling at 20 meters per second (about 45 mph) at the time of the accident.

> **QUESTION**   If the stationary car was not stationary, but slowly rolling in the direction of the total momentum, how would the calculated speed of the other car change if the final momentum remains the same?

Assuming that the cars stick together after the collision simplifies the analysis but is not required. Conservation of momentum applies to all types of collisions. Even if the two cars do not stick together, the original velocity can be determined if the velocity of each car just after the collision can be determined. The cars do not even have to be going in the same initial direction. If the cars suffer a head-on collision, we must be

> **ANSWER**   Because the rolling car accounts for part of the total momentum before the collision, the other car had less initial momentum and therefore a lower speed.

careful to include the directions of the momenta, but the procedure of equating the total momenta before and after the accident remains the same.

Because momentum is a vector, this procedure can also be used in understanding two-dimensional collisions such as occur when cars collide while traveling at right angles to each other. The total vector momentum must be conserved.

Physics on Your Own   Visit a billiards parlor and try some momentum experiments. Ask yourself if momentum is conserved in each situation. (Keep things simple by not putting any additional spin on the ball.)

a. A billiard ball strikes a stationary one head-on.
b. A billiard ball strikes two others placed one behind the other.
c. A head-on collision occurs between billiard balls with equal but opposite velocities.
d. A ball with a smaller mass (for instance, a snooker ball) collides head-on with a stationary billiard ball.
e. A billiard ball collides head-on with a stationary ball with a smaller mass.
f. A billiard ball experiences a glancing blow with a stationary one so that the two balls no longer travel along a straight line.

Studying physics in a billiards parlor improves both your physics and your game.

Airplanes, Balloons, and Rockets

Conservation of momentum also applies to flight. If we look only at the airplane, momentum is certainly not conserved. It has zero momentum before take-off and its momentum changes many times during a flight.

But if we consider the system of the airplane plus the atmosphere, momentum is conserved. In the case of a propeller-driven airplane, the interaction occurs when the propeller pushes against the surrounding air

In flight a Cessna 172's propeller blades push the air backward to go forward.

Figure 5-4 A fire extinguisher provides the impulse for a hallway rocket. Why does your author move backward?

molecules, increasing their momenta in the backward direction. This is accompanied by an equal change of the airplane's momentum in the forward direction. If we could ignore the air resistance, the airplane would continually gain momentum in the forward direction.

> **QUESTION**   Why doesn't the airplane continually gain momentum?

Release an inflated balloon and it takes off across the room. Is this similar to the propeller-driven airplane? No, because the molecules in the atmosphere are not necessary. The air molecules in the balloon rush out, acquiring a change in momentum toward the rear. This is accompanied by an equal change in momentum of the balloon in the forward direction. The air molecules do not need to push on anything; the balloon can fly through a vacuum.

This is also true of rockets and explains why they can be used in space flight. Rockets acquire changes in momentum in the forward direction by expelling gases at very high velocities in the backward direction. By choosing the direction of the expelled gases, the resulting momentum changes can also be used to change the direction of the rocket. An interesting classroom demonstration of this is often done using a modified fire extinguisher as the source of the high-velocity gas as shown in Figure 5-4.

> **Physics on Your Own**   You can make a popular model rocket (available in toy stores) work by filling it with water and then pumping it up with air. Place a piece of cardboard behind the rocket before you fire it. Notice where the water goes. What happens if you don't use any water?

Jet airplanes are intermediate between propeller-driven airplanes and rockets. Jet engines take in air from the atmosphere, heat it to high temperatures, and then expel it at high velocity out the back of the engine. The fast-moving gases impart a change in momentum to the airplane as they leave the engine. Although the gases do not push on the atmosphere, jet engines require the atmosphere as a source of oxygen for combustion.

Angular Momentum

There is another kind of momentum. An object orbiting a point has a *rotational* quantity of motion that is different from linear momentum. This new quantity is called **angular momentum** and is represented by the letter

> **ANSWER**   As the airplane pushes its way through the air, it hits air molecules, giving them impulses in the forward direction. This produces impulses on the airplane in the backward direction. In straight, level flight at a constant velocity the two effects cancel.

L. The size of the angular momentum in this example is equal to the object's linear momentum multiplied by the radius *r* of its circular path.

$$L = mvr$$

angular momentum = linear momentum × radius

A spinning object also has angular momentum because it is really just a large collection of tiny particles, each of which is revolving around the same axis. The total angular momentum of all the particles—the sum of the individual angular momenta—is equal to the angular momentum of the spinning object. If we were to do this summation, we would find that the angular momentum of the rotating object is equal to the product of its rotational inertia *I* (Chapter 3) and its rotational speed *ω*

$$L = I\omega$$

angular momentum = rotational inertia × rotational speed

which is analogous to the expression for the linear momentum.

The Earth has both types of angular momenta: the angular momentum due to its annual revolution around the Sun and that due to its daily rotation on its axis.

Conservation of Angular Momentum   Σ

The angular momentum of a system does not change under certain circumstances. The law of **conservation of angular momentum** is analogous to the conservation law for linear momentum. The difference is that the interaction that changes the angular momentum is a torque rather than a force.

> If the net external torque on a system is zero, the total angular momentum of the system does not change.

conservation of angular momentum

Note that the net external force need not be zero for angular momentum to be conserved. There can be a net external force acting on the system as long as the force does not produce a torque. This is the case for projectile motion because the force of gravity can be considered to act at the object's center of mass. Therefore, even though a thrown baton follows a projectile path, it continues to spin with the same angular momentum around its center of mass. There is no net torque on the baton.

There are some interesting situations where the angular momentum of a spinning object is conserved but the object changes its rotational speed. Near the end of a performance, many ice skaters go into a spin. The spin usually starts out slowly and then gets faster and faster. This might appear to be a violation of the law of conservation of angular momentum, but is, in fact, a beautiful example of its validity.

Angular momentum is the product of the rotational inertia and the rotational speed and, in the absence of a net torque, remains constant. Therefore, if the rotational inertia decreases, the rotational speed must increase. This is exactly what happens. The skater usually begins with arms extended. As the arms are drawn in toward the body, the rotational inertia of the body decreases because the mass of the arms is now closer

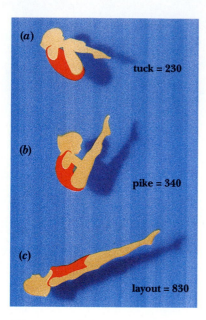

Figure 5-5   Relative values of the rotational inertia for the (a) tuck, (b) pike, and (c) layout positions.

Figure 5-6   In the game of tetherball the ball speeds up as the cord winds around the post, conserving angular momentum.

to the axis of rotation. This requires that the rotational speed increase. To slow the spin, the skater reverses the procedure by extending the arms to increase the rotational inertia.

The same principle applies to the flips and twists of gymnasts and springboard divers. The rate of rotation and hence the number of somersaults that can be completed depends on the rotational inertia of the body as well as the angular momentum and height generated during the take-off. The drawings in Figure 5-5 give the relative values of the rotational inertia for the tuck, pike, and layout positions. The more compact tuck has the smallest rotational inertia and therefore has the fastest rotational speed.

> **QUESTION**   Imagine you are executing a running front somersault when you suddenly realize that you are not turning fast enough to make it around to your feet. What can you do?

> **Physics on Your Own**   Sit on a rotating stool, start yourself spinning slowly with your arms extended, and draw your arms in toward your body. Why does your rotational speed increase? You can increase the effect by holding a large mass in each hand.

Another example is provided in the game of tetherball shown in Figure 5-6. As the ball winds around the pole, the force due to the tension in the rope is directed toward the edge of the pole. Therefore, there is a negligible net torque, and the angular momentum of the ball is almost conserved. As the rope gets shorter the radius of its motion continually gets shorter. Because its angular momentum (mvr) remains constant (as well as its mass), its speed must increase.

A similar thing happens as the Earth moves along its orbit. It is continually being attracted toward the Sun. Because the gravitational force acts toward a single point, there is no net torque on the Earth. And since there's no net torque, angular momentum must be conserved. As Kepler discovered, the orbit of the Earth about the Sun is not a circle, but an ellipse. Thus, the Earth is not always the same distance away from the Sun. This means that when the Earth is closer to the Sun, its speed must be larger in order to keep the angular momentum constant. Similarly, the Earth's speed must be smaller when it is farther away from the Sun.

The Solar System began as a huge cloud of gas and dust that had a very small rotation as part of its overall motion around the center of the Galaxy. As it collapsed under its mutual gravitational attraction, it rotated faster and faster in agreement with conservation of angular momentum. This explains why the planets all revolve around the Sun in the same direction and why the rotation of the Sun itself is also in this direction.

> **ANSWER**   You can tighten your tuck to reduce your rotational inertia. Because angular momentum is conserved, you will rotate faster.

Angular Momentum—A Vector

Like linear momentum, angular momentum is a vector quantity. The conservation of a vector quantity means that both the size and direction are constant. There are some interesting consequences of conserving the direction of angular momentum. The direction of the angular momentum is chosen to lie along the axis, or axle, of rotation. If you curl the fingers of your *right hand* along the direction of motion, your thumb points along the axis in the direction of the angular momentum (Fig. 5-7).

One important application of this principle is the use of a gyroscope for guidance in airplanes and spacecraft. A gyroscope is simply a disc that is rotating rapidly about an axle. The axle is mounted so that the mounting can be rotated in any direction without exerting a torque on the rotating disc (Fig. 5-8). Once the gyroscope is rotating, the axle maintains its direction in space no matter what the orientation of the spacecraft.

> **QUESTION**   Assume that you are at the North Pole holding a rapidly spinning gyroscope that has its angular momentum vector pointing straight up. Which way will it point if you transport it to the South Pole without exerting any torques on it?

Figure 5-7   If you curl the fingers of your right hand along the direction of motion, your thumb points along the axis in the direction of the angular momentum.

Figure 5-8   A spinning gyroscope maintains its direction in space even when its mount is rotated.

> **ANSWER**   It will point toward the ground. Remember this is the same direction (directly toward the North Star) as before. You have changed your orientation because your feet must point toward the center of the spherical Earth.

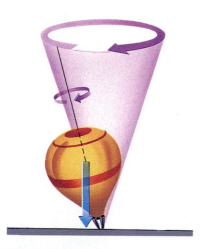

Figure 5-9   Can you make the pencil stand on its point without holding on to it?

Physics on Your Own   Push a pencil through the center of a circular piece of cardboard so that it looks like Figure 5-9. Can you make the pencil stand on its point without holding on to it? Try spinning it. Why does this work? What happens if you try this without the cardboard? Explain your observations.

A spinning top has angular momentum, but it is not usually constant. When the top's center of mass is not directly over the tip, the gravitational force exerts a torque on the top. If the top were not spinning, this torque would simply cause it to topple over. But when it is spinning, the torque produces a change in the angular momentum which causes the angular momentum to change its direction, not its size. The spin axis of the top (and hence its angular momentum) traces out a cone as shown in Figure 5-10. We say that the top *precesses*. The friction of the top's contact point with the table produces another torque. This torque reduces the size of the angular momentum and eventually causes the top to slow down and topple over.

A similar situation exists with the Earth. Because the shape of the Earth is irregular, the gravitational forces of the Sun and Moon on the Earth produces torques on the spinning Earth. These torques cause the Earth's spin axis to precess. This precession is very slow, but does cause the direction of our North Pole to sweep out a big cone in the sky once every 25,780 years (Fig. 5-11). Thus, Polaris (the pole star) is not always the North Star.

Figure 5-10   A top precesses due to the torque produced by the top's weight.

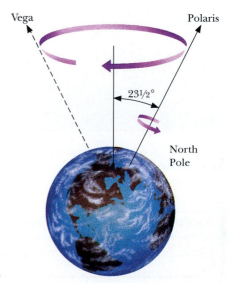

Figure 5-11   The precession of the Earth's axis causes the North Pole to follow a circular path among the stars.

Falling Cats

Cats have the amazing ability of landing on their feet regardless of their initial orientation. Modern strobe photographs have shown that the cat does not acquire a rotation by kicking off. The cat's initial angular momentum is zero. Because the force of gravity acts through the cat's center of mass, it produces no torque and the angular momentum remains zero.

The cat rotates by turning the front and hind ends of its body in different directions. The entire cat has zero angular momentum as long as the angular momenta of the two parts are equal and opposite. Even though these angular momenta are the same size, the amount of rotation can be different, because it depends on the rotational inertia of that part of the body. The cat adjusts the rotational inertia by retracting and extending its legs. The strobe photographs show this complex sequence.

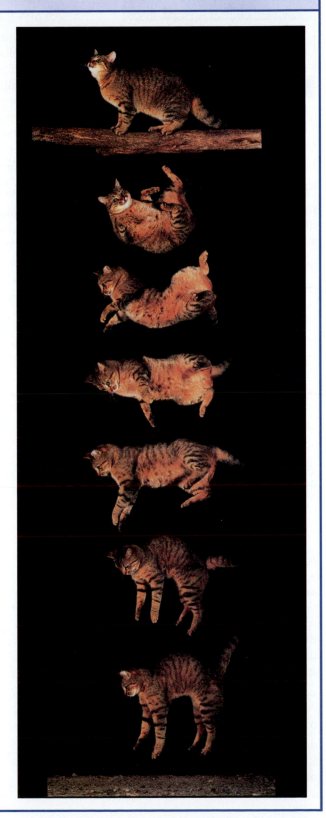

Strobe photos of a falling cat.

CHAPTER

5

REVISITED

The rule—called the conservation of linear momentum—is valid in both directions of time. If we know the velocities and masses of all objects at any time, we can back up the equations to see where the objects came from, and we can go forward to see where they are headed.

SUMMARY

The linear momentum of an object is defined as the product of its inertial mass and its velocity, $\mathbf{p} = m\mathbf{v}$. Momentum is a vector quantity that has the same direction as the velocity. The units for momentum are kilogram-meters per second (kg · m/s).

The momentum of an object changes if its velocity or its mass changes. This change is produced by an impulse, a net force acting on the object for a certain time $\mathbf{F}\Delta t$. Impulse is a vector quantity with the same direction as the force; this is also the direction of the change in momentum. There are many ways of producing a particular change in momentum by changing the strength of the force and the time during which it acts.

The momentum of a system is the vector sum of all the momenta of the system's particles. Assuming that there is no net external force acting on the system, the total momentum does not change. This generalization is known as the law of conservation of momentum. Conservation of momentum applies to many systems from balloons to billiard balls.

For an object orbiting a point, its angular momentum is defined as the product of its linear momentum and the radius of its circular path. For an extended object, its angular momentum is the product of its rotational inertia and its rotational speed.

The angular momentum of a system is conserved if no net external torque acts on the system. External forces may act on the system as long as these forces do not produce a net torque. Even though angular momentum is conserved, the rotational speed can change if the rotational inertia changes.

The conservation of a vector quantity means that both its size and direction are constant. Similarly, a change in angular momentum can be a change in size or direction or both. Conservation of angular momentum can be used to analyze problems ranging from the motion of tops to that of gymnasts and cats.

KEY TERMS

angular momentum: A quantity giving the rotational momentum. For an object orbiting a point, it is the product of the linear momentum and the radius of the path. For an extended body, it is the product of the rotational inertia and the rotational speed.

conservation of angular momentum: If the net external torque on a system is zero, the total angular momentum of the system does not change.

conservation of momentum: If the net external force on a system is zero, the total linear momentum of the system does not change.

conserved: This term is used in physics to mean that a number associated with a physical property does not change; it is invariant.

impulse: The product of the force and the time during which it acts. This vector quantity is equal to the change in momentum.

linear momentum: A vector quantity equal to the product of an object's mass and its velocity.

CONCEPTUAL QUESTIONS

1. What does it mean in physics to say that something is conserved?

2. Under what conditions is mass conserved?

3. How is linear momentum defined?

4. How is impulse defined?

5. Why are supertankers so hard to stop? To turn?

6. Which has the greater momentum, a supertanker docked at a pier or a motorboat pulling a water skier?

7. State Newton's second law in terms of momentum.

8. State Newton's first law in terms of momentum.

9. How does padding dashboards in automobiles make them safer?

10. How does the padding (or air pockets) in the soles of running shoes reduce the forces on your legs?

11. An astronaut training at the Craters of the Moon in Idaho jumps off a platform in full space gear and hits the surface at 5 meters per second. If later, on the Moon, the astronaut jumps from the landing vehicle and hits the surface at the same speed, will the impulse be larger, smaller, or the same as that on Earth?

12. Why is skiing into a wall of deep powder less hazardous to your health than skiing into a wall of bricks? Assume in both cases that you have the same initial speed and come to a complete stop.

13. Assume that a friend jumps from the roof of a garage and lands on the ground. How will the impulses the ground exerts on your friend compare if the landing is on grass or on concrete?

14. Why does an egg break when it is dropped onto the kitchen tile but not when it lands on the living room carpet?

15. Sometimes street fighters carry a roll of pennies in their hands to increase the effectiveness of their punches. What is the physics behind this?

16. Explain why the 12-ounce boxing gloves used in amateur fights hurt less than the 6-ounce ones used in professional fights.

17. Two people are playing catch with a ball. Describe the momentum changes that occur for the ball, the people, and the Earth. Is momentum conserved at all times?

18. Describe the momentum changes that occur when you dribble a basketball.

19. Which produces the larger impulse: a force of 3 newtons acting for 5 seconds or a force of 4 newtons acting for 4 seconds?

20. Which produces the larger change in momentum: a force of 3 newtons acting for 3 seconds or a force of 4 newtons acting for 2 seconds?

21. How can you explain the recoil that occurs when a rifle is fired?

*22. How might you design a rifle that does not recoil?

23. A student who recently studied the law of conservation of momentum decides to propel a go-cart by having a fan blow on a board as shown in the figure. This idea won't work. Why?

Question 23.

*24. If the force exerted on a rocket by the expelled gases from its engine is constant, does the momentum of the rocket increase at a constant rate? What happens to the momentum of the entire system (rocket plus expelled gases)?

25. Under what conditions is linear momentum conserved?

26. The momentum of a freely falling ball is not conserved. Why is this not a violation of the law of conservation of momentum?

27. Your teacher runs across the front of the classroom with a momentum of 200 kilogram-meters per second and foolishly jumps onto a giant skateboard. The skateboard is initially at rest and has a mass equal to your teacher's. If you ignore friction with the floor, what is the total momentum of your teacher and the skateboard before and after the landing?

28. A friend is standing on a giant skateboard that is initially at rest. If you ignore frictional effects with the floor, what is the momentum of the skateboard if your friend walks to the right with a momentum of 150 kilogram-meters per second?

29. The figure shows two air-track gliders held together with a string. A spring is tightly compressed between the gliders and released by burning the string. The mass of the glider on the left is twice that of the glider on the right and they are initially at rest. What is the total momentum of both gliders after the release?

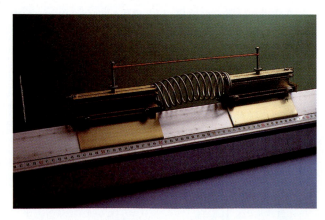

Question 29.

30. If the glider on the right in the preceding question has a speed of 1 meter per second after the release, how fast will the glider on the left be moving?

31. Explain why people who try to jump from rowboats onto docks often end up getting wet.

32. An astronaut in the Space Shuttle pushes off a wall to float across the room. What effect does this have on the motion of the Shuttle?

33. Sometimes a star "dies" in an enormous explosion known as a supernova. What happens to the total momentum of such a star?

*34. During a 4th of July celebration, a rocket is launched from the ground and explodes at the top of its arc. If we ignore air resistance, what happens to the total momentum of all of the rocket's fragments?

35. An astronaut is floating in the center of a space station with no translational motion relative to the station. Is it possible for the astronaut to move to the floor?

*36. During his last trip, Al the Astronaut happened on an enormous bag of gold coins floating in space. He quickly brought his spaceship to a halt, put on his space suit, tied a rope around his waist, and pushed off in the direction of the gold. But problems developed; as he reached the gold coins, the rope broke. Devise a way of getting Al back to his spaceship before his oxygen runs out. Although Al cares most about his life, the creative problem solver can get Al back alive with money.

37. Why does a small helicopter have a rotor on the tail?

38. Why does a helicopter with two sets of blades not need a rotor on the tail?

Question 38.

39. How does a rotary lawn sprinkler work?

40. What do you feel when you turn on an electric drill or a handheld mixer?

41. In which of the following positions would a diver have the smallest rotational inertia for performing a front somersault: tuck, pike, or layout?

42. Why is it possible for a high diver to execute more front somersaults in the tuck position than in the layout position?

43. Why do figure skaters spin faster when they pull in their arms?

Question 43.   Bob Martin/Tony Stone Images

44. An astronaut "floating" in the Space Shuttle has an initial rotational motion but no initial translational motions relative to the Shuttle. Why does the astronaut continue to rotate?

45. Where does a gymnast get the angular momentum to perform a full twist in the middle of a double somersault?

46. An astronaut is suspended in the center of the cargo bay in the Space Shuttle with no translational motion and no rotational motion relative to the Shuttle. How is it possible for the astronaut to change orientations?

Question 46.   NASA

47. A cat that is held upside down and dropped with no initial angular momentum manages to land on its feet. Does the cat need to acquire any angular momentum to do this? Explain your reasoning.

*48. Some people have proposed powering cars by extracting energy from large rotating flywheels mounted in the car. There is a problem with this suggestion if only one flywheel is used. What is the problem and how can it be remedied?

*49. If you sight down the inside of the barrel of a rifle, you see long spiral grooves. When the bullet travels down the barrel, these grooves cause the bullet to spin. Why would we want the bullet to spin?

*50. A billiard ball without spin hits perpendicular to the cushion and bounces back perpendicular to the cushion. However, if the ball is spinning about the vertical, it bounces off to one side and spins at a slower rate. What is the force that causes the change in the ball's linear momentum? Its angular momentum?

51. An air-hockey puck is whirling on the end of a string that passes through a small hole in the center of the table as shown in the figure. What happens to the speed of the puck as the string is slowly pulled down through the hole?

Question 51.

*52. Assume that you step onto a spinning merry-go-round. What happens to your angular momentum, the angular momentum of the merry-go-round, and the total angular momentum?

53. A gyroscope that points horizontally at the North Pole is transported to the South Pole while it continues to spin. Which way does it point?

54. A gyroscope is oriented so that it points toward the North Star when it is in Seattle. If it is carried to the Equator while it continues to spin, which way will it point?

55. The rotational axis of the Earth's spin is tilted 23½ degrees relative to the Earth's axis of revolution about the Sun. The North Pole is tilted toward the Sun on June 22. Which pole is tilted toward the Sun on December 22?

*56. Assume that you are sitting on a freely rotating stool holding a bicycle wheel with its axle vertical so that it rotates in a clockwise direction when viewed from above. What happens when you turn the wheel over?

57. What causes a spinning top to precess?

58. Why is the star Polaris not always located directly above the Earth's geographic North Pole?

*59. A piece of plastic with a very special shape is known as a "rattle-back." If you spin it in one direction, it will stop, rattle, and then spin in the other direction. How might this be possible?

*60. Some tops turn over after they have started spinning. Why doesn't this violate the law of conservation of momentum?

EXERCISES

1. What is the momentum of a 900-kg sports car traveling down the road with a speed of 30 m/s?

2. Calculate the approximate momentum of a world-class sprinter.

3. Does a quarterback with a mass of 100 kg running at 8 m/s have a larger or smaller momentum than a defensive tackle with a mass of 150 kg moving forward at 6 m/s?

4. Which has the larger momentum, an 18-wheeler (mass = 30,000 kg) traveling at 8 km/h (5 mph) or a minivan (mass = 2400 kg) traveling at 105 km/h (65 mph)?

5. What average net force is needed to accelerate a 1400-kg car to a speed of 108 km/h in a time of 7 s?

6. What average net force is needed to accelerate an 80-kg sprinter to a speed of 9 m/s in a time of 1 s? How does this compare with the sprinter's weight?

7. What impulse is needed to stop a 1200-kg car traveling at 30 m/s?

8. A 140-kg football player is running at 8 m/s. What impulse is needed to stop him?

9. A 1400-kg car has a speed of 30 m/s. If it takes 7 s to stop the car, what are the impulse and the average force acting on the car?

10. A pitcher throws a baseball (mass = 145 g) at 45 m/s (100 mph). What is the average force exerted on the catcher's glove if it requires 0.2 s for the ball to stop?

11. A coach is hitting pop flies to the outfielders. If the baseball (mass = 145 g) stays in contact with the bat for 0.05 s and leaves the bat with a speed of 70 m/s, what is the average force acting on the ball?

12. A 150-grain 30-06 bullet has a mass of 0.01 kg and a muzzle velocity of 900 m/s. If it takes 1 ms (millisecond) to travel down the barrel, what is the average force on the bullet?

13. A 30-06 rifle fires a bullet with a mass of 10 g at a velocity of 800 m/s. If the rifle has a mass of 4 kg, what is its recoil speed?

14. A woman with a mass of 50 kg runs at a speed of 8 m/s and jumps onto a giant skateboard with a mass of 25 kg. What is the combined speed of the woman and the skateboard?

15. A 2-kg ball traveling to the right with a speed of 4 m/s collides with a 5-kg ball traveling to the left with a speed of 2 m/s. What is the total momentum of the two balls before they collided? After they collided?

16. A 3-kg ball traveling to the right with a speed of 4 m/s collides with a 4-kg ball traveling to the left with a speed of 3 m/s. What is the total momentum of the two balls before and after the collision?

17. A 2-kg ball traveling to the right with a speed of 4 m/s overtook and collided with a 4-kg ball traveling in the same direction with a speed of 2 m/s. What was the total momentum before and after the collision?

18. A 1200-kg car traveling north at 12 m/s was rear-ended by a 2000-kg truck traveling at 26 m/s. What was the total momentum before and after the collision?

19. If the truck and car in the preceding problem lock bumpers and stick together, what is their speed immediately after the collision?

20. In a railroad hump yard an empty boxcar with a mass of 30 metric tons and a speed of 3 m/s (6 mph) collides with a stationary boxcar containing 30 metric tons of wheat. If the boxcars couple together, what speed do they have just after the collision?

21. A child with a mass of 40 kg is riding on a merry-go-round. If the child has a speed of 3 m/s and is located 1.5 m from the center of the merry-go-round, what is the child's angular momentum?

22. A 1400-kg car is traveling at 30 m/s around a curve with a radius of 100 m. What is the angular momentum of the car?

23. The planet Mercury follows an elliptical orbit that brings it as close as 46 million km to the Sun and as far as 70 million km from the Sun. At both of these locations the velocity of Mercury makes a right angle to the direction to the Sun. If Mercury's speed is 38 km/s when it is farthest from the Sun, how fast is it moving when it is closest to the Sun?

*24. Does Mars or the Earth have the larger angular momentum? The radius, speed, and mass of Mars are 1.5, 0.8, and 0.11 times those of Earth, respectively.

6

Energy is a powerful commodity. People and countries that know how to obtain and utilize energy are generally the wealthiest and most powerful. But what is energy? And what does it mean to say that energy is conserved? (See p. 150 for the answer to this question.)

Energy

Windmills are being developed for power generation to help alleviate the energy crisis.

In the preceding chapter we came to understand momentum by considering the effect of a force acting for a certain time. If, instead, we look at an object's motion after the force has acted for a certain distance, we find another quantity that is sometimes invariant, the energy of motion. This is, however, only one form of a more general and much more profound invariant known as energy.

The total energy of an isolated system does not change.

conservation of energy

Because of the energy crisis afflicting our society the word *energy* has probably appeared more often in the news than any other scientific term. In addition, it is one of the most fundamental and far-reaching concepts in the physics world view. It took nearly 300 years to fully develop the ideas of energy and energy conservation. In view of its importance and popularity, you might think that it would be easy to give a precise definition of energy. Not so.

What Is Energy?

Nobel laureate Richard Feynman in his *Lectures on Physics* captures the essential character of energy and its many forms when he discusses the law of conservation of energy. Feynman says

> there is a certain quantity, which we call energy, that does not change in the manifold changes which nature undergoes. That is a most abstract idea, because it is a mathematical principle; it says that there is a numerical quantity which does not change when something happens. It is not a description of a mechanism, or anything concrete; it is just a strange fact that we can calculate some number and when we finish watching nature go through her tricks and calculate the number again, it is the same. (Something like the bishop on a red square, and after a number of moves—details unknown—it is still on some red square. It is a law of this nature.) Since it is an abstract idea, we shall illustrate the meaning of it by an analogy.
>
> Imagine a child, perhaps "Dennis the Menace," who has blocks which are absolutely indestructible, and cannot be divided into pieces. Each is the same as the other. Let us suppose that he has 28 blocks. His mother puts him with his 28 blocks into a room at the beginning of the day. At the end of the day, being curious, she counts the blocks very carefully, and discovers a phenomenal law—no matter what he does with the blocks, there are always 28 remaining! This continues for a number of days, until one day there are only 27 blocks, but a little investigating shows that there is one under the rug— she must look everywhere to be sure that the number of blocks has not changed. One day, however, the number appears to change—there are only 26 blocks. Careful investigation indicates that the window was open, and upon looking outside, the other two blocks are found. Another day, careful count indicates that there are 30 blocks! This causes considerable consternation, until it is realized that Bruce came to visit, bringing his blocks with him, and he left a few at Dennis' house. After she has disposed of the extra blocks, she closes the window, does not let Bruce in, and then everything is going along all right, until one time she counts and finds only 25 blocks.

However, there is a box in the room, a toy box, and the mother goes to open the toy box, but the boy says "No, do not open my toy box," and screams. Mother is not allowed to open the toy box. Being extremely curious, and somewhat ingenious, she invents a scheme! She knows that a block weighs three ounces, so she weighs the box at a time when she sees 28 blocks, and it weighs 16 ounces. The next time she wishes to check, she weighs the box again, subtracts sixteen ounces and divides by three. She discovers the following:

$$\left(\begin{array}{c} \text{number of} \\ \text{blocks seen} \end{array} \right) + \frac{(\text{weight of box}) - 16 \text{ ounces}}{3 \text{ ounces}} = \text{constant}$$

There then appear to be some new deviations, but careful study indicates that the dirty water in the bathtub is changing its level. The child is throwing blocks into the water, and she cannot see them because it is so dirty, but she can find out how many blocks are in the water by adding another term to her formula. Since the original height of the water was 6 inches and each block raises the water a quarter of an inch, this new formula would be:

$$\left(\begin{array}{c} \text{number of} \\ \text{blocks seen} \end{array} \right) + \frac{(\text{weight of box}) - 16 \text{ ounces}}{3 \text{ ounces}}$$

$$+ \frac{(\text{height of water}) - 6 \text{ inches}}{\frac{1}{4} \text{ inch}} = \text{constant}$$

In the gradual increase in the complexity of her world, she finds a whole series of terms representing ways of calculating how many blocks are in places where she is not allowed to look. As a result, she finds a complex formula, a quantity which *has to be computed,* which always stays the same in her situation. . . .

The analogy has the following points. First, when we are calculating the energy, sometimes some of it leaves the system and goes away, or sometimes some comes in. In order to verify the conservation of energy, we must be careful that we have not put any in or taken any out. Second, the energy has a large number of *different forms,* and there is a formula for each one. . . . If we total up the formulas for each of these contributions, it will not change except for energy going in or out.

It is important to realize that in physics today, we have no knowledge of what energy is. We do not have a picture that energy comes in little blobs of a definite amount. It is not that way. However, there are formulas for calculating some numerical quantity, and when we add it all together it gives "28"— always the same number. It is an abstract thing in that it does not tell us the mechanisms or the *reasons* for the various formulas.*

Energy of Motion

The most obvious form of energy is the one an object has because of its motion. We call this quantity of motion the **kinetic energy** of the object. Like momentum, kinetic energy depends on the mass and the motion of

*R.P. Feynman, R.B. Leighton, and M. Sands. *The Feynman Lectures on Physics.* Vol. I, pp. 4-1 and 2. Reading, MA: Addison-Wesley, 1963.

the object. But the expression for the kinetic energy is different from that for momentum. The kinetic energy *KE* of an object is

$$KE = \tfrac{1}{2}mv^2$$

kinetic energy = ½ mass × speed squared

where the factor of ½ makes the kinetic energy compatible with other forms of energy that we will study later.

Notice that the kinetic energy of an object increases with the square of its speed. This means that if an object has twice the speed, it has four times the kinetic energy; if it has three times the speed, it has nine times the kinetic energy; and so on.

> **QUESTION** What happens to the kinetic energy of an object if its mass is doubled while its velocity remains the same?

The units for kinetic energy, and therefore for all types of energy, are kilograms multiplied by (meters per second) squared (kg · m²/s²). This energy unit is called a **joule** (rhymes with tool). Kinetic energy differs from momentum in that it is not a vector quantity because an object has the same kinetic energy regardless of its direction as long as its speed does not change.

A typical textbook dropped from a height of 10 centimeters (about 4 inches) hits the floor with a kinetic energy of about 1 joule (J). A sprinter crosses the finish line with as much as 5000 joules of kinetic energy.

COMPUTING | KINETIC ENERGY

We can use this relationship to calculate the kinetic energy of a 70-kilogram (154-pound) person running at a speed of 8 meters per second. Putting these values into the equation, we have

$$KE = \tfrac{1}{2}mv^2 = \tfrac{1}{2}(70\ \text{kg})(8\ \text{m/s})^2 = (35\ \text{kg})(64\ \text{m}^2/\text{s}^2) = 2240\ \text{J}$$

> **QUESTION** An object with a speed of 4 meters per second has a kinetic energy of 20 joules. How much kinetic energy would it have if its speed were reduced to 2 meters per second?

> **ANSWER** Since the kinetic energy is directly proportional to the mass, the kinetic energy doubles.

> **ANSWER** Cutting the speed in half reduces the kinetic energy by a factor of 4. Therefore, the kinetic energy would be 5 joules.

Conservation of Kinetic Energy

The search for invariants of motion often involved collisions. In fact, early in their development the concepts of momentum and kinetic energy were often confused. Things became much clearer when these two were recognized as distinct quantities. We have already seen that momentum is conserved during collisions. Under certain, more restrictive, conditions, kinetic energy is also conserved before and after collisions.

Consider the collision of a billiard ball with a hard wall. Obviously the kinetic energy of the ball is not constant. At the instant the ball reverses its direction, its speed is zero and, therefore, its kinetic energy is zero. As we will see, even if we include the kinetic energy of the wall and the Earth, the kinetic energy of the system is not conserved.

However, if we don't concern ourselves with the details of what happens during the collision and only look at the kinetic energy before and after the collision, we find that the kinetic energy is nearly conserved. During the collision the ball and the wall distort, resulting in internal frictional forces that reduce the kinetic energy slightly. But once again we ignore some things to get at the heart of the matter. Let us assume we have "perfect" materials and can ignore these frictional effects. In this case the kinetic energy of the ball after it leaves the wall equals its kinetic energy before it hit the wall. Collisions in which kinetic energy is conserved are known as **elastic** collisions.

elastic collisions conserve kinetic energy

Actually, many atomic and subatomic collisions are perfectly elastic. A larger scale approximation of an elastic collision would be one between air-hockey pucks that have magnets stuck on top so that they repel each other.

Figure 6-1   The collision-ball apparatus demonstrates the conservation of momentum and kinetic energy.

Physics on Your Own   The apparatus shown in Figure 6-1 may look familiar. It has five balls of equal mass suspended in a line so that they can swing along the direction of the line. If you pull on a ball at one end and release it, the last ball at the other end leaves with the velocity of the original. Knowing that momentum and kinetic energy are conserved, can you predict what will happen if you pull back two balls and release them together? Try this experiment if you have one of these toys.

Collisions in which kinetic energy is lost are known as **inelastic** collisions. The loss in kinetic energy shows up as other forms of energy, primarily in the form of heat, which we discuss in Chapter 9. Collisions in which the objects move away with a common velocity are never elastic. We will see this in the billiard ball example presented at the end of this section.

The outcomes of collisions are determined by the conservation of momentum and the extent to which kinetic energy is conserved. We know that the collisions of billiard balls are not perfectly elastic because we hear them collide. (Sound is a form of energy and therefore carries off some of the energy.)

Physics on Your Own   Determine the relative elasticities of balls made of various materials by dropping them from a uniform height onto a very hard surface, such as a concrete floor or a thick steel plate. The more elastic the material, the closer the ball will return to its original height. Is a rubber ball more or less elastic than a glass marble? Do your results agree with the everyday use of the word *elastic*?

COMPUTING   CONSERVATION OF KINETIC ENERGY Σ

Another example of nearly elastic collisions occurs between very hard objects such as billiard balls. Consider the head-on collision of a moving billiard ball with a stationary billiard ball [Fig. 6-2(a)]. The moving ball stops and the stationary one acquires the initial velocity of the moving ball [Fig. 6-2(b)]. Suppose that the mass of each ball is 0.5 kilogram and the initial velocity is 4 meters per second. The total momenta before and after the collision are

$$p(\text{before}) = m_1 v_1 + m_2 v_2 = (0.5 \text{ kg})(4 \text{ m/s}) + 0 = 2 \text{ kg} \cdot \text{m/s}$$

$$p(\text{after}) = m_1 v_1 + m_2 v_2 = 0 + (0.5 \text{ kg})(4 \text{ m/s}) = 2 \text{ kg} \cdot \text{m/s}$$

Therefore, momentum is conserved. Likewise, the kinetic energies before and after the collision are

$$KE(\text{before}) = \tfrac{1}{2}m_1 v_1^2 + \tfrac{1}{2}m_2 v_2^2 = \tfrac{1}{2}(0.5 \text{ kg})(4 \text{ m/s})^2 + 0 = 4 \text{ J}$$

$$KE(\text{after}) = \tfrac{1}{2}m_1 v_1^2 + \tfrac{1}{2}m_2 v_2^2 = 0 + \tfrac{1}{2}(0.5 \text{ kg})(4 \text{ m/s})^2 = 4 \text{ J}$$

Therefore, kinetic energy is also conserved.

But this is not the only possibility that conserves momentum. Another is for the two balls to both move in the forward direction, each with one-half of the initial velocity [Fig. 6-2(c)]. Both possibilities conserve the total momentum of the two-ball system. Yet when we do the experiment, the first possibility is the one that we always observe.

Why does one occur and not the other? The first possibility conserves kinetic energy, but the second does not. To see that the second doesn't conserve kinetic energy, we again calculate the kinetic energy after the collision.

$$KE(\text{after}) = \tfrac{1}{2}m_1 v_1^2 + \tfrac{1}{2}m_2 v_2^2$$

$$= \tfrac{1}{2}(0.5 \text{ kg})(2 \text{ m/s})^2 + \tfrac{1}{2}(0.5 \text{ kg})(2 \text{ m/s})^2 = 2 \text{ J}$$

One-half of the kinetic energy would be lost in this case.

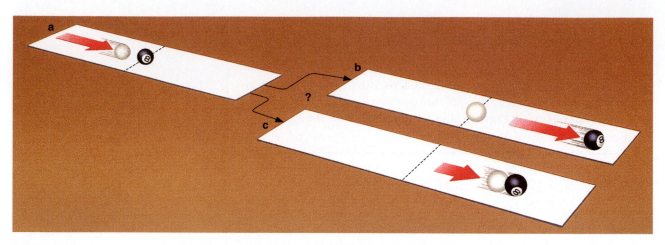

Figure 6-2   (a) A moving billiard ball collides head-on with a stationary one. Which possibility occurs? (b) The moving ball stops and the stationary ball takes on the initial velocity. (c) The balls move off together with one-half the initial velocity.

Changing Kinetic Energy

A box sliding along a frictionless, horizontal surface has a certain kinetic energy because it is moving. A net force on the box can change its speed and thus its kinetic energy. If you push in the direction it is moving, you increase the box's kinetic energy. Pushing in the opposite direction slows the box down and decreases its kinetic energy (Fig. 6-3).

When a force acts for a certain time, it produces an impulse that changes the momentum of the object. In contrast, the *distance* through

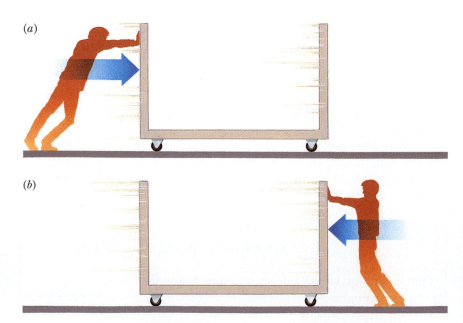

Figure 6-3   (a) A force in the direction of motion increases the kinetic energy. (b) A force opposite the direction of motion decreases the kinetic energy.

which the force acts determines how much the kinetic energy changes. The product of the force *in the direction of motion F* and the distance moved *d* is known as **work** *W.*

$$W = Fd$$

work = force × distance moved

From the definition of work, we conclude that the units of work are newton-meters. If we substitute our previous expression for a newton $(kg \cdot m/s^2)$, we find that a newton-meter $(N \cdot m)$ equals a $kg \cdot m^2/s^2$, which is the same as the units for energy; that is, a joule. Therefore, the units of work are the same as those of energy. In the English system the units of work are foot-pounds.

Stopping Distances for Cars

The data for stopping distances given in driver's manuals do not seem to have a simple pattern, other than the faster you drive, the longer it takes to stop. The driver's manual for the state of Montana, for example, states that it takes at least 186 feet to stop a car traveling at 50 mph and only 65 feet for a car which is traveling at 25 mph. Why is the stopping distance ⅓ as much when the speed is ½ as much? Furthermore, these tables make no mention of the size of the car.

A car's kinetic energy depends on its speed and its mass. When the brakes are applied, the brake pads slow the wheels, which, in turn, apply a force on the highway. The reaction force of the highway on the car does work on the car, reducing its kinetic energy to zero. To stop the car, the total work done on the car must be equal to its initial kinetic energy.

This frictional force between the tires and the road depends on the car's mass and whether it is rolling or skidding. (Skidding greatly reduces the frictional force.) The force of friction does not change very much for various tire designs or road surfaces providing the roads are not wet or icy. Since both the frictional force and kinetic energy are proportional to the car's mass, the stopping distance is independent of the car's mass. So how far the car travels after the brakes are applied depends almost entirely on its speed.

The distances in driver's manual tables include the distance traveled during approximately 1 second of reaction time in addition to the distance required for the brakes to stop the car. At 50 mph the car is traveling at 74 feet per second, so the distance required to stop the

car once the brakes are applied is 186 feet − 74 feet = 112 feet.

Because the car's kinetic energy changes as the square of its speed, a car that is traveling at 25 mph has only ¼ the kinetic energy of one traveling at 50 mph. Because it only requires ¼ as much work to stop the car, the force must only act through ¼ the distance. A car traveling at 25 mph can stop in ¼ × 112 feet = 28 feet. During the 1-second reaction time the car travels an additional ½ × 74 feet = 37 feet, so that the total stopping distance is 28 feet + 37 feet = 65 feet. The stopping distances for other speeds are given in the table.

Stopping Distances for Automobiles Traveling at Selected Speeds				
Speed		**Stopping Distance**		
(mph)	(ft/s)	Reaction (ft)	Braking (ft)	Total (ft)
10	15	15	5	20
20	29	29	18	47
30	44	44	40	84
40	59	59	72	131
50	74	74	112	186
60	88	88	161	249
70	103	103	220	323
80	117	117	287	404
90	132	132	363	495
100	147	147	448	595

Newton's second law and our expressions for acceleration and distance traveled can be combined with the definition of work to show that the work done on an object is equal to the change in its kinetic energy

$$W = \Delta KE$$

work done = change in kinetic energy

If the net force is in the same direction as the velocity, the work is positive and the kinetic energy increases. If the net force and velocity are oppositely directed, the work is negative and the kinetic energy decreases.

Forces That Do No Work

The meaning of *work* in physics is different from the common usage of the word. Commonly, people talk about "playing" when they throw a ball and "working" when they study physics. The physics definition of work is quite precise—work occurs when the product of the force and the distance is nonzero. When you throw a ball, you are actually doing work on the ball; its kinetic energy is increased because you apply a force through a distance. Although you may move pages and pencils as you study physics, the amount of work is quite small.

Similarly, if you hold a suitcase above your head for 30 minutes, you would probably claim it was hard work. According to the physics definition, however, you did not do any work on the suitcase; the 30 minutes of straining and groaning did not change the suitcase's kinetic energy. A table could hold up the suitcase just as well. Your body, however, is doing physiological work because the muscles in your arms do not lock in place but rather twitch in response to nerve impulses. This work shows up as heat (as evidenced by your sweating) rather than as a change in the kinetic energy of the suitcase.

There are other situations in which a net force does not change an object's kinetic energy. If the force is applied in a direction perpendicular to its motion, the velocity of the object changes but its speed doesn't. Therefore, the kinetic energy does not change. The definition of work takes this into account by stating that it is only the force that acts along the direction of motion that can do work.

Often, a force is neither parallel nor perpendicular to the velocity of an object. Because force is a vector, we can think of it as having two components, one that is parallel and one that is perpendicular to the motion as illustrated in Figure 6-4. The parallel component does work but the perpendicular one does not do any work.

It takes no work to hold a cheerleader in the air.

Consider an air-hockey puck moving in a circle on the end of a string attached to the center of the table shown in Figure 6-5. Because the speed is constant, the kinetic energy is also constant. The force of gravity is balanced by the upward force of the table. These vertical forces cancel and do no work. The centripetal force that the string exerts on the puck is not canceled, but does no work because it acts perpendicular to the direction of motion.

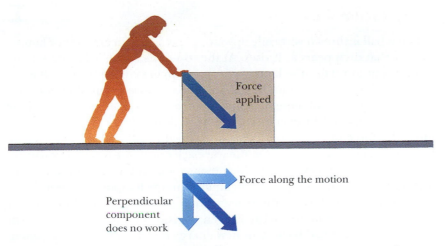

Figure 6-4   Any force can be replaced by two perpendicular component forces. Only the component along the direction of motion does work on the box.

If the orbit of the Earth were a circle with the Sun at the center, the gravitational force on the Earth would do no work and the Earth would have a constant kinetic energy and therefore a constant speed. However, since the orbit is an ellipse, the force is not always perpendicular to the direction of motion (Fig. 6-6) and therefore does work on the Earth. During one-half of each orbit a small component of the force acts in the direction of motion, increasing the kinetic energy and the speed. During the other half of each orbit, the component is opposite the direction of motion, and the kinetic energy and speed decrease.

Figure 6-5   When the force is perpendicular to the velocity, the force does no work.

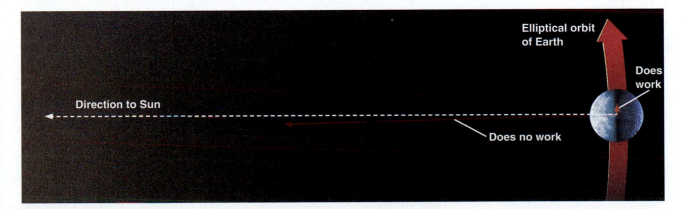

Figure 6-6   The gravitational force of the Sun does work on Earth whenever the force is not perpendicular to the Earth's velocity. The elliptical nature of the orbit has been exaggerated to show the parallel component.

Gravitational Energy   Σ

When a ball is thrown vertically upward, it has a certain amount of kinetic energy that disappears as it rises. At the top of its flight, it has no kinetic energy, but as it falls, the kinetic energy reappears. If we believe that energy is an invariant, we must be missing one or more forms of energy.

The loss and subsequent reappearance of the ball's kinetic energy can be understood by examining the work done on the ball. As the ball rises, the force of gravity performs work on the ball, reducing its kinetic energy until it reaches zero. On the way back down, the force of gravity increases the ball's kinetic energy by the same amount it lost on the way up. Rather than simply saying that the kinetic energy temporarily disappears, we can retain the idea of the conservation of energy by defining a new form of energy. Kinetic energy is then transformed into this new form and later transformed back. This new energy is called **gravitational potential energy.**

We have some clues about the expression for this new energy. Its change must also be given by the work done by the force of gravity, and it must increase when the kinetic energy decreases, and vice versa. Therefore, we define the gravitational potential energy of an object at a height h above some zero level as equal to the work done by the force of gravity on the object as it falls to height zero. The gravitational potential energy *GPE* of an object near the surface of the Earth is then given by

$$GPE = mgh$$

gravitational potential energy = force of gravity × height

As an example, we can calculate the gravitational potential energy of a 6-kilogram ball located 0.5 meter above the level that we choose to call zero.

$$GPE = mgh = (6 \text{ kg})(10 \text{ m/s}^2)(0.5 \text{ m}) = 30 \text{ J}$$

Notice that only the vertical height is important. Moving an object 100 meters sideways does not change the gravitational potential energy because the force of gravity is perpendicular to the motion and therefore does no work on the object.

> **QUESTION** What is the change in gravitational energy of a 50-kilogram person who climbs a flight of stairs with a height of 3 meters and a horizontal extent of 5 meters?

The amount of gravitational potential energy an object has is a relative quantity. Its value depends on how we define the height; that is, what height we take as the zero value. We can choose to measure the height from any place that is convenient. The only thing that has any physical sig-

> **ANSWER** The change in gravitational potential energy is *GPE* = *mgh* = (50 kg)(10 m/s^2)(3 m) = 1500 J. The horizontal extent has no effect on the answer.

nificance is the *change* in gravitational potential energy. If a ball gains 20 joules of kinetic energy as it falls, it must lose 20 joules of gravitational potential energy. It does not matter how much gravitational potential energy it had at the beginning; it is only the amount lost that has any meaning in physics.

Conservation of Mechanical Energy

The sum of the gravitational potential and kinetic energies is conserved in some situations. This sum is called the **mechanical energy** of the system.

$$ME = KE + GPE = \tfrac{1}{2}mv^2 + mgh$$

mechanical energy = kinetic energy + gravitational potential energy

When frictional forces can be ignored and the other nongravitational forces do not perform any work, the mechanical energy of the system does not change. The simplest example of this circumstance is free fall (Fig. 6-7). Any decrease in the gravitational potential energy shows up as an increase in the kinetic energy, and vice versa.

Let's use the conservation of mechanical energy to analyze an idealized situation originally discussed by Galileo. Galileo released a ball that rolled down a ramp, across a horizontal track, and up another ramp as shown in Figure 6-8. He remarked that the ball always returned to its original height. From an energy point of view, he gave the ball some initial gravitational potential energy by placing it at a certain height above the horizontal ramp. As the ball rolled down the first ramp, its gravitational potential energy was transformed into kinetic energy. Going up the opposite ramp merely reversed this process; the ball's kinetic energy was transformed back into gravitational potential energy. The ball continued to move until its kinetic energy was entirely transformed back to gravitational potential energy, which occurred when it once again reached its original height. This result is independent of the slope of the ramp on the right. During the horizontal portion of the ball's trip, the gravitational potential energy remained constant. Hence the kinetic energy did not change and the speed remained constant, in agreement with Newton's first law.

Figure 6-7   As a ball falls, its mechanical energy is conserved; any loss in gravitational potential energy shows up as a gain in kinetic energy.

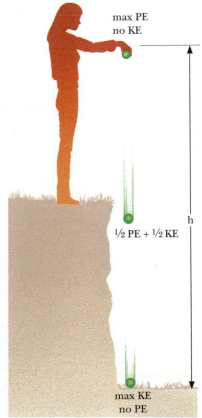

max PE
no KE

½ PE + ½ KE

h

max KE
no PE

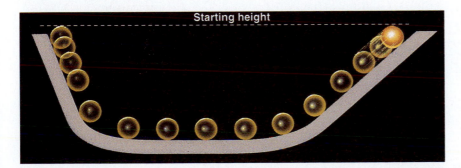

Starting height

Figure 6-8   Galileo's two-ramp experiment can be analyzed in terms of the conservation of energy.

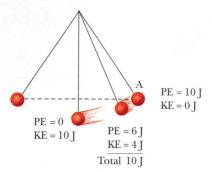

Figure 6-9   The total mechanical energy (kinetic plus gravitational potential) remains equal to 10 joules.

The pendulum bob shown in Figure 6-9 cyclically gains and loses kinetic and gravitational potential energy. Note that the force exerted by the string does no work. Therefore, if we ignore frictional forces, the total mechanical energy is conserved. Suppose that at the beginning, the bob is at rest at point A with a gravitational potential energy of 10 joules. (We have chosen the zero for gravitational potential energy to be at the lowest point of the bob's path.) Since the kinetic energy is zero, the total mechanical energy is the same as the gravitational potential energy; that is, 10 joules.

The bob is released. As it swings down, it loses gravitational potential energy and gains kinetic energy. The gravitational potential energy is zero at the lowest point of the swing, and so the mechanical energy is all kinetic and equal to 10 joules. The bob continues to rise up the other side until all of the kinetic energy is transformed back to gravitational potential energy. Because the gravitational potential energy depends on the height, the bob must return to its original height. And the motion repeats.

> **QUESTION**   Suppose the bob was released at twice the height. What would the maximum kinetic energy be?

Even when the bob is someplace between the highest and lowest points of the swing, the total mechanical energy is still 10 joules. If at this point, we determine from the height of the bob that it has 6 joules of gravitational potential energy, we can immediately declare that it has 4 joules of kinetic energy. Because we know the expression for the kinetic energy, we can calculate the speed of the bob at this point.

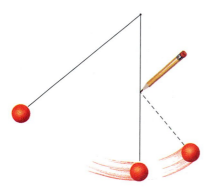

Figure 6-10   How high will the pendulum bob swing?

> **Physics on Your Own**   Construct a simple pendulum. If you let it swing freely, it returns to its original height. If you place a pencil in the path of the string as shown in Figure 6-10, does the bob still swing to the original height? Do your results depend on the vertical height of the pencil's position? Try placing the pencil near the lowest point of the swing.

Roller Coasters

Imagine trying to determine the speed of a roller coaster inside the thrilling loop-the-loop as it traverses the track's hills and valleys. Determining the speed at any spot using Newton's second law is difficult because the forces are continually changing size and direction. We can, however, use conservation of mechanical energy to determine the speed of an object without knowing the details of the net forces acting on it, providing we can ignore the frictional forces. If we know the mass, speed, and

> **ANSWER**   It would be twice as big; that is, 20 J.

height of the roller coaster at some spot, we can calculate its mechanical energy. Now determining its speed at any other spot is greatly simplified. The height gives us the gravitational potential energy. Subtracting this from the total mechanical energy yields the kinetic energy from which we can obtain the roller coaster's speed.

Suppose the roller coaster ride were designed like the one in Figure 6-11, and upon reaching the top of the lower hill, your car almost comes to rest. Assuming there are no frictional forces to worry about (not true in real situations), is there any possibility that you can get over the higher hill? The answer is no. Your gravitational potential energy at the top of the hill is nearly equal to the mechanical energy. This energy is not enough to get you over the higher hill. You will gain speed and thus kinetic energy as you coast down the hill, but as you start up the other hill you will find that you cannot exceed the height of the original hill.

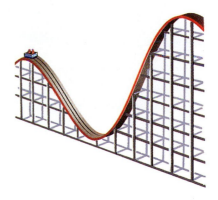

Figure 6-11   The car does not have enough gravitational potential energy to coast over the higher hill.

> **QUESTION**   Is there any way that the roller coaster car can make it over the second hill when starting on top of the first hill?

Our use of the conservation of mechanical energy is limited because we ignore losses due to frictional forces and, in many cases, this is not realistic. These frictional forces do work on the car and thus drain away some of the car's mechanical energy. If we know the magnitude of these

The mechanical energy of a roller coaster on a loop-the-loop is conserved if the frictional forces can be neglected.

> **ANSWER**   Yes. If your car has some kinetic energy at the top of the first hill, it might be possible to make it over the second hill. You would need enough kinetic energy to equal or exceed the extra gravitational potential energy required to climb the second hill.

frictional forces, however, and the distances through which they act, we can calculate the energy transformed to other forms and allow for the loss of mechanical energy.

Where does the energy go that is thus drained away? The answer to this question is still ahead of us. To maintain the notion that energy is an invariant, we will have to search for other forms of energy.

Other Forms of Energy

We can identify other places where kinetic energy is temporarily stored. For example, a moving ball can compress a spring and lose its kinetic energy (Fig. 6-12). While the spring is compressed, it stores energy much like the ball in the gravitational situation. If we latch the spring while it is in the compressed state, we can store its energy indefinitely as elastic potential energy. Releasing it at some future date will transform the spring's elastic potential energy back into the ball's kinetic energy.

> **QUESTION**   When a ball is hung from a vertical spring, it stretches the spring. As it drops, it loses gravitational potential energy, but this does not all show up as kinetic energy. What happens to the gravitational potential energy?

Physics Update

The physics of bungee jumping involves primarily the conversion of gravitational potential energy into the elastic energy of a stretched cord. Originating on Pentecost Island in the Pacific, the practice of a person jumping from a high place harnessed to a flexible attachment was introduced to Western culture in 1979 by the Oxford University Dangerous Sport Club. An all-important parameter, the amount by which the cord stretches at the bottom of the fall, should be accurately known in order to avert death.

Many objects or materials when distorted by a force hold some elastic potential energy as a result of the distortions. A floor "gives" when we jump on it. Our kinetic energy on impact is transformed, in part, to elastic potential energy of the floor. As the floor springs back to its original shape, we regain some of the kinetic energy. In these cases, some of the energy (maybe most) is lost to the dissipating effects of the distortion.

> **ANSWER**   The gravitational potential energy is converted to kinetic energy and to elastic potential energy of the spring. At the bottom, it is all elastic potential energy.

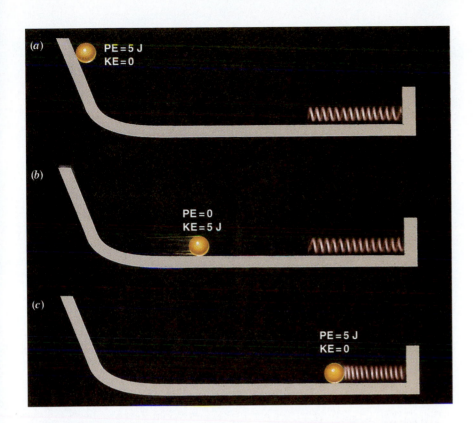

Figure 6-12   The gravitational potential energy (a) is converted to kinetic energy (b), which is then converted to elastic potential energy of the spring (c).

The gravitational potential energy of the water behind Grand Coulee Dam is converted to electric energy.

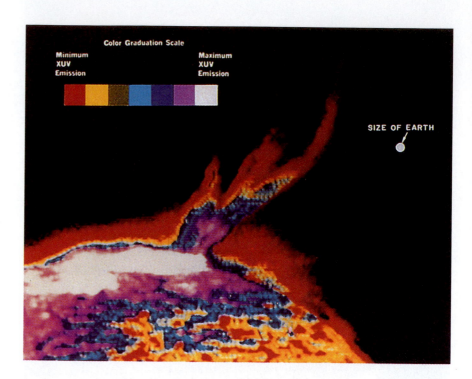

The Sun is a great storehouse of nuclear potential energy.

We have described a gravitational potential energy that is associated with the gravitational force. Other potential energies are associated with other forces in nature. The elastic potential energy in a spring is due to electromagnetic (electric and magnetic) forces. There are also other forms of electromagnetic potential energy. Chemical energy is really just a potential energy associated with the electromagnetic force. We will see in Chapter 25 that nuclear potential energy is associated with the nuclear force.

The transformation of these various forms of potential energy to kinetic energy is what powers our civilization. The gravitational potential energy of the water behind dams powers hydroelectric plants; that of a weight runs grandfather clocks. Most of the energy we use every day is the result of releasing the chemical potential energy of fuels. Nuclear power plants are designed to release nuclear potential energy. Nuclear potential energy is the ultimate source of the energy we receive from the Sun.

We have defined a variety of potential energies to explain the temporary losses of energy. However, this explanation doesn't apply to situations that include friction. To understand the physics of friction, consider the following. We start a box sliding across the floor with a given amount of kinetic energy. As it slows down and comes to a halt, its kinetic energy decreases and finally reaches zero. We might imagine that this energy was stored in some form of "frictional potential energy." If this were the case, we could somehow release this frictional potential energy and the box would move across the floor, continually gaining kinetic energy. This does not happen. As we will discuss in Chapter 9, when frictional forces act, some of the energy changes form and appears as thermal energy.

Is Conservation of Energy a Hoax?

It may seem strange that whenever we discover a situation in which energy does not appear to be conserved, we invent a new form of energy. How can the conservation law have any validity if we keep modifying it whenever it appears to be violated? In fact, the whole procedure would be worthless if it were not internally consistent, meaning that the total amount of energy stays the same no matter what sequence of changes takes place. This idea of internal consistency places a rather strong constraint on the law of conservation of energy.

Imagine a rotating wheel that has 10 joules of kinetic energy. If we extract this energy by stopping the wheel with a brake, we end up with 10 joules of thermal energy as shown in Figure 6-13(a). Imagine that another time, we convert the 10 joules of kinetic energy to electric energy by turning a generator to charge a battery, then using the battery's chemical energy to produce the electricity to run a heater [Fig. 6-13(b)]. We still get 10 joules of thermal energy. This result is reassuring. If the result were different, we would not have a valid scientific principle. Conservation of energy is a useful concept because the mathematical expressions for the various kinds of energy are independent of the specific situations in which they occur.

Conservation of energy is one of the most powerful principles in the physics world view, partly because of its generality; its only restriction is that the system be isolated, or closed. Conservation of energy can be applied to a wide range of problems in physics and in our everyday world.

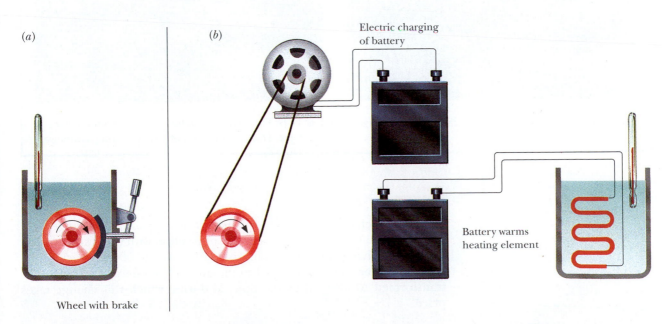

(*a*)

(*b*)

Electric charging of battery

Battery warms heating element

Wheel with brake

Figure 6-13   The two different paths for the transformation of the kinetic energy of the wheel yield the same amount of thermal energy.

Exponential Growth

America's energy usage has grown at a rate of about 5 percent per year for the past few decades. At first glance this sounds relatively harmless, but don't be fooled. Anything that grows in proportion to its current size gets out of control. This kind of growth is called *exponential growth* and, whether it's a bank account, energy usage, or population, it follows the same pattern. This pattern can be illustrated with a simple story of a creative mathematician.

Legend has it that a mathematician in ancient India invented the game of chess. The Ruler of India was so pleased that the mathematician was allowed to choose his own reward. This clever fellow shunned the obvious bounties of gold and jewels and, instead, requested grains of wheat. His plan was to put one grain on the first square of the chessboard, and then double the number of grains on each successive square. So there would be two grains of wheat for the second square, four grains for the third square, and so on. You can determine the number on each square with a simple calculator. Just start with the first grain and keep multiplying by two, 64 times.

At the end of the first row, there are 128 grains on the eighth square (Figure A). But that's not very many. By the end of the second row, the mathematician only has enough wheat to make a few loaves of bread. But what happens if we continue this pattern of doubling? The photo in Figure B is only symbolic, as the number of grains on the 64th square is about 9 billion billion! And the grand total is twice this! This is roughly 500 times the amount of wheat grown in the entire world in a year's time, and probably more wheat than has been grown during the entire history of humankind.

The pattern is very important. The doubling process starts out very slowly—a gentle, rather innocent, growth that eventually gets totally out of hand.

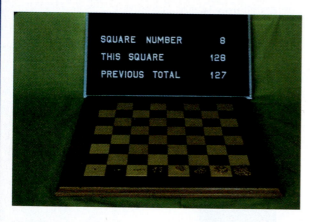

Figure A. The eighth square has 128 grains of wheat.

Figure B. The 64th square has 9 billion billion grains of wheat. The photograph, of course, is only symbolic.

Power

In previous chapters we discussed how various quantities change with time. For example, speed is the change of position with time, and acceleration is the change of velocity with time. The change of energy with time is called **power.** Power P is equal to the amount of energy transferred or transformed ΔE divided by the time Δt during which this change takes place.

$$\text{power} = \frac{\text{energy transformed}}{\text{time taken}} \qquad\qquad P = \frac{\Delta E}{\Delta t}$$

Exponential Growth (Cont.)

The legend ends with the mathematician being beheaded around the 35th square when the granaries of India were exhausted.

There are many examples of exponential growth in nature. Consider the compound interest that you earn in a savings account. Suppose that the bank offers a guaranteed interest rate of 5% per year. The growth pattern is the same as for the chess example. Your money will double after a certain number of years and double again each time this number of years elapses. The "doubling time" is obtained by dividing 70 by the percentage rate. For your savings account this would be 70/(5/year), or 14 years. If you put $1 in the bank today, it will double every 14 years. In 100 years your ancestors will collect $128. If they choose to leave the money in the bank for their ancestors—say, for another 400 years—your initial one dollar becomes $64 billion! (While bankers are not worried about 500-year-old accounts, they are well aware of this pattern and have rules about inactive accounts.)

The growth of the world's population is roughly exponential. It was approximately 5.9 billion when this book went to press in July 1997. Most people are not worried because the population is growing at what seems like a very low rate, about 1.7% per year. But at this rate our population doubles in 41 years. So the population will be 11.8 billion around 2038 and 23.6 billion by 2079.

Exponential growth can have very serious consequences because the big changes come on very fast. We can look at another example to illustrate the nature of the problem. Bacteria are rather curious in that they multiply by dividing. Assume you have a colony of bacteria in which each bacterium divides exactly in two at the end of each minute. Therefore, the doubling time is one minute. Let's further assume that the bottle the bacteria are in will be exactly full at the end of one hour; that is, it will take 60 doublings to fill the bottle. When will the bottle be half-filled? In 59 minutes. If you were a bacterium, when would you realize that you were running out of space? Since you're reading this essay, assume that at 55 minutes you've recognized that the growth is exponential and that there will be a major problem. At this time the bottle is only 3% full. How successful would you be in convincing your fellow bacteria that there's only 5 minutes left?

Imagine that during this debate an adversary argues that with proper government funding the colony can find more space, and, in fact, they do. After a great deal of expense and effort, they locate three new bottles. The resources have quadrupled. How much longer can the growth continue? The answer is only two more minutes!

The point of this discussion is to understand our current usage of energy. Clearly, we cannot let our usage grow exponentially. Although there are sizable variations in the estimates of the world's nonrenewable energy resources, all estimates are rather inconsequential if exponential growth continues. Does it really make a difference if we have underestimated these resources by a factor of two, or even four? No. If we continue to increase our use of these resources by even a small percentage each year, they will quickly be exhausted. During the next doubling time, we will use as much energy as has been used by humankind up to the present.

Source: Bartlett, A.A. "Forgotten Fundamentals of the Energy Crisis," *The Physics Teacher* **46** 876 (1978).

Power is measured in units of joules per second, a metric unit known as a *watt* (W). One watt of power would raise a 1-kilogram mass (with a weight of 10 newtons) a height of 0.1 meter each second.

The English unit for electric power is the watt, but a different English unit is used for mechanical power. A horsepower is defined as 550 foot-pounds per second. This definition was proposed by the Scottish inventor James Watt because he found that an average strong horse could produce 550 foot-pounds of work during an entire working day. One horsepower is equal to 746 watts.

We get electric energy from our local "power" company. But power is not an amount of energy. The energy transformed during a period of

Human Power

A human can generate 1500 watts (2 horsepower) for very short periods of time, such as in weightlifting. The maximum average human power for an 8-hour day is more like 75 watts (0.1 horsepower). Each person in a room generates thermal energy equivalent to that of a 75-watt light bulb. That's one of the reasons why crowded rooms warm up!

Achieving human-powered flight has been a dream for centuries. People failed because it has been difficult to create enough aerodynamic lift with the power output that's humanly possible. In 1979, an American team led by Paul MacCready designed, built, and flew a "winged bicycle" that convincingly demonstrated that human-powered flight is possible. Professional cyclist Bryan Allen crossed the English Channel in the *Gossamer Albatross*. During this flight he generated an average power output of 190 watts (0.25 horsepower).

Russian weightlifter Andrei Chemerkin lifts 260 kilograms to win the gold medal in the 1996 Summer Olympic Games.

The human-powered aircraft *Gossamer Albatross*, which successfully flew across the English Channel in 1979.

time is given by the power multiplied by the time this power is expended. Power companies usually bill us for the amount of energy we use, not the rate of consumption.

COMPUTING POWER

A compact car traveling at 27 meters per second (60 mph) on a level highway experiences a frictional force of about 300 newtons due to the air resistance and the friction of the tires with the road. Therefore, the car must obtain enough energy by burning gasoline to compensate for the work done by the frictional forces each second.

$$\Delta E = W = Fd = (300 \text{ N})(27 \text{ m}) = 8100 \text{ J}$$
$$P = \frac{\Delta E}{\Delta t} = \frac{8100 \text{ J}}{1 \text{ s}} = 8100 \text{ W}$$

This means that the power needed is 8100 watts, or 8.1 kilowatts. This is equivalent to a little less than 11 horsepower.

How much electrical energy does a motor running at 1000 watts for 8 hours require?

$$\Delta E = P\Delta t = (1000 \text{ W})(8 \text{ h}) = 8000 \text{ Wh}$$

This is usually written as 8 kilowatt-hours (kWh). Although this doesn't look like an energy unit, it is—(energy/time) × time = energy. The energy used by the motor in 1 hour is

$$\Delta E = P\Delta t = (1000 \text{ W})(1 \text{ h}) = (1000 \text{ J/s})(3600 \text{ s}) = 3,600,000 \text{ J}$$

In other words, 1 kilowatt-hour = 3.6 million joules.

> **QUESTION**   How much energy is required to leave a 75-watt yard light on for 8 hours?

SUMMARY

Energy is an abstract quantity that is conserved whenever a system is closed. Independent of the kinds of transformations that take place within a system, the total amount of energy remains the same.

Kinetic energy is the energy of motion and is defined to be one-half the mass times the speed squared, $\frac{1}{2}mv^2$. If the kinetic energy before a collision is the same as that after the collision, the collision conserves kinetic

---

ANSWER $\Delta E = P\Delta t = (75 \text{ W})(8 \text{ h}) = 600 \text{ Wh} = 0.6 \text{ kWh}$

We really don't know what energy *is* but we know the many forms it takes and we have an accounting system for determining the amount of energy in a system. When we say that energy is conserved, we are acknowledging that while energy changes from one form into other forms, the total amount of energy in the system stays the same, and, because of this, we are able to keep track of the energy.

energy and is said to be elastic. Kinetic energy is transformed into other forms of energy in inelastic collisions.

Work is equal to the product of the force in the direction of motion and the distance traveled; that is, $W = Fd$. If the force is perpendicular to the velocity of the object, or if the object does not move, no work is done by the force. The change in kinetic energy of an object is equal to the work done on the object.

The gravitational potential energy is equal to the work done by the force of gravity when an object falls through a height h; that is, $GPE = mgh$. The location for the zero value of gravitational potential energy is arbitrary because only the change in gravitational potential energy has any physical meaning. If gravity is the only force that does work on an object, the total mechanical energy (kinetic plus gravitational potential) is conserved. Therefore, any loss in gravitational potential energy shows up as a gain in the kinetic energy, and vice versa.

Other forms of potential energy can be associated with the electromagnetic and nuclear forces. However, a potential energy cannot be associated with the frictional force. This force transforms mechanical energy into thermal energy.

Power is the rate at which energy is transformed from one form into another, $P = \Delta E / \Delta t$. A kilowatt-hour is an energy unit because it is power multiplied by time.

KEY TERMS

elastic: A collision or interaction in which kinetic energy is conserved.

gravitational potential energy: The work that would be done by the force of gravity if an object fell from a particular point in space to the location assigned the value of zero.

inelastic: A collision or interaction in which kinetic energy is not conserved.

joule: The SI unit of energy, equal to 1 newton acting through a distance of 1 meter.

kinetic energy: The energy of motion, $\frac{1}{2}mv^2$, where m is the object's mass and v is its speed.

mechanical energy: The sum of the kinetic energy and various potential energies, which may include the gravitational and the elastic potential energies.

power: The rate at which energy is converted from one form to another. Measured in joules per second, or watts.

watt: The SI unit of power, 1 joule per second.

work: The product of the force along the direction of motion and the distance moved. Measured in energy units, joules.

CONCEPTUAL QUESTIONS

1. Under what conditions is energy conserved?

2. Which of the following physical quantities is not necessarily an invariant even if the system is closed: mass, energy, momentum, volume, or angular momentum?

3. What happens to the total energy of a star that undergoes a supernova explosion?

4. A sports car with a mass of 1000 kilograms travels down the road with a speed of 20 meters per second.

Why can't we say that its momentum is smaller than its kinetic energy?

5. How is kinetic energy defined?

6. Can the kinetic energy of an object ever be negative?

7. Which has the greater kinetic energy, a supertanker berthed at a pier or a motorboat pulling a water skier?

8. Two cars have the same mass, but the red car has twice the speed of the blue car. We now know that the red car has _____ times the kinetic energy of the blue car.

9. Assume that a red car has a mass of 1000 kilograms and a white car has a mass of 2000 kilograms. If both cars are traveling at the same speed, which car has the higher kinetic energy?

10. If the red car in the preceding question has twice the speed of the white car, which car has the higher kinetic energy?

11. Compare the kinetic energies of two identical objects under the following conditions:

    a. The velocity of object A is twice that of object B.

    b. Object A is moving north at the same speed that object B is moving west.

12. Compare the kinetic energies of two identical objects under the following conditions:

    a. Object A is traveling in a straight line at the same speed that object B is moving in a circle.

    b. Both objects make 1 revolution per minute along circular paths. A's circle has only one-half the circumference of B's.

13. What will happen if the end balls of the collision-ball apparatus in Figure 6-1 are pulled out the same distances and let go?

14. What will happen if you pull two balls from the same side of the collision-ball apparatus in Figure 6-1 and let them go?

15. How is work defined in physics? How does this definition differ from everyday usage?

16. A bowler lifts a bowling ball from the floor and places it on a rack. If you know the weight of the ball, what else must you know in order to calculate the work she does on the ball?

17. An object has a velocity toward the south. If a force is directed toward the north, will the kinetic energy of the object initially increase, decrease, or stay the same?

18. An object has a velocity toward the east. If a force directed toward the south acts on the object, will the kinetic energy of the object initially increase, decrease, or stay the same?

19. Two objects have different masses but the same kinetic energies. If you stop them with the same retarding force, which one will stop in the shorter distance?

*20. Two objects have different masses but the same linear momenta. If you stop them with the same retarding force, which one will stop in the shorter distance? (*Hint:* Which one has the larger kinetic energy?)

21. An object executes circular motion with a constant speed whenever the net force on it acts perpendicular to its velocity. What happens to the speed if the force is not perpendicular?

22. The tractor of an 18-wheeler performs work on its trailer when the truck is traveling along a level highway with a constant velocity. Why doesn't the trailer continually gain kinetic energy; that is, continually speed up?

*23. We can use Newton's third law to demonstrate that the momentum lost by one object is gained by another. Can you also do this for kinetic energy? Explain why or why not.

24. In what ways do the laws of conservation of momentum and conservation of kinetic energy differ?

25. Give three examples in which a net force does not change the kinetic energy of an object.

Question 16.

26. Is it possible to change an object's momentum without changing its kinetic energy? What about the reverse situation?

27. Which of the following does the most work: a force of 3 newtons acting through a distance of 3 meters or a force of 4 newtons acting through a distance of 2 meters?

28. Which of the following produces the largest change in the kinetic energy: a force of 5 newtons acting through a distance of 3 meters or a force of 4 newtons acting through a distance of 4 meters?

29. Which of the following affect the value for the gravitational potential energy of an object: the strength of gravity, the object's height, the location of the zero value, or the object's mass?

30. What is the gravitational potential energy of a 10-newton ball resting on a shelf 2 meters above the floor?

31. What happens to their gravitational potential energy when firefighters slide down fire poles?

32. How much work is performed in carrying a suitcase across an airport lounge?

33. The kinetic energy of a freely falling ball is not conserved. Why is this not a violation of the law of conservation of mechanical energy?

34. Does the mechanical energy of a falling ball increase, decrease, or stay the same? Explain.

35. A man with a mass of 80 kilograms falls 10 meters. How much mechanical energy does he gain or lose?

36. Which of the following is conserved as a ball falls freely in a vacuum: the ball's kinetic energy, gravitational potential energy, momentum, or mechanical energy?

37. At which point in the swing of an ideal pendulum (ignoring friction) is the gravitational potential energy a maximum?

38. At which point in the swing of an ideal pendulum (ignoring friction) is the kinetic energy a maximum?

39. If we do not ignore frictional forces, what can you say about the height to which a pendulum bob swings on consecutive swings?

40. If we do not ignore air resistance, what happens to the mechanical energy of a falling ball?

41. A block of wood loses 100 joules of gravitational potential energy as it slides down a ramp. If it has 90 joules of kinetic energy at the bottom of the ramp, what can you conclude?

42. A ball dropped from a height of 10 meters only bounces to a height of 5 meters. Are any of the following conserved in this situation: kinetic energy, mechanical energy, or gravitational potential energy? How do you know?

43. Describe the energy transformations that occur as the Earth orbits the Sun in its elliptical orbit.

Question 31.

44. Imagine a giant catapult that could hurl a spaceship to the Moon. Describe the energy transformations that would take place on such a journey.

45. Describe the energy changes that take place when you dribble a basketball.

46. An elephant, an ant, and a professor jump from a lecture table. Assuming no frictional losses, which of the following could be said about their motion just before they hit the floor?

 a. They all have the same kinetic energy.

 b. They all started with the same gravitational potential energy.

 c. They will all experience the same force on stopping.

 d. They all have the same speed.

47. An air-hockey puck is fastened to the table with a spring so that it oscillates back and forth on top of the table. Describe the energy transformations that take place. How would your description change if the puck were suspended from the ceiling by the spring?

48. Magnets mounted on top of air-hockey pucks allow the pucks to "collide" without touching each other. Describe the energy transformations that take place when one puck collides head-on with another.

49. Mountain highways often have emergency ramps for truckers whose brakes fail. Why are these covered with soft dirt or sand rather than being paved?

50. Athletes will sometimes run along the beach to increase their workouts. Why is running on soft sand so tiring?

51. Describe the energy changes that take place when you stop your car by using the brakes.

52. What happens to the chemical potential energy in the batteries used to power electric socks?

53. Why can we not associate a potential energy with the frictional force as we did with the gravitational force?

54. State how you would respond to a friend who maintains that the law of conservation of energy is a hoax because any time a violation is discovered a new form of energy is invented.

55. How does power differ from energy?

56. Give an example that clearly distinguishes the concepts of power and energy.

57. Which of the following is an energy unit: newton, kilowatt, kilogram-meter per second, or kilowatt-hour?

58. Which of the following is not a unit of energy: joule, newton-meter, kilowatt-hour, or watt?

EXERCISES

1. What is the kinetic energy of a 1200-kg sports car traveling down the road with a speed of 25 m/s?

2. What is the kinetic energy of an 84-kg sprinter running at 10 m/s?

3. A 2-kg air-hockey puck with an initial speed of 2 m/s collides head-on with a stationary 1-kg puck. After the collision, the pucks travel in the same direction. The 2-kg puck has a speed of 1 m/s, and the 1-kg puck has a speed of 2 m/s. Are momentum and kinetic energy conserved in this collision?

4. A 2-kg air-hockey puck traveling north at 8 m/s collides with a 3-kg puck traveling north at 3 m/s. If the pucks stick together and have a common velocity of 5 m/s north, are momentum and kinetic energy conserved?

5. A 4-kg toy car with a speed of 5 m/s collides head-on with a stationary 1-kg car. After the collision, the cars are locked together with a speed of 4 m/s. How much kinetic energy is lost in the collision?

6. A 3-kg toy car with a speed of 6 m/s collides head-on with a 2-kg car traveling in the opposite direction with a speed of 4 m/s. If the cars are locked together after the collision, how much kinetic energy is lost?

7. A net force of 0.3 N acts on a 0.4-kg air-hockey puck for a distance of 1.2 m. What is its kinetic energy if it was initially at rest?

8. A 0.5-kg air-hockey puck is initially at rest. What will its kinetic energy be after a net force of 0.4 N acts on it for a distance of 0.8 m?

9. A toy car has a kinetic energy of 6 J. What is its kinetic energy after a frictional force of 0.5 N has acted on it for 4 m?

10. A radio-controlled car has an initial kinetic energy of 6 J. If the drive wheels exert a force of 0.8 N for 5 m, what is the car's new kinetic energy?

11. How much work is performed by the gravitational force on a geosynchronous satellite during one day?

12. A ball on the end of a string travels in a horizontal circle at a constant speed. The circle has a circumference of 3 m, the ball has a speed of 2 m/s, and the centripetal force is 4 N. How much work is done on the ball each time it goes around?

13. How much work does gravity do on a baseball (mass = 145 g) falling from a height of 8 m?

14. How much work does a 70-kg person do against gravity in walking up a trail that gains 500 m in elevation?

15. A woman with a mass of 60 kg climbs a set of stairs that are 3 m high. How much gravitational potential energy does she gain?

16. What is the gravitational potential energy of a ball with a weight of 60 N when it is sitting on a shelf 2 m above the floor? What assumption do you need to make to get your answer?

17. A 2-kg mass is released at the top of a frictionless slide that is 4 m high. What is the kinetic energy of the mass when it reaches the bottom?

18. If a 0.4-kg ball is dropped from a height of 5 m, what is its kinetic energy when it hits the ground?

19. A 1000-kg frictionless roller coaster starts from rest at a height of 20 m. What is its kinetic energy when it goes over a hill that is 12 m high?

20. How much kinetic energy must a high jumper with a mass of 70 kg generate in order to clear the bar at 2.4 m? Assume that the jumper's center of mass is normally at 0.8 m and the center of mass goes under the bar by 0.2 m.

21. What average power does a weightlifter need to lift 300 lb a distance of 4 ft in 1.2 s?

22. If a 60-kg sprinter can accelerate from a standing start to a speed of 10 m/s in 3 s, what average power is generated?

23. If a CD player uses electricity at a rate of 15 W, how much energy does it use during an 8-h day?

24. If a hair dryer is rated at 800 W, how much energy does it require in 3 min?

INTERLUDE

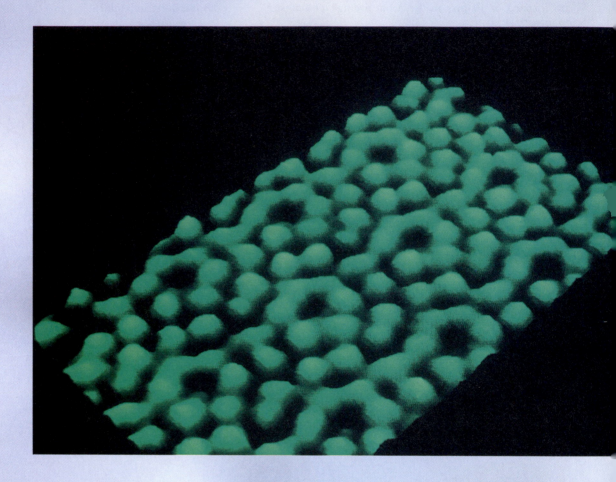

The surface of silicon viewed through a scanning tunneling microscope.

The Search for Atoms

One of the oldest challenges in building a physics world view is the search for the fundamental building blocks of matter. This search began more than 2000 years ago. The first ideas appeared in writings about the Greek philosopher Leucippus, who lived in the fifth century B.C. Leucippus asked a simple question: If you take a piece of gold and cut it in half and then cut one of the halves in half, and so on, will you always have gold? We know what happens initially; one piece of gold yields two pieces of gold, either of these pieces yields two more pieces of gold, and so on. But what if you could continue the process indefinitely? Do you think you would eventually reach an end—a place where either you couldn't cut the piece or, if you could, you would no longer have gold?

Leucippus and his student Democritus felt that this process would eventually stop—that gold has a definite elementary building block. Once you get to this level, further cuts either fail or yield something different. These elementary building blocks were (and still are) known as atoms—from the Greek word for "indivisible."

We know of these two early atomists through the writings of Aristotle in the fourth century B.C. Aristotle disagreed with their atomistic view. Aristotle realized that if matter were made of atoms, one needed to ask, "What is between them?" Presumably nothing. He felt that a void—pure nothingness—between pieces of matter was philosophically unacceptable. Furthermore, the atomistic view held that the atoms were eternal and in continual motion. Aristotle felt this was foolish and in total conflict with everyday experience. He saw a world where all objects wear out and where their natural motion is one of rest. Even if pushed, they quickly slow down and stop when the pushing stops.

> **Although the evidence for atoms has grown tremendously during the last 300 years, it might be surprising to learn that the first proof of the existence of atoms didn't come until 1905.**

As we view the woman's face from closer and closer, we begin to see the building blocks that form the image. Can we do the same for the building blocks of our material world?

Aristotle's arguments were powerful and his world view remained relatively unchanged for the next 19 centuries. Galileo's idea of inertia and the possibility of perpetual motion in the absence of friction did not occur until the 17th century. The technology for making a good vacuum—Aristotle's unacceptable void—did not occur until later. But eventually Aristotle's two impossibilities became possible.

Although the evidence for atoms has grown tremendously during the last 300 years, it might be surprising to learn that the first proof of the existence of atoms didn't come until 1905.

Albert Einstein published a paper showing that atoms in continual, random motion could explain a strange jiggling motion of microscopic objects sus-

pended in fluids observed by a Scottish biologist, Robert Brown, almost 80 years earlier. Not until recently have new electronic microscopes been able to show direct evidence of atoms.

Different combinations and arrangements of a relatively small number of atoms make up the diverse material world around us. What are the properties of these solids, liquids, and gases? How are their macroscopic properties, such as mass, volume, temperature, pressure, and elasticity, related to the underlying microscopic properties? We will examine both macroscopic and microscopic properties and their connections in the next two chapters. This study will then allow us to expand our understanding of one of the most fundamental concepts in the physics world view, the conservation of energy.

Structure of Matter

Hot air allows balloonists to enjoy the scenery from a high vantage point. But what do the macroscopic properties of gases—volume, pressure, and temperature—tell us about the existence of atoms and molecules and their microscopic properties of size, mass, and speed? (See p. 178 for the answer to this question.)

Hot air balloons floating over the Rocky Mountains.

a

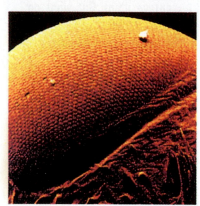

b

c

An electron micrograph of (*a*) a fly's head (27×) and its eye at (*b*) 122× and (*c*) 1240×.

When we talk about the properties of objects, we usually think about their bulk, or **macroscopic,** properties such as size, shape, mass, color, surface texture, and temperature. For instance, a gas occupies a volume, exerts a pressure on its surroundings, and has a temperature. But a gas is composed of molecules that have their own characteristics, such as velocity, momentum, and kinetic energy. These are the **microscopic** properties of the gas. It seems reasonable that there should be connections between these macroscopic and microscopic properties. At first glance, you might assume that the macroscopic properties are just the sum, or maybe the average, of the microscopic ones; however, the connections provided by nature are much more interesting.

We could begin our search for the connections between the macroscopic and the microscopic by examining a surface with a conventional microscope and discovering that rather than being smooth, as it appears to the naked eye, the surface has some texture. The most powerful electron microscope shows even more structure than an optical microscope reveals. But, until recently, instruments could not show us the basic underlying structure of things. In order to gain an understanding of matter at levels beyond which they could observe directly, scientists constructed models of various possible microscopic structures to explain their macroscopic properties.

Building Models

By the middle of the 19th century the body of chemical knowledge had pretty well established the existence of **atoms.** All of the evidence was indirect but was sufficient to create some idea of how atoms combine to form various substances. This description of matter as being composed of atoms is a *model*. It is not a model in the sense of a scale model such as a model railroad or an architectural model of a building, but rather a theory or mental picture.

To illustrate this concept of a model, suppose someone gives you a tin can without a label and asks you to form a mental picture of what might be inside. Suppose you hear and feel something sloshing when you shake the can. You guess that the can contains a liquid. Your model for the contents is that of a liquid, but you do not know if this model actually matches the contents. For example, it is possible (but unlikely) that the can contains an electronic device that imitates sloshing sounds. However, you can use your knowledge of liquids to test your model. You know, for example, that liquids freeze. Therefore, your model predicts that cooling the can would stop the sloshing sounds. In this way you can use your model to help you learn more about the contents. The things you can say, however, are limited. For example, there is no way to determine the color of the liquid or its taste or smell.

Sometimes a model takes the form of an analogy. For example, the flow of electricity is often described in terms of the flow of water through pipes. Similarly, little, solid spheres might be analogous to atoms. We could use the analogy to develop rules for combining these spheres and learn something about how atoms combine to form molecules. However, it is important to distinguish between a collection of atoms that we cannot

Prayer To Become Jubilee People

*F*ather of Time, Mother of Creation, we thank you for the gift of Jubilee that sanctifies both time and space. During this great Jubilee at the dawn of the new millennium, teach us the wisdom of Sabbath rest for your earth and for the land of ourselves.

*T*each us the wisdom of forgiveness of debts for those who cannot pay and for those we refuse to release. Teach us the wisdom of Jubilee justice to remind us that all we have belongs to you and to ensure that everyone has enough.

*T*each us the wisdom of Jubilee liberation that we might free those who are oppressed and languish in captivity.

*S*trengthen our families, parishes, church and nation that we might truly become Jubilee people.

Amen

Courtesy of Tom Cordaro and Pax Christi USA

... late. Sooner or later the ... water and atoms are not

...mematical equation. Physi... structure and behavior ... be very abstract, they can ...re behaves. Suppose that ...und followed by a clunk ...equation for the length of ... the wall before hitting an ...to predict the time delays

...arize and account for the ...behaves. And to be useful, ...new situations. The model ...edict that liquid will flow ...or the structure of matter ...sted by experiment. If the ...belief in the model. If they ...or abandon it and invent a

...group of experimenters ...our modern atomic world ...ooked. We hear only that ...me foolishly trying to turn ...he Aristotelian world view ...s of four elements—earth, ...nuous, it made sense that ...g the proportions of the ...nd heating materials were ...e two of the elements. A ...ding fire changed its char...sed some air even though ...air. Adding water to sugar ...This kind of experimenta...lending, grinding, and so

But the alchemists' role in the building of our world view was more than just finding ways of manipulating materials. By showing that the common materials were not all that existed, the alchemists set the stage for the experimenters who followed. These new experimenters began with a different background of knowledge, a different common sense about the world, and a different expectation of what was possible. They went beyond cataloging ways of manipulating substances and began considering the structures of matter that could be causing these results—they built models. Although it is difficult to pinpoint the transition from

This early 19th-century Japanese woodcut shows an artisan extracting copper from its ore.

Table 7-1 Lavoisier's List of the Elements (1789)

Lavoisier's Name	Modern English Name	Lavoisier's Name	Modern English Name
Lumiere	Light*	Etain	Tin
Calorique	Heat*	Fer	Iron
Oxygene	Oxygen	Manganese	Manganese
Azote	Nitrogen	Mercure	Mercury
Hydrogene	Hydrogen	Molybdene	Molybdenum
Soufre	Sulfur	Nickel	Nickel
Phosphore	Phosphorus	Or	Gold
Carbone	Carbon	Platine	Platinum
Radical muriatique	†	Plomb	Lead
Radical fluorique	†	Tungstene	Tungsten
Radical boracique	†	Zinc	Zinc
Antimoine	Antimony	Chaux	‡Calcium oxide (lime)
Argent	Silver	Magnésie	‡Magnesium oxide
Arsenic	Arsenic	Baryte	‡Barium oxide
Bismuth	Bismuth	Alumina	‡Aluminum oxide
Cobalt	Cobalt	Silice	‡Silicon dioxide (sand)
Cuivre	Copper		

*Not elements
†No elements with these properties have ever been found.
‡These "elements" are really compounds.

alchemy to chemistry, the early chemists were part of the Newtonian age and had a different way of looking at the world.

By the latter half of the 17th century the existence of the four Aristotelian elements was in doubt. They were of little or no help in making sense out of the chemical data that had been accumulated. However, the idea that all matter was composed of some basic building blocks was so appealing that it persisted. This belief fueled the development of our modern atomic model of matter.

The simplest, or most elementary, substances were known as **elements.** These elements could be combined to form more complex substances, the **compounds.** It was not until the 1780s that the French chemist and physicist Antoine Lavoisier and his contemporaries had enough data to draw up a tentative list of elements (Table 7-1). Something was called an element if it could not be broken down into simpler substances.

> **QUESTION**   Could these early chemists know for sure that a substance was an element?

A good example of an incorrectly identified element is water. It was not known until the end of the 18th century that water is a compound of the elements hydrogen and oxygen. Hydrogen had been crudely separated during the early 16th century but oxygen was not discovered until

> **ANSWER**   They never really knew if the substance was an element or if they had not yet figured how to break it down.

1774. When a flame is put into a test tube of hydrogen, it "pops." One day while popping hydrogen, an experimenter noticed some clear liquid in the tube. This liquid was water. This was the first hint that water was not an element. The actual decomposition of water was accomplished at the end of the 18th century by a technique known as electrolysis, by which an electric current passing through a liquid or molten compound breaks it down into its respective elements.

The radically different properties of elements and their compounds are still shocking in the modern world view. Hydrogen, a very explosive gas, and oxygen, the element required for all burning, combine to form a compound that is great for putting out fires! Similarly, sodium, a very reactive metal that must be kept in oil to keep it from reacting violently with moisture in the air, combines with chlorine, a very poisonous gas, to form a compound that tastes great on mashed potatoes—common table salt!

Chemical Evidence for Atoms

Another important aspect of elements and compounds was discovered around 1800. Suppose that a particular compound is made from two elements and that when you combine 10 grams of the first element with 5 grams of the second, you get 12 grams of the compound and have 3 grams of the second element remaining. If you now repeat the experiment, only this time adding 10 grams of each element, you still get 12 grams of the compound, but now have 8 grams of the second element remaining. This result was exciting. It meant that the compound, rather than containing some random mixture of the two elements, has a very definite ratio of their masses. This principle is known as the **law of definite proportions.**

> **QUESTION** How much of the compound would you get if you added only 1 gram of the second element?

This law is difficult to explain using the Aristotelian model, which viewed matter as a continuous, smooth substance. In the continuum model one would expect there to be a range of masses in which the elements could combine to form the compound. An atomic model, on the other hand, provides a simple explanation: The atoms of one element can combine with atoms of another element to form **molecules** of the compound. It may be that one atom of the first element combines with one atom of the second to form one molecule of the compound that contains two atoms. Or it may be that one atom of the first element combines with two atoms of the second element. In any case, the ratio of the masses of the combining elements has a definite value. (Note that the value of the ratio doesn't change with differing amounts. The mass ratio of ten baseballs and ten basketballs is the same as that of one baseball and one

> **ANSWER** Because 10 grams of the first element require 2 grams of the second, 1 gram of the second will combine with 5 grams of the first. The total mass of the compound is just the sum of the masses of the two elements, so 6 grams of the compound will be formed.

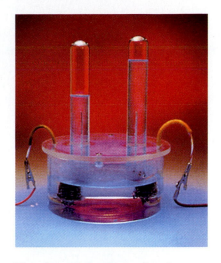

The electrolysis of water breaks up the water molecules into hydrogen and oxygen. Notice that the volume of hydrogen collected in the left-hand tube is twice that of the oxygen collected in the right-hand tube.

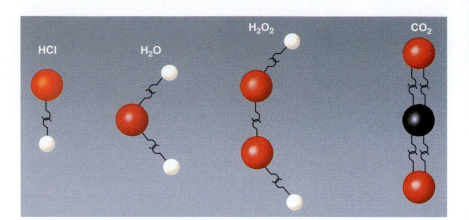

Figure 7-1 Dalton's atomic model has hooks to explain the law of definite proportions.

basketball. It doesn't matter how many balls we have as long as there is one baseball *for each* basketball.)

The actual way in which the elements combined and what caused them to always combine in the same way was unknown. The English mathematician and physicist John Dalton hypothesized that the elements might have hooks (Fig. 7-1) that control how many of one atom combine with another. Dalton's hooks can be literal or metaphorical; the actual mechanism is not important. The essential point of his model was that different atoms have different capacities for attaching to other atoms. Regardless of the visual model we use, atoms combine in a definite ratio to form molecules. One atom of chlorine combines with one atom of sodium to form salt. The ratio in salt is always one atom to one atom.

In retrospect it may seem that the law of definite proportions was a minor step and that it should have been obvious once mass measurements were made. However, it was difficult to see this relationship because some processes did not obey this law. For instance, *any* amount of sugar (up to some maximum) dissolves completely in water. One breakthrough came when it was recognized that this process was distinctly different. The sugar–water solution was not a compound with its own set of properties, but simply a mixture of the two substances. Mixtures had to be recognized as different from compounds and eliminated from the discussion.

Physics Update

New studies confirm that chlorofluorocarbons (CFCs) cause ozone depletion. Because of the "political sensitivity of the ozone-depletion issue," a team of NASA (Langley) and University of California–Irvine scientists undertook satellite measurements of stratospheric hydrogen chloride and hydrogen fluoride. Their four-year study, the Halogen Occultation Experiment (HALOE), shows that CFC emissions, and not natural emissions arising, say, from oceans and biomass burning (sources which can now be accounted for), are chiefly to blame for the recent global buildup of stratospheric chlorine and the consequential depletion of ozone. International agreements limiting CFC emissions have already begun to lessen the growth of chlorine in the stratosphere.

Another complication occurred because some elements can form more than one compound. Carbon atoms, for example, could combine with one or two oxygen atoms to form two compounds with different characteristics. When this happened in the same experiment, the final product was not a pure compound, but a mixture of compounds. This result yielded a range of mass ratios and was quite confusing until the chemists were able to analyze the compounds separately.

Fortunately, the atomic model makes predictions about situations in which two elements form more than one compound. Imagine for a moment that atoms can be represented by nuts and bolts. Suppose a hypothetical molecule of one compound consists of one nut and one bolt, and another consists of two nuts and one bolt. Because there are twice as many nuts for each bolt in the second compound (Fig. 7-2), it has a mass ratio of nuts to bolts that is twice that of the first compound. This prediction was confirmed for actual compounds and provided further evidence for the existence of atoms.

(*a*)   (*b*)

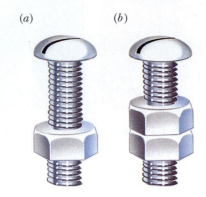

Figure 7-2   A simple model of two compounds: (a) "bolt-mononut" and (b) "bolt-dinut." The ratio of the mass of nuts to the mass of bolts is twice as large for "bolt-dinut."

Physics on Your Own   Collect a number of identical bolts and identical nuts. Assemble a number of "bolt-mononuts"; that is, put one nut on each bolt. Take these apart and measure the total mass of the bolts and the total mass of the nuts. What is the ratio of the mass of nuts to that of bolts? Does this ratio depend on how many bolt-mononuts you built?

Repeat this process with "bolt-dinuts"; that is, put two nuts on each bolt. How does the mass ratio for the bolt-dinuts compare to the mass ratio for the bolt-mononuts?

Masses and Sizes of Atoms

Even with their new information, the 18th-century chemists did not know how many atoms of each type it took to make a specific molecule. Was water composed of one atom of oxygen and one atom of hydrogen, or was it one atom of oxygen and two atoms of hydrogen, or two of oxygen and one of hydrogen? All that was known was that 8 grams of oxygen combined with 1 gram of hydrogen. These early chemists needed to find a way of establishing the relative masses of atoms.

The next piece of evidence was an observation that occurred when gaseous elements were combined; the gases combined in definite *volume* ratios when their temperatures and pressures were the same. This statement was not surprising except that the volume ratios were always simple fractions. For example, 1 liter of hydrogen combines with 1 liter of chlorine (a ratio of 1 to 1), 1 liter of oxygen combines with 2 liters of hydrogen (a ratio of 1 to 2), 1 liter of nitrogen combines with 3 liters of hydrogen (a ratio of 1 to 3), and so on (Fig. 7-3).

It was very tempting to propose an equally simple underlying rule to explain these observations. The Italian physicist Amedeo Avogadro suggested that under identical conditions, each liter of any gas contains the

Figure 7-3   Gases combine completely to form compounds when the ratios of their volumes are equal to the ratios of small whole numbers. One liter of nitrogen combines completely with 3 liters of hydrogen to form 2 liters of ammonia.

same number of molecules. Although it took more than 50 years for this hypothesis to be accepted, it was the key to unraveling the question of the number of atoms in molecules.

> **QUESTION**   Given that oxygen and hydrogen gases are each composed of molecules with two atoms each, how many atoms of oxygen and hydrogen combine to form water?

Once the number of atoms in each molecule was known, the data on the mass ratios could be used to calculate the relative masses of different atoms. For example, an oxygen atom has about 16 times the mass of a hydrogen atom.

To avoid the use of ratios, a mass scale was invented by choosing a value for one of the elements. An obvious choice was to assign the value of 1 to hydrogen since it is the lightest element. However, setting the value of carbon equal to 12 **atomic mass units** (amu) makes the relative masses of most elements close to whole-number values. These values are known as **atomic masses,** and keep the value for hydrogen very close to 1. Even though the values for the actual atomic masses are now known, it is still convenient to use the relative atomic masses.

> **QUESTION**   What is the atomic mass of carbon dioxide, a gas formed by combining two oxygen atoms with each carbon atom?

> **ANSWER**   The observation that 2 liters of hydrogen gas combine with 1 liter of oxygen means that there are two hydrogen molecules for each oxygen molecule. Therefore, there are four hydrogen atoms for every two oxygen atoms. The simplest case would be for these to form two water molecules with two hydrogen atoms and one oxygen atom in each water molecule. This is confirmed by the observation that 2 liters of water vapor are produced.

> **ANSWER**   We have 12 amu for the carbon atom and 16 amu for each oxygen atom. Therefore, 12 amu + 32 amu = 44 amu.

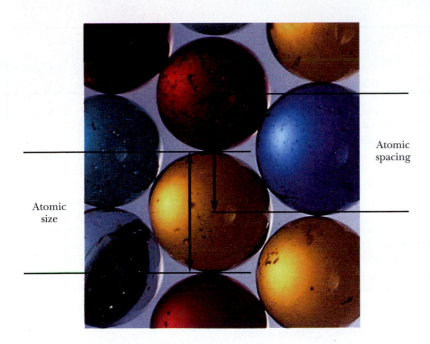

Atomic spacing

Atomic size

Figure 7-4 Assuming that the atoms are "touching" like marbles, the spacing between their centers is the same as their diameters.

The problem of determining the masses and diameters of individual atoms required the determination of the number of atoms in a given amount of material. Diffraction experiments, like those described in Chapter 17 but using X rays, determined the distance between individual atoms in solids to be about 10^{-10} meter, one ten-billionth of a meter.* If we assume that atoms in a solid can be represented by marbles like those in Figure 7-4, the diameter of an atom is about equal to their spacing. This means that it would take 10 billion atoms to make a line 1 meter long. Stated another way, if we imagine expanding a baseball to the size of the Earth, the individual atoms of the ball would only be the size of grapes!

A useful quantity of matter for our purposes is the mole. If the mass of the molecule is some number of atomic mass units, a mole of the substance is this same number of grams. For instance, a mole of carbon is 12 grams. Further experiments showed that a mole of any substance contained the same number of molecules, namely 6.02×10^{23} molecules, a number known as **Avogadro's number.** With this number we can calculate the size of the atomic mass unit in terms of kilograms. Because 12 grams of carbon contain Avogadro's number of carbon atoms, the mass of one atom is

$$m_{\text{carbon}} = \frac{12 \text{ g}}{6.02 \times 10^{23} \text{ molecules}} = 2 \times 10^{-23} \text{ g/molecule}$$

Because one carbon atom also has a mass of 12 atomic mass units, we obtain

$$\frac{2 \times 10^{-23} \text{ g}}{12 \text{ amu}} = 1.66 \times 10^{-24} \text{ g/amu}$$

*For an explanation of this notation, see Appendix B (Numbers Large and Small).

Seeing Atoms

Atoms are so small that they cannot be seen with the most powerful optical microscopes. This limitation is due to the wave properties of light that we will discuss in Chapter 18. The relatively large wavelength of the light scatters from the atoms without yielding any information about the individual atoms. Even the most powerful electron microscope does not allow us to view individual atoms.

A new type of microscope, developed in the early 1980s, provides views of atomic surfaces with a resolution thousands of times greater than is possible with light. The developers of the *scanning tunneling microscope,* or STM, were awarded the Nobel Prize in 1986 (along with the inventor of the electron microscope).

The STM uses a very sharp probe (several atoms across at the tip) placed close to the surface (about one ten-billionth of a meter) to "view" the surface under investigation. As the probe is moved back and forth over the surface, electrons are pulled from the atoms to the probe, creating a current that is monitored by a computer. The current is greatest when the probe is directly over an atom and quite small when the probe is between atoms. By carefully mapping the surface, the structure of the surface can be plotted with a resolution smaller than the atomic dimensions. Computer displays like the one shown here clearly indicate the orderly arrangement of the atoms on the surface of the material. These measurements can also show the location of individual atoms that are deposited onto the surface. Such studies are important in understanding the physics of surfaces as well as catalytic processes used in manufacturing.

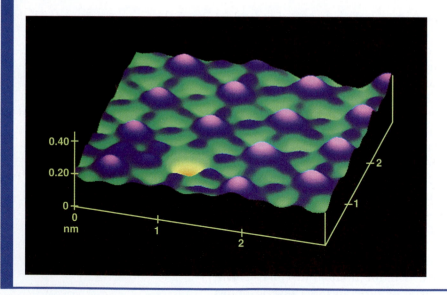

A scanning tunneling microscope image of iodine atoms absorbed on a platinum surface. Note the orderly arrangement of the atoms, and that one of the atoms is missing.

Therefore, 1 atomic mass unit equals 1.66×10^{-27} kilogram, a mass so small that is very hard to imagine. This is the approximate mass of one hydrogen atom. The most massive atoms are about 260 times this value.

The Ideal Gas Model

Many macroscopic properties of materials can be understood from the atomic model of matter. Under many situations the behavior of real gases is very closely approximated by an **ideal gas.** The gas is assumed to be composed of an enormous number of very tiny particles separated by relatively large distances. These particles are assumed to have no internal structure and to be indestructible. They also do not interact with each other except when they collide, and then they undergo elastic collisions much like air-hockey pucks. Although this model may not seem realistic, it follows in the spirit of Galileo in trying to get at essential features. Later we can add the complications of real gases.

For this model to have any validity it must describe the macroscopic behavior of gases. For instance, we know that gases are easily compressed. This makes sense; the model says that the distance between particles is very much greater than the particle size and they don't interact at a distance. There is, then, a lot of space in the gas and it should be easily compressed. This aspect of the model also accounts for the low mass-to-volume ratio of gases.

Since a gas completely fills any container and the particles are far from one another, the particles must be in continual motion. Is there any other evidence that the particles are continually moving? We might ask, "Is the air in the room moving even with all the doors and windows closed to eliminate drafts?" The fact that you can detect an open perfume bottle across the room indicates that some of the perfume particles have moved through the air to your nose.

More direct evidence for the motion of particles in matter was observed in 1827 by the Scottish botanist Robert Brown. In order to view pollen under a microscope without it blowing away, Brown mixed the pollen with water. He discovered that the pollen grains were constantly jiggling. Brown initially thought that the pollen might be alive and moving erratically on its own. However, he observed the same kind of motion with inanimate objects as well.

Brownian motion is not restricted to liquids. Observation of smoke under a microscope shows that the smoke particles have the same very erratic motion. This motion never ceases. If the pollen and water are kept in a sealed container and put on a shelf, you would still observe the motion years later. The particles are in continual motion.

It was 78 years before Brownian motion was rigorously explained. Albert Einstein demonstrated mathematically that the erratic motion was due to collisions between water molecules and pollen grains. The number and direction of the collisions occurring at any time is a statistical process. When the collisions on opposite sides have equal impulses, the grain is not accelerated. But when more collisions occur on one side, the pollen experiences an abrupt acceleration that is observed as Brownian motion.

Pollen grains suspended in a liquid exhibit continual, erratic motion known as Brownian motion.

Pressure

Σ

Let's take a look at one of the macroscopic properties of an ideal gas that is a result of the atomic motions. **Pressure** is the force exerted on a surface divided by the area of the surface; that is, the force per unit area.

$$P = \frac{F}{A}$$

$$\text{pressure} = \frac{\text{force}}{\text{area}}$$

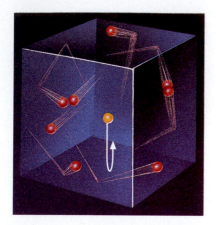

Figure 7-5 Gas particles are continually colliding with each other and with the walls of the container.

This definition is not restricted to gases and liquids. For instance, if a crate weighs 6000 newtons and its bottom surface has an area of 2 square meters, what pressure does it exert on the floor under the crate?

$$P = \frac{F}{A} = \frac{6000 \text{ N}}{2 \text{ m}^2} = 3000 \text{ N/m}^2$$

Therefore, the pressure is 3000 newtons per square meter. The SI unit of pressure [newton per square meter (N/m^2)] is called a pascal (Pa). Pressure in the English system is often measured in pounds per square inch (psi) or atmospheres (atm), where 1 atmosphere is equal to 101 kilopascals or 14.7 pounds per square inch.

Imagine a cubical container of gas in which particles are continually moving around and colliding with each other and with the walls (Fig. 7-5). In each collision with a wall the particle reverses its direction. Assume a head-on collision as in the case of the yellow particle in Figure 7-5. (If the collision was a glancing blow, only the part of the velocity that is perpendicular to the wall would be reversed.) This means that its momentum is also reversed. The change in momentum means that there must be an impulse on the particle and an equal and opposite impulse on the wall (Chapter 5). Our model assumes that an enormous number of particles strike the wall. The average of an enormous number of impulses produces a steady force on the wall that we experience as the pressure of the gas.

We can use this application of the ideal gas model to make predictions that can be tested. Suppose, for example, that we shrink the volume of the container. This means that the particles have less distance to travel between collisions with the walls and should strike the walls more frequently, increasing the pressure. Therefore, decreasing the volume increases the pressure provided that the average speeds of the molecules do not change. Similarly, if we increase the number of particles in the container, we expect the pressure to increase because there would be more frequent collisions with the walls.

Atomic Speeds and Temperature

Presumably, the atomic particles making up a gas have a range of speeds due to their collisions with the walls and with each other. The distribution of these speeds can be calculated from the ideal gas model and a connection made with temperature. Therefore, a direct measurement of these speeds would provide additional support for the model. This is not an easy task. Imagine trying to measure the speeds of a very large group of invisible particles. One needs to devise a way of starting a race and recording the order of the finishers.

One creative approach led to a successful experiment in 1920. The gas leaves a heated vessel and passes through a series of small openings that select only those particles going in a particular direction [Fig. 7-6(a)]. Some of these particles enter a small opening in a rapidly rotating drum, as shown in Figure 7-6(b). This arrangement guarantees that a group of particles start across the drum at the same time. Particles with different

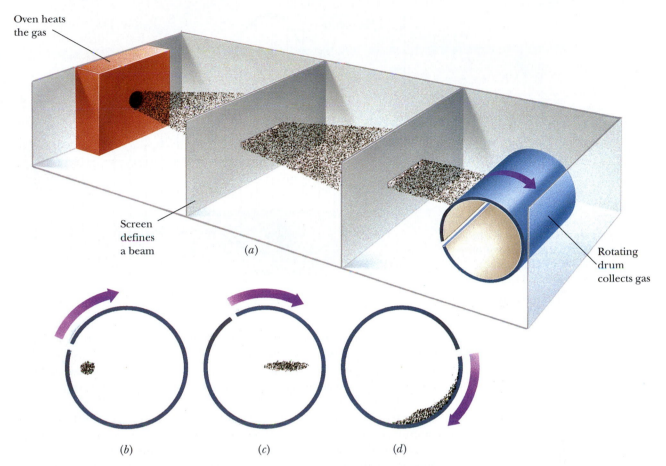

Oven heats
the gas

Screen
defines
a beam

(a)

Rotating
drum
collects gas

(b)   *(c)*   *(d)*

Figure 7-6 (a) Apparatus for measuring the speeds of atomic particles in a gas. (b) A small bunch of particles enters the rotating drum through the narrow slit. (c) The particles spread out as they move across the drum because of their different speeds. (d) The particles are deposited on a sensitive film at locations determined by how much the drum rotates before they arrive at the opposite wall.

speeds take different times to cross the drum and arrive on the opposite wall after the drum has rotated by different amounts. The locations of the particles are recorded by a film of sensitive material attached to the inside of the drum [Fig. 7-6(d)]. A drawing of the film's record is shown in Figure 7-7(a). A graph of the number of particles versus their position along the film is shown in Figure 7-7(b).

Most of the particles have speeds near the average speed, but some move very slowly and some very rapidly. The average speed is typically about 500 meters per second, which means that an average gas particle could travel the length of five football fields in a single second. This high value may seem counter to your experience. If the particles travel with this speed, why does it take several minutes to detect the opening of a perfume bottle on the other side of the room? This delay is due to collisions between the gas particles. On average a gas particle travels a distance of only 0.0002 millimeter before it collides with another gas particle. (This

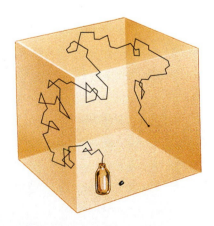

Atomic particles in air travel in zigzag paths because of numerous collisions with air molecules.

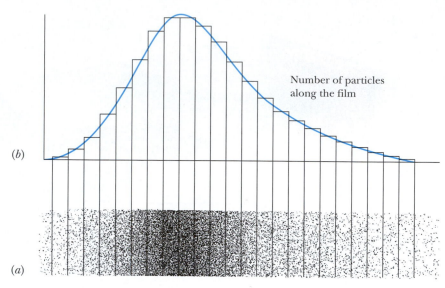

Number of particles along the film

(b)

(a)

Film from inside the drum

Figure 7-7   (a) The distribution of atomic particles as recorded by the film along the circumference of the drum. (b) The distribution of atomic speeds calculated from this experiment.

distance is about one thousand particle diameters.) Each particle makes approximately 2 billion collisions per second. During these collisions the particles can radically change directions, resulting in very zigzag paths. So although their average speed is quite fast, it takes them a long time to cross the room because they travel enormous distances.

When the speeds of the gas particles are measured at different temperatures, something interesting is found. As the temperature of the gas increases, the speeds of the particles also increase. The distributions of speeds for three temperatures are given in Figure 7-8. The calculations based on the model agree with these results. A relationship can be derived that connects temperature, a macroscopic property, with the average kinetic energy of the gas particles, a microscopic property. However, the simplicity of this relationship is only apparent with a particular temperature scale.

Temperature

We generally associate temperature with our feelings of hot and cold; however, our subjective feelings of hot and cold are not very accurate. Although we can usually say which of two objects is hotter, we can't state just how hot something is. To do this, we must be able to assign numbers to various temperatures.

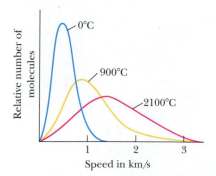

Figure 7-8   The distribution of atomic speeds changes as the temperature of the gas changes.

> **Physics on Your Own**   Place one hand in a pan of hot water and the other in a pan of ice cold water. After several minutes, place both hands in a pan of lukewarm water. Do your hands feel like they are at the same temperature? What does this say about using your body as a thermometer?

Figure 7-9   The height of the liquid in Galileo's first thermometer indicated changes in temperature.

Galileo was the first person to develop a thermometer. He observed that some of an object's properties change when its temperature changes. For example, with only a few exceptions, when an object's temperature goes up, it expands. Galileo's thermometer (shown in Fig. 7-9) was an inverted flask with a little water in its long neck. As the enclosed air got hotter, it expanded and forced the water down the flask's neck. Conversely, the air contracted on cooling and the water rose. Galileo completed his thermometer by marking a scale on the neck of the flask.

Unfortunately, the water level also changed when other local conditions such as atmospheric pressure changed. Galileo's thermometer was replaced by the alcohol-in-glass thermometer that is still popular today. The column is sealed so that the rise and fall of the alcohol is due to its change in volume and not the atmospheric pressure. The change in height is amplified by adding a bulb to the bottom of the column. When the temperature rises, the larger volume in the bulb expands into the narrow tube, making the expansion much more obvious.

In 1701, Newton proposed a method for standardizing the scales on thermometers. He put the thermometer in a mixture of ice and water, waited for the level of the alcohol to stop changing, and marked this level as zero. He used the temperature of the human body as a second fixed temperature, which he called 12. The scale was then marked off into 12 equal divisions.

Shortly after this, the German physicist Gabriel Fahrenheit suggested that the zero point correspond to the temperature of a mixture of ice and salt. Since this was the lowest temperature producible in the laboratory at that time, it avoided the use of negative numbers for temperatures. The original 12 degrees were later divided into eighths and renumbered so that body temperature became 96 degrees.

It is important that the fixed temperatures be reliably reproducible in different laboratories. Unfortunately, neither of Fahrenheit's fixed points could be reproduced with sufficient accuracy. Therefore, the fixed points were changed to those of the freezing and boiling points of pure water at standard atmospheric pressure. To get the best overall agreement with

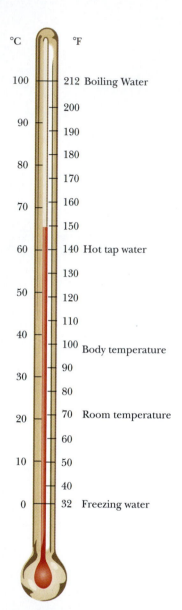

°C   °F

100 ——— 212 Boiling Water

              200

90 ———   190

              180

80 ———   170

70 ———   160

              150

60 ———   140 Hot tap water

              130

50 ———   120

              110

40 ———   100 Body temperature

               90

30 ———   80

20 ———   70 Room temperature

               60

10 ———   50

               40

0 ———   32 Freezing water

Figure 7-10   A comparison of the Fahrenheit and Celsius temperature scales.

Figure 7-11   When the graph of volume versus temperature of an ideal gas is extrapolated to zero volume, the temperature scale reads −273°C.

the previous scale, these temperatures were defined to be 32°F and 212°F, respectively. This is how we ended up with such strange numbers on the **Fahrenheit temperature scale.** On this scale, normal body temperature is 98.6°F.

At the time the metric system was adopted, a new temperature scale was defined with the freezing and boiling points as 0°C and 100°C. The name of this centigrade (one hundred point) scale was changed to the **Celsius temperature scale** in 1948 in honor of the Swedish astronomer Anders Celsius, who devised the scale. A comparison of the Fahrenheit and Celsius scales is given in Figure 7-10. This figure can be used to convert temperatures roughly from one scale to the other.

> **QUESTION**   What are room temperature (68°F) and body temperature (98.6°F) on the Celsius scale?

> **Physics on Your Own**   If you live in an area with a high elevation, measure the boiling point of water. Does it boil at 100°C (212°F)? Does it matter how hard the water boils? What does this say about the possibility of using the boiling point of water as a fixed point for calibrating thermometers?

Assume that we have a quantity of ideal gas in a special container designed to always maintain the pressure of the gas at some constant low value. When the volume of the gas is measured at a variety of temperatures, we obtain the graph shown in Figure 7-11. If the line on the graph is extended down to the left, we find that the volume goes to zero at a temperature of −273°C (−459°F). Although we could not actually do this experiment with a real gas, this very low temperature arises in several theoretical considerations and is the basis for a new, more fundamental temperature scale.

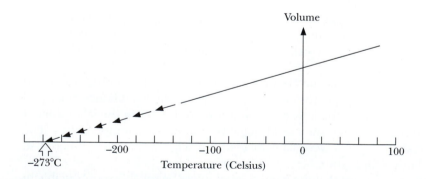

> **ANSWER**   Using Figure 7-10, we see that room temperature is about 20°C and body temperature is about 37°C.

The **Kelvin temperature scale,** also known as the **absolute temperature scale,** has its zero at −273°C and the same size degree marks as the Celsius scale. The difference between the Celsius and Kelvin scales is that temperatures are 273 degrees higher on the Kelvin scale. Water freezes at 273 K and boils at 373 K. (Notice that the degree symbol is dropped from this scale. The freezing point of water is read "273 kelvin," or "273 kay.") The Kelvin scale is named for the British physicist William Thomson, who is more commonly known as Lord Kelvin.

K = C + 273

> **QUESTION**   What is normal body temperature (37°C) on the Kelvin scale?

It would seem that all temperature scales are equivalent and which one we use would be a matter of history and custom. It is true that these scales are equivalent since conversions can be made between them. However, the absolute temperature scale has a greater simplicity for expressing physical relationships. In particular, the relationship between the volume and temperature of an ideal gas is greatly simplified using absolute temperatures. *The volume of an ideal gas at constant pressure is proportional to the absolute temperature.* This means that if the absolute temperature is doubled while keeping the pressure fixed, the volume of the gas doubles.

V proportional to T

The volume of an ideal gas at constant pressure can be used as a thermometer. All we need to do to establish the temperature scale is to measure the volume at one fixed temperature. Of course, thermometers must be made of real gases. But real gases behave like the ideal gas if the pressure is kept low and the temperature is well above the temperature at which the gas liquifies.

This new scale also connects the microscopic property of atomic speeds and the macroscopic property of temperature. *The absolute temperature is directly proportional to the average kinetic energy of the gas particles.* This means that if we double the average kinetic energy of the particles, the absolute temperature of a gas doubles. Remember, however, that the average speed of the gas particles does not double because the kinetic energy depends on the square of the speed (Chapter 6).

T proportional to KE_{ave}

Physics Update

In 1995 cesium atoms were cooled to 700 nanokelvins (a nanokelvin is a billionth of a kelvin), the lowest three-dimensional temperatures ever achieved in an atom. Researchers at the National Institute for Standards and Technology make special laser light patterns called "optical lattices" which trap precooled atoms at regular locations in the pattern. Then, by gradually decreasing the intensity of the laser light, they diminish the interactions between the light and the atom. This causes each atom to occupy a larger region of space. In "expanding" to their new volume, the atoms expend energy, making them colder. Such supercold atoms may someday be employed to make more accurate atomic clocks.

ANSWER   Body temperature is 37 + 273 = 310 K.

QUESTION   What temperature change would double the average speed?

The Ideal Gas Law   Σ

The three macroscopic properties of a gas—volume, temperature, and pressure—are related by a relationship known as the **ideal gas law.** This law states that

$$PV = cT$$

ideal gas law

where P is the pressure, V is the volume, T is the *absolute* temperature, and c is a constant which depends on the amount of gas.

This relationship is a combination of three experimental relationships that had been discovered to hold for the various pairs of these three macroscopic properties. For example, if we hold the temperature of a quantity of gas constant, we can experimentally determine what happens to the pressure as we compress the gas. Or we can vary the pressure and measure the change in volume. This experimentation leads to a relationship known as Boyle's law, which states that the product of the pressure and the volume is a constant. This is equivalent to saying that they are inversely proportional to each other; as one increases, the other must decrease.

PV = constant

In a similar manner, we can investigate the relationship between temperature and volume while holding the pressure constant. The results for a gas at one pressure are shown in Figure 7-11. As stated in the previous section, the volume is directly proportional to the absolute temperature.

$V = T \times$ constant

The third relationship is between temperature and pressure at a constant volume. The pressure is directly proportional to the absolute temperature.

$P = T \times$ constant

Each of these relationships can be obtained from our model for an ideal gas. For example, let's take a qualitative look at Boyle's law. As we decrease the volume, the molecules must hit the sides more frequently and therefore the pressure increases in agreement with the statement of Boyle's law.

QUESTION   How does the ideal gas model explain the rise in pressure of a gas as its temperature is raised without changing its volume?

ANSWER   We would need to quadruple the absolute temperature.

ANSWER   Raising the temperature of the gas increases the kinetic energies of the particles. The increased speeds of the particles not only means that they have larger momenta, but that they also hit the walls more frequently.

Evaporative Cooling

Many of the concepts of the ideal gas model apply to liquids as well as gases. One big difference between liquids and gases is the strength of the attractive forces between the molecules. In our ideal gas model we assumed that these forces could be neglected until the molecules collided. As a consequence a gas expands to fill its container. However, liquids have a definite volume. It is intermolecular forces that hold the molecules in the liquid together.

If we assume a model for liquids that is similar to that for gases, we can begin to understand the evaporation of liquids. Assume that the kinetic energies of the molecules in liquids have a distribution similar to that in gases and that the average kinetic energy of the particles increases with increasing temperature. The intermolecular forces perform work on molecules that try to escape the liquid, allowing only those with large enough kinetic energies to succeed. Therefore, the molecules that leave the liquid have higher than average kinetic energies. With their removal, the average kinetic energy of the molecules left behind is lower, and the liquid is cooled. Water is often carried in canvas bags on desert trips to keep the water cool. The canvas bag is kept wet by water seeping through the canvas. The evaporation of the water from the wet canvas keeps the rest of the water cool.

Evaporative cooling can be demonstrated with a simple experiment. Wrap the end of a thermometer in cotton soaked in room temperature water and observe the temperature as the water evaporates.

Your body uses evaporative cooling to maintain your body temperature on very hot days or during heavy exercise. The evaporating sweat cools our bodies.

Canvas bags like the one shown here keep water cool, even in a hot desert.

If you didn't sweat, you might die of heat prostration! A more effective way of cooling your body is the alcohol rub used to reduce fevers.

SUMMARY

The most elementary substances are elements, which are composed of a large number of very tiny, identical atoms. These atoms combine in definite mass ratios to form molecules according to the law of definite proportions. Observations that gaseous elements combine in definite volume ratios allowed the determination of the relative masses of different atoms. The mass of a carbon atom is set equal to 12 atomic mass units.

Modern experiments show that atoms have diameters of approximately 10^{-10} meter and masses ranging from 1 to more than 260 atomic mass units, where 1 atomic mass unit is equal to 1.66×10^{-27} kilogram.

CHAPTER

7

REVISITED

Observation of the motion of smoke particles provided evidence for the existence of atoms and molecules. Knowledge of the volumes of gases that combined chemically to form other gases helped establish the masses of individual atoms and molecules. The ideal gas law tells us that the average kinetic energy of the gas particles is determined by the absolute temperature.

The Celsius temperature scale is defined with the freezing and boiling points of water as 0°C and 100°C, respectively. The absolute, or Kelvin, temperature scale has its zero point at −273°C and the same size degree as the Celsius scale.

The ideal gas model assumes the gas to be composed of an enormous number of tiny, indestructible spheres with no internal structure, separated by relatively large distances, and interacting only via elastic collisions. The pressure exerted by a gas is due to the average of the impulses exerted by the gas particles on the walls of the container. Most of the particles have speeds near the average speed, but some are moving very slowly and some very rapidly. The average speed is typically about 500 meters per second. The absolute temperature of a gas is proportional to the average kinetic energy of the gas particles.

The ideal gas law states the relationship between the pressure, the volume, and the absolute temperature of a gas, $PV = cT$, where c is a constant that depends on the amount of gas.

KEY TERMS

absolute temperature scale: The temperature scale with its zero point at absolute zero and degrees equal to those on the Celsius scale. This is the same as the Kelvin temperature scale.

atom: The smallest unit of an element that has the chemical and physical properties of that element.

atomic mass: The mass of an atom in atomic mass units.

atomic mass unit: One-twelfth of the mass of a carbon atom. Equal to 1.66×10^{-27} kilogram.

Avogadro's number: 6.02×10^{23} molecules, the number of molecules in 1 mole of any substance.

Celsius temperature scale: The temperature scale with the values of 0° and 100° for the temperatures of freezing and boiling water, respectively.

compound: A combination of chemical elements that forms a substance with its own properties.

element: Any chemical species that cannot be broken up into other chemical species.

Fahrenheit temperature scale: The temperature scale with the values of 32° and 212° for the temperatures of freezing and boiling water, respectively. Its degree is 5/9 that on the Celsius or Kelvin scales.

ideal gas: An enormous number of very tiny particles separated by relatively large distances. The particles have no internal structure, are indestructible, and do not interact with each other except when they collide, and all collisions are elastic.

ideal gas law: $PV = cT$, where P is the pressure, V is the volume, T is the absolute temperature, and c is a constant that depends on the amount of gas.

Kelvin temperature scale: The temperature scale with its zero point at absolute zero and a degree equal to that on the Celsius scale. Also called the absolute temperature scale.

law of definite proportions: When two or more elements combine to form a compound, the ratios of the masses of the combining elements have fixed values.

macroscopic: The bulk properties of a substance such as mass, size, pressure, and temperature.

microscopic: Properties not visible to the naked eye such as atomic speeds or the masses and sizes of atoms.

molecule: A combination of two or more atoms.

pressure: The force per unit area of surface. Measured in newtons per square meter, or pascals.

1. What are the two essential features of a good model?

2. A friend has created a model of how a candy vending machine works. His theory says that a little blue person (LBP) lives inside each vending machine. This person takes your coins and gives you candy in return. Although this little blue person theory may not seem reasonable to you, can you suggest ways of disproving it without opening the machine?

3. Why did the Greek "atomists" believe in atoms?

4. What role did the alchemist play in the development of an atomistic world view?

5. Which of the following is not an element: hydrogen, oxygen, sodium, chlorine, or water?

6. Would you expect sulfur dioxide to be an element or a compound?

7. When the element mercury is heated in air, a red powder is formed and the mass increases. Is this powder an element or a compound? Explain your reasoning.

8. What techniques might you use to show that table salt is a compound?

9. What does the law of definite proportions state about the elements that make up a compound?

10. Give an example that illustrates the meaning of the law of definite proportions.

11. What are the basic differences between mixtures and compounds?

12. Do water and salt form a compound or a mixture?

13. The atomic mass of zinc oxide is 81 atomic mass units. What is the atomic mass of zinc if a molecule of zinc oxide consists of one atom of zinc and one atom of oxygen?

14. The atomic mass of nitrogen dioxide is 46 atomic mass units. What is the atomic mass of nitrogen if a molecule of nitrogen dioxide consists of one atom of nitrogen and two atoms of oxygen?

15. Gold has an atomic mass of 197. How does the mass of a gold atom compare with that of a hydrogen atom?

16. If an atom of titanium is four times as massive as an atom of carbon, what is the atomic mass of titanium?

17. How does the number of molecules in 1 liter of oxygen compare with the number of molecules in 1 liter of carbon dioxide if they are both at the same temperature and pressure?

18. Assume that you have equal volumes of oxygen and hydrogen at the same temperature and pressure. If each molecule of oxygen and hydrogen contains two atoms, how do the numbers of oxygen and hydrogen atoms in the gases compare?

19. If the atomic mass of oxygen is 16 atomic mass units, how much oxygen would be needed to have an Avogadro's number of oxygen atoms?

20. If the atomic mass of oxygen is 16 atomic mass units and there are two oxygen atoms to a molecule, how much oxygen is needed to have an Avogadro's number of oxygen molecules?

21. What are the features of the ideal gas model?

22. If a gas is condensed to form a liquid, the liquid occupies a much smaller volume than the gas. How does the ideal gas model account for this?

23. How can you use the model of an ideal gas to explain the Brownian motion of smoke particles?

24. Some people have proposed that a drunk looking for his keys in the grass is an analogy to Brownian motion. Does this analogy have any validity?

25. What are the units of pressure in terms of length, mass, and time?

26. A cube and a spherical ball are made of the same material and have the same mass. Which exerts the larger pressure on the floor?

27. Describe how a gas exerts a pressure on the walls of its container.

28. Why does the pressure in a tire increase as you add more air?

29. Why does a tire that is low on air flatten out along the ground?

30. If you screw the cap of an empty plastic drinking bottle on tightly while walking in the mountains, why are the sides of the bottle caved in when you return to the valley?

31. If the average speed of a perfume molecule is 500 meters per second, why does it take several minutes before you smell the perfume from a bottle opened across the room?

32. What happens to the average speed of the molecules of a gas as it is heated?

33. Why does an alcohol-in-glass thermometer have a bulb at the bottom?

34. How might you amplify the expansion of a metal rod so that it can be used as a thermometer?

35. Why is body temperature not a good fixed temperature for establishing a temperature scale?

36. What conditions must be imposed before the boiling point of water can be used as a fixed temperature?

37. Two students are sick in bed with fevers. One has a temperature of 39.0°C; the other 100.6°F. Which student has the higher fever?

38. Is a sauna at a temperature of 190°F hotter or colder than one at 90°C?

39. If your mother would only let you go swimming when the temperature was above 70°F, what temperature on the Celsius scale would be required?

40. Where should you set your new Celsius thermostat so that your hot tub stays at a comfortable 102°C?

41. What is the freezing point of water on the Kelvin scale?

42. If liquid nitrogen boils at 77 K, at what Celsius temperature does it boil?

43. What microscopic property of an ideal gas doubles when the absolute temperature is doubled?

44. On which temperature scale does the average kinetic energy of the molecules in an ideal gas double when you double the temperature?

45. How do the average kinetic energies of each gas in a mixture compare?

46. In a mixture of different gases, which molecules have the highest average speed? (*Hint:* Both gases are at the same temperature.)

47. If you heat a gas in a container with a fixed volume, the pressure increases. Use the ideal gas model to explain this.

48. Use the ideal gas model to explain why the pressure of a gas rises as the volume is reduced while keeping the temperature constant.

49. What macroscopic property of an ideal gas doubles when the absolute temperature is doubled while keeping the pressure constant?

50. If the volume of an ideal gas is held constant, what happens to the pressure if the absolute temperature is cut in half?

51. What happens to the temperature of an ideal gas if you reduce its volume to one-third while holding the pressure constant?

52. If you hold the temperature of an ideal gas constant, what happens to its volume when you triple its pressure?

Question 40.

53. Why does the pressure inside the tires increase after a car has been driven?

54. If you put a sealed plastic bottle partially filled with hot tea into the refrigerator, the sides of the bottle will cave in as the tea cools. Why?

55. The water in a canvas water bag placed in front of your car when driving across the desert stays cooler than the surrounding air. Explain this in terms of the average kinetic energies of the water molecules that leave and stay behind.

56. How does an alcohol rub cool your body?

57. Why might hikers get hypothermia during wet weather even when the temperature is above freezing?

58. The temperature of boiling water does not increase even if the heat is turned on high. How can you use the ideal gas model to explain this?

Question 55.

EXERCISES

1. Assume that you have a collection of nuts and bolts and that each nut has a mass of 1 g and each bolt a mass of 2 g. Show that the ratio of the mass of one nut to the mass of one bolt is the same as that of ten nuts to ten bolts.

2. If the ratio of the mass of ten white marbles to ten red marbles is 1.5, what is the mass ratio of one white to one red marble?

3. If 1 g of hydrogen combines completely with 8 g of oxygen to form water, how many grams of hydrogen does it take to combine completely with 24 g of oxygen?

4. If 12 g of carbon combine completely with 16 g of oxygen to form carbon monoxide, how many grams of oxygen does it take to combine completely with 60 g of carbon?

5. Given that 1 g of hydrogen combines completely with 8 g of oxygen to form water, how much water can you make with 8 g of hydrogen and 16 g of oxygen?

6. Given that 12 g of carbon combines completely with 16 g of oxygen to form carbon monoxide, how much carbon monoxide can be made from 48 g of carbon and 48 g of oxygen?

7. Given that the carbon atom has a mass of 12 amu, how many carbon atoms are there in a diamond with a mass of 1 g?

8. Given that the nitrogen molecule has a mass of 28 amu, how many nitrogen molecules are there in 1 g of nitrogen?

9. How many molecules of water are there in 1 L of water? A liter of water has a mass of 1 kg and the mass of a water molecule is 18 amu.

10. A liter of oxygen has a mass of 1.4 g and the oxygen molecule has a mass of 32 amu. How many oxygen molecules are there in a liter of oxygen?

*11. One liter of nitrogen combines with 3 L of hydrogen to form 2 L of ammonia. If the molecules of nitrogen and hydrogen have two atoms each, how many atoms of hydrogen and nitrogen are there in one molecule of ammonia?

*12. One liter of oxygen combines with 1 L of hydrogen to form 1 L of hydrogen peroxide. Given that the molecules of hydrogen and oxygen contain two atoms each, how many atoms are there in one molecule of hydrogen peroxide?

13. Seven grams of nitrogen combine completely with 8 g of oxygen to form nitric oxide. The nitric oxide molecule consists of one nitrogen atom and one oxygen atom. What is the atomic mass of nitrogen?

*14. Seven grams of nitrogen combine completely with 4 g of oxygen to form nitrous oxide. The nitrous oxide molecule consists of two nitrogen atoms and one oxygen atom. What is the atomic mass of nitrogen?

15. About how many atoms would you expect there to be in a cube of material 1 cm on each side?

16. About how many atoms would it take to deposit a single layer on a surface with an area of 1 cm²?

17. A woman with a mass of 50 kg is wearing shoes with heels that have a cross-sectional area of 2 cm². What is the pressure she exerts on the floor when she is standing on one heel?

18. What average pressure does a 1400-kg car exert on the ground if each tire makes contact with 0.02 m² of ground?

19. What happens to the volume of 1 L of an ideal gas when the pressure is tripled while the temperature is held fixed?

20. What happens to the temperature of an ideal gas at 27°C if the volume is forced to double at a constant pressure?

*21. A helium bottle with a pressure of 100 atm has a volume of 2 L. How many balloons can the bottle fill if each balloon has a volume of 1 L and a pressure of 1.25 atm?

*22. When the temperature of an automobile tire is 20°C, the pressure in the tire reads 32 psi on a tire gauge. (The gauge measures the difference between the pressures inside and outside the tire.) What is the pressure when the tire heats up to 40°C while driving? You may assume that the volume of the tire remains the same and that atmospheric pressure is a steady 14 psi.

*23. A volume of 150 cm³ of an ideal gas has an initial temperature of 20°C and an initial pressure of 1 atm. What is the final pressure if the volume is reduced to 100 cm³ and the temperature is raised to 40°C?

*24. An ideal gas has the following initial conditions: $V_i = 500$ cm³, $P_i = 3$ atm, and $T_i = 100$°C. What is its final temperature if the pressure is reduced to 1 atm and the volume expands to 1000 cm³?

States of Matter

Many materials exist as solids, liquids, or gases, each with its own characteristics. However, different materials share many characteristics; for instance, many materials form crystalline shapes in the solid form. What does the shape of a crystal tell us about the underlying structure of the material? (See p. 202 for the answer to this question.)

All matter is composed of approximately 100 different elements. Yet the material world we experience, say in a walk through the woods, holds a seemingly endless variety of forms. This variety arises from the particular combinations of elements and the structures they form, which can be divided into four basic forms, or states: solid, liquid, gas, and plasma.

Many materials can exist in the solid, liquid, and gaseous states if the forces holding the chemical elements together are strong enough that their melting and vaporization temperatures are lower than their decomposition temperatures. Hydrogen and oxygen in water, for example, are so tightly bonded that water exists in all three states. Sugar, on the other hand, decomposes into its constituent parts before it can turn into a gas.

If we continuously heat a solid, the average kinetic energy of its molecules rises and the temperature of the solid increases. Eventually the intermolecular bonds break and the molecules slide over one another (the process called melting) to form a liquid. The next change of state occurs when the substance turns into a gas. In the gaseous state the molecules have enough kinetic energy to be essentially independent of each other. In a plasma individual atoms are literally ripped apart into charged ions and electrons, and the subsequent electrical interactions drastically change the resulting substance's behavior.

Atoms

At the end of the previous chapter we established the evidence for the existence of atoms. It would be natural to ask, "Why stop there?" Maybe atoms are not the end of our search for the fundamental building blocks of matter. They are not. Atoms have structure and we will devote two chapters near the end of the book to developing our understanding of this structure further. For now it is useful to know a little about this structure so that we can understand the properties of the states of matter.

A useful model for the structure of an atom for our current purposes is the "solar system" model developed early in this century. In this model the atom is seen as consisting of a very tiny central nucleus that contains almost all of the atom's mass. This nucleus has a positive electric charge that binds very light, negatively charged electrons to the atom in a way analogous to the Sun's gravitational attraction for the planets. It is the electrons' orbits that define the size of the atom and give it its chemical properties.

The basic force that binds materials together is the electrical attraction between atomic and subatomic particles. As we will see in later chapters, the gravitational force is too weak and the nuclear forces are too short-ranged to have much effect in chemical reactions. We live in an electrical universe when it comes to the states of matter. How these materials form depends on these electric forces. And the form they take determines the properties of the materials.

Density

One characteristic property of matter is its **density.** Unlike mass and volume, which vary from one object to another, density is an inherent property of the material. A ton of copper and a copper coin have drastically different masses and volumes but identical densities. If you were to find an unknown material and could be assured that it was pure, you could go a long way toward identifying it by measuring its density.

Density is defined as the amount of mass in a standard unit of volume and is expressed in units of kilograms per cubic meter.

$$D = \frac{M}{V}$$

$$\text{density} = \frac{\text{mass}}{\text{volume}}$$

For example, an aluminum ingot is 3 meters long, 1 meter wide, and 0.3 meter thick. If it has a mass of 2430 kilograms, what is the density of aluminum? We calculate the volume first and then the density.

$$V = lwh = (3 \text{ m})(1 \text{ m})(0.3 \text{ m}) = 0.9 \text{ m}^3$$

$$D = \frac{M}{V} = \frac{2430 \text{ kg}}{0.9 \text{ m}^3} = 2700 \text{ kg/m}^3$$

Therefore, the density of aluminum is 2700 kilograms per cubic meter. Densities are also often expressed in grams per cubic centimeter. Thus, the density of aluminum is also 2.7 grams per cubic centimeter (g/cm^3). Table 8-1 gives the densities of a number of common materials.

> **QUESTION**   Which has the greater density, 1 kilogram of iron or 2 kilograms of iron?

Physics on Your Own   Determine your density. You can get a pretty good measure of your volume by submerging yourself in a bathtub. Multiply the area of the tub by the difference in the water levels with you in and out of the tub. Your mass can be determined from your weight.

The densities of materials range from the small for a gas under normal conditions to the large for the element osmium. One cubic meter of osmium has a mass of 22,480 kilograms (a weight of nearly 50,000 pounds), about 22 times more than the same volume of water. It is interesting to note that the osmium atom is less massive than a gold atom. Therefore, the higher density of osmium indicates that the osmium atoms must be packed closer together.

> **ANSWER**   They have the same density; the density of a material does not depend on the amount of material.

Table 8-1 Densities of Some Common Materials

Material	Density (g/cm³)
Air*	0.0013
Ice	0.92
Water	1.00
Magnesium	1.75
Aluminum	2.70
Iron	7.86
Copper	8.93
Silver	10.5
Lead	11.3
Mercury	13.6
Uranium	18.7
Gold	19.3
Osmium	22.5

*At 0°C and 1.0 atm.

Density Extremes

Which weighs more: a pound of feathers or a pound of iron? The answer to this junior high school puzzle is, of course, that they weigh the same. The key difference between the two materials is *density;* although the two weigh the same, they have very different volumes. Density is a comparison of the masses of two substances with the same volumes. Obviously, a cubic meter of iron has much more mass than a cubic meter of feathers.

The densities of objects vary over a large range. The densities of common Earth materials pale in comparison to those of some astronomical objects. After a star runs out of fuel, its own gravitational attraction causes it to collapse. The collapse stops when the outward forces due to the pressure in the star balance the gravitational forces. The resultant stellar cores can have astronomically large densities. White dwarf stars are the death stage of most stars. They can have masses up to 1.4 times that of our Sun compressed to a size about that of Earth, with resulting densities a million times larger than the density of water. Neutron stars are cores left after a star explodes and can have densities a billion times larger than white dwarfs. A teaspoon of material from a neutron star would weigh a billion tons on Earth!

A new very low density solid, called *silica aerogel,* has been created at the Lawrence Livermore National Laboratory in California. This solid is made from silicon dioxide and has a density that is only three times

Silica aerogel has an extremely low density, but still can support 1600 times its own weight.

that of air. Because of this very low density, it is sometimes known as "solid smoke." Since silica aerogel is a solid, it holds its shape. In fact, it can support 1600 times its own weight!

The density of interstellar space is very much smaller than that of air; there is about one atom per cubic centimeter, resulting in a density about a billion trillionth that of air at 1 atmosphere of pressure, or about a trillion trillionth (10^{-24}) that of water.

The materials that we commonly handle have densities around the density of water, 1 gram per cubic centimeter. A cubic centimeter is about the volume of a sugar cube. The densities of surface materials on the Earth average approximately 2.5 grams per cubic centimeter. The density at the core of the Earth is about 9 grams per cubic centimeter, making the average density of the Earth about 5.5 grams per cubic centimeter.

QUESTION   If a hollow sphere and a solid sphere are both made of the same amount of iron, which sphere has the greater average density?

ANSWER   The solid sphere has the greater average density because it occupies the smallest volume for a given mass of iron.

Solids

Solids have the greatest variety of properties of the four states of matter. The character of a solid substance is determined by its elemental constituents and their particular structure. This underlying structure depends on the way it was formed. For example, slow cooling often leads to a solidification with the atoms in an ordered state known as a **crystal.**

Crystals grow in a variety of shapes. Their common property is the orderliness of their atomic arrangements. It is this orderliness that distinguishes solids from liquids. The orderliness consists of a basic arrangement of atoms that repeats throughout the crystal, analogous to the repeating geometric patterns in some wallpapers.

The repeating pattern in wallpaper is analogous to the underlying atomic structure in crystals.

> **Physics on Your Own**   Salol, a compound used by pharmacists in medicines, is interesting because its melting point is only a little above room temperature. Melt a pinch of salol in a small glass bowl floating in hot water. As soon as it melts, remove the glass bowl from the hot water and watch the liquid as it solidifies.

The microscopic order of the atoms is not always obvious in macroscopic samples. For one thing there are very few perfect crystals; most samples are aggregates of small crystals. However, macroscopic evidence of this underlying structure does exist. A common example in northern climates is a snowflake. Its sixfold symmetry is evidence of the structure of ice. Another example is mica (Fig. 8-1), a mineral you might find on a hike in the woods. Shining flakes of mica can be seen in many rocks. Larger pieces can be easily separated into thin sheets. The thinness of the sheets seems (at least on the macroscopic scale) to be limitless. It is easy to

The sixfold symmetry exhibited by snowflakes is evidence that ice crystals have hexagonal shapes.

Figure 8-1   Samples of mica exhibit a two-dimensional crystalline structure as evidenced by the ability to peel thin sheets from the larger crystal.

convince yourself that the atoms in mica are arranged in two-dimensional sheets with relatively strong bonds between atoms within the sheet and much weaker bonds between the sheets.

In contrast to mica, ordinary table salt exhibits a three-dimensional structure of sodium and chlorine atoms. If you dissolve salt in water and let the water slowly evaporate, the salt crystals that form have very obvious cubic structures. If you try to cut a small piece of salt with a razor blade, you find that it doesn't separate into sheets like mica but fractures along planes parallel to its faces. (Salt from a salt shaker displays this same structure but the grains are usually much smaller. A simple magnifying glass allows you to see the cubic structure.) Precious stones also have planes in their crystalline structure. A gem cutter studies the raw gemstones very carefully before making the cleavages that produce a fine piece of jewelry.

Physics on Your Own   Examine salt and sugar crystals with a magnifying glass. If you sprinkle salt or sugar crystals on a black piece of paper and carefully separate them, you can see distinct structures that provide clues to the underlying atomic ordering.

There are four major types of bonding within crystals: ionic, covalent, metallic, and an intermolecular bond called the van der Waals attraction. We learned in Chapter 6 that there is a potential energy associated with each force. The stable atomic arrangements are states of lowest potential energy. In general, atoms are electrically neutral. The binding that occurs between atoms is due to changes in the arrangements of their electrons, either through an outright exchange or through changes in their distributions.

Common table salt is an example of **ionic bonding.** The sodium atom literally gives one of its electrons to a chlorine atom, leaving both atoms electrically charged. These charged ions then bond together through their strong electrical attraction.

In **covalent bonding,** the atoms share some of their outer electrons. Instead of being exchanged, the outer electrons orbit both atoms. Depending on how the sharing takes place, radically different properties of the solid can be produced. The most common example of this variation is exhibited by two structures of pure carbon (Fig. 8-2). Diamond is formed when the sharing takes place in a three-dimensional fashion, resulting in a very hard substance that is treasured for its optical brilliance. In graphite, strong sharing occurs in two-dimensional planes, creating sheets of material that are relatively free to move over each other. Because of its slippery nature, graphite is used as a lubricant and as the "lead" in pencils.

The third type, metallic bonding, is the most common: All but 21 of the chemical elements are metals. **Metallic bonding** is very much like covalent bonding except the shared electrons are not confined to a particular group of atoms. In this case, the atoms bind together into a lower energy state, leaving the outer electron(s) relatively free. This results in a crystal structure combined with a "gas" of free electrons. Certain metals,

Figure 8-2 Synthetic diamonds and finely divided graphite are two different crystalline forms of carbon.

Figure 8-3   Dry ice leaves no puddle because the solid evaporates directly to a gas, unlike water ice (right) which normally passes through the liquid state.

such as gold, silver, and copper, are particularly good conductors of electricity. Most of the common metals are relatively easy to mold into many shapes.

The last type of crystalline structure has the weakest bonding, the **van der Waals bonding** (named for Johannes van der Waals, a Dutch physicist). Here the electric bonding force is created by slight imbalances in the electric charge distribution within molecules. If, for example, an electrically neutral molecule had more negative charges on one end than the other, the molecule would attract other similar molecules. This happens in ice. Another common example is "dry ice," the solid state of carbon dioxide. Charge fluctuations within the carbon dioxide molecules produce the attractive forces. Unlike regular ice (the solid form of water), dry ice always *sublimes*, passing directly from its solid state to its gaseous state without going through the liquid state (Fig. 8-3).

With the knowledge that most materials are combinations of more basic elements came the realization that new combinations could be produced. In early times most of the new combinations were probably created more by chance than by design. New combinations produced by mixing molten metals are called **alloys.** The Bronze Age, for example, began when early people discovered that copper and tin could be combined into an alloy that was stronger and more durable than either of its component elements. Brass (a combination of copper and zinc) and steel (iron combined with carbon) followed. Alloys are not new chemical compounds; the metals are mixed in various proportions and cooled.

Unlike alloys, **polymers** (from the Greek words *poly* meaning "many" and *meros* meaning "parts") are combined at the molecular level. Smaller carbon–hydrogen molecules are linked with each other and, in some cases, with other elements into very long chains called macromolecules that have special material properties (Fig. 8-4). In the past 50 years many new materials have been created in the area of polymer chemistry. Synthetic resins such as Bakelite, fibers such as rayon, and plastics such as celluloid were the first polymers made early in this century.

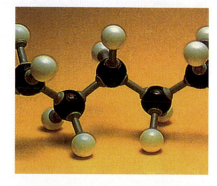

King Tut's mask was easily formed by the artist because of the malleability of the gold.

Figure 8-4   A model for a polymer.

Solid Liquids and Liquid Solids

Many substances exist between the ordinary boundaries of solids and liquids. When materials such as glass or wax cool, the molecules are frozen in space without arranging themselves into an orderly crystalline structure. These solids are amorphous, meaning that they retain some of the properties of liquids. A common example of an amorphous solid is the clear lollipop made by rapidly cooling liquid sugar. The average intermolecular forces in an amorphous material are weaker than those in a crystalline structure.

Despite the solid rigidity of an amorphous material, this form is more like a liquid than a solid because of its lack of order. In addition, the melting points of amorphous materials are not clearly defined. An amorphous material simply gets softer and softer, passing into the fluid state. Another characteristic of these solid liquids is that they actually do flow like a liquid although on time scales that make it hard to detect. Old church windows in Europe are thicker at the bottom than at the top due to centuries of flow.

Other substances are liquids that retain some degree of orderliness, characteristic of solids. Liquid crystals can be poured like regular liquids. They lack positional order, but they possess an orientational order. Small electric voltages can align the rodlike molecules along a particular direction.

Liquid crystals have some interesting applications because polarized light behaves differently, depending on whether it is traveling parallel or perpendicular to the alignment direction. For example, the orientation can be manipulated electrically to produce the numbers in a digital watch or an electronic calculator. You can verify that the light emerging from a liquid crystal display is polarized by looking at the display through polarized sunglasses. Changing the orientation of the sunglasses will vary the intensity of the image.

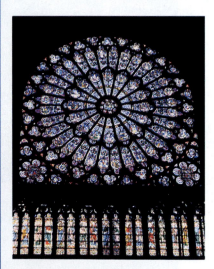

Notre Dame's Rose Window is thicker at the bottom due to a very slow flowing of the glass.

The display on this digital watch is an application of liquid crystal technology.

Approximately 80 percent of the organic chemistry industry is devoted to the production of synthetic polymers, from phonograph records and compact discs to the lining of your frying pan and the soles of your shoes. Nearly 180 pounds of these exotic materials are made each year for each person in the United States. In 1976, plastics surpassed steel as the nation's most widely used material; more plastic is used than steel, aluminum, and copper combined.

Scientists are constantly searching for combinations that yield new properties. Academic departments of "materials science" have been established in universities during the past two decades, much of the work of which involves looking for the special electronic properties of manufactured materials for use in modern circuitry.

Physics Update

Metallic hydrogen has been produced at the Lawrence Livermore Laboratories in a sample of fluid hydrogen. Hydrogen atoms constitute the bulk of the Universe's ordinary matter, so scientists have long sought to understand the properties and states of this simplest of elements. Squeezing hydrogen atoms until they surrender their electrons has been tried ever since Eugene Wigner predicted in 1935 that hydrogen would metallize at sufficiently high pressure. It was long thought that the road to metallic hydrogen lay with crystalline hydrogen rather than with the fluid state. Indeed, solid hydrogen has been crushed in diamond anvil presses up to pressures of 2.5 million atmospheres, but without making hydrogen metallic. Therefore, William Nellis was somewhat surprised when he succeeded at lesser pressures with fluid hydrogen. Besides being of interest to physicists studying the transition from insulator to metal, the formation of metallic hydrogen is of interest to fusion scientists who need to know what hydrogen does at high pressures, and to astronomers who model the interiors of gas giants like Jupiter and Saturn, which are expected to harbor vast reservoirs of metallic fluid hydrogen.

Liquids

When a solid melts, the intermolecular bonds break, allowing the molecules to slide over each other, producing a liquid. **Liquids** fill the shape of the container that holds them much like the random stacking of a bunch of marbles.

The temperature at which a solid melts varies from material to material simply because the bonding forces between molecules are different. Hydrogen molecules are so loosely bound to other hydrogen molecules that hydrogen becomes a liquid at 14 K. Oxygen and nitrogen—the constituents of the air we breathe—melt at 55 K and 63 K, respectively. The fact that ice doesn't melt until 273 K (0°C) tells us that the bonding between water molecules is relatively high.

Water is an unusual liquid. Although water is abundant, few liquids actually occur at ordinary temperatures on Earth. The bonding between the hydrogen atoms and the oxygen atom is especially strong, creating a very strong redistribution of the electrons within the molecule. This redistribution results in a relatively strong bond between the water molecules that requires a high temperature to separate them into the gaseous state.

The intermolecular forces in a liquid create a special "skin" on the surface of the liquid. This can be seen in Figure 8-5 in which a glass has been filled with milk beyond its brim. What is keeping the extra liquid from flowing over the edge?

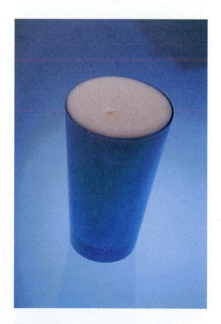

Figure 8-5   A glass filled with milk beyond the brim is evidence of surface tension.

Figure 8-6 A steel needle floats on water because of the surface tension of the water.

Imagine two molecules, one on the surface of a liquid and one deeper into the liquid. The molecule beneath the surface experiences attractive forces in all directions because of its neighbors. The molecule on the surface only feels forces from below and to the sides. This imbalance tends to pull the "surface" molecules back into the liquid.

Physics on Your Own A dry needle with a density much greater than water can actually float on the surface of the water as shown in Figure 8-6. Use tweezers to slowly lower the needle onto the surface.

Surface tension also tries to pull liquids into shapes with the smallest possible surface areas. The shapes of soap films are determined by the surface tension trying to minimize the surface area of the film (Fig. 8-7). If there are no external forces, the liquid forms into spherical drops. In fact, this has been proposed as a way of making nearly perfect spheres by letting liquids cool in space. In the free-fall environment of an orbiting Space Shuttle the force of gravity can be ignored and liquid drops are nearly spherical.

Surface tensions vary among liquids. Water, as you might expect, has a relatively high surface tension. If we add soap or oil to the water, its surface tension is reduced, meaning that the water molecules are not as attracted to each other. It is probably reasonable to infer that the new molecules in the solution are somehow shielding the water molecules from each other.

Figure 8-7 Surface tension minimizes the surface area of this soap film.

Gases

When the molecules separate totally, a liquid turns into a **gas.** (See Chapter 7 for a discussion of an ideal gas.) The gas occupies a volume about one thousand times larger than the liquid. In the gaseous state the molecules have enough kinetic energy to be essentially independent of each other. A gas fills the container holding it, taking its shape and volume. Because gases are mostly empty space, they are compressible and can be readily mixed with each other.

Gases and liquids have some common properties because they are both "fluids." All fluids are able to flow, some more easily than others. The viscosity of a fluid is a measure of the internal friction within the fluid. You can get a qualitative feeling for the viscosity of a fluid by pouring it. Those fluids that pour easily, such as water and gasoline, have low viscosities. Those that pour very slowly, such as molasses, honey, and egg whites, have high viscosities. Glass is a fluid with an extremely high viscosity. In the winter drivers put lower viscosity oils in their cars so the oils will flow better on cold mornings.

The viscosity of a fluid determines its resistance to objects moving through it. A parachutist's safe descent is due to the viscosity of air. Air and water have drastically different viscosities. Imagine running a 100-meter dash in a meter of water!

Honey is a very viscous fluid.

> **QUESTION**   How might you explain the observation that the viscosities of fluids decrease as they are heated?

Plasmas

At around 4500°C all solids have melted. At 6000°C all liquids have been turned into gases. And at somewhere above 100,000°C most matter is ionized into the **plasma** state. In the transition between a gas and a plasma the atoms themselves break apart into electrically charged particles.

Although more rare on Earth than the solid, liquid, and gaseous states, the fourth state of matter, plasma, is actually the most common state of matter in the Universe (over 99 percent). Examples of naturally occurring plasmas on Earth include fluorescent lights and neon-type signs. Fluorescent lights consist of a plasma created by a high voltage that strips mercury vapor of some of its electrons. "Neon" signs employ the same mechanism but use a variety of gases to create the different colors.

Perhaps the most beautiful naturally occurring plasma effect is the aurora borealis, or "northern lights." Charged particles emitted by the Sun and other stars are trapped in the Earth's upper atmosphere to form a plasma known as the Van Allen radiation belts. These plasma particles can interact with atoms of nitrogen and oxygen over both magnetic poles, causing them to emit light as discussed in Chapter 21.

> **ANSWER**   The increased kinetic energy of the molecules means that the molecules are more independent of each other.

The aurora borealis results from the interaction of charged particles with air molecules.

Plasmas are important in nuclear power as well as in the interiors of stars. An important potential energy source for the future is the "burning" of a plasma of hydrogen ions at very high temperatures to create nuclear energy. We will discuss nuclear energy more completely in Chapter 25.

Pressure

A macroscopic property of a fluid—either a gas or a liquid—is its change in **pressure** with depth. As we saw in Chapter 7, pressure is the force per unit area exerted on a surface, measured in units of newtons per square meter (N/m^2), or pascals (Pa).

When a gas or liquid is under the influence of gravity, the weight of the material above a certain point exerts a force downward, creating the pressure at that point. Therefore, the pressure in a fluid varies with depth. You have probably felt this while swimming. As you go deeper, the pressure on your ear drums increases. If you swim horizontally at this depth, you notice that the pressure doesn't change. In fact, there is no change if you rotate your head; the pressure at a given depth in a fluid is the same in all directions.

Consider the thin horizontal section of the fluid shown in Figure 8-8. Since the fluid above this thin section does not move, the net force on it must be zero. Therefore, the section must be exerting an upward force equal to the weight of the entire column of fluid above it. The pressure is just this force per unit area.

Our atmosphere is held in a rather strange container, the two-dimensional surface of the Earth. Gravity holds the atmosphere down so that it doesn't escape. There is no definite top to our atmosphere; it just gets thinner and thinner the higher you go above the Earth's surface.

The air pressure at the Earth's surface is due to the weight of the column of air above the surface. At sea level the average atmospheric pressure is about 101 kilopascals. This means that a column of air that is 1 square meter in cross section and reaches to the top of the atmosphere would weigh 101,000 newtons and have a mass of 10 metric tons. A column of air 1 square inch in cross section would weigh 14.7 pounds; therefore, atmospheric pressure is 14.7 pounds per square inch.

We can use these ideas to describe what happens to atmospheric pressure as we go higher and higher. You might think that the pressure drops to one-half of the surface value halfway to the "top" of the atmosphere. However, this is not true because the air near the Earth's surface is much denser than that near the top of the atmosphere. This means that there is much less air in the top half compared to the bottom half. Because the pressure at a given altitude depends on the weight of the air above that altitude, the pressure changes more quickly near the surface. In fact, the pressure drops to half at about 5500 meters (18,000 feet) and then drops in half again in the next 5500 meters. This means that commercial airplanes flying at a typical altitude of 36,000 feet experience pressures that are only one-fourth those at the surface.

Like fish living on the ocean floor, we land-lovers are generally unaware of the pressure due to the ocean of air above us. Although the at-

Figure 8-8   The thin, horizontal section of the fluid shown in white must support the weight of the fluid above it.

The pressure on a scuba diver increases with depth.

Figure 8-9 Two teams of eight horses could not separate Von Guericke's evacuated half-spheres.

mospheric pressure at sea level might not seem like much, consider the total force on the surface of your body. A typical human body has approximately 2 square meters (3000 square inches) of surface area. This means that the total force on the body is about 200,000 newtons (20 tons!).

> **QUESTION** Why doesn't the very large force on the surface of your body crush you?

An ingenious experiment conducted by a contemporary of Isaac Newton demonstrated the large forces that can be produced by atmospheric pressure. The German scientist Otto von Guericke joined two half-spheres (Fig. 8-9) with just a simple gasket (no clamps or bolts). He then pumped the air from the inside of the sphere, creating a partial vacuum. Two teams of eight horses were unable to pull the hemispheres apart!

In weather reports atmospheric pressure is often given in units of millimeters or inches of mercury. A typical pressure is 760 millimeters (30 inches) of mercury. Since pressure is a force per unit area, reporting pressure in units of length must seem strange. This scale comes from the historical method of measuring pressure. Early pressure gauges were similar to the simple mercury barometer shown in Figure 8-10. A sealed glass tube is filled with mercury and inverted into a bowl of mercury. After inversion, the column of mercury does not pour out into the bowl but maintains a definite height above the pool of mercury. Since the mercury is not flowing, we know that the force due to atmospheric pressure at the bottom of the column is equal to the weight of the mercury column. This means that the atmospheric pressure is the same as the pressure at the bottom of a column of mercury 760 millimeters tall if there is a vacuum above the mercury. Therefore, atmospheric pressure can be characterized by the height of the column of mercury it will support.

Atmospheric pressure also allows you to drink through a straw. As you suck on the straw, you reduce the pressure above the liquid in the straw, allowing the pressure below to push the liquid up. In fact, if you could suck hard enough to produce a perfect vacuum above water, you could use a straw 10 meters (almost 34 feet) long! So although we often talk of sucking on a soda straw and pulling the soda up, in reality we are removing the air pressure on the top of the soda column in the straw and the atmospheric pressure is pushing the soda up.

> **ANSWER** You aren't crushed because the pressure inside your body is the same as outside. Therefore, the inward force is balanced by the outward force.

Figure 8-10 In a mercury barometer the atmospheric pressure is balanced by the pressure due to the weight of the mercury column.

Underwater explorers must use vessels such as this bathysphere at the great depths of the ocean floor.

QUESTION   How high a straw could you use to suck soda?

As you dive deeper in water, the pressure increases for the same reasons as in air. Because atmospheric pressure can support a column of water 10 meters high, we have a way of equating the two pressures. The pressure in water must increase by the equivalent of 1 atmosphere (atm) for each 10 meters of depth. Therefore, at a depth of 10 meters, you would experience a pressure of 2 atmospheres, 1 from the air and 1 from the water. The pressures are so large at great depths that very strong vessels must be used to prevent the occupants from being crushed.

QUESTION   What is the pressure on a scuba diver at a depth of 30 meters (100 feet)?

Sink and Float

Floating is so commonplace to anyone who has gone swimming that it might not have occurred to ask, "Why do things sink or float?" "Why does a golf ball sink and an ocean liner float?" "And how is a hot-air balloon similar to an ocean liner?"

Anything that floats must have an upward force counteracting the force of gravity, since we know from Newton's first law of motion (Chapter 2) that an object at rest has no unbalanced forces acting on it. To understand why things float, therefore, requires that we find the upward **buoyant force** opposing the gravitational force.

The buoyant force exists because the pressure in the fluid varies with depth. To understand this consider the cubic meter of fluid in Figure 8-11. The pressure on the bottom surface is greater than on the top surface, resulting in a net upward force equal to the weight of the fluid. The downward force on the top surface is due to the weight of the fluid above the cube. The upward force on the bottom surface is equal to the weight of the column of fluid above the bottom of the cube. The difference between these two forces is just the weight of the fluid in the cube, as expected. The pressures do not change if the cube of fluid is replaced by a cube of some other material. Therefore, the net upward force is still equal to the weight of the fluid that was replaced. This result is **Archimedes' principle,** named for the Greek scientist who discovered it.

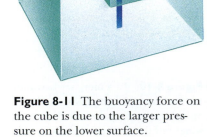

Figure 8-11 The buoyancy force on the cube is due to the larger pressure on the lower surface.

ANSWER   Because soda is mostly water, we assume that it has the same density as water. Therefore, the straw could be 10 meters high—but only if you have very strong lungs. A typical height is more like 5 meters.

ANSWER   The pressure would be (30 m)/(10 m/atm) = 3 atmospheres due to the water plus 1 atmosphere due to the air above the water, for a total of 4 atmospheres.

The buoyant force is equal to the weight of the displaced fluid.

Changing the shape of the steel makes it a "floater." A solid piece of steel with the mass of the ocean liner sinks.

Archimedes' principle

When you place an object in a fluid, it displaces more and more fluid as it sinks lower into the liquid and, therefore, the buoyant force increases. If the buoyant force equals the object's weight before it is fully submerged, the object floats. This occurs whenever the density of the object is less than that of the fluid.

We can change a "sinker" into a "floater" by increasing the amount of fluid it displaces. A solid chunk of steel equal in weight to an ocean liner clearly sinks in water. We can make the steel float by reshaping it into a hollow box. We don't throw away any material; we only change its volume. If we make the volume big enough, it will displace enough water to float.

Ice floats because of a buoyant force. When water freezes, the atoms arrange themselves in a way that actually takes up more volume. As a result the ice has a lower density and floats on the surface. This is fortunate; otherwise, ice would sink to the bottom of lakes and rivers, freezing the fish and plants.

Physics on Your Own   The next time you go swimming or to a hot tub, estimate your density by observing how much of your body floats above the surface of the water.

The buoyant force is present even when the object sinks! For example, any object weighs less in water than in air. You can verify this by hanging a small object by a rubber band. As you lower it into a glass of water, the rubber band is stretched less because the buoyant force helps support the object.

QUESTION   A piece of iron with a mass of 790 grams displaces 100 grams of water when it sinks. What does the iron weigh in air and underwater?

ANSWER   In air the weight is given by $mg = (0.79 \text{ kg})(10 \text{ m/s}^2) = 7.9$ newtons. In water this is reduced by the weight of the displaced water. Therefore, we have 7.9 newtons − 1 newton = 6.9 newtons.

How Fatty Are You?

Exercising does not automatically reduce your weight. One outcome of exercising is the conversion of fatty tissue into muscle without changing your weight. Since healthy people have more muscle, it is important to be able to determine the percentage of body fat. That's a question that just stepping on the bathroom scale won't tell you. However, a 2000-year-old technique developed by Archimedes does work.

Around 250 B.C., Archimedes was chief scientist for King Hieron of Syracuse (now modern Sicily). As the story goes, the king was concerned that his crown was not made of pure gold but had some silver hidden under its surface. Not wanting to destroy his crown to find out if he had been cheated, he challenged his scientist to find an alternative procedure. Everybody knows the legend of Archimedes leaping from his bathtub and shouting, "Eureka, I have found it!"

The key to Archimedes' solution is determining the average density of the crown, or in our case, your body. There is no problem getting your weight. A simple bathroom scale will do. The tricky part is determining your volume. You, like the king's crown, are oddly shaped (sorry!), not matching any of the geometric volumes you studied in school.

Archimedes discovered that an object immersed in water feels an upward buoyant force. If you were to stand on a scale while totally submerged, you would weigh less because the buoyant force supports part of your weight. This buoyant force is equal to the weight of the water your body displaces. From your weights in air and underwater, your volume can be calculated.

Human performance scientists consider the body to be made of fat and "muscle." (Everything but fat—skin, bone, and organs—is grouped as muscle.) From the study of cadavers, the density of human fat is found to be about 90% the density of water, whereas the density of "muscle" is about 110% the density of water. The more fat you have, the lower your average density will be. The percentage of fat for healthy adults should be less than 15% for men and 22% for women. Champion distance runners and bicyclists have about 5% fat.

Bernoulli's Effect

The pressure in a stationary fluid changes with depth but is the same if you move horizontally. If the fluid is moving, however, the pressure can also change in the horizontal direction. Suppose that we have a pipe that has a narrow section like the one shown in Figure 8-12. If we put pressure gauges along the pipe, the surprising finding is that the pressure is lower than in the narrow region of the pipe. If the fluid is not compressible, the fluid must be moving faster in the narrow region. That is, the same amount of fluid must pass by every point in the pipe or it would pile up. Therefore, the fluid must flow faster in the narrow regions. This might lead one to conclude incorrectly that the pressure would be higher in this region. The Swiss mathematician and physicist Daniel Bernoulli stated the correct result as a principle.

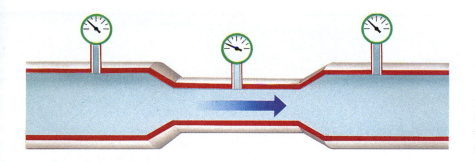

Figure 8-12 The pressure is smaller in the narrow region of the pipe where the velocity of the fluid is greater.

The pressure in a fluid decreases as its velocity increases.

Bernoulli's principle

We can understand **Bernoulli's principle** by "watching" a small cube of fluid flow through the pipe (Fig. 8-13). The cube must gain kinetic energy as it speeds up entering the narrow region. Since there is no change in its gravitational potential energy, there must be a net force on the cube that does work on it. Therefore, the force on the front of the cube must be less than on the back. That is, the pressure must decrease as the cube moves into the narrow region. As the cube of fluid exits from the narrow region, it slows down. Therefore, the pressure must increase again.

There are many examples of Bernoulli's effect in our everyday activities. Smoke goes up a chimney partly because hot air rises but also because of the Bernoulli effect. The wind blowing across the top of the chimney reduces the pressure and allows the smoke to be pushed up. This effect is also responsible for houses losing roofs during hurricanes (or attacks by big bad wolves). When the hurricane reduces the pressure on the top of the roof, the air *inside* the house lifts the roof off.

A fluid moving past an object is equivalent to the object moving in the fluid, so the Bernoulli effect should occur in these situations. A tarpaulin over the back of a truck lifts up as the truck travels down the road, due to the reduced pressure on the outside surface of the tarpaulin produced by the truck moving through the air. This same effect causes your car to be sucked toward a truck as it passes you going in the opposite direction. The upper surfaces of airplane wings are curved so that the air has to travel a farther distance to get to the back edge of the wing. Therefore, the air on top of the wing must travel faster than that on the underside and the pressure on the top of the wing is less, providing lift to keep the airplane in the air.

A tornado caused the difference in air pressures inside and outside this house which tore the roof off.

Figure 8-13 The "cube" of fluid entering the narrow region of the pipe must experience a net force to the right.

The Curve Ball

When a ball moves through air, strange things can happen. Perhaps the most common examples are the curve in baseball, the slice in golf, and the top-spin serve in tennis. Isaac Newton wrote about the unusual behavior of spinning tennis balls, and baseball players and scientists have debated the behavior of the curve ball since the first baseball was thrown over a hundred years ago. The balls naturally follow projectile paths due to gravity (Chapter 3) but it's the extra motion that is the bane of all batters.

Early on, the debate centered on whether or not the curve ball even existed. Scientists, believing that the only forces on the ball are gravity and air resistance (drag), argued that the curve ball must be just an opti-

cal illusion. "Not true!" retorted the baseball players. "It's like the ball rolled off a table just in front of the plate."

When there's a debate about the material world, the best procedure is to devise an experiment; that is, to ask the question of the material world itself. In the early 1940s *Life* magazine commissioned strobe photos of a curve ball and concluded that the scientists were right: The curve ball is an optical illusion. Not to be outdone, *Look* magazine commissioned its own photos and concluded that the scientists (and *Life* magazine!) were wrong.

More recently, three scientists reexamined the question. They found a dark warehouse, a bank of

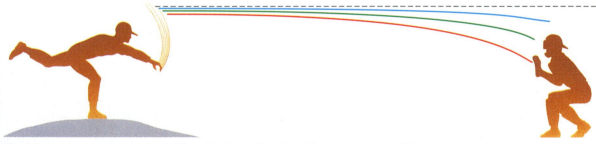

The blue curve represents the ball's path due to its spin without any gravity. The green curve is the ball's path due to gravity without spin. The red curve shows the combined effects of spin and gravity.

Figure 8-14 What happens when you blow over the top of the paper?

Physics on Your Own   Hold a piece of typing paper in front of your face as shown in Figure 8-14. What happens as you blow over the top of the paper? Why?

Physics on Your Own   Some vacuum cleaners can be reversed so that they blow air. If you have access to such a vacuum cleaner, place a ping-pong ball in the stream of air when it is aimed vertically upward. Why does the ball stay in the stream? Can you tilt the airstream without dropping the ball? This project can also be done by blowing through a straw.

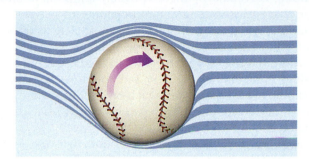

The spinning ball causes the airflow to deflect upward, imparting a downward force on the ball.

strobe lights, and—most importantly—a professional pitcher. After careful analysis the verdict was clear: The ball does indeed curve away from the projectile path, and the deviation is created by the ball's spin. If the ball travels at 75 mph, it takes the ball about half a second to travel the 60 feet to the batter. During this time, the ball rotates about 18 times and deviates from the projectile path by about a foot.

The direction of the deviation depends on the orientation of the spin. The deviation is always perpendicular to the axis of the spin; therefore, spin around a vertical axis moves the ball left or right. Because this would only change the point of contact with a horizontal bat, it is not very effective. Spin around a horizontal axis causes the ball to move up or down. Backspin (the bottom of the ball moving toward the batter) causes the

ball to stay above the projectile path, which helps the batter. The best situation (for the pitcher!) is topspin. This increases the drop as the ball approaches the batter.

The question of when the ball deviates was also answered by this experiment. Batters have claimed for years that the ball travels along its normal path and then breaks at the last moment. Scientists claim that the forces—both gravity and the one caused by the spin— are constant and, thus, the path is a continuous curve. Our study of projectile motion has shown that a ball falls farther during each succeeding second. This is compounded by the extra drop due to the spin. Thus, the vertical speed is much faster near the plate. But the drop is a continuous one; it does not abruptly change.

We are now left with the question of what causes the downward force. The baseball's cotton stitches— 216 on a regulation ball—grab air, creating a layer of air that is carried around the spinning ball. (An insect sitting on the spinning ball would feel no wind—just like dust on a fan's blades is undisturbed by the fan's rotation.) To account for the force we must look at the turbulence, or wake, behind the ball. The airflow over the top of the ball has a larger speed relative to the surrounding air and breaks up sooner than the airflow under the bottom as shown in the figure. This causes the wake behind the ball to be shifted upward. According to Newton's third law, the momentum imparted to the wake in the upward direction causes an equal momentum to be imparted downward on the ball and the curve ball drops.

Let the games begin!

SUMMARY

Density is an inherent property of a substance and defined as the amount of mass in one unit of volume.

Elements combine into substances that can exist in four states of matter: solids, liquids, gases, and plasmas. The transitions between states occur when energy is supplied to or taken from substances. When a solid is heated above its melting point, the molecular bonds break to form a liquid in which the molecules are free to move about. Upon further heating the molecules totally separate to form a gas. In the plasma state the atoms have been torn apart, producing charged ions and electrons. Although rare on Earth, plasma is the most common state in the Universe.

The electric forces between atoms bind all materials together. If the atoms are ordered, a crystalline structure results from one of four major

CHAPTER 8 REVISITED

The crystal's macroscopic shape results from a growth process that adds to the overall structure of the crystal, atom by atom. Study of these crystalline shapes gives scientists clues about the way atoms combine.

types of bonding: ionic, covalent, metallic, or van der Waals. Ionic bonding occurs when atoms exchange one or more electrons. In covalent bonding electrons are shared between atoms, whereas in metallic bonding the electrons form an electron "gas" that is shared by all the atoms. Van der Waals bonding is created by slight (possibly temporary) imbalances in the electric charge distribution of neutral atoms.

Liquids take the shape of their container, and most lack an ordered arrangement of their molecules. The intermolecular forces in a liquid create a surface tension that holds the molecules to the liquid. A gas fills the container holding it, assuming its shape and volume. All gases are compressible and can be readily mixed with each other. The viscosity of a fluid determines how easily it pours and what resistance it offers to objects moving through it.

The pressure in a liquid or gas varies with depth because of the weight of the fluid above that point. At sea level the average atmospheric pressure is about 101 kilopascals (14.7 pounds per square inch).

An object in a fluid experiences a buoyant force equal to the weight of the fluid displaced; therefore, all objects weigh less in water than in air. The buoyant force exists because the pressure in a fluid varies with depth. The pressure on the bottom surface of an object is greater than on its top surface. Objects less dense than the fluid float. The pressure in a moving fluid decreases with increasing speed.

KEY TERMS

alloy: A metal produced by mixing other metals.

Archimedes' principle: The buoyant force is equal to the weight of the displaced fluid.

Bernoulli's principle: The pressure in a fluid varies inversely with its velocity.

buoyant force: The upward force exerted by a fluid on a submerged or floating object. (See Archimedes' principle.)

covalent bonding: The binding together of atoms by the sharing of their electrons.

crystal: A material in which the atoms are arranged in a definite geometric pattern.

density: A property of a material equal to the mass of the material divided by its volume. Measured in kilograms per cubic meter.

gas: Matter with no definite shape or volume.

ionic bonding: The binding together of atoms through the transfer of one or more electrons from one atom to another.

liquid: Matter with a definite volume that takes the shape of its container.

metallic bonding: The binding together of atoms through the sharing of electrons throughout the material.

plasma: A highly ionized gas with equal numbers of positive and negative charges.

polymer: A material produced by linking carbon–hydrogen molecules to form very long macromolecules.

pressure: The force per unit area of surface. Measured in newtons per square meter, or pascals.

solid: Matter with a definite size and shape.

van der Waals bonding: A weak binding together of atoms or molecules due to their electrical attraction.

viscosity: A measure of the internal friction within a fluid.

CONCEPTUAL QUESTIONS

1. What are the four states of matter?

2. Is the average kinetic energy of the molecules in a liquid greater or smaller than in a solid of the same material?

3. Which has a greater density, a tiny industrial diamond used in grinding powders or a 3-carat diamond in a wedding ring?

4. Does the aluminum in a soda can or in an automobile engine have the larger density?

5. Gold and silver have densities of 19.3 and 10.5 grams per cubic centimeter, respectively. If you have equal masses of each, which one will occupy the larger volume?

6. Aluminum and magnesium have densities of 2.70 and 1.75 grams per cubic centimeter, respectively. If you have equal volumes of each, which one will have the larger mass?

7. Why do soda bottles break when the soda freezes?

8. Although the uranium atom is more massive than the gold atom, gold has the larger density. What does this tell you about the two solids?

9. How do crystals differ from amorphous solids?

10. How do true solids and liquid solids differ?

11. Are the crystal structures of ice and table salt the same? How do you know?

12. How does the crystal structure of mica differ from that of table salt?

13. How does the structure of diamond differ from that of graphite?

14. What does the observation that mica can be separated into thin sheets tell you about the crystal structure of mica?

15. From the observation that solids melt to form liquids with the same chemical identity, what can you conclude about the strength of the forces holding atoms together to form molecules relative to those forces holding the molecules together in a solid?

16. Are the intermolecular or the intramolecular forces stronger in a typical liquid? What evidence do you have for your answer?

17. How might you show that wax is a liquid and not a solid?

18. Is there any validity to the claim that liquid crystals are an intermediate state between solids and liquids?

19. What distinguishes alloys from other metals?

20. How do polymers differ from crystalline solids?

21. What evidence do you have to indicate that the molecular bonding in solid oxygen is less than in solid nitrogen?

22. Is the bonding between molecules in liquid nitrogen stronger or weaker than that in liquid oxygen?

23. Why does water bead up when it is spilled on a waxed floor?

24. What shape would you expect a drop of water to take if it is suspended in the air in the Space Shuttle?

25. If you fill a glass with water so the water is level with the top of the glass, you can carefully drop several pennies into the glass without spilling any water. How do you explain this?

26. Why does soapy water bead up less than plain water on a countertop?

27. How does a gas differ from a plasma?

28. What state of matter forms the Van Allen belts?

29. Why do janitors worry about the spiked heels on women's shoes damaging floors?

30. What must happen to the area of a tire touching the ground if you reduce the pressure?

31. Why doesn't atmospheric air pressure collapse a balloon?

32. If you fill a can with steam and close the lid tightly, it collapses as it cools. Why?

33. It takes longer to cook potatoes at higher elevations because the boiling point of water decreases as the atmospheric pressure decreases. Using the ideal gas model, can you think of a reason why this might be so?

34. Why can a chef cook beans faster in a pressure cooker than in an ordinary pot?

*35. If you journey to the mile-high city of Denver, there is 1 mile less atmosphere above you. And yet the typical atmospheric pressure given on the daily weather report is the same as on the coast. How can this be?

*36. Salt water is denser than fresh water. This means that the mass of 1 cubic centimeter of salt water is larger than that of 1 cubic centimeter of fresh water. Would a scuba diver have to go deeper in salt water or fresh water to reach the same pressure?

37. Was it possible for the little Dutch boy to hold back the sea by putting his finger in the dike?

38. Why doesn't the pressure crush a scuba diver at a depth of 30 meters?

39. Why can't water be "sucked" to a height greater than 10 meters even with a very good suction pump?

40. Alcohol is less dense than water. Could you drink alcohol or water through a longer straw?

41. How could you use a barometer to measure the height of a mountain pass?

42. Why are concrete dams wider at the bottom than at the top?

43. Why does a loaded freighter sit lower in the water than an empty one?

44. What happens in a submarine that allows it to surface?

45. Spilled gasoline can sometimes be seen as colorful films on the top of rain puddles. What does this tell you about the density of gasoline?

46. Some toys contain two different colored liquids that do not mix. If the blue liquid always stays on top of the red liquid, what can you say about the densities of the liquids?

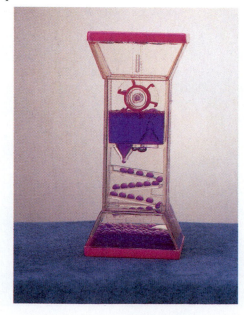

47. Why are you more easily able to sink to the bottom of the swimming pool when you expel as much air as possible from your lungs?

48. What happens to the depth of a scuba diver who breathes deeply?

49. Salt water is slightly denser than fresh water. Will a boat float higher in salt water or fresh water?

50. The densities of mercury and lead are 13.6 and 11.3 grams per cubic centimeter, respectively. Will a block of lead float in mercury? (*Caution:* Do *not* try this experiment; mercury is very toxic!)

*51. An ice cube is floating in a glass of water. Will the water level in the glass rise, go down, or stay the same as the ice cube melts?

*52. A small boat containing a lead ball is floating in an aquarium. Will the water level in the aquarium be higher, lower, or the same after the lead ball is removed from the boat and gently lowered into the water?

53. Why does a propeller have a cross-sectional shape that looks like that of a wing?

54. Why does your car get pulled sideways when a truck passes you going in the opposite direction on a two-lane highway?

55. Why do tennis players put top-spin on serves?

56. How does a curve ball in baseball differ from a "sinker"?

57. Why does the shower curtain get pulled into the shower when you suddenly turn the water on full force?

58. A partial vacuum can be created by installing a pipe at right angles to a water faucet and turning the water on as shown in the figure. What is the physics behind this?

Vacuum created

59. The Green Building at the Massachusetts Institute of Technology (MIT) is a tall tower built on an inverted U-shaped base that is open to the Charles River Basin. Why might the doors in the opening have opened "by themselves" on windy days before revolving doors were installed to correct the design flaw?

60. Why would an aneurysm (a widening of an artery) be especially subject to rupturing?

EXERCISES

1. What is the density of a substance that has a mass of 30 g and a volume of 8 cm^3?

2. An object has a mass of 800 kg and a volume of 0.25 m^3. What is its average density?

3. A solid ball with a volume of 0.2 m^3 is made of a material with a density of 3000 kg/m^3. What is the mass of the ball?

4. What is the mass of a lead sinker with a volume of 2 cm^3?

5. What is the volume of a solid silver medallion with a mass of 80 g?

6. A cube with a mass of 48 g is made from a metal with a density of 6 g/cm^3. What is the volume of the cube and the length of each edge?

7. If 1000 cm^3 of a gas with a density of 0.0009 g/cm^3 condenses to a liquid with a density of 0.9 g/cm^3, what is the volume of the liquid?

8. What is the density of steam if it occupies 900 times the volume of room-temperature water?

*9. Two barometers are made with water and mercury. If the mercury column is 25 in. tall, how tall is the water column?

*10. On a day when a water barometer has a height of 9 m, what is the height of a mercury barometer?

11. Given that atmospheric pressure drops by a factor of 2 for every gain in elevation of 18,000 ft, what is the height of a mercury barometer in an unpressurized compartment of an airliner flying at 36,000 ft?

12. Given that atmospheric pressure drops by a factor of 2 for every gain in elevation of 18,000 ft, what is the height of a mercury barometer in an unpressurized compartment of a surveillance plane flying at 72,000 ft?

*13. Each cubic inch of mercury has a weight of 0.5 lb. What is the pressure at the bottom of a column of mercury 30 in. tall if there is a vacuum above the mercury?

*14. If a cubic meter of water has a mass of 1000 kg, what is the pressure at a depth of 100 m? Is the atmospheric pressure important?

*15. A plastic ball has a volume of 300 cm³ and a mass of 30 g. What fraction of the ball's volume will float above the surface of a lake?

*16. A plastic fishing bobber has a volume of 12 cm³ and a mass of 6 g. How much of the bobber will float above the surface?

17. A metal box is 6 cm wide, 8 cm long, and 3 cm high. If it has a mass of 96 g, how deep will it sink into water?

18. A metal buoy has a volume of 2 m³. How much of the buoy will be submerged if it has a mass of 500 kg?

19. A cubic meter of iron has a mass of 7860 kg, and a cubic meter of water has a mass of 1000 kg. What will the cube of iron weigh underwater?

20. A ball has a volume of 5 cm³ and a mass of 30 g. What will the ball weigh underwater?

Thermal Energy

This infrared photograph of a dolphin's tail flukes shows the temperature variations, with white being the hottest. The dolphins use the tail flukes to get rid of the heat generated by muscle activity and to cool blood flowing to the reproductive organs. What factors control the rate at which the radiation is emitted or absorbed and the resulting temperature changes? (See page 225 for the answer to this question.)

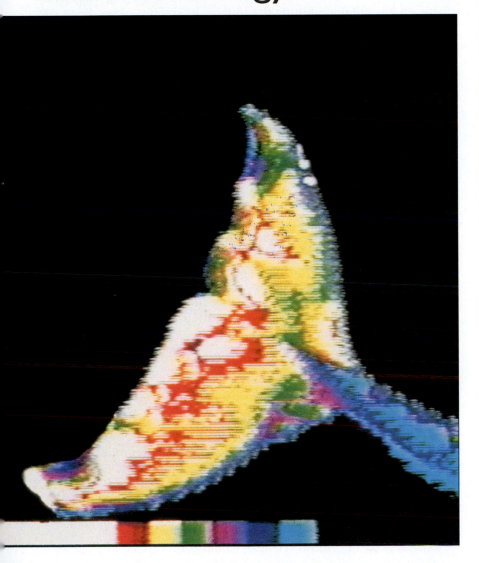

If we examine any system of moving objects very carefully, or if we look at it for long enough, we find that the law of the conservation of mechanical energy fails. A pendulum bob swinging back and forth does, in fact, come to rest. Its original mechanical energy disappears.

Other examples show the same thing. Rub your hands together. You are doing work—applying a force through a distance—but clearly your hands do not fly off with some newly found kinetic energy. Similarly, take a hammer and repeatedly strike a metal surface. The moving hammer has kinetic energy but upon hitting the surface, its kinetic energy disappears. What happens to the energy? It is not converted to potential energy as happened in Chapter 6 because the energy doesn't reappear. So either the kinetic energy truly disappears and total energy is not conserved, or it is transferred into some form of energy that is not a potential energy.

There are similarities in the examples given above. When you rub your hands together, they feel hot. The metal surface and the hammer also get hotter when they are banged together. The pendulum bob is not as obvious; the interactions are between the bob and the surrounding air molecules and between the string and the support. But closer examination shows that, once again, the system gets hotter.

At first glance it is tempting to suggest that temperature, or maybe the change in temperature, could be equated with the lost energy. However, neither of these ideas works. If the same amount of energy is expended on a collection of different objects, the resulting temperature increases are not equal. Suppose, for example, that we rub two copper blocks together. The temperature of the copper blocks increases. If we repeat the experiment by expending the same amount of mechanical energy with two aluminum blocks, the change in temperature will not be the same. The temperature change is an indication that something has happened, but it is not equal to the lost energy.

The Nature of Heat

Early ideas about the nature of heat centered on the existence of a fluid that was transferred between objects at different temperatures. Temperature was considered to be a measure of the concentration of this fluid, and heat was a measure of the quantity of the fluid in the object. When something was brought near a fire, for example, some of this fluid was transferred from the fire to the object. Presumably, the fluid was either massless, or nearly so, as experimenters could not detect any changes in the mass of an object as it was heated.

An 18th-century British scientist, Count Rumford, conducted an experiment that established a different explanation. At the time, he was in charge of boring cannons at a military arsenal in Munich and was struck by the enormous amount of heat produced during the boring process. Rumford decided to investigate this. He placed a dull boring tool and a brass cylinder in a barrel filled with cold water. The boring tool was forced against the bottom of the cylinder and rotated by two horses. These are the results described by Rumford:

Rumford investigated the nature of heat while boring cannons.

and at the end of 2 hours and 30 minutes it (the water in the barrel) actually boiled!
It would be difficult to describe the surprise and astonishment expressed by the
countenances of the by-standers, on seeing so large a quantity of cold water heated,
and actually made to boil without any fire.

Rumford showed that large quantities of heat could be produced by mechanical means without fire, light, or chemical reaction. (This is a large-scale version of the simple hand-rubbing experiment.) The importance of his experiment was the demonstration that the production of heat seemed inexhaustible. As long as the horses turned the boring tool, heat was generated without any limitation. He concluded that anything that could be produced without limit could not possibly be a material substance. Heat was not a fluid, but something generated by motion.

In our modern physics world view, **heat** is the *flow of energy* between two objects due to a difference in temperature. We measure the amount of energy gained or lost by an object by the resulting temperature change in the object. By convention, 1 **calorie** is defined as the amount of heat that raises the temperature of 1 gram of water by 1°C. In the English system, the unit of heat, called a **British thermal unit** (Btu), is the amount of energy needed to change the temperature of 1 pound of water by 1°F. One British thermal unit is equal to approximately 252 calories.

heat = flow of energy

QUESTION   How many calories are required to raise the temperature of 8 grams of water by 5°C?

ANSWER   To raise the temperature of 1 gram by 5°C requires 5 calories. Therefore, 8 grams requires 5 calories/gram × 8 grams = 40 calories.

Mechanical Work and Heat

Σ

The Rumford experiment used the *work* supplied by the horses to raise the temperature of the water, clearly demonstrating an equivalent way of "heating" the water. The water got hotter *as if* it were heated by a fire, but there was no fire.

There is a close connection between work and heat. Both are measured in energy units, but neither resides in an object. In Chapter 6 we saw that work was a measure of the "flow" of energy from one form to another. For example, the gravitational force does work on a freely falling ball, causing its kinetic energy to increase—potential energy changes to kinetic energy. Similarly, heat does not reside in an object, but is a flow of

From the Brewery to Physics

James Prescott Joule

Fame doesn't always go hand in hand with achievement, not even in the world of science. A case in point is James Prescott Joule. Even today, when his name is mentioned, it's usually as a unit of energy, and not with reference to the man or his work.

Joule's work was inventive and extraordinarily precise. Working in his home, Joule carried out a series of experiments that established the first convincing evidence for one of the great laws of physics: the conservation of energy.

James Joule was born in Salford, England, on December 24, 1818. He was born with delicate health, a spinal deformity, and indications of hemophilia. As a young man he was nervous and awkward, both in movement and speech. Despite these health problems, Joule was fond of the outdoors, and could often be found with his younger brother conducting simple experiments in electricity, sound, and weather.

Joule's family owned a large and successful brewery—a profitable business that allowed Joule to attend the best schools and freed him to pursue a wide variety of interests. He was tutored in his early education by the brilliant young chemist John Dalton. And at the age of 19, Joule explored the nature of electromagnets, which led to the publishing of his first scientific paper.

His most important work began in an attempt to improve the efficiencies of steam engines. In order to measure efficiency, Joule measured the heat developed by electric currents. This led him to design and build highly accurate thermometers, capable of measuring to within one 200th of a degree. This, in turn, led to his remarkable experiments in energy transformation.

As discussed in this chapter, Joule measured the temperature rise of the water as a function of the mechanical energy imparted to paddle wheels submerged in water. From a series of such experiments, he firmly established the equivalence of mechanical and thermal energy. His extraordinarily precise experiments offered solid proof of a relationship others had already surmised.

The Royal Society (a prestigious organization of British scientists dating back to the time of Newton), however, was not impressed. Luckily, William Thomson (also known as Lord Kelvin) was. With his help, Joule's paper was eventually published, and he and Thomson became close friends and collaborators.

In 1861, the noise complaint of a neighbor ended Joule's work on a powerful new type of engine, leaving him greatly bothered by the embarrassment of the incident. Joule then worked in secret until his finances ran out in 1875. By then Joule's inventive spirit was also exhausted, and chronic illness began to plague him. He died in 1889, a tired man, leaving the further development of what is now known as *thermodynamics* to his protégé Lord Kelvin.

Adapted from an essay by Steven Janke for Pasco Scientific. *Reference:* Crowther, T.B. *Men of Science.* New York, 1936.

energy into or out of an object, increasing the internal energy of the object. This internal energy is sometimes known as **thermal energy,** and the area of physics that deals with the connections between heat and other forms of energy is called **thermodynamics.**

Although Rumford's experiment hinted at the equivalence between mechanical work and heat, James Joule uncovered the precise equivalence 50 years later. Joule's experiment used a container of water with a paddle wheel arrangement like that shown in Figure 9-1. The paddles are connected via pulleys to a weight. As the weight falls, the paddle wheel turns and the water's temperature goes up. The potential energy lost by the falling weight results in a rise in the temperature of the water. Since Joule could raise the water temperature by heating it or by using the falling weights, he was able to establish the equivalence between the work done and the heat transferred. Joule's experiment showed that 4.2 joules of work are equivalent to 1 calorie of heat.

There are other units of energy. The calorie used when referring to the energy content of food is not the same as the calorie defined here. The food Calorie (properly designated by the capital C to distinguish it from the one used in physics) is equal to 1000 of the physics calories. A piece of pie rated at 400 Calories is equivalent to 400,000 calories of thermal energy, or nearly 1.7 million joules of mechanical energy.

1 calorie = 4.2 joules

> **QUESTION**   Since joules and calories are both energy units, do we need to retain both of them?

Temperature Revisited

If we bring two objects at different temperatures into contact with each other, there is an energy flow between them with energy flowing from the hotter object to the colder. We know from the structure of matter (Chapter 7) that the molecules of the hotter object have a higher average kinetic energy. Therefore, on the average, the faster moving particles of the hotter object lose some of their kinetic energy when they collide with the slower ones of the colder object. The average kinetic energy of the hotter object's particles decreases and that of the colder object's particles increases until they become equal. On a macroscopic scale the temperature changes for each object; the hotter object's temperature drops and the colder object's temperature rises. The flow of energy stops when the two objects reach the same temperature, a condition known as **thermal equilibrium.** Atomic collisions still take place, but, on the average, the particles do not gain or lose kinetic energy.

Let's assume that we have two objects, labeled A and B, that cannot be placed in thermal contact with each other. How can we determine whether they would be in thermal equilibrium if we could bring them to-

> **ANSWER**   No. However, both are presently used for historical reasons. The Europeans are much further along than the Americans in converting from Calories to kilojoules in the labeling of food.

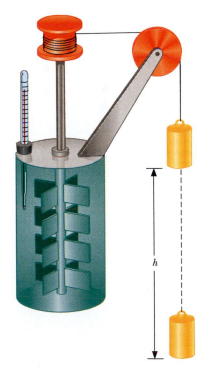

Figure 9-1   Joule's apparatus for determining the equivalence of work and heat. The decrease in the gravitational potential energy of the falling mass produces an increase in the energy of the water.

gether? Let us also assume that we have a third object, labeled C, that can be placed in thermal contact with A and that A and C are in thermal equilibrium. If C is now placed in thermal contact with B and if B and C are also in thermal equilibrium, then we can conclude that A and B are in thermal equilibrium. This is summarized by the statement of the **zeroth law of thermodynamics.**

zeroth law of thermodynamics

> If objects A and B are each in thermal equilibrium with object C, then A and B are in thermal equilibrium with each other.

Although this statement might seem to be so obvious that it is not worth elevating to the stature of a law, it plays a very fundamental role in thermodynamics because it is the basis for the definition of temperature. Two objects in thermal equilibrium have the same temperature. On the other hand, if two objects are not in thermal equilibrium, they must have different temperatures. The zeroth law was developed later in the history of thermodynamics but labeled with a zero because it is more basic than the other laws of thermodynamics.

Heat, Temperature, and Internal Energy

Heat and temperature are not the same thing. Heat is a flow of energy, whereas temperature is a macroscopic property of the object. Two objects can be at the same temperature (the same average atomic kinetic energy) and yet transfer vastly different amounts of energy to a third object. For example, a swimming pool of water and a coffee cup of water at the same temperature can melt very different amounts of ice.

Physics on Your Own   Mix equal amounts of water at different temperatures to see if the equilibrium temperature is midway between the hot and cold temperatures. Styrofoam cups isolate the system from the surroundings pretty well. The losses to the surroundings can be further minimized if one temperature is above room temperature and the other is below. Next try unequal amounts of water and predict the final temperature.

When we consider the total microscopic energy of an object—such as translational and rotational kinetic energies, vibrational energies, and the energy stored in molecular bonds—we are talking about the **internal energy** of the object. There are two ways of increasing the internal energy of a system. One way is to heat the system; the other is to do work on the system. The law of conservation of energy tells us that the total change in the internal energy of the system is equal to the change due to the heat added to the system plus that due to the work done on the system. This is called the **first law of thermodynamics** and is really just a restatement of the law of conservation of energy.

> The increase in internal energy of a system is equal to the heat added plus the work done on the system.

first law of thermodynamics

This law sheds more light on the nature of internal energy. Let's assume that if 10 calories of heat are added to a sample of gas, its temperature rises by 2°C. If we add the same 10 calories to a sample of the same gas that has twice the mass, we discover that the temperature rises by only 1°C (Fig. 9-2). Adding the same amount of heat does not produce the same rise in temperature. This makes sense because the larger sample of gas has twice as many particles and therefore each particle receives only half as much energy on the average. The average kinetic energy, and thus the temperature, should increase by half as much. An increase in the temperature is an indication that the internal energy of the gas has increased, but the mass must be known to say how much it increases.

Absolute Zero

The temperature of a system can be lowered by removing some of its internal energy. Since there is a limit to how much internal energy can be removed, it is reasonable to assume that there is a lowest possible temperature. This temperature is known as **absolute zero** and has a value of −273°C, the same temperature used to define the zero of the Kelvin scale.

The existence of an absolute zero raised the challenge of experimentally reaching it. The feasibility of doing so was argued extensively during the first three decades of the 20th century and it was eventually concluded that it was impossible. This belief is formalized in the statement of the **third law of thermodynamics.**

> Absolute zero may be approached experimentally but can never be reached.

third law of thermodynamics

There appears to be no restriction on how close experimentalists can get, only that it cannot be reached. Small systems in low-temperature laboratories have reached temperatures closer than a few billionths of a degree to absolute zero.

A substance at absolute zero has the lowest possible internal energy. Originally, it was thought that all atomic motions would cease at absolute zero. The development of quantum mechanics (Chapter 23) showed that all motion does not cease; the atoms sort of quiver with the minimum possible motion. In this state the atoms are packed closely together. Their mutual binding forces arrange them into a solid block.

Specific Heat

Suppose we have the same number of molecules of two different gases and each gas is initially at the same temperature. If we add the same amount of heat to each gas, we find that the temperatures do not rise by the same amount. Even though the gases undergo the same change in

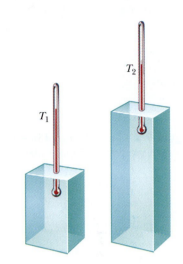

Figure 9-2 Adding equal amounts of heat to different amounts of a material produces different temperature changes.

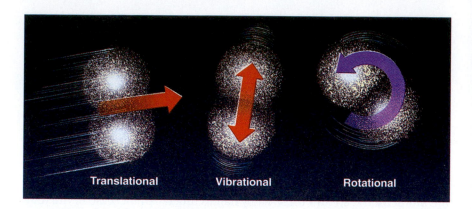

Figure 9-3 Three forms of internal energy for a diatomic molecule.

their internal energies, their molecules do not experience the same changes in their average translational kinetic energies. Some of the heat appears lost. Actually the heat is transformed into other forms of energy. If the gas molecules have more than one atom, part of the internal energy is transformed into rotational kinetic energy of the molecules and part of it can go into the vibrational motion of the atoms (Fig. 9-3). Only a small fraction of the increase in internal energy for most real gases goes into increasing the average kinetic energy that shows up as an increase in temperature.

The amount of heat it takes to increase the temperature of 1 gram of any material by 1°C is known as the **specific heat** of the material. Specific heat is an intrinsic property of the material and does not depend on the size or shape of objects made from the material. By definition, the specific heat of water is numerically 1; that is, 1 calorie raises the temperature of 1 gram of water by 1°C. The specific heat for a given material in a particular state depends slightly on the temperature, but is usually assumed to be constant. The specific heats of some common materials are given in Table 9-1. Notice that the SI units for specific heat are joules per kilogram-kelvin. These are obtained by multiplying the values in calories per gram-degree Celsius by 4186. Note also that the value for water is quite high compared with most other materials.

> **QUESTION**   What is the rise in temperature when 20 calories are added to 10 grams of ice at −10°C?

When we bring two different materials into thermal contact with each other, they reach thermal equilibrium but don't experience the same changes in temperature because of their different specific heats and masses. However, conservation of energy tells us that the heat lost by the hotter object is equal to the heat gained by the colder object. (We're assuming that no energy is "lost" to the environment.)

> **ANSWER**   This is the same as adding 2 calories to each gram. Because ½ calorie is required to raise the temperature of 1 gram of ice by 1°C, the 2 calories will raise its temperature by 4°C.

Table 9-1
Specific Heats for Various Materials

Material	Specific Heat (cal/g·°C)	Specific Heat (J/kg·K)
Solids		
Aluminum	0.215	900
Copper	0.092	385
Diamond	0.124	519
Gold	0.031	130
Ice	0.50	2090
Silver	0.057	239
Liquids		
Ethanol	0.75	3140
Mercury	0.033	138
Water	1.00	4186
Gases		
Air	0.24	1000
Helium	1.24	5190
Nitrogen	0.25	1040
Oxygen	0.22	910

COMPUTING   Sᴘᴇᴄɪꜰɪᴄ Hᴇᴀᴛ

The specific heat c is obtained by dividing the heat Q added to the material by the product of the mass m and the resulting change in temperature ΔT.

$$c = \frac{Q}{m\Delta T}$$

For example, if it requires 11 calories to raise the temperature of an 8-gram copper coin 15°C, we can calculate the specific heat of copper.

$$c = \frac{Q}{m\Delta T} = \frac{11 \text{ cal}}{(8 \text{ g})(15°\text{C})} = 0.092 \text{ cal/g·°C}$$

Note that this agrees with the entry in Table 9-1.

We can rearrange our definition of specific heat to obtain an expression for the heat required to change the temperature of an object by a specific amount. For instance, suppose that you have a cup of water at room temperature that you want to boil. How much heat will this require? Let's assume that the cup contains ¼ liter of water at 20°C and that we can ignore the heating of the cup itself. The mass of the water is 250 g and the boiling point of water is 100°C at 1 atmosphere of pressure. Therefore, the temperature change is 80°C and we have

$$Q = cm\Delta T = \left(1 \frac{\text{cal}}{\text{g·°C}}\right)(250 \text{ g})(80°\text{C}) = 20{,}000 \text{ cal} = 20 \text{ kcal}$$

This 20 kilocalories of energy must be supplied by the stove or microwave oven.

specific heat

$$= \frac{\text{heat added}}{\text{mass} \times \text{temperature change}}$$

The specific heats of the materials on the surface of the Earth account for the temperature extremes lagging behind the season changes. The first day of summer in the Northern Hemisphere usually occurs on June 21. On this day the soil receives the largest amount of solar radiation because it is the longest day of the year and the sunlight arrives closest to the vertical. And yet the hottest days of summer occur several weeks later. It takes time for the ground to warm up because it requires a lot of energy to raise its temperature each degree.

Change of State

We continue our investigation of internal energy by continually removing energy from a gas and watching its temperature. If we keep the pressure constant, the volume and temperature of the gas decreases rather smoothly until the gas reaches a certain temperature. At this temperature there is a rapid drop in volume and *no* change in temperature. Drops of liquid begin to form in the container. As we continue to remove energy from the gas, more and more liquid forms but the temperature remains the same. When all of the gas has condensed into liquid, the temperature

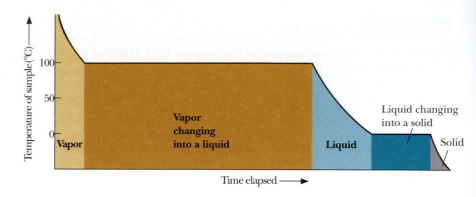

Figure 9-4 A graph of the temperature of water versus time as thermal energy is removed from the water. Notice that the temperature remains constant while the steam condenses to liquid water and while the liquid water freezes to form ice.

drops again (Fig. 9-4). The change from the gaseous state to the liquid state (or from the liquid to the solid), or vice versa, is known as a **change of state.**

While the gas was condensing into a liquid, energy was continually leaving the system, but the temperature remained the same. Most of this energy came from the decrease in the electric potential energy between the molecules as they got closer together to form the liquid. This situation is analogous to the release of gravitational potential energy as a ball falls toward the Earth's surface. The energy that must be released or gained per unit mass of material is known as the **latent heat.** The values of the latent heat for melting and vaporization are given in Table 9-2.

Physics on Your Own   Folklore has it that hot water freezes faster than cold water. Investigate this by placing equal amounts of hot and cold water in identical containers in your freezer or outside on a cold night. Is there any truth to the folklore? Does your answer depend on the type of container or how hot and cold the water is?

The same processes occur when you heat a liquid. If you place a pan of water on the stove, the temperature rises until the water begins to boil. The temperature then remains constant as long as the water boils. It

Table 9-2 Melting Points, Boiling Points, and Latent Heats for Various Materials

Material	Melting Point (°C)	Latent Heat (melting) (kJ/kg)	(cal/g)	Boiling Point (°C)	Latent Heat (vaporization) (kJ/kg)	(cal/g)
Nitrogen	−210	25.7	6.14	−196	199	47.5
Oxygen	−218	13.8	3.3	−183	213	50.9
Water	0	334	79.8	100	2,257	539
Aluminum	660	396	94.6	2467	10,900	2600
Gold	1064	63	15	2807	1,710	409

doesn't matter whether the water boils slowly or rapidly. (Since the rate at which foods cook depends only on the temperature of the water, you can conserve energy by turning the heat down as low as possible while still maintaining a boil.) During the change of state, the additional energy goes into breaking the electric bonds between the water molecules and not into increasing the average kinetic energy of the molecules. Each gram of water requires a certain amount of energy to change it from liquid to steam without changing its temperature. In fact, this is the same amount of energy that must be released to convert the steam back into liquid water. Furthermore, the temperature at which steam condenses to water is the same as the boiling point. The melting and boiling points for some common substances are given in Table 9-2.

The melting of snow and ice in Glacier National Park is a slow process because of the latent heat required to change the ice to liquid water.

> **QUESTION**   At the boiling temperature, what determines whether the liquid turns into gas or the gas turns into liquid?

A similar change of state occurs when snow melts. The snow does not suddenly become water when the temperature rises to 0°C (32°F). Rather, at that temperature, the snow continues to take in energy from the surroundings, slowly changing into water as it does. Incidentally, we are fortunate that it behaves this way; otherwise, we would have gigantic floods the moment the temperature rose above freezing! The latent heat required to melt ice explains why ice can keep a drink near freezing until the last of the ice melts.

On nights when the temperature is predicted to drop below freezing, owners of fruit orchards in California and Florida turn on sprinklers to keep the fruit from freezing. As the water freezes, heat is given off that maintains the temperature of the fruit at 0°C, a temperature above that where the fruit freezes. Once the ice is completely frozen, the fruit is still protected because ice does not conduct heat very well. The ice serves as a "sweater" for the fruit. However, if the air temperature drops too low, the fruit will be ruined.

> **Physics on Your Own**   Determine the number of calories needed to melt 1 gram of ice. Place some ice in a styrofoam cup of water after you have measured the mass of ice, the mass of water, and the temperature of the water. To minimize the loss of thermal energy to the surroundings, place a piece of styrofoam (or another cup) over the top while the ice melts. Does the loss of thermal energy to the surroundings cause your value to be too high or too low?

> **ANSWER**   If heat is being supplied, the liquid will boil to produce additional gas. However, if heat is being removed, some of the gas will condense to form additional liquid.

Thermal energy is transported along the branding iron by conduction because the brand is hotter than the handle.

Conduction

Thermal energy is transported from one place to another via three mechanisms: conduction, convection, and radiation. Each of these is important in some circumstances and can be ignored in others.

If temperature differences exist within a single, isolated object such as a branding iron held in a campfire, thermal energy will flow until thermal equilibrium is achieved. We say that the thermal energy is conducted through the material. **Conduction** takes place via collisions between the particles of the material. The molecules and electrons at the hot end of the branding iron collide with their neighbors, transferring some of their kinetic energy, on the average. This increased kinetic energy is passed along the rod via collisions until the end in your hand gets hot.

The rate at which energy is conducted varies from substance to substance. Solids, with their more tightly packed particles, tend to conduct thermal energy better than liquids and gases. The mobility of the electrons within materials also affects the thermal conductivity. Metals such as copper and silver are good **thermal conductors** as well as good electrical conductors. Conversely, electrical insulators such as glass and asbestos are also good **thermal insulators.** You could hold a rod made from asbestos for a very long time without the end getting hot. The ceramic bowl in Figure 9-5 can have regions at drastically different temperatures.

The differences in the conductivity of materials explain why aluminum and wooden benches in a football stadium do not feel the same on a cold day. Before you sit on either bench, they are at the same temperature. When you sit down, some of the thermal energy in your bottom flows into the bench. Since the wooden bench does not conduct the heat very well, the spot you are sitting on warms up and feels more comfortable. On the other hand, the aluminum bench continually conducts heat away from your bottom, making your seat feel cold.

Figure 9-5   The left end of this ceramic dish is ice cold while the right end is very hot. This can occur because ceramic is a very poor conductor of thermal energy.

The rate at which thermal energy is conducted through a slab of material depends on many physical parameters besides the type of material. You might correctly guess that it depends on the area and thickness of the material. A larger area allows more thermal energy to pass through, and a greater thickness allows less. The temperatures on the two sides of the slab also make a difference. Experimentation tells us that it is only the difference in temperature that matters. The greater the difference, the greater the flow. Table 9-3 gives the thermal conductivity of a variety of common materials.

In our everyday lives we are more concerned with reducing the transfer of thermal energy than increasing it. We wear clothing to retain our body heat and we insulate our houses to reduce our heating and air-conditioning bills. Figure 9-6 shows regions of a roof after a snowstorm. The unmelted patches exist where there is better insulation or, in the case of an unheated porch or garage, where there is little or no temperature difference. Table 9-3 allows us to compare the heat loss through slabs of different materials of the same size and thickness for the same difference in temperatures.

Examination of Table 9-3 reveals that static air is a pretty good insulator. This insulating property of air means that porous substances with many, small air spaces are good insulators, and it explains why the goose down used in sleeping bags keeps you so warm. It also explains how fishnet long underwear keeps you warm. The air trapped in the holes keeps your body heat from being conducted away.

Snow contains a lot of air space between the snowflakes, which makes snow a good thermal insulator. Mountaineers often dig snow caves to escape from severe weather. Likewise, snow-covered ground does not freeze as deep as bare ground.

Table 9-3
Thermal Conductivities for Various Materials

Material	Conductivity (W/m · °C)
Solids	
Silver	428
Copper	401
Aluminum	235
Stainless steel	41
Building Materials	
Polyurethane foam	0.024
Fiberglass	0.048
Wood	0.08
Window glass	0.8
Concrete	1.1
Gases	
Air (stationary)	0.026
Helium	0.15

A camper keeps warm inside a snow cave in Idaho.

Figure 9-6   Melting snow patterns reveal differences in thermal conduction. For example, the old garage has been converted to living space but it appears not to have been insulated.

Figure 9-7   Glider pilots search for convective thermals to gain altitude.

QUESTION   Why do people who spend time outdoors in cold weather wear many layers of clothing?

Convection

Thermal energy can also be transferred in fluids by **convection.** In convection the energy is transported by the movement of the fluid. This movement could be forced, as in heating systems or the cooling system in an automobile, or it could happen because of the changes that occur in the density of the fluid when it is heated or cooled. As the gas near the flame of a candle is heated, it becomes less dense and rises due to the buoyant force (Chapter 8).

Convection in the Earth's atmosphere plays a fundamental role in our global climate as well as our daily weather. Convection currents arise from the uneven heating of the Earth's surface. Glider pilots, hang-glider fliers, and birds of prey (such as hawks and eagles) use convection currents called thermals to provide them with the lift they need to keep aloft (Fig. 9–7).

Local winds near a large body of water can be caused by temperature differences between the water and the land. The specific heat of water is much greater than that of rock and soil. (Convection currents in the water also moderate the changes in the water temperature.) During the

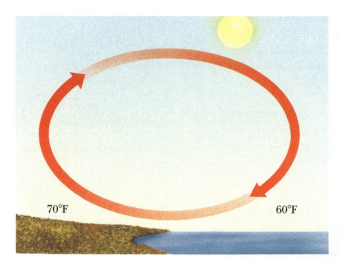
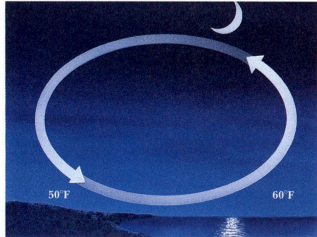

Figure 9-8   The difference in temperature between the land and the water causes breezes to blow (a) onshore during the morning and (b) offshore during the evening.

ANSWER   In addition to the flexibility of adding and removing layers to get the required insulation, the air spaces between the layers contribute to the overall insulation.

morning, the land warms up faster than the water. The hotter land heats the air over it, causing the air to rise. The result is a pleasant "sea breeze" of cooler air coming from the water [Fig. 9-8(a)]. During the evening, the land cools faster, reversing the convection cycle [Fig. 9-8(b)].

> **QUESTION**   What role does convection play in bringing a pot of water to a boil?

Radiation

The third mechanism for the transfer of thermal energy involves electromagnetic waves. As we will see in Chapter 21, these waves can travel through a vacuum, and thus radiation is still effective in situations where the conduction and convective processes fail. With radiation, some of the energy of the object is converted to electromagnetic radiation, as we will see in Chapter 22. The electromagnetic **radiation** then travels through space and is converted back to thermal energy when it hits other objects. The Sun's thermal energy is transmitted to the Earth and the other planets via radiation. Most of the heat that you feel from a cozy fire is transferred by radiation, especially if the fire is behind glass or in a stove.

All objects emit radiation. Although radiation from objects at room temperature is not visible to the human eye, it can be viewed with special "night glasses" or recorded on infrared sensitive film. When colors are artificially added, we can distinguish the different temperatures of objects as shown in Figure 9-9.

As the temperature of the object rises, more and more of the radiation becomes visible. Objects such as the heating coils on kitchen stoves glow with a red-orange color. Betelgeuse, the red star marking the right shoulder of the constellation Orion, has a surface temperature of approximately 3000 kelvin. As an object gets hotter, the color shifts to yellow and then white. Our Sun appears white (above the Earth's atmosphere) with a temperature of 5800 kelvin. The hottest stars appear blue and have temperatures exceeding 8000 kelvin.

The radiation that Earth receives from the Sun is typical of an object at 5800 kelvin. If nothing else happened, the Earth would continue to get hotter and hotter until life as we know it could not exist. However, the Earth also radiates. Its temperature rises until it radiates as much energy into space as it receives; it reaches an equilibrium.

A phenomenon known as the *greenhouse effect* can have a major effect on the equilibrium condition. Visible light easily passes through the windows of a greenhouse or a car, heating up the interior. However, the infrared radiation given off by the interior does not readily pass through the glass and is trapped inside. Only when the temperature reaches a high value is equilibrium established. This is one reason why we are

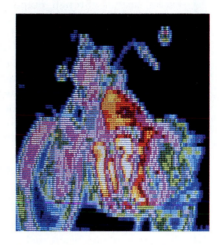

Figure 9-9   This infrared photograph of a motorcycle shows the different temperatures of its parts.

> **ANSWER**   As the flame or heating element warms up the water near the bottom of the pan, it becomes less dense and rises. This circulation causes all of the water to warm up at the same time.

Wind Chill

The meteorologist on television announces the temperatures for the day and then adds that it's going to feel even colder because of the wind. If the air is a certain temperature, why does it matter whether the wind is blowing?

Your body is constantly producing heat that must be released to the environment to keep your body from overheating. The primary way that your body gets rid of excess heat is through evaporation. For each liter (about 1 quart) of water that evaporates, roughly 600 kilocalories of heat are absorbed from your body. While most of us correctly associate this mechanism with sweating, a surprising 25% of the heat lost by evaporation in a resting individual is due to the evaporation of water from the linings of our lungs into the air we exhale. Vigorous activity can produce sweating at a rate in excess of 2 liters per hour, removing 1200 kilocalories per hour, a rate tens of times larger than for a resting individual.

Another form of heat loss is due to convection of air away from the body. Even with no wind, the air leaves your skin on a cold day because its density changes as it is warmed by your body. A third form of heat loss is radiation loss. If your body is warmer than the surrounding objects such as the walls in a room, your body radiates energy to the walls. This is why you feel cold some mornings even though the air in the room has been heated to normal room temperature. The walls take some time to warm up and you will continue to radiate to them until they warm up.

The wind near your body greatly alters the effectiveness of these heat transfers. In stationary air, the layer of air next to your skin becomes warm and moist, reducing the further loss of heat to this layer of air. However, if there is a wind, the wind brings new air to your skin that is colder and drier. Warming and adding moisture to this new air requires additional heat from your body.

In the mid-1940s a single index was created—the *windchill factor*—to express the cooling effects for various ambient temperatures and wind speeds in terms of an "equivalent" temperature with no wind. We can use the table shown here to find the windchill temperature for a thermometer reading of 25°F on a day when the wind is blowing at 10 mph. Look along the top of the table until you locate the 25°F and then move down this column to the row labeled 10 mph along the left-hand side of the table. The entry at the intersection of this row and this column is the equivalent temperature. Therefore, the cooling effects are equivalent to a temperature of 9°F on a calm day.

Although there continues to be debate about the details of the numbers in various windchill tables, it is generally agreed that wind makes you feel colder.

MPH	Equivalent Temperature* of Wind Chill Index (°F)																
Calm	35	30	25	20	15	10	5	0	−5	−10	−15	−20	−25	−30	−35	−40	−45
5	33	27	21	16	12	7	1	−6	−11	−15	−20	−26	−31	−35	−41	−47	−54
10	21	16	9	2	−2	−9	−15	−22	−27	−31	−38	−45	−52	−58	−64	−70	−77
15	16	11	1	−6	−11	−18	−25	−33	−40	−45	−51	−60	−65	−70	−78	−85	−90
20	12	3	−4	−9	−17	−24	−32	−40	−46	−52	−60	−68	−76	−81	−88	−96	−103
25	7	0	−7	−15	−22	−29	−37	−45	−52	−58	−67	−75	−83	−89	−96	−104	−112
30	5	−2	−11	−18	−26	−33	−41	−49	−56	−63	−70	−78	−87	−94	−101	−109	−117
35	3	−4	−13	−20	−27	−35	−43	−52	−60	−67	−72	−83	−90	−98	−105	−113	−123
40	1	−4	−15	−22	−29	−36	−45	−54	−62	−69	−76	−87	−94	−101	−107	−116	−128
45	1	−6	−17	−24	−31	−38	−46	−54	−63	−70	−78	−87	−94	−101	−108	−118	−128
50	0	−7	−17	−24	−31	−38	−47	−56	−63	−70	−79	−88	−96	−103	−110	−120	−128

Region labels: Cold, Very cold, Bitterly cold, Extremely cold

*Wind speeds greater than 40 mph have little additional chilling effect.

warned to never leave pets or children in a car with the windows rolled up on a hot day.

A similar thing happens with the Earth. The atmosphere is transparent to visible light, but the water vapor and carbon dioxide in the atmosphere tend to block the infrared radiation from escaping, causing the Earth's temperature to increase. The high surface temperatures on Venus are due to the greenhouse effect of its thick atmosphere. It is feared that increases in the carbon dioxide concentration in the Earth's atmosphere (due to such things as the burning of fossil fuels) will cause global warming that in turn will cause unwanted changes in the Earth's climate. Such alterations in the climate could change the types of crops that will grow and melt the polar ice caps, flooding coastal cities!

Thermal Expansion Σ

All objects change size as they change temperature. When the temperature increases, nearly all materials expand. But not all materials expand at the same rate. Solids, being most tightly bound, expand the least. All gases expand at the same rate, following the ideal gas equation developed in Chapter 7. Each material's characteristic **thermal expansion** is reflected in a number called its *coefficient of expansion*. The coefficient of expansion gives the fractional change in the size of the object per degree change in temperature.

Expansion slots allow bridges to change length with temperature changes without damage.

COMPUTING THERMAL EXPANSION Σ

Because the coefficient of thermal expansion tells us how much a unit length of material will expand as the temperature is raised 1°C, the expansion for a particular object is given by

$$\Delta L = \alpha L \Delta T$$

where ΔL is the change in length, α is the coefficient of thermal expansion, L is the original length, and ΔT is the change in temperature. There is a similar expression for the volume expansion of liquids.

As an example, the coefficient of expansion for steel is 0.000011 meter for each meter of length for each degree Celsius rise in temperature. This means that a bridge that is 50 meters long expands by 0.00055 meter, or 0.55 millimeter, for each degree of temperature increase. If the temperature increases by 40°C from night to day, the bridge expands by 22 millimeters (almost 1 inch).

We can also obtain this answer using the relationship for thermal expansion.

$$\Delta L = \alpha L \Delta T = \frac{0.000011}{°C}(50 \text{ m})(40°C) = 0.022 \text{ m}$$

Physics Update

A new material shrinks when heated over a wide temperature range, from 0.3 K up to 1050 K. Unlike most materials, which expand when heated, the zirconium tungstate compound made by an Oregon State–Brookhaven collaboration exhibits a negative thermal expansion in three dimensions. Previously known shrinking materials have done so only over a small temperature range, or, like some bakeware ceramics, shrink in one dimension but actually expand in the other two dimensions. An important role for the material would be as a component in composite materials where it is desirable to keep thermal expansion to a minimum.

Freezing Lakes

Life as we know it depends on the unique thermal expansion properties of water. All materials change size when their temperatures change. Since density is the ratio of mass to volume and since the mass of an object does not change when heated or cooled, a change in size means a change in the object's density.

As stated in the chapter, most objects expand when heated and contract when cooled. Water does both! Over most of its liquid range, water behaves as expected, decreasing in volume as its temperature decreases. As the water is cooled below 4°C, however, it expands!

This unusual property affects the way lakes freeze. While cooling toward 4°C, the surface water becomes more dense and therefore sinks (Chapter 8), cooling the entire lake. However, once the entire lake becomes 4°C, the surface water expands as it cools further and becomes less dense. Therefore, the cooler water floats on the top and continues to cool until it freezes. Lakes freeze from the top down. However, since ice is a good thermal insulator, most lakes do not freeze to the bot-

tom. If water were like most other materials, the very cold water would sink and lakes would freeze from the bottom up, creating a challenging evolutionary problem for all aquatic and marine life.

Figure 9-10 A bimetallic strip is used in thermostats to control furnaces.

Thermal expansion has many consequences. Civil engineers avoid the possibility of a bridge buckling by including expansion slots and by mounting one end of the bridge on rollers. The gaps between sections of concrete in highways and sidewalks allow the concrete to expand and contract without breaking or buckling. The romantic "clicketty clack" of train rides is due to expansion joints between the rails.

> **QUESTION**   Why are telephone wires higher in winter than in summer?

We use the differences in the thermal expansions of various materials to our advantage. Some thermostats are constructed of two different metal strips bonded together face to face as shown in Figure 9-10. Since the metals have different coefffficients of expansion, they expand by different amounts, causing the bimetallic strip to bend. Placing electric contacts in appropriate places allows the thermostat to function as an electric switch to turn a furnace, heater, or air conditioner on and off at specified temperatures.

> **ANSWER**   The wires expand with the hotter temperatures in summer and therefore hang lower.

> **QUESTION** How does running hot water on a jar lid loosen it?

Physics Update

One of the great mysteries of water, its tendency to shrink when warmed, has been successfully modeled for the first time. As cold water is heated, it reaches a minimum volume—and therefore a maximum density—at around 4°C. No theoretical model has been able to explain this "density anomaly." This shortcoming compromises the accuracy of molecular-scale models of proteins and other systems involving water. Now researchers at Texas Tech University have proposed an explanation for the density anomaly by looking beyond neighboring molecules in the liquid and focusing on more distant "second neighbors." In all ten known forms of ice, an H_2O molecule is surrounded by its closest neighbors in the same way. Using a simple model in which an oxygen atom and its second neighbors arrange themselves on a one-dimensional array, the Texas Tech researchers obtained density curves with a temperature and pressure behavior similar to that of water.

SUMMARY

The law of conservation of mechanical energy fails whenever frictional effects are present. Often the transformation of mechanical energy to thermal energy is accompanied by temperature changes that produce observable changes in the object. Heat and temperature are not the same thing. Heat is a flow of energy, and temperature is a macroscopic property of the object. The number of calories required to raise the temperature of 1 gram of a substance by 1°C is known as its specific heat.

The first law of thermodynamics tells us that the total change in the internal energy of a system is the sum of the heat added and the work done on the system. This is just a restatement of the law of conservation of energy. Performing 4.2 joules of work on a system is equivalent to adding 1 calorie of thermal energy. Part of this energy increases the average kinetic energy of the atoms; the absolute temperature is directly proportional to this average kinetic energy. Other parts of this energy break the bonds between the molecules and cause substances to change from solids to liquids to gases. At higher temperatures, molecules, atoms, and even nuclei break apart.

There is a limit to how much internal energy can be removed from an object, and thus there is a lowest possible temperature—absolute zero, or

CHAPTER

9

REVISITED

The rate at which an object radiates energy is determined by the difference in temperature between the object and its surroundings, the surface area of the object, and characteristics of the surface. The change in temperature of the object depends primarily on the amount of energy radiated away, the mass of the object, and the specific heat of the material.

> **ANSWER** The metal expands more than the glass, and the lid pulls away from the jar.

−273°C—the same as the zero on the Kelvin scale. A substance at absolute zero has the lowest possible internal energy.

The temperature of a substance does not change while it undergoes a physical change of state. The energy that is released or gained per gram of material is known as the latent heat.

The natural flow of thermal energy is always from hotter objects to colder ones. In the process called conduction, thermal energy is transferred by collisions between particles; in convection the transfer occurs through the movement of the particles; and in radiation the energy is carried by electromagnetic waves.

KEY TERMS

absolute zero: The lowest possible temperature; 0 K, −273°C, or −459°F.

British thermal unit: The amount of heat required to raise the temperature of 1 pound of water by 1°F.

calorie: The amount of heat required to raise the temperature of 1 gram of water by 1°C.

change of state: The change in a substance between solid and liquid or liquid and gas.

conduction, thermal: The transfer of thermal energy by the collisions of the atoms or molecules within a substance.

conductor, thermal: A material that allows the easy flow of thermal energy. Metals are good conductors.

convection, thermal: The transfer of thermal energy in fluids by means of currents such as the rising of hot air and the sinking of cold air.

heat: A flow of energy due to a difference in temperature.

insulator, thermal: A material that is a poor conductor of thermal energy. Wood and stationary air are good thermal insulators.

internal energy: The total microscopic energy of an object, which includes its atomic and molecular translational and rotational kinetic energies, vibrational energy, and the energy stored in the molecular bonds.

joule: The SI unit of energy, equal to 1 newton acting through a distance of 1 meter.

latent heat: The amount of heat required to melt (or vaporize) 1 gram of a substance. The same amount of heat is released when 1 gram of the same substance freezes (or condenses).

radiation: The transport of energy via electromagnetic waves.

specific heat: The amount of heat required to raise the temperature of 1 gram of a substance by 1°C.

thermal energy: Internal energy.

thermal equilibrium: A condition in which there is no net flow of thermal energy between two objects. This occurs when the two objects obtain the same temperature.

thermal expansion: The expansion of a material when heated.

thermodynamics: The area of physics that deals with the connections between heat and other forms of energy.

thermodynamics, first law of: The increase in internal energy of a system is equal to the heat added plus the work done on the system.

thermodynamics, third law of: Absolute zero may be approached experimentally but never be reached.

thermodynamics, zeroth law of: If objects A and B are each in thermodynamic equilibrium with object C then A and B are in thermodynamic equilibrium with each other. All three objects are at the same temperature.

CONCEPTUAL QUESTIONS

1. A block of wood falls from a chair and lands on the floor. What happens to the kinetic energy of the block?

2. What happens to the sound energy from your stereo speakers?

3. What was the purpose of the experiment that Count Rumford conducted with the cannons?

4. What evidence did Rumford have that heat was not a fluid?

5. What was the purpose of Joule's experiment with the paddle wheel?

6. How is a calorie defined?

7. What effect would it have on Joule's experiment if the masses fell rapidly and hit the floor with a substantial speed?

8. What would you expect to find if you measure the temperature of the water at the top and bottom of Niagara Falls? Explain your reasoning.

9. What does the zeroth law of thermodynamics tell us about measuring the temperature of an object?

10. Give an example that illustrates the meaning of the zeroth law of thermodynamics.

11. Can two objects be in thermal equilibrium if they are not touching? Explain.

12. Could two objects be touching, but not be in thermal contact?

13. How do the units of heat and temperature differ?

14. Is it ever correct to talk about the flow of temperature from a hot object to a colder object? If so, give an example; if not, state why.

15. What is the difference between heat and temperature?

16. How do heat and internal energy differ?

17. How do the internal energies of a cup of water and a gallon of water at the same temperature compare?

18. Which of the following does not affect the amount of internal energy an object has: its temperature, the amount of material, its state, the type of material, or its shape?

19. Under what conditions is the first law of thermodynamics valid?

20. Compare the heat added to a system with the work done on the system when there is no change in internal energy of the system.

21. Give an example that illustrates the meaning of the first law of thermodynamics.

22. Give an example that illustrates the meaning of the third law of thermodynamics.

23. How is specific heat defined?

24. Describe an experiment that you could perform to determine the specific heat of a metal ball.

25. Does it take more thermal energy to raise the temperature of 5 grams of water or 5 grams of ice by 6°C?

26. Ice and air have the same specific heats. Would 100 calories of heat raise the temperature of 1 liter of ice by the same amount as 1 liter of air? Explain your reasoning.

27. Why do climates near the coasts tend to be more moderate than in the middle of the continent?

28. Why does the coldest part of winter occur during late January and February when the shortest day is near December 21?

29. Given that the melting and freezing temperatures of water are identical, what determines whether a mixture of ice and water will freeze or melt?

30. If you make the mistake of removing ice cubes from the freezer with wet hands, the ice cubes will stick to your hands. Why does the water on your hands freeze rather than the ice cubes melt?

31. Why can an iceberg survive for several weeks floating in seawater that's above freezing?

32. The boiling point for liquid nitrogen at atmospheric pressure is 77 K. Is the temperature of an open container of liquid nitrogen higher, lower, or equal to 77 K?

33. What would happen to a pot of water on the stove if there were no latent heat of vaporization required for converting water to steam?

34. Why is steam at 100°C more dangerous than water at 100°C?

35. If you put an unwrapped steak in the freezer, it freezes and then over time dries out. Why?

36. In Washington, D.C., the weather report sometimes states that the temperature is 95°F and the humidity is 95 percent. Why does the high humidity make it so uncomfortable?

37. An old hiker's adage is, "If your feet get cold, put on a hat." What is the physics behind this?

38. Why would putting a rug on a tiled bathroom floor make it more comfortable for bare feet?

39. Which of the following does not determine the rate at which heat is conducted through a slab of material: the type of material, the difference in temperature on the two sides, the thickness of the slab, or the average temperature of the slab?

40. How does a metal rod conduct thermal energy from the hot end?

41. Which of the following is the best insulating material: static air, glass, wood, or brick?

42. Which of the following is the best thermal conductor: ceramic, water, wood, or copper?

43. Why is it possible to hold a lighted match until the flame burns quite close to your fingers?

44. Why might a cook put large aluminum nails in potatoes before baking them?

45. Why can you quickly bring a pan of water to a boil when water is a poor conductor of heat?

46. When pilots fly under clouds they often experience a downdraft. Why might you expect this?

47. In northern climates drivers often encounter signs that read "BRIDGE FREEZES BEFORE ROADWAY." Why does this occur?

48. A metal spoon is placed in one of two identical cups of hot coffee. Which cup will be hotter after a few minutes?

49. Earth satellites orbit the Earth in a very good vacuum. Would you expect these satellites to cool off when they enter the Earth's shadow?

50. Suppose you made a videotape in a café using infrared-sensitive equipment. How could you tell which coffee mugs were full?

51. A Thermos bottle is usually constructed from two nested glass containers with a vacuum between them as shown in the figure. The walls are usually silvered as well. How does this construction minimize the loss of thermal energy?

vacuum

silvered

Questions 51 and 52.

52. Will a Thermos bottle (shown in the figure) keep something cold as well as it keeps it hot?

53. Why might a glass dish taken from the oven and put into cold water shatter?

54. The metal roof on a wooden shed makes noises when a cloud passes in front of the Sun. Why?

55. Which of the following does not affect the change in length of a bridge: its length, its cross-sectional area, the type of construction materials used, the change in the temperature?

56. Give an example that clearly illustrates the concept of a coefficient of thermal expansion.

*57. Suppose that the column in an alcohol-in-glass thermometer is not uniform. How would the spacing between the degrees on a wide portion of the thermometer compare with those on a narrow portion?

*58. When a mercury thermometer is first put into hot water, the level of the mercury drops slightly before it begins to climb. Why?

EXERCISES

1. How much heat is required to raise the temperature of 400 g of water from 20°C to 30°C?

2. If the temperature of 500 g of water drops by 8°C, how much heat is released?

*3. How much work does it require to push a crate with a force of 200 N across a floor a distance of 6 m? How many calories of thermal energy are produced by the friction?

*4. How many joules of gravitational potential energy are converted to kinetic energy when 100 g of lead shot falls from a height of 50 cm? How many calories are produced if none of the kinetic energy is converted to other forms?

5. A typical jogger burns up food energy at the rate of about 40 kJ per minute. How long would it take to run off a piece of cake if it contains 400 Calories?

*6. A physics student foolishly wants to lose weight by drinking cold water. If he drinks 1 L (1000 cm³) of water at 10°C below body temperature, how many Calories will it take to warm the water up?

7. What is the change in the internal energy of a system if 15 J of work are done on the system and 6 J of heat are removed from the system?

8. During a process, 20 J of heat are transferred into a system, while the system itself does 12 J of work. What is the change in the internal energy of the system?

9. If the internal energy of an ideal gas increases by 150 J when 220 J of work are done to compress it, how much heat is released?

10. When an ideal gas was compressed, its internal energy increased by 180 J and it gave off 150 J of heat. How much work was done on the gas?

11. Eighty grams of water at 70°C are mixed with an equal amount of water at 30°C in a completely insulated container. The final temperature of the water is 50°C.

   a. How much heat is lost by the hot water?

   b. How much heat is gained by the cold water?

   c. What happens to the total amount of internal energy of the system?

*12. If 200 g of water at 100°C are mixed with 300 g of water at 50°C in a completely insulated container, what is the final equilibrium temperature?

13. It takes 150 cal to raise the temperature of a metallic cross from 20°C to 30°C. If the cross has a mass of 60 g, what is the specific heat of the metal?

14. If it takes 3400 cal to raise the temperature of a 500-g statue by 44°C, what is the specific heat of the material used to make the statue?

15. An iron weight has a mass of 2 kg. If it takes 3680 J to raise its temperature by 4°C, what is the specific heat of iron?

16. A brass monkey has a mass of 0.25 kg. What is the specific heat of brass if it takes 1900 J to increase its temperature by 20°C?

17. How many calories will it take to raise the temperature of a 50-g gold chain from 20°C to 100°C?

18. How many calories would it take to raise the temperature of a 400-g aluminum pan from 293 K to 373 K?

19. How many joules would it take to raise the temperature of 5 kg of ice from −5°C to 0°C?

20. How much energy would be required to raise the temperature of a 10-g gold wedding band from room temperature (20°C) to body temperature (37°C)?

21. How much heat would it take to melt a 1-kg block of ice?

*22. How much heat is required to convert 200 g of ice at −5°C to water at +5°C?

23. What is the change in length of a metal rod with an original length of 2 m, a coefficient of thermal expansion of 0.00002/°C, and a temperature change of 20°C?

24. A steel railroad rail is 24.4 m long. How much does it expand during a day when the low temperature is 50°F (18°C) and the high temperature is 100°F (38°C)?

Available Energy

The energy of the hot water and steam from a geyser can be used to run engines, yet the internal energy of a pail of room-temperature water, while quite high, isn't as useful as an energy source. Why is this true and what would it take to extract this internal energy and make it available for performing useful work? (See p. 247 for the answer to this question.)

Old Faithful Geyser in Yellowstone National Park, Wyoming. Hot water from geysers can be used as an energy source.

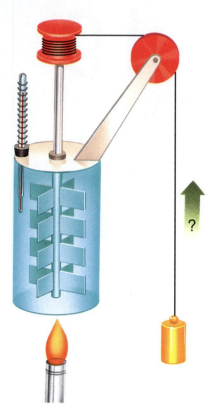

Why won't the weight rise when we heat the water?

Mechanical energy can be completely converted into the internal energy of an object. This is clearly demonstrated every time an object comes to rest due to frictional forces. In the building of a scientific world view, the belief in a symmetry often leads to interesting new insights. In this case, it seems natural to ask if the process can be reversed. Is it possible to recover this internal energy and get some mechanical energy back? The answer is yes, but it is not an unqualified yes.

Imagine that we try to run Joule's paddle-wheel experiment (Chapter 9) backward. Suppose we start with hot water and wait for the weight to rise up from the floor. Clearly, we don't expect this to happen. The water can be heated to increase its internal energy, but the hot water will not rotate the paddle wheel. Water can be very hot and thus contain a lot of internal energy, but mechanical work does not spontaneously appear.

The first law of thermodynamics doesn't exclude this; it places no restrictions on which energy transformations are possible. As long as the internal energy equals or exceeds that needed to raise the weight, there would be no violation of the first law if some of the internal energy were used to do this. And yet it doesn't happen. The energy is there, but it is not available. Apparently the availability of energy depends on the form that it takes.

This aspect of nature, unaccounted for so far in the development of our world view, is addressed in the second law of thermodynamics. It's a subtle law that has a rich history, resulting in three different statements of the law. The first form deals with heat engines.

Heat Engines

We can extract some of an object's internal energy under certain circumstances. Heat naturally flows from a higher temperature region to one of lower temperature. A hot cup of coffee left on your desk cools off as some of its thermal energy flows to the surroundings. The coffee continues to cool until it reaches thermal equilibrium with the room. Energy flows out of the hot region because of the pressure of the cold one. But no work—no mechanical energy—results from this flow.

Many schemes have been proposed for taking part of the heat and converting it to useful work. Any device that does this is called a **heat engine.** The simplest and earliest heat engine is traced back to Hero of Alexandria. About A.D. 50, Hero invented a device similar to that shown in Figure 10-1. Heating the water-filled container changes the water into steam. The steam, escaping through the two tubes, imparts angular momentum to the container, causing it to rotate. From our modern perspective we might judge this to be more of a toy than a machine. The importance of this device, however, was that mechanical energy was in fact obtained—the container rotated. Apparently, Hero did do something useful with his heat engine; there are stories that he used pulleys and ropes to open a temple door during a religious service. (Probably much to the shock of the worshippers!)

Figure 10-1   A modern version of Hero's heat engine rotates under the action of the escaping steam.

Physics on Your Own   Make the Hero heat engine illustrated in Figure 10-2. Punch holes large enough for small straws on opposite sides of a 1-pound coffee can. Put a piece of straw in each hole and bend and tape them so that they both point in the clockwise direction. Fill the can about one-quarter full of water and seal the lid with tape. Suspend the can over a heating element as shown. The escaping steam will cause the can to rotate rapidly. *Be careful* of the hot water and steam.

Figure 10-2   A coffee can steam engine.

The Industrial Revolution began with the move away from animal power toward machine power. The first machines were steam engines used to pump water out of mines in England. Figure 10-3 shows the essential features of the most successful of the early steam engines. It was invented by James Watt in 1769 and had a movable piston in a cylinder. The flow of heat was not a result of direct thermal contact between objects but the transfer of steam that was heated by the hot region and cooled by the cold region. By alternatively heating and cooling the steam in the cylinder, the piston was driven up and down, producing mechanical work.

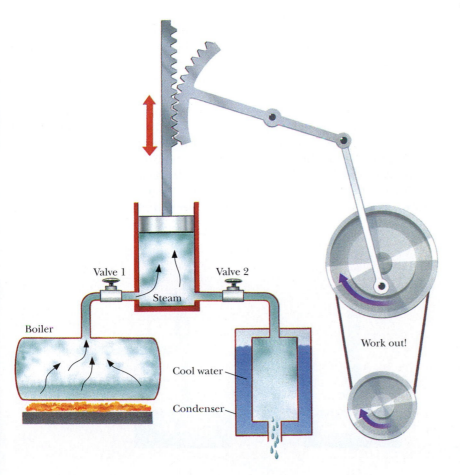

Figure 10-3   The essential features of Watt's steam engine. Opening valve 1 lets steam into the chamber raising the piston. Closing valve 1 and opening valve 2 reduces the pressure allowing the piston to fall.

The steam locomotive is a heat engine.

The internal combustion engine in a Ford Probe is an example of a modern heat engine.

The opening of the American West was helped by another heat engine, "the iron horse," or steam locomotive. A more modern heat engine is the internal combustion engine used in automobiles. Replacing the wood and steam, gasoline explodes to produce the high-temperature gas. The explosions move pistons, allowing the engine to extract some of this energy to run the automobile. The remaining hot gases are exhausted to the cooler environment.

Although there are many types of heat engine, all of them can be represented by the same schematic diagram. To envision this, recall that all heat engines involve the flow of energy from a hotter region to a cooler one. Figure 10-4(a) shows how we represent this flow. A heat engine is a device placed in the path of this flow to extract mechanical energy. Figure 10-4(b) shows heat flowing from the hotter region; part of this heat is converted to mechanical work and the remainder is exhausted to the

Figure 10-4 (a) A schematic showing the natural flow of thermal energy from a higher temperature region to a lower temperature region. (b) The white square represents the many types of heat engine. The heat engine converts part of the heat from the hot region to mechanical work and exhausts the remaining heat to the cold region.

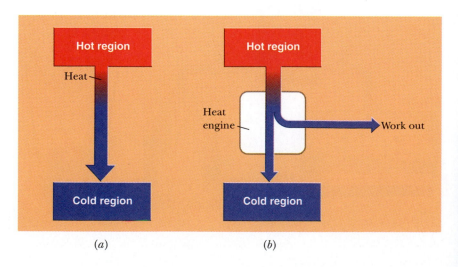

colder region. You can verify that heat is thrown away by an automobile's engine by putting your hand near (but not directly on) the exhaust pipe. The exhausted gases are hotter than the surrounding air. Without the cool region the flow would stop and there would be no energy extracted.

> **QUESTION**   How much work does a heat engine perform if it extracts 100 joules of energy from a hot region and exhausts 60 joules to a cold region?

Ideal Heat Engines

The first law of thermodynamics requires that the sum of the mechanical work and the exhausted heat be equal to the heat extracted from the hot region. But exhausted heat means wasted energy. When engines were developed at the beginning of the Industrial Revolution, engineers asked: What kind of engine would get the maximum amount of work from a given amount of heat?

In 1824, the French scientist Sadi Carnot published the answer to this question. Carnot found the best possible engine by imagining the whole process as a thought experiment. He assumed his engine would use idealized gases and that there would be no frictional losses due to parts rubbing against each other. His results were surprising. Even under these ideal conditions the heat engine must exhaust some heat.

Carnot's work led to one version of the **second law of thermodynamics** and an understanding of why internal energy cannot be completely converted to mechanical energy.

> It is impossible to build a heat engine to perform mechanical work that does not exhaust heat to the surroundings.

heat engine form of the second law of thermodynamics

That is, even the best theoretical heat engine cannot convert 100 percent of the incoming heat into work.

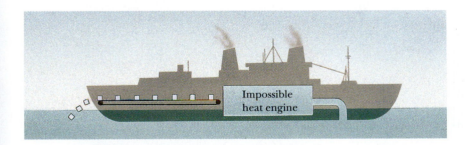

Impossible heat engine

An impossible engine in an ocean liner extracts energy from ocean water, makes ice cubes, and propels the ship.

> **ANSWER**   Conservation of energy requires that the work be equal to 40 joules, the difference between the input and the output.

Stated slightly differently, the fact that the engine must throw away heat means that no heat engine can run between regions at the same temperature. This is unfortunate, as it would be a boon to civilization if we could "reach in" and extract some internal energy from a single region. For example, think of all the energy that would be available in the oceans. If we could build a single-temperature heat engine, it could be used to run an ocean liner. Its engine could take in ocean water at the front of the ship, extract some of the water's internal energy, and drop ice cubes off the back. Notice that this hypothetical engine isn't intended to get something for nothing; it only tries to get out what is there. But the second law says that this is impossible.

Perpetual Motion Machines

Since the beginning of the Industrial Revolution inventors and tinkerers have tried to devise a machine that would run forever. This search was fueled by the desire to get something for nothing. A machine that did some task without requiring energy would have countless applications in society, not to mention the benefits that would accrue to its inventor. No such machine has ever been invented; in fact, the only "successful" perpetual motion machines have been devices that were later shown to be hoaxes.

In the 17th century, English physician Robert Fludd proposed the device shown in Figure 10-5. Fludd wanted to turn the waterwheel to move the millstone and at the same time return the water to the upper level. It didn't work. Machines like Fludd's violate the first law of thermodynamics by trying to get more energy out of a device than is put into it. Every machine is cyclic by definition. After doing something, the machine returns

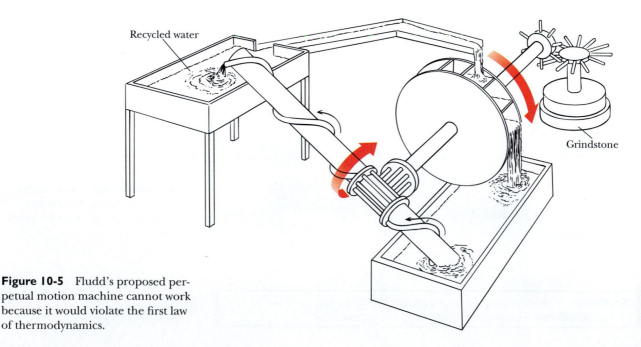

Figure 10-5 Fludd's proposed perpetual motion machine cannot work because it would violate the first law of thermodynamics.

to its original state to start the whole process over again. Fludd's scheme was to have the machine start with a certain amount of energy (the potential energy of the water on the upper level), do some work, and then return to its original state. But if it did some work, some energy was "spent." Returning the machine to its original state would require getting something for nothing, a violation of the first law.

Other perpetual motion machines have been less ambitious. They were simply supposed to run forever without extracting any useful output. They also didn't work. Even if we ask nothing of such a machine except to run—and thus stay within the confines of the first law of thermodynamics—it still will not run forever. These machines failed because they were attempting to violate the second law of thermodynamics.

To illustrate this, suppose we invent such a machine as a thought experiment. Figure 10-6 is a diagram of one possibility. A certain amount of energy is initially put into the system and, somehow, the system is started. The paddle wheel turns in the water and, as the Joule experiment showed us, produces an increase in the temperature of the water. Mechanical energy has been converted to internal energy as indicated by the water's temperature rise. Heat from the hot water flows to the pad below the water. The pad is a heat engine designed to capture this energy and convert it to mechanical energy. The mechanical energy, in turn, is used to drive gears that rotate the paddles.

Does the machine continue to run? Notice that this is a perfect machine and we can assume that there are no frictional losses. So if it fails, it does so because of some fundamental reason. Such engines have been built—they have all failed. The problem lies with the heat engine in the pad. Any heat engine, no matter how good, returns less mechanical energy than the thermal energy it receives. Therefore, our machine runs down and we have failed. We do not fail because we try to create energy but because we cannot use the energy that we have. Once the energy is converted to internal energy it is no longer fully available.

Even in a simpler machine without a heat engine, there would still be moving parts. Any rubbing of moving parts increases the machine's internal energy. All of this internal energy would have to be converted back into mechanical energy if the machine were to run forever. The second law tells us that this cannot happen.

All machines and heat engines, no matter how complicated, cannot avoid the constraints imposed by the first and second laws of thermodynamics.

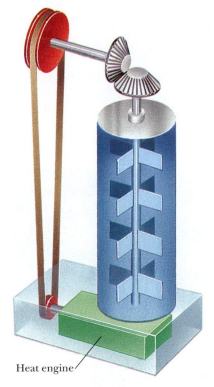

Heat engine

Figure 10-6 This proposed perpetual motion machine would violate the second law of thermodynamics.

Real Engines

All engines can be rated according to their efficiencies. In general, efficiency can be defined as the ratio of the output to the input. In the case of heat engines the **efficiency** η of an engine is equal to the ratio of the work W produced divided by the heat Q extracted from the hot region.

$$\eta = \frac{W}{Q}$$

$$\text{efficiency} = \frac{\text{work out}}{\text{heat in}}$$

Carnot's ideal heat engine has the maximum theoretical efficiency. This efficiency can be expressed as a simple relationship using Kelvin temperatures.

efficiency of an ideal heat engine

$$\eta = 1 - \frac{T_c}{T_h}$$

The Carnot efficiency depends only on the temperatures of the two regions. It is larger (that is, closer to 1) if the temperature of the exhaust region T_c is low or that of the hot region T_h is high. The largest efficiency occurs when the temperatures are as far apart as possible. Usually this efficiency is multiplied by 100 so that it can be stated as a percentage. Real engines produce more waste heat than Carnot's ideal engine, but his relationship is still important because it sets an upper limit for their efficiencies.

Today steam engines are used primarily to drive electric generators in power plants. Although these engines are certainly not Carnot engines, their efficiencies can be increased by making their input temperatures as high as possible and their exhaust temperatures as low as possible. Due to constraints imposed by the properties of the materials used in their construction, these engines cannot be run much hotter than 550°C. Exhaust temperatures cannot be much lower than about 50°C. Given these constraints, we can calculate the maximum theoretical efficiency of the steam engines used in electric generating plants. Remembering to convert the temperatures from Celsius to Kelvin, we have

$$\eta = 1 - \frac{T_c}{T_h} = 1 - \frac{323 \text{ K}}{823 \text{ K}} = 0.61 = 61\%$$

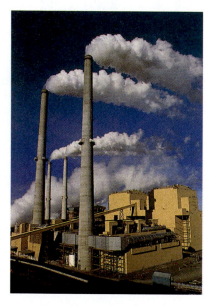

A coal-fired electrical generating plant.

Actual steam engines have efficiencies closer to 50%. In addition to the efficiency of the engine, we must consider the efficiency of the boiler for converting the chemical energy of the fuel into heat. Typical oil- or coal-fired power plants have overall efficiencies of about 40% or less. Nuclear power plants use uranium as a fuel to make steam. Safety regulations require that they run at lower temperatures, so they are 5 to 8% less efficient than the oil- or coal-fired plants. In other words, nuclear plants exhaust more waste heat for each unit of electricity generated.

> **QUESTION**   What is the maximum theoretical efficiency for a heat engine running between 127°C and 27°C?

Not all engines are heat engines. Those that don't go through a thermodynamic process are not heat engines and can have efficiencies closer to 100%. Electric motors are close to being 100% efficient if you look only at the electric motor. However, the overall efficiency of the electric motor

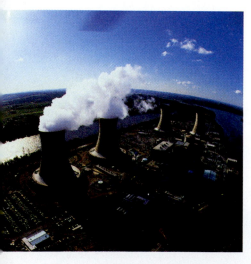

Three Mile Island nuclear power plant.

> **ANSWER**   Being careful to convert these temperatures to the Kelvin scale, we have
>
> $$\eta = 1 - \frac{T_c}{T_h} = 1 - \frac{300 \text{ K}}{400 \text{ K}} = 1 - 0.75 = 0.25 = 25\%$$

must include the energy costs of generating the electricity and the energy losses that occur in the electric transmission lines. Because most electricity is produced by steam engines, the overall efficiency is quite low.

Physics Update

Internal combustion engines, on the molecular level, convert chemical energy into mechanical motion by burning single molecules. No such motor has yet been made artificially, but examples abound in biology, where the "fuel" is adenosine triphosphate (ATP), the basic energy-carrying molecule in cells. Two examples of "protein motors" are myosin, which slides past actin filaments to produce muscle contraction, and kinesin, which transports material in cells. The size of the kinesin motor is 12 nanometers, 50 times smaller than the smallest transistor now made. Marcelo Magnasco of Rockefeller University and NEC Research Institute has developed a general framework for such motors that describes the relationship between their state of motion and the rate at which they consume chemical fuel. Magnasco's description paves the way for a fundamental, physics-based understanding of motor proteins and provides insights into designing artificial ones.

Physics on Your Own   Call your local power company and ask them for the efficiency of their fossil-fuel-fired power plants. If they also use nuclear power, ask them how their efficiencies for these plants compare with those of the fossil-fuel-powered plants.

Running Heat Engines Backward

There are devices that extract heat from a cooler region and deposit it into a hotter one. These are essentially heat engines running backward. **Refrigerators** and air conditioners are common examples of such devices. By reversing the directions of the arrows in Figure 10-4(b), we produce the schematic of the process shown in Figure 10-7.

Refrigerators move heat in a direction opposite to its natural flow. The inside of your refrigerator is the cold region and your room is the hot region. The refrigerator removes heat from the inside and exhausts it to the warm region outside. If you put your hand by the base of the refrigerator, you can feel the heat being transferred to the kitchen. Similarly, if you walk by the external part of an air conditioner, you feel heat being expelled outside the house. In both cases the warm region gets hotter and the cool gets colder.

Refrigerators require work to move energy from a lower temperature region to a higher temperature region. Therefore, the heat delivered to the higher temperature region is larger than that extracted from the lower temperature region. The extra heat is the amount of work done on the system. In fact, this process leads to an equivalent statement of the **second law of thermodynamics.**

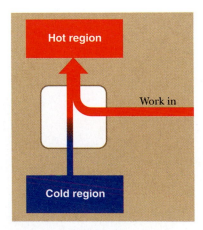

Figure 10-7 A refrigerator uses mechanical work to transfer heat from a colder region to a hotter one.

refrigerator form of the second law
of thermodynamics

> It is impossible to build a refrigerator that can transfer heat from a lower temperature region to a higher temperature region without using mechanical work.

Heat pumps can be used to cool houses in summer and heat them in winter. They can extract energy from the outside air even in cold weather.

A fairly recent device, called a **heat pump,** is used in many homes to both cool during the summer and heat in winter. It is simply a reversible heat engine. In the summer it functions as an air conditioner by extracting heat from inside your house. In the winter it runs in reverse, extracting energy from the outside air and putting it into your house.

It might seem strange to be able to extract heat from the cold air. (Incidentally, this also happens in your refrigerator.) Remember that heat and temperature are not the same. Even on the coldest days, the outside air still has enormous amounts of internal energy. The heat pump just transfers some of this internal energy. Actually, there is an economical limit. If the outside temperature is too low, it takes more energy to run the heat pump than it does to use electric baseboard heaters. This situation can be corrected if a warmer region such as well water is available.

> **QUESTION**   How much energy does an air conditioner exhaust if it requires 200 joules of mechanical work to extract 1000 joules of energy from a house?

Order and Disorder

Whether it is a heat engine or a refrigerator, the second law of thermodynamics essentially says that we can't break even. If we are trying to get mechanical energy from a thermal source, we have to throw some heat away. If we want to cool something, we have to do work to counteract the natural flow of energy. But these two forms of the second law offer little insight into why this is so. The reason lies in the microscopic behavior of the many-particle systems that make up matter. To understand the second law of thermodynamics, we need to look at the details of such systems.

We begin by looking at systems in general. Any system consists of a collection of parts. Assuming that these parts can be shifted around, the system has a number of possible arrangements. We will examine systems and determine their organization because, as we will see, it is their organizational properties that lie at the heart of the second law. An organizational property can be such a thing as the position of objects in a box or the height of people in a community.

Suppose your system is a new deck of playing cards. The cards are arranged according to suits and within suits according to value. The organizational property in this case is the position of the card within the deck. If somebody handed you a card, you would have no trouble deciding where it came from. If the deck were arranged by suit but the values

> **ANSWER**   Conservation of energy requires that the exhaust equal 1200 joules, the sum of the inputs.

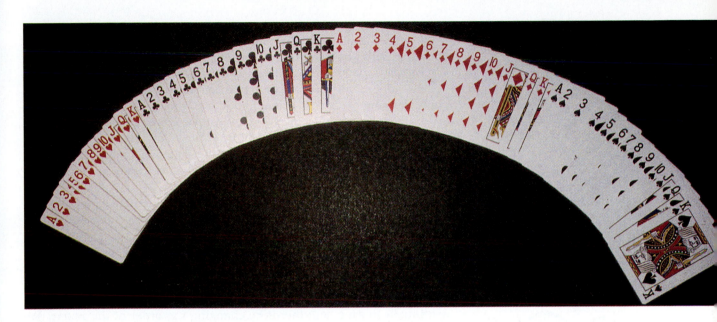

A new deck of cards has a high degree of order.

within the suits were mixed, you would not be able to pinpoint the location of the card but could say from which fourth of the deck it came. In a completely shuffled deck, you would have no way of knowing the origin of the card. The first arrangement is very organized, the second less so, and the last one very disorganized. We call a system that is highly organized an **ordered system** and one that shows no organization a **disordered system.**

Another way of looking at the amount of organization is by asking how many equivalent arrangements are possible in each of these situations. When the deck is arranged by suit and value, there is only one arrangement. When the values are shuffled within suits, there are literally billions of ways the cards could be arranged within each suit. When we completely shuffle the deck, the number of possible arrangements becomes astronomically large (8×10^{67}). The more disordered system is the one with the larger number of equivalent arrangements.

> **QUESTION**   How many different arrangements can you make with three colored blocks: one *red*, one *blue*, and one *green*?

A simple system of three coins on a tray can model microscopic order. An obvious organizational property of this system is the number of heads (or tails) facing up. How many different arrangements are there? If the

> **ANSWER**   The arrangements are rbg, rgb, bgr, brg, grb, and gbr. Therefore, there are six possible arrangements.

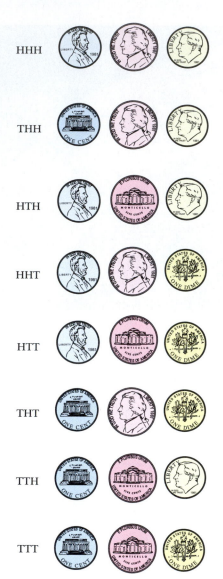

HHH

THH

HTH

HHT

HTT

THT

TTH

TTT

Figure 10-8 The eight possible arrangements of three different coins.

coins are different, there are eight different arrangements for the three coins as shown in Figure 10-8. Including the location of each coin in our considerations would increase the number of arrangements.

In real situations the macroscopic property—for example, the total energy—doesn't depend on the actual properties of a particular atom. To apply our coin analogy to this situation, we would examine the different arrangements without identifying individual coins. In other words, we should simply count the number of heads and tails. How many different arrangements are there now? Four. Of the eight original arrangements, two groups of three are now indistinguishable as shown in Figure 10-9. Each group has three *equivalent states*. The arrangements with all heads or all tails occur in only one way and thus have only one state each. Arrangements with only one state have the highest order. The arrangements that occur in three ways have a lower order.

Understanding the microscopic form of the second law of thermodynamics depends on realizing that the order of real systems is constantly changing. This dynamic nature of systems occurs because macroscopic objects are composed of an immense number of atomic particles that are continually moving, and therefore changing the order of the system. We can use our coins to understand this.

To model the passage of time, we shake the tray to flip the three coins. If we repeatedly start our system with a particular high-order arrangement, we observe that after one flip, the amount of order in the system usually changes. Each of the eight states has an equal probability of occurring. However, the probability of a low-order arrangement occurring is not equal to that of a high-order arrangement, simply because there are more low-order possibilities. Of the eight possible states, only two yield high-order arrangements. The chance is only 2 in 8, or 25%, of obtaining a high-order arrangement. In the other 75% of the time, we expect to obtain two heads and a tail or two tails and a head.

Physics on Your Own   Flip three coins 80 times. How many times do you get all heads? All tails? Two heads and one tail? Two tails and one head? Are your numbers consistent with the predicted odds? Would you expect the consistency to improve if you threw the coins 8000 times?

If we increase the number of coins, the probabilities change. With four coins, for example, the number of possible states grows to 16. Adding one additional coin doubles the total number of combinations because we are adding a head or a tail to each of the previous combinations. But again, there are only two arrangements with the highest order, those having all heads or all tails. The probability now of obtaining a high-order arrangement is 2 in 16, or about 12%. As the number of elements in the system increases, the probability of obtaining a high-order arrangement gets smaller. Conversely, the probability of obtaining a low-order arrangement increases with the size of the system.

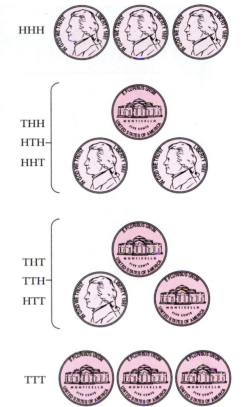

> **QUESTION**   What is the probability of obtaining all heads or all tails with five coins?

For example, if we have 8 coins, the probability of obtaining the highest order distribution (all heads or all tails) is less than 1%. If we increase the number of coins to 21, then we only have one chance in a million of getting all heads or all tails. If the number of coins increases to that of the number of air molecules in a typical bedroom (10^{27}), the probability decreases to 1 in 10^{82}.

As another example, let's consider the air molecules in your room. Presumably any air molecule can be anywhere in the room. A high-order arrangement might have all the molecules in one small location; a low-order arrangement would have them spread uniformly throughout the available space in one of many equivalent states. The number of low-order arrangements is astronomically higher than the number of high-order ones.

Entropy   Σ

We now introduce a new concept, called **entropy**, as a measure of a system's organization. A system that has some recognizable order has low entropy. The more disorganized the system, the higher its entropy. As with gravitational potential energy, we are only concerned with changes in entropy; the actual numerical value of the entropy does not matter.

We argued in the last section that the order of a system of many particles tends to decrease with time. Therefore, the entropy of the system tends to increase. This tendency is expressed by another version of the **second law of thermodynamics.**

Figure 10-9 The four possible arrangements of three indistinguishable coins.

> The entropy of a system tends to increase.

entropy form of the second law of thermodynamics

The word *tends* needs to be stressed. There is nothing that says the entropy *must* increase. It happens because of the overwhelming odds in its favor. It is still *possible* for all the air molecules to be in one corner of your room. (That would leave you gasping for air if you weren't in that corner!) Nothing prohibits this from happening. Fortunately, however, although this arrangement has the same chance of occurring as any one of the others, the likelihood of it actually happening is vanishingly small.

This entropy version looks so different from the forms of the second law developed for heat engines and refrigerators that you might think that they are different laws. The connection between them is that the motion associated with internal energy has a low-order arrangement,

> **ANSWER**   Adding the additional coin doubles the total number of possible combinations to 32. Therefore, we have 2 chances out of 32, or 6%, of obtaining all tails or all heads.

Arrow of Time

The concept of entropy gives an insight into the character of time. Imagine a motion picture of a football game played backward. It looks silly because the order of the events doesn't match our experiences. There is a direction to time. But why do things happen in a particular sequence? According to most of the laws of nature we have established, other sequences of events *could* happen: The parabolic curve of the thrown football is the same regardless of which direction the film is played. Yet the game—played backward—is unreal. Even a repeating, cyclic event like a child on a swing has clues telling you whether the movie is running forward or backward. If the child were not pumping, the swinging dies down because of frictional effects. Left without an input of energy, everyday motions stop when the macroscopic energy gets transformed into internal energy.

In principle, a stationary pendulum bob could transfer some of its internal energy into swinging the bob. This would (simply) require that the bob's atomic particles all move in the same direction at the same time. They are all moving but in random directions. It could happen that at some time they could all be moving in the same direction. It could happen, but it doesn't. The entropy form of the second law of thermodynamics tells us why: All systems tend toward disorder. This fact of nature gives a direction to time.

In the ghost town of Castle, Montana, the population is low but the entropy is high.

The direction of time is implicit in many events. Bright new paint has a radiant color because of the particular molecules and their particular arrangement. As time passes, there is a mixing of the molecules with the air and a rearranging of the molecules in the paint. Paint fades. Structures crumble; their original shapes are very ordered, but as time passes, they deteriorate. Order is lost. The culprit is chance and the consequence is increasing entropy. Time has a direction because the Universe has a natural tendency toward disorder.

whereas macroscopic motion requires a high degree of organization. Atomic motions are random; at any instant, particles are moving in many directions with a wide range of speeds. If we could take a series of snapshots of these particles and scramble the snapshots, we would be unable to distinguish them; one snapshot is just like another. There appears to be no order to the motions or positions of the particles.

The macroscopic motion of an object, on the other hand, gives organization to the motions of the object's atoms. Although the particles are randomly moving in all directions on the atomic level, they also have the macroscopic motion of the object. All the atoms in a ball falling with a certain kinetic energy are essentially moving in the same direction (Fig. 10-10). When the ball strikes the ground, the energy is not lost; the collective motions are randomized. The atoms move in all directions with equal likelihood.

Although it is certainly physically possible for all the particles to once again move in a single direction at some future time (without somebody picking it up), it is so unlikely that we never expect to see it happen. If it did happen, however, some of the internal energy of the ball's particles would then be available to do some mechanical work. This would be very bizarre; at some moment, a ball initially at rest would suddenly fly off with some kinetic energy.

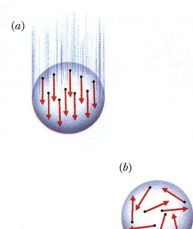

(a)

(b)

Figure 10-10 (a) The motions of atoms in a falling ball have a high degree of order. (b) The motions after the ball hits the ground are random with a low degree of order.

Physics Update

In any computational process, energy is dissipated whenever information is destroyed. The lesson is: to dissipate less energy, don't discard information. One way to do this would be to build logic circuits that run in reverse. That is, the circuits (and the computers using them) would return to their original states at the end of the computation cycle after having performed the required calculations. This idea has been promoted by Rolf Landauer and Charles Bennett of IBM. In principle such circuits, performing reversible operations, would dissipate less energy than irreversible circuits (operating with no regard for the retention or destruction of information) performing irreversible operations. Ralph Merkle of Xerox said that the use of reversible circuitry, entailing also the use of reversible software, was not yet a priority for computer architects but would be early in the next century when the problems of heat dissipation (requiring large heat sinks) and energy consumption become more pressing. Already, Merkle said, computer systems account for 5% of all commercial electricity use in the United States; this might double by the turn of the century.

Decreasing Entropy

Although the overall entropy of a system tends to increase, isolated pockets of activity can show a decrease in entropy; that is, an increase in order. But these increases in order are paid for with an overall decrease in the order of the Universe. Life, for example, is a glorious example of increasing order. It begins as a simple fertilized egg and evolves into a complicated human being. Clearly, order is increasing. But this increase is accomplished through the use of energy. To bring about the increase in human order, a great deal of energy becomes less available for performing useful work. The order of the environment decreases accordingly.

As a simpler example, imagine that you have two buckets of water, one hot and the other cold. If you place a heat engine between the buckets, you can do some work. After the heat engine operates for a while, the buckets reach the same temperature. The system now has a higher entropy. You can decrease the entropy by heating one bucket and cooling the other. But you can only do this by increasing the entropy of the Universe. We saw this with refrigerators. You can make the hot region hotter and the cold region colder but only after mechanical work is put in. That use of work lowers the availability of the energy in the Universe to do mechanical work.

Entropy and Our Energy Crisis

Save energy! That should be the battle cry these days. But why do we have to *save* energy? According to the first law of thermodynamics, energy is neither created nor destroyed. The total amount of energy is constant.

Quality of Energy

With the exceptions of a geothermal source of hot water heating a home or a windmill pumping water out of a well, most energy that we use in our lives has been converted from its original form to the form we use. An obvious example is the electrical energy that lights our homes. This energy often comes from a fossil fuel that is burned to heat water to form steam, which turns a generator to produce the electricity that lights our homes.

It is not possible to convert, or even transfer, energy without "losing" some of the energy in the process. (By *losing energy* we mean that some of the energy is transformed to some form other than the one desired.) Even charging a battery—storing energy for later use—requires more energy than gets stored in the battery. A more obvious loss is that due to frictional forces. If we use a waterwheel to turn an electric generator, some of the input energy is transformed to thermal energy by the frictional forces in the rubbing of the mechanical parts. Losses also occur when electrical energy is transported from one place to another. A final loss comes from the thermal bottleneck described by the second law of thermodynamics. Whenever we convert thermal energy to *any* other form of energy, the second law of thermodynamics tells us that we always lose some energy to other forms: It is impossible to make the conversion with 100% efficiency.

All conversions of energy are not equivalent. Some energy conversions are less efficient, and thus more costly to our world's energy budget. To obtain an accurate measure of the efficiency of the energy you use, you need to go back to the original source and calculate the overall efficiency of delivering it to your home and its final use. Suppose, for example, that the initial conversion is 50% efficient. For every unit of energy you start with you have to give up half a unit. If the process of transporting the energy to your home is also 50% efficient, the amount that you get to use is one-half of one-half, or one-fourth, of the original. To calculate the total efficiency you multiply the efficiencies of each stage together.

The table shows the relative efficiencies of two methods of heating water in your home. The efficiency of each step is shown as well as the overall efficiency through that step. Notice that the lowest efficiency occurs during the generation of electricity in the coal-fired plant. The extra steps of making and transporting the electricity reduce the efficiency of electric heat by a factor of 2½ compared to heating with natural gas. Therefore, the quality of natural gas for heating is much higher than that of electricity produced from coal.

Efficiency for Heating Water		
	Step Efficiency %	**Accumulative Efficiency %**
Electric		
Mining of coal	96	96
Transportation of coal	97	93
Generation of electricity	33	31
Electrical transmission	85	26
Heating	92	24
Gas		
Production of natural gas	96	96
Transportation	97	93
Heating	64	60

Gas water heaters are more efficient than electric ones.

There is no need to conserve energy; it happens naturally. The crux of the matter is really the second law. There are pockets of energy in our environment that are more valuable than others. Given the proper conditions, we can use this energy to do some useful work for society. But if we later add up all the energy, we still have the same amount. The energy used to drive our cars around town is transformed via frictional interactions and exhaust, the consumption of food results in our body temperatures being maintained as well as moving us around, and so on. All the energy is present and accounted for.

Water naturally flows downhill. The water is still there at the bottom of the hill, but it is less useful. If all the water is at the same level, there is no further flow. Similarly, if all the energy in the Universe is spread out uniformly, we can get no more "flow" from one pocket to another.

The real energy issue is the preservation of the valuable pockets of energy. We can only burn a barrel of oil once. When we burn it, the energy becomes less useful for doing work and the entropy of the Universe goes up. The second law doesn't tell us how fast entropy should increase. It does not tell us how fast we must use up our pockets of energy; it only says that it *will* occur. The decision of finding an acceptable rate is left to us. The battle cry for the future should be, "Slow down the increase in entropy!"

SUMMARY

Mechanical energy can be completely converted into internal energy, but it is not possible to recover all of this internal energy and get the same amount of mechanical energy back. The energy is there, but it is not available.

We can extract some of an object's internal energy using a heat engine. Part of the heat is converted to mechanical work and the remainder is emitted as exhaust. The first law of thermodynamics requires that the sum of the mechanical work and the exhaust heat be equal to the heat extracted. The second law of thermodynamics says that we cannot build a heat engine without an exhaust. As a consequence it is impossible to completely convert heat to mechanical energy.

People have tried to construct two kinds of perpetual motion machines. The first kind violates the first law of thermodynamics by trying to get more energy out of a device than is put into it. The perpetual motion machines of the second kind do not try to do any useful work but just try to run forever. They fail because they violate the second law of thermodynamics.

The efficiency of an ideal heat engine—a Carnot engine—depends on the temperatures of the input heat source and the exhaust region. The maximum theoretical efficiency increases as the difference in these two temperatures increases and is given by $\eta = 1 - T_c/T_h$, where T_h is the input temperature and T_c is the exhaust temperature in kelvin. The efficiencies of real engines are always less than this.

Refrigerators and air conditioners are essentially heat engines running backward. These devices require work to move energy from a lower

CHAPTER

10

REVISITED

Performing useful work with the internal energy in an object requires that the energy be able to flow from the object to a region at a lower temperature. Regardless of the amount of internal energy in an object, if there is no colder region nearby, the energy is useless for running an engine. On the other hand, the colder the exhaust region, the larger the fraction of the internal energy that can be used.

temperature region to a higher temperature region. An equivalent form of the second law says that we cannot build a refrigerator that moves energy from a colder region to a hotter region without work being done.

The third form of the second law says that the entropy of an isolated system tends to increase. Entropy is a measure of the order of a system. Entropy increases as the order within a system decreases. The entropy of a system can decrease but the decrease is paid for with an overall increase in the entropy of the Universe. Life is an example of decreasing entropy (increasing order).

Our energy crisis is really related to the using up of the pockets of energy in our environment that are more valuable than others. We still have the same amount of energy as we've always had.

KEY TERMS

disordered system: A system with an arrangement equivalent to many other possible arrangements.

efficiency: The ratio of the work produced to the energy input. For an ideal heat engine, the Carnot efficiency is given by $1 - T_c/T_h$.

entropy: A measure of the order of a system. The second law of thermodynamics states that the entropy of an isolated system tends to increase.

heat: A flow of energy due to a difference in temperature.

heat engine: A device for converting heat into mechanical work.

heat pump: A reversible heat engine that acts as a furnace in winter and an air conditioner in summer.

ordered system: A system with an arrangement belonging to a group with the smallest number (possibly one) of equivalent arrangements.

refrigerator: A heat engine running backward.

thermodynamics, second law of: There are three equivalent forms: (1) It is impossible to build a heat engine to perform mechanical work that does not exhaust heat to the surroundings; (2) It is impossible to build a refrigerator that can transfer heat from a lower temperature region to a higher temperature region without using mechanical work; and (3) The entropy of a system tends to increase.

CONCEPTUAL QUESTIONS

1. What does a heat engine do?

2. What restrictions does the first law of thermodynamics place on the operation of a heat engine?

3. Why is it not possible to run an ocean liner by taking in seawater at the bow of the ship, extracting internal energy from the water, and dropping ice cubes off the stern?

4. One possible end to the Universe is for it to reach thermal equilibrium; that is, it would have a uniform temperature. Why could this be called the "heat death" of the Universe? Would this temperature be absolute zero?

5. State the heat engine form of the second law of thermodynamics in your own words.

6. Give an example that clearly illustrates the meaning of the heat engine form of the second law of thermodynamics.

7. Does the following statement agree with the second law of thermodynamics? No engine can transform its entire heat input into work.

8. Would it be possible to design a heat engine that produces no thermal pollution?

9. It is possible to float heat engines on the ocean and extract some of the internal energy of the water. What conditions must be present for this to occur?

10. A hurricane can be thought of as a heat engine that converts thermal energy from the ocean to mechanical motion of its winds. Use this idea to explain why the wind speeds decrease as the hurricane moves away from the Equator.

11. Many people have tried to build perpetual motion machines. What restrictions does the first law of thermodynamics place on the possibility of building a perpetual motion machine?

12. What restrictions does the second law of thermodynamics place on the possibility of building a perpetual motion machine?

13. Explain how the following simplified statements of the first and second laws of thermodynamics are consistent with the versions given in this chapter.

First: You cannot get ahead.

Second: You cannot even break even.

14. A student proposes to run an automobile without using any fuel. He says that he will build a windmill on top of the car. The car's motion will cause the windmill to rotate and generate electricity. The electricity will run a motor, maintaining the car's motion, which in turn causes the windmill to rotate. Can you see anything wrong with his proposal?

15. What does it mean to say that the human body is a heat engine with an efficiency of 20 percent?

16. Why are nuclear power plants less efficient than coal-fired plants?

17. What happens to the efficiency of an ideal heat engine as its input temperature is increased while its exhaust temperature is held fixed?

18. If the input temperature of an ideal heat engine is fixed, what happens to its efficiency as its exhaust temperature is increased?

19. How does the efficiency of an idealized automobile engine depend on the operating temperature of the engine and the temperature of its surroundings?

20. How would you expect the efficiency of a heat pump to depend on the temperature of the surrounding air?

21. Why is saying that the efficiency of a heat engine must be less than one equivalent to the statement of the heat engine form of the second law of thermodynamics?

*22. With his paddle-wheel apparatus James Joule determined that 4.2 joules of mechanical work were equivalent to 1 calorie of heat. If he had used a heat engine and had measured the heat flowing into the engine and the work done by the engine, what would he have determined about the relative sizes of the joule and the calorie?

23. Would it be possible to cool a room by opening the door of the refrigerator?

24. An air-conditioner mechanic is testing a unit by running it on the workbench in an isolated room. What happens to the temperature of the room?

25. State the refrigerator form of the second law of thermodynamics in your own words.

26. Give an example that clearly illustrates the meaning of the refrigerator form of the second law of thermodynamics.

27. Does the following statement agree with the refrigerator form of the second law of thermodynamics? The natural direction for the flow of heat is from hotter objects to colder objects.

28. The coefficient of performance for a refrigerator is defined as the ratio of the heat extracted from the colder system to the work required. Will this number be greater than, equal to, or less than one for a good refrigerator?

29. Can one side ever have an advantage when making the call on a flip of a coin?

30. You watch a friend flipping coins and notice that heads has come up four times in a row. Does this mean that it is more likely that a tail will come up on the next throw?

*31. What sums of three dice have the highest order?

*32. What sums of two dice have the highest and lowest order?

33. Why does the sum of two dice equal 7 more often than any other number?

34. Why do "boxcars" (a pair of 6's) occur so rarely when throwing two dice?

35. State the entropy form of the second law of thermodynamics in your own words.

36. Give an example that clearly illustrates the meaning of the entropy form of the second law of thermodynamics.

37. What happens to the entropy of an egg as it is scrambled?

38. What happens to the entropy of the Universe as the orange liquid diffuses into the clear liquid?

39. A cold piece of metal is dropped into an insulated container of hot water. After the system has reached an equilibrium temperature, has the entropy of the Universe increased or decreased?

40. What happens to the entropy of the Universe as an ice cube melts in water?

41. What happens to the entropy of a human as it grows from childhood to adulthood? Is this consistent with the second law of thermodynamics?

*42. Describe a system in which the entropy is decreasing. Is this system isolated from its surroundings?

43. A ringing bell is inserted into a large glass of water. The bell and the water are initially at the same temperature and are insulated from their surroundings. Eventually the bell stops vibrating and the water comes to rest.

   a. What happens to the mechanical energy of the bell?

   b. What happens to the temperature of the system?

   c. What happens to the entropy of the system?

44. Why doesn't the bell in the previous question start ringing when it is removed from the water? Would it help to warm the water first?

45. Why is it that a coin doesn't acquire thermal energy from its surroundings and leap up off a table?

*46. Are Mexican jumping beans a violation of the second law of thermodynamics?

47. Imagine that you could film the motion of the gas molecules in the room. Would you be able to tell whether the film was running forward or backward? Would it make a difference if air were being released from a balloon?

*48. Describe a situation involving motion that could be filmed and run backward without viewers realizing that it was being played in reverse.

Question 41.

49. Which of the following statements explains why we are currently experiencing a worldwide energy crisis?

   a. The amount of energy in the world is decreasing rapidly.

   b. The entropy of the world is increasing rapidly.

   c. The entropy of the world is decreasing rapidly.

50. How does slowing down the increase in entropy help solve the world's energy crisis?

51. Why is heating water on a gas stove more efficient than heating it on an electric stove?

52. Since childhood we've been told to turn out the lights when we leave a room. Does this really reduce the electric bill for a house with electric heating?

53. Which results in the larger increase in the entropy of the Universe; heating a liter of room-temperature water to boiling using natural gas or using electricity?

EXERCISES

1. What input energy is required if an engine performs 200 kJ of work and exhausts 400 kJ of heat?

2. How much work is performed by a heat engine that takes in 1000 J of heat and exhausts 600 J?

3. An engine takes in 7000 cal of heat and exhausts 4000 cal of heat each minute it is running. How much work does the engine perform each minute?

4. A heat engine requires an input of 9 kJ per minute to produce 3 kJ of work per minute. How much heat must the engine exhaust per minute?

5. What is the efficiency of a heat engine that does 40 J of work for every 200 J of heat it takes in?

6. An engine takes in 4000 cal of heat for every 1000 cal of work produced. What is its efficiency?

7. An engine takes in 8000 cal of heat and exhausts 5000 cal each minute it is running. How much work does the engine do each minute? What is its efficiency?

8. An engine takes in 600 cal of heat and exhausts 450 cal of heat each second it is running. How much work does the engine do each minute? What is its efficiency?

9. An engineer has designed a machine to produce electricity by using the difference in the temperature of ocean water at different depths. If the surface temperature is 20°C and the temperature at 50 m below the surface is 12°C, what is the maximum efficiency that this machine could produce?

10. A heat engine takes in 1000 J of energy at 1000 K and exhausts 600 J at 500 K. What are the actual and maximum theoretical efficiencies of this heat engine?

11. An ideal heat engine has a theoretical efficiency of 40% and an exhaust temperature of 27°C. What is its input temperature?

12. What is the exhaust temperature of an ideal heat engine that has an efficiency of 50% and an input temperature of 400°C?

13. A heat engine operating at an efficiency of 20% performs work at a rate of 20 kW. What is the input power to the engine?

*14. The thermal input power to an engine is 1000 W at 600 K. What is the minimum power of the engine's exhaust if the exhaust temperature is 400 K?

15. How much work is required by a refrigerator that takes in 1000 J from the cold region and exhausts 1500 J to the hot region?

16. A refrigerator uses 500 J of work to remove 1800 J of heat from a room. How much heat does it exhaust?

17. How much work per second (power) is required by a refrigerator that takes 800 J of thermal energy from a cold region each second and exhausts 1000 J to a hot region?

18. What is the power of the exhaust of an air conditioner that uses 1200 W to remove heat from a room at a rate of 4800 W?

19. Show that four coins can be arranged in 16 different ways.

20. Show that the combination of four coins with the lowest order (two heads and two tails) is the one with the largest number of arrangements.

21. What is the probability of rolling a total of 7 with two dice?

22. What is the probability of rolling a sum of 10 with two dice?

*23. The total number of possible states for three dice is $6 \times 6 \times 6 = 216$. What is the probability of throwing a sum equal to 5?

*24. The total number of possible states for three dice is $6 \times 6 \times 6 = 216$. What is the probability of throwing a sum equal to 6?

INTERLUDE

Why do the riders not fall out of the roller coaster car as it executes the loop-the-loop?

Universality of Motion

A set of rules—Newton's laws—correctly describes the motion of ordinary objects. These rules are the foundation of the classical physics world view and they match our commonsense notions about the behavior of the material world. Are Newton's laws valid for all situations and for all regions of the Universe?

As we developed these rules, we presumed that all observers were standing still on the Earth's surface. In Chapter 1 when describing the speed of a ball dropping from a height, we simply said, "The ball is falling at 30 meters per second." It was unnecessarily cumbersome at that time to say, "The ball is falling at 30 meters per second *relative to the surface of the Earth.*" Is this extra qualification ever needed? To answer this question we need to ask how the motion would appear to somebody moving relative to the Earth's surface. We need to ask if that person—say somebody riding in a train or standing in a freely falling elevator—would develop the same rules of motion. If the rules are different, it is possible that the laws of motion are not universal. The Universe might have an enormous number of different rules—one set for each different point of view. On the other hand, if we can deduce that these laws are universal, the payoffs in terms of our world view are large. In science the fewer the rules, the more beautiful the world view. When a rule is universal, such as Newton's law of gravitation (Chapter 4), we believe it is a more fundamental aspect of nature.

As we try to establish the universality of the laws of motion, we find that the price for this new level of understanding is a restructuring of our ideas about the concepts of space and time. The person primarily responsible for this restructuring was Albert Einstein, who, in addition to his stature among physicists, captured the popular imagination as no other scientist ever has. The mere mention of his name conjures up such images as time as the fourth dimension, people growing older slowly, and warped space. His popularity was due to the seemingly bizarre ideas that he brought to our world view. In fact, Einstein didn't like the publicity. He wished to be left alone in the solitude of his work. But his ideas were too shocking not to create a stir.

In 1905, this quiet man was a clerk in a Swiss patent office. During that year he published four papers; two were on the subject we now call the special theory of relativity. Interestingly, it was work in another area of physics that resulted in his being awarded the Nobel Prize in 1921. His ideas revolutionized the physics world view and propelled him into the center of scientific activity for the next half century. How this person—who could not get a university teaching job when he graduated—created such revolutionary ideas is a fascinating story in itself.

Some people may feel that Einstein's ideas have little or no connection with reality—that they are a fantasy-based creation resulting from some mathematical trickery. This perception couldn't be further from the truth.

Although the ideas of relativity had their beginnings in a realm of the physical world beyond our everyday experience, they produced profound changes in the very foundations of our world view. But the ideas of relativity didn't start with Einstein. As early as Galileo, questions were being asked about the absolute nature of position, speed, and acceleration.

> **Although the ideas of relativity had their beginnings in a realm of the physical world beyond our everyday experience, they produced profound changes in the very foundations of our world view. But the ideas of relativity didn't start with Einstein. As early as Galileo, questions were being asked about the absolute nature of position, speed, and acceleration.**

11

Everybody has been told that the Earth rotates on its axis once each day and yet it appears that the Sun, Moon, and stars all go around the Earth. What evidence do we have to support the idea that it is the Earth that is really moving? (See p. 275 for the answer to this question.)

Classical Relativity

The apparent motion of the stars is due to the Earth's rotation.

Do observers moving relative to each other agree on the description of the motion of an object? Most of us feel that they would not. Consider, for example, the situation where one observer is unfortunate enough to be in a freely falling elevator and the other is standing safely on the fifth floor. How do the two observers describe the motion of an apple that is "dropped" by the observer in the elevator?

The observer in the elevator sees the apple suspended in midair [Fig. 11-1(a)]. It has no speed and no acceleration. The observer on the fifth floor sees the apple falling freely under the influence of gravity. It has a constant downward acceleration and therefore is continually gaining speed [Fig. 11-1(b)].

Is there something fundamentally different about these descriptions, or are the differences just cosmetic? And most important, do the differences mean that the validity of Newton's laws of motion is in question? Are the laws valid for the observer in the elevator? If not, the consequences for our physics world view could be serious.

A Reference System

We see motion when something moves relative to other things. Imagine sitting in an airplane in straight, level flight at a constant speed. As far as the activities inside the plane are concerned, you don't think of your seat as moving. From your point of view, the seat remains in the same spot relative to everything else in the plane.

The phrase *point of view* is too general. Since all motion is viewed relative to other objects, we need to agree on a set of objects that are not moving relative to each other and that can therefore be used as the basis for

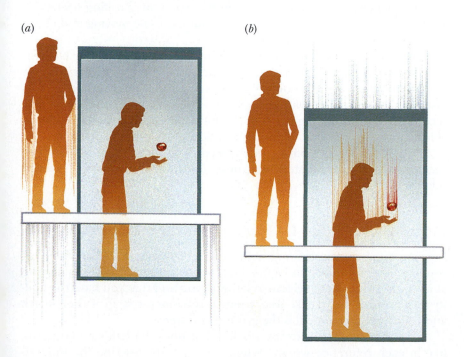

(a) (b)

Figure 11-1 The dropped apple appears suspended in midair as viewed by the elevator passenger (a) and as falling freely by the observer standing on the floor (b).

Which car is moving?

detecting and describing motion. This collection of objects is called a **reference system.**

One common reference system is the Earth. It consists of such things as the houses, trees, and roads that we see every day. This reference system appears to be stationary. In fact, we are so convinced that it is stationary that we occasionally get tricked. If while sitting in a car waiting for a traffic light to change, the car next to you moves forward, you occasionally experience a momentary sensation that your car is rolling backward. This illusion occurs because you expect your car to be moving and everything outside the car to be stationary.

It doesn't matter if you are in a moving car or sitting in your kitchen; both are good reference systems. Consider your room as your reference system. To describe the motion of an object in the room, you measure its instantaneous position with respect to some objects in the room and record the corresponding time with a clock. This probably seems reasonable, and quite obvious. But complications—and interesting effects—arise when the same motion is described from two different reference systems. We begin by studying these interesting effects in classical relativity.

Motions Viewed in Different Reference Systems

Imagine that you are standing next to a tree and some friends ride past you in a truck, as shown in Figure 11-2. Suppose that the truck is moving at a very high, constant velocity relative to you and that you have the ability to see inside the truck. Your friends' reference system is their truck and your reference system is the ground and trees.

One of your friends drops a ball. What does the ball's motion look like in each reference system? When your friends describe the motion,

Figure 11-2   Your friends move by you at a very high, constant velocity and drop a ball.

they refer to the walls and floor of the truck. They see the ball fall straight down and hit the truck's floor directly below where it was released (Fig. 11-3).

You describe the motion of the ball in terms of the ground and trees. Before the ball is released, you see it moving horizontally with the same velocity as your friends. Afterwards, the ball has a constant horizontal component of velocity, but the vertical component increases uniformly. That is, you see the ball follow the projectile path shown in Figure 11-4.

The ball's path looks quite different when viewed in different reference systems. Galileo asked if observers could decide whose description was "correct." He concluded that they couldn't. In fact, each observer's description was correct. We can understand this by looking at the explanations that you and your friends give for the ball's motion.

We begin by examining the horizontal motion. Your friends, observing that the ball doesn't move horizontally, conclude that the net horizontal force is zero. On the other hand, you do see a horizontal velocity. But since it is constant, you also conclude that the net horizontal force is zero.

What about vertical forces? Your friends see the ball exhibit free fall with an acceleration of 10 (meters per second) per second. The vertical component of the projectile motion that you observe is also free-fall motion with the same acceleration. Each of you concludes that there is the same net constant force acting downward.

Although you disagree with your friends' description of the ball's path, you agree on the forces involved. Any experiments that you do in

Figure 11-3   From your friends' point of view, they are at rest and see you moving. In their system, they see the ball fall vertically.

Figure 11-4   From the ground you see the ball follow a projectile path.

your reference system will yield the same forces that your friends find in their system. In both cases the laws of motion explain the observed motion.

inertial reference system

We define an **inertial reference system** as one in which Newton's first law (the law of inertia) is valid. Each of the systems above was assumed to be an inertial reference system. In fact, any reference system that has a constant velocity relative to an inertial system is also an inertial system.

The principle that the laws of motion are the same for any two inertial reference systems is called the **Galilean principle of relativity.** Galileo stated that if one were in the hold of a ship moving at a constant velocity, there would be no experiment this person could perform that would detect the motion. This means that there is no way to determine which of the two inertial reference systems is "really" at rest. There seems to be no such thing in our Universe as an absolute motion in space; all motion is relative.

principle of relativity

> The laws of physics are the same in all inertial reference systems.

The principle of relativity says that the laws of motion are the same for your friends in the truck as they are for you. A very important consequence is that the conservation laws for mass, energy, and momentum are valid in the truck system as well as the Earth system. If your friends say that momentum is conserved in a collision, you will agree momentum is conserved even though you do not agree on the values for the velocities of each object.

Physics on Your Own   Conduct some physics experiments the next time you ride in a car, train, bus, or airplane. During the constant velocity portions of the trip, try tossing a ball into the air, playing catch with a friend, using a plumb line or a carpenter's level, observing the surface of a beverage, using an equal-arm balance, or weighing something. Do you notice any differences in the results of these experiments and those performed in your house?

Comparing Velocities

Is there any way that you and your friends in the truck can reconcile the different velocities that you have measured? Yes. Although you each see different velocities, you can at least agree that each person's observations make sense within their respective reference system. When you measure the velocity of the ball moving in the truck, the value you get is equal to the *vector* sum of the truck's velocity (measured in your system) and the ball's velocity (measured relative to the truck).

Suppose your friends roll the ball on the floor at 2 meters per second due east and the truck is moving at a velocity of 3 meters per second due east relative to your system. In this case, the vectors point in the same direction, so you simply add the speeds to obtain 5 meters per second due east, as shown in Figure 11-5(a). If, instead, the ball rolls due west at 2 meters per second, you measure the ball's velocity to be 1 meter per second due east.

> **QUESTION**   What do you observe for the velocity of the ball if it is rolling eastward at 2 meters per second while the truck is moving westward at 4 meters per second?

Although this rule works well for speeds up to millions of kilometers per hour, it fails for speeds near that of light, about 300,000 kilometers per second (186,000 miles per second). This is certainly not a speed that we encounter in our everyday activities. The fantastic, almost unbelievable, effects that occur at speeds approaching that of light are the subject of our next chapter.

Accelerating Reference Systems

Let's expand our discussion of your friends in the truck. This time, suppose that their system has a constant forward acceleration relative to your reference system. Your friends find that the ball doesn't land directly

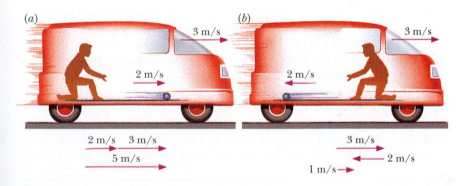

Figure 11-5   The velocity of the ball relative to the ground is the vector sum of its velocity relative to the truck and the truck's velocity relative to the ground.

> **ANSWER**   The ball is moving 2 meters per second westward.

Figure 11-6 In an accelerating system your friends see the ball fall toward the back of the truck.

beneath where it was released but falls toward the back of the truck, as shown in Figure 11-6. In your reference system, however, the path looks the same as before. It is still a projectile path with a horizontal velocity equal to the ball's velocity at the moment it was released. The ball stops accelerating horizontally when it is released but your friends continue to accelerate. Thus, the ball falls behind.

> **QUESTION** Where would the ball land if the truck was slowing down?

As before, the descriptions of the ball's motion are different. But what about the explanations? Your explanation of the motion—the forces involved, the constant horizontal velocity, and the constant vertical acceleration—doesn't change. But your friends' explanation does change; the law of inertia does not seem to work anymore. The ball moves off with a horizontal acceleration. In their reference system, they would have to apply a horizontal force to make an object fall vertically, a contradiction of the law of inertia. Such an accelerating system is called a **noninertial reference system.**

There are two ways for your friends to explain the motion. First, they can abandon Newton's laws of motion. This is a radical move requiring a different formulation of these laws for each type of noninertial situation. This is intuitively unacceptable in our search for universal rules of nature.

Second, they can keep Newton's laws by assuming that there is a horizontal force acting on the ball. But this would indeed be strange; there would be a horizontal force in addition to the usual vertical gravitational force. This also poses problems. In inertial reference systems we can explain all large-scale motion in terms of gravitational, electric, or magnetic forces. The origin of this new force is unknown and, furthermore, its size and direction depend on the acceleration of the system. We know, from your inertial reference system, that the strange new force your friends

> **ANSWER** It would land forward of the release point since the ball continues moving with the horizontal velocity it had when released, whereas the truck is slowing down.

seem to experience is due entirely to their accelerated motion. Forces that arise in accelerating reference systems are called **inertial forces.**

If inertial forces seem like a way of getting around the fact that Newton's laws don't work in accelerated reference systems, you are right. They do not exist; they are invented to preserve the Newtonian world view in reference systems where it does not apply. In fact, another common label for these forces is *fictitious forces*.

If you are in the accelerating system, these fictitious forces seem real. We have all felt the effect of being in a noninertial system. If your car suddenly changes its velocity—speeding up, slowing down, or changing direction—you feel pushed in the direction opposite the acceleration. When the car accelerates rapidly, we often say that we are being pushed back into the seat.

> **QUESTION**   What is the direction and cause of the fictitious force you experience when you suddenly apply the brakes in your car?

> **Physics on Your Own**   Fasten a helium-filled balloon on a string to the seat in your car. Observe which way it floats when the car speeds up, moves at a constant velocity, and slows down. Describe the inertial forces in each case.

Realistic Inertial Forces

If you were in a windowless room that suddenly started accelerating relative to an inertial reference system, you would know something happened. You would feel a new force. Of course, in this windowless room, you wouldn't have any visual clues to tell you that you were accelerating; you would only know that some strange force was pushing in a certain direction.

This strange force would seem very real. If you had force measurers set up in the room, they would all agree with your sensations. This experience would be rather bizarre; things initially at rest would not stay at rest. Vases, ashtrays, chairs, and even people would need to be fastened down securely or they would move.

This situation occurs whenever we are in a noninertial reference system. Imagine riding in an elevator accelerating upward from the Earth's surface. You would experience a fictitious force opposite the acceleration in addition to the gravitational force. In this case, the inertial force would be in the same direction as gravity and you would feel "heavier." You can even measure the change by standing on a bathroom scale.

> **ANSWER**   Assuming that you are moving forward, the inertial force acts in the forward direction, "throwing" you toward the dashboard. It arises because of the car's acceleration in the backward direction due to the braking.

If the elevator stands still or moves with a constant velocity, a bathroom scale indicates your normal weight [Fig. 11-7(a)]. Because your acceleration is zero, the net force on you must also be zero. This means that the force $\mathbf{F}_s$ exerted on you by the scale must balance the gravitational force $\mathbf{F}_g$ exerted on you by the Earth and therefore $\mathbf{F}_s$ is equal and opposite to $\mathbf{F}_g$. Since the size of the gravitational force is equal to the mass m times the acceleration due to gravity g, we sometimes say that you experience a force of 1 "g."

If the elevator accelerates upward [Fig. 11-7(b)], you must experience a net upward force as viewed from the ground. Since the gravitational force does not change, the force $\mathbf{F}_s$ exerted on you by the scale must be larger than the gravitational force $\mathbf{F}_g$. This change in force would register as a heavier reading on the scale. You would also experience the effects on your body. Your stomach "sinks" and you feel heavier.

If the upward acceleration is equal to that of gravity, the net upward force on you must have a magnitude equal to $\mathbf{F}_g$. Therefore, the scale must exert a force equal to twice $\mathbf{F}_g$ and the reading shows this. You experience a force of 2 g's and feel twice as heavy. Astronauts experience maximum forces of 3 g's during launches of the Space Shuttle. During launches of the Apollo missions to the Moon, the astronauts experienced up to 6 g's. When pilots eject from jet fighters, the forces approach 20 g's for very short times.

Figure 11-7(c) shows the situation as seen from the ground when the elevator accelerates in the downward direction. In the elevator the inertial force is upward and subtracts from the gravitational force. You feel lighter; we often say that our stomachs are "up in our throats."

If the downward acceleration is equal to that of gravity, you feel *weightless*. You and the elevator are both accelerating downward at the ac-

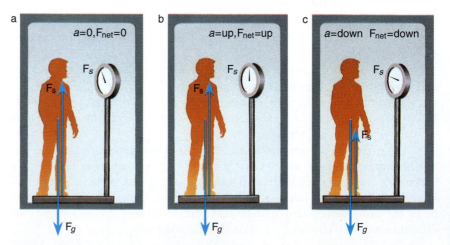

Figure 11-7   The weight of the elevator passenger $\mathbf{F}_s$ registered on the dial depends on the acceleration and the force of gravity. (a) $\mathbf{F}_s$ is equal to the force of gravity $\mathbf{F}_g$ when the elevator has no acceleration, (b) $\mathbf{F}_s$ is larger than $\mathbf{F}_g$ when the elevator accelerates upward, and (c) $\mathbf{F}_s$ is smaller than $\mathbf{F}_g$ when the elevator accelerates downward.

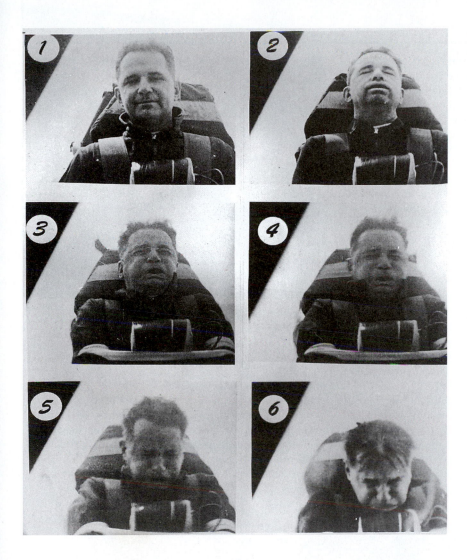

This sequence of photographs taken during the experiments before the first space flight shows the effects of inertial forces during large accelerations.

celeration due to gravity. The bathroom scale does not exert any force on you. You appear to be "floating" in the elevator, a situation sometimes referred to as "zero g."

QUESTION If you are traveling upward in the elevator and slowing down in order to stop at a floor, will the scale read heavier or lighter?

If somehow your elevator accelerates in a sideways direction, the extra force is like the one your friends felt in the truck; the inertial force is horizontal and opposite the acceleration. During the takeoff of a commercial jet airplane, passengers typically experience horizontal accelerations of ¼ g.

ANSWER Because you are accelerating in the downward direction while slowing down, the inertial force is upward and the scale will read lighter.

Living in Zero G

Astronauts in space stations orbiting Earth experience weightlessness—"zero g"—because the station continuously falls toward the Earth. They float about the station and can do gymnastic maneuvers involving a dozen somersaults and twists. They can release objects and have them stay in place suspended in the air. Of course, when they try to move a massive object, they still experience the universality of Newton's second law; being weightless does not mean being massless. Even when they leave the space station to go for a space "walk," the effect is the same: They float along with the space station.

Although they experience the sensation of being weightless, the gravitational force on them is definitely *not* zero; at an altitude of a few hundred kilometers, the gravitational force is only a few percent lower than at the Earth's surface. In the accelerating, noninertial reference system of the space station, the gravitational force and the inertial force cancel each other, producing the sensation of weightlessness. Any experiment they could devise inside the space station, however, yields the same result; gravity appears to have been turned off.

Although a pleasant experience for a while, living in "zero g" can create problems over long periods of time because our bodies have evolved in a gravitational field. Astronauts report puffiness in the face, presumably from body fluids not being held down by gravity. Scientists also report that changes occur in astronauts' hearts due to the lower stress levels—sort of the reverse

Astronaut Shannon Lucid reads a book in the Mir Space Station.

of exercise. For longer periods of living in zero g, it is expected that bone growth may be impaired.

All of these issues will need to be addressed in the next decade as the United States and Russia develop plans to send astronauts and cosmonauts to Mars. Such a trip will require several years, much longer than the record 439 days for living in zero g held by the Russians.

Astronauts working in the Spacelab science module in Atlantis' cargo bay.

Figure 11-8 The floating cork can be used as a detector of inertial forces.

Centrifugal Forces

A rotating reference system—such as a merry-go-round—is also noninertial. If you are on the merry-go-round, you feel a force directed outward. This fictitious force is opposite the centripetal force we discussed in Chapter 3 and is called the **centrifugal force.** It is only present when the system is rotating. As soon as the ride is over, the centrifugal force disappears.

centrifugal force

If the merry-go-round were to spin very fast, you would get thrown off. Consider the carnival ride that spins you in a huge cylinder [Fig. 11-9(a)]. As the cylinder spins, you feel the centrifugal force pressing you against the wall. When the cylinder reaches a large enough rotational speed, the floor drops out from under you. You don't fall, however, because the centrifugal force pushing you against the wall increases the frictional force with the wall enough to prevent you from sliding down the wall. If you try to "raise" your arms away from the wall, you feel the force pulling them back to the wall.

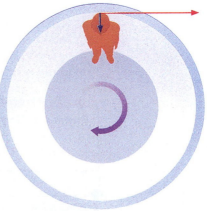

Figure 11-9 Although the people in the rotating cylinder feel forces pushing them against the wall, an inertial observer says that the wall must push on the people to make them go in circles.

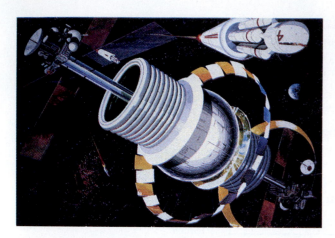

Artificial gravity can be created by rotating a space station.

Somebody looking into the cylinder from outside (an inertial system) sees the situation shown in Figure 11-9(b); the only force is the centripetal one acting inward. Your body is simply trying to go in a straight line and the wall is exerting an inward-directed force on you, causing you to go in the circular path. This real force causes the increased frictional force.

> **QUESTION**   What is the net force on someone standing on the floor of the rotating space station as viewed from their reference system?

An artificial gravity in a space station can be created by rotating the station. A person in the station would see objects "fall" to the floor and trees grow "up." If the space station had a radius of 1 kilometer, a rotation about once every minute would produce an acceleration of 1 g near the rim. Again, viewed from a nearby inertial system, the objects don't fall, they merely try to go in straight lines. Living in this space station would have interesting consequences. For example, climbing to the axis of rotation would result in "gravity" being turned off.

> **Physics on Your Own**   Grow some plants on a phonograph turntable rotating at 78 rpm. Normally plants grow up; that is, the direction opposite gravity. Which direction is "up" on the turntable?

> **ANSWER**   The net force would be zero because the person is at rest relative to the floor. The pilot of an approaching spaceship would see a net centripetal force acting on the person in the space station.

The Earth: A Nearly Inertial System   Σ

The Earth is moving. This is probably part of your commonsense world view because you have heard it so often. But what evidence do you have to support this statement? To be sure that you are really a member of the moving-Earth society, point in the direction that the Earth is moving right now. This isn't easy to do.

We do not feel our massive Earth move, and it seems more likely that it is motionless. But in fact it is moving at a very high speed. A person on the Equator travels about 1700 kilometers per hour due to the Earth's rotation. The speed due to the Earth's orbit around the Sun is even larger, 107,000 kilometers per hour (67,000 mph!).

What was it that led us to accept this idea that the Earth is moving? If we look at the Sun, Moon, and stars, we can agree that *something* is moving. The question is this: Are the heavenly bodies moving with the Earth at rest, or are the heavenly bodies at rest with the Earth moving? The Greeks believed that the motion was due to the heavenly bodies traveling around a fixed Earth located in the center of the Universe. This scheme is called the **geocentric model.**

They assumed that the stars were fixed on the surface of a huge celestial sphere with the Earth at its center (Fig. 11-10). This sphere rotated on an axis through the North and South Poles, making one complete revolution every 24 hours. You can easily verify that this model describes the motion of the stars by observing them during a few clear nights.

The Sun, Moon, and planets were assumed to orbit the Earth in circular paths at constant speeds. When this theory did not result in a model that could accurately predict the positions of these heavenly bodies, the Greek astronomers developed an elaborate scheme of bodies moving around circles that were in turn moving on other circles, and so on. Although this geocentric model was fairly complicated, it described most of the motions in the heavens.

This brief summary doesn't do justice to the ingenious astronomical picture developed by the Greeks. The detailed model of heavenly motion developed by Ptolemy in A.D. 150 resulted in a world view that was accepted for 1500 years. Ptolemy's theory was so widely accepted because it was a relatively simple, commonsense picture that predicted the positions of the Sun, Moon, planets, and stars quite accurately. It was also very comforting for philosophic and religious reasons to know that the Earth (and consequently earthlings!) was at the center of the Universe.

In the 16th century a Polish scientist named Copernicus began a radical revolution in our celestial world view. He abandoned the idea of a fixed Earth; he claimed to be able to explain the heavenly motions by having the Earth rotate about an axis once every 24 hours while revolving about the Sun once a year. Only the Moon remained as a satellite of Earth; the planets were assumed to orbit the Sun. As his proposal put the Sun in the center of the Universe, it is called the **heliocentric model** (Fig. 11-11).

How does one choose between two competing views? One criterion—simplicity—doesn't help here. Although the basic model of Copernicus

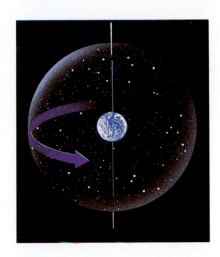

Figure 11-10   The geocentric view of the Universe has the Earth at its center.

Figure 11-11 In the heliocentric model of the Solar System, the planets orbit the Sun.

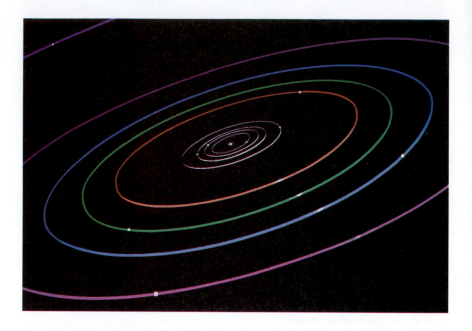

was simpler to visualize than that of Ptolemy, it required about the same mathematical complexity to achieve the same degree of accuracy in predicting the positions of the heavenly bodies.

A second criterion is whether one model can explain more than the other. Here Copernicus was the clear winner. His model predicted the order and relative distances of the planets, explained why Mercury and Venus were always observed near the Sun, and included some of the details of planetary motion in a more natural way. It would seem that the Copernican model should have quickly replaced the older Ptolemaic model.

But the Copernican model appeared to fail in one crucial prediction. Copernicus' model meant that the Earth would orbit the Sun in a huge circle. Therefore, observers on Earth would view the stars from vastly different positions during the Earth's annual journey around the Sun. These different positions would provide different perspectives of the stars and thus they should be observed to shift their positions relative to each other on an annual basis. This shift in position is called *parallax*.

You can demonstrate parallax to yourself with the simple experiment shown in Figure 11-12. Hold a finger in front of your face and look at a distant scene with your left eye only. Now look at the same scene with only your right eye. Since your eyes are not in the same spot, the two views are not the same. You see a shifting of your finger relative to the distant

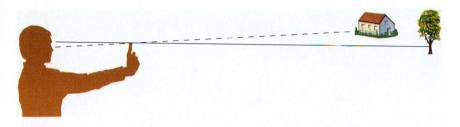

Figure 11-12   The position of the finger changes relative to the background when viewed by the other eye.

scene. Notice also that this effect is more noticeable when your finger is close to your face. As you move your finger farther and farther away, the effect becomes smaller and smaller.

Unfortunately for Copernicus, the stars were not observed to shift as predicted; they did not exhibit parallax. Undaunted by the null results, Copernicus countered that the stars were too far away and the instruments too crude to measure this effect. Although his counterclaim was a possible explanation, the lack of observable parallax was a strong argument against his model and delayed its acceptance. It is interesting to note that the biggest stellar parallax is so small that it was not observed until 1838—300 years later.

There was another problem with Copernicus' model. Copernicus developed these ideas before Galileo's time and did not have the benefit of Galileo's work on inertia or inertial reference systems. Since it was not known that all inertial reference systems are equivalent, most people ridiculed the idea that the Earth could be moving: After all, one would argue, if a bird were to leave its perch to catch a worm on the ground, the Earth would leave the bird far behind!

For these reasons, the ideas of Copernicus were not accepted for a long time. In fact, 70 years later, Galileo was being censured for his heretical stance that the Earth does indeed move. Galileo's arguments centered on his idea that a person in the hold of a ship moving at a constant velocity would not be able to detect the ship's motion. Most people of the time were still convinced that a moving Earth would play havoc on such simple everyday events as falling balls or birds catching worms.

One of the reasons that it took thousands of years to accept the motion of the Earth is that the Earth is very nearly an inertial reference system. Were the Earth's motion undergoing large accelerations, the effects would have been indisputable. Even though the inertial forces are very small, they do provide evidence for the Earth's motion.

Copernicus' critics argued that if the Earth were moving, birds would be left behind.

Noninertial Effects of the Earth's Motion   Σ

A convincing demonstration of the Earth's rotation was given by the French physicist J. B. L. Foucault around the middle of the 19th century. He showed that the plane of swing of a pendulum appears to rotate. Foucault's demonstration is very popular in science museums; almost everyone has a large pendulum with a sign saying that it shows the rotation of the Earth. But how does this show that the Earth is rotating?

First, we must ask what would be observed in an inertial system. In the inertial system, the only forces on the swinging bob are the tension in the string and the pull of gravity; both of these act in the plane of swing.

A Foucault pendulum shows that the Earth rotates.

So in an inertial system, there is no reason for the plane to change its orientation.

The noninertial explanation is simplest with a Foucault pendulum on the North Pole as shown in Figure 11-13. The plane of the pendulum rotates once every 24 hours. That is, if you start it swinging along a line on the ground, some time later the pendulum will swing along a line at a slight angle to the original line. In 12 hours it will be along the original line again (the pendulum's plane is half-way through its rotation) and finally, after 24 hours, the pendulum will once again be realigned with the original line. If you lie on your back under the pendulum and observe its motion with respect to the distant stars, you would see that the plane of the pendulum remains fixed relative to them. It is the Earth that is rotating.

The weight of objects on the Earth is affected by the Earth's rotation. A person on the Equator is traveling along a circular path, but a person on the North Pole is not. The person on the North Pole feels the force of gravity; the person on the Equator feels the force of gravity plus the fictitious centrifugal force. The effect of this centrifugal force is small; it is only one-third of 1 percent of the gravitational force. That means if we transported a 1-newton object from the North Pole to the Equator, its weight would be 0.997 newton.

You can experience the effects of the Coriolis force by playing catch while riding on a merry-go-round.

Physics Update

Changes in the Earth's rotation rate occur at the level of several parts in 10^8. Using laser ranging (bouncing radar waves off the Moon or satellites) and very long baseline interferometry, length-of-day measurements can detect 0.03-millisecond changes. Jean Dickey, a Jet Propulsion Laboratory geophysicist, cities three main types of change: a linear increase owing to tidal dissipation; larger, irregular variations, on the scale of decades, owing to core–mantle interactions; and shorter-term (seasonal) changes from the angular momentum exchange between crust and atmosphere.

Top views

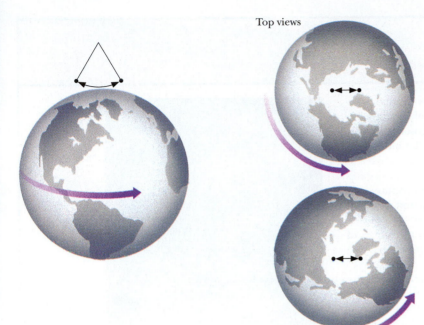

Figure 11-13 A Foucault pendulum at the North Pole appears to rotate relative to the ground once in 24 hours.

Another inertial force in a rotating system, known as the **Coriolis force,** is the force you feel when you move along a radius of the rotating system. If, for example, you were to walk from the center of a merry-go-round to its edge, you would feel a force pushing you in the direction opposite the rotation.

From the ground system, the explanation is straightforward. A point on the outer edge of a rotating merry-go-round has a larger speed than a point closer to the center because it must travel a larger distance during each rotation. As you walk toward the outer edge, the floor of the merry-go-round moves faster and faster. Your inertial tendency is to keep the same speed relative to the ground system. The merry-go-round moves out from under you, giving you the sensation of being pulled in the opposite direction. If you move inward toward the center of the merry-go-round, the direction of this inertial force is reversed. The Coriolis force is more complicated than the centrifugal force in that it depends on the velocity of the object in the noninertial system as well as the acceleration of the system.

QUESTION If you drop a ball from a great height, it will experience a Coriolis force. Will the ball be deflected to the east or the west?

ANSWER This situation is analogous to walking toward the center of the merry-go-round. Therefore, the ball will be deflected in the direction of the Earth's rotation; that is, to the east.

Planetary Cyclones

The atmospheres of the gaseous planets—Jupiter, Saturn, Uranus, and Neptune—are very unlike the Earth's atmosphere. The atmospheres are composed primarily of hydrogen molecules with a much smaller amount of helium. All of the other gases comprise less than 1% of the atmospheres. Yet the colors provided by these gases (for instance, the clouds on Jupiter and Saturn are composed of crystals of frozen ammonia and those on Uranus and Neptune are composed of frozen methane) give us some visual clues about the effects of the Coriolis force on the large-scale motions in these planetary atmospheres.

The most famous cyclone in the Solar System is the Great Red Spot on Jupiter, which was first observed more than 300 years ago. It is a giant, reddish oval that is about 26,000 kilometers across the long dimensions—about the size of two Earths side by side. Since the Great Red Spot is located in Jupiter's southern hemisphere, it might be expected to rotate in the clockwise direction. However, it is observed to rotate *counterclockwise* with a period of 6 days. Therefore, the Coriolis effect tells us that the Great Red Spot must be a high-pressure storm rather than the low-pressure regions typical of hurricanes and cyclones on Earth. Jupiter also has three "white ovals" that were first observed in 1938 and have diameters about 10,000 kilometers across.

When *Voyager 2* flew by Neptune in the fall of 1989, planetary scientists were surprised and pleased to observe a Great Dark Spot. It is located in Neptune's southern hemisphere, is about 10,000 kilometers across, and rotates counterclockwise with a period of 17 days. *Voyager* also observed a few small storms on the

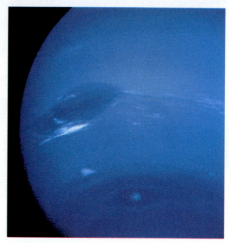

Voyager 2 discovered this Great Dark Spot on Neptune during its flyby in the fall of 1989.

order of 5000 kilometers across in Saturn's atmosphere but none in Uranus' atmosphere.

Although no one knows the origins of these planetary storms, scientists are able to explain their long lives. Hurricanes on Earth die out rather quickly when they travel across land areas. Although each of these planets has a "rocky" core, the cores are relatively small compared to the planet's size. The resultant thickness of the atmospheres contributes to the long lifetimes of the storms. Another factor is size. Larger storms are more stable and last longer.

The Great Red Spot on Jupiter is a high-pressure cyclonic storm that has lasted for at least 300 years.

Air moving poleward from the Equator is traveling east faster than the land beneath it and veers to the east (turns right in the Northern Hemisphere and left in the Southern Hemisphere)

Air moving toward the Equator is traveling east slower than the land beneath it and veers to the west (turns right in the Northern Hemisphere and left in the Southern Hemisphere)

Coriolis effect. Air moving north or south is deflected by the rotation of the Earth.

   The Coriolis force acts on anything moving along the Earth's surface and deflects it toward the right in the Northern Hemisphere and toward the left in the Southern Hemisphere. This reversal was experienced by British sailors during World War I. During a naval battle near the Falkland Islands (50° south latitude) they noticed that their shells were landing about 100 meters to the left of the German ships. The Coriolis corrections that were built into their sights were correct for the Northern Hemisphere, but were in the wrong direction for the Southern Hemisphere!

   The Coriolis force also causes large, flowing air masses in the Northern Hemisphere to be deflected to the right. As the air flows in toward a low-pressure region, it is deflected to the right and therefore circulates counterclockwise as viewed from above. The result is that hurricanes in the Northern Hemisphere rotate counterclockwise. The circulation pattern is reversed for hurricanes in the Southern Hemisphere and for high-pressure regions in the Northern Hemisphere. Figure 11-14 shows a hurricane in the Northern Hemisphere as seen from one of NASA's satellites. Folklore has it that the Coriolis force causes toilets and bathtubs to drain counterclockwise in the Northern Hemisphere, but its effects on this scale are so small that other effects dominate.

   Even if the Earth was not rotating, it would still not be an inertial reference system. Although the Earth's orbital velocity is very large, the change in its velocity is small. The acceleration due to its orbit around the Sun is about one-sixth that of its daily rotation on its axis. In addition, the Solar System orbits the center of the Milky Way Galaxy once every 250 million years with an average speed of 1 million kilometers per hour. The associated inertial forces are about 100 million times smaller than those due to rotation. The Galaxy has an acceleration within the local group of galaxies, and so on.

Figure 11-14 Hurricanes in the Northern Hemisphere turn counter-clockwise as seen from above. This image of Hurricane Fran was taken from GEOS-8 less than 7 hours before the eye went ashore at Cape Fear, North Carolina.

In terms of our daily lives, the Earth is very nearly an inertial reference system. Any system that is moving at a constant velocity relative to its surface is, for most practical purposes, an inertial reference system.

Physics on Your Own   You can experience the Coriolis force by playing catch with a friend while riding on a merry-go-round. Notice how the ball appears to curve in the air when you expect it to go straight.

About 100,000 light years

Our Sun's position

The Earth is located in one of the spiral arms of the Milky Way Galaxy.

SUMMARY

All motion is viewed relative to some reference system, the most common being the Earth. An inertial reference system is one in which the law of inertia is valid. Any reference system that has a constant velocity relative to an inertial reference system is also an inertial reference system.

The principle that the laws of motion are the same for any two inertial reference systems is called the Galilean principle of relativity. Observers moving relative to each other report different descriptions for the motion of an object, but the objects obey the same laws of motion regardless of reference system. For example, measurements in one inertial reference system yield the same forces as measurements in any other inertial reference system.

Observers in different reference systems can reconcile the different velocities they obtain for an object by adding the relative velocity of the reference systems to that of the object. However, this procedure breaks down for velocities near that of light.

In a reference system accelerating relative to an inertial reference system, the law of inertia does not work without the introduction of fictitious forces that are due entirely to the accelerated motion. Centrifugal and Coriolis forces arise in rotating reference systems and are examples of inertial forces. The Earth is a noninertial reference system, but its accelerations are so small that we often consider the Earth to be an inertial reference system.

CHAPTER 11 REVISITED

The most direct evidence is provided by the Foucault pendulum. The plane in which the pendulum swings stays fixed relative to the distant stars and rotates relative to the ground, demonstrating that the Earth is rotating. The effect of the Coriolis force on storm systems is also evident. This force only occurs in a rotating system and causes hurricanes to rotate in opposite directions on either side of the Equator.

KEY TERMS

centrifugal force: A fictitious force arising when a reference system rotates (or changes direction). It points away from the center, in the direction opposite to the centripetal acceleration.

Coriolis force: A fictitious force that occurs in rotating reference systems. It is responsible for the direction of winds in hurricanes.

Galilean principle of relativity: The laws of motion are the same in all inertial reference systems.

geocentric model: A model of the Universe with the Earth at its center.

heliocentric model: A model of the Universe with the Sun at its center.

inertial force: A fictitious force that arises in accelerating (noninertial) reference systems. Examples are the centrifugal and Coriolis forces.

inertial reference system: Any reference system in which the law of inertia (Newton's first law of motion) is valid.

noninertial reference system: Any reference system in which the law of inertia (Newton's first law of motion) is not valid. An accelerating reference system is noninertial.

reference system: A collection of objects not moving relative to each other that can be used to describe the motion of other objects. See inertial and noninertial reference systems.

CONCEPTUAL QUESTIONS

1. *Alice in Wonderland* begins with Alice falling down a deep, deep rabbit hole. As she falls, she notices that the hole is lined with shelves and grabs a jar of orange marmalade. Upon discovering that the jar is empty, she tries to set it back on a shelf—a difficult task since she is falling. She is afraid to drop the jar for it might hit somebody on the head. What would really happen to the jar if Alice had dropped it? Describe its motion from Alice's reference system and from the reference system of someone sitting on a shelf.

*2. Imagine riding in a glass-walled elevator that goes up the outside of a tall building at a constant speed of 20 meters per second. As you pass a window washer, he throws a ball upward at a speed of 20 meters per second. Assume, furthermore, that you drop a ball out a window at the same instant.

   a. Describe the motion of each ball from the point of view of the window washer.

   b. Describe the motion of each ball as you perceive it from the reference system of the elevator.

3. How is an inertial reference system defined?

4. What condition must be imposed on the velocities of two reference systems for them both to be inertial reference systems?

5. An object is thrown horizontally near the surface of the Earth. As viewed from a reference system that is stationary relative to the Earth, the path of the object is that of a projectile. What is the shape of the path as viewed from a reference system moving at the same horizontal velocity as the object? Explain your reasoning.

6. Assume that you are riding in a windowless train on perfectly smooth tracks. (You can't feel any motion.) Imagine that you have a collection of objects and measuring devices in the train. What experiment could you do to prove that the train is moving horizontally at a constant velocity?

7. Where will the ball in the figure land according to the observer in the train? Will an observer on the ground agree?

8. What would an observer on the ground (standing next to the tracks) say about the horizontal velocity of the ball in the figure while it is falling?

9. What would the observer in the train say about the horizontal forces acting on the ball in the figure?

10. What would an observer on the ground say about the horizontal forces acting on the ball in the figure?

11. What value would an observer on the ground obtain for the acceleration of the ball in the figure?

12. What value would an observer on the train obtain for the acceleration of the ball in the figure?

13. What would our friend in the truck say about the validity of the laws of conservation of momentum and energy?

14. When you throw a ball vertically up in a car moving with a constant velocity, it lands back in your hand. Why?

15. Assume that you are driving down a straight road at a constant speed. A small ball is tied to a string hanging from the rear-view mirror. Which way will the ball swing when you apply the brakes? Explain your reasoning.

*16. Assume that you are driving down a straight road at a constant speed. A helium-filled balloon is tied to a string that is pinned to the front seat. Which way will the balloon swing when you apply the brakes? Explain your reasoning.

17. Where will the ball in the figure land as seen by the observer on the train? Does an observer on the ground agree?

18. An observer in the train notices that the ball in the figure falls in a straight line that is slanted toward the

Questions 7–12. A train is traveling along a straight, horizontal track at a constant velocity of 50 km/h. An observer in the train holds a ball directly over a white spot on the floor of the train and drops it.

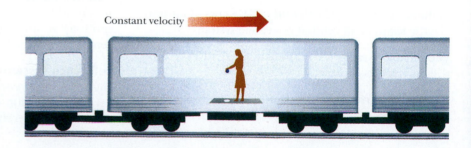

Constant velocity

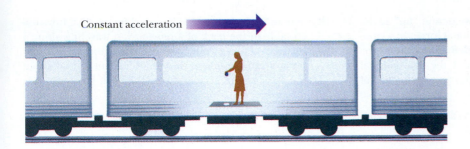

Constant acceleration

Questions 17–22. A train is traveling along a straight, horizontal track with a constant acceleration in the forward direction. An observer in the train holds a ball directly over a white spot on the floor of the train. At the instant the speed is 50 km/h, she drops the ball.

back of the train. Is the acceleration of the ball along this line greater than, equal to, or less than the usual acceleration due to gravity?

19. What would an observer on the ground obtain for the horizontal speed of the ball in the figure while it is falling?

20. What would an observer in the train obtain for the horizontal speed of the ball in the figure while it is falling?

21. What would the observer in the train say about the horizontal forces acting on the ball in the figure?

22. What would an observer on the ground say about the horizontal forces acting on the ball in the figure?

23. A person with a weight of 180 pounds takes a ride in the elevator that goes up the side of the Space Needle in Seattle. Much to the amusement of the other passengers, this person stands on a bathroom scale during the ride. During the time the elevator is accelerating upward, is the reading on the scales greater than, equal to, or less than 180 pounds?

24. During the time the elevator in the preceding question is moving upward with a constant speed, is the reading on the scales greater than, equal to, or less than 180 pounds?

25. Assume you are standing on a bathroom scale while an elevator slows down to stop at the top floor. Will the reading on the scale be greater than, equal to, or less than the reading after the elevator stops?

26. What happens to the reading in the preceding question when the elevator starts down?

27. If a child weighs 200 newtons standing at rest on Earth, would she weigh more, less, or the same if she were in a spaceship accelerating at 10 (meters per second) per second in a region of space far from any celestial objects?

28. A room is being accelerated through space at 10 (meters per second) per second relative to the "fixed stars." It is far away from any massive objects. If a woman weighs 700 newtons when she is at rest on Earth, will she weigh more, less, or the same in the room?

29. If you were allowed to leave your tray down while the DC-9 accelerates down the runway, why would objects slide off the tray?

30. What happens to the surface of a drink if you hold the drink while the Boeing 777 accelerates down the runway?

*31. Assume that you weigh a book on an equal-arm balance while an elevator is stopped at the ground floor. Would you get the same result if the elevator were accelerating upward? Explain your reasoning.

*32. Assume that a meterstick balance is balanced with a 20-gram mass at 40 centimeters from the center and a 40-gram mass at 20 centimeters from the center. Will it remain balanced if it is in an elevator accelerating downward?

33. On the way to the movies in your car, you are disappointed that your date is sitting next to the door. Up ahead you notice a sharp corner to the right and decide to take advantage of the COD (Come Over, Dear) corner. The seat is slick and your date slides over beside you. Describe the motion of your date from the reference system in the car and one fastened to the roadway.

34. Why would the pilot of an acrobatic airplane have a tendency to black out when pulling out of a vertical descent?

35. In an inertial reference system, we define "up" as the direction opposite to the gravitational force. In a non-inertial reference system, up is defined as the direction opposite to the vector sum of the gravitational force and the inertial force. Which way is up in each of the following cases?

    a. An elevator accelerates downward with an acceleration *smaller* than that of free fall.

    b. An elevator accelerates upward with an acceleration *larger* than that of free fall.

    c. An elevator accelerates downward with an acceleration *larger* than that of free fall.

36. Using the definition of "up" in the previous question, which way is up for each of the following situations?

    a. A child rides near the outer edge of a merry-go-round.

    b. A dining car going around a curve accelerates to the right with an acceleration equal to that of gravity.

    c. A skier skies down a hill with virtually no friction.

37. Which way is "up" for astronauts orbiting Earth in the Space Shuttle?

38. What happened to the astronauts' sense of up and down as the Apollo spacecraft passed the point in space where the gravitational attraction of the Earth and Moon on the spacecraft are equal?

39. A ball is thrown vertically upward from the center of a moving railroad flatcar. Where, relative to the center of the car, does the ball land in each of the following cases?

    a. The flatcar moves at a constant velocity.

    b. The velocity of the flatcar increases.

    c. The velocity of the flatcar decreases.

    d. The flatcar travels in a circle at constant speed.

40. You and a friend are rolling marbles across a horizontal table in the back of a moving van traveling along a straight section of interstate highway. You roll the marbles toward the side of the van. What can you say about the velocity and acceleration of the van if you observe the marbles

    a. head straight for the wall?

    b. curve toward the front of the truck?

41. Why don't drinks spill when a jetliner changes directions?

42. Would it be possible to take a drink at the top of a loop-the-loop on a roller coaster? Explain.

43. For a science project a student raises some fast-growing plants on a turntable rotating at 78 rpm. Draw a side view of the experiment showing how the plants will look.

44. A student hangs two pendulums from the outer edge of a turntable rotating at 78 rpm. Draw a side view of the apparatus.

45. Why does the mud fly off the tires of your pickup as you drive down the interstate?

46. The Red Cross uses centrifuges to separate the various components of blood. Which component goes to the bottom of the test tube?

47. Why is the centrifugal force called a fictitious force, but the centripetal force is considered to be real?

48. Give an example which clearly distinguishes between *centripetal* and *centrifugal* forces.

49. A large cylindrical spaceship is drifting through space and rotating about the axis of the cylinder. Which way is "up" for the occupants of the spaceship?

50. A space station, far from any large masses, can be spun so that people inside the station feel the effects of an "artificial gravity." Why does this work?

51. A large cylindrical spaceship is at rest relative to the "fixed" stars. It is rotating counterclockwise about the axis of the cylinder. An astronaut standing on the outside wall of the cylinder releases a ball. Which way will the ball travel as viewed by someone at rest relative to the "fixed" stars?

52. Which way will the ball in the preceding question travel as viewed by the astronaut?

53. How do the geocentric and heliocentric models account for our observations of the daily motion of the stars?

54. The Moon orbits the Earth once a month, and the Earth orbits the Sun once a year. Draw the yearly path of the Moon as you might see it from a point high above the Sun.

55. What evidence do you have to support the belief that the Earth rotates on its axis?

56. What evidence do you have to support the belief that the Earth orbits the Sun?

57. Would a Foucault pendulum rotate at the Equator?

58. Why are there no hurricanes on the Equator?

59. Assuming that the Earth is a perfect sphere and that the gravitational force is constant over the surface, would your weight at the Equator be greater, smaller, or the same as that at the North Pole?

*60. It is known that the Earth is bigger around the Equator than around the poles. How does this equatorial bulge support the idea that the Earth is rotating?

EXERCISES

1. A spring gun fires a ball horizontally at 15 m/s. It is mounted on a flatcar moving in a straight line at 30 m/s. Relative to the ground, what is the horizontal speed of the ball when the gun is aimed

   a. forward?

   b. backward?

2. A moving van is traveling along a straight stretch of interstate at 120 km/h. A water pistol can fire a stream of water at a speed of 50 km/h. What is the speed of the water relative to the ground if the pistol is aimed backward? What if the pistol is aimed forward?

3. A child can throw a ball at a speed of 50 mph. If the child is riding in a bus traveling at 20 mph, what is the speed of the ball relative to the ground if the ball is thrown

   a. forward?

   b. backward?

4. A train is traveling along a straight, horizontal track at a constant speed of 60 km/h. If a ball is fired forward with a speed of 100 km/h relative to the train, what is its speed relative to the ground? What if the ball is fired backward?

5. An observer measures the acceleration due to gravity in an elevator near the surface of the Earth. What would the value and direction be if the elevator

   a. accelerates upward at 6 m/s^2?

   b. travels upward with a constant speed of 6 m/s?

6. What would an observer measure for the value and direction of the acceleration due to gravity in an elevator near the surface of the Earth if the elevator

   a. accelerates downward at 4 m/s^2?

   b. accelerates downward at 14 m/s^2?

7. What is the maximum total force exerted on a 60-kg astronaut by her seat during the launch of a Space Shuttle?

8. What was the maximum total force exerted on an 80-kg astronaut by his seat during the launch of one of the Apollo missions to the Moon?

9. An elevator is being accelerated downward with an acceleration equal to one-half that of gravity. If a person who weighs 900 N when at rest on the Earth steps on a bathroom scale in this elevator, what will the reading be?

10. A child weighs 400 N standing on Earth. What is the weight of the child in an elevator accelerating upward at 0.25 g?

11. A monkey with a mass of 15 kg rides on a bathroom scale in an elevator that is accelerating upward at ½ g. What does the scale read?

12. What does the scale read if a 6-kg cat lies on a bathroom scale in an elevator accelerating downward at 0.2 g?

13. A room is being accelerated through space at 3 m/s^2 relative to the "fixed stars." It is far away from any massive objects. If a man weighs 900 N when he is at rest on Earth, how much will he weigh in the room?

14. A woman with a weight of 600 N on Earth is in a spacecraft accelerating through space a long way from any massive objects. If the acceleration is 4 m/s^2, what is her weight in the ship?

*15. A cylindrical space station with a radius of 40 m is rotating so that points on the walls have speeds of 20 m/s. What is the acceleration due to this artificial gravity at the walls?

*16. What is the centrifugal acceleration on the equator of Mars given that it has a radius of 3400 km and a rotational period of 24.6 h? How does this compare to the acceleration due to gravity on Mars of 3.7 m/s^2?

Special Relativity

At various times in our lives, we have all had impressions of the passage of time. An hour in a dentist's chair seems much longer than two hours watching a good movie. But what is time? If we develop a foolproof way of measuring time, will all observers in the Universe accept our measurements? (See p. 299 for the answer to this question.)

Time and space are central in the theory of special relativity, and they take on new roles.

When observers in different inertial reference systems describe the same events, their reports don't match. In the framework of classical relativity they disagree in their descriptions of the paths and on the values of an object's velocity, momentum, and kinetic energy. On the other hand, they agree on relative positions, lengths, time intervals, accelerations, masses, and forces. Even the laws of motion and the conservation laws are the same.

We never asked, or even thought to ask, whether some of these were actually the same for all reference systems or whether we had just assumed them to be the same. In classical relativity we assumed that the concepts of length, time, and mass were the same. But are they really the same?

Albert Einstein asked this question. He reexamined the process of describing events from different reference systems with an emphasis on the concepts of space and time. This led to the development of the **special theory of relativity.**

Einstein arrived at the special theory of relativity by setting forth two postulates, or conditions, that are assumed to be true. He then examined the effects of these postulates on our basic concepts of space and time. The predictions of special relativity were then compared with actual experimental measurements. The theory had to agree with nature to have any validity.

Albert Einstein (1879–1955).

The First Postulate

The **first postulate** is related to the question of whether there exists an absolute space—some signpost in the Universe from which all motion can be regarded as absolute. This postulate says that there is no absolute space; any inertial reference system is just as good as any other. Einstein's first postulate is a reaffirmation of the Galilean principle of relativity.

The laws of physics are the same in all inertial reference systems.

first postulate of special relativity

As we discussed in the previous chapter, Galileo argued that a traveler in the hold of a ship moving with a constant velocity could not conduct experiments that would determine whether the ship was moving or at rest. However, the Galilean principle of relativity came into question near the end of the 19th century. A theory by the Scottish physicist James Clerk Maxwell describing the behavior of electromagnetic waves, such as light and radio, yielded unexpected results.

In Newton's laws, reference systems moving at constant velocities are equivalent to each other. If, however, one system accelerates relative to another, the systems are not equivalent. Because Newton's laws depend on acceleration and not on the velocity, acceleration of a reference system can be detected, but its velocity cannot.

In Maxwell's theory, however, the *velocity* of the electromagnetic waves appears in the equations rather than their acceleration. According to the classical ideas, the appearance of a velocity indicated that inertial systems were not equivalent. In principle, you could merely turn on a flashlight and measure the speed of light to determine how your reference system was moving.

Maxwell's equations and the Galilean principle of relativity were apparently in conflict. It seemed that the physics world view could not accommodate both. During his studies, Einstein had developed a firm belief that the principle of relativity must be a fundamental part of any physical theory. At the same time, he wasn't ready to abandon Maxwell's new ideas about light. He felt that the conflict could be resolved and that both the principle of relativity and Maxwell's equations could be retained.

Meanwhile, others were pursuing different options. If there were an absolute reference system in the Universe, it should be possible to find it. The key seemed to lie in the behavior of light.

Searching for the Medium of Light   Σ

Although it was well established in the late 1800s that light was a wave phenomenon, nobody knew what substance was waving. Sound waves move by vibrating the air, ocean waves vibrate water, and waves on a rope vibrate the rope. What did light vibrate? It was assumed that there must be some medium through which light traveled. This medium was called the **ether.**

But if space were filled with such a medium, it should be detectable. From their knowledge of the behavior of other waves, scientists were convinced that the ether had to be fairly rigid. Therefore, as the Earth passed through this ether in its annual journey around the Sun, it should be slowed down by friction. However, no such slowing down was detected. How could the ether be rigid and yet so intangible that the Earth could pass through it without slowing down?

> **QUESTION**   If there were such a thing as an ether wind, wouldn't people feel it?

Two American physicists, A. A. Michelson and E. W. Morley, tried to detect the ether with an experiment that raced two light beams in perpendicular directions as shown in Figure 12-1. They reasoned that the Earth's annual motion around the Sun should create an ether "wind" on Earth much like a moving car creates a wind for the passengers. This ether wind would affect the speed of light differently along the two paths and the race would not end in a tie. They calculated that their experiment was sensitive enough to measure a speed relative to the ether as small as one-hundredth of the Earth's orbital speed. Although the experiment was conducted at many times of the year and in many different orientations, the results were always the same—every race ended in a tie!

Not finding the ether wind with such a straightforward experiment was shocking. Physicists were receiving conflicting information. First,

> **ANSWER**   The ether's effects were predicted to be very small because the Earth did not appear to experience any frictional drag.

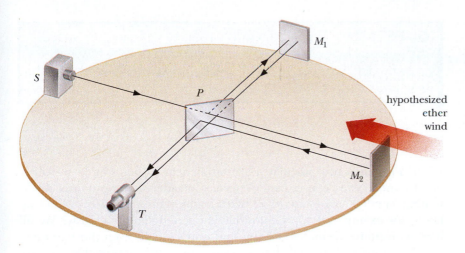

Figure 12-1   A race between two light beams in perpendicular directions was supposed to detect the hypothesized ether. In this experiment, a light beam from the source *S* is split by the partially silvered mirror *P* and travels two different paths to the telescope *T*.

there *must* be an ether wind and, second, there *must not* be an ether wind. The problem was in the first message; light does not require a medium. It can travel through a vacuum. It is a wave that doesn't wave anything.

The Second Postulate

It is difficult (if not impossible) to re-create a creative process. Although Einstein mentioned the failure to find the ether in his 1905 paper, years later he indicated that his primary motivation in formulating the **second postulate** was his deep belief in the principle of relativity. He could reconcile the apparent contradiction between the principle of relativity and Maxwell's equations with his second postulate because it eliminated the possibility of using the speed of light to distinguish between inertial reference systems.

> The speed of light in a vacuum is a constant value regardless of the speed of the source or the speed of the observer.

second postulate of special relativity

At first glance, the second postulate might seem like a rather innocent statement. But consider the situation of our friends in the truck from the previous chapter. We agreed that the velocity of an object measured relative to the ground was different from the velocity measured relative to the truck—the difference depended on how fast the truck was moving relative to the ground. Einstein's second postulate says that this doesn't happen with light.

If our friends move toward us and turn on a flashlight, we might expect that we would measure the speed of light to be greater than that from a flashlight on the ground. We find, however, that we get the same speed. It doesn't matter that the flashlight is moving relative to us. Even if we move very rapidly toward the flashlight, the results would be the same. Regardless of any relative motion, any measurement of the speed of light yields the same value: 300,000 kilometers per second (186,000 miles per second).

QUESTION   If we were communicating with an alien space-ship approaching Earth at 20% of the speed of light, at what speed would we receive their signals and at what speed would they receive ours?

Simultaneous Events

When Einstein's two postulates are applied to rather simple measurements, unexpected consequences occur. Consider the question of determining whether two events took place at the same time. How would we know, for example, if two explosions happened simultaneously? We all have an intuitive feeling about this and don't usually even think to question it. Einstein cautioned that we must not simply accept this intuitive feeling. We should look very carefully at how we determine the validity of such statements.

To determine the simultaneity of two events, we must receive some type of signal indicating that each event occurred. To be specific, let's determine if two paint cans exploded at the same time. If the two cans are in the same place, as in Figure 12-2, we can agree that they exploded simultaneously if the light from the two explosions arrived together. The signals traveled the same distance and their simultaneous arrival means that the explosions occurred at the same time. The simultaneity of events at a single location does not present a problem.

If the paint cans are not in the same place, we have to be more careful. The signals could arrive together even though the explosions occurred at different times. The results depend on the distances to each explosion. Clearly, the easiest case occurs when the observer is the same distance from each event; then the simultaneous arrival of the signals indicates that the events occurred simultaneously.

Einstein had no quarrel with this method of determining simultaneity. His concern was whether *all* observers would agree on the simultane-

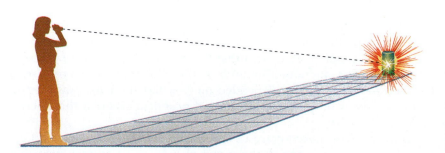

Figure 12-2 The simultaneity of events at a single location presents no problem.

ANSWER   Because radio and light behave the same, it would not matter which type of signal we used. In either case, the second postulate tells us that both observers would receive the signals at the speed of light, not at 120% of this speed, as would be predicted intuitively.

ity; he concluded that they wouldn't. He claimed that two observers moving relative to each other at a constant velocity cannot always agree on whether two events happen at the same time. This statement probably seems incredible. You might say, "How can two people see the same physical events and disagree on their simultaneity? They really did happen at the same time . . . didn't they?"

Einstein would say that there is no such thing as universal agreement about simultaneity. To understand this, let's return to your friends in the truck. Assume that the paint cans are located equal distances to the right and left of one of your friends and that the truck is moving with constant velocity to the right relative to you on the ground. Assume that at the moment when you were also equal distances from the paint cans, the cans exploded as shown in Figure 12-3(a). You know that they exploded simultaneously because the signals arrived at your eyes simultaneously and you can verify that you are equal distances from the paint marks on the ground.

How would this apply to one of your friends in the truck? During the time it took the signals to reach him, he approached the right-hand signal and receded from the left-hand one, as shown in Figure 12-3(b). The signal from the right-hand explosion, therefore, reached his eyes before the left-hand one. Both of you agree that the arrival of the signals at his eyes was not simultaneous.

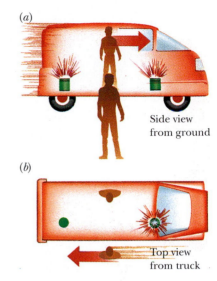

Figure 12-3   (a) The paint cans explode when they are equal distances from each observer. While the ground observer claims the events were simultaneous, the observer in the truck (b) claims the can on the right exploded first.

> **QUESTION**   Why do you both agree that the signals arrived at your friend's eyes at different times?

Your friend concludes that the explosions were not simultaneous. He reports, "I'm standing here in the middle of the truck. I can tell by the paint marks on the floor of the truck that the explosions happened equal distances from me, but the signals did not reach my eyes simultaneously. Clearly, the one that reached me first came from the explosion that happened first."

"Well," you might counter, "I understand why you think that. You were moving and that's why you reached a different conclusion."

"I'm not moving!" retorts your friend.

According to the first postulate, his motion is no more certain than yours. From his point of view, he is standing still and you are moving. There is nothing either of you can do to determine who is *really* moving. From his point of view, *you* falsely concluded that the events were simultaneous because *you* moved to the left and thus shortened the distance that the left-hand signal traveled to your eyes.

How do we get ourselves out of this predicament? Einstein concluded that we don't. You and your friends are both correct. You each believe that you have the correct answer and that the other is moving and therefore has been fooled. There is no way to resolve the conflict other than to admit that simultaneity is relative.

simultaneity is relative

> **ANSWER**   Different observers agree on the simultaneity (or nonsimultaneity) of events at a single location, in this case, at your friend's eyes.

Synchronizing Clocks

Since our concepts of motion are very fundamental to the physics world view, disagreements in simultaneity could result in a radical revision. For example, when we discussed the motion of a ball and talked about the ball being 20 meters above the ground at a time of 3 seconds, we were claiming that the ball was at this position at the same instant that the hands on the clock indicated 3 seconds. That is, the two events occurred simultaneously.

Remember that disagreements only occur when the events are at different locations. Rather than trying to determine distant events with a single clock located near you, you could set up a series of clocks distributed throughout space. Then each event could be recorded on a clock at that location, and there would be no problem with the simultaneity of the event and the clock reading.

However, for this to work, all of the clocks must be synchronized. But how do we know that they are synchronized? Even if they are synchronized in one reference system, will they be synchronized in all inertial reference systems?

To answer these questions we need to examine the process of synchronizing clocks in different places. It might be tempting to suggest that we follow the procedure used for years in war movies. The soldiers rendezvous to synchronize their wristwatches and then disperse. Clearly, this method worked quite satisfactorily for them, but we have no guarantee that the watches remain synchronized. We don't know, for example, if the motion of a clock affects its timekeeping ability.

One way of synchronizing separated clocks is illustrated in the sequence of strobe drawings in Figure 12-4. A flashbulb is mounted on top of a pole located midway between the two clocks. Initially, the clocks are preset to the same time and are not running. They are designed to start when light signals hit photocells mounted on their roofs. After the flash (a), the light expands in a sphere centered on the top of the pole (b and c). The light signals are detected as they arrive at each clock (d), starting the clocks simultaneously. The two clocks are now synchronized (e).

Let's now attempt to synchronize clocks in two different inertial systems. We assume that each system has the same setup as that used in Figure 12-4 and that the clocks are located along a line parallel to the direction of *relative* motion. Pretend you are located in the lower system of Figure 12-5 and see the upper system moving to the right with constant velocity. The flashbulb goes off as the two poles meet (b). You see the light expanding in a sphere about the pole in your inertial system (c). (It does not matter which bulb flashes, or even if both flash, because you measure the speed of light to be a constant independent of the motion of the source.) Since you see the upper system moving to the right, the left-hand clock moves toward the light signal and starts first (d). The two clocks in your system start simultaneously (e). Notice, however, that it takes some additional time for the light signal to catch up with the right-hand clock (h) in the upper system and it starts after the left-hand clock.

You report that your clocks are synchronized, but that the clocks in the other system are not synchronized. Because the upper system was moving to the right during the time the light signal was en route, the

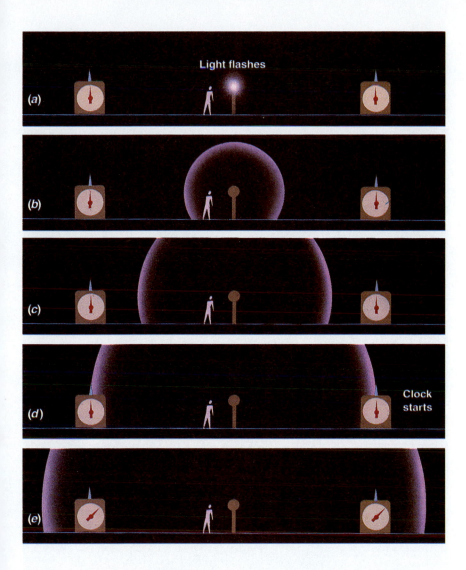

Figure 12-4 Strobe drawings illustrating a method of synchronizing two separated clocks in a single inertial system.

clock on the left moved toward the signal, while the one on the right moved away from it. You observe that the light traveled a shorter distance to the left-hand clock and it was, therefore, started before the right-hand one.

What would the observer viewing the events from the other inertial system say? Let's repeat the analysis assuming that you are now in the upper system. From this point of view you observe the lower system moving to the left as shown in Figure 12-6. Once again you see the light signal expand in a sphere centered on the top of the pole in your system (c). The two clocks in your system start simultaneously (e). You see the right-hand clock in the lower system approach the light signal and start first (d). Only later is the left-hand clock in the lower system started (h). You conclude that the clocks in your system are synchronized, but the clocks in the lower system are not synchronized.

All observers conclude that the clocks in their own reference system are synchronized and the clocks in all other reference systems are not

(Text continues on page 290.)

Figure 12-5 An attempt to synchronize clocks in two inertial systems as viewed by an observer in the lower system.

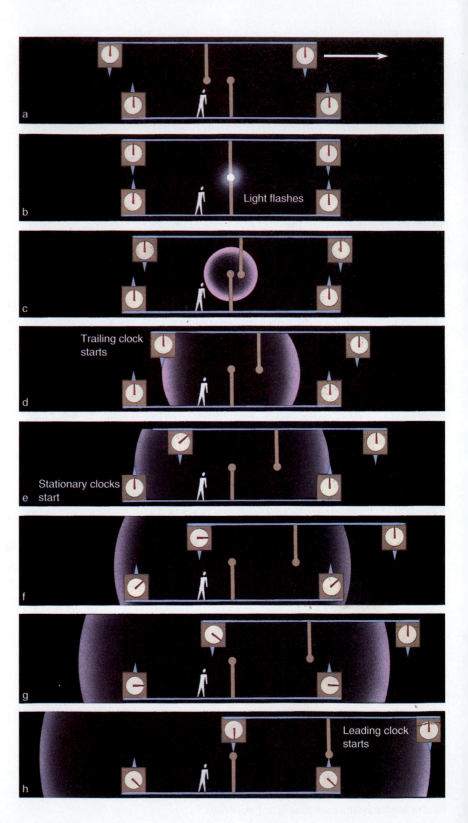

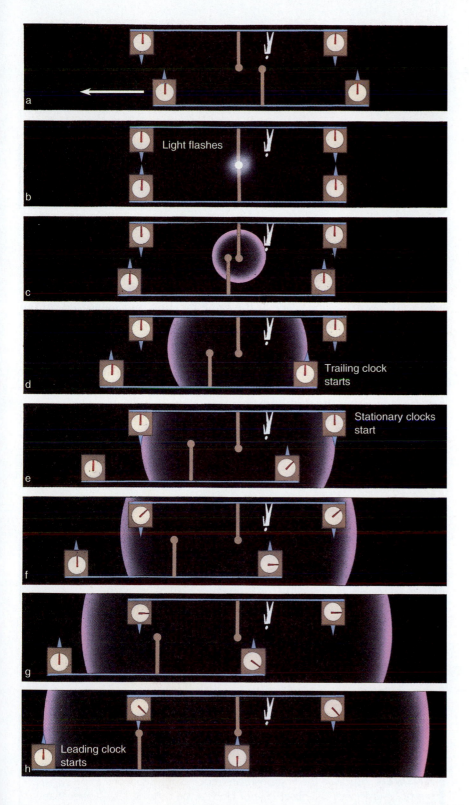

Figure 12-6 The same attempt to synchronize the clocks shown in Figure 12-5 but as viewed by an observer in the upper system.

synchronized. This conflict cannot be resolved. The first postulate says that the two inertial reference systems are equivalent; no experiments can be performed to determine which observer is "really" at rest.

The equivalence can be made more apparent by noting that in each case it was the trailing clock that moved toward the light signal and thus started early. We can summarize the situation by observing that the *trailing clocks lead*.

trailing clocks lead

> **QUESTION**   A conductor on board a fast-moving train verifies that all clocks in the train are synchronized. What do observers on the ground say about this?

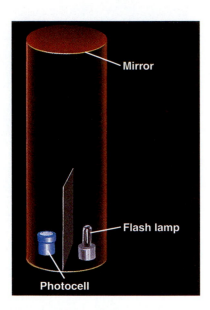

Figure 12-7   A light clock.

Time Varies

Can observers in different inertial systems agree on the time interval between two events taking place at the same location and measured on the same clock? To examine this question consider the very unusual but legitimate clock shown in Figure 12-7. Clocks keep time by counting some regular cycle. In this clock, the cycle is initiated by firing a flashbulb at the bottom. The light signal travels to the mirror at the top of the cylinder and is reflected downward. The photocell receiving the signal initiates a new cycle by firing the flashbulb again.

Imagine an identical clock in your friends' truck. The light that strikes the top mirror and returns to the photocell must be that portion of the flash that left at an angle to the right of the vertical. Therefore, it travels the larger distance shown in Figure 12-8. Because the speed of light is a constant, the time for the round trip must be longer. The time interval between flashes is longer for the moving clock; that is, time is dilated. The moving clock runs slower.

> **QUESTION**   Assume that both you and your friends are carrying clocks. If you determine that your friends' clock is running 10% slow, what will your friends say about your clock?

trailing clocks lead

Your friends' report is different. The light signal in their clock travels straight up and down, whereas the signal in your clock takes the longer path. They claim that your time is dilated. Note that each of you agree that moving clocks run slower. This equivalence is in agreement with the first postulate.

> **ANSWER**   They find that the clock in the caboose is ahead of the clock in the engine.

> **ANSWER**   Because the first postulate requires the situations to be symmetric, your friends will observe your clock to be running 10% slower than theirs.

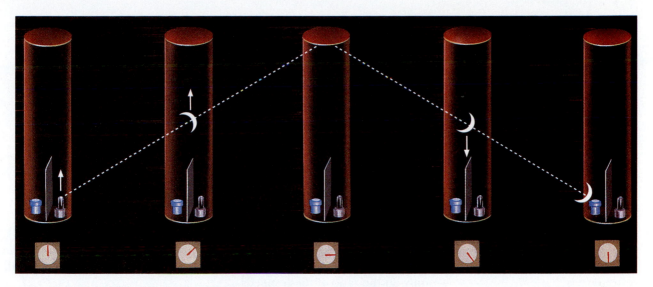

Figure 12-8 The light in the moving clock travels farther and therefore the clock runs slower.

The implications of this are startling. According to the first postulate, *all* clocks in the moving system must run at the same rate. This statement applies to physical, chemical, and even biological clocks. Thus, pulse rates will be lower, biological aging will be slower, the pitch of musical notes will be lower, and so on. Time itself changes when viewed from different inertial systems.

At first glance, it may seem like Einstein discovered the fountain of youth. By traveling at a high velocity, clocks would run slower and we would live longer. Unfortunately, this isn't the case. Within our own inertial system everything is normal. Our biological clocks run at their normal pace and we age normally. Nothing changes.

We should also note that we have not invented a time machine that will allow us to go back into history. Although we can make moving clocks run very slowly by giving them very high speeds, we cannot make them run backward. If such were possible, a person moving relative to you could conceivably see your death before your birth! Obviously, this would play havoc with our ideas of cause and effect.

Experimental Evidence for Time Dilation

The size of the effects predicted by the special theory of relativity increases with speed. The time interval in the moving system is equal to the time interval in the rest system multiplied by an adjustment factor. The relativistic adjustment factor is called gamma (γ) and is given by

$$\gamma = \frac{1}{\sqrt{1 - \left(\frac{v}{c}\right)^2}}$$

the relativistic adjustment factor

Table 12-1	The value of the adjustment factor for various speeds	
Speed	**Adjustment Factor**
The fastest subsonic jet plane	1.0000000000006
Three times the speed of sound	1.000000000005
One-half the speed of light	1.15
80% of the speed of light	1.67
99% of the speed of light	7.09
99.99% of the speed of light	70.7
99.9999% of the speed of light	707
The speed of light	Infinite

In this expression, v is the relative speed of the inertial systems and c is the speed of light. Notice that the value of the adjustment factor depends only on the ratio of these speeds.

In Table 12-1 we have calculated the values for the adjustment factor for different speeds of the moving system. As you can see from the first two entries in the table, a clock moving at ordinary speeds relative to an observer is slowed by a seemingly negligible amount. For instance, a clock moving at three times the speed of sound would have to travel for 63 centuries before it lost 1 second relative to a clock at rest!

An experiment to detect the slowing of a clock during a transcontinental flight would need to detect differences of a few billionths of a second. However, modern atomic clocks are sensitive to such small time differences. Jet planes, each with several atomic cesium clocks, were flown in opposite directions around the Earth. Two experiments—one in 1971 and one in 1977—confirmed the predictions.

The graph of the adjustment factor versus speed drawn in Figure 12-9 shows that the effects become infinitely large as the speed approaches that of light. An early verification of time dilation at these large speeds involved the behavior of subatomic particles known as *muons*. These muons are created high in our atmosphere by collisions of particles from outer space with air molecules. Time dilation can be tested with these fast-moving muons because they are radioactive; they spontaneously break up into other particles. This radioactive decay provides us with a simple but very accurate clock.

> **QUESTION**   Assuming that you measure the radioactivity of muons flying past you, would you expect the muons to live for a longer or a shorter time due to the relativistic effects?

Knowing the number of muons present at a high elevation and the characteristics of the radioactive "clocks," one can predict quite accurately, in the absence of any relativistic effects, how many muons should

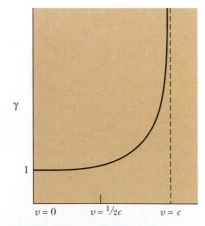

Figure 12-9   A graph of the adjustment factor versus the ratio of the speed of the system or object to the speed of light.

> **ANSWER**   Because the muons are moving, their radioactive decays should be slowed down as viewed from Earth. Therefore, they will live longer than if they were at rest in the laboratory or if you were moving along with them.

reach the Earth's surface before disintegrating. Experiments yielded a much greater number of muons at sea level than predicted. In fact, the number of muons agrees with calculations that assume the decay time for the muons is dilated as predicted.

Length Contraction Σ

The existence of time dilation suggests that space travel to distant galaxies is possible. Although the galaxies are enormous distances from Earth, space explorers traveling fast enough could complete the trips within their (time-dilated) lifetimes.

An examination of such a trip leads us to another startling consequence of Einstein's ideas. Consider a trip to our nearest neighbor star, Proxima Centauri, which is 42 trillion kilometers from Earth. Even light traveling at its incredible speed takes 4.4 years to make the trip. A spaceship capable of traveling at 99% of the speed of light would make the trip in about 4.5 years according to clocks on Earth. However, according to Table 12-1, clocks inside the spaceship will record that the trip takes one-seventh as long, or about 0.64 years.

But there is a catch. Both the space travelers and their Earth-bound friends agree that their relative velocity is 99% of the speed of light. How, then, can the space travelers reconcile making this trip in only 0.64 years? The distance to Proxima Centauri must be *contracted.* The amount of contraction is just right to compensate for the time dilation. Our space travelers measure a distance that is one-seventh as long. Measurements of space, like those of time, change with relative motion.

To see why this happens, we follow Einstein's advice and carefully consider how we measure the length of something. If you are at rest relative to a stick, there is no problem measuring its length. You simply measure it with a ruler, or mark the position of the two ends on the floor and measure the distance between the marks.

What if the stick is moving? Again, you could mark the floor at each end of the stick as it passes by. But you have to be careful. Clearly, you could get a variety of lengths if you mark one end first and the other end at various times later. To obtain the correct length, you must mark the position of the two ends *simultaneously;* for example, by exploding paint cans at each end.

What will a person at rest relative to the stick (Fig. 12-10) think of this measurement? She agrees with your procedure, but says that the paint

Figure 12-10 According to the observer on the ground, the paint can at the front of the pole explodes before the can at the back. Therefore, the two paint splashes on the ground are closer together.

The Twin Paradox

The prediction of time dilation is usually greeted with disbelief. Surely people in different inertial systems can stop their experiment, come together, and compare clocks. They should be able to resolve the question of which clock is really running slower. This feeling was ingeniously expressed in a hypothetical situation that led to an apparent paradox, called the *twin paradox*. Suppose that twins decide to do a time-dilation experiment. One twin gets in a spaceship and flies away from Earth. The spaceship travels out to a distant star and returns to Earth.

The twin on Earth observes that clocks in the spaceship run slower than on Earth. Therefore, the twin in the spaceship ages more slowly. The twin should return from the journey at a younger age than the one who stayed at home. Meanwhile, the twin in the spaceship observes that the clocks on Earth are running slower. So the Earth-bound twin will age slower and should be the younger at the reunion. Thus, we have a paradox. How can each twin be younger than the other?

The paradox arises because we assumed that everything was symmetric, that there was no way of deciding who was taking a trip. But the situation is not symmetric. It becomes clear that the twin in the spaceship is taking the trip as soon as the spaceship accelerates to leave or turns around to return. The inertial forces that arise during the acceleration give it away.

It is sometimes thought that the special theory of relativity can only be applied to situations involving inertial systems; that is, where there is no acceleration. This is not true; there are several ways of getting the correct answer within the framework of special relativity. All solutions agree that the twin in the spaceship is younger at the reunion than the twin who stayed on Earth.

Imagine making such a trip. Suppose that the journey takes 40 years as measured by clocks on Earth, but only 10 years elapse on the spaceship's clocks. On your return you would find that society's technology and institutions have jumped ahead by 40 years. It is possible that you would return and be younger than your children. The social consequences of this *family* shock could be even more mind-boggling than the expected *future* shock that you would experience.

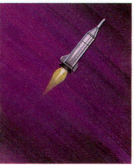

One twin makes a journey to a distant star while the other remains at home. Upon meeting, the twin who made the journey is younger than the twin who stayed at home.

cans didn't explode simultaneously. She claims that the paint can at the front exploded earlier. By the time the can at the back exploded, the back end of the stick had moved closer to the first mark. Thus, the length you measured is shorter than hers. That is, the length of a moving stick is contracted.

moving sticks are shorter

Consider another way to measure the length of the moving stick. You could measure the stick's velocity and record the elapsed time between the passing of the front and back ends. Again, the observer on the moving stick disagrees with your results. She says that your clocks are running slower and, therefore, the elapsed time is shorter. Once again, your measurement yields a contracted length.

The moving length is equal to the length measured at rest *divided* by the relativistic adjustment factor. It is important to note that this length contraction only occurs along the direction of the relative motion. Lengths along the direction perpendicular to this motion are the same in the two inertial systems.

> **QUESTION** Assume that both you and your friends in the truck are carrying metersticks pointing along the direction of relative motion. If your friends measure your stick to be ½ meter long, what length would you measure for your friends' stick?

Spacetime

Einstein's ideas changed the role time plays in our world view. In the Newtonian world view, we considered motion by looking at the spatial dimensions and looking *independently* at time. Einstein demonstrated that time is not a separate quantity but rather is intimately connected to the spatial dimensions. When space changes, there is a corresponding change in the time. Time is now truly the fourth dimension.

The theory of special relativity must be self-consistent. All observers must find that events obey the laws of physics. As we have seen, they do not have to agree on their particular measurements, but they must be able to make sense of the events within their own reference system.

A hypothetical situation illustrates this point. Suppose that an ingenious student claims that she can fit a 10-meter pole into a 6-meter-long barn (Fig. 12-11). Knowing about length contraction, she proposes to

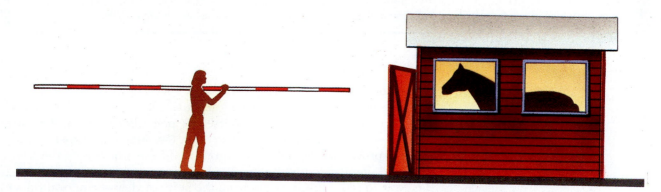

Figure 12-11 The observer on the ground tries to put a 10-meter-long pole in a 6-meter-long barn.

> **ANSWER** The first postulate requires that the situations be symmetric. *Each* observer says that the other's stick is contracted. Therefore, you would also measure a length of ½ meter.

Relatively Modest

Albert Einstein.

Albert Einstein, one of the greatest physicists of all times, was born in Ulm, Germany, in 1879. As a child he showed little intellectual promise, reportedly leading one instructor in the highly disciplined German school system to say, "You will never amount to anything, Einstein." Following a vacation in Italy, Einstein resumed his education at the Swiss Federal Polytechnic School. While there, Einstein attended very few lectures, and passed courses through the excellent lecture notes taken by a friend.

In 1905, at the age of 26, Einstein published four scientific papers that revolutionized physics. One of these papers, on the interpretation of the photoelectric effect, drastically changed our concept of light and earned Einstein the Nobel Prize in 1921. The second paper of this series dealt with Brownian motion, the random motions of macroscopic particles suspended in a fluid that provided evidence for the existence of atoms. The remaining two papers were the brilliant foundation of what is now known as the special theory of relativity. Later in that same year Einstein earned his Ph.D. In 1915, Einstein published his general theory of relativity, a theory of gravity that relates gravitational forces to a warping of spacetime.

Following academic appointments in Switzerland and Czechoslovakia, Einstein accepted a special position created for him at the Kaiser Wilhelm Institute in Berlin. This appointment provided an appreciable allowance and relieved him of teaching duties, allowing him to devote all of his time to research. Shortly after Hitler came to power, Einstein left Germany to accept a position at the Institute for Advanced Study at Princeton, where he remained until he died. Although Einstein was a pacifist, he was persuaded by fellow physicist Leo Szilard to write a letter to President Franklin D. Roosevelt urging him to initiate a program to develop a nuclear bomb.

Einstein's work on the photoelectric effect changed our views of light and, eventually, led to the development of a whole new field of physics known as quantum mechanics (see Chapter 23). Later, however, he became deeply disturbed by the ideas of quantum mechanics. In particular, he could never accept the probabilistic view of nature that is central to the highly successful quantum theory. Einstein once said, "God does not play dice with nature." The last few decades of his life were devoted to an unsuccessful search for a unified theory that would combine the gravitational and electromagnetic forces into a single force.

Adapted from Serway, R.A. *Physics for Scientists and Engineers*, updated 3rd ed. Philadelphia: Saunders, 1992.

---

propel the pole into the barn at 80% of the speed of light because the adjustment factor is five-thirds, giving a moving length of 6 meters. Just enough to fit into the barn! Of course, the pole will only be in the barn for an instant because it is moving very fast. Our ingenious experimenter plans to prove that the pole was entirely in the barn by closing the front door and simultaneously opening the back door.

Now consider this situation from the point of view of a person riding on the pole (Fig. 12-12). The pole is 10 meters long, but the barn is moving and is contracted to 3.6 meters! Clearly, he is not going to agree that the pole was ever entirely in the barn, not even for an instant.

There is no paradox, however. The rider does not agree that the back door was opened at the same time that the front door was closed. Recalling that trailing clocks lead, he says that the back door opened before the

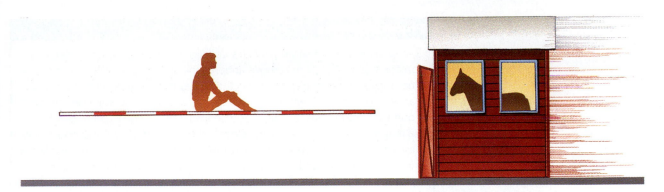

Figure 12-12   The observer on the pole tries to put a 10-meter-long pole in a 3.6-meter-long barn.

front door closed. In fact, careful calculations of this situation verify the self-consistency. The time interval between these events is just enough to allow the "extra" 6.4 meters to pass through.

Relativistic Laws of Motion

As we did in our study of the classical ideas of motion, we now expand our considerations beyond describing motion to consider the laws of motion. Many approaches can be taken to develop laws of motion that are consistent with the ideas of special relativity, although we don't have an entirely free hand. The new laws must have a structure that is logical and internally consistent, and the predictions of the laws must agree with the results of experiments. Furthermore, the new formulations must reduce to the older ones (Newton's laws) when the velocities are small because we know Newton's laws work for small velocities.

The form of Newton's second law regarding momentum carries over into special relativity providing that the expression for the momentum is modified so that $p = \gamma mv$. This is the classical formula multiplied by the adjustment factor. Notice that this expression reduces to the classical one for small speeds because the adjustment factor is very close to 1 in this case. With this modification, the second law is written as

$$F = \frac{\Delta p}{\Delta t} = \frac{\Delta(\gamma mv)}{\Delta t}$$

relativistic form of Newton's second law

Careful analysis of symmetrical collisions of identical balls in different inertial reference systems demonstrates that conservation of momentum is still valid provided this relativistic form for momentum is used.

A force acting for a time still produces the same impulse and therefore the same change in the relativistic momentum. However, since the adjustment factor increases with speed, the acceleration decreases and goes to zero as the object approaches the speed of light. This means that a material object cannot be accelerated to a speed equal to or greater than that of light. Nothing can go faster than the speed of light.

The speed of light is the speed limit of the Universe.

Physics Update

Gerald Gabrielse of Harvard and his colleagues can adjust their tabletop antiproton trap at CERN to remove antiprotons until only a single one remains. Measuring a single antiproton allows the effects of special relativity to become manifest. The rate at which the circulating antiproton completes an orbit around the trap, known as the cyclotron frequency, is equal to the product of the magnetic field strength and its electric charge divided by the product of its mass and gamma, the relativistic adjustment factor. The researchers applied a brief radio pulse to add energy to the antiproton, increasing its velocity and consequently gamma, according to special relativity. As the antiproton dissipated this energy in the trap, the researchers observed the expected increase in cyclotron frequency.

In separate measurements of a single proton in their trap, they confirmed that the cyclotron frequency of antiprotons and protons is identical to one part in a billion, confirming that proton and antiproton masses are equal to a new level of precision.

The law of conservation of energy is also valid in special relativity. If we calculate the work done by a force acting on an object that is initially at rest and equate this work to the kinetic energy of the object as we did in classical physics, we arrive at the expression

relativistic kinetic energy

$$KE = \gamma mc^2 - mc^2$$

The relativistic kinetic energy of an object is equal to the difference between two terms. The second term mc^2 is the energy of the particle at rest. Therefore, it is known as the **rest-mass energy** E_o. This is the origin of Einstein's famous mass–energy equation

mass–energy relationship

$$E_o = mc^2$$

Because the kinetic energy is the additional energy of the object because of its motion, the first term γmc^2 is identified as the total energy of the particle. This expression tells us that the total energy of an object increases with speed. In fact, since the adjustment factor approaches infinity as the speed approaches that of light, the energy also approaches infinity.

Notice also that even when the object is at rest, it has an amount of energy mc^2 stored in its mass. Mass is another form of energy. Thus, the law of conservation of energy must be modified once more to include a new form of energy, mass–energy. This relationship produced a major change in the physics world view. It tells us that mass can be converted into energy and energy can be converted into mass. We will discuss this more fully when we discuss nuclear reactors and the properties of subatomic particles.

The authors of some popular books and articles about special relativity make the statement that mass increases with speed. These authors define a relativistic mass $m = \gamma m_o$, where m_o is called the *rest mass* and is the mass measured in a system at rest relative to the object. This statement does not change any of the mathematics, but does change the interpreta-

tion of some of the expressions. The modern view is that mass is an invariant, and that the introduction of a relativistic mass is unnecessary and sometimes leads to errors.

SUMMARY

The ideas contained in the special theory of relativity are based on two postulates: (1) The laws of physics are the same in all inertial reference systems, and (2) the speed of light in a vacuum is a constant, regardless of the speed of the source or the observer.

As a consequence of adopting these two postulates, observers in different inertial reference systems cannot agree on the simultaneity of events at different places nor the synchronization of separated clocks. These observers do agree that trailing clocks lead.

Time seems normal in one's own reference system, but is dilated when viewed from another reference system. Both observers see the other's clocks running slower. The time interval in the moving system is equal to the interval in the rest system multiplied by the relativistic adjustment factor,

$$\gamma = \frac{1}{\sqrt{1 - \left(\frac{v}{c}\right)^2}}$$

Observers in two different inertial systems agree that objects in the other's system are shorter along the direction of the relative velocity. However, both observers agree on the relative speed of the two systems.

Time intervals are measured in comparison to some kind of periodic motion. Even though all observers in our own inertial reference system will agree with our foolproof method for measuring time, observers moving relative to us will not accept these time measurements. There is no absolute time; time measurements depend on the particular observer.

The conservation laws are valid in special relativity if we introduce the ideas of relativistic momentum and energy and consider mass to be another form of energy. The speed of light is the speed limit of the Universe.

KEY TERMS

ether: The hypothesized medium through which light traveled.

first postulate of special relativity: The laws of physics are the same for all inertial reference systems.

rest-mass energy: The energy associated with the mass of a particle. Given by $E_o = mc^2$, where c is the speed of light.

second postulate of special relativity: The speed of light in a vacuum is a constant regardless of the speed of the source or the speed of the observer.

special theory of relativity: A comprehensive theory of space and time that replaces Newtonian mechanics when velocities get very high.

CONCEPTUAL QUESTIONS

1. If you were located in a spaceship traveling with a constant velocity somewhere in the Galaxy, could you devise experiments to determine your speed? If so, what kinds of experiments?

2. Why did Maxwell's equations appear to be in conflict with the Galilean principle of relativity?

3. Does the first postulate require that the laws of physics be the same in all frames of reference?

4. Does the first postulate require that the speed of light in a vacuum be the same in all inertial reference systems?

5. If an observer in the train in the figure holds a ball directly over a white spot on the floor and drops it, where will the ball land relative to the white spot?

6. If an observer in the train in the figure drops a ball and measures its acceleration, will she obtain a value equal to, greater than, or less than the usual acceleration due to gravity?

7. A spaceship heading for Earth sends a signal by turning a light on and off. What value would a person in the spaceship obtain for the velocity of the light signal? What is the velocity of the light measured by someone on Earth?

8. A very massive star is approaching Earth at 10% of the speed of light when it suddenly explodes as a supernova. With what speed does the flash of light leave the supernova? How fast does it approach Earth?

9. An observer in the train in the figure turns on a light in the caboose and measures the time it takes to get to the engine. (Assume that the light is traveling in a vacuum.) Knowing the length of the train, he is able to calculate the speed of light. Will he obtain a speed less than, greater than, or equal to c?

10. If an observer on the ground uses her own instruments to measure the speed of the light in the previous question, will she obtain a value less than, greater than, or equal to c?

Questions 5, 6, 9, 10, 17, 18, and 41–44. A train is traveling along a straight, horizontal track at a constant speed that is only slightly less than that of light.

11. Can observers in two different inertial systems agree on the simultaneity of events at a single location?

12. Can observers in different inertial systems agree on the simultancity of events at different locations?

13. Is it possible for an observer to reverse the order of birth of twins? If so, how must the observer be moving?

14. Two observers moving in opposite directions claim that events taking place at two different locations occurred in different orders. Which event really happened first?

*15. Suppose you had a row of clocks along a line perpendicular to the direction of relative motion. Would observers in both reference systems agree on the synchronization of these clocks? Explain.

*16. Two events occur at different locations along a line perpendicular to the direction of relative motion. Will observers in both reference systems agree on the simultaneity of these events?

17. An observer in the train in the figure determines that firecrackers go off simultaneously in the engine and in the caboose. What will an observer on the ground say about this?

18. An observer on the ground reports that as the midpoint of the train in the figure passes her, simultaneous flashes occurred in the engine and caboose. What would an observer in the train say about this?

19. As a friend passes you at a very high speed to the right, he explodes a firecracker at each end of his skateboard. These explode simultaneously from his point of view. Which one explodes first from your point of view? How must a third person be moving for her to have observed the other firecracker explode first?

20. Two lights on lamp posts flash simultaneously as seen by an observer on the ground. How would you have to be moving in order to see the right-hand light flash first? The left-hand light flash first?

21. Does the principle of relativity require that all observers see the same sequence of events?

22. Does the principle of relativity require that all clocks behave the same?

23. If a woman conceives a child while on board the first ship sent to colonize the fourth planet out from a neighboring star, how long would you expect it to be before the baby is born according to clocks in the ship? According to clocks back on Earth?

24. If a musician plays middle C on a clarinet while traveling at 85% of the speed of light in a spaceship, will passengers in the ship hear a lower note, a higher note, or the same note?

25. In what sense is it correct to say that time stands still for a person traveling at the speed of light?

26. How fast would a person have to travel to go backward in time?

27. Can a person live long enough to make a trip to a star that is 200 light-years away? A light-year is the distance that light travels in 1 year.

28. Superman wants to travel back to his native Krypton for a visit, a distance of 3,000,000,000,000 meters. (It takes light 10,000 seconds to travel this distance.) If Superman can hold his breath for 1000 seconds and travel at any speed less than that of light, can he make it?

29. On average, an isolated neutron at rest lives for 17 minutes before it decays. If neutrons are moving relative to you, will you observe that they have a longer, a shorter, or the same average life?

30. In an experiment to measure the lifetime of muons moving through the laboratory, scientists obtained an average value of 8 microseconds before a muon decayed into an electron and two neutrinos. If the muons were at rest in the laboratory, would they have a longer, a shorter, or the same average life?

31. A warning light in the engine of a fast-moving train flashes once each second. Will an observer on the ground measure the time between flashes to be greater than, less than, or equal to 1 second?

32. A warning light on the ground flashes once each second. Will an observer in a fast-moving train measure the time between flashes to be greater than, less than, or equal to 1 second?

33. Peter volunteers to serve on the first mission to visit Alpha Centauri. Even traveling at 80 percent of the speed of light, the round trip will take a minimum of 10 years. When Peter returns from the trip, how will his age compare with that of his twin brother Paul, who remained on Earth?

34. Is it physically possible for a 30-year-old college professor to be the natural parent of a 40-year-old student?

35. What does the special theory of relativity say about the possibility of the event described in the following limerick?

> There was a young lady named Bright
> Who could travel much faster than light.
> She went away one day
> In a relative way
> And returned on the previous night.

36. According to the special theory of relativity, a twin who makes a long trip at a high speed can return to Earth

at a younger age than the twin that remains at home. Is it possible for one twin to return before the other is born? Explain.

37. In *A Connecticut Yankee in King Arthur's Court*, Mark Twain chronicles the adventures of a New England craftsman who in 1879 is suddenly transported back in time to Camelot in the year 528. What does the special theory of relativity say about this possibility? What effect would such a trip have on our beliefs about cause and effect?

38. In the TV series *The Voyagers*, two heros used a time device to travel back in time to observe (and preserve) various historic events, such as Lindberg's flight across the Atlantic Ocean in the *Spirit of St. Louis*. What does the special theory of relativity say about this possibility?

39. Suppose a meterstick zips by you at a speed only slightly less than the speed of light. If you measure the length of the meterstick as it goes by, would you determine it to be longer, shorter, or equal to 1 meter long?

40. Is it possible for length contraction to occur without time dilation?

41. An observer on the ground and an observer in the train in the figure each measure the distance between two posts located along the tracks. Does the observer on the ground obtain a longer, shorter, or the same length as the observer in the train?

42. An observer on the ground and an observer in the train in the figure each measure the length of the train. Does the observer on the ground obtain a longer, shorter, or the same length as the observer in the train?

43. An observer on the ground and an observer in the train in the figure each measure the width of the train. Does the observer on the ground obtain a longer, shorter, or the same width as the observer in the train?

44. An observer on the ground and an observer in the train in the figure each measure the distance between the rails. Does the observer on the ground obtain a longer, shorter, or the same distance as the observer in the train?

45. Do length contraction and time dilation occur at speeds of 100 mph?

46. Why don't we notice relativistic effects in our daily lives?

47. In view of the fact that clocks run slower and metersticks are shorter in a moving system, how is it possible for an observer in a moving system to obtain the same speed for light as we do in our system?

*48. In this chapter we discussed the attempt to synchronize clocks in two different inertial systems. In that discussion we did not mention time dilation or length contraction. Do either of these effects change our discussion?

49. An observer in a rapidly moving train claims that the engine came out of the tunnel at the same time as the caboose entered it.

   a. Would an observer on the ground agree? If not, which event would the observer say happened first?

   b. According to this observer, which is longer, the train or the tunnel?

   c. Are your answers consistent with each other?

50. An observer on the ground claims that the engine of a rapidly moving train come out of a tunnel at the same time as the caboose entered.

   a. Would an observer in the train agree? If not, which event would the observer say happened first?

   b. According to this observer, which is longer, the train or the tunnel?

   c. Are your answers consistent with each other?

51. What happens to the acceleration of a proton as it is accelerated closer and closer to the speed of light?

52. An electron is accelerated close to the speed of light by a constant force. Is the acceleration of the electron constant?

53. How much energy would it take to accelerate an object to the speed of light?

54. Identify each of the following energy expressions.

   a. mc^2 c. $\gamma mc^2 - mc^2$

   b. γmc^2 d. $\frac{1}{2}mv^2$

55. Does the idea included in $E_o = mc^2$ violate the law of conservation of energy?

56. Why is it not correct to claim that "matter can neither be created nor destroyed"?

57. If mass is a form of energy, is it correct to claim that a spring has more mass when compressed than when relaxed?

58. An artist is making a metallic statue of Einstein. Does the mass of the statue change as the metal cools? If so, does it get larger or smaller?

59. Under what conditions would you use the special theory of relativity instead of Newton's three laws of motion that we studied in Chapter 2?

60. Why do scientists believe in the special theory of relativity?

61. Special relativity establishes an upper limit for all speeds. Does it do the same for momenta and kinetic energies?

EXERCISES

1. If it takes light 8.3 min to reach the Earth from the Sun, how far away is the Sun?

2. If it takes light 4.4 years to reach the Earth from the nearest star system, how far is it to the star system?

3. When Mars is closest to the Earth, it is approximately 56 million km away. If the radio telescope at Arecibo, Puerto Rico, bounces a radio signal from Mars' surface, how long will it take the radio signal to make the round trip?

4. How long would it take a radio signal to reach a space probe in orbit about Pluto when Pluto is 6 trillion m from Earth?

5. What is the size of the adjustment factor for a speed of $0.2c$?

6. What is the size of the adjustment factor for a speed of 40% that of light?

7. The average lifetime of a muon is 2.2 millionths of a second when measured at rest. What would you expect the average lifetime to be if the muons were traveling at 50% of the speed of light? At 99%?

8. The lifetime of neutrons measured at rest relative to the neutrons is 920 s. What is the lifetime of neutrons traveling at 99% of the speed of light?

9. How many minutes would a clock traveling at 10% of the speed of light lose each day?

*10. A clock is traveling in a jet plane at three times the speed of sound. What fraction of a second would it lose in one year?

11. If a twin travels at a speed of $0.99c$ for 2 years measured on clocks within the spacecraft, how much younger will she be than her sister who remained on Earth?

*12. In 1969, James A. McDivitt remained in orbit around the Earth for 10 days at a speed of 28,500 km/h. How much younger was he when he returned to Earth than he would have been if he had stayed home? The adjustment factor for this speed is approximately 1.0000000003.

13. How long is a 150-m rocket traveling at 80% of the speed of light?

14. How long would a meterstick be if it were traveling by you at a speed of $0.9999c$?

15. What is the distance to the nearest star system measured by an observer in a rocket ship traveling to the star system with a speed of $0.95c$? The distance is 42 trillion km as measured by an observer on Earth.

16. The diameter of our Galaxy is thought to be about 6×10^{14} km. What would the pilot of an interstellar spaceship determine this distance to be if the ship is traveling at $0.98c$?

17. According to Newton's second law, it would require a force of 9.5 N acting for a year to accelerate a 1-kg mass to a speed of $0.99c$. What force is required according to special relativity?

18. What is the difference in the impulse needed to accelerate a 1-kg mass to 50% of the speed of light according to special relativity and Newton's second law?

19. How fast would an electron have to be moving for its kinetic energy to be equal to its rest-mass energy?

20. By what factor does the total energy of a particle increase when the speed doubles from $0.3c$ to $0.6c$?

Newton's work on motion and gravity led to our understanding of the motion of an apple in free fall. Einstein's work on space and gravity led to our understanding of black holes. What are the connections between these two views of gravity? (See p. 317 for the answer to this question.)

General Relativity

Newton's apple falling into an Einsteinian black hole.

The rules of motion are identical in any reference system moving at a constant velocity. But constant relative to what? We need some place to start; some *first* inertial system with which to compare all the others. Our commonly used inertial system, the Earth, is really accelerating as it rotates on its axis and revolves around the Sun. The Sun, the Galaxy, and even the local cluster of galaxies are accelerating. There seems to be no place that is at rest. Without something at rest, all motion seems truly relative. Since all systems moving at a constant velocity are equivalent, we can say arbitrarily that one system is at rest and the others moving, or that all are moving. This reciprocity implies that there is no absolute velocity. But there is something unsettling about this. We want to ask, "What's really at rest?"

We can detect accelerations. If an observer is accelerating relative to an inertial observer, we cannot switch their roles. In noninertial situations people feel new forces—the inertial forces we discussed in Chapter 11— and therefore know that they are accelerating.

For about a decade after Einstein's publication of his theory of special relativity, he worked on generalizing his ideas. The outcome—his **general theory of relativity**—deals with the roles of acceleration and gravity in our attempts to find our place in the Universe.

Gravity and Acceleration   Σ

What is the acceleration of a ball dropped in an accelerating spaceship?

principle of relativity

Armed with his deep relativistic philosophy, Einstein started by expanding the principle of relativity, to include all areas of physics and noninertial, or accelerating, reference systems.

> The laws of physics are the same in all reference systems.

Einstein chose a windowless elevator to perform a thought experiment that would explore the consequences of the principle of relativity. Our modern version of Einstein's elevator is a spaceship. Imagine a spaceship very far from any stars that is accelerating relative to some "fixed" stars at 10 (meters per second) per second. The astronauts feel an inertial force equivalent to the gravitational force they would feel on Earth. If they release a ball, it falls freely with an acceleration of 10 (meters per second) per second. However, an observer in an inertial system outside the spaceship would give a different explanation: The ball continues in the forward direction with the velocity it had at the time of release. The floor accelerates toward the ball at 10 (meters per second) per second, making it seem as if the ball were falling.

QUESTION   What would the astronauts observe if they release two balls with different masses?

ANSWER   The two balls would appear to fall with the same acceleration. This is easiest to see from outside the spaceship. The two balls move side by side with the same velocity while the floor accelerates toward them.

The fact that the astronauts can attribute their motion to gravitational effects is only possible because the mass that appears in Newton's second law (Chapter 2) is identical to the mass that appears in the universal law of gravitation (Chapter 4). It might seem surprising that these two notions of masses are not automatically the same, but recall that they arise in different physical circumstances. It makes sense to question whether they are the same. Newton's second law gives a relationship between applied force and the resulting acceleration. This idea of mass depends on the inertial properties of mass and is therefore called the **inertial mass.** The universal law of gravitation refers to the strength of the attractive force between two objects. This mass is known as the **gravitational mass.**

If the masses are different, objects should reflect this difference. But it took a very clever experiment to search for the very small differences that might occur between these two ideas of mass. An accurate test of the equality of inertial and gravitational mass was performed in 1889 by a Hungarian scientist Baron Roland von Eötvös. We can understand the results by examining a straightforward approach that would work if sensitive enough equipment were available. Imagine dropping balls in a long vacuum tube. The force on each ball is determined by its gravitational mass, whereas the acceleration of each ball depends on this force and its inertial mass. Therefore, if there is a difference between inertial and gravitational mass, balls made of different materials will not experience the same accelerations. We need only detect a difference in the arrival times of the balls. Eötvös found that the inertial and gravitational masses differ by no more than three parts per billion. More recently a team of scientists from Moscow State University has confirmed these results with even greater accuracy.

gravitational mass = inertial mass

As a result of the equality of inertial and gravitational mass, any experiment using material objects would not be able to reveal to the astronauts whether the force is due to the gravitational attraction of a nearby mass or the accelerated motion of their spaceship. Believing that *all* motion is relative, Einstein felt that the astronauts could not make any distinction between the two alternatives. He formalized his belief as the **equivalence principle.**

equivalence principle

> Constant acceleration is completely equivalent to a uniform gravitational field.

> **QUESTION**   What is the path of a ball dropped in a train that is accelerating uniformly in the forward direction?

> **ANSWER**   The ball will fall in a straight line angled toward the back of the train. The acceleration of the train forward is equivalent to a gravitational force in the backward direction. The effective gravitational force is the vector sum of this inertial force and the usual gravitational force (Fig. 13-1).

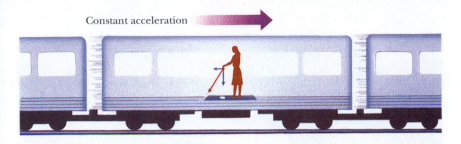

Constant acceleration

Figure 13-1   The effective gravitational force is the vector sum of the inertial force and the usual gravitational force. It causes the ball to fall along a straight line angled toward the back of the train.

Physics on Your Own   Ride in an elevator and interpret your sensations in terms of changes in the strength of gravity. You may find it interesting to stand on a bathroom scale while performing this experiment.

Gravity and Light

Before Einstein there seemed to be a way for the astronauts to distinguish between gravitational and inertial forces. According to the ideas of that time, the astronauts would only have to shine a flashlight across their windowless ship and observe the path of the light by placing frosted glass at equal intervals across the spaceship as shown in Figure 13-2. With the ship at rest on a planet the beam of light would pass straight across the room because the gravitational field would have no effect on it. However, in an accelerating spaceship the astronauts could see the light bend. While the light travels across the ship, the ship accelerates upward,

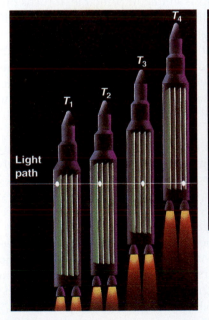

Light path

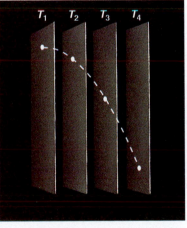

Figure 13-2   While the light travels across the spaceship, the spaceship accelerates upward, causing the light to intersect the frosted glass closer and closer to the floor. The path relative to the ship is a parabola just like a falling ball that is projected horizontally on Earth.

Figure 13-3 The path of starlight is bent as it passes the Sun. This effect has been greatly exaggerated in the drawing.

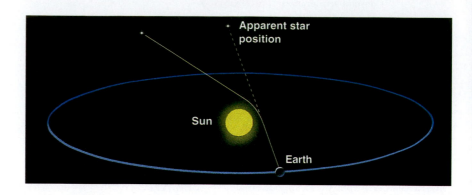

light is bent by gravity

making the differences in the vertical positions on adjacent screens get larger and larger.

Einstein agreed that light would bend in the accelerating spaceship but disagreed that light would be unaffected in the ship in the presence of gravity. Since he firmly believed in the equivalence principle, the astronauts should not be able to tell any difference between the two situations. Einstein was forced to conclude that light is bent by gravity.

The only way to decide which reasoning agrees with nature is to do experiments. A first impulse might be to try to test in the laboratory whether light is bent by gravity. However, the deflection of the beam would be much, much smaller than the diameter of an atom. To get a measurable deflection, we need a longer path, stronger gravity, or both.

Einstein suggested that his idea could be tested using the Sun's gravitational field. He theorized that the Sun's gravity would bend a star's light from its original direction, causing us to see the star in a slightly different position as illustrated in Figure 13-3.

Although Einstein's experiment is simple in principle, it turns out to be tricky in practice. The Sun's brilliance makes the experiment difficult to perform except when the sunlight is blocked by the Moon. The stars near the Sun can be photographed during a total solar eclipse and their positions compared with those taken at other times. British expeditions to Africa and Brazil in 1919 obtained photographs like the one shown schematically in Figure 13-4. Displacements of the stars consistent with the predicted values were observed and Einstein was immediately showered with fame.

Recent experiments have used the same idea but avoided the problem of waiting for solar eclipses. Since radio and light are two different forms of electromagnetic radiation, astronomers have been able to use radio telescopes whenever the Sun is near the line of sight to a celestial object that emits radio signals. All of the resulting measurements are in agreement with the predictions of general relativity.

Further confirmation of the bending of light by gravitational fields has occurred in the last few years with observations of very bright objects near the visible edges of the Universe. When this light passes near a massive galaxy on its way to Earth, the gravitational field of the galaxy can act as a *gravitational lens,* producing multiple images of the object.

Figure 13-4 The positions of the stars as seen during an eclipse. The open circles show their positions in the absence of the Sun. This effect has been greatly exaggerated in the drawing.

Black Holes

A bizarre astronomical object has been hypothesized that dramatically illustrates the bending of light by extremely massive objects. Stars collapse due to their own gravitational attraction after their source of fuel is exhausted. It has been suggested that stars much more massive than our Sun should collapse to a size so small that the increased gravity near the star would prohibit anything—including light—from escaping. This object is called a *black hole* because there is no light coming from this region of space. Light inside the black hole is bent back on itself and never escapes.

The idea of a black hole illustrates something on the fringes of the physics world view. There are astronomers who believe such objects exist and others who don't (although the number of skeptics is shrinking).

Because even light can't escape from a black hole, we have to search for more indirect ways of "seeing" a black hole. The key to finding a black hole is the influence its gravitational field has on nearby objects. The majority of stars in the Universe occur in groups of two or more that are bound together by their mutual gravitational attraction. In many of these binary star systems, one of the stars is not visible. Some of these invisible stars could be black holes. The mass of the unseen star can be determined by examining the behavior of the visible companion. The visible star would orbit the black hole along an elliptical path because the black

Newton's apple falling into an Einsteinian black hole.

hole would continually exert a centripetal force on the star.

Present theories of stellar evolution indicate that only black holes can occur at the end of stellar evolution for masses larger than about three or four times the mass of the Sun. Although the determination of the masses of the unseen stars is not very accurate, there are several cases with large enough masses to be black holes. Since the existence of invisible stars with large masses is very weak evidence, stronger collaborating evidence is needed. Although light from a black hole cannot escape, light from events taking place near the black hole should be visible. It is believed that mass from the companion star should be captured by the black hole's powerful gravitational field. As the mass falls into the black hole, it should emit X rays.

Several candidates for black holes were detected by the Uhuru satellite launched in December of 1970, the most famous being Cygnus X-1, the first X-ray source in the constellation Cygnus. Current evidence indicates that Cygnus X-1 is very compact, with a mass about six times that of the Sun, and is probably a black hole. Two other binary star systems are also good prospects for black holes. There are also indications that super massive black holes (with masses up to 100 million solar masses) are found at the centers of galaxies.

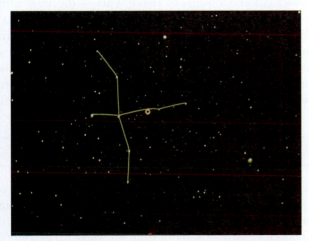

The circle indicates the location of Cygnus X-1, which is believed to be a black hole.

Physics Update

In 1919 Arthur Eddington observed the deflection of starlight as it grazed the Sun. The measurement, in agreement with the prediction of general relativity, helped to make Einstein world famous. Performing a new version of this test, astronomers from Harvard, MIT, and the Haystack Observatory have used radio telescopes in Massachusetts and California to measure the deflection of radio waves coming from the extragalactic object 3C279 as they passed near the Sun. The use of coordinated but widely spaced telescopes produces a much more accurate measurement than is possible with a single radio telescope. The ratio of measured to predicted deflection was 0.9998, with an uncertainty of 0.0008.

Gravity and Time

Einstein's work in general relativity also showed that time is altered by a gravitational field. To see that this is plausible, consider the following thought experiment. Identical clocks are placed at the center and near the edge of a very rapidly rotating merry-go-round, as shown in Figure 13-5. The clock near the edge is moving at a high velocity. We know from special relativity that a moving clock runs slower; the faster it moves, the slower it runs. But the clock near the edge also experiences a large acceleration, and the equivalence principle tells us that this acceleration is equivalent to a gravitational field. Therefore, clocks must run slower in a gravitational field. The stronger the gravitational field, the slower the clocks run.

clocks run slower in stronger gravitational fields

This effect is very small. We need large gravitational fields or extremely sensitive clocks (or both) to test the prediction. As we mentioned in the previous chapter, atomic clocks were flown in jets to detect time differences between in-flight clocks and clocks "stationary" on Earth. In one experiment the in-flight clocks were 47 billionths of a second ahead of the ground-based clocks after 15 hours of flying. Part of the time difference (−6 billionths of a second) was due to the time dilation effects of special relativity, since the Earthbound and in-flight clocks had different speeds. The remainder of the difference (+53 billionths of a second) was due to the different accelerations of the two sets of clocks and the weaker gravitational field experienced by the in-flight clocks. The combination of special and general relativistic effects accounted for the time differences observed.

> **QUESTION**   Why is the first time difference negative?

The effect due to the much larger gravitational field of the Sun can also be measured using naturally occurring "clocks." Atoms give off light that vibrates at a well-defined frequency. The frequency determines the color of the light. Therefore, any slowing of time will show up as a shift in

> **ANSWER**   The clocks in the airplanes are moving faster than those on the ground and are therefore slowed by time dilation.

color. We will discuss this shift in much more detail in later chapters. For now, let us just say that such shifts have been confirmed for light from our Sun and from very compact stars called white dwarfs. Because this effect causes visible light to shift toward the red colors, it is known as the **gravitational redshift.**

In 1960, two American physicists, R. V. Pound and G. A. Rebka, performed a laboratory experiment in which they examined "light" from radioactive nuclei after it traveled up a 22-meter (72-foot) tower. The detected radiation showed the predicted redshift. Their accomplishment was remarkable because the shift was only a few parts in a million billion (10^{15}).

The gravitational redshift must be taken into account in the operation of the Global Positioning System (GPS). In the GPS system, the location of a GPS receiver is determined by measuring its distances from three or more satellites. These distances are determined from timing signals broadcast by the satellites. In order to provide a positioning accuracy of a few meters, the rates of the clocks in the satellites must be adjusted to take general relativity into account. This practical application of general relativity can be used to locate lost hunters and land commercial aircraft.

Gravity and Space

The special theory of relativity has shown us that there is a very intimate relationship between space and time. Since gravity affects time, we should also expect it to affect space. To understand this we once again return to our merry-go-round, but this time we place a meterstick along a radius and another along the rim, as shown in Figure 13-6. There is no length contraction for the meterstick along the radius because it is perpendicular to the relative velocity. However, the meterstick along the rim is contracted. The contraction increases as the distance from the center of the merry-go-round increases and as the angular velocity increases. Invoking the equivalence principle and equating acceleration with gravity, we conclude that gravity alters space.

We have already encountered evidence for the distortion of space when we discussed the bending of light by a gravitational field. To make this more explicit, consider three space stations in orbit around the Sun. When astronauts in each space station take sightings to obtain the angle between the other two, they discover that the sum of the angles of the triangle formed by the three spaceships is greater than 180°, as illustrated in Figure 13-7.

The astronauts are faced with a dilemma. It is customary to define a straight line by the path of light in a vacuum. However, then our example does not obey the rules of ordinary Euclidean geometry wherein the sum of the angles in any triangle is supposed to be 180°.

Non-Euclidean geometries were already familiar to mathematicians. The rules of Euclidean geometry apply to flat planes, but they do not apply to curved surfaces. Surveyors know that they cannot use Euclidean geometry on the surface of the Earth unless they restrict themselves to small regions. The jogs in the section lines on U.S. Forest Service maps are due to the spherical geometry of the Earth.

Figure 13-5 The clock near the edge of the merry-go-round has a larger acceleration and runs slower. According to the equivalence principle, clocks in gravitational fields must run slower.

Figure 13-6 The meterstick near the edge of the merry-go-round has a larger acceleration and is shorter. According to the equivalence principle, sticks in gravitational fields are shorter.

Figure 13-7 The sum of the angles in this triangle as measured by the astronauts in the space stations is greater than 180˚.

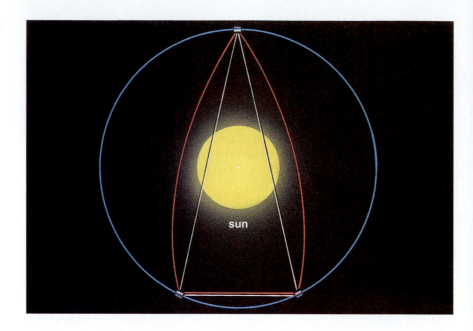

As an example, consider two lines starting out at right angles to each other at the North Pole. These are lines of longitude that differ by 90°. If we extend these lines "straight," they eventually cross the Equator. Figure 13-8 shows that the triangle formed by these two lines and the Equator has three 90° angles for a total of 270°. Although the lines of longitude are not straight in three-dimensional space, they are straight to someone on the surface of the Earth.

> **QUESTION**   Another axiom in Euclidean geometry is that parallel lines never cross. Is this true on the surface of a sphere?

Warped Spacetime

The Newtonian world view considered space to be flat (Euclidean) and completely independent of matter. Objects naturally travel in straight lines in this space. The addition of matter introduced forces that caused objects to deviate from these natural paths. The matter interacted with the objects but did not affect space. Time was independent of space.

The Einsteinian world view begins with a four-dimensional flat spacetime. The addition of matter warps this spacetime. The matter does not act directly on objects, but changes the geometry of space. The objects travel in "straight" lines in this four-dimensional spacetime. Although these lines are straight in four dimensions, the paths that we view in three-

Figure 13-8   The sum of the three interior angles for triangles drawn on spheres is larger than 180˚. In this case the sum is 270˚.

> **ANSWER**   No. The lines of longitude described above are perpendicular to the Equator, and can therefore be considered to be parallel. But they cross at both poles.

dimensional space are not necessarily straight. This situation is analogous to that in which the shadow of a straight stick is projected onto the surface of a sphere. The shadow is not necessarily straight, as shown in Figure 13-9.

We can see what is meant by replacing the gravitational field by a warped spacetime by considering the following situation. Imagine that you are looking down into a completely darkened room where somebody is rolling bowling balls that glow in the dark along the floor. Assume that you record the paths of the balls and your record looks like Figure 13-10.

What reasons can you give to explain the pattern that emerges from these paths? You might suggest that the balls are attracted to an invisible mass that is fixed in the center of the floor [Fig. 13-11(a)]. Or you might suggest that the floor has a dip in the center [Fig. 13-11(b)].

In his classic book, *Flatland,* author and mathematician Edwin A. Abbott tells a bizarre tale of an inhabitant of a two-dimensional world. This fellow was given the "pleasure" of going to another dimension. Once he went into the third dimension, the shape of his normally flat world became obvious to him. Things that were incomprehensible in Flatland—like looking inside a closed figure—became trivial in the third dimension. When he returned to Flatland, he tried to convince his fellow inhabitants of his newly gained insight. They thought he was insane with his strange talk of "up."

Are we similarly doomed to never understand four-dimensional spacetime? If our space has certain shapes that are obvious in the next dimension, can we deduce them without stepping into that additional dimension? Much as two-dimensional creatures on the surface of a sphere

Figure 13-9   The shadow of a straight stick projected onto a sphere is not necessarily straight.

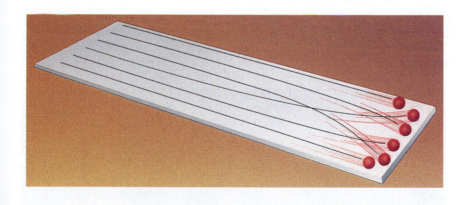

Figure 13-10   The paths of the glowing bowling balls as seen in the dark.

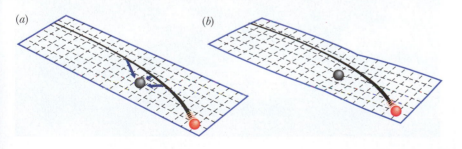

Figure 13-11   Two possible explanations for the paths in Figure 13-10. (a) The bowling balls are attracted by a mass. (b) The mass causes the floor to sag.

can examine their geometry to determine that their space is not flat, we can examine the geometry of our space to learn about the geometry of spacetime.

Physics on Your Own   Test the ideas of warped space using a water bed. Roll tennis balls across the bed and notice that they go in fairly straight lines. Place something heavy like a bowling ball in the center of the bed and roll tennis balls by it. Notice that the balls are "attracted" to the bowling ball, and that the attraction varies with distance.

Further Tests of Relativity

One of Einstein's early calculations of the effect of mass on space and time concerned the orbit of the planet Mercury. As the ability to measure astronomical distances improved in the 1800s, it was discovered that Mercury did not retrace the same path during each revolution (Fig. 13-12). This anomaly results in the long axis of Mercury's elliptical path rotating, or precessing, around the Sun. The effect is extremely small, about 5600 seconds of arc each century. Astronomers were quick to realize that this discrepancy might be explained as the result of the gravitational forces the other planets exert on Mercury. After doing the calculations, however, there was a small remainder of 43 seconds of arc per century that could not be explained. Einstein's theory of general relativity accounts for this remainder.

In 1974 two astronomers discovered a very special star system that has provided an extraordinarily precise verification of general relativity. The system consists of two very compact, dense neutron stars that orbit

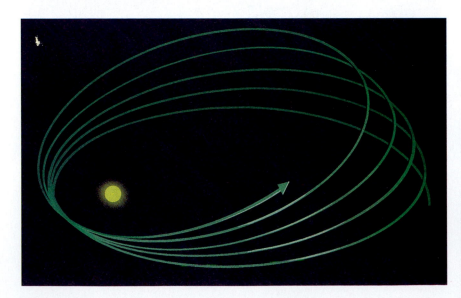

Figure 13-12   The long axis of Mercury's orbit precesses. The shape of the orbit and its precession have been greatly exaggerated in the drawing.

Gravitational Waves

When Einstein proposed his general theory of relativity in 1916, he also postulated the existence of gravitational waves. In many ways gravitational waves are analogous to electromagnetic waves, which we will discuss in Chapter 21. The acceleration of electric charges produces electromagnetic waves that we experience as light, radio, radar, TV, and X rays. Gravitational waves result from the acceleration of masses. Both types of wave carry energy through space, travel at the speed of light (3×10^8 m/s), and decrease in intensity as the inverse square of the distance from the source.

Since electromagnetic waves are so easily detected by a \$10 radio and our eyes, why have gravitational waves not been detected? The first reason is that the gravitational force is 10^{43} times weaker than the electromagnetic force. Even in the most favorable cases gravitational waves are weaker by this same factor. In addition, the detectors are less sensitive in intercepting gravitational waves by at least another factor of 10^{43}.

The detection of gravitational waves is one of the most fundamental challenges in modern physics. Joseph Weber, a physicist at the University of Maryland, pioneered the efforts to detect gravitational waves. He used a large solid cylinder, 2 meters long and weighing several tons, as a detector. A passing gravitational wave causes vibrations in the length of the cylinder due to the differences in the forces on atoms in various parts of the cylinder. The detection system is amazingly sensitive; it can detect changes in the length of the cylinder of 2×10^{-16} meter, about one-fifth the radius of a proton. However, a cylinder can vibrate for many other reasons, such as passing trucks. To eliminate these vibrations Weber uses two cylinders placed a thousand miles apart and requires that both vibrate together.

Although there may be many sources of gravitational waves, only a few should emit strong enough signals and be close enough to be within the range of current instruments. The details of the various processes are not well understood, so the calculations are rough estimates. When a supernova occurs, a very large burst of gravitational waves should be given off. This process may give off a large enough pulse to be detected, but is only expected to happen once every 15 years within our Galaxy. Gravitational waves should also be given off strongly from a pair of neutron stars or black holes orbiting each other at close range.

In 1991 the U.S. Congress approved the construction of the Laser Interferometer Gravitational-Wave Observatory (LIGO). LIGO will consist of identical facilities in Washington and Louisiana, widely separated locations to eliminate extraneous vibrations. Each facility will look for changes in the distance between pairs of mirrors 4 kilometers apart. The distances will be measured by splitting a laser beam into two beams that travel at right angles to each other, bounce off mirrors, and are then recombined. When the beams recombine, very slight changes in the distances to the distant mirrors alter the brightness of the light. This observatory should ultimately be a million times more sensitive than Weber's experiment.

Although we do not have direct evidence of gravitational waves, there is very strong indirect evidence for the existence of gravitational waves. In 1974 Joseph Taylor and Russell Hulse discovered a pair of neutron stars orbiting each other at close range. Neutron stars have masses slightly larger than the mass of our Sun but are only about 10 kilometers in diameter. At these very great densities, electrons and protons combine to form neutrons, so the star consists almost entirely of neutrons. The two neutron stars orbit each other every 8 hours and reach orbital speeds of 0.13% of the speed of light. One of the neutron stars is a pulsar that gives a pulse of radiation every 59 milliseconds. This acts like a clock that is as good as any atomic clock that we have on Earth. This clock allows very, very precise timings of the orbits, and almost 20 years of measurements indicate that the orbital period is decreasing. In fact, it is decreasing at precisely the rate expected from the loss of energy due to gravitational radiation. The 1993 Nobel Prize in physics was awarded to Taylor and Hulse for this work.

Most physicists believe that gravitational waves will be detected directly; it is only a matter of continuing to develop the technology for improving the sensitivity of the detectors.

Joseph Weber's gravitational wave detector.

each other every 8 hours. The long axis of their elliptical orbits precesses very much faster than that of Mercury. After 20 years of measurements, the precession has been determined to an accuracy of 1 part in 4 million and agrees with the predictions of general relativity.

A modern test of general relativity determines the orbits of the planets by bouncing radar signals from their surfaces. The general theory of relativity says that light will travel farther because of the spacetime curvature. This prediction has been checked for the orbits of the inner planets, Mercury and Venus, and the recorded delays agree with the predictions.

> **QUESTION**   Why do physicists believe in the general theory of relativity?

Einstein in Perspective

Let's take a moment to notice a "crack" in the old physics world view. Once we accepted special relativity, we could have predicted trouble for Newton's law of gravity. Consider two masses held apart by spring balances as shown in Figure 13-13. Assume that they are far enough away from everything else that we need only worry about the force of gravity between them. Now suppose two observers read the forces indicated by the spring balances. One observer is at rest relative to the masses, and the other has a large relative velocity. Since this is a static situation, there should be no disagreement on the scale readings.

But here's the problem. The observers disagree on the distance between the two objects and, therefore, on the strength of the gravitational force. So once again we find different views. If you are a Newtonian, you would say, "Fine, the moving observer has a way of telling absolute motion." If you are a relativist, you might say, "Something is wrong. No one observer should be special. The law of gravity must be wrong."

Once we accepted the equivalence of energy and mass, we might have predicted that light would be bent by a gravitational field. This was predicted before Einstein but went unnoticed by the scientific community. It took the genius of Einstein to fit everything together and to inspire the eclipse expeditions to look for this effect.

Figure 13-13   Two masses in outer space are held apart by springs.

> **ANSWER**   Physicists believe in general relativity because its predictions agree with all of the experimental tests.

SUMMARY

Einstein's generalization of his ideas on relativity led to an expanded principle of relativity, which included a new concept of gravity. In general relativity, accelerations can be replaced by gravitational fields. The equivalence principle requires that the mass that appears in Newton's second law (inertial mass) be identical to the mass that appears in the universal law of gravitation (gravitational mass).

The equivalence of acceleration and gravitation leads to a number of consequences. Light and other electromagnetic waves are bent by gravitational fields. Astronomical observations of the deflection of light passing near massive bodies are consistent with the predicted values.

The general theory of relativity also predicts that time is slowed as the strength of the gravitational field increases. Supersensitive atomic clocks flown in very fast jets have detected the predicted time differences between in-flight clocks and clocks on Earth. This difference is confirmed by other experiments such as measuring gravitational redshifts.

Space and time form a four-dimensional spacetime that is warped by the presence of matter. Objects travel in straight lines in this four-dimensional spacetime, but the paths that we view in three-dimensional space are not necessarily straight.

CHAPTER

13

REVISITED

There are actually very few connections between the Newtonian world view of gravity and the Einsteinian explanation except that they both explain the motions of objects near massive objects. Because Einstein's explanation includes a distortion of space, it predicts that light is bent by a gravitational field, something not predicted by Newton.

KEY TERMS

equivalence principle: Constant acceleration is completely equivalent to a uniform gravitational field.

general theory of relativity: An extension of the special theory of relativity to include the concept of gravity.

gravitational mass: The property of a particle that determines the strength of its gravitational interaction with other particles. Measured in kilograms.

gravitational redshift: The decrease in the frequency of electromagnetic waves due to a gravitational field.

inertial mass: An object's resistance to a change in its velocity. Measured in kilograms.

inertial reference system: Any reference system in which the law of inertia (Newton's first law of motion) is valid.

noninertial reference system: Any reference system in which the law of inertia (Newton's first law of motion) is not valid. An accelerating reference system is noninertial.

spacetime: A combination of time and three-dimensional space that forms a four-dimensional geometry.

CONCEPTUAL QUESTIONS

1. How do the reference systems included in the general and special theories of relativity differ?

2. Would it be correct to state that special relativity is a special case of general relativity? Explain.

3. Does the statement of the principle of relativity used for general relativity include the statements for special and classical relativity?

4. How do the statements of the principle of relativity differ in special and general relativity?

5. How does the inertial mass of an object differ from its gravitational mass? Give an example that clearly illustrates this difference.

6. Create a thought experiment to distinguish between a gravitational and an inertial force if gravitational

mass were not the same as inertial mass. Explain your reasoning.

7. Suppose two teams of astronauts who think they are accelerating through space are actually sitting on the surfaces of Earth and Mercury. The gravitational field of Mercury is much smaller than that of Earth. Which team thinks it has the larger acceleration? Which one thinks it has the larger velocity?

8. Two spaceships are traveling through space with different accelerations. If the passengers believe that they are actually sitting on planets, which ship is sitting on the planet with the larger acceleration due to gravity?

9. The prolonged absence of gravity (and exercise) weakens the hearts of astronauts. How might an artificial gravity be created in a space station to avoid this problem?

10. Inhabitants of a large, rotating cylindrical space station measure the acceleration "due to gravity" at the floor and at the ceiling of their living room. Which is larger and why? Compare this with similar measurements performed on Earth.

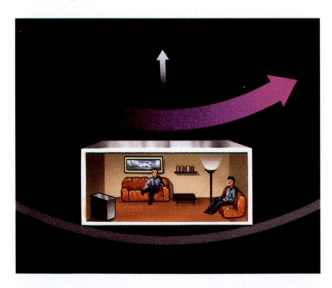

11. State the equivalence principle in your own words and give an example that illustrates its meaning.

12. What does the equivalence principle say is equivalent?

13. Can you distinguish between a uniform gravitational field and a constant acceleration? If so, how?

14. Is the Earth's gravitational field equivalent to a constant acceleration?

15. Why do we usually not notice the bending of light?

16. Does the bending of a high-speed bullet differ from that of a beam of light?

17. If our Sun had the same mass but only half the diameter, would the positions of the stars on photographs taken during an eclipse be any different? If so, how?

18. If our Sun had twice the mass, would the positions of the stars on photographs during an eclipse be any different? If so, how?

19. If you were worried about growing old, would you prefer to live on the top or ground floor of a skyscraper?

20. What would happen to the color of the light emitted by atoms near the surface of a shrinking star?

21. What does the equivalence principle say about the rate at which clocks run?

22. Would you expect a clock on the Equator to run faster or slower than one at the North Pole?

23. Would clocks on board the Space Shuttle gain or lose time during a launch?

24. Would clocks in a space station orbiting Earth gain or lose time compared with clocks on Earth?

25. A train has a large, uniform acceleration along a long, straight track. Would there be a change in the color of a beam of light from a laser as it travels from the caboose to the engine?

26. If we were to shoot a laser beam down a hole drilled through the center of the Earth, would there be a change in the color of the laser light?

27. Assume that there are three space stations in orbit around a massive star and that observers in each station take sightings to the other two to obtain the angle between them. Does general relativity tell us that the sum of the three angles will be less than, greater than, or equal to 180°?

28. Can parallel lines cross in general relativity?

29. Two people stand 1000 meters apart along the Equator. As they walk due north at the same speed, they notice that they are continually getting closer together. Explain this observation in terms of forces and in terms of curved space.

30. How would you explain the observation that the Earth orbits the Sun if the gravitational attraction of the Sun is replaced by a warping of space?

31. You would feel weightless in a freely falling elevator. Would a horizontal cannon shoot straight across the elevator?

32. Would it be possible to measure the distance to a black hole by bouncing radar signals off it?

33. Would light bend as it passes a black hole?

*34. Under what conditions would light travel in circles?

35. What are three possible sources of gravity waves?

36. Why are gravity waves so hard to detect?

*37. What are the key differences and similarities in the special and general theories of relativity?

38. Now that we have the general theory of relativity, should we throw away Newton's law of universal gravitation?

39. Optical lenses work by bending light to form images. Explain how a large galaxy might act as a gravitational lens for a distant source of light.

EXERCISES

1. Imagine a spaceship that is far from any large masses so that the effects of gravity are negligible. This spaceship has a velocity of 6000 km/s and an acceleration in the forward direction of 8 m/s². What is the acceleration of a ball relative to the spaceship after it is released in this spaceship?

2. A spaceship is resting on the surface of Mars where the acceleration due to gravity is 40% of that on Earth. If the astronauts think that they are accelerating through space, what would their acceleration be?

3. Supernova 1987A exploded at a distance of 1.61×10^{21} m from Earth. How long did it take the gravity waves produced by this explosion to travel to Earth? How long would it take the light from the explosion to arrive?

4. The center of our Galaxy is 2.84×10^{20} m from Earth. If gravity waves are produced at this moment near the center of the Galaxy, how long will it be before we could detect them?

5. If light could somehow continually travel perpendicular to a gravitational field with a strength of 10 N/kg, how far would the light bend in 1 s?

6. How far would light bend in traveling across the United States, a distance of approximately 5000 km?

7. The sum of the angles of a triangle drawn on the surface of a sphere is greater than 180°. What is the largest possible sum? What does this triangle look like?

8. The ratio of the circumference to the diameter of a circle drawn on a flat plane is about 3.14. What is the value of this ratio if the circle is the Equator of the Earth? In this case the center of the circle is the North Pole, as shown in the figure.

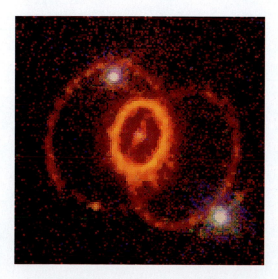

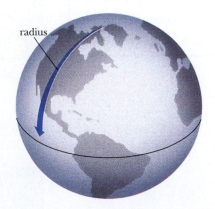

radius

INTERLUDE

Sculling on Lake Powell creates interesting wave patterns.

Waves—Something Else That Moves

Imagine standing near a busy highway and trying to get the attention of a friend on the other side. How could you signal your friend? You might try shouting first. You could throw something across the highway, you could make a loud noise by banging two rocks together, you could shine a flashlight at your friend, and so on.

Signals can be sent by one of two types of methods. One includes ways in which material moves from you to your friend—such as throwing a pebble. The other type includes ways in which energy moves across the highway without any accompanying material. This second type represents phenomena that we usually call **waves.**

The study of waves has greatly expanded the physics world view. Surprisingly, however, waves do not have a strong position in our commonsense world view. It's not that wave phenomena are uncommon, but rather that many times the wave nature of the phenomena is not recognized. Plucking a guitar string, for example, doesn't usually invoke images of waves traveling up and down the string. But that is exactly what happens. The buzzing of a bee probably does not generate thoughts of waves either. We will discover interesting examples of waves in unexpected situations.

Waves are certainly common enough—we grow up playing with water waves and listening to sound waves—but most of us do not have a good intuitive understanding of the behavior of waves. Ask yourself a few questions about waves: Do they bounce off materials? When two waves meet, do they crash like billiard balls? Is it meaningful to speak of the speed of a wave? When speaking of material objects, the answers to these questions seem obvious, but when speaking of waves, the answers require closer examination.

We study waves for two reasons. First, because they are there; studying waves adds to our understanding of how the world works. The second reason is less obvious. As we delve deeper and deeper into the workings of the world, we reach limits beyond which we cannot observe phenomena directly. Even the best imaginable magnifying instrument is too weak to allow direct observation of the subatomic worlds. Our search to understand these worlds yields evidence only by indirect methods. We must use our common experiences to model a world we cannot see. In many cases the modeling process can be reduced to asking whether the phenomenon acts like a wave or like a particle.

To answer the question of whether something acts like a wave or a particle we must expand our commonsense world view to include waves. After you study such common waves as sound waves, we hope you will be ready to "hear" the harmony of the subatomic world.

> **It's not that wave phenomena are uncommon, but rather that many times the wave nature of the phenomena is not recognized.**

Water drops falling onto the surface of water produce waves that move outward as expanding rings. But what is moving outward? Does the wave disturbance carry energy or momentum? What happens when two waves meet? How does wave motion differ from particle motion?

(See p. 344 for the answer to this question).

Vibrations and Waves

Circular waves are formed by falling water drops.

If you stretch or compress a spring and let go, it vibrates. If you pull a pendulum off to one side and let it go, it oscillates back and forth. Such vibrations and oscillations are very common motions in our everyday world. If these vibrations and oscillations affect surrounding objects or matter, a wave is often generated. Ripples on a pond, musical sounds, laser light, exploding stars, and even electrons all display some aspects of wave behavior.

Although many of us have little difficulty visualizing a ball falling or an electron moving in a wire, some of us have a great deal of trouble visualizing a wave. And yet, waves are responsible for many of our everyday experiences. Fortunately, nature has been kind; nearly all waves have the same characteristics. Once you understand one type, you will know a great deal about the others.

We begin our study of waves with the source of all waves, simple vibrations. We then examine common waves, such as waves on a rope, water waves, and sound waves, and later progress to more exotic examples, such as radio, television, light, and even "matter" waves.

Simple Vibrations

If you distort an object and release it, elastic forces restore the object to its original shape. In returning to its original shape, however, the inertia of the displaced portion of the object causes it to overshoot, creating a distortion in the opposite direction. Again, restoring forces attempt to return the object to its original shape and, again, the object overshoots. This back-and-forth motion is what we commonly call a **vibration,** or an **oscillation.** For all practical purposes, the labels are interchangeable.

A mass hanging on the end of a vertical spring exhibits a very simple vibrational motion. Initially, the mass stretches the spring so that it hangs at the position where its weight is just balanced by the upward force of the spring, as shown in Figure 14-1. This position—called the **equilibrium position**—is analogous to the undistorted shape of an object. If you pull downward (or push upward) on the mass, you feel a force in the opposite direction. The size of this restoring force increases with the amount of stretch or compression you apply. If the applied force is not too big, the restoring force is proportional to the distance the mass is moved from its equilibrium position. If the force is too large, the spring will be permanently stretched and not return to its original length. In the discussion that follows we assume that the stretch of the system is not too large. Many natural phenomena obey this condition, so little is lost by imposing this constraint.

Imagine pulling the mass down a short distance and releasing it as shown in Figure 14-2(a). Initially there is a net upward force which accelerates the mass upward. As the mass moves upward, the net force decreases in size (b), becoming zero when the mass reaches the equilibrium position (c). Because the mass has inertia, it overshoots the equilibrium position. The net force now acts downward (d) and slows the mass to zero speed (e). Then the mass gains speed in the downward direction (f). Again, the mass passes the equilibrium position (g). Now the net force is

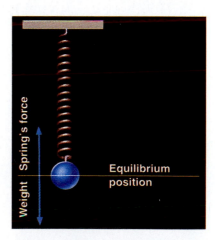

Figure 14-1   At the equilibrium position, the upward force due to the spring is equal to the weight of the mass.

Figure 14-2 A time sequence showing one complete cycle for the vibration of a mass on a spring. The clocks show that equal time intervals separate the images.

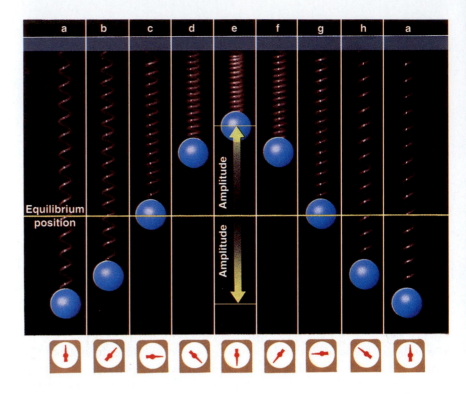

once again upward (h) and slows the mass until it reaches its lowest point (a). This sequence [Figs. 14-2(a through a)] completes one **cycle.**

Actually, a cycle can begin at any position. It lasts until the mass returns to the original position *and* is moving in the same direction. For example, a cycle might begin when the mass passes through the equilibrium point on its way up (c) and end when it next passes through this point on the way up. Note that the cycle does not end when the mass passes through the equilibrium point on the way down (g). This motion is known as periodic motion, and the length of time required for one cycle is known as the **period** *T.*

period is the time to complete one cycle

Because of energy conservation (Chapter 6), the mass travels the same distance above and below the equilibrium position. This distance is marked in Figure 14-2 and is known as the **amplitude** of the vibration. The amplitude decreases and eventually the motion dies out because of the frictional effects that convert mechanical energy into thermal energy.

We might guess that the time it takes to complete one cycle would change as the amplitude changes, but experiments show that the period remains essentially constant. It is fascinating that the period and frequency are *not* affected by the amplitude of the motion. (Again, we have to be careful not to stretch the system "too much.") This means that a vibrating guitar string always plays the same frequency regardless of how hard the string is plucked.

We can describe the time dependence of the vibration equally well by giving its **frequency** *f,* the number of cycles that occur during a unit of time. Frequency is often measured in cycles per second, or hertz (Hz). For example, concert A (the note that orchestras use for tuning) has a frequency of about 440 hertz, household electricity oscillates at 60 hertz, and your favorite FM station broadcasts radio waves near 100 million hertz.

There is a simple relationship between the frequency f and the period T; one is the reciprocal of the other.

$$f = \frac{1}{T}$$

$$T = \frac{1}{f}$$

$$\text{frequency} = \frac{1}{\text{period}}$$

$$\text{period} = \frac{1}{\text{frequency}}$$

To illustrate this relationship, let us calculate the period of a spring vibrating at a frequency of 4 hertz.

$$T = \frac{1}{f} = \frac{1}{4\text{ Hz}} = \frac{1}{4\text{ cycles/s}} = \frac{1}{4}\text{s}$$

This calculation shows that a frequency of 4 cycles per second corresponds to a period of ¼ second. This makes sense since a spring vibrating four times per second should take ¼ of a second for each cycle. (When we state the period, we know it refers to one cycle and don't write "second per cycle.")

QUESTION   What is the period of a mass that oscillates with a frequency of ten times per second?

Although the period for a mass oscillating on the end of a spring does not depend on the amplitude of the oscillation, we might expect the period to change if we switch springs or masses. The stiffness of the spring and the size of the mass do change the rate of vibration.

The stiffness of a spring is determined by how much force it takes to stretch it by a unit length. For moderate amounts of stretch or compression, this value is a constant known as the **spring constant** k. In SI units this constant is measured in newtons per meter. Larger values correspond to stiffer springs.

In trying to guess the relationship between spring constant, mass, and period, we would expect the period to decrease as the spring constant increases since a stiffer spring means more force and therefore a quicker return to the equilibrium position. Furthermore, we would expect the period to increase as the mass increases since a larger mass will slow the motion because of its inertia.

ANSWER   Because the period is the reciprocal of the frequency, we have
$$T = \frac{1}{(10\text{ Hz})} = 0.1\text{ s}$$

period of a mass on a spring

COMPUTING PERIOD OF A MASS ON A SPRING

The mathematical relationship for the period of a mass on a spring can be obtained theoretically, and is verified by experiment:

$$T = 2\pi \sqrt{\frac{m}{k}}$$

where $\pi = 3.14$.

As an example, consider a 0.2-kilogram mass hanging from a spring with a spring constant of 0.8 newton per meter.

$$T = 2\pi \sqrt{\frac{m}{k}} = 2\pi \sqrt{\frac{0.2 \text{ kg}}{0.8 \text{ N/m}}} = 6.28 \sqrt{\frac{1}{4} \text{s}^2} = 3.14 \text{ s}$$

Therefore, this mass–spring combination oscillates with a period of 3.14 seconds, or a frequency of 0.32 hertz.

QUESTION What is the period of a 0.1-kilogram mass hanging from a spring with a spring constant of 0.9 newton per meter?

The Pendulum

The pendulum is another simple system that oscillates. Students are often surprised to learn (or to discover by experimenting) that the period of oscillation does not depend on the amount of swing. To a very good approximation, large- and small-amplitude oscillations have the same period if we keep their amplitudes less than 30°. This amazing property of pendula was first discovered by Galileo when he was a teenager sitting in church watching a swinging chandelier. (Clearly he was not paying attention to the service.) Galileo tested his hypothesis by constructing two pendula of the same length and swinging them with different amplitudes. They swung together, verifying his hypothesis.

Let's consider the forces on a pendulum when it has been pulled to the right as shown in Figure 14-3. The component of gravity acting along the string is balanced by the tension in the string. Therefore, the net force is a component of gravity at right angles to the string and directed toward the lower left. This restoring force causes the pendulum bob to accelerate toward the left. Although the restoring force on the bob is zero at the lowest point of the swing, the bob passes through this point (the equilibrium position) because of its inertia. The restoring force now points toward the right and slows the bob.

Because the gravitational force is proportional to the mass of the bob, we would expect that the motion would not depend on the mass. This prediction is true and can be verified easily by making two pendula of the

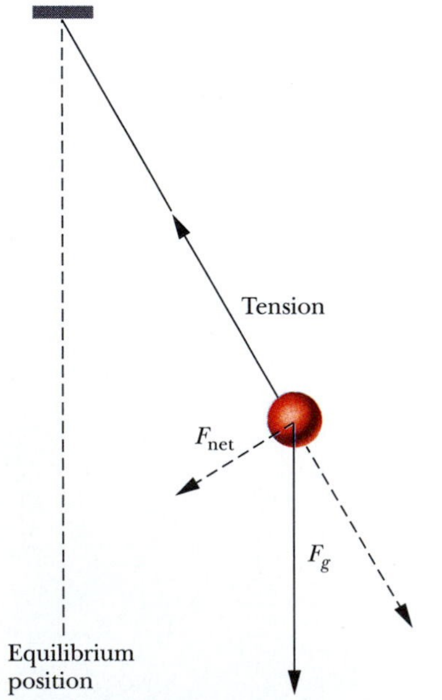

Figure 14-3 The net force on the pendulum bob accelerates it toward the equilibrium position.

ANSWER 2.09 seconds.

same length with bobs of the same size made out of different materials so they have different masses. The two pendula will swing side by side.

> **QUESTION**   Why do we suggest using different materials?

We also know from our experiences with pendula that the period depends on the length of the pendulum; longer pendula have longer periods. Therefore, the length of the pendulum can be changed to adjust the period.

> **Physics on Your Own**   Construct a simple pendulum and investigate the validity of some of the statements made in this section. Does the period depend on the amplitude of the swing, the length of the string, or the mass of the bob?

A strobe photograph of a pendulum taken at 20 flashes per second. Note that the pendulum bob moves the fastest at the bottom of the swing.

Because the restoring force for a pendulum is a component of the gravitational force, you might expect that the period depends on the strength of gravity much as the period of the mass on the spring depends on the spring constant. This hunch is correct and can be verified by taking a pendulum to the Moon, where the acceleration due to gravity is only one-sixth as large as on Earth.

COMPUTING PERIOD OF A PENDULUM Σ

The period of a pendulum is given by

$$T = 2\pi \sqrt{\frac{L}{g}}$$

As an example, consider a pendulum with a length of 10 meters.

$$T = 2\pi \sqrt{\frac{L}{g}} = 2\pi \sqrt{\frac{10 \text{ m}}{10 \text{ m/s}^2}} = 6.3 \sqrt{1 \text{ s}^2} = 6.3 \text{ s}$$

Therefore, this pendulum would oscillate with a period of 6.3 seconds.

> **QUESTION**   What would you expect for the period of a 1.7-meter pendulum on the Moon?

period of a pendulum

> **ANSWER**   If we use the same type of material, the size has to be different to get different masses. Different sizes might also affect the period. When doing an experiment, it is important to keep all but one factor constant.

> **ANSWER**   6.3 seconds.

A replica of an early mechanical clock.

Clocks

Keeping time is a process of counting time intervals. So it is reasonable that periodic motions have been important to timekeepers. Devising accurate methods for keeping time has kept many scientists, engineers, and inventors busy throughout history. The earliest methods for keeping time depended on the motions in the heavens. The day was determined by the length of time it took the Sun to make successive crossings of a north–south line and was monitored with a sundial. The month was determined by the length of time it took the Moon to go through its phases. The year was the length of time it took to cycle through the seasons and was monitored with a calendar, a method of counting days.

As science and commerce advanced, the need grew for increasingly accurate methods of determining time. An early method for determining medium intervals of time was to monitor the flow of a substance such as sand in an hourglass or water in a water clock, but neither of these was very accurate and because they were not periodic, they had to be restarted for each time interval. It is interesting to note that Galileo kept time with a homemade water clock in many of his early studies of falling objects.

The next generation of clocks took on a different character, employing oscillations as their basic time-keeping mechanism. Galileo's determination that the period of a pendulum does not depend on the amplitude of its swing led to Christian Huygens' development of the pendulum clock in 1656, ten years after Galileo's death. One of the difficulties Huygens encountered was to develop a mechanism for supplying energy to the pendulum to maintain its swing.

The development of clocks that would keep accurate time over long periods was spurred by seafarers. To determine longitude requires measuring the positions of prominent stars and comparing these positions with their positions as seen from Greenwich, England, *at the same time.* Since pendulum clocks did not work on board swaying ships, several cash prizes were offered for the design and construction of suitable clocks. Beginning in 1728, John Harrison, an English instrument maker, developed a series of clocks that met the criteria, but he was not able to collect his money until 1765. One of Harrison's clocks had an accuracy of a few seconds in five months at sea.

Any periodic vibration can be used to run clocks. Grandfather clocks utilize pendulums to regulate the hands and are powered by hanging weights. Mechanical watches have a balance wheel fastened to a spring. Electric clocks use 60-hertz alternating electrical current. Digital clocks utilize the vibrations of quartz crystals or resonating electric circuits.

Modern time is kept with atomic clocks, which use the frequencies of atomic transitions (see Chapter 22) and are extremely insensitive to such changes in the clocks' environment as pressure and temperature. Atomic clocks are accurate to a few seconds in a million years.

Modern time is kept by extremely accurate atomic clocks operated by the National Institute of Standards and Technology.

Physics on Your Own   Make a list of the different types of mechanisms used to run clocks. How would you adjust the natural frequency of each clock to make it run at the correct rate?

Physics Update

Scientists at the University of Tokyo have achieved the highest peak speed for a computer performing a scientific calculation: 1.08 terraflops (short for trillion floating point operations per second). With their special-purpose GRAPE-4 machine, Junichiro Makino and Makoto Taiji perform simulations of the complex interactions among astronomical objects such as stars and galaxies. This type of simulation, referred to as an *N*-body problem because the behavior of each of the *N* objects is affected by all the other objects, is particularly computation intensive. Fortunately, two advances have made possible ever-larger simulations. One is computer speed, which is up by a factor of 100 over the past ten years for the fastest computers. Another is improved algorithms for conducting efficient calculations. GRAPE-4 reaches its record speeds using 1692 processor chips, each performing at rates of 640 megaflops. The Tokyo researchers hope to achieve petaflops (10^{15} operations per second) by the turn of the century with a suite of 20,000 processors each operating at 50 gigaflops.

Resonance

We discovered with the mass on a spring and the pendulum that each system had a distinctive, natural frequency. The natural frequency of the pendulum is determined by its length and the acceleration due to gravity. Pulling the bob back and releasing it produces a vibration at this particular frequency.

A child on a swing is an example of a life-size pendulum. If the child does not "pump" her legs and if no one pushes her, the amplitude of the swing continually decreases and the child comes to rest. As every child knows, however, the amplitude can be dramatically increased by pumping or pushing. A less obvious fact is that the size of the effort—be it from pumping or pushing—is not important, but its timing is crucial. The inputs must be given at the natural frequency of the swing. If the child pumps at random times, the swinging dies out.

This phenomenon of a large increase in the vibrational amplitude when a periodic force is applied to a system at its natural frequency is called **resonance.** Resonance can also be achieved by using impulses at other special frequencies, but each of these has a definite relationship to the resonant frequency. For example, if you push the child on the swing every other time, you are providing inputs at one-half of the resonant frequency; every third time gives inputs at one-third of the resonant frequency; and so on. Each of these frequencies causes resonance.

If kids pump at the right frequencies, they can increase the amplitudes of their motions.

> **QUESTION** What happens to the amplitude of the swing if you push at twice the natural frequency?

> **ANSWER** In this case you would be pushing twice for each cycle. One of the pushes would negate the other and the swing would stop.

> **Physics on Your Own**   The next time you go to a playground, experiment with the phenomenon of resonance. Try pushing or pumping a swing at a frequency other than its natural one. What happens? If you're the one on the swing, what happens to the resonant frequency when you stand up?

More complex systems also have natural frequencies. If someone strikes a spoon on a table, you are not likely to mistake its sound for that of a tuning fork. The vibrations of the spoon produce sounds that are characteristic of the spoon. All objects have natural frequencies at which they vibrate. The factors that determine these vibrational frequencies are rather complex. In general, the dominant factors are the stiffness of the material, the mass of the material, and the size of the object.

Resonance can have either good or bad effects. Although your radio receives signals from many stations simultaneously, it only plays one station at a time. The radio can be tuned so that its resonant frequency matches the broadcast frequency of your favorite station. Tuning puts the radio in resonance with one particular broadcast frequency and out of resonance with the frequencies of the competing stations. On the other hand, if the radio has an inferior speaker with one or two strong resonant frequencies, it will distort the sounds from the radio station by not giving all frequencies equal amplification.

Suppose you had a collection of pendula of different lengths, as shown in Figure 14-4. Notice that two of these pendula have the same length, and thus the same natural frequency. The pendula are not independent because they are all tied to a string. The motion of one of them is felt by all the others through pulls by the string. If you start the left-hand pendulum swinging, its back-and-forth motion creates a tug on the common string with a frequency equal to its natural, or resonant, frequency. Pendula with different frequencies jiggle a little bit, but are not affected very much. However, the pendulum with the same frequency resonates with the input frequency, drastically increasing its amplitude. In exactly the same way, objects resonate when input frequencies are the same as any of their natural frequencies.

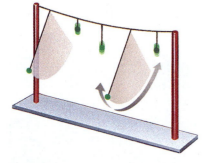

Figure 14-4 Pendulums with the same natural frequency resonate with each other.

Waves: Vibrations That Move

Most **waves** begin with a disturbance of some material. Some disturbances, such as the clapping of hands, are one-time, very abrupt events; whereas others, such as the back-and-forth vibration of a guitar string, are periodic events. The simplest wave is a single pulse that moves outward as a result of a single disturbance.

Imagine a long row of dominoes lined up as shown in Figure 14-5. Once a domino is pushed over, it hits its neighbor, and its neighbor hits its neighbor, and so on, sending the disturbance along the line of dominoes. The key point is that "something" moves along the line of dominoes—from the beginning to the end—but it is not any individual domino.

Tacoma Narrows Bridge

Resonant effects can sometimes have disastrous consequences. In 1940, a new bridge across one of the arms of Puget Sound in the state of Washington was opened to traffic. It was a suspension bridge with a central span of 850 meters (2800 feet). Since the bridge was designed for two lanes, it had a width of only 12 meters (40 feet). Within a few months after it opened, early morning winds in the Sound caused the bridge to oscillate in standing wave patterns that were so large in amplitude that the bridge failed structurally and fell into the water below, as shown in the figure.

But why did this bridge fail when other suspension bridges are still standing (including the bridge that now spans the Sound at the location of the original)? The bridge was long and narrow, and particularly flexible. Motorists often complained about the vertical oscillations and nicknamed the bridge "Galloping Gertie." However, the amplitudes of the vertical oscillations were relatively small until that fateful morning. The wind was blowing along the arm of the Sound (perpendicular to the length of the bridge) at moderate to high velocities but was not near gale force. One might speculate that fluctuations in the wind speed matched the natural frequency of the bridge, causing it to resonate. However, the wind was reasonably steady and wind fluctuations are normally quite random. Furthermore, the forces would be horizontal and the oscillations were vertical.

The best explanation involves the formation and shedding of vortices in the wind blowing past the bridge. *Vortices* are like the eddies that you get near the ends of the oars when you row a boat. The vortices rotate in opposite directions in the wind blowing over and under the bridge. As each vortex is shed, it exerts a vertical impulse on the bridge. Therefore, if the frequency of vortex formation and shedding is near the natural frequency of the bridge for vertical oscillations, a standing wave will form just like those on a guitar string. (The frequency does not have to match exactly; it only needs to be close. How close depends on the details of the bridge construction.)

The bridge would have been fine except for another unfortunate circumstance. Besides the vertical standing wave, it is also possible to set up torsional, or twisting, standing waves on the bridge. Normally, the frequencies of the two standing waves are quite different. But for the Tacoma Narrows Bridge the two frequencies were fairly close (8 per minute for the vertical motion compared with 10 per minute for the twisting motion). This allowed some of the energy from the vertical motion to be transferred to the twisting motion that eventually led to the mechanical failure of the bridge.

The Tacoma Narrows bridge collapsed when winds set up resonant vibrations.

Figure 14-5 A wave pulse travels along a line of dominoes.

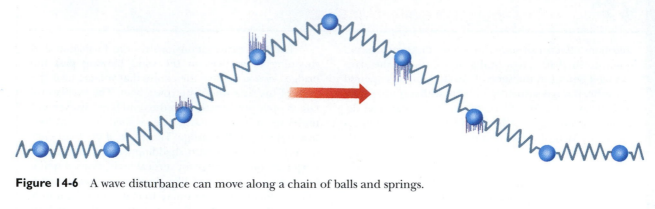

Figure 14-6   A wave disturbance can move along a chain of balls and springs.

Actually, the domino example is not completely analogous to what happens in most situations involving waves because there is no mechanism for restoring the dominoes in preparation for the next pulse. We can correct this omission by imagining a long chain of balls connected by identical springs as shown in Figure 14-6. Striking the center ball causes it to move up and down. The springs on each side of the ball exert restoring forces, causing it to vibrate. These springs also exert forces on the neighboring balls, moving them from their equilibrium positions. And these, in turn, experience restoring forces, causing them to vibrate and exert forces on their neighbors. In this way the wave disturbance moves outward from its original location.

In a similar manner, a pebble dropped into a pond depresses a small portion of the surface. Each vibrating portion of the surface generates disturbances in the surrounding water. As the process continues, the disturbance moves outward in circular patterns, such as those shown in Figure 14-7.

This type of disturbance, or pulse, occurs in a number of common, everyday events. A crowd transmits single pulses when a small group begins pushing. This push spreads outward through the crowd much like a ripple moves over a pond's surface. Similarly, the disturbance produced by a clap sends a single sound pulse through the air. Other examples of nonrecurrent waves include tidal bores, tidal waves, explosions, and light pulses emitted by supernovas (exploding stars).

Although a wave moves outward from the original disturbance, there is no overall motion of the material. As the wave travels through the medium, the particles of the material vibrate about their equilibrium positions. Although the wave travels down the chain, the individual balls of the chain return to their original positions. The wave transports energy rather than matter from one place to another. The energy of an undisturbed particle in front of the wave is increased as the wave passes by and then returns to its original value. In a real medium, however, some of the energy of the wave is left behind as thermal energy in the medium.

There are two basic types of waves. A wave in which the vibration of the medium is perpendicular to the motion of the wave is called a **transverse wave.** Waves on a rope are transverse waves. A wave in which the vibration of the medium is along the same direction as the motion of the

Figure 14-7   Water drops produce disturbances that move outward in circular patterns.

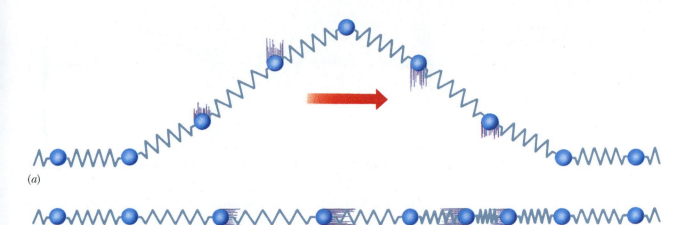

(a)

Figure 14-8 In a transverse wave (a) the medium moves perpendicular to the direction of propagation of the wave, whereas in a longitudinal wave (b) the medium's motion is parallel to the direction of propagation.

wave is called a **longitudinal wave.** Both types of wave can exist in the chain of balls. If a ball is moved vertically, a transverse wave is generated [Fig. 14-8(a)]. If the ball is moved horizontally, the wave is longitudinal [Fig. 14-8(b)].

Transverse waves can only move through a material that has some rigidity; transverse waves cannot exist within a fluid because the molecules simply slip by each other. Longitudinal waves, on the other hand, can move through most materials because the materials can be compressed and have restoring forces.

> **QUESTION** Is it possible to have transverse waves on the surface of water?

One-Dimensional Waves

Because all waves have similar properties, we can look at waves that are easy to study and then make generalizations about other waves. Imagine a clothesline tied to a post as in Figure 14-9. A flick of the wrist generates a single wave pulse that travels away from you. On an idealized rope, the wave pulse would maintain its shape. On a real rope, the wave pulse slowly spreads out. We will ignore this spreading in our discussion. The wave's speed can be calculated by dividing the distance the pulse travels by the time it takes.

> **ANSWER** Transverse surface waves are possible because the force of gravity tends to restore the surface to its flat equilibrium shape.

Figure 14-9 While pieces of the rope vibrate up and down, the wave moves along the rope.

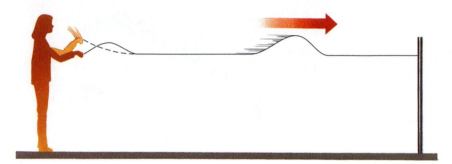

The speed of the wave can be changed. If you pull harder on the rope, the pulse moves faster; the speed increases as the tension in the rope increases. The speed also depends on the mass of the rope; a rope with more mass per unit length has a slower wave speed. Surprisingly, the amplitude of the pulse does not have much effect on the speed.

Probing the Earth

Imagine drawing a circle to represent the Earth. Further, imagine drawing a dot to show how far the interior of the Earth has been explored by direct drilling and sampling techniques. Where would you place the dot? The dot should be placed on the original circle. The Earth is about 6400 kilometers (4000 miles) in radius and we have only drilled into its interior about 12.2 kilometers, less than 0.2% of the distance to the center. Therefore, we must learn about the interior of the Earth using indirect means such as looking at signals from explosions and earthquakes.

Three kinds of wave are produced in an earthquake. One type travels along the surface and the other two travel through the Earth's interior; one of the interior waves is a longitudinal wave and the other is a transverse wave. These waves move outward in all directions from the earthquake site and are received at numerous earthquake monitoring sites around the Earth. The detection of these waves and their arrival times provide clues about the Earth's interior.

Two major things happen to the waves: First, partial reflections occur at boundaries between distinctly different regions. Second, the waves change speed as the physical conditions—such as the elasticity and density—change. Changing a wave's speed usually results in the wave changing direction, a phenomenon known as refraction, which we will discuss in Chapter 17. As the waves go deeper into the Earth, they speed up, causing them to change direction.

The longitudinal waves are called the primary, or P-waves, and are created by the alternating expansion and compression of the rocks near the source of the shock. This push–pull vibration can be transmitted through solids, liquids, and gases. P-waves move with the highest speeds and therefore are the first to arrive at a seismograph station. P-waves move at about 5 kilometers per second (11,000 mph!) near the surface and speed up to about 7 kilometers per second toward the base of the upper crust.

The secondary, or S-waves, are transverse waves. In this case, the rock movement is perpendicular to the direction the wave is traveling. S-waves travel through solids but cannot propagate through liquids and gases because fluids lack rigidity.

We wouldn't be able to infer much about the Earth's interior if only one signal arrived at each site. There would be a number of paths that could account for the characteristics and timing of the signal. Fortunately, many sites receive multiple signals that allow large computers to piece the information together to form a model of the Earth's interior. Information that is not received is also important. After an earthquake there are many sites that do not receive any transverse signals. This tells us that they are located in a shadow region behind a liquid core as shown in the figure.

*Reference: Levin, H., *The Earth Through Time.* Philadelphia: Saunders, 1992.

These observations make sense if we consider the vibrations of a small portion of the rope. The piece of rope is initially at rest and moves as the leading edge of the pulse arrives. How fast the rope returns to its equilibrium position determines how the pulse passes through the region (and hence, the speed of the pulse). The more massive the rope, the more sluggishly it moves. Also, if the rope is under a larger tension, the restoring forces on the piece of rope are larger and cause it to return to its equilibrium position more quickly.

When a pulse hits the end that is attached to the post, it "bounces" off and heads back. This reflected pulse has the same shape as the incident pulse, but is inverted as shown in Figure 14-10. If the incident pulse is an "up" pulse (a **crest**), the reflected pulse is a "down" pulse (a **trough**). If the end of the rope is free to move up and down, the pulse still reflects but no inversion takes place.

There is no inversion of the front and back of the wave pulse. To observe this you would generate a pulse that is not symmetrical. The pulses

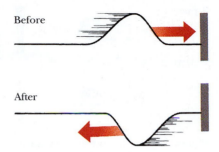

Figure 14-10 A wave pulse is inverted when it reflects from a fixed end. Note that the steep edge leads on the way in and on the way out.

Cross-section of the Earth showing paths of some waves produced by an earthquake. The path of *P*-waves (black) that penetrate the core are sharply bent. *S*-waves (red) end at the core, although some portion may be converted to *P*-waves, traverse the core, and emerge in the mantle again as *P*- and *S*-waves. Both *P*- and *S*-waves may also be reflected back into the Earth at the surface. Transverse waves from an earthquake are not detected in the shadow of the Earth's liquid core.

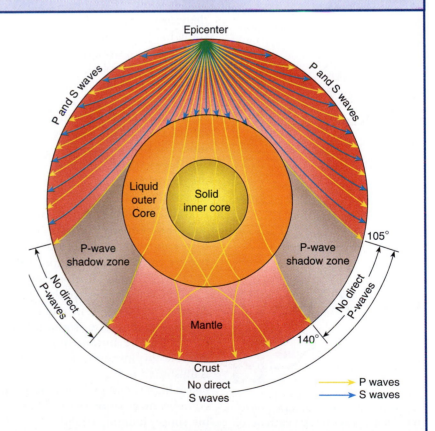

Before

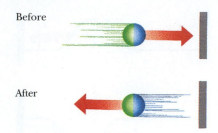

After

Figure 14-11 In contrast to a wave, the blue half of the ball leads on the way in and trails after reflection from the wall.

(a)

(b)

(c)

(d)

(e)

(f)

Figure 14-12 The two wave pulses on the rope pass through each other as if the other were not present.

shown in Figure 14-10 are steeper in front than in back. The steeper edge is away from you when the pulse moves down the rope and toward you when the reflected pulse returns. The leading edge continues to lead.

These inversions contrast with the behavior of a ball when it "reflects" from a wall. If the ball is not spinning, the top of the ball remains on top, but the leading edge is interchanged. Figure 14-11 shows that the blue half leads before the collision, whereas the green half leads afterwards.

> **Physics on Your Own**   You can investigate the behavior of waves on a long clothesline by tying one end to a rigid post. (As an alternative, you can use a long Slinky on a floor.) Do small pulses and big pulses travel away from you at the same speed? Do the shapes of the pulses remain the same? Do reflected waves pass through waves you send down the line? Is a crest inverted when it reflects from the post?

Superposition

Suppose you send a crest down the rope and when it reflects as a trough, you send a second crest to meet it. An amazing thing happens when they meet. The waves pass through each other as if the other were not there. This is shown in Figure 14-12. Each pulse retains its own shape, clearly demonstrating that the pulses are not affected by the "collision." A similar thing happens when you throw two pebbles into a pond. Even though the wave patterns overlap, you can still see a set of circular patterns move outward from *each* splash (Fig. 14-7).

In contrast, imagine what would happen if two particles—say, two Volkswagens—were to meet. Particles don't exhibit this special property. It would certainly be a strange world if waves did not pass through each other. Two singers singing at the same time would garble each other's music, or the sounds from one might bounce off those from the other.

During the time the wave pulses pass through each other, the resulting disturbance is a combination of the individual ones; it is a **superposition** of the pulses. As shown in Figure 14-13, the distance of the medium from the equilibrium position, the **displacement,** is the algebraic sum of the displacements of the individual wave pulses. If we consider displacements above the equilibrium position as positive and those below as negative, we can obtain the shape of the resultant disturbance by adding these numbers for each location along the rope.

If two crests overlap, the disturbance is bigger than either one alone. A crest and a trough produce a smaller disturbance. If the crest and the trough are the same size and have symmetrical shapes, they completely cancel for the instant during total overlap. A high-speed photograph taken at this instant yields a picture of a straight rope. This phenomenon is illustrated in Figure 14-12(d). This is not as strange as it may seem. If we take a high-speed photograph of a pendulum just as it swings through the equilibrium position, it would appear that the pendulum was not moving, but simply hanging straight down. In either case, longer exposures would blur, showing the motion.

Figure 14-13   This time sequence shows that the superposition of two wave pulses yields shapes that are the sum of the individual shapes.

Periodic Waves   Σ

A rope moved up and down with a steady frequency and amplitude generates a train of wave pulses. All the pulses have the same size and shape as they travel down the rope. The drawing in Figure 14-14 shows a **periodic wave** moving to the right. New effects emerge when we examine periodic waves. For one thing, unlike the single pulse, periodic waves have a frequency. The frequency of the wave is the oscillation frequency of any piece of the medium.

An important property of a periodic wave is the distance between identical positions on adjacent wave pulses, called the **wavelength** of the periodic wave. This is the smallest distance for which the wave pattern repeats. It might be measured between two adjacent crests, or two adjacent troughs, or any two identical spots on adjacent pulses as shown in Figure 14-15. The symbol for wavelength is the Greek letter lambda λ.

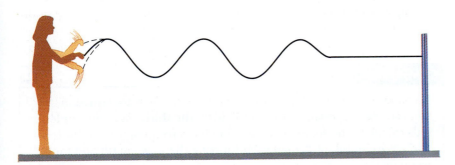

Figure 14-14   A periodic wave on a rope can be generated by moving the end up and down with a constant frequency.

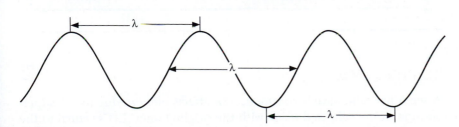

Figure 14-15   The wavelength of a periodic wave is the distance between any two identical spots on the wave.

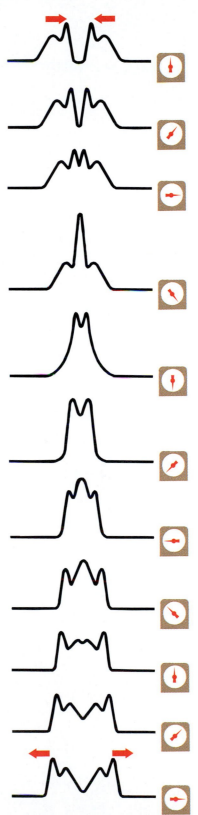

The speed of the wave can be determined by measuring how far a particular crest travels in a certain time. In many situations, however, the speed is too fast, the wavelength too short, or the amplitude too small to allow us to follow the motion of a single crest. We then use an alternative procedure.

Suppose you take a number of photographs at the same frequency as the vertical vibration of any portion of the rope. You would find that all of the pictures look the same. During the time the shutter of the camera was closed, each portion of the rope went through a complete cycle, ending in the position it had during the previous photograph. But this means that each crest moved from its original position to the position of the crest in front of it. That is, the crest moved a distance equal to the wavelength λ. Since the time between exposures is equal to the period T, the wave's speed v is

$$v = \frac{\lambda}{T}$$

It is usually easier to measure the wave's frequency than its period. Because the frequency is just the reciprocal of the period, we can change the equation to read

speed = wavelength × frequency

$$v = \lambda f$$

Although we developed this relationship for waves on a rope, there is nothing special about these waves. This relationship holds for all periodic waves, such as radio waves, sound waves, and water waves.

COMPUTING Sᴘᴇᴇᴅ ᴏғ ᴀ Wᴀᴠᴇ

If you know any two of the three quantities in the wave equation, you can use this relationship to calculate the third. As an example, let's calculate the speed of a wave that has a frequency of 40 hertz and a wavelength of ¾ meter. Multiplying the wavelength and the frequency gives us the speed.

$$v = \lambda f = (\text{¾ m})\,(40\ \text{Hz}) = 30\ \text{m/s}$$

QUESTION   If water waves have a frequency of 5 hertz and wavelength of 8 centimeters, what is the wave speed?

Standing Waves

When a periodic wave is confined, new effects emerge due to the superposition of the reflected waves with the original ones. Let's return to the example of a periodic wave moving down a rope toward a rigid post.

ANSWER $v = \lambda f = (8\ \text{cm})(5\ \text{Hz}) = 40\ \text{cm/s}$

Figure 14-16 A strobe drawing of a standing wave on a rope shows how the shape of the rope changes with time. The shape does not move to the left or right.

When the periodic wave reflects from the post, it superimposes with the wave heading toward the post. The complete pattern results from the superposition of the original wave and reflections from both ends. In general, we get a complicated pattern with a small amplitude, but certain frequencies cause the rope to vibrate with a large amplitude. Figure 14-16 shows multiple images of a resonating rope. Although the superimposing waves move along the rope, they produce a resonant pattern that does not move along the rope. Because the pattern appears to stand still (in the horizontal direction), it is known as a **standing wave.**

It might seem strange that two identical waves traveling in opposite directions combine to produce a vibrational pattern that doesn't travel along the rope. We can see how this happens by using the superposition principle to find the results of combining the two traveling waves. Let's start at a time when the crests of the traveling wave moving to the right (the blue line in Fig. 14-17) line up with the crests of the wave moving to the left (yellow line). Adding the displacements of the two traveling waves yields a wave which has the same basic shape, but which has twice the amplitude. This is shown by the black line in Figure 14-17(a).

Figure 14-17(b) shows the situation a short time later. The blue wave has moved to the right, and the yellow wave has moved the same distance to the left. The superposition at this time produces a shape that still looks like one of the traveling waves but is not as high. A short time later the crests of one wave line up with the troughs of the other. At this time the two waves cancel each other and the rope is straight. Although the rope is straight at this instant, some parts of the rope are moving up while others are moving down. The remaining drawings in Figure 14-17 show how this pattern changes as time progresses during the rest of the cycle.

Notice that some portions of the rope do not move. Even though each traveling wave by itself would cause all pieces of the rope to move, the waves interfere to produce no motion at these points. Such locations are known as **nodes** and are located on the vertical lines indicated by N's in Figure 14-17. The positions on the rope that have the largest amplitudes are known as **antinodes** and are on the lines marked by A's. Notice that the nodes and antinodes alternate and are equally spaced.

There is a relationship between the shape of the resonant pattern, or standing wave, and the moving periodic waves that superimpose to create it. The "wavelength" of the standing wave is equal to the wavelength of the underlying periodic wave. This can be seen in Figure 14-17.

Unlike the pendulum or the mass on a spring where there was only one resonant frequency, periodic waves that are confined have many different resonant frequencies. Strobe photographs of the standing wave with the lowest, or **fundamental, frequency** show that the rope has shapes

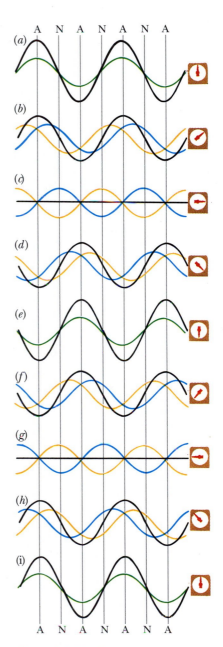

Figure 14-17 This set of strobe drawings shows how two traveling waves (the blue and yellow lines) combine to form a standing wave (the black line). Only the black line would be visible in an actual photograph.

Figure 14-18 The shapes of a rope oscillating as a standing wave of the lowest frequency.

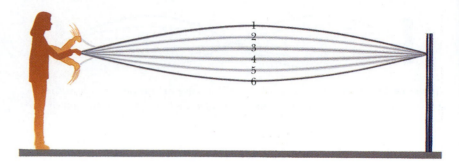

like those drawn in Figure 14-18. The images show how the shape of the rope changes during one-half cycle. At position 1 the rope has the largest possible crest. The displacement continually decreases until it becomes flat and the rope is straight (between positions 3 and 4). The rope's inertia causes it to overshoot and form a trough that grows in size. At the end of the half cycle, the rope is in position 6 and beginning its upward journey. The process then repeats.

> **QUESTION**   How does the distance between adjacent nodes or antinodes compare with the wavelength?

This pattern has the lowest frequency and thus the longest wavelength of the resonant modes. Notice that one-half wavelength is equal to the length of the rope, or the wavelength is twice the length of the rope. Because this is also the wavelength of the traveling waves, the longest resonant wavelength on a rope with nodes at each end is twice the length of the rope.

If we increase the frequency of the traveling wave slightly, the amplitude quickly decreases. The vibrational patterns are rather indistinct until the next resonant frequency is reached. This new resonant frequency has twice the frequency of the fundamental and is known as the second **harmonic.** Six shapes of the rope for this standing wave pattern are shown in Figure 14-19. When the rope is low on the left side, it is high on the right, and vice versa. Because the rope has the shape of a full wavelength, the wavelength of the traveling waves is equal to the length of the rope. Note that this pattern has one more node and one more antinode than the fundamental.

We can continue this line of reasoning. A third resonant frequency can be reached by again raising the frequency of the traveling wave. This third harmonic frequency is equal to three times the fundamental frequency. This standing wave has three antinodes and four nodes including the two at the ends of the rope. Other resonant frequencies occur at whole-number multiples of the fundamental frequency.

> **ANSWER**   Because one antinode is up when the adjacent ones are down, each antinodal region corresponds to a crest or a trough. Therefore, the distance between adjacent antinodes or adjacent nodes is one-half wavelength.

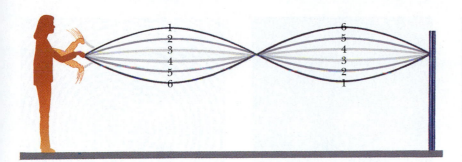

Figure 14-19 The shapes of a rope for a standing wave with the second resonant frequency.

Remember that the product of the frequency and the wavelength is a constant. This is a consequence of the fact that the velocities of all waves on this rope are the same. Therefore, if we increase the frequency by some multiple while keeping the velocity the same, the wavelength must decrease by the same multiple. The fundamental wavelength is the largest and its associated frequency is the smallest. As we march through higher and higher frequencies, we get shorter and shorter wavelengths.

> **QUESTION** How does the wavelength of the third harmonic compare with the length of the rope?

Interference

Standing waves on a rope are an example of the superposition, or **interference,** of waves in one dimension. If we use a two-dimensional medium, say the surface of water, we can generate some new effects.

Suppose we use two wave generators to create periodic waves on the surface of water in a ripple tank like the one in Figure 14-20(a). Because the two waves travel in the same medium, they have the same speed. We also assume that the two sources have the same frequency and that they are **in phase;** that is, both sources produce crests at the same time, troughs at the same time, and so on. The superposition of these waves creates the interference pattern shown in the photograph and drawing of Figure 14-20. The bright regions are produced by the crests, whereas the dark regions are produced by the troughs.

In some places crest meets crest to form a supercrest and, one-half period later, trough meets trough to form a supertrough. This meeting point is a region of large amplitude; the two waves form antinodal regions. In other places, crest and trough meet. Here, if the two waves have about the same amplitude, they cancel each other, resulting in little or no amplitude; the two waves form nodal regions.

> **ANSWER** The wavelength of the third harmonic is one-third the length of the fundamental wavelength. Because the fundamental wavelength is twice the length of the rope, the wavelength of the third harmonic would be two-thirds the length of the rope.

(a)

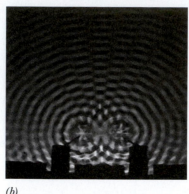
(b)

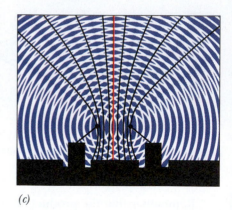
(c)

Figure 14-20   (a) A lightbulb above a ripple tank produces light and dark lines on the floor due to the water waves. (b) The interference pattern produced by two point sources of the same wavelength and phase. (c) The locations of the nodal lines in this pattern are shown in blue; the central antinodal line is shown in red.

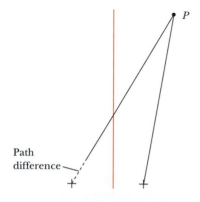

Figure 14-21   Whether the region at *P* is a nodal or antinodal region depends on the difference in the path lengths from the two sources.

Because of the periodic nature of the waves, the nodal and antinodal regions have fixed locations. These stationary interference patterns can be observed only if the two sources emit waves of the same frequency; otherwise one wave continually falls behind the other and the relationships between the two waves change. The two wave sources do not have to be in phase; there can be a time delay between the generation of crests by one source and the other as long as the time delay is constant. For simplicity, we usually assume that this time delay is zero; that is, the two sources are in phase.

The regions of crests and troughs lie along lines. One such antinodal line lies along the perpendicular to the midpoint of the line joining the two sources. This is the vertical red line in Figure 14-20(c). This central antinodal line is the same distance from the two sources. Therefore, crests generated a the same time at the two sources arrive at the same time at this midpoint to form supercrests. Similarly, two troughs arrive together, creating a supertrough.

Consider a point *P* off to the right side of the central line, as shown in Figure 14-21. Although nothing changes at the sources, we get a different result. Crests from the two sources no longer arrive at the same time. Crests from the left-hand source must travel a greater distance and therefore take longer to get to *P*. The amount of delay depends on the difference in the two path lengths.

If the point *P* is chosen such that the distance to the sources differs by an amount equal to one-half the wavelength, when waves from the two sources meet, crests overlap with troughs, and vice versa. The waves cancel. There are many points that have this path difference. They form nodal lines that lie along each side of the central line as shown by the blue lines in Figure 14-20(c). An antinodal line occurs when the path lengths differ by one wavelength; the next nodal line when the paths differ by 1½ wavelengths, and so on.

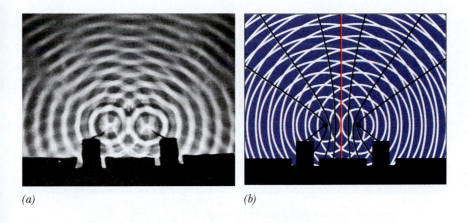

Figure 14-22 The nodal lines (blue) are more widely spaced for longer wavelengths. The central antinodal line (red) remains in the same location. Compare this with Figure 14-20.

> **QUESTION** Would the central line still be an antinodal line if the two sources were completely out of phase; that is, if one source generates a crest at the same time as the other generates a trough?

The photograph in Figure 14-22 shows the interference pattern for water waves with a longer wavelength than those in Figure 14-20. The nodal lines are now more widely spaced for the same source separation. Therefore, longer wavelengths produce wider patterns. Actually, the width of the pattern depends on the relative size of the source separation and the wavelength. As the ratio of the wavelength to the separation gets bigger, the nodal lines spread out. If the wavelength is much larger than the separation, the pattern is essentially that of a single source, whereas if it is much smaller, the nodal lines are so close together that they cannot be seen.

Diffraction

In the photographs in Figure 14-23 periodic water waves move toward a barrier. We see that the waves do not go straight through the opening in the barrier, but spread out behind the barrier. This bending of a wave is called **diffraction** and is definitely not a property of particles. If a BB gun is fired through an opening in a barrier, the pattern it produces is a precise "shadow" of the opening if we assume that no BBs bounce off the opening's edges.

The amount of diffraction depends on the relative sizes of the opening and the wavelength. If the wavelength is much smaller than the opening, very little diffraction is evident. As the wavelength gets closer to the size of the opening, the amount of diffraction gets bigger. In Figure

> **ANSWER** The central line would now be a nodal line because crests and troughs would arrive at the same time.

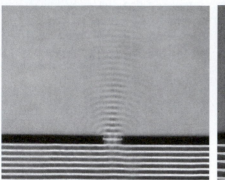

Figure 14-23 Ripple tank patterns of water waves moving upward and passing through a narrow barrier. Note that the amount of diffraction increases as the wavelength gets longer.

14-23(c) the opening and the wavelength are approximately the same size and the diffraction is evident.

Notice that diffraction produces nodal and antinodal lines similar to those observed in the interference patterns from two point sources. In this case, there is a broad central antinodal region with nodal lines on each side and the diffraction pattern is created by different portions of the wave interfering with themselves. The spacing of these lines is determined by the ratio of the wavelength and the width of the opening.

Physics on Your Own   Watch for diffraction effects when ocean waves strike offshore rocks. Are the "shadows" very sharp?

CHAPTER

14

REVISITED

When waves move in a medium, the medium oscillates in place. No material is transported from one location to another; it is the disturbance that moves. Unlike particles, when two waves pass through the same region at the same time, the individual disturbances add together. Afterwards, each wave retains its own identity.

SUMMARY

Vibrations and oscillations are described by the length of time required for one cycle, the period T (or its reciprocal, the frequency f), and the amplitude of the vibration, the maximum distance the object travels from the equilibrium point. When vibrations are small, the period is independent of the amplitude. The pendulum and a mass hanging on a spring are examples of systems that vibrate.

All systems have a distinctive set of natural frequencies. A simple system such as a pendulum has only one natural frequency, whereas more complex systems have many natural frequencies. When a system is excited at a natural frequency, it resonates with a large amplitude.

Waves are vibrations moving through a medium; it is the wave (energy) that moves through the medium, not the medium itself. Transverse waves vibrate perpendicular to the direction of the wave, whereas longitudinal waves vibrate parallel to the direction of the wave. The speed of a periodic wave is equal to the product of its frequency and its wavelength.

Waves pass through each other as if the other were not there. When they overlap, the shape is the algebraic sum of the displacements of the individual waves. When a periodic wave is confined, resonant patterns known as standing waves can be produced. Portions of the medium that do not move are called nodes, whereas portions with the largest amplitudes are known as antinodes. The fundamental has the lowest frequency and the longest wavelength.

Two identical periodic wave sources with a constant phase difference produce an interference pattern consisting of large-amplitude, antinodal regions and zero-amplitude, nodal regions. The spacing of the interference pattern depends on the relative size of the source separation and the wavelength.

Waves do not go straight through openings or around barriers, but spread out. This diffraction pattern contains nodal and antinodal regions and depends on the relative sizes of the opening and the wavelength.

KEY TERMS

amplitude: The maximum distance from the equilibrium position that occurs in periodic motion.

antinode: One of the positions in a standing wave or interference pattern where there is maximal movement; that is, the amplitude is a maximum.

crest: The peak of a wave disturbance.

cycle: One complete repetition of a periodic motion. It may start anyplace in the motion.

diffraction: The spreading of waves passing through an opening or around a barrier.

displacement: In wave (or oscillatory) motion, the distance of the disturbance (or object) from its equilibrium position.

equilibrium position: A position where the net force is zero.

frequency: The number of times a periodic motion repeats in a unit of time. It is equal to the inverse of the period.

fundamental frequency: The lowest resonant frequency for an oscillating system.

harmonic: A frequency that is a whole-number multiple of the fundamental frequency.

in phase: Two or more waves with the same wavelength and frequency that have their crests lined up.

interference: The superposition of waves.

longitudinal wave: A wave in which the vibrations of the medium are parallel to the direction the wave is moving.

node: One of the positions in a standing wave or interference pattern where there is no movement; that is, the amplitude is zero.

oscillation: A vibration about an equilibrium position or shape.

period: The shortest length of time it takes a periodic motion to repeat. It is equal to the inverse of the frequency.

periodic wave: A wave in which all the pulses have the same size and shape. The wave pattern repeats itself over a distance of one wavelength and over a time of one period.

resonance: A large increase in the amplitude of a vibration when a force is applied at a natural frequency of the medium or object.

spring constant: The amount of force required to stretch a spring by one unit of length. Measured in newtons per meter.

standing wave: The interference pattern produced by two waves of equal amplitude and frequency traveling in opposite directions. The pattern is characterized by alternating nodal and antinodal regions.

superposition: The combining of two or more waves at a location in space.

transverse wave: A wave in which the vibrations of the medium are perpendicular to the direction the wave is moving.

trough: A valley of a wave disturbance.

vibration: An oscillation about an equilibrium position or shape.

wave: The movement of energy from one place to another without any accompanying matter.

wavelength: The shortest repetition length for a periodic wave. For example, it is the distance from crest to crest or trough to trough.

CONCEPTUAL QUESTIONS

1. If the net force on a mass oscillating at the end of a vertical spring is zero at the equilibrium point, why doesn't the mass stop there?

2. If the restoring force on a pendulum is zero when it is vertical, why doesn't it quit swinging at this point?

3. If a mass hanging from a vertical spring is above the equilibrium point, which way does the net force point?

4. If a pendulum is to the right of vertical, which way does the restoring force on the pendulum bob point?

5. How could you increase the period of a mass on a spring?

6. How can you increase the period of a pendulum?

7. The amplitudes of real pendula decrease because of frictional forces. How does the period of a real pendulum change as it dies down?

8. Assume that you pull the mass on the spring 1 centimeter from the equilibrium position, let go, and measure the period of the oscillation. Would you expect the period to be larger, the same, or smaller if you pulled the mass 2 centimeters from the equilibrium position?

9. What is the frequency of the hand on a clock that measures the seconds?

10. What is the frequency of the hand on a clock that measures the minutes?

11. What is the period of the hand on a clock that measures the seconds?

12. What is the period of the hand on a clock that measures the minutes?

13. Suppose your grandfather clock runs too fast. If the mass on the pendulum can be moved up or down, which way would you move it to adjust the clock? Explain your reasoning.

*14. Does the natural frequency of a swing depend on whether you are standing up or sitting down?

15. Why do soldiers "break step" before crossing a suspension bridge?

16. Why is it not desirable for a radio loudspeaker to have resonant frequencies in the audio range? (This is a problem with some inexpensive AM radios.)

17. Which of the following multiples of the fundamental frequency will not produce resonance for a child's swing: ⅓, ½, 1, or 2?

18. A short pendulum has a natural frequency of 4 hertz. Which of the following driving frequencies would not produce resonance for this pendulum: 8 hertz, 4 hertz, 2 hertz, or 1 hertz?

19. What is being transported along a clothesline when a wave moves from one end to the other?

20. What actually moves across the surface of a lake when you drop a rock into it?

21. Sonar devices use underwater sound to explore the ocean floor. Would you expect sonar to be a longitudinal or a transverse wave? Explain.

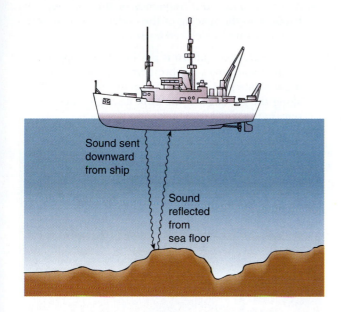

Sound sent downward from ship

Sound reflected from sea floor

22. Both transverse waves and longitudinal waves can exist in a long aluminum rod. Explain how you would produce each type.

23. Is it possible to send a small and a large pulse down a rope and observe one overtake the other? Explain your reasoning.

24. Is it possible for a shout to overtake a whisper?

25. Which of the following properties affect the speed of waves along a rope: amplitude of the pulse, shape of the pulse, tension in the rope, and the mass per unit length of the rope?

26. What can you do to increase the speed of waves on a rope?

27. When a trough reflects from the free end of a rope, does it return as a trough or a crest? Does this change if the end is fixed?

28. When a crest reflects from the free end of a stiff rod, does it return as a crest or a trough? How does this change if the end is fixed?

29. A pulse in the shape of a crest is sent from left to right along a stretched rope. A trough travels in the opposite direction so that the pulses meet in the middle of the rope. Would you expect to observe a crest or a trough arrive at the right-hand end of the rope?

30. How could you demonstrate that sound waves pass through each other without affecting each other?

31. If shapes a and b in the figure correspond to idealized wave pulses on a rope, what shape is produced when they completely overlap?

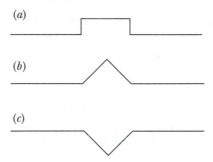

(*a*)

(*b*)

(*c*)

32. Repeat the previous question for shapes a and c.

33. In the following list of properties of periodic waves, which one is independent of the others: frequency, wavelength, speed, amplitude?

34. Two waves with the same speed have different frequencies. Which one has the longer wavelength?

35. When you decrease the frequency of a wave, what happens to its wavelength?

36. When you increase the period of a wave, what happens to its wavelength?

37. What happens to the wavelength of water waves as you decrease the frequency of putting your finger into the water?

38. What happens to the wavelength of waves on a rope if you move your hand up and down more rapidly?

39. If the frequency of a periodic wave is doubled while the velocity remains the same, what happens to the wavelength?

40. What happens to the wavelength of a periodic wave if the period is doubled while the velocity remains the same?

41. If the speed of a periodic wave doubles while the period remains the same, what happens to the wavelength?

42. What happens to the wavelength of a periodic wave if both the speed of the wave and the frequency are cut in half?

43. How far does a wave travel during one period?

44. How long does it take a wave to travel a distance equal to its wavelength?

45. How many nodes are there when a rope fixed at both ends vibrates in its fourth harmonic?

46. How many antinodes are there when a rope vibrates in its third harmonic if both ends of the rope are fixed?

47. How much higher is the frequency of the third harmonic on a rope than that of the fundamental?

48. How much higher is the frequency of the fourth harmonic on a rope than that of the second?

49. How does the wavelength of the fourth harmonic on a rope compare with the length of the rope?

50. How does the wavelength of the fourth harmonic on a rope compare with that of the second?

51. Assume that a rope has one fixed end and one free end. What is the wavelength of the fundamental? Assume that there is an antinode at the free end.

52. Can both the second and third harmonics exist for a rope with one fixed and one free end? Assume that there is an antinode at the free end.

53. A longitudinal standing wave can be established in a long, aluminum rod by stroking it with rosin on your fingers. If the rod is held at its midpoint, what is the wavelength of the fundamental standing wave? Assume that there are antinodes at each end of the rod and a node where the rod is held.

54. What is the wavelength of the fundamental for the rod in the preceding question if it is held midway between the center and one end?

55. Two point sources produce waves of the same wavelength and are in phase. At a point midway between the sources, would you expect to find a node or an antinode?

56. Two point sources produce waves of the same wavelength and are completely out of phase (that is, one produces a crest at the same time as the other produces a trough). At a point midway between the sources, would you expect to find a node or an antinode?

57. What happens to the spacing of the antinodal lines in an interference pattern when the two sources are moved farther apart?

58. As you increase the frequency of the sources, what happens to the spacing of the nodal lines in an interference pattern produced by two sources?

*59. An interference pattern is produced in a ripple tank. As the two sources are brought closer together, does the separation of the locations of maximum amplitude along the far edge of the tank decrease, increase, or remain the same?

*60. As the frequency of the two sources forming an interference pattern in a ripple tank increases, does the separation of the locations of minimum amplitude along the far edge of the tank increase, decrease, or remain the same?

61. What happens to the spacing of the antinodal lines in a diffraction pattern when the width of the opening is made larger?

62. As you decrease the frequency of the source, what happens to the spacing of the nodal lines in a diffraction pattern?

EXERCISES

1. If a mass on a spring takes 3 s to complete one cycle, what is its period?

2. A mass on a spring bobs up and down over a distance of 20 cm from the top to the bottom of its path twice each second. What are its period and amplitude?

3. A bob vibrates up and down on a vertical spring. Calculate its frequency if the period of the motion is 0.5 s.

4. A Foucault pendulum with a length of 9 m has a period of 6 s. What is its frequency?

5. If a mass on a spring has a frequency of 5 Hz, what is its period?

6. If a very long pendulum has a frequency of 0.2 Hz, how long does it take the pendulum to complete one cycle?

7. What is the period of a 0.4-kg mass suspended from a spring with a spring constant of 40 N/m?

8. A boy with a mass of 50 kg is hanging from a spring with a spring constant of 200 N/m. With what frequency does the boy bounce up and down?

9. A pendulum has a length of 10 m. What is its period?

10. A girl with a mass of 40 kg is swinging from a rope with a length of 2.5 m. What is the frequency of her swinging?

11. You observe the wave shape shown in the figure traveling along a rope. Draw the shape you expect to see reflected if the end of the rope is free.

12. Draw the shape you expect to see reflected from the fixed end of a rope if it initially looks like the shape in the figure.

*13. The highly idealized wave pulses shown in the figure at a time equal to zero have the same amplitudes and travel at 1 cm/s. Draw the shape of the rope at 2, 4, 5, and 8 s.

*14. Work the preceding problem, but change the rectangular pulse from a crest to a trough.

15. A periodic wave on a string has a wavelength of 50 cm and a frequency of 2 Hz. What is the speed of the wave?

16. If the breakers at a beach are separated by 5 m and hit shore with a frequency of 0.3 Hz, with what speed are they traveling?

17. What is the distance between adjacent crests of ocean waves that have a frequency of 0.2 Hz if the waves have a speed of 2 m/s?

18. Sound waves in iron have a speed of about 5100 m/s. If the waves have a frequency of 300 Hz, what is their wavelength?

19. If the wavelength of standing waves on a guitar string is 1.6 m and the speed of the waves is 410 m/s, what is the frequency of the note being played?

20. What is the period of waves on a rope if their wavelength is 0.4 m and their speed is 2 m/s?

21. A rope is tied between two posts separated by 2 m. What possible wavelengths will produce standing waves on the rope?

22. A 2-m-long rope is tied to a very thin string so that one end is essentially free. What possible wavelengths will produce standing waves on this rope?

*23. What is the fundamental frequency on a 4-m rope that is tied at both ends if the speed of the waves is 20 m/s?

*24. The speed of sound waves in aluminum is 5100 m/s. What is the fundamental frequency for standing waves in a 2-m aluminum rod if it is held at its center?

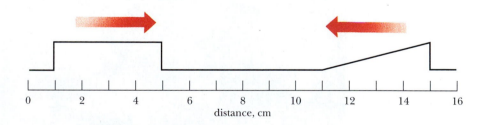

distance, cm

Question 13.

Sounds are all around us. Some are pleasant and some irritate and distract us. How do the sounds of music differ from those sounds we call noise and why do musicians use different size instruments? (See p. 370 for the answer to this question.)

Sound and Music

The West Chester University marching band.

hen we think of sound, we generally think of signals traveling
through the air to our ears. But sound is more than this. Sound also
travels through other media. For instance, old-time Westerns showed cow-
boys and Indians listening for the "iron horse" by putting their ears to the
rails or the ground; two rocks clapped together underwater are easily
heard by swimmers below the surface; a fetus inside a mother's womb can
be examined with ultrasound; and the voices of people talking in the next
room are often heard through the walls. Sounds can be soothing and musical, but they can also be irritating, or even painful.

Sound is a wave phenomenon. For example, we talk of the pitch, or
frequency, of sounds, which is definitely a wave characteristic. But what
other evidence do we have? The conclusive evidence is that sound exhibits superposition, something that we know distinguishes waves from
particles.

Sound

A vibrating object produces disturbances in the surrounding air. When it
moves outward, the sound wave pushes the air molecules away, creating a
compression. When it moves in the opposite direction, a partial void, or
rarefaction, is created near the object. Pressure differences cause the air
molecules to rush back into the region only to get pushed out again.
Thus, the air molecules vibrate back and forth near the object's surface as
illustrated by the tuning fork in Figure 15-1. The compressions and

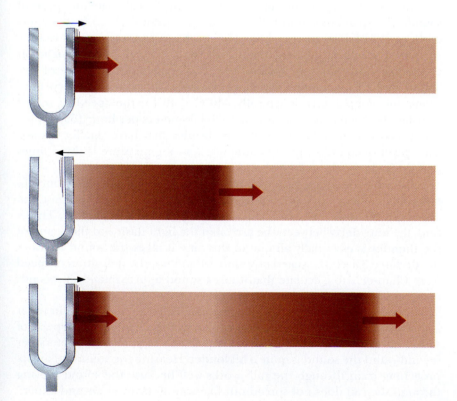

Figure 15-1 Sound is a longitudinal wave in which the air molecules
vibrate along the direction the wave
is traveling, producing compressions
and rarefactions that travel through
the air.

rarefactions travel away from the vibrating surface as a sound wave. Because the vibrations of the air molecules are along the same direction as the motion of the wave, sound is a **longitudinal wave.**

sound is a longitudinal wave

> **Physics on Your Own**   Investigate the behavior of longitudinal waves with a Slinky. For example, fix one end and send a compression wave pulse down the Slinky, bouncing it off the end. If you send two wave pulses down the Slinky, the first pulse will reflect and collide with the trailing one. Do these pulses superimpose the same way that transverse pulses do? Can you generate standing waves on the Slinky similar to those discussed in Chapter 14?

As with all waves, it is energy—not mass—that is transported along the path. The individual molecules of the air are not moving from one place to another; they simply vibrate back and forth. It is the disturbance that moves across the room when you talk. This disturbance, or wave, moves with a certain speed.

Speed of Sound

speed of sound = 343 m/s

Echoes demonstrate that sound waves reflect from surfaces, and that they move with a finite speed. You can use this phenomenon to measure the speed of sound in air. If you know the distance to the reflecting surface and the time it takes for the echo to return, you can calculate the speed of sound. The speed of sound is 343 meters per second (1125 feet per second), or 1235 kilometers per hour (767 mph), at room temperature.

Experiments have shown that the speed of sound does not depend on the pressure, but does depend on the temperature and the type of gas. Sound is slower at lower temperatures. At the altitude of jet airplanes, where the temperature is typically −40°C (−40°F), the speed of sound drops to 310 meters per second, or 1020 kilometers per hour (690 mph). The speed is higher for gases with molecules that have smaller masses. The speed of sound in pure helium at room temperature is three times that in air.

Knowing the speed of sound in air allows you to calculate your distance from a lightning bolt. Since the speed of light is very, very fast, the time light takes to travel from the lightning to you is negligible. Therefore, the time delay between the arrival of the light flash and the sound of the thunder is essentially all due to the time it takes the sound to travel the distance. Given the speed of sound, we can use the definition of speed from Chapter 1 to calculate that it takes sound approximately 3 seconds to travel 1 kilometer (5 seconds per mile).

Sound waves also travel in other media. The speed of sound in water is about 1500 meters per second, much faster than in air. The speed of sound in solids is usually quite a bit higher—as high as 5000 meters per second—and the sound is quite a bit louder. Hearing the sounds of an approaching train through the rails works well because the sound moving through the rail does not spread out like sound waves in air and experiences less scattering along the path. You can easily experience this phenomenon in the classroom. Have a friend scratch one end of a meterstick while you hold the other end next to your ear. The effect is striking.

Sonar (sound navigation ranging) uses the echoes of sound waves in water to determine the distances to underwater objects. These sonar devices emit sounds and measure the time required for the echo to return. Sonar is used to determine the depth of the water, search for schools of fish, and locate submarines.

> **QUESTION**   If it takes the thunder 9 seconds to reach you, how far away is the lightning?

> **Physics on Your Own**   Hook an elastic band to your eyetooth and pluck it. The sound is much louder to you than to a nearby friend because you hear the sound that travels through your jawbone, but your friend hears only the sound traveling through air. What happens when you stretch the band?

Hearing Sounds

Sound perception is a complex phenomenon whose study involves a wide variety of sciences including physiology, psychology, and the branch of physics called acoustics. Experiments with human hearing, for example, clearly show that our perception is often different from the simple interpretations of the measurements taken by instruments. Our perception of pitch depends mainly on frequency but is also affected by other properties, such as loudness. And loudness depends on the amplitude of the wave (as well as the response of our ears). When our ear–brain systems tell us that one sound is twice as loud as another, instruments show that the power output is nearly eight times greater.

Our ears intercept sound waves from the air and transmit their vibrations through internal bone structures to special hairlike sensors. The ear canal acts as a resonator, greatly amplifying frequencies near 3000 hertz. This amplified sound wave moves the eardrum, which is located at the end of the ear canal as shown in Figure 15-2. The eardrum is connected to three small bones in our middle ear. When sound reaches the middle ear, it has been transformed from a wave in air to a mechanical wave in the bones. These bones then move a smaller oval window inside the ear. The leverage advantage of the bone structure and the concentration of the pressure vibration onto a smaller window further amplify the sound, increasing our ability to hear faint sounds. The final transformation of mechanical sound waves to nerve impulses takes place in the inner ear.

> **ANSWER**   The lightning bolt must have been 3 kilometers away, or a little less than 2 miles.

Figure 15-2 The structure of the human ear.

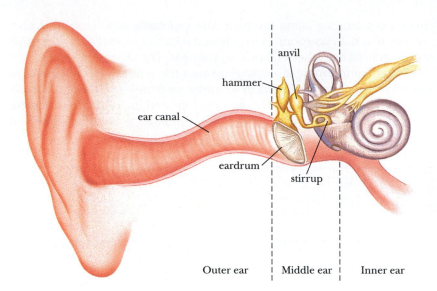

Outer ear | Middle ear | Inner ear

The pressure vibrations in the fluid of the inner ear resonate with different hairlike sensors depending on the frequency of the sound.

The range of frequencies that we can hear clearly depends on the resonant structures within our ears. When a frequency is too high or too low, the sound wave is not amplified like those within the audible range. The audible range is normally from 20 hertz to 20,000 hertz, although it varies with age and the individual. The sensitivity of our ears varies over this range, being very insensitive near both ends. As we get older, our ability to hear higher frequencies decreases.

Physics Update

Acoustic time-reversal mirrors (TRMs) are devices that record a sound wave emanating from a source and generate a new one that behaves as if the original were traveling backwards in time. Previously TRMs had been rigorously tested for sound propagating through fluids such as water or air. For example, shouting "too" at the device would yield a reversed acoustic wave (sounding something like "oot") that converges backward toward the speaker's mouth. The principle has now been shown to be valid in solids by a team at the University of Paris. TRMs have potential applications for detecting tiny metallic defects in airplanes and for locating and destroying kidney stones. For example, shining ultrasound waves through a patient with a kidney stone produces distinctive echoes from the stone. The TRM would record the echoes and then generate a reversed wave, sending back sound energy that would travel back to the stone, shattering it.

The Recipe of Sounds

Suppose you are in a windowless room but can hear sounds from the outside. You would have no trouble identifying most of the sounds you hear. A bird sounds different from a foghorn, a trumpet different from a baritone. Why is it that you can recognize these different sources of sound?

They might be producing different notes; a bird sings much higher than a foghorn. They may make different melodies; some birds have a rather complex sequence of notes in their call, whereas the foghorn produces one continuous note.

What if all the sound makers outside your room are restricted to a single note? Do you suppose that you could still identify the different sources? Most likely. When each source produces the note, it is accompanied by other higher, resonant frequencies. You actually hear a superposition of these frequencies. The reason you can distinguish among the sources is that each sound has a unique combination of intensities of the various harmonics (Chapter 14). One sound might be composed of a strong fundamental and weaker, even higher harmonics, and another may have a particularly strong second harmonic. Each sound has a particular recipe of resonant frequencies that combine to make the total sound.

The vibrating strings in a piano, violin, guitar, or banjo have their own combinations of the fundamental frequency and higher harmonics. The relative intensities of these harmonics depend primarily on the way the string was initially vibrated and on the vibrational characteristics of the body of the instrument. The initial part of the sound is called the attack and its character is determined, in part, by how the sound is produced. Bowing, for example, produces a different sound from plucking or striking. The manner in which the various components of the sound decay also differs from one instrument to another. The wave patterns produced by various sounds are illustrated in Figure 15-3.

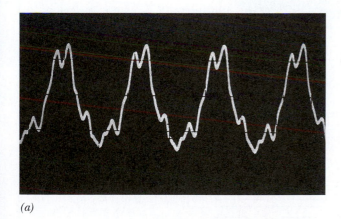

(a)

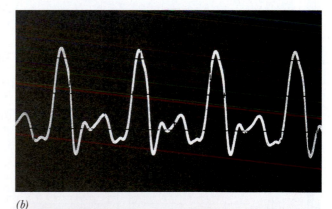

(b)

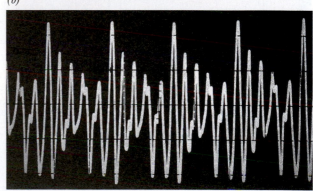

(c)

Figure 15-3 Wave patterns of (a) a violin, (b) a trumpet, and (c) a bassoon. These photographs show only general characteristics; the details depend on the placement of the microphone.

Animal Hearing

Animals have different ranges of sensitivity for hearing than humans. Dogs and bats, for example, have hearing ranges that extend to ultrasonic frequencies, frequencies above those that we can hear.

Unlike humans, most animals use their hearing as an aid in gathering food and escaping danger. Bats squeak at ultrasonic frequencies and detect the echoes from small flying insects (Fig. A); robins cock their heads in early spring as they listen for the very faint sounds of worms in the ground, and owls have two different types of ears providing them with binocular hearing for finding mice moving through grassy fields.

Animal researchers have found that homing pigeons and elephants hear very, very low-frequency sounds, infrasonic frequencies. Low-frequency sounds do not attenuate as rapidly, and therefore travel much farther than high-frequency sounds. In the case of pigeons, hearing the sounds of skyscrapers swaying in the wind in distant cities may provide them with navigational bearings. There is evidence that elephants communicate with each other over distances of miles using a low-frequency rumble. Figure B displays the frequency ranges that some animals can hear.

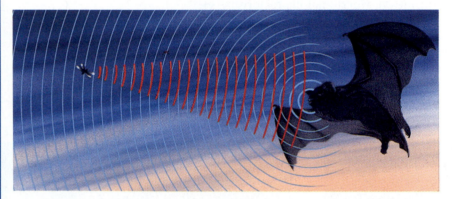

Figure A
The bat emits sound waves (shown in blue) that reflect off the insect and return (shown in red) to the bat giving away the insect's position.

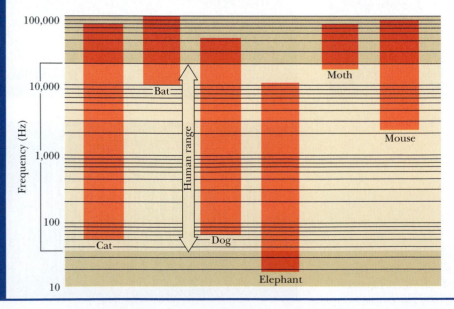

Figure B
The frequency ranges of hearing for some animals.

Our voices have the same individual character. We recognize different voices because of their particular recipe of harmonics. Research has shown that this recipe changes under emotional stress. Some people have suggested that these changes in the recipe of sound of a voice can be used in lie detection. Electronic devices can now decompose the composite waveform—the superposition of all the harmonics—and evaluate the relative strength of each harmonic. This technique spots changes in higher harmonics that are virtually undetectable in the composite waveform.

Physics Update

Digital versatile discs (DVDs) appeared in consumer products in early 1997. The same size as conventional compact discs (CDs), DVDs hold about 14 times more data because of a combination of innovations. For example, both types of disc encode digital data as tiny pits on thin plastic platters, but for DVDs the pits are smaller (0.4 versus 0.83 micrometers [micro = 10^{-6}] for CDs), the tracks of pits are closer together (0.7 versus 1.6 micrometers), and the laser light used to read data has a wavelength of 635 to 650 nanometers (nano = 10^{-9}) rather than the 780 nanometers used for CDs; all of these factors allow data to be crowded in more densely. Furthermore, the DVD consists of two layers, each of which can hold data. With a capacity of 4.7 gigabytes (giga = 10^{9}), the DVD provides a variety of multimedia products, even movies, at least in compressed form. And all of this is without resorting to blue-light lasers (still under development), whose shorter wavelengths would permit even more data to be compressed into a given area.

How does music differ from other sounds? This question is difficult because what is music to one person might be noise to you. In general, music can be defined as that collection of periodic sounds that is pleasing to the ear.

Most cultures divide the totality of musical frequencies into groups known as octaves. A note in one octave has twice the frequency of the corresponding note in the next lower octave. For example, the pitches labeled C in ascending adjacent octaves have frequencies of 262, 524, and 1048 hertz, respectively. In Western cultures, most music is based on a scale that divides the octave into 12 steps. It may seem strange that an octave spans 12 notes, since the word *octave* derives from the Latin word meaning "eight." An octave contains seven notes and five sharps and flats. Indian and Chinese music have different divisions within their octaves.

Through the years, people have created instruments to produce sounds that were pleasing to them. Nearly all of these instruments involve the production of standing waves. Although there is an enormous variety of instruments, most of them can be classified as string, wind, or percussion instruments.

Stringed Instruments   Σ

When a string vibrates, it moves the air around it, producing sound waves. Because the string is quite thin, not much air is moved and, consequently, not much sound is produced. In acoustic stringed instruments, this lack

Loudest and Softest Sounds

The intensity of sound is measured in terms of the sound energy that crosses a square meter of area in one second and is measured by a sound level meter. At a reference value of 1000 hertz, the faintest sounds that can be heard by the human ear have intensities a little less than one-trillionth (10^{-12}) of a watt per square meter. This results from a variation in pressure of about 0.3 billionths of an atmosphere and corresponds to a displacement of the air molecules of approximately one-billionth of a centimeter, which is less than the diameter of the molecule! The ear is a very sensitive instrument.

On the loud side the ear cannot tolerate sound intensities much greater than one watt per square meter without experiencing pain. This corresponds to a variation in pressure of about one-thousandth of an atmosphere, one million times bigger than for sounds that can be barely heard. The displacements of the molecules are also one million times larger.

A source of sound that is transmitting at 100 watts in a spherically symmetric pattern is painful to our ears at a distance of 1 meter and theoretically audible at about 3000 kilometers (2000 miles!), if we assume no losses in moving through the air. Psychoacoustic scientists report that when the intensity of sound is increased about 8 times, people report a doubling in the loudness of the sound.

Because of the very large range of sensitivity of the ear as well as our perceptual scale, the scale scientists use to distinguish different sound levels is based on multiples of 10. A sound that has 10 times as much intensity as another sound has a level of 10 decibels, or 10

dB's (pronounced "dee bees") higher than the first sound. Since every increase of 10 decibels corresponds to a factor of 10 increase in intensity, an increase of 20 decibels means that the intensity increases by $10 \times 10 = 100$ times. Sound level meters use this same nonlinear scale. The table gives some representative values of sound intensity.

Decibel levels for some common sounds

Source	Decibels	Energy Relative to Threshold	Sensation
Nearby jet taking off	150	1 quadrillion	
Jackhammer	130	10 trillion	
Rock concert, automobile horn	120	1 trillion	Pain
Police siren	110	100 billion	
Power lawn mower, loud music	100	10 billion	
Screaming baby	90	1 billion	Endangers hearing
Traffic on a busy street	80	100 million	Noisy
Vacuum cleaner	70	10 million	
Conversation	60	1 million	
Library	40	10,000	Quiet
Whisper	30	1000	Very quiet
Breathing	10	10	Just audible
	0	1	Hearing threshold

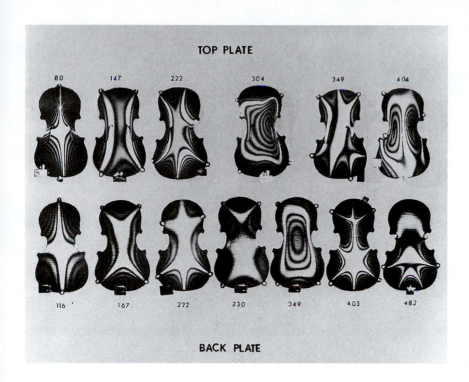

TOP PLATE

80   147   222   304   349   404

116   167   222   230   349   403   482

BACK PLATE

Figure 15-4 The vibrations in the body of the violin are made visible by holographic techniques.

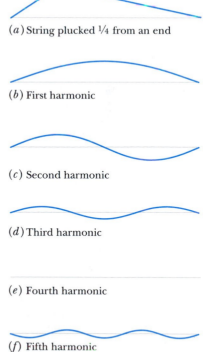

(*a*) String plucked ¼ from an end

(*b*) First harmonic

(*c*) Second harmonic

(*d*) Third harmonic

(*e*) Fourth harmonic

(*f*) Fifth harmonic

Figure 15-5 The shape of the plucked string (a) is the superposition of the first five harmonics (b–f). Note that the fourth harmonic does not contribute to this particular shape.

of volume is solved by mounting the vibrating string on a larger body. The vibrations are transmitted to the larger body, which can move more air and produce a louder sound. Figure 15-4 shows a variety of vibrations produced in a violin's body by its vibrating strings. The design of the instrument produces variations in the instrument's vibrational patterns and thus changes the character of the sound produced. There was something special about the way Antonio Stradivari made his violins that made their sound more pleasing than others.

Modern electric guitars do not use the body of the instrument to transmit the vibrations of the strings to the air. The motions of the vibrating strings are converted to an oscillating electric signal by a crystal mounted on the bridge. This signal is then amplified and sent to the speakers.

Whether acoustic or electric, plucking a guitar string creates vibrations. This initial distortion causes waves to travel in both directions along the string and reflect back and forth from the fixed ends. The pluck generates a unique set of standing wave frequencies (Chapter 14). The initial shape of the string is equivalent to a superposition of many waves. Figure 15-5 shows the shape of a plucked string at the moment of release and the contributions of the first five harmonics. (For this particular shape, the fourth harmonic is zero everywhere and does not contribute.)

These harmonic waves travel back and forth on the string, creating standing waves with nodes at the two ends of the string. The first three standing waves are shown in Figure 15-6. Since the distance between nodes is one-half wavelength, the wavelength of the fundamental, or first

Figure 15-6 The first three standing waves on a guitar string. The wavelengths of these harmonics are $2L$, L, and $2L/3$, respectively.

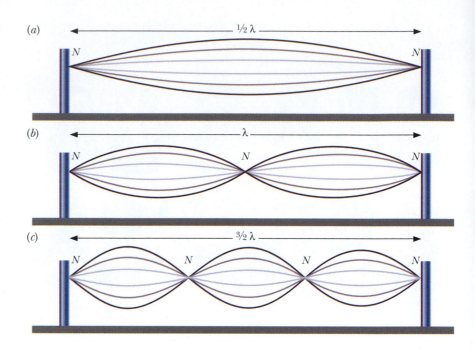

harmonic (a), is twice as long as the string, or $\lambda_1 = 2L$. The wavelength of the second harmonic (b) $\lambda_2 = L$, the length of the string, and therefore half as long as that of the fundamental. The wavelength of the third harmonic (c) $\lambda_3 = 2L/3$, one-third of the fundamental wavelength.

The frequencies of the various harmonics can be obtained from the relationship for the wave's speed, $v = \lambda f$. As long as the vibrations are small, we can assume that the speeds of the different waves on the plucked string are all the same. Therefore, $v = \lambda_1 f_1 = \lambda_2 f_2 = \lambda_3 f_3 = \ldots$ Because the second harmonic has half the wavelength, it must have twice the frequency. The third harmonic has one-third the wavelength and three times the frequency. The harmonic frequencies are whole-number multiples of the fundamental frequency.

harmonic wavelengths λ_1, $\lambda_1/2$, $\lambda_1/3, \ldots$

speed = wavelength × frequency

harmonic frequencies f_1, $2f_1$, $3f_1, \ldots$

> **QUESTION**   What is the wavelength corresponding to the third harmonic on a 60-centimeter-long wire?

There is a simple way to verify that more than one standing wave is present on a vibrating string. By lightly touching the string in certain places, particular standing waves can be damped out and thus make others more obvious. For example, suppose you touch the center of the vibrating guitar string. We can see from Figure 15-6(a) that the fundamental will be damped out since it has an antinode at the center. The second harmonic, however, has a node in the middle [Fig. 15-6(b)], so it is

> **ANSWER**   It must be one-third the length of the fundamental wavelength. Because the fundamental wavelength is twice the length of the string, we obtain 2(60 cm)/3 = 40 cm.

unaffected by your touch. The third harmonic is damped because it also has an antinode in the middle of the string. In fact, all odd-numbered harmonics are damped out, and all even-numbered ones remain.

When you do this experiment, you hear a shift in the lowest frequency. With the initial pluck the fundamental is the most prominent frequency. After touching the middle of the string, the fundamental is gone, so the lowest frequency is now that of the second harmonic, a frequency twice the original frequency. In musical terms we say that the note shifts upward by one octave.

> **Physics on Your Own** Investigate the harmonics produced on a guitar string. Pluck the string and listen for the fundamental (first) harmonic. While the string is still vibrating, gently touch the string at its midpoint. You should be able to hear the second harmonic. Find the higher harmonics by plucking the string again and gently touch it ⅓, ¼, ⅕ . . . of the way from one end.

Music, of course, consists of many notes. The string we have been discussing can play only one note at a time since the fundamental frequency determines the musical note. Instruments must be able to play many different frequencies to be useful. To play other notes we must have more strings, or a simple way of changing the vibrational conditions on the string. Most stringed instruments use similar methods for achieving different notes. Pianos, harps, and harpsichords have many strings. Different notes are produced by striking or plucking different strings.

> **QUESTION** What are the different ways to raise the fundamental frequency of a note played on a guitar string?

A guitar usually has six strings, some more massive than others to extend the range of frequencies. Still, only six strings aren't enough to produce many interesting songs. A guitarist must be able to change the note produced by each string. By fingering the string the guitarist shortens the vibrating portion of the string, creating new conditions for standing waves. The new fundamental wavelength is now twice this *shortened* length and therefore smaller than before. This smaller wavelength produces a higher frequency.

The guitarist can change the speed of the waves by changing the tension of the string. An increase in the tension increases the speed and therefore increases the frequency. It takes too long to change the tension in the middle of a song so changes in tension are only used to tune the instruments. An exception is the "washtub bass" shown in Figure 15-7.

> **ANSWER** To raise the fundamental frequency of a note on a guitar string, you could increase the tension, finger the string, or use a string with less mass per unit length.

Guitarist Eric Clapton changes notes by pushing the strings against the frets to shorten the lengths that vibrate.

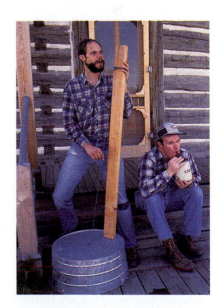

Figure 15-7 The note played by this one-string bass is changed by varying the tension (and thus the wave velocity) in the string.

Figure 15-8 A drawing of a longitudinal sound wave. The dark regions represent compressions (high density); the lighter regions represent rarefactions (low density).

Wind Instruments

Σ

Wind instruments—such as clarinets, trumpets, and organ pipes—are essentially containers for vibrating columns of air. Each has an open end that transmits the sound and a method for exciting the air column. With the exception of the organ pipe, all wind instruments also have a method for altering the fundamental frequency. In many ways wind instruments are analogous to the stringed instruments. A spectrum of initial waves is created by a disturbance. The standing waves that are generated are governed by the instrument; all other frequencies are quickly damped out. The sound we hear is a combination of the frequencies of these standing waves.

But there are three main differences between wind and string instruments. First, unlike vibrating strings, the vibrational characteristics of air cannot be altered to change the speed of the waves. Only the length of the vibrating air column can be changed. Second, a string has a node at each end, but there is an antinode near the open end of the wind instrument. At an antinode the vibration of the air molecules is a maximum. A node occurs at the closed end, where there is no vibration of the air molecules. Finally, the disturbance in wind instruments produces longitudinal waves in the air column instead of the transverse waves of the stringed instruments. A longitudinal wave at a single moment is represented in Figure 15-8.

Drawing longitudinal standing waves is difficult because the movement of the air molecules is along the length of the pipe. If we draw them in this manner for even one period, crests and troughs overlap and produce a very confusing drawing. When illustrating longitudinal waves in a wind pipe, we normally draw them *as if* the air movements were transverse. This should not be overly confusing as long as you remember that these drawings are basically graphs of the displacements of the air molecules. In Figure 15-9 two curves are drawn to represent the range of dis-

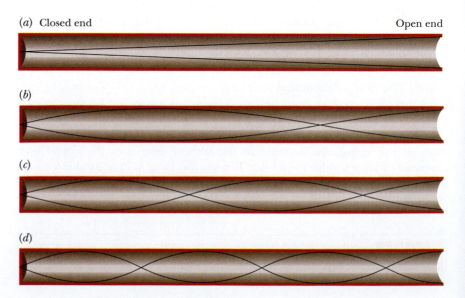

(a) Closed end   Open end

(b)

(c)

(d)

Figure 15-9 The first four standing waves in a closed organ pipe illustrated as transverse displacements. Note that there is a node at the closed end and an antinode at the open end.

placement of the standing waves. Note that the curves meet at the closed end of the pipe. There is a node at that spot, indicating that the air molecules near the wall do not move.

Consider a closed organ pipe—one that is closed at one end and open at the other. The largest wavelength that produces a node at the closed end and an antinode at the open end is four times the length of the pipe, or $4L$. This is illustrated in Figure 15-9(a). The next-smaller wavelength [Fig. 15-9(b)] is four-thirds the length of the pipe, or $4L/3$; and the wavelength in Figure 15-9(c) is four-fifths the length of the pipe, or $4L/5$.

As with the stringed instruments we can generate a relationship among the various frequencies by examining the relationship between the wave's speed, its wavelength, and its frequency; $v = \lambda f$. The possible wavelengths are 1, ⅓, ⅕, ⅐, . . . times the fundamental. Because the wave speeds are constant, the corresponding frequencies are 1, 3, 5, 7, . . . times the fundamental.

The turbulence caused by blowing into the pipe creates a wide range of possible frequencies. As with the stringed instruments, the surviving standing waves are those that match the conditions at the boundaries of the instrument and "fit" into the instrument. The closed organ pipe, for example, does not have resonant frequencies that are 2, 4, 6, . . . times the fundamental; the even harmonics are missing because these frequencies would not produce a node at the closed end and an antinode at the open end. On the other hand, the open organ pipe (one that is open at both ends) has all harmonics.

> **QUESTION** What is the wavelength corresponding to the fifth harmonic in a 50-centimeter-long closed organ pipe?

> **Physics on Your Own** Create a soft-drink bottle instrument. The bottles produce musical notes when you blow across their tops. You can adjust the frequencies of the notes by filling the bottles with different amounts of water, thus changing the length of the vibrating column of air above the water.

Percussion Instruments

Percussion instruments are characterized by their lack of the harmonic structure of the string and wind instruments. Percussionists employ a wide range of instruments, including drums, cymbals, bells, triangles, gongs, and xylophones. Although all of these resonate at a variety of frequencies, the higher frequencies are not whole-number multiples of the lowest frequency.

> **ANSWER** It must be one-fifth the length of the fundamental wavelength. Because the fundamental wavelength is four times the length of the pipe, we obtain 4(50cm)/5 = 40 cm.

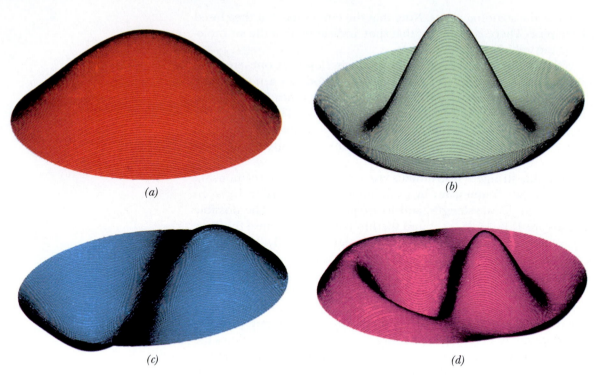

Figure 15-10   A circular drumhead can vibrate in many different modes with non-harmonic frequencies.

Each of the percussion instruments behaves in a different way. The restoring force in a drum is provided by the tension in the drumhead. The two-dimensional standing-wave patterns are produced by the reflection of transverse waves from the edges of the drum and are characterized by nodal lines. These nodal lines are the two-dimensional analogs of the nodal points in vibrating strings.

The nodal lines for a circular membrane are either along a diameter or circles about the center. Of course, there is always a nodal line around the edge of the drumhead. Figure 15-10(a) shows the fundamental mode in which the center of the drumhead moves up and down symmetrically. Figure 15-10(c) shows a mode with an antinodal line along a diameter. When one-half of the drumhead is up, the other half is down. This mode has a frequency 1.593 times that of the fundamental. The second symmetric mode is shown in Figure 15-10(b) and has a frequency 2.295 times the fundamental.

Beats   Σ

When we listen to two steady sounds with nearly equal frequencies, we hear a periodic variation in the volume. This effect is known as **beats** and should not be confused with the rhythm of the music that you might dance to; beats are the result of the superposition of the two waves. Because the two waves have different frequencies, there are times when they add together and times when they are out of step and cancel. The result is a periodic cancellation and reinforcement of the waves that is heard as a periodic variation in the loudness of the sound.

This is illustrated by the drawings in Figure 15-11. It is important to realize that these drawings do not represent strobe pictures of the waves. The horizontal line represents time, not position. The drawings represent the variation in the amplitude of the sound at a single location; for instance, as it reaches one of your ears. Figures 15-11(a) and (b) show each wave by itself, while Figure 15-11(c) shows the superposition of these two waves. Your ears hear a frequency that is the average of the two frequencies and that varies in amplitude with a beat frequency. The loudness you hear is related to your ear's response and the amplitude of the combined sound wave. For our purposes, the variation of the amplitude in the drawing is an adequate indication of the variation in the loudness you perceive.

We can obtain the beat frequency by examining Figure 15-11. If the time between vertical lines is 1 second, wave (a) has a frequency of 10 hertz and wave (b) 11 hertz. At the beginning of a second a crest from wave (a) cancels a trough from wave (b), and the sound level is zero. Wave (a) has a lower frequency and continually falls behind. At the end of 1 second it has fallen an entire cycle behind, and the waves once again cancel. The difference of 1 hertz in their frequencies shows up as a variation in the sound level that has a frequency of 1 hertz. This same process is valid for any frequencies that differ by 1 hertz. For example, the beat frequency produced when two sound waves of 407 hertz and 408 hertz are combined is also 1 hertz.

If the frequencies differ by 2 hertz, it only takes ½ second for the lower frequency wave to fall one cycle behind. Therefore, this pattern happens two times per second, or with a beat frequency of 2 hertz. This reasoning can be generalized to show that the beat frequency is equal to the difference in the two frequencies.

beat frequency $= \Delta f$

Piano tuners employ this beat phenomenon when tuning pianos. The tuner produces the desired frequency by striking a tuning fork, and

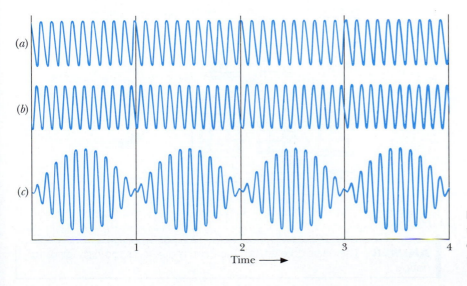

(a)

(b)

(c)

1   2   3   4

Time ⟶

Figure 15-11   The superposition of two sound waves of different frequencies produces a sound wave (c) that varies in amplitude.

adjusts the piano wire's tension until the beats disappear. Modern electronics have now made it possible for a tone-deaf person to make a living tuning pianos.

> **QUESTION**   Why don't we hear beats when adjacent keys on a piano are hit at the same time?

> **Physics on Your Own**   Play the same note on two strings of a guitar. As you adjust the tension of one of them, can you hear beats?

Doppler Effect   Σ

What we hear is not necessarily the sound that was originally produced, even when only a single source is involved. Recall the sound of a car as it passes you. This variation in sound is especially noticeable with racing cars and some motorcycles since they emit distinctive roars. Imagine standing by the side of a road as a car passes you with its horn blasting constantly. Two things about the horn's sound change. First, as the car ap-

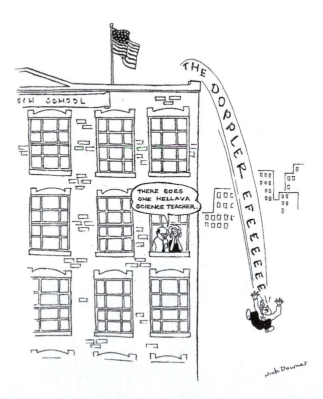

> **ANSWER**   The beats do exist, but the beat frequency is too high for us to notice.

proaches, its sound gets progressively louder; as it leaves, the sound gets quieter. The volume changes simply because the wave spreads out as it moves away from the source. If you are far from the source, the energy of the sound waves intercepted by your ears is smaller.

The second change in the sound of the horn might not be as obvious. The frequency *that you hear* is not the same as the frequency that is actually emitted by the horn. The frequency you hear is shifted upward as the car approaches you and downward as the car recedes from you. This shift in frequency due to the motion is called the **Doppler effect,** after the Austrian physicist and mathematician Christian Doppler.

The pitch of the sound we hear is determined by the frequency with which crests (or troughs) hit our ears. Our ears are sensitive to the frequency of a wave, not to the wavelength. Figure 15-12 shows a two-dimensional drawing of sound waves leaving a tuning fork. A small portion of the sound is intercepted by the ear. If the tuning fork and the ear are stationary relative to the air, the frequency heard is the same as that emitted.

When the tuning fork is moving, the listener hears a different frequency. Because the tuning fork moves during the time between the generation of one crest and the next, the sound waves crowd together in the forward direction and spread out in the backward direction as shown in Figure 15-13. If the tuning fork moves toward the listener, the ear detects the sound waves that are crowded together. The frequency with which the waves hit the ear is therefore higher than was actually emitted by the source. Similarly, if the tuning fork moves away from the listener, the observed frequency is lower.

Notice, however, that in both Figures 15-12 and 15-13 the spacings do not change with the separation of the source and receiver. Because the shift does not depend on the distance, the Doppler-shifted frequency does not change as the source gets closer or farther away.

> **QUESTION** If you were flying a model airplane on a wire so that it traveled in circles about you, would you hear a Doppler shift?

We will see in later chapters that the Doppler effect also occurs for light. The observation that the light from distant galaxies is shifted toward lower frequencies (their colors are redshifted) tells us that they are moving away from our Galaxy and that the Universe is expanding.

The Doppler effect also occurs when the receiver moves and the source is stationary. If the listener moves toward a stationary tuning fork, her ear intercepts wave crests at a higher rate and she hears a higher frequency. If she recedes from the tuning fork, the crests must continually catch up with her ear, which therefore intercepts the crests at a reduced rate, and she consequently hears a lower frequency. Notice that the shift in the frequency is still constant as long as the velocities are constant. Of

> **ANSWER** Because the airplane is not moving toward or away from you, you would not hear a Doppler shift. However, someone standing off to the side watching you would hear a Doppler shift.

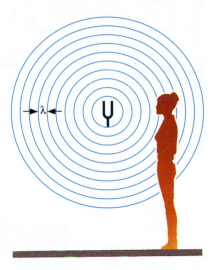

Figure 15-12 If the source and the ear are stationary relative to the air, the ear hears the same frequency as emitted by the source.

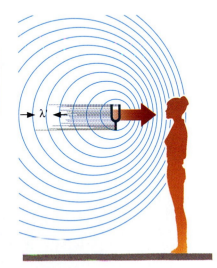

Figure 15-13 If the source of sound moves toward the right, the waves bunch up on the right-hand side and spread out on the left-hand side. The person hears a frequency that is higher than that emitted by the source.

Breaking the Sound Barrier

A common misunderstanding is that the sonic boom occurs when the aircraft "breaks" the sound barrier. The boom actually continues as long as the plane flies at supersonic speeds. However, at a given location, the loud noise is heard only when the edge of the pressure cone reaches the ear as illustrated in the figure. This short, loud noise results from the accumulated superposition of the wave crests generated by the plane.

The first person to break the sound barrier was Chuck Yeager on October 14, 1947. He flew a Bell XS-1 rocket plane at 12,800 meters over Edwards Air Force Base in California at a speed of 1080 kilometers per hour (670 mph). This speed is known as Mach 1.015 since it is 1.015 times the speed of sound at this altitude. The supersonic Concorde cruises at speeds up to Mach 2.2 (1450 mph), while the Russian MIG-25B has been tracked on radar at speeds up to Mach 3.2 (2110 mph).

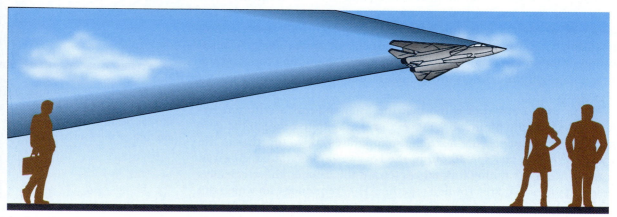

A plane traveling faster than the speed of sound produces a shock wave that is heard after the plane passes overhead. The couple on the right has not yet heard the sonic boom or any other sound from the plane.

course, the loudness of the signal decreases as the distance to the tuning fork increases.

When a wave bounces off a moving object, it experiences a similar Doppler shift in frequency due to successive crests (or troughs) having longer or shorter distances to travel before reflecting. By monitoring these frequency shifts, the speed of objects toward or away from the original source can be determined. Because the Doppler effect occurs for all kinds of waves, this technique is used in many situations, from catching speeding motorists using radar (an electromagnetic wave) to monitoring the movement of dolphins using sound waves in water.

Shock Waves

When a source of waves moves faster than the speed of the waves in the medium, the next crest is generated in front of the leading edge of the previous crest. This causes the expanding waves to superimpose and form the conical pattern shown in Figure 15-14. The amplitude along the cone's edge can become very large because the waves add together with their crests lined up. The edge of the cone is known as a **shock wave** because it arrives suddenly and with a large amplitude.

Shock waves are common in many media. When speedboats go much faster than the speed of the water waves, they create shock waves commonly known as wakes. The Concorde, a supersonic plane, travels much faster than the speed of sound in air and therefore produces shock waves. Some people are concerned about the effects of the sonic boom when the edge of the pressure cone reaches the ground.

> **QUESTION** Would you expect a spacecraft traveling to the Moon to produce a shock wave during its entire trip?

The return of a Space Shuttle to Earth produces a double sonic boom. One boom is produced by the noise; the other is produced by the engine housings near the rear of the spacecraft. Listen for this the next time you watch a Shuttle return on television.

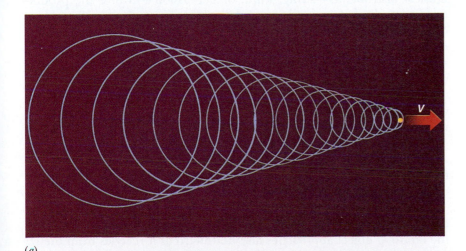

(a)

(b)

Figure 15-14 (a) When the source moves faster than the speed of sound, the waves form a cone known as the shock wave. (b) The presence of the shock wave indicates that ducks can swim faster than the speed of the water waves.

> **ANSWER** No. Because there is no air between the Earth and the Moon, the spacecraft would not produce any sound.

C H A P T E R

15

R E V I S I T E D

By studying the sounds produced by instruments such as guitars, pianos, and organs, we learn that musical sounds often have combinations of frequencies that are whole-number multiples of the lowest frequency. The size of the body of the instrument determines the range of frequencies that it amplifies.

S U M M A R Y

Sound is a longitudinal wave that travels through a variety of media. The speed of sound is 343 meters per second in air at room temperature, four times faster in water, and more than ten times faster in solids. In a gas, the speed of sound depends on the temperature (slower at lower temperatures) and the type of gas (faster in molecules with less mass).

Ears detect sound waves and send electric signals to the brain. The range of frequencies that can be detected depends on the resonant structures in the animals' ears. The audible range in humans is roughly 20–20,000 hertz, although it varies with age and the individual.

You can recognize different sources of sound because each source produces a unique combination of intensities of the various harmonics—its own recipe of harmonics.

Music generally differs from other sounds in its periodicity. Nearly all musical instruments—string, wind, or percussion—involve the production of standing waves. In many of these instruments the standing waves, or harmonics, are whole-number multiples of the fundamental. The frequency of the second harmonic is twice the fundamental frequency. The string's actual shape is a superposition of all the standing waves and looks quite different from those of the individual harmonics.

Different notes are generally produced by changing the standing-wave conditions. The relationship $v = \lambda f$ tells us that if the speed or the wavelength is changed, the frequency must change. A smaller wavelength gives a higher frequency when the speed is held constant.

Two steady sounds with nearly equal frequencies superimpose to form beats, a variation in the loudness of the sound that has a frequency equal to the difference in the frequencies.

The frequency of the sound from a moving object is shifted in frequency according to the Doppler effect. Frequency is shifted upward as the emitting object approaches and downward as it recedes from the receiver. The shift in the frequency is constant as long as the velocity is constant. The Doppler effect occurs for all kinds of waves.

When a source of waves moves faster than the speed of the waves in the medium, a conical shock wave is formed. When the edge of the cone created by supersonic aircraft reaches us, we hear a sonic boom.

K E Y   T E R M S

amplitude: The maximum distance from the equilibrium position that occurs in periodic motion.

antinode: One of the positions in a standing wave or interference pattern where there is maximal movement; that is, the amplitude is a maximum.

beats: A variation in the amplitude resulting from the superposition of two waves that have nearly the same frequencies. The frequency of the variation is equal to the difference in the two frequencies.

Doppler effect: A change in the frequency of a periodic wave due to the motion of the observer, the source, or both.

frequency: The number of times a periodic motion repeats in a unit of time. It is equal to the inverse of the period.

fundamental frequency: The lowest resonant frequency for an oscillating system.

fundamental wavelength: The longest resonant wavelength for a standing wave.

harmonic: A frequency that is a whole-number multiple of the fundamental frequency.

longitudinal wave: A wave in which the vibrations of the medium are parallel to the direction the wave is moving.

node: One of the positions in a standing wave or interference pattern where there is no movement; that is, the amplitude is zero.

period: The shortest length of time it takes a periodic motion to repeat. It is equal to the inverse of the frequency.

periodic wave: A wave in which all the pulses have the same size and shape. The wave pattern repeats itself over a distance of one wavelength and over a time of one period.

shock wave: The characteristic cone-shaped wave front that is produced whenever an object travels faster than the speed of the waves in the surrounding medium.

standing wave: The interference pattern produced by two waves of equal frequency traveling in opposite directions. The pattern is characterized by alternating nodal and antinodal regions.

transverse wave: A wave in which the vibrations of the medium are perpendicular to the direction the wave is moving.

wavelength: The repetition length for a periodic wave. For example, it is the distance from crest to crest or trough to trough.

CONCEPTUAL QUESTIONS

1. How could you infer that sound waves are longitudinal?

2. What is the evidence that sound is a wave phenomenon?

3. Does a 220-hertz sound wave move faster, slower, or at the same speed as a 440-hertz sound wave?

4. Which of the following has the most effect on the speed of sound in air: amplitude, frequency, wavelength, or temperature?

5. Is the speed of sound faster in air or water?

6. Why does the voice of a clown who has inhaled helium sound so high pitched?

7. Why do you see the lightning before you hear the thunder?

8. When the time interval between seeing the lightning and hearing the thunder is short, the thunder is loud. Why?

9. What is an echo?

10. Even though you may be far away from an orchestra, the tuba and the flute do not sound "out of step" with each other. What does this tell us about sound waves?

11. How much more intense is a rock concert than a police siren?

12. If earplugs are advertised to reduce sound levels by 10 decibels, by how much do they reduce the intensity of the sound?

13. What observable property of a sound wave is determined by its amplitude?

14. What property of a sound wave determines its loudness (or intensity)?

15. What determines the pitch of a sound wave?

16. Which is longer—the wavelength of infrasound or ultrasound?

17. Why would the different hairlike sensors in your ear resonate with different frequency sounds?

18. What kind of hearing problem would be caused by a hole in the eardrum?

19. Which harmonic is one octave higher in frequency than the fundamental?

20. What change in frequency is required to move up two octaves on the musical scale?

21. What can you do to raise the frequency of a note on a guitar string?

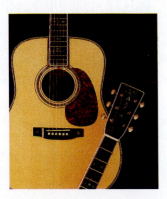

22. How can you increase the speed of the waves on a guitar string?

23. Two standing waves are created on a guitar string. Will the frequency of the one with the shorter wavelength be higher than, lower than, or the same as the frequency of the one with the longer wavelength?

24. How does the wavelength of the fundamental on a violin string compare with the length of the string?

25. The frequency of the third harmonic on a guitar string is _____ times that of the fundamental.

*26. Where on a guitar string of length L would you place your finger to damp out the third harmonic? Do these places differ in their effect on other harmonics?

27. Does increasing the tension of the string affect the wavelength of the fundamental on a guitar string?

28. Does plucking the string harder affect the fundamental frequency of a guitar string?

*29. Why is the mass per unit length of the strings on a bass larger than that on a violin?

30. Does the second or the third harmonic on a guitar string have traveling waves with the higher speed, or are the speeds the same?

31. Would you expect to find nodes or antinodes at the ends of a guitar string?

32. Would you expect to find a node or an antinode at each end of an organ pipe that's closed at one end?

33. How does the wavelength of the fundamental standing wave for a closed organ pipe compare with the length of the pipe?

34. How long is the wavelength of the fundamental in an open organ pipe compared with the length of the pipe?

35. If you close one end of an open organ pipe, what happens to the wavelength of the fundamental?

36. What happens to the frequency of the fundamental when you open the closed end of a closed organ pipe?

37. Would you expect to find a node or an antinode near the open end of a trumpet?

*38. Which of the following musical instruments does not have even harmonics: guitar, open organ pipe, closed organ pipe, or harp?

39. What happens to the wavelength of an organ pipe if you saw off a bit of the end of the pipe?

40. What happens to the fundamental frequency of an organ pipe as the temperature in the room increases?

41. What would happen to the wavelength of an organ pipe if you fill it with helium instead of air?

*42. What would happen to the frequency of an organ pipe if it were filled with helium instead of air?

43. The fundamental wavelength for a bar on a xylophone that is suspended at points midway between the ends and the middle of the bar is _____ the length of the bar.

44. What is the fundamental wavelength for standing waves on a rod held at its center?

45. Why does middle C played on an oboe sound different from middle C played on a piano?

46. How do we recognize people by their voices?

47. What causes the phenomenon known as beats?

48. As the frequencies of two waves get closer together, what happens to the beat frequency?

*49. The two wires corresponding to one key on a piano are out of tune. If we increase the tension of the wire producing the lower frequency, will the beat frequency increase, decrease, or stay the same?

50. You suspect that a tuning fork has been damaged and its fundamental frequency slightly changed. How can you use another tuning fork to check your hypothesis?

51. Describe the sounds you would hear if a train passed you with its whistle blowing.

52. Would you expect to hear a lower, the same, or a higher frequency as a tuning fork moves away from you? Explain your reasoning.

53. An automobile sounding its horn is moving toward you at a constant speed. How does the frequency you hear compare with that heard by the driver?

54. Which of the following properties of the wave does not change in the Doppler effect: wavelength, speed, or frequency?

55. If a wind is blowing from a siren toward your ear, the frequency of the siren will change. Will you hear a higher or a lower frequency? Why?

*56. Does the frequency of the detected radar change as a police car approaches a stationary wall at a steady speed?

57. What does the observation of a wake tell you about the speed of a boat compared with the speed of water waves?

58. Explain why a sonic boom sounds much like an explosion.

*59. What are some of the effects that you should consider in the design of a concert hall?

EXERCISES

1. The musical note "middle C" has a frequency of 262 Hz. What is its period of vibration?

2. What is the frequency of a tuning fork vibrating with a period of 0.0008 s?

3. What is the wavelength of a musical note with a frequency of 524 Hz?

4. How long is a wavelength of infrasound with a frequency of 3 Hz?

5. What frequency would you need to produce a sound wave in room-temperature air with a wavelength of 1 m?

6. What frequency of sound has a wavelength of 5 m in air?

7. What is the longest wavelength that can be heard by a normal ear?

8. What is the shortest wavelength that the average human can hear?

9. You observe that the delay between a lightning flash and the thunder is 8 s. How far away is the lightning?

10. If the "shot heard around the world" could actually travel around the world, how long would it take? (Assume that the circumference of the Earth is 40,000 km.)

11. If the sonar signal sent straight down from a boat takes 0.4 s to return, how deep is the lake?

12. How deep is a school of fish if the sonar echo is received 0.06 s after it was sent?

13. What is the intensity of a typical conversation compared to a typical whisper?

14. If ear protectors can reduce the sound intensity by a factor of 100,000, by how many decibels is the sound level reduced?

15. If the fundamental frequency of a guitar string is 196 Hz, what is the frequency of the fourth harmonic?

16. If the fundamental frequency of a guitar string is 220 Hz, what harmonic frequencies are possible?

17. If the fundamental frequency of a 80-cm-long guitar string is 500 Hz, what is the speed of the traveling waves?

18. What is the fundamental frequency for a 60-cm banjo string if the speed of waves on the string is 470 m/s?

19. What harmonic frequencies are possible in a closed organ pipe that is 1.2 m long?

*20. What frequencies are possible in an open organ pipe that is 1.2 m long?

21. What length of closed organ pipe is required to produce the note B_4 with a frequency of 493.88 Hz?

22. How long is an open organ pipe with a fundamental frequency of 493.88 Hz?

23. A tuning fork has been damaged and its frequency slightly changed. What could its altered frequency be if it produces two beats per second with a tuning fork that is known to vibrate at 440 Hz?

24. What is the beat frequency of two piano wires that vibrate at 440 Hz and 444 Hz?

An image produced by a kaleidoscope has symmetries due to the arrangement of its mirrors.

The Mystery of Light

Throughout history our knowledge, attitudes, and values have been reflected in our sciences and in our arts. The most obvious example of this parallelism is light. Light and one of its characteristics, color, have followed parallel trends in science and art. In the Middle Ages, for example, light was sacred and mysterious; artists placed it in some heavenly plane. When St. John the Evangelist said, "God is light," there was an implicit message that light was too holy to be studied and analyzed. During this time bright colors were used in mosaics and stained glass windows, but not studied.

Light, as seen in these early paintings, had a surrealistic behavior. It wasn't until the Renaissance that artists studied the effects of light on objects, how it illuminates some sides of an object but not others. During the late 1600s, when light was shown to take a finite time to travel through space, paintings were just beginning to show objects casting shadows on the ground.

Still the scientific questions remained: What is light? Does light behave as a wave or as a stream of particles? It was known that light travels from the Sun to Earth and that space is a very, very good vacuum. If light is a wave, how can it travel through a vacuum where there is nothing to wave? The history of this debate about the wave or particle nature of light goes back at least

> **The next time you look at a star, think about the starlight that reaches your eyes. It probably began its journey hundreds or thousands of years ago, in many cases long before civilization began on Earth.**

A prism spreads light out into its component colors.

to Newton's time. Newton, perhaps influenced by his successful work on apples and moons, believed that light was a stream of very tiny particles. His contemporary, the Dutch scientist Christian Huygens, believed that light was a wave. The controversy was not resolved until 100 years later by a French scientist. (However, the question of the nature of light reemerged during the 20th century with the development of quantum mechanics.)

And what is the origin of the colors? Prior to Newton's work with clear glass prisms, color was thought to be something emanating from objects. A ruby's redness was only due to the characteristics of the ruby. As we will see, this is partially correct; the character of the light shining on the ruby also plays a role.

The next time you look at a star, think about the starlight that reaches your eyes. It probably began its journey hundreds or thousands of years ago, in many cases long before civilization began on Earth. When it arrives at your eye, its journey is over. The light is absorbed in your retina, producing an electrical signal that travels to your brain. However, this light is carrying a much more complex message than simply the location of the star in the heavens. With the exception of a few artificial satellites within the Solar System, and a variety of cosmic particles striking the Earth, our entire knowledge of the cosmos comes from the information carried by light.

We continue to build our physics world view by studying this mysterious phenomenon in the next three chapters; then we will use our newly gained knowledge of light to probe the structure of matter at the atomic level . . . and beyond.

Physics Update

Isotope effects in sonoluminescence have been observed by Seth Putterman and Robert Hiller at UCLA. Sonoluminescence is a mysterious phenomenon in which acoustic energy is changed into light energy; high-frequency sound waves are absorbed by tiny bubbles in water. The bubbles, oscillating wildly, re-emit the energy in the form of tiny, focused light bursts. Many things about sonoluminescence are still unknown, such as the nature of the light-emitting process or why the light pulses are so short. The UCLA work has established one new fact: Substituting heavy water (one of the hydrogen atoms has a neutron in addition to the usual proton) for ordinary water as the liquid medium causes the sonoluminescence spectrum to dramatically shift from ultraviolet toward red wavelengths. This result seems to represent yet a new mystery. According to the researchers, "The shift is remarkably large, especially in view of the small difference in chemical and elastic properties between light and heavy water."

Light

The early-morning surface of a quiet New Hampshire lake acts like a mirror to produce an inverted image of the shoreline.

CHAPTER

16

The trees in this photograph are not really growing in both directions. And when swords are thrust through a "lovely assistant" or the haunted mansion has spooks sitting next to you as you traverse its dark innards, the response is, "It's done with mirrors." But how does light produce these illusions that are so convincing? (See p. 397 for the answer to this question.)

Although light played a central role in the histories of religion, art, and science and is so common to our everyday experience, it is actually quite elusive. Even the act of seeing has confused people. We talk of looking *at* things—of looking *into* a microscope, of sweeping our glance *around* the room—as if seeing were an active process, much like beaming something in the direction of interest. This notion goes back to some early ideas about light in which rays were supposedly emitted by the eyes and found the objects seen. The idea is still common, perhaps perpetuated in part by the Superman stories. In these stories Superman supposedly has the ability to emit powerful X rays from his eyes, enabling him to see through brick walls, or in the modern movie version, through clothes.

Seeing is actually a rather passive activity. What you see depends on the light that enters your eyes and not on some mysterious rays that leave them. The light is emitted whether or not your eyes are there to receive it. You simply point your eyes toward the object and intercept some of the light. In fact, light passing through clean air is invisible. If a flashlight is shined across a room, you don't see the light passing through the air. You only see the light that strikes the wall and bounces back into your eyes. (Sometimes you can see the beam's path because part of the light scatters from dust or smoke particles in the air and is sent toward your eyes.)

Shadows

One of the earliest studies of light was how it moved through space. By observing shadows and the positions of the light sources and the objects causing the shadows, it is easy to deduce that light travels in straight lines. In drawings illustrating the paths of light it is convenient to use the idea of **light rays.** Because there are an infinite number of paths, we draw only enough to illustrate the general behavior.

A point source emits light in all directions. An opaque object near the source blocks some of the rays from reaching a screen, producing a shadow like the one shown in Figure 16-1. The shadow has the same shape as the cross section of the object but is larger.

Most sources of light are not points, but extend over some space. However, we can think of each small portion of the source as a point source casting its own sharp shadow. All these point-source shadows are superimposed on the screen. The darkest region is where all of the shadows overlap. This is known as the **umbra** (Fig. 16-2). Surrounding the umbra is the **penumbra,** where only some of the individual shadows overlap.

Rather than looking at the shadow, imagine standing in the shadow looking back at the source. If your eye is located in the umbra, you will not be able to see any portion of the source; the object blocks out all of the light. If your eye is in the penumbra, you will be able to see part of the light source.

Shadows that we observe on Earth are not totally black, devoid of light. We can, for example, see things in shadows. This is due to the light that scatters into the shadow from the atmosphere or from other objects. On the Moon, however, the shadows are much darker because there is no

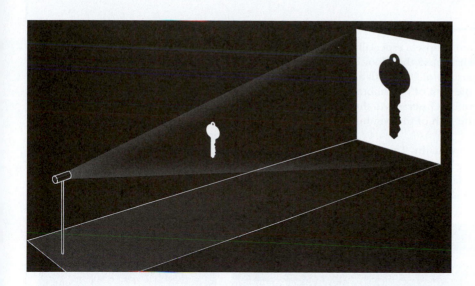

Figure 16-1   The shadow produced by a point source of light is very sharp.

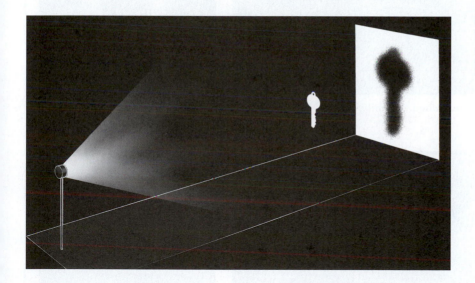

Figure 16-2   The shadow produced by an extended source of light has a dark central umbra surrounded by a lighter penumbra.

lunar atmosphere. Astronauts exploring the Moon's surface have to be careful. Stepping into their own shadows can be dangerous because the extreme blackness of the shadow would hide everything within it—sharp rocks, uneven terrain, even a deep hole.

Pinhole Cameras

Light that strikes most objects leaves in a great many directions. (This must be true since we can see the object from many different viewing directions.) If we place a piece of photographic film in front of a vase

Eclipses

The most spectacular shadows are eclipses, especially total solar eclipses. During a solar eclipse the Moon's shadow sweeps a path across a portion of the Earth as shown in Figure A. If you are in the path of the umbra, the Sun is totally obscured (Figure B). Observers to the side of the umbra's path but in the path of the penumbra see a partial eclipse.

One of Aristotle's arguments that the Earth is a sphere involved the shadow during lunar eclipses (Figure C). Here, the Earth's shadow falls onto the face of the Moon. Since the shape of the shadow is always circular and since the only solid that always casts a circular shadow is a sphere, Aristotle correctly concluded that the Earth must be a sphere.

Few people experience a solar eclipse, while the Earth's population on half the Earth can see a lunar eclipse. In order to see a solar eclipse, observers must be directly in the shadow of the Moon. This only covers a small portion of the Earth's surface. During a lunar eclipse, we observe the Earth's shadow on the Moon (Figure D). Anyone who can see the Moon can therefore see this eclipse.

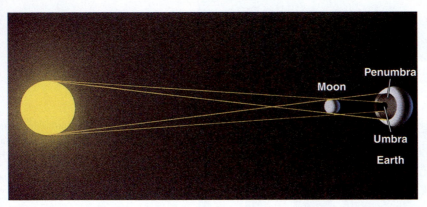

Figure A   A total solar eclipse occurs when the umbra of the Moon's shadow falls on the Earth.

Figure B   A total eclipse of the Sun.

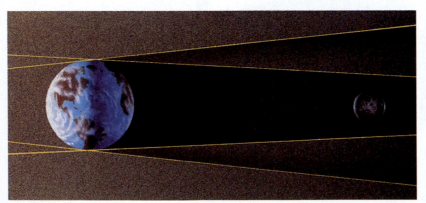

Figure C   During a lunar eclipse, the Moon is in the Earth's shadow.

Figure D   During a lunar eclipse, we can see part of the Earth's circular shadow.

of flowers as in Figure 16-3(a), the film will be completely exposed leaving no record of the scene. Light from one part of the vase hits the film at the same place as light from many other parts of the vase and flowers. Every spot on the film receives light from virtually every spot that faces the film.

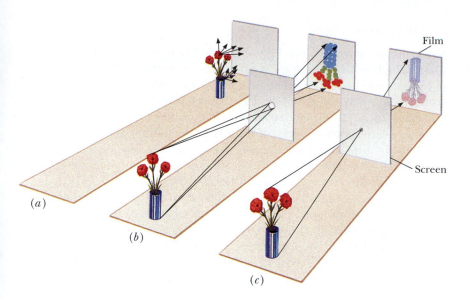

(a)

(b)

(c)

Film

Screen

Figure 16-3 (a) Each part of the film receives light from many parts of the vase and flowers and no image is recorded. (b) The screen restricts the light so that each part of the film receives light from a small portion of the scene. (c) Reducing the size of the hole produces a sharper but dimmer image.

We can get an image by controlling which light rays hit the film. A screen with a small hole in it is placed between the vase and the film as shown in Figure 16-3(b). Now only the light from a small portion of the vase reaches a given region of the film. Making the hole smaller, as in Figure 16-3(c), further reduces the portion of the vase exposed to a given spot on the film. If the hole is made small enough, a recognizable image of the vase is formed on the film. This technique can be used to make a pinhole camera by enclosing the film in a light-tight box with a hole that can be opened and shut. The photograph in Figure 16-4 was taken with a pinhole camera made from a shoe box.

The problem with pinhole cameras is the very small amount of light that reaches the film. Exposure times generally must be very long, requiring that the scene be relatively static. We will see in Chapter 17 that this problem was eventually overcome by using a lens.

Pinhole cameras were used before the invention of film. If you add another hole to the box, as illustrated in Figure 16-5, you can see the image on the back wall. During solar eclipses, some observers make use of this process to watch the partial phases of the eclipse safely. On a grander scale an entire room can be converted into a pinhole camera known as a *camera obscura*. Smaller versions were used by Renaissance artists to help them draw landscapes and portraits; they traced the images formed on the back wall.

Figure 16-4 A photograph made with a pinhole camera. The exposure time was 2 seconds.

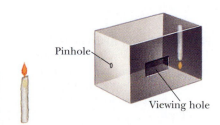

Pinhole

Viewing hole

Figure 16-5 A pinhole in one end of the box produces an inverted image on the opposite end.

Physics on Your Own   Convert your room into a camera obscura by covering the windows with black plastic garbage bags. Make a hole about the size of a quarter in one of the bags. On a sunny day you should get a good image of the outdoors on the wall opposite the window. Experiment with the size of the hole to obtain the best picture.

Figure 16-6 A beam of light striking a rough surface is scattered in many directions.

Reflections

Some things emit their own light—a candle, a light bulb, and the Sun, to mention a few. But we see most objects because they reflect some of the light that hits them. The incident light, or the light striking the object, is scattered in many directions by the relatively rough surface of the object, a process known as **diffuse reflection.** If the surface of the object is very smooth, a light beam reflects off it much like a ball rebounds from a wall. Presumably then, when light hits rougher surfaces, the same thing happens. But with rough surfaces, different portions of the incident light reflect in many directions as shown in Figure 16-6.

By looking at the reflection of a thin beam of light (a good approximation to a "ray" of light) on a smooth surface, we can discover a rule of nature. Figure 16-7(a) shows the reflections of light beams hitting a mirror at three different angles. Clearly, the angle between the incident and the reflected ray is different in each situation. However, if we examine only one case at a time, we notice that the angles the incident and reflected rays make with the surface of the mirror are equal. If we measure the angles of the incident ray and the reflected ray for many such situations, we discover that the two angles are always equal.

> **QUESTION** Assume that you are stranded on an island. Where would you aim a mirror to signal a searching aircraft with sunlight?

In our illustration of the law, we used the angles made with the reflecting surface. It is more convenient to consider the angles between the rays and the **normal,** an invisible line that touches the reflecting surface at the spot where the rays hit the surface and is perpendicular to the surface [Fig. 16-7(b)]. The reflected ray lies in the same plane as the normal and incident ray. Using these angles we state the **law of reflection.**

law of reflection

> The angle of reflection is equal to the angle of incidence.

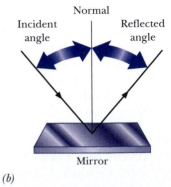

Figure 16-7 (a) Reflections of three thin beams of light hitting a mirror at different angles from the left. (b) For each one the angle of reflection is equal to the angle of incidence.

(a)   *(b)*

> **ANSWER** The normal to the mirror would have to be directed at the point midway between the Sun and the aircraft. Luckily, this is easier to do than it might seem.

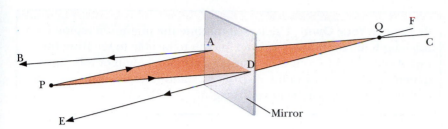

Figure 16-8 A ray diagram showing that the location of the image *Q* is determined by the crossing of the two lines of sight. The image *Q* is as far behind the mirror as the object *P* is in front.

Flat Mirrors

When we look at smooth reflecting surfaces, we don't see light rays; we see images. We can see how images are produced by looking at the paths taken by light rays. In Figure 16-8, we locate the image *Q* of a single point *P* in front of a flat mirror. Light leaves the point *P* in all directions; some of the light strikes the mirror at point *A,* then reflects and travels in the direction of *B.* An eye at *B* would receive the light coming along the direction *AB.*

Remember that seeing is a passive activity. Our eye–brain system only records the direction from which the light arrives. We do not know how far away the light originated, but we do know that it came from someplace along the line from *B* to *C.* However, we can say the same thing about the light that arrives from another direction, say at *E.* The eye perceives this light as coming from someplace along the line *EF.* Because the only place that lies on both lines is *Q,* our brain says that the light originated at point *Q,* This is the location of the image. We see an image of *P* located behind the mirror at point *Q.* After the light reflects from the mirror, it has all of the properties it would have had if the object had actually been at *Q.*

The image has a definite location in space. Figure 16-8 shows that the point *Q* is located the same distance behind the mirror as the point *P* is in front of the mirror. A straight line drawn between *P* and *Q* is normal to the mirror. These observations allow us to locate the image quickly. For example, in Figure 16-9 we can locate the image of the pencil by first locating the image of its tip and then the image of its eraser. The size of the image is the same as that of the object (although it is far away from you and looks smaller). Note that the pencil doesn't have to be directly in front of the mirror. However, the mirror must be between the entire image and the observer's eyes as illustrated in Figure 16-10.

> **QUESTION**   If one wall of a living room is completely covered with mirrors, how much larger does the room look?

> **ANSWER**   Because the image of the wall opposite the mirrors is as far behind the mirrors as the real wall is in front of them, the room will appear to be twice as big.

Figure 16-9 The image of a pencil formed by a flat mirror is located behind the mirror.

Figure 16-10 A flat mirror forms an image of an object even if the object is not located directly in front of the mirror.

(a)

(b)

Figure 16-11   (a) The woman's head appears to sit on the table. How is this done? (b) The same scene with both mirrors removed.

Physics on Your Own   Use tape to outline the minimum region on a full-length mirror that is required for you to be able to see from the top of your head to your toes. Does the size or location of this region depend on how far from the mirror you stand? How does the height of this region compare with your height?

A magician's trick illustrates the realism of an image by presenting the audience with a live, talking head on a table as in Figure 16-11(a). Like many other illusions, the trick lives up to the cliche that "it's done with mirrors." But where is the person's body? Your eye–brain system is tricked by the images. You think that the table is an ordinary one with legs and a wall behind it. You "know" this because you can see the wall between the table's legs. Figure 16-11(b) shows the set-up with the mirrors removed. The table has mirrors between its legs so that the walls you see under the table are really images of the side walls.

Multiple Reflections

When a light beam reflects from two or more mirrors, we get interesting new optical effects from the multiple reflections. If the two mirrors are directly opposite each other such as in some barbershops and hair salons, we get an infinite number of images, with successive images being farther and farther away from the object. To see how this works, remember that the light appearing to come from an image has the same properties as if it actually came from the object. Each mirror forms images of everything in front of it, *including* the images formed by the other mirror. Each of the mirrors forms an image of the object. The light from these images forms an additional set of images on the opposing mirrors, and on and on, as illustrated in Figure 16-12.

QUESTION   Would the opposing mirrors in Figure 16-12 allow you to see the back of your head?

Physics on Your Own   Use two mirrors to make a periscope that will allow you to see around a corner. How are the images formed?

In Figure 16-13(a) we have placed a figure of a lion in front of two mirrors that form a right angle. Figure 16-13(b) shows the paths taken by the rays that entered the camera. Notice that there are three images; one image is formed by each mirror and then each mirror forms an image of

ANSWER   Yes, provided the mirrors are tilted a bit so that your head doesn't get in the way of your view. The second image in a mirror will be of the back of your head.

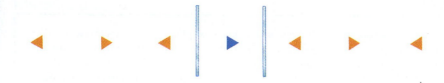

Figure 16-12   Two opposing parallel mirrors form an infinite number of images. Notice the reversals in the orientations of the triangles.

(a)

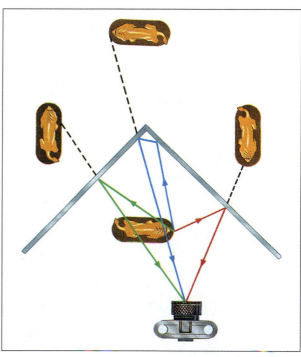

(b)

Figure 16-13   (a) Two mirrors at right angles to each other produce three images of the lion. (b) The red, green, and blue lines show sample paths taken by rays that enter the camera lens.

these images. (Remember that a mirror does not have to actually extend between the image and the object. However, it is often useful to imagine that each mirror is extended.) If the angle between the mirrors is precisely 90°, these latter two images overlap to form a single image beyond the corner.

Physics on Your Own   Examine the image of your face formed by two mirrors at right angles to each other. What do you see when you touch one of your cheeks? How does the image differ from the image you see in a flat mirror?

If the angle between the mirrors is made smaller, the overlapping images beyond the corner separate. When the angle between the mirrors reaches 60°, we once again get overlapping images beyond the corner and a total of five images as shown in Figure 16-14.

Figure 16-14   When the mirrors form an angle of 60°, five images of the lion are produced.

Retroreflectors

An interesting consequence of having two mirrors at right angles is that an incoming ray (in a plane perpendicular to both mirrors) is reflected back parallel to itself, as shown by the three rays in the figure. This works for all rays if we add a third mirror to form a "corner" reflector, like putting mirrors on the ceiling and the two walls in the corner of a room. These *retroreflectors* are used in the construction of bicycle reflectors so that the light from a car's headlights is reflected back to the car driver and not off to the side as a single mirror would. Examination of many reflectors reveals a surface covered with holes in the shape of the corners of cubes. Cloth reflectors used on clothing and the surfaces of STOP signs are often covered with a layer of reflective beads. The surfaces in the regions between the beads work like corner reflectors.

An outer space application of retroreflectors involves an experiment to accurately measure the distance to the Moon. The Apollo astronauts placed panels of retroreflectors on the Moon to allow scientists on Earth to bounce a laser beam off the Moon and receive the reflected signal back on Earth. The time for the beam to make the round trip, multiplied by the speed of light, yields the distance. (Remember, speed × time = distance.) Ordinary mirrors would not have worked because the astronauts could not have aimed them well enough to send the beam back to Earth. Additionally, the Moon's wobbly rotation would send the reflected beam in many directions. Since retroreflectors work for all incident angles, they could be simply laid on the Moon's surface. We now know that the Earth–Moon distance increases by 3 to 4 centimeters per year.

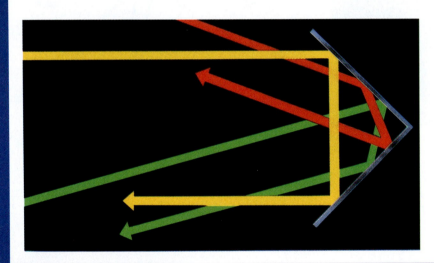

Each of the three rays is reflected back parallel to itself.

The curved surfaces of fun-house mirrors produce interesting images.

Curved Mirrors

Fun-house mirrors are interesting because of the distortions they produce. The distortions are not caused by a failure of the law of reflection, but result from the curvature of the mirrors. Some distortions are desirable. If the distortion is a magnification of the object, we can see more detail by looking at the image.

Cosmetics mirrors and some rearview mirrors on cars are simple curved mirrors that don't produce bizarre distortions, but do change the image size. A cosmetics mirror uses the concave side—the reflecting surface is on the inside of the sphere—to generate a magnified image of your face. The convex reflecting surface—the outside of the sphere—always produces a smaller image but has a bigger field of view. Convex mirrors are quite often used on cars and trucks and on "blind" street corners because they provide a wide-angle view.

Cylindrical mirrors are used to concentrate sunlight at this solar farm.

QUESTION   Sometimes mirrors are installed in stores to inhibit shoplifting. Are these concave or convex mirrors?

Figure 16-15 shows the essential geometry for a concave spherical mirror; the reflecting surface is the inside of a portion of a sphere. The line passing through the center of the sphere *C* and the center of the mirror *M* is known as the **optic axis.** Light rays parallel to the optic axis are reflected back through a common point *F* called the **focal point.** The focal

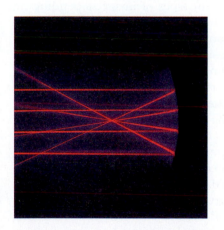
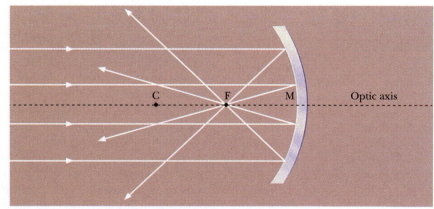

Figure 16-15   The focal point *F* of a spherical concave mirror lies along the optic axis midway between the center of the sphere *C* and the center of the mirror *M.* Rays parallel to the optic axis are focused at the focal point *F.*

ANSWER   The mirrors must be convex to provide a wide view of the store.

Convex mirrors are used to obtain wide-angle views.

point is located halfway between the mirror and the center of the sphere. The distance from the mirror to the focal point is known as the **focal length** and is equal to one-half of the radius *R* of the sphere.

Concave mirrors can be used to focus light. Some solar collectors use them to concentrate sunlight from a large area onto a smaller heating element. Because the Sun is very far away, its rays are essentially parallel and a concave spherical mirror can focus the sunlight at the focal point. A cylindrical mirror focuses sunlight to a line instead of a point. A pipe containing a fluid can be placed along this line to carry away the thermal energy. If the heat energy is used to generate electricity, the higher temperature makes the process more efficient.

Light rays are reversible. The law of reflection is still valid when the incident and reflected rays are reversed. So the shape that focuses parallel rays to a point will take rays from that point and send them out as parallel rays. This idea is used in automobile headlights. The bulb is placed near the focal point of the mirror, producing a nearly parallel beam.

The optic axis of a convex mirror also passes through the center of the sphere *C*, the focal point *F*, and the center of the mirror *M*, but in this case, the focal point and the center of the sphere are on the back side of the mirror. (Again, the focal point is halfway between the center of the sphere and the center of the mirror.) Rays parallel to the optic axis are now reflected *as if they came from* the focal point behind the mirror, as shown in Figure 16-16. Convex mirrors always produce images that are erect and smaller than the object.

Images Produced by Mirrors

A concave mirror can form two different types of image depending on how close the object is to the mirror. Imagine walking toward a very large concave mirror. At a large distance from the mirror, you will see an image of your face that is inverted and reduced in size (Fig. 16-17). This image is formed by light from your face reflecting from the mirror and converging to form an image in front of the mirror. This image is known as a **real image,** because the rays reflect from the mirror and converge to form the image. The rays then diverge, behaving as if your face were actually at the image location.

As you walk toward the mirror, the image of your face moves toward you and gets bigger. As you pass the center point, the image moves behind you. When your face is closer to the mirror than the focal point, the image is similar to that in a flat mirror except that it is magnified (Fig. 16-18). As with the flat mirror, the reflected rays diverge after reflecting and the image is located behind the mirror's surface. In this case, there can be no light at the location of the image because it is formed behind the mirror. The light only appears to come from the location of the image. This second type of image is called a **virtual image.**

The essential difference between a real and a virtual image is whether the light actually comes from the image location or only appears to come from there. If the rays diverge upon reflecting, they never come together to form a real image. They will, however, appear to originate from a

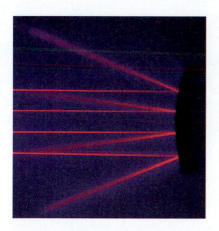
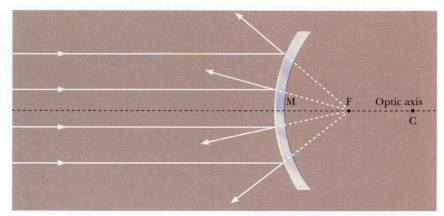

Figure 16-16 The focal point *F* of a spherical convex mirror lies along the optic axis midway between the center of the sphere *C* and the center of the mirror *M*. Rays parallel to the optic axis are reflected as if they come from the focal point *F*.

common location behind the mirror. Reflected rays that converge to form a real image can be seen on a piece of paper placed at the image location because the light actually converges at that location. However, if you put a piece of paper at the location of a virtual image, you get nothing since there is no light there.

Locating the Images Σ

There is a simple way of locating an image without measuring any angles by looking at a few special rays. Light leaves each point on the object in all directions; those rays that strike the mirror form an image. Although any of these rays can be used to locate the image of a point, three are easy to draw and, therefore, useful in locating the image. Because all rays from a given point on the object are focused at the same place for a real image (or appear to come from the same point for a virtual image), we need only find the intersection of any two of them. The third one can be drawn as a check. (For a spherical mirror these three rays do not usually meet at a point. However, they give the approximate location of the image if the object is small enough that the special rays strike the mirror near the optic axis.)

The three rays that are useful in these ray diagrams are shown in Figure 16-19. The easiest ray to draw is the red one lying along a radius of the sphere. It strikes the mirror normal to the surface and reflects back on itself. Another easy ray is the blue one that approaches the mirror parallel to the optic axis. It is reflected back toward the focal point. The third ray (shown in green) is a reverse version of the second one; a ray passing through the focal point reflects back parallel to the optic axis. (This ray does not actually need to pass through the focal point. If the object is closer than the focal point, the ray still lies along the line from the focal point to the mirror.)

Figure 16-17 An inverted image is formed when the head of your author is located outside the focal point of a concave mirror. Note that both the face and the floor tiles are inverted.

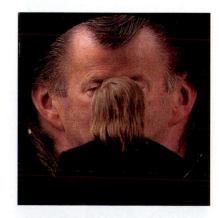

Figure 16-18 A magnified, erect image is formed when the head is located inside the focal point of a concave mirror.

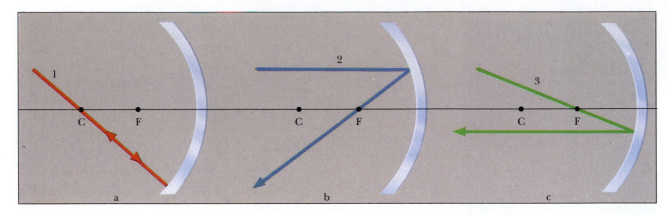

Figure 16-19   The three light rays used in drawing ray diagrams for concave mirrors.

If the mirror is small, some of these special rays may not strike the actual surface of the mirror. For the purposes of the ray diagram, we extend the mirror because a larger mirror with the same focal length would produce the same image. In fact, the mirror could be so small that none of the three easily drawn rays hit the mirror.

The descriptions of these rays can be abbreviated as follows:

rays for curved mirrors

1. Along radius—back on itself.
2. Parallel to optic axis—through focal point.
3. Through focal point—parallel to optic axis.

To illustrate the use of these ray diagrams, consider an object located inside the focal point of a concave mirror [Fig. 16-20(a)]. Figure 16-20(b) shows the three special rays that we use to locate the image of the tip of the candle. The rays are color-coded to correspond to the descriptions given above. Because the base of the candle is on the optic axis, we know that the image of the base is also on the optic axis. So finding the location of the tip of the candle gives the image location and orientation. For the case illustrated in Figure 16-20, we can see that a virtual image is produced that is erect and magnified.

(a)

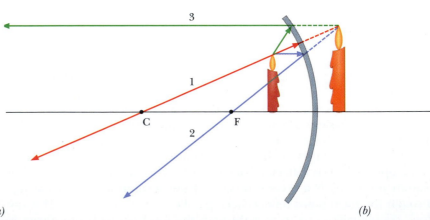

(b)

Figure 16-20   The image of a candle inside the focal point of a concave spherical mirror is virtual, erect, and magnified.

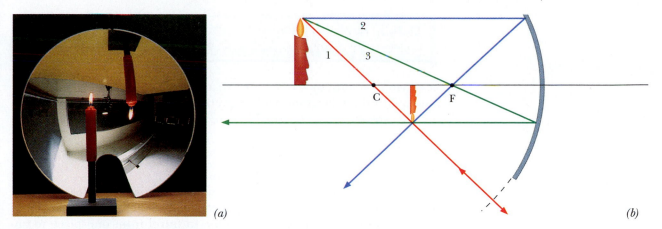

(a)   (b)

Figure 16-21   The image of a candle outside the focal point of a concave spherical mirror is real, inverted, and may be larger or smaller than the object.

As the candle is moved away from the mirror, the image size and the distance of the image behind the mirror increase. As the candle approaches the focal point, the image becomes infinitely large and infinitely far away. You can verify this by drawing a couple of ray diagrams.

When the candle is beyond the focal point, a real image is formed [Fig. 16-21(a)]. The ray diagram in Fig. 16-21(b) shows that the reflected rays do not diverge as in the previous case, but come together, or converge. These rays actually cross at some point in front of the mirror to form an image.

Fortunately, locating images formed by convex mirrors is the same process used for images formed by concave ones. A ray diagram showing how to locate the image is given in Figure 16-22(b). The same three rays

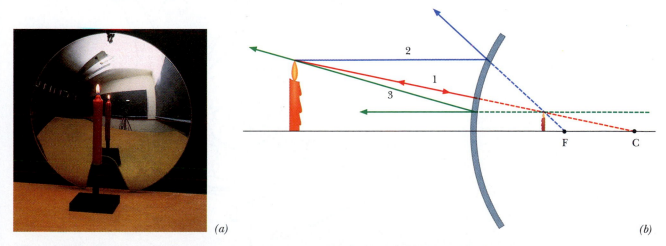

(a)   (b)

Figure 16-22   The image of a candle in front of a convex spherical mirror is always virtual, erect, and reduced in size. Ray 2 reflects as if it came from the focal point and ray 3 starts toward the focal point.

> **QUESTION**   Where would you place an object to get an image at the same location?

are used. You must only remember that the focal point is now behind the mirror. With ray diagrams you can verify that images formed by convex mirrors are always erect, virtual, and reduced in size.

Speed of Light $\boxed{\Sigma}$

In addition to traveling in a straight line in a vacuum, light moves at a very, very high speed. In fact, people originally thought that the speed of light was infinite—that it took no time to travel from one place to another. It was clear to these observers that light travels very much faster than sound; lightning striking a distant mountain is seen much before the thunder is heard.

Because of its very great speed, early attempts to measure the speed of light failed. One interesting attempt was made by Galileo when he tried to measure the speed of light by sending a light signal to an assistant on a nearby mountain. The assistant was instructed to uncover a lantern upon receiving Galileo's signal. Galileo measured the time that elapsed between sending his signal and receiving that of his assistant. Knowing the

> **ANSWER**   At the center of the sphere. This can be checked with a ray diagram.

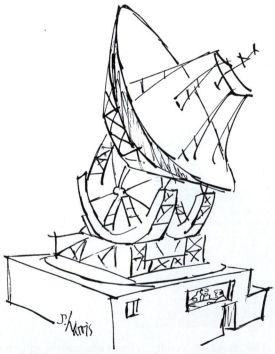

"AS I UNDERSTAND IT, THEY WANT AN IMMEDIATE ANSWER. ONLY TROUBLE IS, THE MESSAGE WAS SENT OUT 3 MILLION YEARS AGO."

distance to the mountain, he was able to calculate the speed of light. Upon repeating the experiment with a more distant mountain, however, he found the same elapsed time! Had the speed of light increased? No. Galileo correctly concluded that the elapsed time was due to his assistant's reaction time. Therefore, the time it took light to travel the distance was either zero or much smaller than he was able to measure.

About 40 years later, in 1675, the Danish astronomer Ole Roemer made observations of the moons of Jupiter that showed that light had a finite speed. Roemer found that the period of revolution of a moon around Jupiter was shorter during the part of the year when Earth approached Jupiter and longer when Earth receded from Jupiter. He expected the period to be constant like that of our Moon about Earth. He correctly concluded that the period of the Jovian moon was constant and that the variations observed on Earth were due to the varying distance between Earth and Jupiter. When Earth approaches Jupiter, the light emitted at the beginning of the period must travel farther than that emitted at the end of the period. Therefore, the light emitted at the end of the period arrives sooner than it would if Earth and Jupiter remained the same distance apart. This difference in distance makes the period appear shorter. Once the radius of the Earth's orbit was determined, the speed of light could be calculated.

The first nonastronomical measurement of the speed of light was performed by the French physicist Hippolyte Fizeau in 1849. He sent light through the gaps in the teeth of a rotating gear to a distant mirror. The mirror was oriented to send the light directly back (Fig. 16-23). At moderate speeds of rotation, the returning light would strike a tooth. But at a certain rotational speed, the light would pass through the next gap. Knowing the speed of the gear and the distance to the mirror, Fizeau was able to calculate the speed of light to a reasonable accuracy.

The speed of light has been measured many, many times. As the methods improved, the uncertainty in its value became less than one meter per second. In 1983, an international commission set the speed of light in a vacuum to exactly 299,792,458 meters per second and used it with atomic clocks to define the length of the meter. The speed of light is usually rounded off to 3×10^8 meters per second (186,000 miles per second). If light traveled in circles, it could go around the Earth's Equator

speed of light

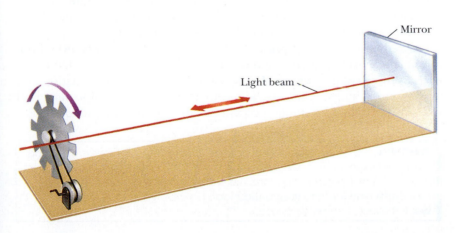

Figure 16-23 A schematic drawing of Fizeau's apparatus for measuring the speed of light.

7.5 times in 1 second. It's no wonder that early thinkers thought light traveled at an infinite speed. Although the speed of light is finite, it can be considered infinite for many everyday experiences.

> **QUESTION**   Given that radio waves travel at the speed of light, would you expect there to be longer than normal pauses in a conversation between two people talking via radio from the two coasts? How about in a conversation with astronauts on the Moon?

Color

Color is all around us and adds beauty to our lives. Color and the way we perceive it is an area of research involving many disciplines, including physics, chemistry, physiology, and psychology. One example of our body's role in detecting color is the fact that there is no such thing as white light. What we perceive as white is really the summation of many different colors reaching our eyes. The color of an object is determined by the color, or colors, of the light that enters our eyes and the way that this is interpreted by our brain.

In addition to the additive effect within our brain, there is also the issue of what reaches our eyes. If you have ever tried to match the color of two pieces of clothing under different kinds of lighting, you know it can be difficult. They may match very well in the store but be quite different in sunlight. The colors we perceive are determined by two factors: the color present in the illuminating light and the colors reflected by the object. A red sweater is red because the pigments in the dye absorb all colors except red. When viewed under white light, the sweater looks red. If the illuminating light does not contain red, the sweater will appear black, because all of the light is absorbed. A brightly colored box and dice look different under different illuminating lights (Fig. 16-24).

Although most lights give off all colors, the colors do not have the same relative intensities found in sunlight. Fluorescent lights are often brighter in the blue region and therefore highlight blues. At the risk of taking some of the romance out of candlelit dinners, we note that the warm glow of your date is due to the yellow-red light from the candles and may have nothing to do with your date's feelings toward you.

Often an object that appears to be a single color reflects several different colors. Although different colors may enter our eyes, we do not see each of these colors. Our eye–brain systems process the information and we perceive a single color sensation at each point. The color perceived may appear to have nothing in common with the component colors. For

> **ANSWER**   Because light can travel around the world 7.5 times in 1 second, we would expect it to travel across the United States and back in a small part (about $\frac{1}{30}$) of 1 second. Therefore, the pauses would seem normal. The round-trip time for the signal to the Moon is a little less than 3 seconds; this would produce noticeable pauses.

(a)   *(b)*

Figure 16-24   (a) A brightly colored box and dice illuminated with yellow light. (b) The same box and dice illuminated with white light.

instance, if red and green are reflected by the object, it will appear yellow. This behavior is in sharp contrast with the sense of hearing. Our ear–brain combination can hear and distinguish many different sounds coming from the same place at the same time.

The additive effects of color can be demonstrated by placing colored filters in front of spotlights or slide projectors and allowing the colored beams from each one to overlap. One combination that produces most of the colors that we perceive is red, green, and blue light beams. Figure 16-25 illustrates the colors that would be seen on the screen. For instance, red and green yield yellow, blue and green yield cyan (a bluish green), and red and blue yield magenta (a reddish purple). All three colors together produce white!

Two colors that add together to produce white light are called **complementary colors.** This process is illustrated by positive and negative color film of the same scene; the colors on one are the complements of those on the other as illustrated in Figure 16-26.

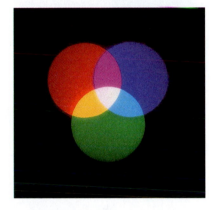

Figure 16-25   The addition of colored light produces new colors.

Figure 16-26   The colors in the positive image on the film are the complements of those in the negative image.

Figure 16-27 The array of colors on your television come from the addition of three basic colors—red, green, and blue.

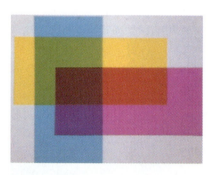

Figure 16-28 Filters produce different results than mixing colored lights.

By using dimmer switches to vary the brightness of each beam, we can generate a very wide range of colors. This process is the basis of color television. Examine a color TV screen with a magnifying glass and you will see that it is covered with arrays of red, green, and blue dots or lines like those in Figure 16-27. A similar process is used to print color photographs in magazines.

> **QUESTION** What color would you expect to see if you remove some of the blue light from white light?

It may seem strange to find that mixing certain colors yields white; from childhood we have learned that mixing many different colored paints together does not yield white, but rather a dark brownish color. Mixing paints is different from mixing colored lights. Mixing paints is a *subtractive* process. When white light strikes a red object, the pigment in the object subtracts all of the colors except the red and reflects the red back to the viewer. Likewise, a red filter subtracts all of the colors except the red that passes through it. Each additional color pigment or filter subtracts out more colors from the incident light. Therefore, if we overlap magenta, cyan, and yellow filters, we obtain a neutral gray (ideally it would be black) as illustrated in Figure 16-28.

This brief coverage of color perception allows us to answer the questions, Why is the sky blue? and, Why is the Sun yellow? The Sun radiates light that is essentially white. Since it appears yellow, we can assume that some of the complementary color has somehow been removed. The complement of yellow is blue—the color of the sky. The molecules in the atmosphere are more effective in scattering blue light than red light. As the sunlight passes through the atmosphere, more and more of the blue end of the spectrum is removed, leaving the transmitted light with a yellowish color (Fig. 16-29). When we look away from the Sun, the sky has a bluish cast because more of the blue light is scattered into our eyes. This effect is

Figure 16-29 The sky is blue and the Sun yellow because air scatters more blue light than red light.

> **ANSWER** This leaves red and green behind, creating a yellow color.

enhanced when the rays have a longer path through the atmosphere: The Sun turns redder near sunrise and sunset. It also increases with increased numbers of particles in the air (such as dust) and is more pronounced under these conditions. The additional dust in the air during harvest time produces the spectacular harvest moons. Although much less romantic, the same effect results from the air pollution near urban industrial sites.

These ideas also account for the color of water. Because water absorbs red light more than the other colors, the water takes on the color that is complementary to red; that is, cyan. Consequently, underwater photographs taken without artificial lighting look bluish green.

> **QUESTION**   If red light were scattered more than blue light, what color would the Sun and sky appear?

SUMMARY

Light is seen only when it enters our eyes. Since we know that light travels in straight lines, we can use rays to understand the formation of shadows and images.

When light reflects from smooth surfaces, it obeys the law of reflections, which states that the angles the incident and reflected rays make with the normal to the surface are equal. The reflected ray lies in the same plane as the normal and the incident ray.

Mirrors produce real and virtual images. Light converges to form real images which can be projected, whereas light only appears to come from virtual images. The virtual image formed by a flat mirror is located on a normal to the mirror at the same distance behind the mirror as the object is in front of the mirror. The sizes of the image and the object are the same.

Images formed by spherical mirrors can be located by drawing three special rays: (1) along the radius—back on itself; (2) parallel to the optic axis—through the focal point; and (3) through the focal point—parallel to the optic axis. The focal point is located halfway between the surface and the center of the sphere.

Light travels through a vacuum at 299,792,458 meters per second.

The additive effects of color mean that red and green lights add to yield yellow, blue and green lights yield cyan, and red and blue lights yield magenta. All three colors produce white. Mixing paints is a subtractive process. The sky is blue and the Sun is yellow because the molecules in the atmosphere preferentially scatter blue light, leaving its complement.

CHAPTER

16

REVISITED

When light reflects from a smooth mirror or clean piece of glass, its direction is changed, giving the viewer false information about the location of the source of the light. For instance, reflections of the walls of a box make the box appear to be empty when, in fact, a rabbit is hiding behind the mirrors. This is the basis of many ancient visual illusions and, consequently, the origin of the popular saying.

> **ANSWER**   The sky would appear red due to the scattered light, and the Sun would appear cyan because of the removal of more of the red end of the spectrum.

KEY TERMS

complementary color: For lights, two colors that combine to form white.

diffuse reflection: The reflection of rays from a rough surface. The reflected rays do not leave at fixed angles.

focal length: The distance from a mirror to its focal point.

focal point: The location at which a mirror focuses rays parallel to the optic axis or from which such rays appear to diverge.

law of reflection: The angle of reflection (measured relative to the normal to the surface) is equal to the angle of incidence. The incident ray, the reflected ray, and the normal all lie in the same plane.

light ray: A line that represents the path of light in a given direction.

normal: A line perpendicular to a surface or curve.

optic axis: A line passing through the center of a curved mirror and the center of the sphere from which the mirror is made.

penumbra: The transition region between the darkest shadow and full brightness. Only part of the light from the source reaches this region.

real image: An image formed by the convergence of light.

umbra: The darkest part of a shadow where no light from the source reaches.

virtual image: The image formed when light only appears to come from the location of the image.

CONCEPTUAL QUESTIONS

1. A professor shines a light beam across the front of a lecture hall. Why can you see the light on the wall, but not in the air?

2. Why can you usually see the beam of an airport beacon pass through the air?

3. Why can we see things in shadows?

*4. When you look at a crescent moon on a clear night, you can often see the rest of the face of the Moon very dimly lit. How does this happen?

5. Which of the following will cast a shadow that has an umbra but no penumbra: the Sun, a light bulb, a campfire, or a point source of light?

*6. Will the shadow of a hand produced by the beam from a slide projector have an umbra and a penumbra? Explain.

7. A point source of light produces a shadow of a ball on a wall. What happens to the size of the shadow as the ball is moved closer to the source?

8. A spotlight produces a shadow of a ball on a wall. What happens to the size of the umbra as the ball is moved farther away from the light?

*9. Under what conditions would the shadow of a ball on a screen not have an umbra? What does this have to do

with the observation that some solar eclipses are not total for any observer on Earth?

10. During some solar eclipses the angular size of the Moon is smaller than that of the Sun. What would observers on Earth see if they stood directly in line with the Sun and Moon?

11. What happens to the image produced by a pinhole camera when you move the back wall farther from the pinhole?

12. What effect(s) does enlarging the hole in a pinhole camera have on the image?

13. When the incident ray IO reflects from the mirror in the figure, the reflected ray lies along the line O_____?

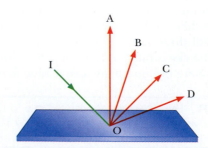

14. Which letter corresponds to the location of the image of the object O in the figure?

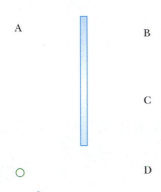

15. How do the size and location of your image change as you walk toward a flat mirror?

16. If a 0.8-meter-tall child stands 0.5 meter in front of a vertical plane mirror, how tall will the image of the child be?

*17. If you want to just be able to see the top of your head, where should you position the top of a flat mirror relative to the height of the top of your head?

*18. How does the height of the shortest mirror in which a person can see her entire body compare with her height? Does your answer depend on how far she stands from the mirror?

19. If you stand 1 meter in front of a flat mirror, how far away from you is your image?

20. Where is the image of the arrow shown in the figure located? Shade in the region where an observer could see the entire image.

*21. What is the magnification of a flat mirror? What is its focal length?

*22. If you walk toward a flat mirror at a speed of 1.2 meters per second, with what speed does your image move?

*23. It is often stated that a mirror reverses left and right. If this is correct, why doesn't it also reverse up and down? What does a mirror reverse?

*24. A flat piece of glass reflects part of the light and allows the rest to pass through. How might you use a flat pane of glass and appropriate lighting to put the image of a king on a throne and remove it from the throne by simply turning the light on and off?

25. How many images would be formed by two mirrors that form an angle of 45°?

26. Why do the images produced by two opposing mirrors appear to get progressively smaller?

27. If rays of light parallel to the optic axis converge to a point after leaving the mirror, what kind of mirror is it?

28. Why are the back surfaces of automobile headlights curved?

29. Can the image produced by a convex mirror ever be larger than the object?

*30. The image produced by a convex mirror is always closer to the mirror than the object. Then why is it that the convex mirrors used on cars and trucks often have the warning "Caution: Objects Are Closer Than They Appear" printed on them?

31. What are the size and location of the image of your face when you hold your face very close to a concave mirror? How do the size and location change as you move away from the mirror?

32. What are the size and location of the image of your face when you hold your face very close to a convex mirror? How do the size and location change as you move away from the mirror?

33. What type of mirror would you use to produce a magnified image of your face?

34. What type of mirror should a solar engineer use to concentrate light to boil water?

35. What is the fundamental difference between a real image and a virtual one?

36. What kind of image is formed by a bathroom mirror?

37. Can both real and virtual images be projected onto a screen?

38. Can both real and virtual images be photographed? Explain.

39. A concave mirror is used to form a real image of a candle. If the upper half of the mirror is covered by a piece of paper, what happens to the image?

40. What happens to the image if all of a convex mirror is covered up except for a hole at the center?

41. What happens to the real image produced by a concave mirror if you move the object to the location of this image?

*42. Argue that the reversible nature of light rays means that you can interchange an object and a real image.

43. Light from the nearest star (other than our Sun) takes 4.3 years to reach Earth. What can we say about what is happening on this star at this moment?

44. Astronomers claim that looking at distant objects is the same as looking back in time. In what sense is this true?

45. Why does the arrival of the sound from the bass drum in a distant band not correspond to the blow of the drummer?

46. Without asking, how could you tell whether you are talking to astronauts on the Moon or on Mars?

47. What color is produced by the overlap of a blue spotlight and a green spotlight? Does the illuminated surface make a difference?

48. If you remove some of the green light from white light, what color would you expect to see?

49. If a material can reflect only red light, what color will it appear under white light? Under blue light?

50. A substance is known to reflect red and blue light. What color would it have when it is illuminated by white light? By red light?

51. Some yellow objects actually absorb yellow light but reflect red and green light. If we shine yellow light on such a yellow object, what color will it appear to our eyes?

52. An actress wears a blue dress. How could you use spotlights to make the dress appear to be black?

53. The color we call brown is not a pure color, but a mixture of yellow with a bit of red and green. If we shine red light on a brown object, what color will it be?

54. A Crest toothpaste tube viewed under white light has a red C on a white background. What would you see if you used red light?

55. What color do you expect to get if you mix magenta and cyan paints?

56. What pigments would an artist mix to get a yellow color?

57. What do we mean by the term *complementary colors?*

58. A lens for a spotlight is coated so that it does not transmit yellow light. If the light source is white, what color is the spot?

59. What is the color of the sky as seen from the Moon, where there is no atmosphere?

60. How would the color of sunlight change if the atmosphere were much more dense?

EXERCISES

1. A 5-cm-diameter ball is located 40 cm from a point source and 80 cm from a wall. What is the size of the shadow on the wall?

2. A 5-cm-diameter ball is located 50 cm in front of a pinhole camera. If the film is located 10 cm from the pinhole, what is the size of the image on the film?

3. Use a ruler and a protractor to verify that the image produced by a flat mirror is as far behind the mirror as the object is in front.

4. Use a compass and a protractor to verify that the three rules for drawing ray diagrams for spherical mirrors satisfy the law of reflection.

5. A telescope mirror is part of a sphere with a radius of 6 m. What is the focal length of the mirror?

6. What is the radius of the spherical surface that would produce a mirror with a focal length of 4 m?

7. An object is located midway between the focal point and the center of a concave spherical mirror. Draw a ray diagram to locate its image. Is the image real or virtual, erect or inverted, magnified or reduced in size?

8. An object is located three times the focal length from a concave spherical mirror. Draw a ray diagram to locate its image. Is the image real or virtual, erect or inverted, magnified or reduced in size?

9. A 4-cm-tall object is placed 75 cm from a concave mirror with a focal length of 25 cm. Where is the image located? What is the size of the image?

10. Where is the image of a 5-cm-tall object located 60 cm from a convex mirror with a focal length of 30 cm?

11. A convex mirror has a focal length of 30 cm. What is the location and magnification of the image of an object located 15 cm from the mirror?

12. Repeat the previous exercise for a concave mirror.

13. If Galileo and his assistant were 15 km apart, how long would it take light to make the round trip? How does this time compare with reaction times of about 0.2 s?

14. Approximately how long would it take a telegraph signal to cross the United States from the east coast to the west coast? (Telegraph signals travel at about the speed of light.)

15. At a time when the Sun is 150 million km from Earth, how long does it take sunlight to reach the Earth?

16. The average distance from the surface of the Earth to the surface of the Moon is 376,000 km. How long after a large flash occurs on the Moon's surface would it take for observers on Earth to see it?

17. How far does light travel in one year? This distance is known as a light-year and is a commonly used length in astronomy.

18. How far does light travel in one nanosecond; that is, in one-billionth of a second?

Refraction of Light

When light travels from one transparent material to another, some interesting visual effects are created. Fish in an aquarium look bigger and a tree that has fallen into a lake looks bent. Is it ever possible that the light cannot travel from one transparent material to another? (See p. 425 for the answer to this question.)

Refraction of light in the atmosphere causes the oval shape of the setting sun.

W hen light strikes a transparent material, it usually changes direction. This change accounts for many interesting effects ranging from the apparent distortion of objects to the beauty of an afternoon rainbow. This bending of light that occurs at the surface of a transparent object is called **refraction.**

Refraction can be studied by looking at the paths the light takes as the incident angle is varied, as shown in Figure 17-1. As in reflection the angles are measured with respect to the normal to the surface. In this case, the normal is extended into the material and the angle of refraction is measured with respect to the extended normal. The amount of bending is zero when the angle of incidence is zero. That is, light incident along the normal to the surface is not bent. As the angle of incidence increases relative to the normal, the amount of bending increases; the angle of refraction differs more and more from the angle of incidence.

Index of Refraction

The amount of bending that occurs when light enters the material depends on the incident angle and an optical property of the material called the **index of refraction.** (We will refine the definition of the index of refraction in the next chapter.) A mathematical relationship can be

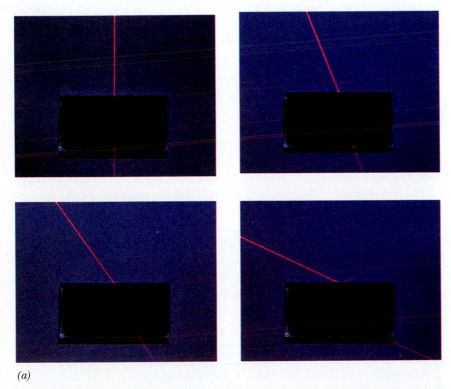

(a)

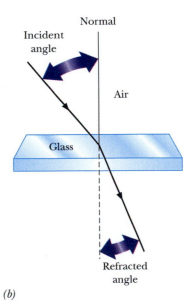

(b)

Figure 17-1   The amount of refraction depends on the angle of incidence. Notice that some of the incident light is reflected and some is refracted.

written that predicts the refracted angle given the incident angle and the type of material. This rule, called *Snell's law,* is not as simple as the rule for reflection since it involves trigonometry. A simpler way to express the relationship is to construct a graph of the experimental data. Of course, although graphs are easier to use, they often have the disadvantage of being less general. In this case, a graph has to be made for each substance. The graph in Figure 17-2 gives the angle of refraction in air, water, and glass for each angle of incidence in a vacuum. Although the curves for water and glass have similar shapes, light is refracted more on entering glass than water.

If no refraction takes place, the index of refraction is equal to 1. You can see from the graph that very little bending occurs when light goes from a vacuum into air; the index of refraction of air is only slightly greater than 1. Because the index of refraction of air is very close to 1, air and vacuum are nearly equivalent. Therefore, we will use the graph in Figure 17-2 for light entering water or glass from either air or a vacuum. The index of refraction for water is 1.33, and for different kinds of glass it varies from 1.5 to 1.9. The curve for glass on the graph is drawn for an index of 1.5. The index of refraction for diamond is 2.42. A larger index of refraction means more bending for a given angle of incidence. For example, the graph indicates that light incident at 50° has an angle of refraction of 31° in glass and 35° in water. Thus, the light is bent 19° going into glass (index = 1.5) and only 15° going into water (index = 1.33).

> **QUESTION**   What is the angle of refraction for light incident on glass at 30°? How much does the ray bend?

Light entering a transparent material from air bends *toward* the normal. What happens if light originates in the material and exits into the air? Experiments show that the paths of light rays are reversible. The photographs in Figure 17-1 can be interpreted as light inside the glass passing upward into the air. (If this were really the case, however, there would also be a faint reflected beam in the glass.) This example shows that when light moves from a material with a higher index of refraction to a lower one, the light leaving the material is bent *away from* the normal. Because of the reversibility of the rays, you can still use the graph in Figure 17-2 to find the angle of refraction; simply reverse the labels on the two axes.

> **QUESTION**   If a ray of light in water strikes the surface at an angle of incidence of 40°, at what angle does it enter the air?

> **ANSWER**   The graph in Figure 17-2 gives an angle of refraction of approximately 20°. Therefore, the ray bends 30° − 20° = 10° from its original direction.

> **ANSWER**   Locate the 40° angle on the *vertical* axis of the graph in Figure 17-2 and move sideways until you encounter the curve for water. Then moving straight down to the horizontal axis, we obtain an angle of 58°.

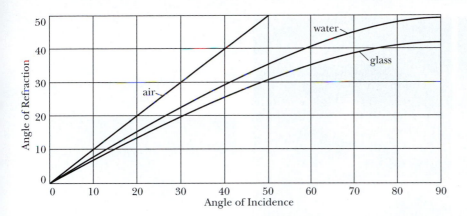

Figure 17-2 This graph shows the relationship between the angle of refraction and the angle of incidence for light entering air, water, and glass from a vacuum.

Another consequence of this reversibility is that light passing through a pane of glass that has parallel surfaces continues in its original direction after emerging. The glass has the effect of shifting the light sideways as shown in Figure 17-1.

The refraction of light produces interesting optical effects. A straight board partially in water appears bent at the surface. The photograph of a meterstick in Figure 17-3 illustrates this effect. Looking from the top, we see that the portion of the meterstick in the water appears to be higher than it actually is.

This phenomenon can also be seen in the photographs of identical coins, one underwater as shown in Figure 17-4(a) and the other in air as shown in Figure 17-4(b). Even though the coins are the same distance from the cameras, the one underwater appears closer and larger. The drawing in Figure 17-4(c) shows some of the rays that produce this illusion. This effect also makes fish appear larger—although never as large as the unlucky fisherman would like you to believe.

> **QUESTION**   If you keep your stamp collection under thick pieces of glass for protection, will they appear to have their normal sizes?

Let's examine the reason for the coin's appearing larger when it is in the water. Is it due to the image being closer, or is the image itself bigger? It is fairly straightforward to see that the increase in size is due to the *image* being closer. To see that the image hasn't increased in size, we need to remind ourselves that rays normal to the surface are not refracted. Therefore, if we use vertical rays to locate the images of all points on the rim of the coin, each image will be directly above the corresponding point on the rim. This means that the image has the same size as the coin.

> **ANSWER**   No. Just like the coin in water, the stamps appear to be closer and are therefore apparently larger in size.

Figure 17-3 A straight pencil appears to be bent at the surface of the water.

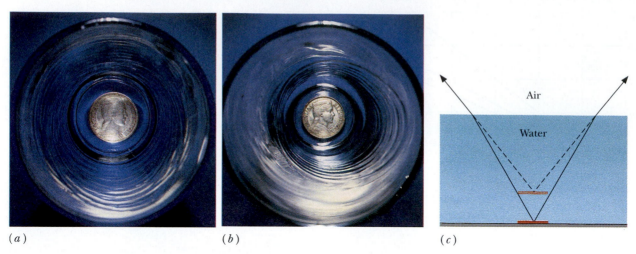

(a)   (b)   (c)

Figure 17-4   A coin underwater (a) appears closer than an identical coin in air (b). (c) Some of the rays that produce the image.

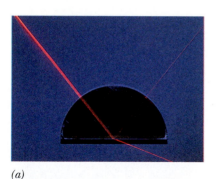

(a)

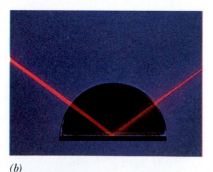

(b)

Figure 17-5   (a) Light traveling from glass into air at the lower surface bends away from the normal. (b) When the incident angle is larger than some critical angle, the light is totally reflected.

Total Internal Reflection

There are situations where light can't pass between two substances even if they are both transparent. This occurs at large incident angles when the light strikes a material with a lower index of refraction such as from glass into air as shown at the lower surface in Figure 17-5. At small angles of incidence, both reflection and refraction take place. The refracted angle is larger than the incident angle as shown in Figure 17-5(a). As the incident angle increases, the refracted angle increases even faster. At a particular incident angle, the refracted angle reaches 90°. Beyond this incident angle—called the **critical angle**—the light no longer leaves the material. The light is totally reflected as shown in Figure 17-5(b). This is called **total internal reflection.**

The critical angle can be found experimentally by increasing the incident angle and watching for the disappearance of the emerging ray. Since the graph in Figure 17-2 works for both directions, we can find the critical angle by looking for the angle of refraction for an incident angle of 90°. Using Figure 17-2 with the labels on the axes reversed indicates that the critical angle for our glass is about 42°. The critical angle for diamond is only 24°.

QUESTION   What is the critical angle for water?

ANSWER   The graph in Figure 17-2 shows that the angle of refraction in water never exceeds 49°, so this is the critical angle.

Mirages

Normal atmospheric conditions are altered when the surface of the Earth is very warm or very cold, producing some interesting optical effects. When the surface is cold, such as over cold water, the density of the atmosphere decreases more rapidly than usual. This bends light rays traveling near the surface so that even terrestrial objects shift in position. For example, an object floating on the water may appear in midair, an effect known as *looming*, shown in Figure A.

A layer of very warm air near the ground can cause the opposite effect. If the air near the ground is warmed, the index of refraction is low near the ground.

This can cause the light near the ground to be bent upward. The combination of the direct and refracted light produces *mirages* like the one shown in Figure B. This is why motorists see "wet" spots on the road in the distance when the weather is particularly hot; the mirage of the sky looks like water on the road.

Mirages can often be seen along dark walls facing the Sun. Have a friend hold a bright object near the wall. Stand next to the wall about 10 meters from your friend and look with one eye. You may need to move your eye toward or away from the wall to see the mirage.

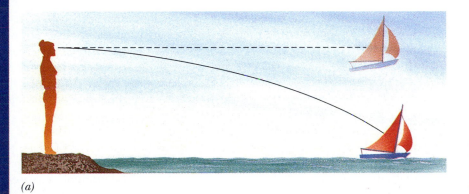

(a)

Looming occurs when the colder air near the surface causes the light rays to bend downward.

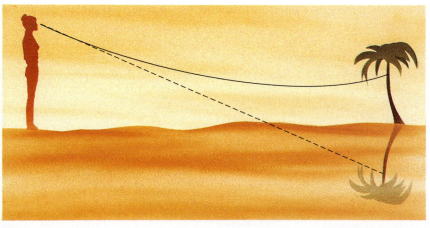

(b)

Mirages occur when the hot air near the surface causes the light rays to bend upward.

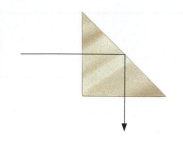

Figure 17-6 A prism acts like a flat mirror when the light is totally internally reflected.

This total internal reflection has many applications. For example, a 45° right prism can act as a mirror, as shown in Figure 17-6. When the light beam hits the back surface, if the incident angle of 45° is greater than the critical angle, the beam is totally reflected. This reflecting surface has many advantages over ordinary mirrors. It doesn't have to be silvered, is easier to protect than an external surface, and is also more efficient for reflecting light.

Another application of this principle is to "pipe" light through long narrow fibers of solid plastic or glass as shown in Figure 17-7. Light enters the fiber from one end. Once inside, the light doesn't escape out the side because the angle of incidence is greater than the critical angle. The rays finally exit at the end of the fiber, because there the incident angles are smaller than the critical angle. Fiber-optics applications are found in photography, medicine, telephone transmissions, and even decorative room lights.

Atmospheric Refraction

We live at the bottom of an ocean of air. Light that reaches us travels through this air and is modified by it. The Earth's atmosphere is not uniform. Under most conditions the density of the atmosphere decreases with increasing altitude. As you might guess, the index of refraction depends on the density of a gas because the less dense the gas, the more like a vacuum it becomes. We conclude, therefore, that the index of refraction of the atmosphere gradually decreases the higher we go.

Refraction occurs whenever there is any change in the index of refraction. When there is an abrupt change as at the surface of glass, the change in the direction of the light is abrupt. But when the change is gradual, the path of a light ray is a gentle curve. The gradual increase in the index of refraction as light travels into the lower atmosphere means that light from celestial objects such as the Sun, Moon, and stars bends toward the vertical. Figure 17-8 shows that this phenomenon makes the object appear higher in the sky than its actual position. Astronomers must correct for atmospheric refraction to get accurate positions of celestial objects.

This shift in position is zero when the object is directly overhead and increases as it moves toward the horizon. Atmospheric refraction is large enough that you can see the Sun and Moon before they rise and after

Light pipe

Figure 17-7 Light may be "piped" through solid plastic or glass rods using total internal reflection.

they set. Of course, without knowing where the Sun and Moon should be, you are not able to detect this shift in position. You can, however, see distortions in their shapes when they are near the horizon as shown in the photographs in Figure 17-9. Since the amount of refraction is larger closer to the horizon, the apparent change in position of the bottom of the Moon is larger than for the top. This results in a shortening of the diameter of the Moon in the vertical direction and gives the Moon an elliptical appearance.

There are other changes in the atmosphere's index of refraction. Because of the atmosphere's continual motion, there are momentary changes in the density of local regions. Stars get their twinkle from this variation. As the air moves, the index of refraction along the path of the star's light changes and the star appears to change position slightly and vary in brightness and color; that is, twinkle. Planets do not twinkle as much because they are close enough to Earth to appear as tiny discs. Light from different parts of the disc averages out to produce a steadier image.

Dispersion

Although the ancients knew that jewels produced brilliant colors when sunlight shone on them, they were wrong about the origin of the colors. They thought the colors were part of the jewel. Newton used a prism to show that the colors don't come from jewels but rather from light itself. He used a prism to demonstrate that the colors are already present in sunlight. When sunlight passes through a prism, the light refracts and is split up into a spectrum of colors ranging from red to violet, a phenomenon known as **dispersion** (Fig. 17-10). To eliminate the idea that the colors are somehow produced by the prism, Newton did two experiments. He took one of the colors from a prism and passed it through a second prism,

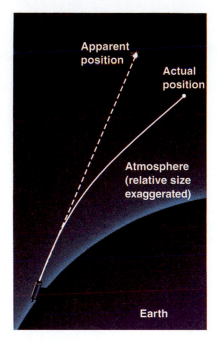

Figure 17-8 Atmospheric refraction changes the apparent positions of celestial objects, making them appear higher in the sky.

Figure 17-9 The atmospheric distortion of the setting Moon increases as the Moon gets closer to the horizon, giving the Moon an elliptical shape.

Figure 17-10 A prism separates white light into the colors of the rainbow.

demonstrating that no new colors were produced. He also recombined the colors and obtained white light. His experiments showed conclusively that white light is a combination of all colors. The prism just spreads them out so that the individual colors can be seen.

The name ROY G. BIV is a handy mnemonic for remembering the order of the colors produced by a prism or those in the rainbow: red, orange, yellow, green, blue, indigo, and violet. (Indigo is included mostly for the mnemonic; people can seldom distinguish it from blue or violet.)

The light changes direction as it passes through the prism because of refraction at the faces of the prism. Dispersion tells us that the colors have slightly different indices of refraction in glass. Violet light is refracted more than red and, therefore, has a larger index. The brilliance of a diamond is due to the small critical angle for internal reflection and the separation of the colors due to the high amount of dispersion.

Rainbows

Sometimes after a rain shower, you get to see one of nature's most beautiful demonstrations of dispersion, a rainbow. Part of its appeal must be that it appears to come from thin air. There seems to be nothing there but empty sky.

In fact, the rainbow is formed by water droplets in the air. You can verify this by making your own rainbow. Turn your back to the Sun and spray a fine mist of water from your garden hose in the direction opposite the Sun. Each color forms part of a circle about the point directly opposite the Sun (Fig. 17-11). The angle to each of the droplets along the circle of a given color is the same. Red light forms the outer circle and blue light the inner one.

> **Physics on Your Own**   Produce a "rainbow" by shining a flashlight at a quart jar filled with water. Covering the flashlight with an index card with a slit cut in it produces a narrow vertical beam that makes the colors more prominent.

If you are willing to get wet, it is possible to see a complete circular rainbow. Near noon on a sunny day, spray the space around you with a fine mist. Looking down, you will find yourself in the center of a rainbow. A circular rainbow can sometimes be seen from an airplane.

> **QUESTION**   If you see a rainbow from an airplane, where do you expect to see the shadow of the airplane?

Figure 17-11 A rainbow's magic is that it seems to appear out of thin air. Notice the secondary rainbow on the right.

> **ANSWER**   Because the center of the rainbow is always directly opposite the Sun, the shadow of the airplane will be at the center of the rainbow.

A rainbow formed in the spray from a sprinkler hose.

Rainbows result from the dispersion of sunlight by water droplets in the atmosphere. The dispersion that occurs as the light enters and leaves the droplet separates the colors that compose sunlight. Figure 17-12 shows the paths of the red and violet light. The other colors are spread out between these two according to the mnemonic ROY G. BIV. Each droplet disperses all colors. Your eyes, however, are only in position to see one color coming from a particular droplet. For instance, if the droplet is located such that a line from the Sun to the droplet and a line from your eyes to the droplet form an angle of 42°, the droplet appears red (Fig. 17-13). If this angle is 40°, the droplet appears violet. Intermediate angles yield other colors. Whether or not you believe there is a pot of gold at the end of the rainbow, you will never be able to get there to find out. As you move, the rainbow "moves." In your new position you see different droplets producing your rainbow.

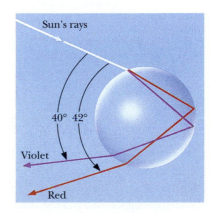

Figure 17-12 Dispersion of sunlight in a water droplet separates the sunlight into a spectrum of colors.

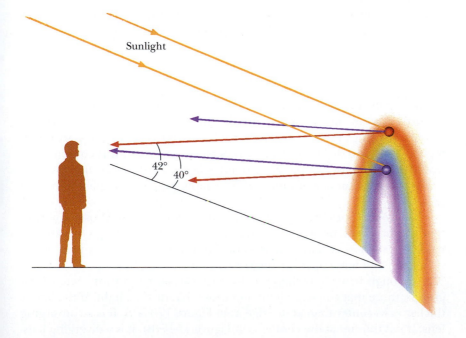

Figure 17-13 The color of each water droplet forming the rainbow depends on the viewing angle.

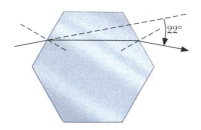

Figure 17-14 Light passing through alternate surfaces of a hexagonal ice crystal changes direction by at least 22°.

Figure 17-15 A photograph of the 22° halo and its associated sundogs taken with a fisheye lens. A building was used to block out the Sun's direct rays.

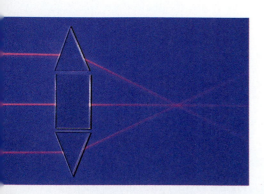

Figure 17-16 Two prisms and a rectangular block form a primitive lens.

If viewing conditions are good, you can see a secondary rainbow that is fainter and larger than the first. It is centered about the same point, but the colors appear in the reverse order. This rainbow is produced by light that reflects twice inside the droplets.

Halos

Sometimes a large halo can be seen surrounding the Sun or Moon. These halos and other effects, such as *sundogs* and various arcs, are caused by the refraction of light by ice crystals in the atmosphere.

Atmospheric ice crystals have the shape of hexagonal prisms. Each one looks like a slice from a wooden pencil that has a hexagonal cross section. Light hitting the crystal is scattered in many different directions depending on the angle of incidence, which face it enters, and which face it exits. Light entering and exiting alternate faces, as shown in Figure 17-14, has a minimum angle of scatter of 22°. Although light is scattered at other angles, most of the light concentrates near this angle.

To see a ray of light that has been scattered by 22°, you must look in a direction 22° away from the Sun. Light scattering this way from crystals randomly oriented in the atmosphere forms a 22° halo around the Sun as shown in Figure 17-15. The random nature of the orientations ensures that at any place along the halo there will be crystals that scatter light into your eyes. Dispersion in the ice crystal produces the colors in the halo.

Occasionally one also sees "ghost" suns located on each side of the Sun at the same height as the Sun, as seen in Figure 17-15. These sundogs are produced by ice crystals which have axes that are oriented vertically. These crystals can only refract light into your eyes when they are located along or just outside the halo's circle at the same altitude as the Sun.

An even larger but dimmer halo at 46° exists but is less frequently seen. It is formed by light passing through one end and one side of the crystals. Other effects are produced by light scattering through other combinations of faces in crystals with particular orientations.

Lenses

Unlike a pane of glass, when light enters a material with entrance and exit surfaces that are not parallel, the direction of the light beam changes. Two prisms and a rectangular block can be used to focus light as shown in Figure 17-16. However, most other rays passing through this combination would not be focused at the same point. The focusing can be improved by using a larger number of blocks or by shaping a piece of glass to form a lens.

We see the world through lenses. This is true even for those of us who don't wear glasses, because the lenses in our eyes focus images on our retinas. Other lenses extend our view of the universe—microscopes for the very small and telescopes for the very distant.

Although many lens shapes exist, they can all be put into one of two groups: those that converge light and those that diverge light. If the lens is thicker at its center than at its edge as in Figure 17-17(a), it is a converging lens. If it is thinner at the center as in Figure 17-17(b), it is a diverging lens.

> **QUESTION** Lenses in eyeglasses are made with one convex surface and one concave surface. How can you tell if the lenses are converging or diverging?

Lenses have two focal points—one on each side. A converging lens focuses incoming light that is parallel to its optic axis at a point on the other side of the lens known as the *principal* **focal point** (Fig. 17-18). The distance from the center of the lens to the focal point is called the **focal length.** We can find the other focal point by reversing the direction of the light and bringing it in from the right-hand side of the lens. The light then focuses at a point on the left-hand side of the lens that we refer to as the "other" focal point in drawing ray diagrams.

For a diverging lens, incoming light that is parallel to the optic axis appears to diverge from a point on the same side of the lens (Fig. 17-19). This point is known as the *principal focal point,* and the focal point on the other side is known as the "other" one. You can show by experiment that the two focal points are the same distance from the center of the lens if the lens is thin. A lens is considered to be thin if its thickness is very much less than its focal length.

The shorter the focal length, the "stronger" the lens; that is, the lens focuses light parallel to the optic axis at a point closer to the lens.

> **Physics on Your Own** Determine the focal length of a magnifying glass or other converging lens by focusing sunlight to the smallest possible spot. Turn the lens around to see if you get the same distance on the other side.

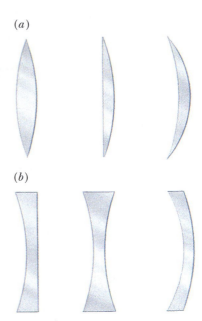

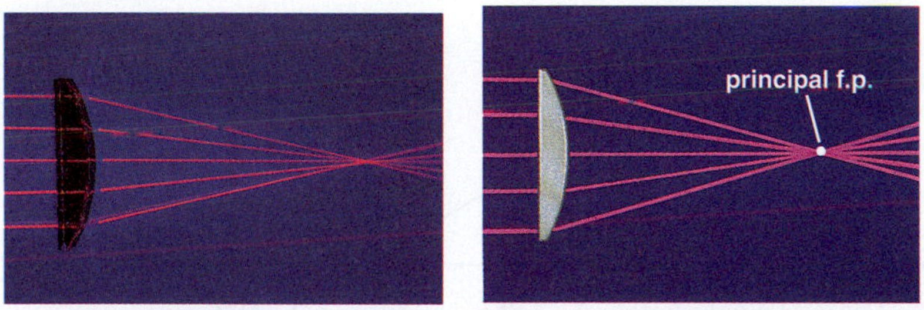

(a)

(b)

Figure 17-17 (a) Converging lenses are thicker at the middle. (b) Diverging lenses are thinner at the middle.

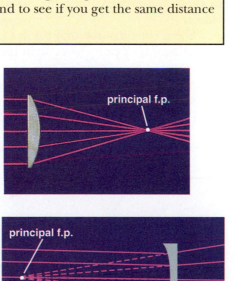

principal f.p.

principal f.p.

Figure 17-18 Light parallel to the optic axis of a converging lens is focused at the principal focal point.

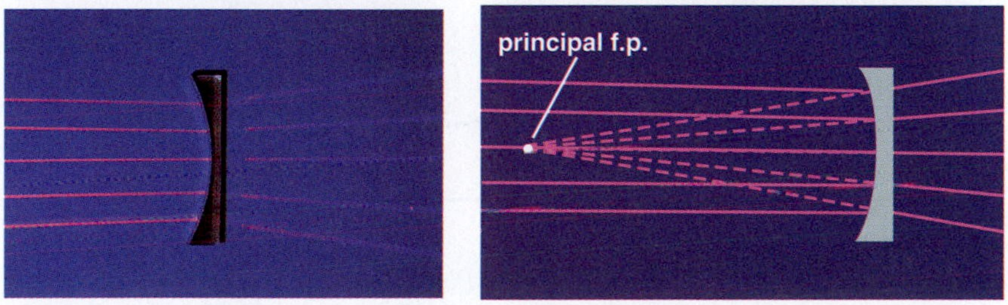

Figure 17-19 Light parallel to the optic axis of a diverging lens appears to come from the principal focal point.

> **ANSWER** Check to see if they are thicker at the center than at the edges. If they are thicker at the center, they are converging.

Images Produced by Lenses   Σ

The same ray-diagramming techniques used for curved mirrors in the previous chapter will help us locate the images formed by lenses. Again, three of the rays are easily drawn without measuring angles. The intersection of any two determines the location of the image. Figure 17-20 shows the three rays.

First, a ray passing through the center of the lens continues without deflection. Next, for a converging lens, a ray parallel to the optic axis passes through the principal focal point; and third, a ray coming from the direction of the other focal point leaves the lens parallel to the optic axis [Fig. 17-20(a)]. (The optic axis passes through the center of the lens and both focal points.) Notice that the second and third rays are opposites of each other. For a diverging lens, the second ray comes in parallel to the optic axis and leaves as if it came from the principal focus, and the third ray heads toward the other focal point and leaves parallel to the optic axis [Fig. 17-20(b)].

These rays are very similar to the ones used for mirrors. There are two main differences. The first ray passes through the center of the lens and not the center of the sphere as it did for mirrors, and there are now two focal points instead of one. We can still give abbreviated versions of these rules.

rays for lenses

1. Through center—continues.
2. Parallel to optic axis—through (from) principal focal point.
3. Through (toward) other focal point—parallel to optic axis.

In these abbreviated rules, the words in parentheses refer to diverging lenses.

These rules assume that the lens is thin. The first rule neglects the offset that takes place when a light ray passes through parallel surfaces of

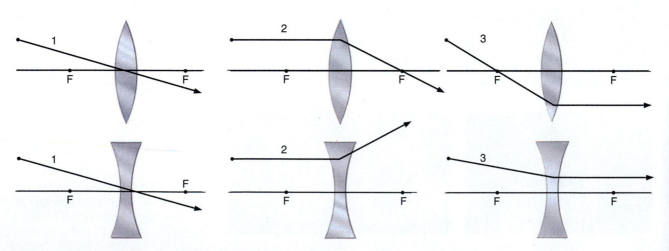

Figure 17-20 The three rays used in drawing ray diagrams for (a) converging and (b) diverging lenses.

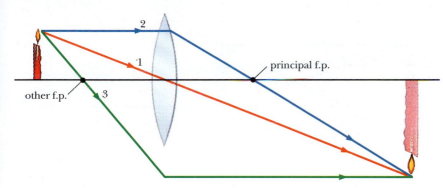

Figure 17-21   The image formed by a candle located outside the focal point of a converging lens is real, inverted, and may be magnified or reduced in size.

glass at other than normal incidence (see Fig. 17-1). For the purposes of drawing these rays, the bending of the light is assumed to take place at a plane perpendicular to the optic axis and through the center of the lens. For a lens with spherical surfaces these three rays do not usually meet at a point. However, they give the approximate location of the image if the object is small enough that the rays strike the lens near the optic axis.

We can apply these rays to locate the image of a candle that is located on the optic axis outside the focal point of a converging lens. The ray diagram in Figure 17-21 shows that the image is located on the other side of the lens and is real and inverted. (See Chapter 16 for a discussion of the types of image.) Whether it is magnified depends on how far it is from the focal point. As the candle is moved away from the lens, the image moves closer to the principal focal point and gets smaller.

If the candle is moved inside the focal point as illustrated in Figure 17-22, the image appears on the same side of the lens. This is the arrangement that is used when a converging lens is used as a magnifying glass. The lens is positioned such that the object is inside the focal point, producing an image that is virtual, erect, and magnified.

A diverging lens always produces a virtual image as shown in Figure 17-23. The image changes location and size as the object is moved, but the image remains erect and virtual.

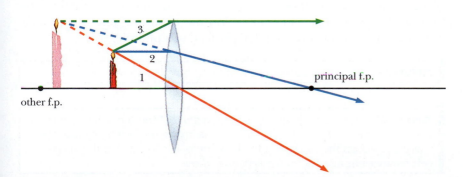

Figure 17-22   The image formed by a candle located inside the focal length of a converging lens is virtual, erect, and magnified.

Figure 17-23   The image formed by a diverging lens is virtual, erect, and reduced in size.

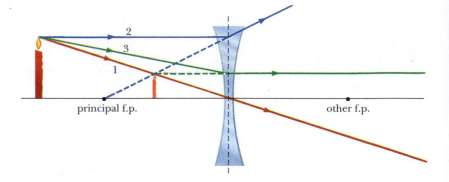

> **QUESTION**   Is the lens used in a slide projector converging or diverging?

Notice that one of the rays in Figure 17-21 does not pass through the lens. This isn't a problem because there are many other rays that do pass through the lens to form the image. Ray diagramming is just a geometric construction that allows you to locate images, a process that can be illustrated with an illuminated arrow and a large-diameter lens as drawn in Figure 17-24. A piece of paper at the image's location allows the image to be easily seen. If the lens is then covered with a piece of cardboard with a hole in it, the image is still in the same location, the same size, and in focus. The light rays from the arrow that form the image are those that pass through the hole. The image is not as bright since less light now forms the image. The orange lines illustrate the paths of some of the other rays.

> **QUESTION**   What happens to the image in Figure 17-24 if you cover up the bottom half of the lens?

Cameras

We saw in the last chapter that pinhole cameras produce sharp images if the pinhole is small. The amount of light striking the film, however, is quite small. Very long exposure times are needed, which means that the objects in the scene must be stationary. The amount of light reaching the film can be substantially increased (and the exposure time substantially reduced) by using a converging lens instead of a pinhole.

> **ANSWER**   It must be converging as it forms a real image on the screen.

> **ANSWER**   Although it is very tempting to say that the top half of the image disappears, this is not correct. All parts of the image are formed by the rays that pass through the top half of the lens. The only effect is that the image gets dimmer because half of the light is blocked.

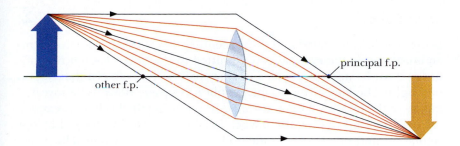

Figure 17-24   The orange rays form the image. The black ones are used because they are easy to draw.

The essential features of a simple camera are shown in Figure 17-25. This camera has a single lens at a fixed distance from the film. The distance is chosen so that the real images of faraway objects are formed at, or at least near, the film. These cameras are usually not very good for taking close-up shots, such as portraits, because the images are formed beyond the film and are therefore out of focus at the film. More expensive cameras have an adjustment that moves the lens relative to the film to position (focus) the image on the film.

QUESTION   If the focal length of the lens in a simple camera is 50 millimeters, how far is it from the lens to the film for a subject that is very far from the camera?

Ideally, all light striking the lens from a given point on the object should be focused to a given point on the film. However, real lenses have a number of defects, or **aberrations,** so that light is not focused to a point but is spread out over some region of space.

A lens cannot focus light from a white object to a sharp point because of dispersion. A converging lens focuses violet light at a point closer to the lens than it does red light. This chromatic aberration produces images with colored fringes. Since the effect is reversed for diverging lenses and since the amount of dispersion varies with material, lens designers minimize the chromatic aberration by combining converging and diverging lenses made of different types of glass.

A spherical lens (or a spherical mirror for that matter) does not focus all light parallel to the optic axis to a sharp point. Light further from the optic axis is focused at a point closer to the lens than light near the optic axis. This spherical aberration is usually corrected by using a combination of lenses. It is also reduced by using a diaphragm to reduce the effective diameter of the lens. Although this sharpens the image, it also reduces the amount of light striking the film. New techniques for reducing spherical aberration by grinding lenses with nonspherical surfaces and by making lenses in which the index of refraction of the glass changes with the distance from the optic axis have been developed.

ANSWER   If the objects are effectively at infinity, the light from each point will be focused at a distance equal to the focal length. Thus, the film should be about 50 millimeters from the center of the lens.

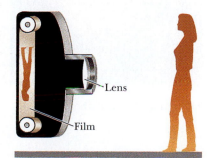

Figure 17-25   The essential features of a simple camera.

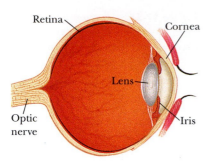

Schematic drawing of the human eye.

Our Eyes

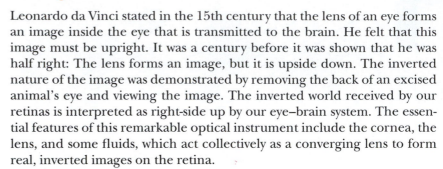

Leonardo da Vinci stated in the 15th century that the lens of an eye forms an image inside the eye that is transmitted to the brain. He felt that this image must be upright. It was a century before it was shown that he was half right: The lens forms an image, but it is upside down. The inverted nature of the image was demonstrated by removing the back of an excised animal's eye and viewing the image. The inverted world received by our retinas is interpreted as right-side up by our eye–brain system. The essential features of this remarkable optical instrument include the cornea, the lens, and some fluids, which act collectively as a converging lens to form real, inverted images on the retina.

Physics on Your Own   You can observe that the real image formed on the retina of your eye is inverted. Punch a pinhole in an index card and hold it about 15 centimeters (6 inches) in front of one eye while looking at a bright light source. Hold the head of the pin in line with the hole and midway between the hole and your eye. You should be able to see the inverted shadow of the pinhead on your retina.

When you look at a distant object, nearby objects are out of focus. Only distant objects form sharp images on the surface of the retina. The nearby objects form images that would be behind the retina and, therefore, the images on the retina are fuzzy. This phenomenon occurs because the locations of images of objects at various distances depend on the distances between the lens and the objects and on the focal length of the lens. The lens in the eye changes its shape and thus its focal length to accommodate the different distances.

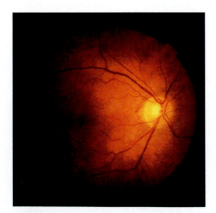

The retina of your author's eye.

Physics Update

The sharpest pictures yet of the living human retina have been produced by University of Rochester researchers, allowing scientists to see the individual cells called cones for the first time in human patients. The Rochester team took their pictures by flashing low-power yellow laser light into the dilated pupils of patients fitted with special lenses, and they captured the images using a special electronic camera with higher sensitivity and resolution than ordinary film. The retina, the screenlike membrane that lines the inside of the eyeball, captures light signals and sends them to the brain. Only 3 micrometers wide, cone cells detect color and facilitate daytime vision. With current instruments, ophthalmologists can only see structures 10 micrometers or larger. The ability to resolve cones in patients could aid in the treatment of age-related macular degeneration, the leading cause of blindness in the elderly.

Opticians measure the strength of lenses in diopters. The *diopter* is equal to the reciprocal of the focal length measured in meters. For example, a lens with a focal length of 0.2 meter is a 5-diopter lens. In this case a larger diopter means that the lens is stronger. Converging lenses have positive values for their diopters, and diverging lenses have negative values. Diopters have the advantage that two lenses placed together have a diopter equal to the sum of the two individual ones.

In the relaxed eye of a young adult all the transparent materials have a total "power" of +60 diopters. Most of the refraction (+40 diopters) is due to the outer element of the eye, the cornea, but the relaxed lens contributes +20 diopters.

The eye can vary the strength of the lens from a relaxed value of +20 diopters to a maximum of +24 diopters. When the relaxed eye views a distant object, the +60 diopters produce an image at 1.7 centimeters (0.7 inches), which is the distance to the retina in a normal eye. The additional +4 diopters allow the eye to view objects as near as 25 centimeters (10 inches) and still produce sharp images on the retina.

The ability of the eye to vary the focal length of the lens decreases with age as the elasticity of the lens decreases. A 10-year-old eye may be able to focus as close as 7 centimeters (+74 diopters), but a 60-year-old eye may not be able to focus any closer than 200 centimeters (6½ feet). An older person often wears bifocals when the eyes lose their ability to vary the focal length.

The amount of light entering the eye is regulated by the size of the pupil. As with the ear, the range of intensities that can be viewed by the eye is very large. From the faintest star that can be seen on a dark, clear night to bright sunlight is a factor of approximately 10^{10} in intensity.

	Sphere	Cylinder	Axis
R	−6.50	+3.25	089
L	−5.75	+2.75	074
R	+2.00	Bifocals	
L	+2.00		

The prescription for your author's new eyeglasses. The spherical and cylindrical corrections are given in diopters. Axis is the number of degrees the axis of the cylinder is rotated from the vertical. The bifocal correction is added to the others.

> **Physics on Your Own**   Another common visual defect is *astigmatism*. When some of the refracting surfaces are not spherical, the image of a point is spread out into a line. Use the pattern in Figure 17-26 to check for astigmatism in your eyes. Lines along the direction in which images of point sources are spread remain sharp and dark, but the others become blurred. Are your two eyes the same?

Magnifiers   Σ

It has been known since the early 17th century that refraction could bend light to magnify objects. The invention of the telescope and microscope produced images of regions of the Universe that until then had been unexplored. Galileo used the newly discovered telescope to see Jupiter's moons and the details of our Moon's surface. The English scientist Robert Hooke spent hours peering into another unexplored world with the aid of the new microscope.

The size of the image on the retina depends on the object's actual size and on its distance away. The image of a dime held at arm's length is much larger than that of the Moon. What really matters is the angular size of the object; that is, the angle formed by lines from your eye to opposite sides of the object. The angular size of an object can be greatly increased

Figure 17-26   A test pattern for astigmatism. If you see some lines blurred while other lines are sharp and dark, you have some astigmatism.

Eyeglasses

Our optical system is quite amazing. The lens in our eyes can change its shape, altering its focal length to place the image on the retina. Sometimes, however, the eye is too long or too short and the images are formed in front of or behind the retina. When the eye is too long, the images of distant objects are formed in front of the retina as shown in Figure A. Such a person has *myopia* (nearsightedness) and can see things that are close, but has trouble seeing distant objects. When the eye is too short, the person has *hyperopia* (farsightedness) and has trouble seeing close objects. Distant objects can be imaged on the retina, but close objects form images behind the retina (Figure B).

Our knowledge of light and refraction allows us to devise instruments—eyeglasses—that correct these deficiencies. The nearsighted person wears glasses with diverging lenses to see distant objects (Fig. A). The farsighted person's sight is corrected with converging lenses (Fig. B).

Even those people with perfect vision early in life lose some of the lens's range, particularly the ability to shorten the focal length. These people have trouble creating a focused image on their retina for objects that are close. Many older people wear bifocals when their eyes lose the ability to shorten the focal length. The upper portion of the lens is used for distant viewing, while the lower portion is used for close work or reading. When people work at intermediate distances such as looking at computer screens, they sometimes wear trifocals.

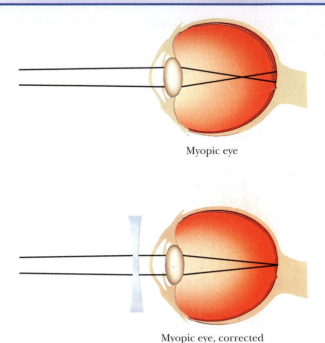

Myopic eye

Myopic eye, corrected

Figure A
The myopic eye forms the images of distant objects in front of the retina. This is corrected with diverging lenses.

by bringing it closer to your eye. However, if you bring it closer than about 25 centimeters (10 inches), your eye can no longer focus on it and its image is blurred. You can get both an increased angular size and a sharp image by using a converging lens as a magnifying glass. When the object is located just inside the focal point of the lens, the image is virtual, erect, and has nearly the same angular size as the object. Moreover, as shown in Figure 17-27, the image is now far enough away that the eye can focus on it and see it clearly.

Physics on Your Own   Place a drop of water on a piece of transparent foil used in wrapping food and use it to magnify the print in this text. Do drops of other clear liquids yield higher magnifications?

The difficulty of making glasses to correct vision is increased when a person has astigmatism. This occurs when some of the refracting surfaces are not spherical and the image of a point is spread out into a line. This visual defect is corrected by adding a cylindrical curvature to the spherical curvature of the lens. You can check your (or a friend's) glasses for astigmatism by looking through them in the normal way but holding the glasses at a distance from your head. When you rotate the lens about a horizontal axis, you will see background distortion if the lens has an astigmatic correction.

Many people wear contact lenses to correct their vision. Contact lenses have created some interesting challenges in correcting for astigmatism and the need to wear bifocals. When correcting for astigmatism, it is very important that the cylindrical correction has the correct orientation. Contact lenses can be weighted at one place on the edge to keep the lens oriented correctly. People who need bifocals can sometimes be fitted with different corrections in each eye; one eye is used for close vision and the other is used for distant vision. The eye–brain system switches from one eye to the other as the situation demands.

Finally, a recent medical procedure uses laser technology to correct nearsightedness. In this procedure, the laser is used to make radial cuts in the cornea of the eye to reduce its curvature.

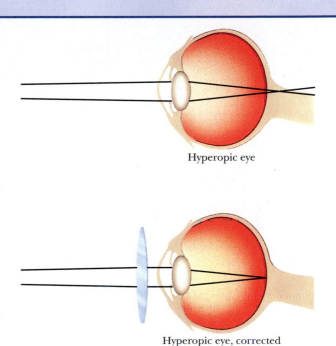

Hyperopic eye

Hyperopic eye, corrected

Figure B
The hyperopic eye forms images of close objects in back of the retina. This is corrected with converging lenses.

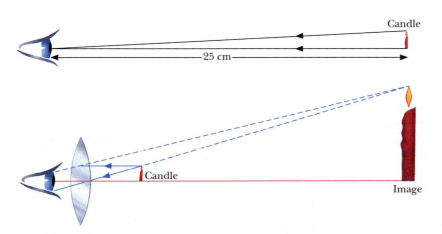

Candle

25 cm

Candle

Image

Figure 17-27 A candle viewed (a) at a distance of 25 centimeters and (b) through a magnifying glass.

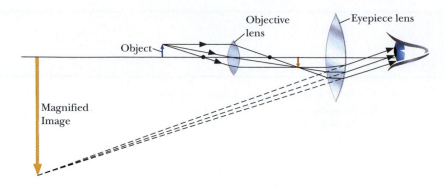

Figure 17-28   Schematic of a compound microscope.

An even higher magnification can be achieved by using two converging lenses to form a compound microscope, as shown in Figure 17-28. The object is located just outside the focal point of the objective lens. This lens forms a real image that is magnified in size. The eyepiece then works like a magnifying glass to further increase the angular size of this image.

Telescopes

There are many varieties of telescope. A simple one using two converging lenses is known as a **refracting telescope,** or refractor. Figure 17-29 shows that this type of telescope has the same construction as a compound microscope except that now the object is far beyond the focal point of the objective lens. Like the microscope, the refractor's lens produces a real, inverted image. Although the image is much smaller than the object, it is much closer to the eye. The eyepiece acts like a magnifying glass to greatly increase the angular size of the image.

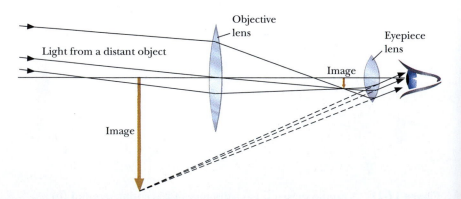

Figure 17-29   Schematic of an astronomical telescope.

The Hubble Space Telescope

At a cost of $1.6 billion, the Hubble Space Telescope was placed in an orbit 575 kilometers (357 miles) above the Earth by the Space Shuttle *Discovery* in April 1990. Its 15-year-long mission is to provide astronomers with observations of the Universe without the disturbances caused by the Earth's atmosphere. The atmosphere absorbs most of the radiation reaching us from space, except for two broad bands in the radio region and around the visible region. A variety of experiments were planned that ranged from viewing distant, faint objects to accurately measuring the positions of stars. The design specifications indicated that the Hubble Space Telescope would be able to see objects seven times farther away than could be observed from the Earth's surface.

However, these experiments were seriously hampered by a defect in the telescope's primary mirror. While the error in its shape was only 0.002 millimeters (about 1/40th the thickness of a human hair) at the edge of the 2.4-meter diameter mirror, it caused light from the edge of the mirror to focus 38 millimeters beyond light from the center. This left the telescope with an optical defect—spherical aberration—that created fuzzy halos around images of stars, and blurred images of extended objects like galaxies and giant clouds of gas and dust. Only 15 percent of the light from a star was focused into the central spot compared to the design value of 70 percent. However, because this type of aberration is well known, computer enhancement was used to sharpen images of brighter objects, producing some rather remarkable views and some very good science, but the technique did not work for faint objects.

Fortunately, the Hubble Space Telescope was designed to be serviced by the Space Shuttle and new optics were designed to compensate for the error. In December 1993, the Space Shuttle *Endeavour* docked with the Hubble Space Telescope to repair the optics and replace a number of mechanical and electrical components that had failed or required scheduled replacement. The repairs required five space walks involving two astronauts each. The total repair mission had a cost of $700 million. The repaired Hubble Space Telescope is able to see much farther into space. This should allow astronomers to determine the size and age of the Universe and to study very faint objects that should yield information about the early history of the Universe.

The Hubble Space Telescope.

Hubble Deep Field

This Hubble Space Telescope image looks farther into the Universe than ever before. It shows a vast array of galaxies that are important to our understanding of the Universe.

Figure 17-30   Schematic of prism binoculars.

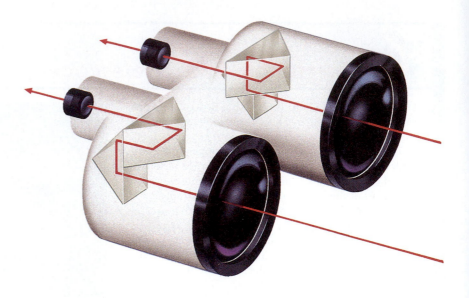

The magnification of a telescope is equal to the ratio of the focal lengths of the objective and the eyepiece. To get high magnification, the focal length of the objective needs to be quite long. Binoculars were designed to provide a long path length in a relatively short instrument. The diagram in Figure 17-30 shows that this is accomplished by using the internal reflections in two prisms to fold the path.

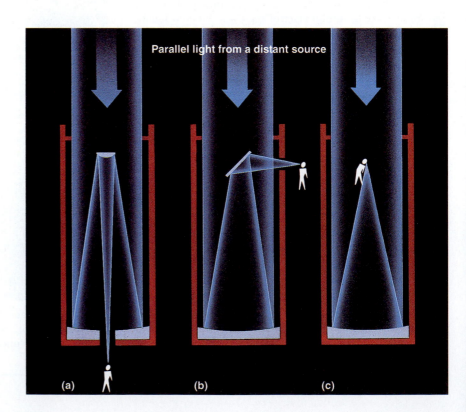

Parallel light from a distant source

(a)   (b)   (c)

Figure 17-31   Schematics of a (a) Cassegrain reflector, (b) Newtonian reflector, and (c) prime focus telescope.

Large-diameter telescopes are desirable because they gather a lot of light, allowing us to see very faint objects or to shorten the exposure time for taking pictures. The problem, however, is making a large-diameter glass lens. It is very difficult, if not impossible, to make a piece of glass of good enough quality. Also, a lens of this diameter is so thick that it sags under its own weight. Therefore, most large telescopes are constructed with concave mirrors as objectives and are known as **reflecting telescopes,** or reflectors. The use of a concave mirror to focus the incoming light has several advantages: The construction of a mirror requires grinding and polishing only one surface rather than two; a mirror can be supported from behind; and, finally, mirrors do not have the problem of chromatic aberration. Figure 17-31 illustrates several designs for reflecting telescopes.

The world's largest refractor has a diameter of 1 meter (40 inches), whereas the largest reflector has a diameter of 6 meters (236 inches). This is just about the limit for a telescope with a single objective mirror; the costs and manufacturing difficulties are not worth the gains. Telescope makers have recently built telescopes in which the images from many smaller mirrors are combined to increase the light-gathering capabilities.

SUMMARY

When light strikes a transparent material, part of it reflects and part refracts. The amount of refraction depends on the incident angle and the index of refraction of the material. Light entering a material of higher index of refraction bends toward the normal. Because the refraction of light is a reversible process, light entering a material with a smaller index of refraction bends away from the normal. For light in a material with the larger index of refraction, total internal reflection occurs whenever the angle of incidence exceeds the critical angle.

The refraction of light at flat surfaces causes objects in or behind materials of higher indices of refraction to appear closer, and therefore larger. The apparent locations of celestial objects are changed by refraction in the atmosphere.

White light is separated into a spectrum of colors because the colors have different indices of refraction, a phenomenon known as dispersion. Rainbows are formed by dispersion in water droplets. Each color forms part of a circle about the point directly opposite the Sun. Halos are caused by the refraction of sunlight in ice crystals.

Ray diagrams can be used to locate the images formed by lenses. The rays are summarized by (1) through center—continues, (2) parallel to optic axis—through (from) principal focal point, and (3) through (toward) other focal point—parallel to optic axis.

Cameras and our eyes contain converging lenses that produce real, inverted images. Converging lenses can be used as magnifiers of objects located inside the focal points. Lenses can be combined to make microscopes and telescopes.

CHAPTER

17

REVISITED

The most obvious consequence of the passage of light between two transparent materials is that the direction of the light changes at the interface. This may produce virtual images that are closer, making the fish look bigger and the tree look bent. It is possible for light to get "trapped" if it is in the material with the larger index of refraction and if the angle of incidence is larger than the critical angle.

KEY TERMS

aberration: A defect in a mirror or lens causing light rays from a single point to fail to focus at a single point in space.

critical angle: The minimum angle of incidence for total internal reflection to occur.

dispersion: The spreading of light into a spectrum of colors.

focal length: The distance from the center of a lens to its focal point.

focal point: The location at which a lens focuses rays parallel to the optic axis or from which such rays appear to diverge.

index of refraction: An optical property of a substance that determines how much light bends upon entering or leaving it.

normal: A line perpendicular to a surface or curve.

optic axis: A line passing through the center of a lens and both focal points.

reflecting telescope: A type of telescope using a mirror as the objective.

refracting telescope: A type of telescope using a lens as the objective.

refraction: The bending of light that occurs at the interface between transparent media.

total internal reflection: A phenomenon that occurs when the angle of incidence of light traveling from a material with a higher index of refraction into one with a lower index of refraction exceeds the critical angle.

CONCEPTUAL QUESTIONS

1. A narrow beam of light emerges from a block of glass in the direction shown in the figure. Which arrow best represents the path of the beam within the glass?

2. A mirror is lying on the bottom of a fish tank that is filled with water. If IN represents a light ray incident on the top of the water, which possibility in the figure best represents the outgoing ray?

3. Swimmers with normal vision cannot focus on distant objects underwater. How do goggles correct this problem?

4. Suppose that you are lying on the bottom of a swimming pool looking up at a ball that is suspended 1 meter above the surface of the water. Does the ball appear to be closer, farther away, or still 1 meter above the surface?

5. Even though water is transparent and colorless, you are able to see a drop of water. Why?

6. Why do clear streams look so shallow?

7. For what incident angles is light totally reflected at a water–air surface?

8. Is the critical angle greater at a water–air surface or a glass–air surface?

9. Telephone companies are using "light pipes" to carry telephone signals between various locations. Why does the light stay in the pipe?

10. In arthroscopic surgery the surgeon makes a very small incision in the skin through which the instruments are inserted. How does the surgeon see inside the body during the operation?

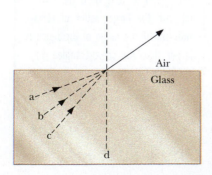

Question 1.

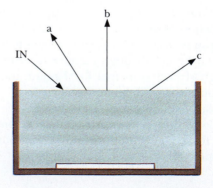

Question 2.

11. We observe stars twinkle when viewed from Earth, but astronauts in the Space Shuttle do not observe this. Why?

12. How does the refraction of light in the atmosphere affect the appearance of the Sun or Moon as it approaches the horizon?

*13. Two stars toward the south are observed to be separated by 5° along an east–west arc. Will this separation appear larger, smaller, or the same when the stars are near the horizon?

*14. In the absence of an atmosphere a star moves across the sky from horizon to horizon at a constant speed. How does the star appear to move in the presence of an atmosphere?

15. How is the time of sunrise affected by atmospheric refraction?

16. How does the presence of an atmosphere affect the length of day and night?

17. Assume that you are trying to spear fish from a boat and spy a fish about 2 meters from the boat. Should you aim high, low, or directly at the fish?

*18. If you were going to send a beam of light to the Moon when it is just above the horizon, would you aim high, low, or directly at the Moon?

19. Why is a diamond more brilliant than a clear piece of glass having the same shape?

20. Does reflected light exhibit dispersion?

21. On a hot day travelers in the desert may "see" a pond even though there is nothing but sand. Draw a ray diagram showing how this can happen.

22. Sometimes mariners report seeing an island floating in the air. Draw a diagram showing how this can happen.

23. How does the order of the colors in the primary rainbow compare with that in the secondary rainbow?

24. If your line of sight to a water droplet makes an angle of 41° with the direction of the sunlight, what color would the raindrop appear to be?

25. Why is the shadow of your head in the center of a rainbow?

26. What might you say to a friend who claims to have seen a rainbow in the east shortly after sunrise?

27. At what time of day might you expect to see the top of a rainbow rise above the horizon?

28. In what direction would you look to see a rainbow at noon in the summer?

29. Would you expect to see a rainbow produced by the Moon?

30. When and where would you expect to see a rainbow that has mostly red and orange colors?

31. What causes halos?

32. At what angle would a halo around the Moon appear?

33. Where would you look for sundogs?

34. Where would you look for the 46° halo?

35. What type of lens would you use to construct an overhead projector?

36. Can a prism be used to form an image?

37. Where does a ray arriving parallel to the optic axis of a converging lens go after passing through the lens?

38. What path does a ray take after passing through a converging lens if it passes through a focal point before it enters the lens?

39. A converging lens is used to form a sharp image of a candle. What effect does covering the upper half of the lens with paper have on the image?

40. How does covering all but the center of a lens affect the image of an object?

41. Two lenses with identical shapes are made from glasses with different indices of refraction. Which one has the shorter focal length? Why?

*42. An air pocket in a rectangular block of glass has the shape of the left-hand lens in Figure 17-20(a). Would this "air lens" cause light parallel to the optic axis to converge or diverge?

43. How might you convince a friend that the image formed by a camera is a real image?

44. What kind of image is formed on the retina of an eye?

45. Why are the backgrounds usually out of focus in close-up photos of people's faces?

46. What kind of lens would you place in front of the lens of a simple camera to turn it into a close-up camera for taking pictures of small objects?

47. What is the purpose of the pupil in the eye?

48. What are the purposes of diaphragms in cameras?

49. Why does a mirror not exhibit chromatic aberration?

*50. When a single converging lens is used to focus white light, the image has a colored fringe due to chromatic aberration. Describe the changes in the color of the fringe as a screen is moved through the focal point.

51. Are reading glasses used by older people converging or diverging? Explain.

52. Why do many older people wear bifocals?

53. Sometimes when a person develops a cataract, the lens of the eye is surgically removed and replaced with a

plastic one. Would you expect this lens to be converging or diverging?

54. Stamp and coin collectors often wear special glasses that allow them to see the details of the stamps and coins. Are the lenses in these glasses converging or diverging?

55. How far from a magnifying glass do you place an object to view it?

56. Does the magnification of a magnifying glass increase or decrease as its focal length decreases?

57. How does the length of a refracting telescope compare with the focal lengths of the lenses?

*58. What differences would there be if binoculars were made with mirrors instead of prisms?

*59. The figure shows the words MAGNESIUM DIOXIDE viewed through a solid plastic rod. Why does MAGNESIUM appear upside down, while DIOXIDE is right-side up?

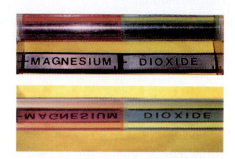

EXERCISES

1. If light in air is incident at 20°, at what angle is it refracted in water? In glass?

2. Light in air is incident on a surface at an angle of 50°. What is its angle of refraction in glass? In water?

3. If light in water is incident at 30°, at what angle is it refracted in air?

4. Light from the bottom of a swimming pool is incident on the surface at an angle of 20°. What is the angle of refraction?

5. Use Figure 17-2 to estimate the angle of refraction for light in air incident at 20° to the surface of glass with an index of refraction of 1.9.

6. Use Figure 17-2 to estimate the critical angle for glass with an index of refraction of 1.9.

7. A prism made of glass with an index of refraction of 1.5 has the shape of an equilateral triangle. A light ray is incident on one face at an angle of 48°. Use a protractor and Figure 17-2 to find the path through the prism and out an adjacent side. What is the exit angle?

8. A prism made of plastic with an index of refraction of 1.33 has the shape of a cube. A light ray is incident on one face at an angle of 70°. Use a protractor and Figure 17-2 to find the path through the prism and out an adjacent side. What is the exit angle?

9. Use a ray diagram to show that an object on the other side of a very thick piece of glass appears to be closer than it actually is.

*10. Use a protractor and Figure 17-2 to show that an object in water appears to be ¾ as deep as it actually is. (Note that the index of refraction for water is ⁴⁄₃.)

11. An object is located midway between the other focal point and the center of a converging lens. Draw a ray diagram showing how you locate the image. Estimate the magnification of the image from your diagram.

12. If an object is three focal lengths from a converging lens, what are the location and magnification of the image?

13. The focal length of a converging lens is 30 cm. Where is the image of an object placed 60 cm from the center of this lens?

14. Where is the image of an arrow placed 60 cm from a diverging lens with a focal length of 30 cm?

*15. Over what range of positions can an object be located so that the image produced by a converging lens is real and magnified?

*16. Show that a diverging lens cannot produce a magnified image of a real object.

17. Draw a ray diagram to find the image of an object located inside the focal point of a diverging lens. Is the image real or virtual? Erect or inverted? Magnified or reduced in size?

18. Where is the image of an object located at the focal point of a diverging lens?

19. How many diopters are there for a converging lens with a focal length of 0.2 m?

20. If a lens has a focal length of 25 cm, how many diopters does it have?

A Model for Light

Our perceptions of this audio cassette are enhanced by viewing it between crossed Polaroid filters to observe the stress patterns in the plastic.

CHAPTER

18

Light is one of our most common phenomena and our knowledge of light is responsible for understanding the beautiful effects in soap bubbles and for technological advances into extending human perception. But even with these advances, light is still elusive and difficult to understand. What is light? (See p. 448 for the answer to this question.)

In the previous two chapters we learned a great deal about light by simply observing how it behaves, but we did not ask, "What is light?" This question is easy to ask, but the answer is elusive. For example, we can't just say, "Let's look!" What we "see" is the stimulation caused by light entering our eyes, not light itself. To understand what light is we need to look for analogies, to ask ourselves what things behave like light. In effect, we are building a model of a phenomenon that we can't observe directly. This same problem occurs quite often in physics since many components of nature are not directly observable. In the case of light we need a model that accounts for the properties that we studied in the previous two chapters. When we look at the world around us, we see that there are two candidates: *particles* and *waves*.

So the question becomes, does light behave as if it were a stream of particles or a series of waves? Newton thought that light was a stream of particles, but other prominent scientists of his time thought that it behaved like waves. Because of Newton's great reputation, his particle model of light was the accepted theory during the 18th century. However, many of the early observations could be explained by a particle *or* a wave model. Scientists continually looked for new observations that could distinguish between the two theories. We will examine the experimental evidence to see whether light behaves like particles or waves. The process of deciding is as important as the answer.

Reflection

We know from such common experiences as echoes and billiards that both waves and particles can bounce off barriers. But this fact is not enough. Our model for the behavior of light must agree with our conclusion that the angle of reflection is equal to the angle of incidence. A particle model for light can account for this if the reflecting surface is frictionless and perfectly elastic. A wave model of light also has no problem accounting for the law of reflection. The photograph in Figure 18-1(a) shows a straight wave pulse on the surface of water striking a

Figure 18-1 (a) The horizontal white line is a straight water wave pulse moving toward the top of the picture. The part that has already hit the black barrier is reflected toward the right. (b) The corresponding ray diagram has rays perpendicular to the wave fronts and shows that the angle of reflection is equal to the angle of incidence.

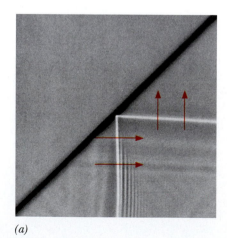

(a)

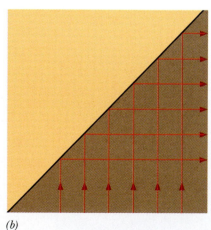

(b)

smooth, straight barrier. The wave was initially moving toward the top of the picture and is being reflected toward the right. We see that the angle between the incident wave and the barrier is equal to that between the reflected wave and the barrier. In the corresponding ray diagram [Fig. 18-1(b)], we draw the rays perpendicular to the straight wave fronts; that is, in the direction the wave is moving.

Observing that light reflects from surfaces gives us no clues as to its true nature; both particles and waves reflect according to the law of reflection.

Physics on Your Own   Observe the reflection of particles and waves in bathtubs and pool halls. Can you devise an experiment that will verify the law of reflection for both?

Refraction

We learned a great deal in the previous chapter about the behavior of light when it passes through boundaries between different transparent materials. Newton thought that the particles of light experienced a force as they passed from air into a transparent material. This inferred force would occur only at the surface, act perpendicular to the surface, and be directed into the material. This force would cause the particles to bend toward the normal as shown in Figure 18-2. In this scheme the light particles would also experience this force upon leaving the material. Because the force acts into the material, it now has the effect of bending the particles away from the normal. Furthermore, when Newton calculated the dependence of the amount of refraction on the angle of incidence, his answer agreed with the curves in Figure 17-2. So the model gives not only the correct qualitative results but also the correct quantitative results.

Water waves also refract. The photograph in Figure 18-3 shows the refraction of water waves. The boundary that runs diagonally across the

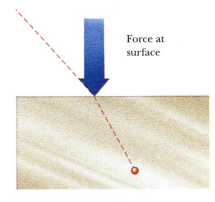

Figure 18-2   A force perpendicular to the surface would cause light particles to be refracted.

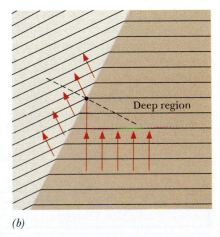

Deep region

(a)   *(b)*

Figure 18-3   (a) Water waves refract when passing from deep to shallow water. (b) The corresponding ray diagram shows that the waves refract toward the normal.

photograph separates the shallow region on the left from the deeper region on the right. Once again the numerical relationship between the angle of refraction and the angle of incidence is in agreement with the law of refraction.

There is an important difference between the wave and particle predictions; the speeds of the particles and waves do not change in the same way. Figures 18-2 and 18-3 both correspond to light entering a substance with a higher index of refraction. In the particle model, particles bend toward the normal because they speed up. Thus, the particle model predicts that the speed of light should be faster for substances with higher indices of refraction.

In waves the opposite is true. Because the crests in Figure 18-3 are continuous across the boundary, we know the frequency of the waves doesn't change. However, the wavelength does change; it is shorter in the shallow region to the left of the boundary. Because $v = \lambda f$ (Chapter 14), a decrease in the wavelength means a decrease in the speed of the wave. This means that the speed of the waves in the shallow region is smaller, and therefore the wave model predicts that the speed of light should be slower for substances with higher indices of refraction.

light travels slower in materials

Because the two models predict opposite results, we have a way of testing them; the speed of light can be measured in various materials to see which model agrees with the results. The speed of light in a material substance was not measured until 1862, almost two centuries after the development of the two theories. The French physicist Jean Foucault measured the speed of light in air and water and found the speed in water to be less. This dealt a severe blow to the particle model of light and consequently caused a modification of the physics world view.

$n = c/v$

The American physicist Albert Michelson improved on Foucault's measurements and found a ratio of 1.33 for the speed of light c in vacuum to light's speed v in water. This value is equal to the **index of refraction** n of water as predicted by the wave model, and thus, $n = c/v$. This gives us a way of predicting the speed of light in any material once we know its index of refraction. The speed of light in a substance is equal to its speed in a vacuum divided by the index of refraction, $v = c/n$. Because the speed of light in a vacuum is the maximum speed, the indices of refraction of substances must be greater than 1.

In the previous chapter we discovered that different colors have slightly different indices of refraction within a material, which results in dispersion. Because the index of refraction is related to the speed, we can now conclude that different colors must have different speeds in a material.

QUESTION   Does red or blue light have the slower speed in glass?

ANSWER   Because blue light is refracted more than red, it has the higher index of refraction. Therefore, blue light has a slower speed.

Interference

Although the speed of light in materials provided definitive support for the fact that light behaved like a wave, there was other supporting evidence early on in the debate. Newton's main adversary in this debate was largely unsuccessful at convincing others of the importance of the evidence. We now examine some of the other properties of waves that differ from those of particles and show that light exhibits these properties.

We begin with interference. The interference of two sources of water waves is shown in the photograph and drawing in Figure 14-20. Attempts to duplicate these results with two lightbulbs, however, fail. But if light is a wave phenomenon, it should exhibit such interference effects. Light from the two sources does superimpose to form an interference pattern, but the pattern isn't stationary. Because the phase between the lightbulbs varies rapidly, the pattern blurs out and the region looks uniformly illuminated. Stationary patterns are only produced when the two sources have the same wavelength and a constant phase difference.

The interference of two light sources was first successfully demonstrated by Thomas Young in 1801 (over a half century before Foucault's work on the speed of light in materials), when he let light from *one* pinhole impinge on two other pinholes. Passing the light through the first pinhole produced light at the second pinholes that was reasonably in phase. Then this light passed through pinholes or thin slits to produce a constant interference pattern on the screen [Fig. 18-4(a)]. The interference pattern shown in Figure 18-4(b) consists of colored bars where light

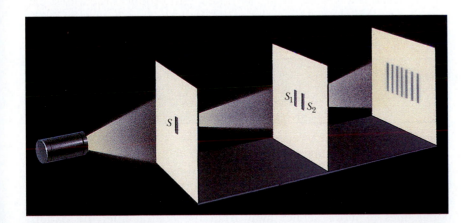

(a)

(b)

Figure 18-4   (a) A schematic of Young's experiment that demonstrated the interference of light. (b) The interference pattern produced by white light incident on two slits.

Figure 18-5 The two-slit interference pattern produced by red light (a) is a wider than that produced by blue light (b).

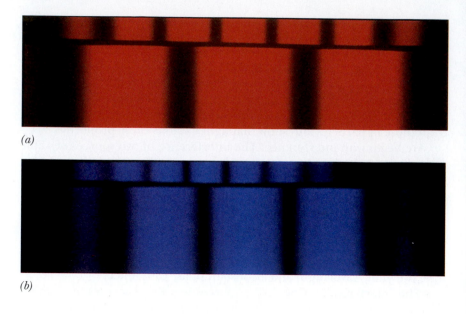

(a)

(b)

of each color interferes constructively. These antinodal regions have large amplitudes and appear bright. Modern versions of this experiment are in complete agreement with the wave model.

If the slits are illuminated with a single color, the pattern looks much more like the one you would expect to find along the far edge of the ripple tank in Figure 14-20. As the color of the light is changed, the pattern on the screen changes size (Fig. 18-5). Red light produces the widest pattern, and violet light produces the narrowest one. The sizes produced by other colors vary in the same order as the colors of the rainbow.

> **QUESTION** What happens to the width of one of these interference patterns if the distance between the two slits is increased?

Recall that in looking at the two-source interference patterns in Chapter 14 the nodal lines on each side of the central line were created by a difference in the path lengths from the two sources that was equal to one-half wavelength (Fig. 14-21). Increasing the wavelength causes these nodal lines to move farther away from the central line; the pattern widens. Therefore, the shifting of the dark regions on our screen with color signals a change in wavelength. We conclude that the color of light depends on its wavelength, with red being the longest and violet the shortest.

> **ANSWER** We learned in Chapter 14 that the spacing of the nodal lines depends on the ratio of the wavelength to the slit separation. Therefore, the wider spacing will produce a narrower pattern.

Measurements of the interference pattern and the separation of the slits can be used to calculate the wavelength of the light. Experiments show that visible light ranges in wavelength from 400 to 750 nanometers (nm), where a nanometer is 10^{-9} meter. It takes more than 1 million wavelengths of visible light to equal 1 meter. Knowing the speed of light to be 3.0×10^8 meters per second, we can calculate the corresponding frequency range to be roughly $(4.0–7.5) \times 10^{14}$ hertz, using the relationship that $c = \lambda f$.

Physics on Your Own   Perform Young's two-slit experiment by tapping the slits to one end of the cardboard tube from a roll of paper towels. To make the two slits, blacken a microscope slide (or other piece of glass) with a candle flame. Scratch two viewing slits on the blackened side by holding two razor blades side by side and drawing them across the slide. Cut a slit about 2 millimeters wide in a piece of paper and tape it to the other end of the tube. Aim this end of the tube at a light source and look through the end with the two slits to see the interference pattern (Fig. 18-6).

You can vary the wavelength of the light by using pieces of red and blue cellophane as filters. The spacing can be varied by putting a third razor blade between the first two but raised so that it doesn't make a slit.

Physics Update

The first direct image of the surface of a star other than our Sun was reported by Andrea Dupree of Harvard–Smithsonian. The surface of the star, Betelgeuse, had been indirectly imaged earlier using speckle interferometry, in which many brief exposures are added up to make a composite image. Dupree's pictures, made with the Hubble Space Telescope, confirm previous suspicions that Betelgeuse's surface exhibits a giant bright spot. According to Dupree, the spot is 2000 K warmer than its surroundings and this might be indicative of a new physical phenomenon at work in some stellar atmospheres.

Figure 18-6   A device for viewing two-slit interference.

Diffraction Limits

The wave nature of light and the size of the viewing instrument limit how small an object we can see, even with the best instruments. Because of the wave nature of light, there is always some diffraction. The image is spread out over a region in space; the region is larger for longer wavelengths and smaller openings.

Consider two objects with small *angular* sizes (they could be very big, but so far away that they look small) separated by a small angular distance. Each of these will produce a diffraction pattern when its light passes through a small opening such as in our eye or a telescope. The diagram and photograph in (a) correspond to the case where the diffraction patterns can still be clearly distinguished. In (c) the overlap is so extensive, you cannot resolve the individual images. The limiting case is shown in (b); the separation of the centers of the two patterns is less than the width of the central maximum of either pattern. At this limit, this means that the first minimum of each pattern lies on top of the maximum of the other pattern.

This minimum angular separation can be calculated mathematically. The light coming from two point sources can be resolved if the angle between the objects is greater than 1.22 times the ratio of the wavelength of the light to the diameter of the aperture. Imagine viewing an object from a distance of 25 centimeters, the nominal distance of closest vision. If we use an average pupil size of 5 millimeters and visible light with a wavelength of 500 nanometers, we calculate that you can distinguish a separation of 0.03 millimeter, about the radius of a human hair. A similar calculation tells us that we should be able to distinguish the headlights of an oncoming car at a distance of 10 kilometers.

From this relationship you can see why astronomers want bigger telescopes. With a larger mirror, the resolving angle is smaller and they can distinguish more detail in distant star clusters. For instance, the resolving angle of a 5-meter telescope in visible light is about 0.02 arcsecond, where 1 arcsecond is 1/3600 of a degree. In practice, this resolution is never obtained because turbulence in the Earth's atmosphere limits the resolution to about 1 arcsecond. This is one of the major reasons for placing the Hubble Space Telescope in orbit.

Diffraction effects also place lower limits on the sizes of objects that can be examined under an optical microscope since the details to be observed must be separated by more than the diffraction limits set by the microscope.

Diffraction

Young's experiment points out another aspect of waves. His interference pattern was only possible because light from one pinhole overlapped that from the other. Light spreads out as it passes through the pinholes. In other words, light exhibits diffraction just like the water waves of Figure 14-23.

The photographs in Figure 18-7 were taken of the diffraction pattern produced by light passing through a narrow slit. The slit was wider in Figure 18-7(b)! But this difference in the patterns makes sense if you examine the photographs of water waves in Figure 14-23. Contrary to what common sense might tell us, the narrower slit produces the wider pattern. These patterns are in agreement with the wave model and contrary to what we would expect if light traveled only in straight lines. In the case of particles, the wider slit would produce the wider pattern, not the narrower one.

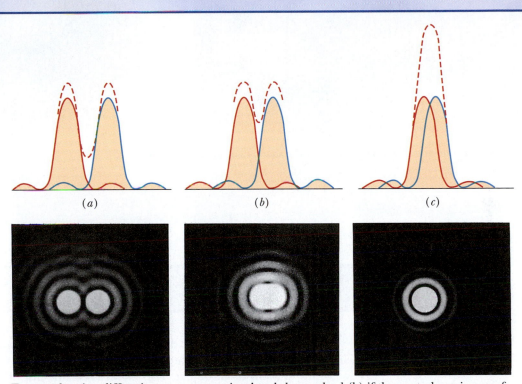

Two overlapping diffraction patterns can just barely be resolved (b) if the central maximum of each pattern lies on the first minimum of the other. If the patterns are closer, they appear to be a single object (c).

Figure 18-7   Diffraction patterns produced by red light incident on (a) a narrower and (b) a wider slit. The narrower slit produced the wider pattern.

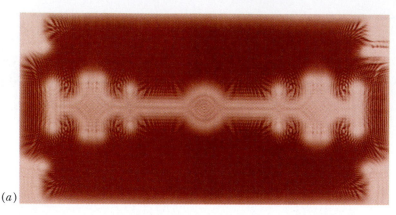

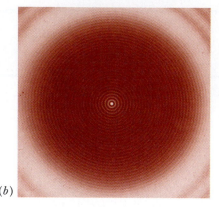

(a)

(b)

Figure 18-8   Photographs of diffraction patterns in the shadows of (a) a razor blade and (b) a penny.

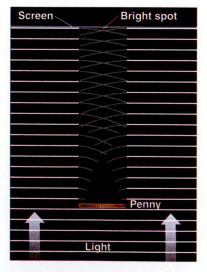

Figure 18-9   Diagram illustrating how wave interference and diffraction produce a bright spot at the center of the penny's shadow.

QUESTION   Does red light or blue light produce the wider diffraction pattern?

Water waves also show that diffraction takes place around the edges of barriers. To observe this effect with light, we should look along the edge of the shadow of an object. This effect is usually not observed with light because most shadows are produced by broad sources of light. The resulting shadows have smooth changes from umbra through penumbra to full brightness. Therefore, we should look at shadows produced by point sources to eliminate the penumbra.

The photographs in Figure 18-8 were created by putting photographic paper in the shadows of a razor blade and a penny. We used red light from a laser because the light waves are in phase and can easily be made to approximate a point source. You can see that the edges of the shadows show the effects of the interference of diffracted light. These effects can only be explained by a wave model of light.

Notice the center of the shadow of the penny. Even though the penny is solid, the shadow has a bright spot in the center. Can you explain this bright spot using the wave model? The diagram in Figure 18-9 shows how the light diffracts around the edge of the penny. The light coming from each point on the edge travels the same distance to the center of the shadow. These waves arrive in phase and superimpose to form the bright spot. This is added evidence for the wave model of light.

ANSWER   The width of the diffraction pattern depends on the ratio of the wavelength to the width of the slit. Because red light has the longer wavelength, it would produce the wider pattern.

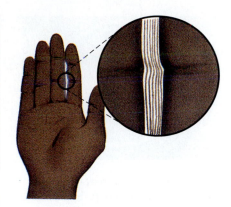

Physics on Your Own   Look at a lightbulb through the slit between two fingers held in front of your eyes as in Figure 18-10. How does the image of the lightbulb change as you slowly change the spacing between your fingers? Notice the dark, vertical lines in the image. What happens to those lines as your fingers move closer together?

To observe two-dimensional diffraction patterns, look at a distant light through a fine mesh screen or a piece of cloth held at arm's length.

Thin Films Σ

Figure 18-10   Light viewed through the thin slit between two fingers exhibits diffraction effects.

Soap bubble, oil slicks, and, in fact, thin films of any transparent material exhibit beautiful arrays of color under certain conditions—an effect not of dispersion but of interference.

Consider a narrow beam of red light incident on a thin film as shown in Figure 18-11. At the first surface, part of the light is reflected and part is transmitted. The same thing happens to the transmitted part when it attempts to exit the film; part is transmitted and part is reflected, and so on. Considering just the first two rays that leave the film on the incident side, we can see that they interfere with each other. Because of the thickness of the film, ray 2 lags behind ray 1 when they overlap on the incident side of the film. Ray 2 had to travel the extra distance within the film. At a certain thickness, this path difference is such that the crests of one ray line up with the troughs from the other and the two rays cancel. In this case little or no light is reflected. At other thicknesses, the crests of one ray line up with the crests from the other and the two rays reinforce. In this case light is reflected from the surface.

Because different colors have different wavelengths, a thickness that cancels one color may not be the thickness that cancels another. If the film is held vertically (Fig. 18-12), the pull of gravity causes the film to vary in thickness, being very thin at the top and increasing toward the bottom. In fact, when the film breaks, it begins at the top where it has been stretched too thin. If the film varies in thickness, and the incident light is white, different colors will be reflected for different thicknesses, producing the many colors observed in soap films.

Notice that the very thin region at the top of the soap film in Figure 18-12 has no light reflecting from the film. At first this seems confusing. Because there is essentially no path difference between the front and back surfaces, we might expect light to be reflected. However, the first ray is inverted when it reflects; that is, crests are turned into troughs, and vice versa. This process is analogous to the inversion that takes place when a wave pulse is reflected from the fixed end of a rope (Chapter 14). Light waves are inverted when they reflect from a material with a higher index of refraction. At the back surface the rays reflect from air (a lower index of refraction than the soap) and no inversion takes place.

Because the ray reflected from the front surface is inverted while the ray reflected from the back surface is not inverted, a crest from the front

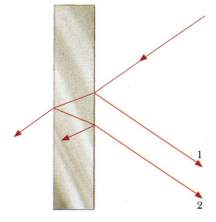

Figure 18-11   Light incident on a thin film is reflected and transmitted at each surface. Rays 1 and 2 interfere to produce more or less light depending on the thickness of the film and the wavelength of the light in the film.

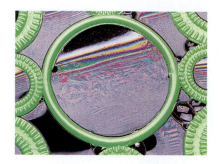

Figure 18-12   Different colors are reflected from different thicknesses of the thin film. The very thin region at the top of the soap film appears black because the reflected light is inverted at the front surface but not at the back surface.

surface will overlap a trough from the back surface if the thickness of the film is much smaller than a wavelength. The two reflected rays will cancel and no light will be reflected. If the light is normal to the surface and the thickness of the film is increased until it is one-quarter wavelength thick, the light will now be strongly reflected. The ray that reflects from the back surface must travel an extra one-half wavelength and these crests and troughs will be delayed so that they now line up with those reflected from the front surface. It is important to note that we are referring to the wavelength of the light *in the film*. Since the frequency of the light is the same in the air and in the film while the speed is reduced, the wavelength in the film is equal to that in the air divided by the index of refraction, $\lambda_f = \lambda / n$.

> **QUESTION**   Why does part of the thin film in Figure 18-12 appear white when white is not one of the rainbow colors?

This colorful phenomenon also occurs after a rain has wetted the highways. Oil dropped by cars and trucks floats on tope of the puddles. Sunlight reflecting from the top surface of the oil and from the oil–water interface interferes to produce the colors. Again, variations in the thickness of the oil slick produce the array of colors.

The colors on the puddle are caused by the interference of light reflected from the top and bottom of a thin oil film floating on the surface.

> **Physics on Your Own**   Observe the colored patterns produced by soap bubbles or when a thin film of oil is placed on the surface of water.

Interference also occurs for the transmitted light. Light passing directly through the thin film (Fig. 18-13) can interfere constructively or destructively with light that is reflected twice within the film. The effects are complementary to those of the reflected light. When the thickness of the film is chosen to minimize the reflection of a certain color of light, the transmission of that color is maximized. This process is a consequence of the conservation of energy. The light must go someplace.

Some modern office buildings have windows with thin films to reduce the amount of light entering the offices. The visors in the helmets of space suits are coated with a thin film to protect the eyes of the astronauts. Thin films are also very important to lens makers because the proper choice of material and thickness allows them to coat lenses so that they do not reflect certain colors or conversely so that they do not transmit certain colors. It is common to coat the lenses in eyeglasses with a thin film to stop the transmission of ultraviolet light. Lenses in high-quality telescopes and binoculars are coated to reduce the reflection (and therefore enhance the transmission) of visible light, increasing the brightness of images.

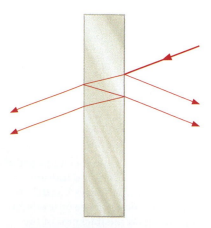

Figure 18-13   The light transmitted through a thin film also displays interference effects.

> **ANSWER**   We observe white when most of the colors overlap.

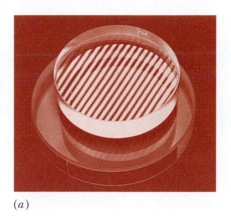

(a)

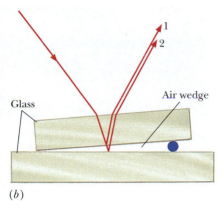

(b)

Figure 18-14   (a) Interference lines are produced by light reflecting from the top and bottom surfaces of an air wedge between two pieces of glass. (b) A drawing of the side view of the air wedge showing the interference of one ray.

An interesting example of thin-film interference was observed by Newton (even though he did not realize that it supported the wave model of light). When a curved piece of glass such as a watch glass is placed in contact with a flat piece of glass, a thin film of air is formed between the two. An interference pattern is produced when the light reflects from the top and bottom of the air gap. The interference pattern that is formed by a wedge-shaped air gap is shown in Figure 18-14. Such patterns are commonly used to test the surface quality of lenses and mirrors.

> **QUESTION**   Would you expect to find a dark spot or a bright spot when you look at the reflected light from an air gap of almost zero thickness?

The thin film on the visor protects the astronaut's eyes.

Polarization

We have established that light is a wave phenomenon, but we have not discussed whether it is transverse or longitudinal; that is, whether the vibrations take place perpendicular to the direction of travel or along it. We will determine this by examining a property of transverse waves that does not exist for longitudinal waves and then see if light exhibits this behavior.

Transverse waves traveling along a horizontal rope can be generated so that the rope vibrates in the vertical direction, the horizontal direction, or, in fact, in any direction in between. If the vibrations are only in one direction, the wave is said to be *plane polarized,* or often just **polarized.** This property becomes important when a wave enters a medium in which various directions of polarization are not treated the same. For instance, if our rope passes through a board with a vertical slit cut in it (Fig. 18-15), the wave passes through the slit if the vibration is vertical, but not if it is horizontal.

> **ANSWER**   The light waves will be inverted when they reflect off the second interface, but not the first one. Therefore, you would see a dark spot in the reflected light and a bright spot in the transmitted light.

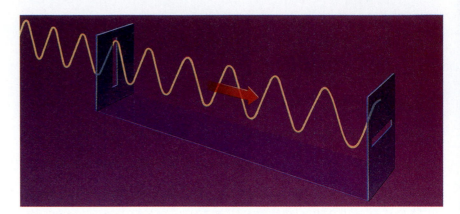

Figure 18-15   Waves with vertical polarization pass through a vertical slit, but not through a horizontal one.

What happens if the slit is vertical, but the polarization of the wave is someplace between vertical and horizontal? Imagine that you are looking along the length of the rope and we represent the polarization of the wave by a two-headed arrow along the direction of vibration. This polarization can be imagined as a superposition of a vertical vibration and a horizontal vibration, as illustrated in Figure 18-16. The vertical slit allows the vertical component to pass through while blocking the horizontal component. Therefore, the wave has a vertical polarization after it passes

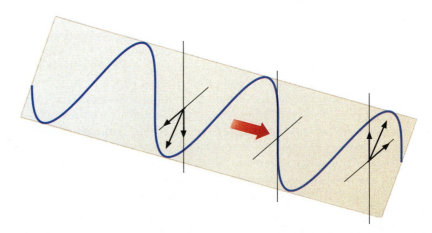

Figure 18-16   The plane-polarized wave can be broken up into two perpendicular component waves; one in the vertical plane and one in the horizontal plane. The displacements of these component waves are shown at two sample locations.

through the vertical slit. The amplitude of the transmitted wave is equal to the amplitude of the vertical component of the incident wave.

Determining whether light can be polarized is a little difficult because our eyes are unable to tell if light is polarized or not. Nature, however, provides us with materials that polarize light. The mere existence of polarized light demonstrates that light is a transverse wave. If it were longitudinal, it could not be polarized. Commercially available Polaroid filters consist of long, complex molecules that have their long axes parallel. These molecules pass light waves with polarizations perpendicular to their long axes, but absorb those parallel to it.

light is a transverse wave

We can use a piece of Polaroid (or the lens from Polaroid sunglasses) to analyze various light sources to see if they are polarized. If the light is polarized, the intensity of the transmitted light will vary as the Polaroid is rotated. This simple procedure shows that common light sources such as ordinary incandescent lamps, fluorescent lights, candles, and campfires are unpolarized. However, if we examine the light reflected from the surface of a lake, we find that it is partially polarized in the horizontal direction. Light reflected from nonmetallic surfaces is often partially polarized in the direction parallel to the surface. This is the reason why Polaroid sunglasses have their axis of polarization in the vertical direction. Boaters know that Polaroid sunglasses remove the glare from the surface of water and allow them to see below the surface.

(a) *(b)*

Photographs taken through a plate-glass window (a) with and (b) without a polarizing filter on the camera lens. Notice that the polarizing filter eliminates the glare.

Holography

Wouldn't it be fantastic to have a method for catching waves, preserving the information they carry, and then at some later time playing them back? In fact, high-fidelity equipment routinely records a concert and plays it back so well that the listeners can hardly tell the difference. Similarly, high-fidelity photography should record the light waves coming from a scene and then play them back so well that the image is nearly indistinguishable from the original scene.

Conventional photography, however, does not do this. While modern chemicals and papers have created images with extremely fine resolution, nobody is likely to mistake a photograph for the real object—because a photographic image is two dimensional, while the actual scene is three dimensional. This loss of depth means that you can view the scene in the photograph from only one angle, or perspective. You cannot look around objects in the foreground to see objects in the background. The objects in the scene do not move relative to one another as you move your point of view. Viewing a truly high-fidelity photograph should be like viewing a scene through a window.

Holography is a photographic method that produces a three-dimensional image that has virtually all of the optical properties of the scene. This process was conceived by Dennis Gabor in 1947. He received a Nobel Prize in 1971 for this work. Gabor chose the word *holo-gram* to describe the three dimensionality of the image by combining the Greek roots *holo* (complete) and *gram* (message).

Although Gabor was able to make a hologram out of a flat transparency, the lack of a proper light source prevented him from making one of an object with depth. With the invention of the laser in 1960, interest in holography was renewed. We will postpone the discussion of laser operation until Chapter 23. At the moment we need only know that it is a device that produces light with a single color and a constant phase relationship.

Figure A shows the essential features of a set-up for making holograms. Notice that there is no lens between the object and the film. Light from the object reflects onto all portions of the film. Although the information about the object is carried by this *object* beam, if this were all that happened, the film would be completely exposed and virtually no information about the object would be recorded.

Another portion of the beam from the laser (*reference* beam) illuminates the film directly. Once again, if this were all that happened, the film would not record the scene. However, the light from the laser has a single well-defined wavelength (color) with the crests lined up. Therefore, the light reflected from the object and the light in the reference beam produce an interference

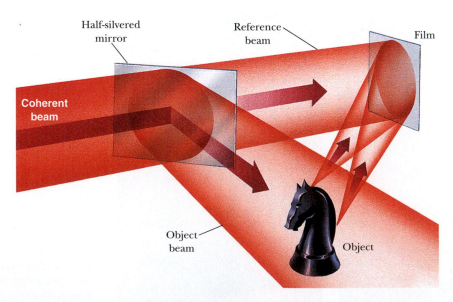

Figure A   Making a hologram.

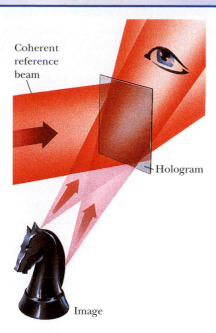

Figure B A hologram is viewed by placing it in the reference beam. The light passing through the film produces a virtual image at the location of the original scene.

pattern that is recorded in the film. This interference pattern contains the three-dimensional information about the scene.

The hologram is viewed by placing it back in a reference beam. The pattern in the film causes the light passing through it to be deflected so that it appears to come from the original scene (Figure B).

Because information from each point in the scene is recorded in all points of the hologram, a hologram contains information from all perspectives covered by the film. Therefore, viewing a hologram is just like viewing the original scene through a window. The photographs in Figure C were taken of a single hologram. Note that the relative locations of the chessmen change as the camera position changes. This is what we would see if we were looking at the actual chessboard.

Because information from each point in the scene is recorded in all points of the hologram, a hologram can be broken and each piece will produce an image of the entire scene. Of course, something must be lost. Each piece will only allow the scene to be viewed from the perspective of that piece. This is analogous to looking at a scene through a window which has been covered except for a small hole.

Recent advances in holography have made it possible to display holograms with ordinary sources of white light instead of the much more expensive lasers. It is now possible to hang holograms in your home as you would paintings. In fact, they have become so inexpensive that they have appeared on magazine covers. Holograms can also be made in the shape of cylinders so that you can walk around the hologram and see all sides of the scene. In some of these, individuals in the scene move as you walk around the hologram. Holographic movies are possible, and holographic television should be possible in the future.

In addition to being a fantastic art medium, holography has found many practical applications. It can be used to measure the three-dimensional wear patterns on the cylinder walls of an engine to an accuracy of a few ten-thousandths of a millimeter (a hundred-thousandth of an inch), perform nondestructive tests of the integrity of machined parts or automobile tires, study the shape and size of snowflakes while they are still in the air, store vast quantities of information for later retrieval, and produce three-dimensional topographical maps.

Figure C The relative positions of the chessmen in the holographic image change as the point of view changes.

445

Physics Update

A digital holographic storage system, one actually integrated with a computer hard drive, has been developed by scientists at Stanford. In such a system data is converted into light patterns. The light waves enter a refractive medium, where they bring about microscopic rearrangements of electric charge, which in turn affect the local index of refraction. To read out the data, a reference laser beam is sent into the medium; the refracted beam, bearing the decoded data, is detected with a charge-coupled device. Data can be stacked up in the hologram by recording at several angles. By home-computer standards, the Stanford results so far are modest: total storage capacity of 163 kilobytes and a data transfer rate of 6.3 million bits per second. The researchers believe future hologram performance should be much better: terabytes (tera $= 10^{12}$) of storage and transfers above 1 gigabits per second (giga $= 10^{9}$).

Physics on Your Own   Examine a variety of light sources through the lens of Polaroid sunglasses to see if the sources are polarized. Looking away from the direction of the Sun, check to see if skylight is polarized. If you have two sets of glasses, you can verify that the lenses are polarizing by rotating one with respect to the other to see if they block out light.

Until now we have discussed materials in which one direction passes the light and the other absorbs it. Some materials allow both components of polarization to pass through but the orientations have different speeds. This can have the effect of rotating the plane of polarization. The cello-

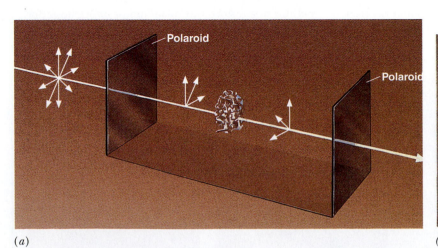

(a)

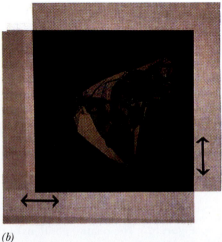

(b)

Figure 18-17   (a) Cellophane or transparent tape rotates the plane of polarization. (b) A pattern formed by different thicknesses of cellophane viewed between two pieces of Polaroid.

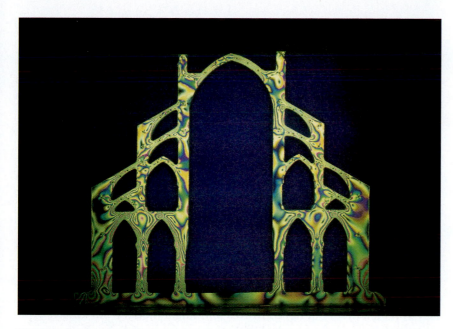

Viewing a plastic model of a structure through crossed pieces of Polaroid shows the stress patterns in the structure. This photograph shows a model of the flying buttresses of the Notre Dame cathedral.

phane on cigarette packages and some types of transparent adhesive tape have this property.

You can make an interesting display by crumbling a cigarette wrapper and looking at it between two Polaroids, as in Figure 18-17(a). The Polaroid on the side near the light polarizes the light striking the display. When this light is viewed with the second Polaroid, the different thicknesses of cellophane have different colors, as shown in Figure 18-17(b). The amount of rotation depends on the thickness of the cellophane and the wavelength of the light. Certain thicknesses rotate certain colors by just the right amount to pass through the second Polaroid. Others are partially or completely blocked. The colors of each section change as either Polaroid is rotated.

Glasses and plastics under stress rotate the plane of polarization; the greater the stress the more the rotation. Plastic models of structures, such as cathedrals, bones, or machined parts, can by analyzed to discover where the stress is greatest.

Physics on Your Own   Place a piece of Polaroid (or the lens from Polaroid sunglasses) over the lens of an empty slide projector. Place a jar of Karo syrup in the light beam and view the transmitted light through another piece of Polaroid. Different path lengths through the syrup produce different amounts of rotation of the plane of polarization.

Looking Ahead

We have not finished our search to understand light. Notice, for example, that although we established that the wave model of light explains all the phenomena we have observed so far, we still have not said what it is that is waving! We know that light travels from the distant stars to our eyes through incredible distances in a vacuum that is much better than any we can produce on Earth. What is in the vacuum that could be the medium for waves?

We will see in Chapter 22 that understanding the nature of light was the key to our modern understanding of atoms. This reexamination of the model for light occurred at the beginning of this century and began with Einstein asking many of the same questions that we addressed here.

C H A P T E R

18

R E V I S I T E D

Light is probably the most fascinating and elusive phenomenon in nature. The answer to the question, "What is light?" actually changes with the techniques used to examine the question. In this chapter we found very good evidence to support the conclusion that light is a wave phenomenon. In future chapters we'll return to this question and delve further into the mystery of light.

S U M M A R Y

Developing a theory of light involves building a model from known experiences that can be compared with the behavior of light. Both particle and wave models are able to account for the law of reflection and the law of refraction. However, only a wave model can correctly account for the speed of light in transparent materials. The index of refraction n is the ratio of the speed of light c in a vacuum (300 million meters per second) to its speed v in the material; that is, $n = c/v$. Dispersion of light indicates that different colors have different speeds in a material.

Other properties of waves that differ from those of particles further support the fact that light is a wave phenomenon. Interference of two coherent light sources produces a stationary pattern. If different colors are used, the pattern changes size, demonstrating that colors have different wavelengths. Red light has the largest wavelength and produces the widest pattern, and violet light produces the narrowest one. Visible light ranges in wavelength from 400 to 750 nanometers.

Light exhibits diffraction, which is the spreading out of a wave as it passes through narrow openings and around the edges of objects. The width of the diffraction pattern depends on the ratio of the wavelength to the size of the opening. The narrower the opening, the wider the pattern. Because red light has the longest visible wavelength, it would produce the widest pattern for a given opening.

Thin films of transparent materials exhibit beautiful arrays of colors due to the interference of light rays reflecting from the two surfaces. A ray that reflects from a material with a higher index of refraction is inverted. Different wavelengths are strongly reflected or transmitted at different thicknesses of film.

Light exhibits polarization, demonstrating that it is a transverse wave. Polarizing materials pass light waves with polarizations perpendicular to one axis, but absorb those parallel to it. Common light sources are usually unpolarized. However, light reflected from the surface of a lake or glass is partially polarized parallel to the surface.

KEY TERMS

diffraction: The spreading of waves passing through an opening or around a barrier.

dispersion: The spreading of light into a spectrum of colors. The variation in the speed of a periodic wave due to its wavelength or frequency.

index of refraction: An optical property of a substance that determines how much light bends upon entering or leaving it. The index is equal to the ratio of the speed of light in a vacuum to its speed in the substance.

interference: The superposition of two or more waves.

normal: A line perpendicular to a surface or curve.

polarized: A property of a transverse wave when its vibrations are all in a single plane.

reflection, law of: The angle of reflection (measured relative to the normal to the surface) is equal to the angle of incidence. The incident ray, the reflected ray, and the normal all lie in the same plane.

refraction: The bending of light that occurs at the interface between two transparent media. It occurs when the speed of light changes.

CONCEPTUAL QUESTIONS

1. Does Newton's idea that a light beam consists of tiny particles correctly predict that light beams travel in straight lines? Explain.

2. Explain how Newton's particles of light could explain that the angle of incidence equals the angle of reflection.

3. Argue that the law of reflection would not hold for particles rebounding from a surface that is not frictionless or not perfectly elastic.

4. How does the particle theory of light account for the diffuse reflection of light?

5. How does Newton's idea of light particles explain the law of refraction?

6. Explain how Newton's idea of light particles predicts that the speed of light in a transparent material will be faster than in a vacuum.

7. How would the photograph in Figure 18-3 change if it represented particles or waves entering a material with a lower index of refraction?

8. Do you expect the speed of light in glass to be slower than, faster than, or the same as that in water?

9. What property of a wave remains the same when the wave crosses a boundary (for example, water waves going from deep to shallow water)?

*10. Does the amplitude of a light wave increase, decrease, or stay the same upon reflection from a transparent material?

11. What property of light waves determines its brightness?

12. Which of the following phenomena does not show a difference between the wave theory and particle theory of light: reflection, refraction, interference, diffraction, or polarization?

13. Explain how measurements of the speed of light in transparent materials can distinguish between the wave and particle models of light.

14. Explain how the wave model for light predicts that the speed of light in a transparent material will be slower than the speed in a vacuum.

15. What is the physical difference between red and blue light?

16. How does the low speed of light in diamonds affect their brilliance?

17. Why do we not notice any dispersion when white light passes through a window pane?

18. What does the dispersion of light tell us about the speeds of various colors of light in a material?

19. Is the index of refraction of a transparent material a constant that does not depend on the properties of the light?

20. Will the converging lens in the figure focus blue light or red light at a closer distance to the lens?

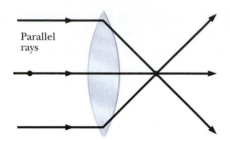

Parallel rays

21. Red light is used to form a two-slit interference pattern on a screen. As the two slits are moved farther apart, does the separation of the bright bands on the screen decrease, increase, or remain the same?

22. What happens to the separation of the bright bands in a two-slit interference pattern as the slits are brought closer together?

23. Would orange light or blue light produce the wider interference pattern? Why?

24. We observe that the two-slit interference pattern produced by blue light is narrower than that produced by red light. What does this tell us about red and blue light?

25. Why don't we notice interference patterns when we turn on two lights in a room?

26. What determines whether two light beams tend to cancel or reinforce each other?

27. If light and sound are both sound phenomena, why can we hear sounds around a corner but not be able to see around corners?

28. What happens to the separation of the bright bands in a two-slit interference pattern as the slits are made narrower but their separation remains the same?

29. Approximately how narrow should a slit be for the diffraction of visible light to be observable?

30. Is the wavelength of visible light most nearly the same size as an apple, a pencil lead, a human hair, or an atom?

31. Blue light is used to form a single-slit diffraction pattern on a screen. As the slit is made wider, does the separation of the bright bands on the screen decrease, increase, or remain the same?

32. Would a slit with a width of 200 nanometers or 400 nanometers produce a wider diffraction pattern?

33. Would orange light or blue light produce the wider diffraction pattern? Why?

34. What does the observation that the single-slit diffraction pattern produced by red light is wider than that produced by blue light tell us about red and blue light?

35. What causes the colors seen in soap bubbles?

36. If a thin film varies uniformly in thickness, why don't we observe the rainbow colors in order?

37. Assume that you have the thinnest film that strongly reflects red light. Would you need to make the film thinner or thicker to completely reflect blue light?

38. Assume that you have the thinnest film that strongly reflects blue light. If you can make the film thinner, is it possible to transmit the blue light strongly?

39. A thin film strongly reflects orange light. Will it still reflect orange light when it is put into water?

*40. A thin, transparent film strongly reflects yellow light. What does the film do when it is applied to a glass lens that has a higher index of refraction than the film?

41. What evidence do you have that light is a transverse rather than a longitudinal wave?

42. If all of the labels have come off the sunglasses in the drug store, how could you tell which ones were polarized?

43. How could you use Polaroid sunglasses to tell if light from the sky is polarized?

44. How could you tell if the light from a rainbow is polarized?

*45. Two pieces of Polaroid with their axes at right angles to each other completely block out the light from an unpolarized source. What would happen to the amount of transmitted light if you placed a third piece of Polaroid between the crossed Polaroids so that its axis was at 45° to each of them?

*46. Suppose we make automobile windshields and headlights out of Polaroid material to reduce the problem of glare from oncoming traffic. How would you orient the axes of polarization so that you could see with your own headlights and still block out the lights of an approaching automobile?

47. Is it better to use red light or blue light to minimize diffraction effects while photographing tiny objects through a microscope?

48. Why can't an ordinary microscope using visible light be used to observe individual molecules?

49. Why are the diffraction effects of your eyes more important during the day than at night?

50. What happens to the diffraction limit of a telescope if you double the diameter of its mirror?

51. How would you distinguish a hologram from a flat transparency?

*52. How would a holographic puzzle differ from an ordinary one?

53. What kind of light is required to make a hologram of a three-dimensional object?

54. What kind of light is required to display a hologram?

55. Why is it not possible to make a hologram of a three-dimensional object with a source of white light?

*56. If a hologram that was produced with red light is illuminated with red light, the image produced by the interference pattern in the film has the same size and location relative to the film that it had when the hologram was made. What would happen to the size and location if you illuminated this hologram with blue light?

57. Why is light sometimes described using the wave model and sometimes using the particle model?

EXERCISES

1. What is the speed of light in glass with an index of refraction of 1.7?

2. What is the speed of light in water?

3. The speed of light in diamond is 1.24×10^8 m/s. What is the index of refraction for diamond?

4. Zircon is sometimes used to make fake diamonds. What is its index of refraction if the speed of light in zircon is 1.6×10^8 m/s?

5. If it takes light 7 ns to travel 1 m in an optical cable, what is the index of refraction of the cable?

6. If an optical cable has an index of refraction of 1.5, how long will it take a signal to travel between two points on opposite coasts of the United States separated by a distance of 5000 km?

7. The red light from a helium–neon laser has a wavelength of 633 nm. What is its frequency?

8. What is the frequency of the yellow light with a wavelength of 590 nm that is emitted by sodium lamps?

9. What is the wavelength of light that has a frequency of 5.0×10^{14} Hz?

10. What is the wavelength of the radio signal emitted by an AM station broadcasting at 1090 kHz? Radio waves travel at the speed of light.

11. What is the wavelength of the red light from a helium–neon laser when it is in glass with an index of refraction of 1.7? The wavelength in a vacuum is 633 nm.

12. A transparent material is known to have an index of refraction equal to 2.2. What is the wavelength of light in this material if it has a wavelength of 550 nm in a vacuum?

13. Light from a sodium lamp with a wavelength in a vacuum of 590 nm enters diamond in which the speed of light is 1.24×10^8 m/s. What is the wavelength of this light in diamond?

14. What is the wavelength of light in water if it has a frequency of 6.6×10^{14} Hz?

15. What is the thinnest film that will strongly reflect light with a wavelength of 600 nm in the film?

16. What is the thinnest soap film that will strongly reflect red light from a helium–neon laser? The wavelength of this light is 633 nm in air and 470 nm in soapy water.

*17. What is the thinnest film that will strongly reflect light with a frequency of 5×10^{14} Hz if the index of refraction of the film is 1.5?

*18. What is the thinnest film (but not zero thickness) that will strongly transmit light with a frequency of 5×10^{14} Hz if the index of refraction of the film is 1.5?

INTERLUDE

The hairs on this girl's head repel each other because they have the same charge.

An Electrical and Magnetic World

If our common experiences are any guide, electrical and magnetic effects appear insignificant. They don't seem to be a common property of all matter, nor do they usually seen very strong. Magnets are used to lift a couple of paper clips, and static electricity lifts bits of paper.

If we ignore the electricity supplied by power companies and occurring in lightning, electricity seems only to give us occasional annoying shocks when we scuff our feet on rugs or slide across vinyl seat covers. Furthermore, the effects are intermittent; shocks usually only occur in winter or when it is especially dry. Even then, they only occur with a few materials; we get shocked when we touch metal doorknobs, but not when we touch wooden doors.

Magnetic properties seem more permanent. A piece of metal that is magnetic in the winter is still magnetic in the summer. But again, these properties don't seem to be universal. Only a few materials are magnetic, and the magnetic forces don't seem very strong.

Most people held this attitude in Newton's time. The investigation of electrical and magnetic properties of matter was not considered important. Gravity reigned supreme as the governing force in the Universe. All objects are attracted by the gravitational force. It holds objects to the surface of the Earth, it holds our life-supporting atmosphere around the Earth, and it even controls the motion of the planets. But in fact, gravity is the *weakest* force. It is the electric and magnetic forces that control the structure of matter on the atomic level and therefore the structure of matter in the macroscopic world.

> **It is the electric and magnetic forces that control the structure of matter on the atomic level and therefore the structure of matter in the macroscopic world.**

We are electrical creatures. What tastes good or bad, what's medicine and what's poison, and the contractions of our running muscles or our vocal cords are all controlled by electrical interactions. Even our thoughts are due to electrical signals. Magnets and magnetism seem more removed, but images of our brains can be taken using magnetic techniques. As we will learn in Chapter 21, the seemingly separate phenomena of electricity and magnetism are intimately connected.

The next three chapters follow a path similar to the one followed by the early pioneers in this area. We first look at electricity and magnetism separately. Then we look for connections between the two. As we develop the connections you will find yourself returning to components of the physics world view that we developed in earlier chapters. All of this discussion will set the stage for the study of the basic building block of nature, the atom.

We see electric effects when sparks are made. On very dry days, we may get shocked when we touch a metal doorknob after walking across a carpet. If you look closely, you can see a spark jump between your hand and the doorknob. Although we may be surprised by them, the sparks do not hurt us. On the other hand, nature produces very long and dangerous sparks during lightning storms. What determines the length of the sparks? (See p. 472 for the answer to this question.)

Electricity

A lightning storm produces very high voltages, creating very large and dangerous electric currents.

The early Greeks knew that amber—a fossilized tree sap currently used in jewelry—had the interesting ability to attract bits of fiber and hair after it was rubbed with fur. This was one way of recognizing an object that was electrified. In modern terminology, we say the object is **charged.** This doesn't explain what **charge** is, but is a handy way of referring to this condition.

In 1600, the English scientist William Gilbert published a pioneering work, *De Magnete,* in which he pointed out that this electric effect was not an isolated property of amber, but a much more general property of matter. Materials such as gems, glass, and sealing wax could also be charged. Rubbing is the most common way of charging an object. In fact, both objects become charged.

After Gilbert, many experimenters joined in the activity of investigating electricity. However, the question of what happens when an object gains or loses this electrical property remained unanswered. A modern response might include words like *electron* or *proton.* But we have to be careful that we are not simply substituting new words or phrases for old ones. To answer this question fully and thus expand our world view, we need to carefully examine our electric world.

Electrical Properties

In an effort to explain electricity, Gilbert proposed the existence of an electric fluid in certain types of objects. He suggested that rubbing an object removed some of this fluid, leaving it in the region surrounding the object. Bits of fiber were attracted to the object by the "draft" of the fluid returning to the object. Although many other electrical phenomena still

The amber used in jewelry will attract pieces of paper when rubbed with fur.

could not be explained with this idea, it was the beginning of attempts to model the underlying, invisible processes that caused the electrical effects.

Little progress was made for more than a century. In the 1730s it was shown that charge from one object would be transferred to a distant object if they were connected by metal wires, but not if they were connected by silk threads. Materials that are able to transfer charge are known as **conductors;** those that cannot are called nonconductors, or **insulators.** It was discovered that metals, human bodies, moisture, and a few other substances are conductors.

The discovery that moisture is a conductor explains why electrical effects vary from day to day. You normally experience bigger shocks in the winter when the humidity is naturally low. On more humid days any charge that you get by scuffing your feet is quickly dissipated by the moisture in the air surrounding your body.

Many charged objects have to be suspended from insulators, such as silk threads or plastic bases, or they quickly lose their charge to the earth, or ground. In fact, we speak of **grounding** an object to ensure that it is not charged. Other objects hold their charges without being insulated. When we charge a plastic rod by rubbing it, the charge stays on the rod even if we hold it. But our bodies are conductors. Why doesn't the charge flow to ground through our bodies? It stays on the rod because the rod is an insulator; charge generated at one end remains there. The charge can be removed by moving our hands along the charged end. As we touch the regions that are charged, the charges flow through our bodies to ground. A metal rod cannot be charged in this way, because metal conducts the charge to our hands. A metal rod can be charged if it is mounted on an insulating stand or if we hold it with an insulating glove; that is, the rod must be insulated from its surrounding.

Even the flow of liquid through pipes is enough of a rubbing motion to charge the liquid. Fuel trucks tend to build up charge as they dispense fuel. If this charge becomes large enough, sparking may occur and cause

Fuel trucks are grounded to remove the charge built up by the fuel flowing through the hoses.

a fire or explosion. To avoid this danger a conducting wire is connected from the truck to the ground to allow the charge to flow into the ground.

Physics on Your Own   Notice the sparks when you do one or more of the following in a dark room. (1) Pull apart some clothes that have just come from a clothes dryer. (2) Rapidly pull some transparent tape from the roll. (3) Comb your hair and bring the comb near a metal object.

Two Kinds of Charge

It was also the early 1730s before there was any mention that charged objects could repel one another. Electricity, like gravity, was believed to be only attractive. It may seem strange to us now since both the attractive and repulsive aspects of electricity are easy to demonstrate. If you comb your hair, the comb becomes charged and can be used to attract small bits of paper. After contacting the comb, some of these bits are then repelled by it.

This phenomenon can be investigated more carefully using balloons and pieces of wool. If we rub a balloon with a piece of wool, the balloon becomes charged; it attracts small bits of paper and sticks to walls or ceilings. If we suspend this balloon by a thread and bring uncharged objects near it, the balloon is attracted to the objects [Fig. 19-1(a)]. Everything seems to be an attractive effect.

Charging another balloon in the same way demonstrates the new effect; the two balloons repel one another [Fig. 19-1(b)]. Because we believe that any two objects prepared in the same way are charged in a like manner, we are led to the idea that like charged objects repel one another.

Whenever we charge an object by rubbing it with another, both objects become charged. If we examine the pieces of wool, we find that they are also charged; they each attract bits of paper.

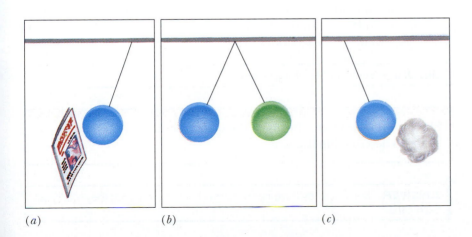

(a)   (b)   (c)

Figure 19-1   A charged balloon is (a) attracted by uncharged objects, (b) repelled by an identically charged balloon, and (c) attracted by the wool used in charging it.

All American Physicist

Benjamin Franklin

Ben Franklin was born on January 17, 1706, in Boston, the 15th of 17 children. Young Benjamin learned to read early, but lack of money cut his formal schooling to two years. He was judged "Excellent" in reading, "Fair" in writing, and "Poor" in arithmetic. This man, who would eventually be regarded as the New World's first physicist, returned home to help make and sell candles and soap.

At 12, Benjamin was apprenticed to his brother James, a printer. His schooling may have ended, but not his education. The young boy sharpened his writing skills, and was soon publishing prolifically in his brother's paper. At 17, although still indentured to his brother, Franklin ran away.

By age 24, Franklin had his own printing shop, and for 37 years published the *Pennsylvania Gazette*. Franklin was a prolific organizer, reformer, and innovator. He was the first to publish a newspaper cartoon, organize a subscription library, establish a volunteer fire department, and start a colonial fire insurance company.

Part of Franklin's creativity found an outlet in scientific experimentation in the fields of thermodynamics, oceanography, meteorology, and, most notably, electricity. Although he had no formal scientific training, he was an astute observer and his experiments were careful, precise, and well described. Perhaps due to his lack of interest in mathematics, the experiments were qualitative.

Franklin was apparently the first to recognize the effectiveness of a pointed conductor for discharging a charged object, or for preventing it from becoming charged. Perhaps surprisingly. Franklin was not the first to suspect the electrical nature of lightning, nor the first to verify the theory experimentally, which he did with his famous kite experiment. Franklin invented the lightning rod, which later saved his own home from destruction.

Adapted from an essay written for Pasco Scientific by Steven Janke.

References: Van Doren, C. *Benjamin Franklin*. New York: Viking Press, 1938. *The Autobiography of Benjamin Franklin*. London: Yale University Press, 1964.

QUESTION   Will the two pieces of wool attract or repel each other?

The piece of wool and the balloon, however, attract each other after being rubbed together [Fig. 19-1(c)]. If they had the same kind of charge, they should repel. We are, therefore, led to the idea that there must be two different kinds of charge and that they attract each other. These experiments can be summarized by stating

Like charges repel; unlike charges attract.

Physics on Your Own   Perform the experiments with the wool and balloons described in the text.

ANSWER   Because they have been charged in a like manner, they will repel each other.

Conservation of Charge

Benjamin Franklin, believing like Gilbert that electricity was a single fluid, thought that an excess of this fluid caused one kind of charged state, whereas a deficiency caused the other. Because he could not tell which was which, he arbitrarily named one positive and the other negative. By convention the charge on a glass rod rubbed with silk or plastic film is positive (Fig. 19-2); that on an amber or rubber rod rubbed with wool or fur is negative.

At first glance these names seem to have no advantage over other possible choices such as black and white or yin and yang. They were, however, more significant. Franklin's use of the fluid model led him to predict that charge should be conserved. The amount of electric fluid should remain the same; it is just transferred from one object to another. If you start with two uncharged objects and rub them together, the amount of excess fluid on one is equal to the deficiency on the other. In other words, the positive and negative charges are equal. Using his system of positive and negative numbers, we can add them and see that the total is still zero.

Although we no longer believe in Franklin's fluid model, his idea about the **conservation of charge** has been verified to a very high precision. It is one of the fundamental laws of physics. In its generalized form, it can be stated

Figure 19-2 The charge on the glass rod is called positive. This is indicated by a few plus signs.

> In an isolated system the total charge is conserved.

conservation of charge

In our modern physics world view, all objects are composed of negatively charged electrons, positively charged protons, and uncharged neutrons. An object is uncharged (or neutral) because it has equal amounts of positive and negative charges, not because it contains no charges. Positively charged objects may have an excess of positive charges or a deficiency of negative charges. We simply call them positive objects since the electrical effects are the same in both situations.

> **QUESTION** Will a negative object have an excess of negative charges or a deficiency of positive charges?

The modern view easily accounts for the conservation of charge when charging objects. The rubbing simply results in the transfer of charges from one object to the other; whatever one object loses, the other gains.

Induced Attractions

Attraction is more common than repulsion because charged objects can attract *uncharged* objects. How do we explain the observation that charged objects attract uncharged objects? Consider a positive rod and

> **ANSWER** It could have either.

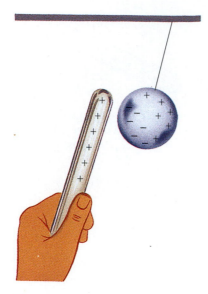

Figure 19-3 A charged rod attracts a neutral metal ball because of the induced separation of charges.

an uncharged metal ball. As the rod is brought near the ball, its positive charges attract the negative charges in the ball and repel the positive charges. Because the charges in the ball are free to move, this results in an excess of negative charges on the near side and an excess of positive charges on the far side of the ball (Fig. 19-3). Because charge is conserved, the negative charge on one side is equal to the positive on the other. Experiments with balloons show that the electric force varies with distance; the force gets weaker as the balloons are moved farther apart. Thus, the ball's negative charge is attracted to the rod more than its positive charge is repelled. These induced charges result in a net attraction of the uncharged ball.

> **Physics on Your Own**   Run a comb through your hair and hold it close to a thin stream of water from a faucet. How do you explain what happens?

Experiments show that if the ball is made of an insulating material, the attraction still occurs. In insulators the charges are not free to move across the object, but there can be motion on the molecular level. Although the molecules are uncharged, the presence of the charged rod might induce a separation of charge within the molecule. Other molecules are naturally more positive on one end and more negative on the other. The positive rod rotates these molecules so that their negative ends are closer to the rod, as shown in Figure 19-4. This once again results in a net attractive force.

A particularly graphic example of the attraction of an uncharged insulator is illustrated in Figure 19-5. A long 2" by 4" is balanced on a watch glass so that the board is free to rotate. When a charged rod is brought near one end, the resulting attractive force produces a torque which rotates the board.

> **QUESTION**   Assuming that the 2" by 4" was attracted by a positive charged rod in the previous example, what direction would it rotate if a negatively charged rod was used?

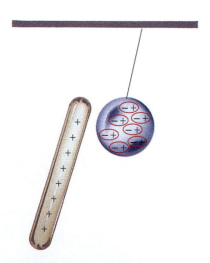

Figure 19-4 An insulating object is attracted because of the separation of charges within the molecules.

We can now return to the charged comb that attracted and then repelled bits of paper. Initially, the bits of paper are uncharged. They are attracted to the comb by the induced charge. When the bits of paper touch the comb, they acquire some of the charge on the comb and are repelled because like charges repel.

> **ANSWER**   It would rotate in the same direction. The interaction between any charged object and an uncharged object is always attractive.

 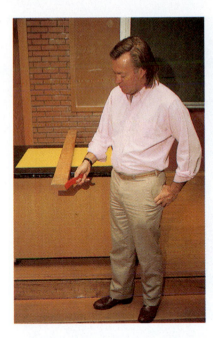 

Figure 19-5   A standard two-by-four balanced on a curved dish can be rotated by the attractive force of a charged rod.

The Electroscope

In most experiments we transfer so few charges that the total attraction, or repulsion, is small compared with the pull of gravity. Detecting that an object is charged becomes difficult unless it is very light—like bits of paper or a balloon. We get around this difficulty with a device called an *electroscope* that gives easily observable results when it is charged. By bringing the object in question near the electroscope, we can deduce that it is charged by the effect it has on the electroscope.

The essential features of an electroscope are a metal rod with a metal ball on top and two very light metal foils attached to the bottom. Figure 19-6 shows a homemade electroscope constructed from a chemical flask. The glass enclosure protects the very light foils from air currents and insulates them from the surroundings.

If the electroscope isn't charged, the foils hang straight down under the influence of gravity. When we touch the ball of the electroscope with a charged rod, some of the rod's excess charge will be shared with the electroscope. The charges flow from the rod onto the electroscope because the charges on the rod repel each other and therefore distribute themselves over the largest region possible, which includes the ball, the metal rod, and the foils. Since the foils have the same charge, they repel each other and swing apart (Fig. 19-7). The separation of the foils indicates that the electroscope is charged. This happens when the electroscope is touched with a positive rod as well as a negative one.

Figure 19-6 An electroscope made from a flask, a metal rod, and two pieces of thin metal foil.

When a charged object is touched to an electroscope, the electroscope takes on the same charge as the object. However, the electroscope only shows the presence of charge; it does not indicate the sign. We can determine the sign by slowly bringing a rod with a known positive or negative charge toward the ball of the charged electroscope and watching the motion of the foils. For example, as a positive rod nears the electroscope, it induces a redistribution of charge in the electroscope. It repels the pos-

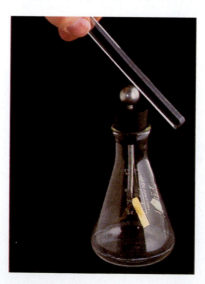

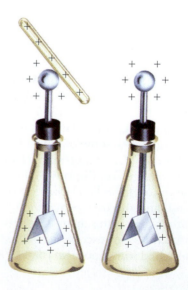

Figure 19-7 When an electroscope is charged by direct contact, it has the same charge as the rod.

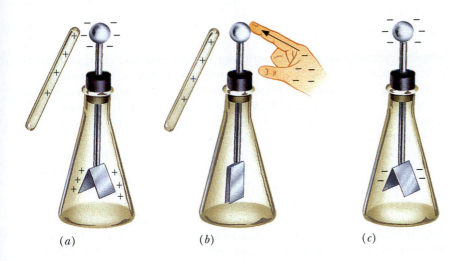

(a)   (b)   (c)

Figure 19-8 (a) The positively charged rod brought near the electroscope attracts negative charges to the ball, leaving the foils with an excess positive charge. (b) Touching the ball with a finger allows negative charges to flow into the electroscope, (c) leaving it with a net negative charge.

itive charge and attracts the negative charge, causing an increase in positive charge (or a decrease in negative charge) on the foils. The foils then move farther apart if they were originally positive or closer together if they were originally negative.

> **QUESTION** When a negative rod is slowly brought near an electroscope that is initially charged, the foils move closer together. What is the charge on the electroscope?

A charged rod will also separate the foils on an uncharged electroscope. As a positive rod is brought near an uncharged electroscope, some of the negative charges in the electroscope are attracted to the ball on top; the ball has a net negative charge and the foils have a net positive charge as shown in Figure 19-8(a). Note that the net charge on the electroscope is still zero since we did not touch it with the rod. We have only moved the charges around in the electroscope by bringing the charged rod near the ball. When the charged rod is removed, the charges redistribute themselves and the foils fall. An analogous phenomenon happens with a negative rod.

Imagine that while the positive rod is near the electroscope, you touch the ball with your finger as shown in Figure 19-8(b). Negative charges are attracted by the large positive charge on the rod and travel from the ground through your body to the electroscope. If you first remove your finger and then the rod, the electroscope is now charged and the foils repel each other. This is known as charging the electroscope by induction and results in it acquiring a charge *opposite* to that on the rod.

> **ANSWER** The negative rod repels negative charges on the electroscope's ball, increasing their concentration on the foils. Because this negative charge reduces the charge on the foils, they must have been positive initially.

The Electric Force

Simple observations of the attraction or repulsion of two charged objects indicate that the electric force depends on distance. For instance, a charged object has more effect on an electroscope as it gets nearer. But we need to be more precise. How does the size of this force vary as the separation between charged objects changes? And how does it vary as the amounts of charge on the objects vary?

In 1785, the French physicist Charles Coulomb measured the changes in the electric force as he varied the distance between two objects and the charges on them. He verified that if the distance between two charged objects is doubled (without changing the charges), the electric force on each of them is reduced to one-fourth the initial value. If the distance is tripled, the force is reduced to one-ninth, and so on. This type of behavior is known as an **inverse-square relationship;** "inverse" because the force gets smaller as the distance gets larger, "square" because the force changes by the square of the factor by which the distance changes.

Coulomb also showed that reducing the charge on one of the objects by one-half reduced the electric force to one-half its original value. Reducing the charge on each by one-half reduced the force to one-fourth the original value. This means that the force is proportional to the product of the two charges.

> **QUESTION**   Coulomb was not able to measure charge directly. What technique could he have used to reduce the charge to one-half its original value?

These two effects are combined into a single relationship known as Coulomb's law.

Coulomb's law

$$F = \frac{kQ_1 Q_2}{r^2}$$

In this equation, the Q's represent the charges on the two objects, r is the distance between their centers, and k is a constant whose value depends on the units chosen for force, charge, and distance.

Each object feels the force due to the other. The forces are vectors and act along the line between the two objects. Each force is directed toward the other if the charges have opposite signs and away from each other if the charges have the same sign (Fig. 19-9). Because the forces are equal, point in opposite directions, and act on different objects, they satisfy Newton's third law of action–reaction pairs.

Because the existence of an elementary, fundamental charge was not known until this century, the unit of electric charge, the **coulomb** (C), was chosen for convenience in use with electric circuits. Using the

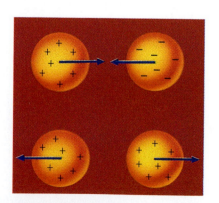

Figure 19-9 The forces on two charged objects are equal in size and opposite in direction, which obeys Newton's third law.

> **ANSWER**   He could have used identical conducting spheres and put the charged sphere into contact with a neutral one. By symmetry each sphere would have one-half the original charge.

coulomb, the value of k in Coulomb's law is determined by experiment to be

$$k = 9 \times 10^9 \ \text{N} \cdot \text{m}^2/\text{C}^2$$

Coulomb's constant

The coulomb is a tremendously large unit for the situations we have been discussing. For instance, the force between two spheres each having 1 coulomb of charge and separated by 1 meter is

$$F = \frac{kQ_1 Q_2}{r^2} = 9 \times 10^9 \frac{\text{N} \cdot \text{m}^2}{\text{C}^2} \frac{1 \ \text{C} \times 1 \ \text{C}}{1 \ \text{m} \times 1 \ \text{m}} = 9 \times 10^9 \ \text{N}$$

This is a force of one million tons!

The charges that we have been discussing are much less. When you scuff your shoes on a carpet, the charges transferred are about a millionth of a coulomb. But even this charge is very large compared with the charge on an electron or proton. The charge on an electron or a proton is now known to be 1.6×10^{-19} coulomb. Conversely, it would take the charge on 6.2×10^{18} protons (or electrons) to equal a charge of 1 coulomb.

Electricity and Gravity

The mathematical form of Coulomb's law is the same as that of the universal law of gravitation discussed in Chapter 4. Therefore, the drawing in Figure 4-5 illustrating the dependence of the inverse-square law on distance also holds for electricity if the Earth is replaced by a charged object.

Professor Sprott is protected from the high-voltage discharge by the metallic cage.

Lightning

Everybody has seen lightning (Fig. A) and has probably been frightened by this spectacular display of energy at some time or other. Lightning is unpredictable, and seems to occur randomly and instantaneously. Everything but the thunder is usually over in less than half a second. And yet the effects are not inconsequential: Each year lightning causes about 80 million dollars of damage in the United States. On the average, lightning kills 85 people a year and injures another 250 in this country alone.

We still have a lot to learn about lightning. For instance, we don't know what causes the initial buildup of charge, what determines the paths the lightning takes, or what triggers the lightning bolts.

We do know that during a storm, clouds develop a separation of electric charge, with the tops of the clouds positively charged and the bottoms negatively charged, as shown in Figure B. The production of a lightning bolt begins when the negative charge on the bottom of the cloud gets large enough to overcome air's resistance to the flow of electricity (Fig. C) and electrons begin flowing toward Earth along a zigzag, forked path at about 60 miles per second! As the electrons flow downward, they collide with air molecules and ionize them, producing more free electrons. Even though the current may be as large as 1000 amperes, this is not the lightning bolt we see.

Meanwhile, as the electrons approach the ground, the ground is becoming more and more positively charged due to the repulsion of electrons in the

(a)

Figure A   Lightning bolts can generate currents up to 200,000 amperes.

There are, however, several important differences between gravity and electricity that govern the role of these two forces in the Universe.

Electricity has two kinds of charge and thus displays both attraction and repulsion, but there is only one kind of mass. Negative mass does *not* exist and therefore the gravitational force is never repulsive. Because of the induced charge distribution possible with two kinds of charge, conducting materials can shield a region from all external electric forces. Suppose we were to build a large metal room. If this room were placed near a large electric charge, the charges in the metal walls would redistribute in such a way as to cancel the effect of the external charge for all ground. This positively charged region moves up through any conducting objects on the ground—houses, trees, and people—into the air (Fig. D). When the downward moving electrons meet the upward flowing positive regions at an altitude of a hundred meters or so, they form a complete circuit (Fig. E) and the lightning begins. In less than a millisecond up to a billion trillion electrons may reach the ground; the current may be as high as 200,000 amperes. While the flow of charge is downward, the point of contact between the cloud charge and the ground charge is flowing up-ward at about 50,000 miles per second. Observers report that the lightning bolt is moving downward because the pathway that is lit is the initial forked pathway coming from the cloud. The upward surge heats the air to 50,000°F. A meter of the lightning's path may shine as bright as a million 100-watt lightbulbs (Fig. F). This rapid heating of the air along the lightning's path also produces a shock wave that we hear as the thunder. What we see as a single lightning bolt is usually composed of several bolts in rapid succession along the same path.

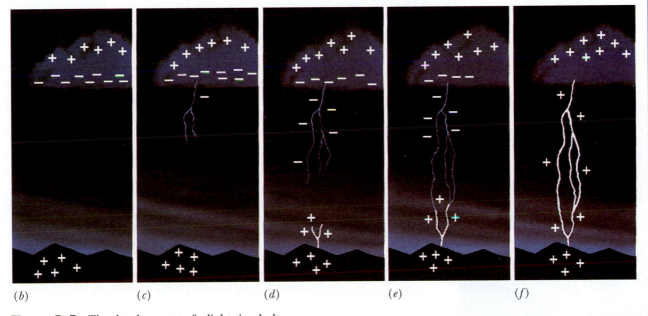

(b)   (c)   (d)   (e)   (f)

Figures B–F   The development of a lightning bolt.

locations in the room. The cancellation only works because there are two kinds of electric charge. This result is not possible for gravitational forces because it is not possible to create a gravitation-free chamber since there is no negative mass. Antigravity spaceships, although they make good material for science fiction writers, are not possible.

Another difference between electricity and gravity lies in the behavior of objects in the respective fields. The gravitational mass is equal to the inertial mass, and therefore objects of all sizes and compositions have the same acceleration near the surface of the Earth. In contrast, the electric charges on objects are *not* proportional to the masses and therefore

The electrical force dominates on the atomic scale, such as in these ice crystals on a strawberry leaf.

charged particles do not all have the same acceleration near a charged object. If we were to release a proton and an electron near a charged object, the electron would have a much larger acceleration since the forces are the same and the electron's mass is much smaller.

Finally, charge comes in one fixed size. Experiments have shown that the charges on all electrons and protons are the same size. The charges on other ordinary subatomic particles are either the same size or integral multiples of this size. Mass also occurs in lumps, but there are many different sizes. All electrons have the same mass but this mass is different from the mass of protons.

A consequence of these differences is the different roles these two forces play in our lives. The electric force is the dominant force in the atomic world. On the other hand, the gravitational force dominates on the macroscopic scale of people, planets, and galaxies. The Earth and Moon would each need 10^{14} coulombs of charge to have an electric force between them equal to their gravitational force. The reason electricity and gravity switch roles in these two domains is that macroscopic objects are essentially uncharged; that is, there are approximately equal numbers of positive and negative charges on most large objects. Even though the total attractive force between all of the positive charges in the Earth and all of the negative charges in the Moon is tremendous, it is equal and opposite to the repulsive force between the negative charges in the Earth and the negative charges in the Moon. The net electric force is essentially zero because of this correlation. However, because there is only one kind of mass, there is nothing to cancel the much weaker gravitational force between the atoms in the Earth and the Moon. So as we move from people to planets and eventually to galaxies, the gravitational force becomes the dominant one.

The gravitational force dominates in the Solar System and beyond, such as in this galaxy.

COMPUTING   GRAVITATIONAL AND ELECTRIC FORCES   Σ

As a numerical example, let's calculate the sizes of the electric and gravitational forces between an electron and a proton in a hydrogen atom. We start by assuming that the electron and proton are separated by a distance of 5.3×10^{-11} meter and use the known masses and charges of the electron and proton.

$$F_g = G \frac{M_e M_p}{r^2} = 6.7 \times 10^{-11} \frac{\text{N} \cdot \text{m}^2}{kg^2} \frac{(9.1 \times 10^{-31} \text{ kg})(1.7 \times 10^{-27} \text{ kg})}{(5.3 \times 10^{-11} \text{ m})^2}$$

$$= 3.7 \times 10^{-47} \text{ N}$$

$$F_e = k \frac{Q_e Q_p}{r^2} = 9 \times 10^9 \frac{\text{N} \cdot \text{m}^2}{\text{C}^2} \frac{(1.6 \times 10^{-19} \text{ C})^2}{(5.3 \times 10^{-11} \text{ m})^2} = 8.2 \times 10^{-8} \text{ N}$$

Therefore, the electric force is more than 10^{39} times stronger than the gravitational force. (That's one thousand, trillion, trillion, trillion times bigger!) Because the two forces change in the same way as the separation of the electron and proton changes, the electric force is always this much stronger.

The Electric Field

Σ

Rather than calculate the forces that an electric charge has on other objects, it is often more convenient to think of the charge as modifying the space around itself and then of the modified space as exerting forces on the objects. This method is analogous to what we did with the gravitational force in Chapter 4.

In Chapter 21 we will learn that these electric fields can take on an identity of their own and, along with their associated magnetic fields, can travel through space as waves. We therefore define an **electric field** in exactly the same way that we defined the gravitational field. By convention the value of the electric field at any point in space is equal to the force experienced by one unit of positive charge placed at that point. (It is assumed that this test charge does not disturb the charges that produce the field.) In fact, electric fields can be measured experimentally, by placing a very tiny test charge at various points in space and determining the force exerted on it at each point. The electric field E at a point is obtained by dividing the force F by the size q of the test charge, $E = F/q$.

electric field = force on unit positive charge

Because force is a vector quantity, the electric field is a vector field; it has a size and a direction at each point in space. You could imagine the space around a positive charge as a "porcupine" of little arrows pointing outward. The arrows farther from the charge would be shorter to indicate that the force is weaker there.

electric field = $\dfrac{\text{force}}{\text{charge}}$

The values for an actual electric field can either be measured with a test probe or calculated from the known distribution of charges. When calculating the value of the field at the point of interest, we can take advantage of the fact that each charge acts independently; the effects simply add. This means that we calculate the contribution of each charge to the field and then add these contributions vectorially.

> **QUESTION**   What does the electric field surrounding a negative charge look like?

We can develop a qualitative feeling for the electric field around charges with photographs such as those in Figure 19-10. They were taken by placing charged rods into an insulating liquid with grass seeds floating on top. These photographs are top views of the surface of the liquid. The rod is positioned vertically into the liquid and thus is pointing into the page. The more visible horizontal lines in the photographs are wires used to charge the rods. The elongated seeds line up in the direction of the electric field. The pattern is most distinct where the field is strongest.

In drawing fields one usually uses a sample of continuous lines rather than little arrows representing the vectors. The lines are drawn such that they point in the direction of the field at each point in space. The relative spacing between neighboring lines represents the strength of the field;

> **ANSWER**   An electric field surrounding a negative charge looks just like the electric field surrounding a positive charge except that all of the arrows are reversed.

Figure 19-10 These patterns of grass seeds show the general features of the electric fields produced by (a) a single positive rod, (b) two rods with equal and opposite charges, and (c) two rods with positive charges. The directions of the fields are indicated in the accompanying diagrams.

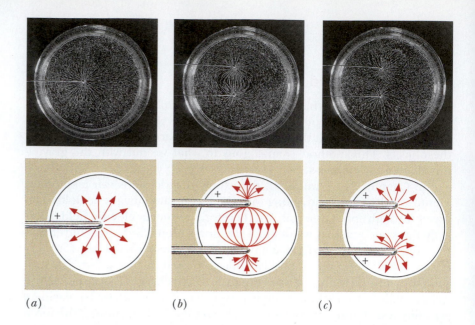

(a)   (b)   (c)

the closer the lines, the stronger the field. The drawing in Figure 19-10(b) shows that each line starts on a positive charge and ends on a negative one and that the lines never cross.

> **QUESTION**   How would the photograph and drawing in Figure 19-10(a) change if the positive charge were replaced by a negative charge?

Electric Potential Σ

Because the gravitational and electric force laws have the same form, the **electric potential energy** of a charged object can be defined for every point in an electric field in the same way it was defined for gravity. For each point, the electric potential energy is equal to the work done in bringing the charged object from some zero reference location to that point. The value of the electric potential energy does not depend on the path, but does depend on the reference point, the point of interest, and the charge on the object (Fig. 19-11).

As with gravitational potential energy, the actual value of the electric potential energy is not important in physical problems; it is only the difference in energy between points that matters. If it requires 10 joules of work to move a charged object from A to B, then the electric potential energy of the object at B is 10 joules more than at A. If A is the zero reference point, the electric potential energy of the object at B is 10 joules.

It is often more convenient to talk about the energy available due to the electric field without reference to a specific charged object. Because an object's electric potential energy depends on its charge, an **electric potential** V can be assigned to the field and defined as the electric potential energy EPE divided by the object's charge q; that is, $V = EPE/q$. This quan-

electric potential energy = work from reference point

electric potential $= \dfrac{\text{electric PE}}{\text{charge}}$

> **ANSWER**   The photograph would not change, but the arrows in the drawing would point in opposite directions.

tity is numerically equal to the work required to bring a positive test charge of 1 coulomb from the zero reference point. The units for electric potential are joules per coulomb, a combination known as a **volt** (V). Because of this, we often speak of the electric potential as a voltage.

Defining the electric potential has the advantage that once we know the electric potential, we can obtain the electric potential energy for any charged object by multiplying the potential by the charge. Once again, it is only the *potential difference* that matters. For instance, a 12-volt battery has an electric potential difference of 12 volts between its two terminals. This means that 1 coulomb of charge moving from one terminal to the other would gain or lose (1 coulomb)(12 volts) = 12 joules of energy.

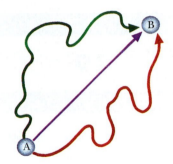

Figure 19-11 The work performed in taking charge from *A* to *B* does not depend on the path.

> **QUESTION** How much work is required to move 3 coulombs of positive charge from the negative terminal of a 12-volt battery to the positive terminal?

If the potential difference, or voltage, between two points is high and the points are close together, the electric field can be strong enough to tear electrons from molecules in the air. The electrons are pulled one way, and the remaining positive ions are pulled the other way. Because electrons have much less mass than ions, they quickly accelerate to high velocities. As they accelerate they collide with other molecules, knocking additional electrons loose. This forms a cascade of electrons that we call a *spark*. Dry air breaks down in this fashion when the electric field reaches about 30,000 volts per centimeter.

> **Physics on Your Own** The next time you have dry weather and you get sparks when touching doorknobs, try holding a key in your hand when you touch metal objects. Why does this hurt less? Estimate your electric potential by estimating the length of the sparks.

Physics Update

Artificial lightning bolts for breaking down and monitoring waste, described by Dan Cohn of MIT and colleagues, represent a potential near-term spinoff of long-term nuclear fusion research. In a pilot-scale research furnace at MIT, a 10,000-degree plasma, created by passing an electric current between a pair of graphite electrodes in a nitrogen-filled gas chamber, has been used to melt waste material (consisting of soil, metals, combustible materials, and sludges) into a lavalike liquid. The liquid solidifies into a stable black glass which can be safely stored or even used as a construction material. The process produces no toxic ash, virtually no dioxin, and less gas emission than traditional incineration techniques. Furthermore, it has the potential of being more economical than present techniques.

> **ANSWER** Each coulomb requires 12 joules of work, so 3 coulombs require 36 joules; $W = QV = (3C)(12V) = 36J$.

Sparks occur when the electric field becomes strong enough to pull electrons from atoms. These electrons are accelerated by the electric field, obtaining sufficient kinetic energy to knock electrons from other atoms. This avalanche of electrons produces the visible sparks that we see. Electric fields of 30,000 volts per centimeter are strong enough to break down atoms in dry air. Because the electric field is proportional to the electric potential difference and inversely proportional to the distance, it takes large voltages to produce long sparks.

S U M M A R Y

Objects become electrically charged or uncharged by transferring charges. An uncharged object has equal numbers of positive and negative charges. Positively charged objects may have an excess of positive charges or a deficiency of negative charges. In an isolated system the total charge is conserved. Charges can flow through conductors such as metal wires, but not through insulators such as silk threads.

In electricity, unlike gravity, there are two different kinds of charge and the direction of the force depends on the relative signs of the charges. Like charges repel; unlike charges attract. A charged object can attract an uncharged object due to an induced separation of charges.

The electric force has the same mathematical form as the gravitational force. The electric force, however, can be repulsive as well as attractive. Electric charge comes in whole-number multiples of one fixed size. All electrons and protons have the same electric charge although opposite in sign.

Electric charges are surrounded by an electric field equal to the force experienced by a unit positive charge. This electric field is a vector field with a size and a direction at each point in space.

The electric potential energy of a charged object in an electric field is equal to the work done in bringing the object from some zero reference location. Only the differences in electric potential energy matter in physical situations. The electric potential equals the electric potential energy divided by the object's charge. This quantity is numerically equal to the work required to bring a positive test charge of one coulomb from the zero reference point. The units for electric potential are joules per coulomb, or volts.

K E Y   T E R M S

charge: A property of elementary particles that determines the strength of its electric force with other particles possessing charge. Measured in coulombs, or in multiples of the charge on the proton.

charged: Possessing a net negative or positive charge.

conductor: A material that allows the passage of electric charge. Metals are good conductors.

conservation of charge: In an isolated system the total charge is conserved.

coulomb: The SI unit of electric charge, the charge of 6.24×10^{18} protons.

electric field: The space surrounding a charged ob-

ject, where each location is assigned a value equal to the force experienced by one unit of positive charge placed at that location.

electric potential: The electric potential energy divided by the object's charge. The work done in bringing a positive test charge of 1 coulomb from the zero reference location to a particular point in space.

electric potential energy: The work done in bringing a charged object from some zero reference location to a particular point in space.

grounding: Establishing an electrical connection to the earth in order to neutralize an object.

insulator: A material that does not allow the passage of electric charge. Ceramics are good insulators.

inverse-square relationship: A relationship in which a quantity is related to the reciprocal of the square of a second quantity; the electric force is proportional to the inverse-square of the distance from the charge.

volt: The SI unit of electric potential.

CONCEPTUAL QUESTIONS

1. A handheld glass rod can be charged by rubbing it with silk or a plastic bag while holding it in your hands. Would you conclude from this that glass is a conductor or an insulator?

2. What evidence supports the conclusion that the material used in making balloons is an insulating material?

3. Why is it not possible to charge just one end of a metal rod?

4. Could you use fur, silk, or wool to charge a metal rod that is held in your hand? Explain.

5. Why is it easier to charge a balloon on a dry day than on a humid day?

6. Why do you sometimes get shocked when you touch a metal doorknob after crossing a carpeted floor?

7. Why do clothes sometimes stick together when removed from a dryer?

8. Before an aircraft is fueled from a truck, a wire from the truck is attached to the aircraft. Why?

9. If silk is used to charge a glass rod, is the charge on the silk positive or negative? Explain.

10. If one rubber rod is charged with fur and another with plastic, they *attract* each other. How would you explain this?

11. Re-create the line of reasoning (with experimental evidence) that leads to the conclusion that there are two kinds of charge.

12. How would you know if you discovered a third kind of charge? What other objects would you expect it to attract and repel?

13. What do we mean when we say that an atom or molecule is neutral?

14. When a neutral water molecule disassociates into an H^+ ion and an OH^- ion, what can we say about the sizes of the charges on the two ions? Explain.

15. In the modern view of electricity, does an object with a positive charge have an excess of positive charge, an excess of negative charge, a deficiency of negative charge, or a deficiency of positive charge?

16. When a charged comb is brought near bits of paper, the bits are first attracted to the comb and then repelled. Describe how the charge on the comb and on the bits of paper changes during this process.

17. When you bring a positive rod near a red rod suspended by a thread, the red rod is attracted to the positive rod. Does this tell us whether the charge on the red rod is positive, negative, or neutral?

18. A green rod is suspended by a thread so that it can rotate freely. A neutral rod attracts the green rod. Does this tell us whether the charge on the green rod is positive or negative?

19. Describe how a charged balloon sticks to a wall.

20. Why does a freshly wiped phonograph record attract dust?

21. Describe how a negatively charged rod attracts (a) an uncharged conducting object and (b) an uncharged insulating object.

22. Why are neutral objects attracted to both negatively and positively charged objects?

23. An electroscope initially has a net negative charge. Why do the foils come together when the electroscope is touched by a human hand?

24. How could you use a negatively charged rod to determine whether a charged electroscope has a negative or a positive charge?

25. Describe how you would use a negatively charged rod to give an electroscope a negative charge.

26. Describe how you would use a negatively charged rod to put a positive charge on an electroscope.

27. Two identical electroscopes, one initially charged and the other initially neutral, are connected by a thin rod. After this, both electroscopes are charged. What can you conclude about the charges on each one?

28. Suppose you find an electroscope with separated

foils. When you bring a negatively charged rod slowly near it, the foils come closer together. What can you conclude about the initial charge on the electroscope?

*29. When Coulomb was developing his law, he did not have an instrument for measuring charge. And yet he was able to obtain spheres with ½, ⅓, ¼ . . . of some original charge. How might he have used a set of identical spheres to do this?

30. The nucleus of a helium atom contains two protons and two uncharged neutrons. If the nucleus is surrounded by two electrons, what is the total charge of the atom?

31. What happens to the electric force between two charged objects if the charge on one of them is changed in sign?

32. If the signs of the charges on two charged objects are both reversed, what happens to the electric force between them?

33. Two charged objects are very, very far from any other charges. If the distance between them is cut in half, what happens to the electric force between them?

34. Assume that you have two identically charged objects separated by a certain distance. How would the force change if the objects were three times as far away from each other?

35. Assume that you have two identically charged objects separated by a certain distance. How would the force change if one object had twice the charge and the other object kept the same charge?

36. Two charged objects are very, very far from any other charges. If the charge on both of them is doubled, what happens to the electric force between them?

37. In what direction does the electric force on a positive charge point when it is located near an isolated negative charge?

38. Two parallel plates have equal but opposite charges. What is the direction of the force on a positive charge placed near the center of the plates?

39. Although the formulas for the electric and gravitational forces have the same form, there are differences in the two. What are some of these differences?

40. What are the similarities between the electric and gravitational forces?

41. Even though electric forces are very much stronger than gravitational forces, gravitational forces determine the motions in the Solar System. Why?

42. When we approach another person, we are not aware of the gravitational and electric forces between us. What are the reasons in each case?

43. What force holds atoms together?

44. What force holds our bodies together, gravitational or electric?

45. Why are the accelerations of all charged objects near a charged sphere not the same?

*46. In this chapter we compared the electric and gravitational forces between an electron and a proton. Why is the result valid for all separations?

47. Why is it relatively safe to remain in your car during a thunderstorm?

48. Why should you run a wire from your satellite dish or your TV antenna to a metal rod driven into the ground?

49. How is the value of the electric field at each point in space defined?

*50. It is not possible for two electric field lines to cross. How do you explain this?

51. Describe the electric field surrounding a ball with a net negative charge.

52. Describe the electric field in the space between two parallel plates containing equal charges of opposite sign.

53. An electron and a proton are released in a region of space where the electric field is vertically upward. How do the forces on the electron and proton compare?

54. How do the accelerations of the electron and proton in the previous question compare?

55. What is the definition of the value of the electric potential energy of a charged object at each point in space?

56. How would the electrical potential energies of an electron and a proton at the same location in a region containing an electric field compare?

57. How is the value of the electric potential at each point in space defined?

*58. If the electric potential at a point is zero, does the electric field also have to be zero?

*59. A proton is released at rest in a uniform electric field. Does the proton's electric potential energy increase or decrease? Does the proton move toward a location with a higher or a lower electric potential?

*60. An electron is released at rest in a uniform electric field. Does the electron's electric potential energy increase or decrease? Does the electron move toward a location with a higher or a lower electric potential?

EXERCISES

1. The nucleus of a certain type of uranium atom contains 92 protons and 143 neutrons. What is the total charge on the nucleus?

2. How many electrons does it take to make 1 C?

3. What is the electric force of attraction between 3 C and −6 C separated by a distance of 3 m?

4. Two small spheres separated by 2 m contain negative charges of 2 C and 3 C. What is the electric force between them? Is it attractive or repulsive?

5. The nucleus of lithium contains 3 protons and 3 neutrons. When 2 electrons are removed from the neutral lithium atom, the remaining electron has an average distance from the nucleus of 0.018 nm. What is the force between the electron and the nucleus at this separation?

6. In an ionic solid like ordinary table salt (NaCl) an electron is transferred from one atom to the other. If the atoms are separated by a distance of 0.1 nm, how strong is the electric force between them?

*7. How much stronger is the electric force between 2 protons than the gravitational force between them?

*8. Calculate the ratio of the electric force to the gravitational force between 2 electrons.

9. What is the electric field 4 m from 8 C of positive charge?

10. What is the electric field at a distance of 2 cm from 4 mC of negative charge?

11. What is the electric field 5.3×10^{-11} m from a proton?

12. What is the electric field at a distance of 0.2 nm from a carbon nucleus containing 6 protons and 6 neutrons?

13. What is the electric field midway between charges of 2 C and 3 C separated by 2 m?

14. What is the electric field midway between an electron and a proton separated by 0.2 nm?

*15. What is the force on a proton located in an electric field of 5000 N/C? What is the proton's acceleration?

*16. How would the values obtained in the previous exercise change if the proton were replaced by an electron?

17. The electric potential energy of an object at point *A* is known to be 60 J. If it requires 12 J of work to move the object to point *B*, what is its potential energy at *B*?

18. The electric potential energies of a charged object at two locations in the laboratory are known to differ by 48 J. How much work does it take to move the object from one location to the other?

19. How much work does a 9-V battery do in pushing 3 mC of charge through a circuit containing one lightbulb?

20. Points *A* and *B* each have an electric potential of +6 V. How much work is required to take 2 μC of charge from *A* to *B*?

21. What was your electric potential relative to a metal pipe if a spark jumped 1.3 cm (0.5 in.) from your finger to the pipe?

22. How far can a spark jump in dry air if the electric potential difference is 6×10^5 V? (This is the reason why high-voltage sources are surrounded by a vacuum or an insulating fluid.)

Electric Charges in Motion

We live in an electric world. It is hard to imagine a world without electricity from batteries and power plants. How does the movement of electrons produce the light that allows us to work and play indoors and at night? (See p. 493 for the answer to this question.)

This night scene of Boathouse Row in Philadelphia is an example of the artistic use of lighting.

New phenomena often enter our world view accidentally, a result of someone with an open mind conducting an experiment. Initially, these phenomena are often just curiosities, new challenges to our understanding of the world. Many of them give us new ways to do things and eventually find widespread use as our understanding of them increases. In some cases, however, a whole new technology is born. The electrical properties we studied in Chapter 19 were such curiosities. In the early days of investigation into these phenomena little use had been made of electricity other than building devices to amaze and shock friends. In the past 200 years electricity has developed into a major technological presence in our society. Today, the average U.S. citizen's annual usage of electric energy is equivalent to the food energy needed to feed more than six people.

An important advance in understanding electricity came with the development of batteries. Batteries could make charges flow continuously as a current rather than in short bursts. This development eventually led to modern electric circuits that have had an obvious impact on our lives. Electric currents can produce heat and light, run motors and radios, operate sound systems and computers, and so on. The list is nearly endless.

An Accidental Discovery

Near the end of the 18th century the Italian anatomist Luigi Galvani announced that an excised frog's leg twitched when he touched it. It was well known from firsthand experiences with static electricity that twitches in humans could result from electric charges. But this was just a leg! Galvani was convinced that he had discovered the secret life force that he believed existed within all animals. Galvani found that a twitch occurred when he touched the frog's leg with a metal object but not when he touched it with an insulator. Furthermore, the metal had to be different from the metal hook holding the other end of the leg.

Another Italian scientist, Alessandro Volta, heard of these results and performed many experiments with very sensitive electroscopes, searching for evidence that electric charge resides in animal tissue. He eventually convinced himself that the electricity was not in the frog's leg but was the result of touching it with the two metals.

Volta tried many different materials and discovered that an electric potential difference was produced whenever two different metals were joined. Some combinations of metals produced larger potential differences than others. One of his demonstrations was to place his tongue between pieces of silver and zinc that were held together at one end. The flow of electricity caused his tongue to tingle. You may have felt a similar sensation when a piece of metal or metal foil touched a filling in your tooth. A more detailed explanation of this process would take us beyond the scope of this book, but it is enough to realize that some chemical process moves charges from one metal to the other, creating a potential difference.

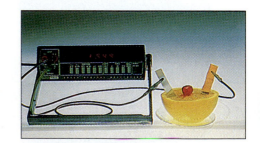

A modern version of Volta's demonstration that zinc and copper rods inserted into a grapefruit produce a potential difference. The cherry is for decorative purposes only.

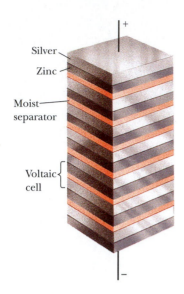

Silver
Zinc
Moist separator
Voltaic cell
+
−

Figure 20-1   A schematic of Volta's battery composed of zinc and silver plates with moist paper between them.

> **Physics on Your Own**   Make a battery by sticking a paper clip and a piece of copper wire into a lemon. You can feel the voltage by placing the tips of the wires to your tongue.

Batteries

Volta used his discovery that two dissimilar metals produce an electric potential difference to make the first battery. He made a single cell by putting a piece of paper that had been soaked in a salt solution between pieces of silver and zinc. He then stacked these cells in a pile like the one shown in Figure 20-1. By doing this he was able to produce a larger potential difference. The potential difference of each cell added to that of the other cells.

A cell's potential difference depends on the choice of metals. Those commonly used in batteries have potential differences of 1½ or 2 volts. A simple 2-volt cell can be made by putting lead and lead oxide plates (known as electrodes) into dilute sulfuric acid (the electrolyte). A 12-volt car battery is made by connecting six of these cells together. The positive electrode of one cell is connected to the negative electrode of the next. The connectors sticking out of the first and last cells (the terminals) are usually marked with a + or − sign to indicate their excess charges.

> **QUESTION**   How many 1½-volt cells does it take to make a 9-volt battery?

Batteries that can be recharged are called *storage batteries*. This recharging is usually accomplished with an electric charger with the negative terminals of the charger and battery connected to each other and the positive terminals connected to each other. The charger forces the current to run backward through the battery, reversing the chemical reactions and refreshing the battery. This electric energy is stored as chemical energy for later use.

Nonrechargeable flashlight batteries (sometimes called *dry cells*) are constructed with a carbon rod down the center, as shown in Figure 20-2. Carbon is a conductor and replaces one of the metals. The rod is surrounded by a moist paste containing the electrolyte. The other electrode is the zinc can surrounding the paste. The cell is then covered with an insulating material. These 1½-volt cells are often used end to end (both pointing in the same direction) to provide the 3 volts used in many flashlights.

The voltage produced by an individual cell depends on the materials used and not on its size. The size determines the total amount of chemicals used and therefore the total amount of charge that can be trans-

A car battery can be recharged with a battery charger.

> **ANSWER**   Because the voltages of the cells add, it takes 6 cells.

(a)

Metal cap
Plastic cover
Seal
Zinc cup
Carbon rod
Electrolyte paste
Outer paper tube

(b)

Figure 20-2   (a) Flashlight batteries. (b) Schematic of a standard flashlight-type dry cell.

ferred. We have seen that the voltage can be increased by placing cells in a row, an arrangement called **series.** Cells (or batteries) can also be placed side by side, or **parallel,** as shown in Figure 20-3. This parallel arrangement does not increase the voltage but does increase the effective size of the battery. The larger amount of chemicals means that the batteries will last longer before they run down.

QUESTION   When might you wish to connect batteries in parallel?

Household electricity has a number of differences from the electricity produced by batteries. The electricity in homes is usually supplied at 120 volts. This electricity is not supplied by batteries but by huge electric generators in dams or power plants burning coal, oil, gas, or nuclear fuel. (We will discuss generators in Chapter 21). In household circuits the flow of charge (current) reverses direction 120 times a second. This is called *60-cycle alternating current* (ac) as opposed to the *direct current* (dc) supplied by batteries.

Because most of the uses of electricity we will discuss in this chapter depend only on the flow of charge and not on its direction, we will simplify our discussions by referring to flashlight batteries. Another advantage of using batteries is that they have low voltages, so you can experiment with some of the ideas mentioned here without fear of being shocked. Household electricity, on the other hand, is dangerous and should not be used in experiments.

ANSWER   This is especially useful when the batteries are difficult to change as in remote radio transmitters on mountaintops.

Figure 20-3   Two cells in parallel have the same voltage as a single cell, but last twice as long.

Physics Update

The development of lighter batteries would please anyone who drives an electric car or uses a laptop computer. Until now, the battery type with the best energy density (stored energy per unit mass) is one that uses a lithium cathode and a lithium compound anode. A new lithium battery developed at Tokyo University uses a lithium metal anode and a cathode consisting of a composite of two organic materials. At peak performance the organic-cathode battery had an energy density of more than 600 watt-hours per kilogram, compared to an equivalent density of about 400 for the most sophisticated commercially available lithium battery. The new battery retained 80 to 90% of its original energy storage capacity even after 100 recharging cycles.

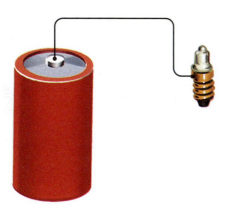

Figure 20-4 An unsuccessful attempt to light a bulb.

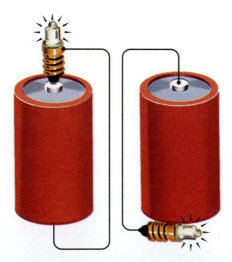

Figure 20-5 Successful ways of lighting a bulb.

Pathways

The invention of batteries provided experimenters with a way of producing a continuous flow of charge, a **current.** Although the properties of interacting electric charges are unchanged from the last chapter, new effects are observed when charges move through conducting pathways in a continuous fashion. The simplest way to learn about these is to experiment with a simple flashlight battery, some wire, and a few flashlight bulbs. The basic properties of household electricity are then simple extensions of these ideas.

Imagine that you have a battery, a wire, and a bulb. From your previous experiences can you predict an arrangement that will light the bulb? A common response is to connect the bulb to the battery as shown in Figure 20-4. The bulb doesn't light. It doesn't matter which end of the battery is used or which part of the bulb is touched. The charges don't flow from one end of the battery to the bulb and cause it to light. Suppose you hold the wire to one end of the battery and the metal tip of the bulb to the other end. Touching the free end of the wire to various parts of the bulb will eventually yield success. Two of the four possible arrangements for lighting the bulb are shown in Figure 20-5.

> **QUESTION**   Would the bulb still light if you turned the battery end to end in the arrangements shown in Figure 20-5?

Comparing the methods for lighting the bulb with the ways that don't light it indicates that we must use two parts of each object: the two ends of the wire, the two ends of the battery, and the two metal parts of the bulb. The two parts of the bulb are the metal tip and the metal around the base (this is threaded on many bulbs). Whenever all six ends are connected in pairs, forming a continuous loop (no matter how you do it), the bulb lights.

> **ANSWER**   Either end of the battery works because reversing the battery does not change the conducting path.

The continuous conducting path (known as a **complete circuit**) allows the electric charges to flow from one end of the battery to the other. If the bulb is to light, the conducting path must go through the bulb. You can verify this by examining a bulb with unfrosted glass, as illustrated in Figure 20-6. The part of the bulb that glows is a thin wire (called the *filament*) that is supported by two thicker wires coming out of the base. The bulb has to be taken apart to see that one of these support wires is connected to the metal tip and the other to the metal side of the bulb. The rest of the bulb is made from insulating materials so that there is a single conducting path through the bulb.

Figure 20-6   Schematic drawing showing the continuous conducting path through a bulb.

Physics on Your Own   Connect a flashlight bulb to a flashlight battery, leaving a gap in the circuit. Test a variety of materials to see if they are conductors by inserting them in the gap as shown in Figure 20-7.

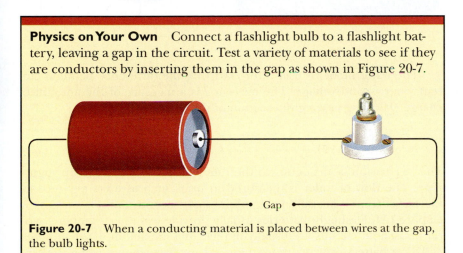

Gap

Figure 20-7   When a conducting material is placed between wires at the gap, the bulb lights.

Combining the concept of a complete circuit with the idea of the conservation of charge leads to the conclusion that electricity flows out one end of the battery and back into the other. The charge that leaves one end of the battery returns to the other end. Charge does not get lost or used up along the way.

> **QUESTION**   How does the information about pathways explain why the bulb in Figure 20-4 doesn't light?

A Water Model

The flow of charge in a complete circuit is analogous to the flow of water in a closed system of pipes. The battery is analogous to a pump, the wires to the pipes, and the bulb (something that transforms energy in the flow to another form) to a paddle wheel that turns when the water is flowing (Fig. 20-8). In a closed system of pipes the water that leaves the pump returns to the intake side of the pump.

> **ANSWER**   Because there is no pathway for the charges to flow back to the other end of the battery, there is no complete circuit.

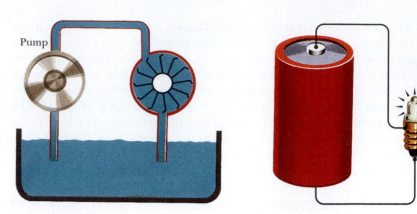

Figure 20-8   The flow of charge is analogous to the flow of water through a closed system.

The water model is useful for clarifying the difference between current, charge, and voltage. The current I is a measure of the amount of charge Q flowing past a given section of the circuit in 1 second,

$$ \text{current} = \frac{\text{charge}}{\text{time}} $$

$$ I = \frac{Q}{t} $$

This expression is analogous to the rate at which water flows through a pipe. The flow of water is measured in units such as liters per second. Electric current is measured in coulombs per second, a unit known as the **ampere** (A). Flashlight batteries usually provide less than 1 ampere, typical household circuits are usually limited to a maximum of 20 amperes, and a car battery provides more than 100 amperes when starting the car.

The voltage between two points in a circuit is a measure of the change in electric potential between these two points. That is, voltage is a measure of the work done in moving a charge through the circuit, which, in turn, is an indication of the force on the electric charge due to the electric field that exists in the wire. You can also think of voltage as being analogous to pressure in our water model.

Although the water model is useful, it is important to realize its limitations. A break in an electric circuit causes the electricity to stop flowing; it doesn't spill out of the end of the wire like water would from a broken pipe. An electric circuit is not analogous to a sprinkler system used to water lawns. A valve blocks the flow of water, whereas a switch puts a gap in the conducting path to stop the flow of electricity.

Another difference is that one or both types of charge may be free to move in a conductor. In liquids and gases, both charges can move. It wasn't until 1879 that experiments determined that only the negatively charged electrons are mobile in metals. For most macroscopic effects, positive charges moving in one direction are equivalent to negative charges moving in the opposite direction. For example, consider two neutral metal spheres. If you move 1 coulomb of positive charge from sphere A to sphere B, B has an excess of 1 coulomb of positive charge. Sphere A has a deficiency of 1 coulomb of positive charge or, what is the equivalent, an excess of 1 coulomb of negative charge. The net charges on the two spheres are identical to moving 1 coulomb of negative charge from B to A. Because of this equivalence, as well as the historic origins of the subject, when discussing the macroscopic effect of current, we adopt the convention of assuming that current is the flow of positive charges.

Resistance

If a bulb is left connected to a battery, the battery runs down in less than a day. (The actual time depends on the particular bulb and battery used.) The bulb has the same brightness for most of this time, but near the end it dims and goes out. However, if you connect just a wire across the ends of the identical battery, it runs down in less than an hour and the wire often gets too hot to touch. Because the battery is drained much faster and something is obviously happening in the wire, we infer that the current is larger through the single wire than through the pathway with the bulb. The bulb offers more **resistance** to the flow of electricity.

This notion of resistance also makes sense according to the water model. More water flows through wide pipes each minute than through narrow pipes. The bulb's filament is a very thin section of wire and should offer more resistance. Our analogy also tells us that long pipes offer more resistance than short pipes. Therefore, we would expect the resistance of wires to increase with length and decrease with diameter. The other factor—not obvious from our model—is that the resistance depends on the type of material used for the wires. These ideas are verified by experiments with wires of different sizes and materials. Table 20-1 compares the resistances of identical wires made of different metals.

Resistance is a result of the interaction of the pathway with the flow of charge. The electrons in a wire feel a net force due to the repulsion of the negative terminal and the attraction of the positive terminal and are accelerated. However, the electrons don't go very far before they bump into atoms, which causes them to lose speed. This impedance to the flow of charge determines the resistance of the wire.

As the electrons move through a wire, they undergo collisions which transfer kinetic energy to the atoms and cause the wire to heat up. If there is enough energy transferred, the wire heats up so much that it glows. Heating coils and elements in stoves, ovens, toasters, baseboard heaters, and lights do this. Resistors are used to control the voltages and currents in circuits used in such devices as radios and curling irons.

Resistance R is defined to be the voltage V across an object divided by the current I through the object

$$R = \frac{V}{I}$$

Resistance is measured in volts per ampere, a unit known as the **ohm** (Ω). This definition is always valid but is most useful when the resistance is constant, or relatively constant. In this case this relationship is known as **Ohm's law.** The resistances of pieces of metal, carbon, and some other substances are approximately constant if they are maintained at a constant temperature. The resistance of the filament in a lightbulb increases as the filament heats up.

Table 20-1

The resistance of 100-meter-long wires with a diameter of 1 millimeter made of different metals

Metal	Resistance (Ohms)
Silver	2.02
Copper	2.16
Gold	3.11
Aluminum	3.59
Tungsten	7.13
Iron	12.7
Nichrome	191

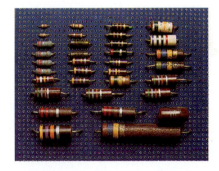

Resistors are used in electronic circuits.

$$\text{resistance} = \frac{\text{voltage}}{\text{current}}$$

COMPUTING OHM'S LAW

Σ

Ohm's law provides us with a relationship between the resistance and the potential difference, or voltage, across the circuit element. For example, suppose we had a 12-volt battery and wanted to produce a current of 1 ampere in a particular circuit. What resistance would the circuit need to have?

$$R = \frac{V}{I} = \frac{12\,V}{1\,A} = 12\,\Omega$$

QUESTION   If a light bulb draws a current of 1/2 ampere when connected to a 120-volt circuit, what is the resistance of its filament?

Ohm's law can be rearranged to find any of the three quantities when the other two are given. Suppose a heating element has a resistance of 9.6 ohms when hot. What current will it draw when connected to 120 volts?

$$I = \frac{V}{R} = \frac{120\,V}{9.6\,\Omega} = 12.5\,A$$

QUESTION   If a 3-volt flashlight bulb has a resistance of 9 ohms, how much current will it draw?

The Danger of Electricity

Electricity is dangerous because our bodies are electric machines. Our muscles constrict when neurons "fire." This reaction is normally triggered by a complex chain of chemical reactions, but can also occur when an electric current is applied across the muscle. This process is the origin of twitches! This situation can become dangerous when the muscle is part of the heart or breathing system. In these cases, the current can also affect the pacing signals, interrupting the normal processes. Currents as small as 50 milliamperes can interrupt breathing in some individuals.

Often people feel that the danger is the voltage. Although high voltages are dangerous, it is the current through a body that is lethal. Ohm's law tells us that the current is equal to the voltage divided by the resis-

ANSWER $R = V/I = (120\,V)(\frac{1}{2}\,A) = 240\,\Omega$.

ANSWER $I = V/R = (3\,V)/(9\,\Omega) = \frac{1}{3}\,A$.

tance of the path. In this case, the body is part of the path. If the conditions are such that the total resistance is low, even a low voltage can cause a dangerously high current.

Batteries and Bulbs

Adding more paths in a circuit increases our understanding of how charges flow through conductors. If you connect a single bulb to a battery and then run an additional wire directly from one end of the battery to the other as shown in Figure 20-9, the bulb goes out, the wire gets very hot, and the battery runs down quickly, which tells us that most of the charge flows through the wire. Actually, a sensitive detector shows that some electricity flows through the path containing the bulb but it is not enough to make it light. In general, the current is larger through the path with the smaller resistance. A path that offers very little resistance is known as a **short circuit.**

We can explore more circuit possibilities by using more bulbs. The brightness of a bulb gives a rough measure of the current through a bulb. Bulbs don't glow at all until the current exceeds a certain value; after that the brightness increases with increased current. Although the relationship is complicated, a simple relationship can be assumed between brightness and current for questions we discuss in this section.

We begin by creating a "standard" to which we can refer; the brightness of a single bulb connected to a single battery will represent a standard current. We will also assume that all batteries and bulbs are identical.

Two bulbs can be connected to a battery so that there is a single path from the battery through one bulb, through the other bulb, and back to the other end of the battery. This arrangement (Fig. 20-10) is known as a **series circuit.** The bulbs are equally bright but less so than our standard. In fact, the total amount of light given off by both bulbs is less than that given off by the standard bulb, and the battery's lifetime is longer than that of the battery in the standard circuit, allowing us to infer that there is a smaller current in the circuit. Therefore, the resistance of two bulbs in series is greater than that of a single bulb.

Of special note is the fact that the two bulbs have the same brightness, telling us that the current through each bulb is the same. Electric charge (or current) is not being used up along the pathway in agreement with the conservation of charge. The same charge flows back into the battery as left it. It is energy that is continually being used (actually transformed) along the pathway. The charges continually gain energy due to the potential difference provided by the battery, and this energy is converted to thermal energy (and light) in the circuit. The wires offer very little resistance compared with the filaments in the bulbs. Most of the electric energy is transformed in the bulbs; the electrons leave half of their energy in each bulb. Again, this makes sense since the two bulbs have the same brightness.

Another consequence of the equal brightnesses is that bulbs cannot be used to determine a direction for the flow of electricity. Everything can

Figure 20-9 When the wire on the left is connected from one end of the battery to the other producing a short circuit, the light goes out.

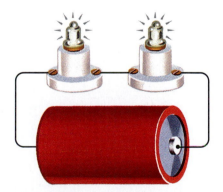

Figure 20-10 Two bulbs wired in series are equally bright, but dimmer than the standard.

Superconductivity

Under normal conditions, all conductors exhibit resistance to the passage of electric current. This makes sense if you visualize the conduction electrons bumping their way through a metal consisting of an array of vibrating atoms. The electrons are accelerated along the wire by the electric potential difference provided by a battery, and then lose their acquired speeds in collisions. The surprise is that the electrical resistance of some materials goes to zero at very low temperatures; they become *superconductors*.

This knowledge is not a new breakthrough: Superconductivity was first discovered by the Dutch physicist Heike Kamerlingh Onnes in 1911. Three years earlier Onnes had developed a process for liquefying helium and, therefore, was able to study the properties of materials at very low temperatures. The first superconductor was solid mercury, which lost its electrical resistance at a critical temperature of 4.15 K. Experimenters searched for metals or alloys that exhibited superconductivity, and new superconductors were found with critical temperatures as high as 23 K.

In 1972 three American physicists, John Bardeen, Leon Cooper, and J. Robert Schrieffer, received the Nobel Prize for explaining superconductivity in 1957. Their theory—called the BCS theory—showed how electrons pair up to create a resonance effect that allows the electron pairs to travel effortlessly through the material. The electrical resistance does not simply drop to a very low value; it drops to zero! Once a current has been established in a superconducting material, it will

A permanent magnet levitates above a superconductor because its magnetic field cannot penetrate the superconducting material.

persist without any applied voltage. Such supercurrents have been observed to last for years, which can have great practical applications in the generation of magnetic fields. Large magnetic fields require very large currents. In ordinary materials, this means the generation of a lot of thermal energy at high cost, and diverting the heat to avoid melting the magnets. In contrast,

be explained equally well assuming a flow of negative charges in one direction (the actual situation in wires), a flow of positive charges in the other direction, or both charges flowing simultaneously in opposite directions (as happens in fluids). In fact, because it is only the motion of charges that matters, one charge moving in both directions also works. In household electricity the negative charges move back and forth with a frequency of 60 hertz.

Two bulbs can also be connected so that each bulb has its own path from one end of the battery to the other, as shown in Figure 20-11. This arrangement is known as a **parallel circuit.** In contrast to the series circuit, the current in one bulb does not pass through the other, which can be seen by disconnecting either bulb and observing that the other is not affected. (Strictly speaking, the addition of the second bulb lowered the brightness of the first bulb slightly. This results from a lowering of the voltage of the battery as it supplies more current. If this were not the case, the bulb would continue to glow when the battery is short-circuited. In the spirit of Galileo, we will ignore this effect and concentrate on the main features.)

once the large currents have been established in superconducting magnets, they can be disconnected from the power supply; no further electrical energy is needed.

Theoretical models for superconductivity predicted an upper limit for the critical temperature of about 30 K. Had this upper limit been true, superconductivity would have remained in the domain of very low temperatures. The only gases with boiling points this low are helium (which is expensive) and hydrogen (which is explosive). In 1986, nearly 80 years after the discovery of the first superconductor, a major new class of superconductors was discovered with much higher critical temperatures. These new superconductors are *ceramics* made from copper oxides mixed with such rare earth elements as lanthanum and yttrium, and have critical temperatures as high as 125 K. This new limit is very significant because this temperature is higher than the boiling point of nitrogen, an abundant gas that is relatively inexpensive to liquefy and very safe to use, allowing liquid nitrogen to be used to keep a material superconducting. A tremendous increase in the use of superconductors will occur if a superconducting material can be made with a critical temperature above room temperature (or at least above those obtained by ordinary refrigeration).

One very important potential application of superconductivity is in the transportation of electrical energy. With conventional transmission lines, a lot of the electricity produced at a distant power plant is lost to resistive heating effects in the wires that carry this energy to our homes. A superconducting transmission line would eliminate these losses. The problem, of course, is that we need to cool these conduits in order to make them superconducting. However, if we can use ordinary refrigeration, the cooling costs will be much less than the costs of resistive losses.

Superconductors have a second property that is very important. They expel magnetic fields when they become superconducting. This also means that magnetic fields cannot penetrate the superconductors and they will, therefore, repel magnets. This is dramatically shown by floating a permanent magnet above a superconductor as shown in the figure. This effect may find applications in improving the magnetic levitation of trains for high-speed transportation. A prototype train has been constructed in Japan using helium-cooled superconducting magnets to levitate the train, as well as to propel it at speeds up to 300 mph.

Many other applications have been proposed. The switching properties of superconducting materials could have a large impact on the field of computer electronics. It might also be possible to construct superconducting generators and motors. These new superconducting materials might also find applications in the field of medical imaging. However, many of these applications will require major technological breakthroughs in fabrication (for instance, ceramics are brittle and, therefore, difficult to form into wires) and in finding materials that will carry larger currents.

Figure 20-11 Two bulbs wired in parallel are equally bright and have the same brightness as the standard.

Figure 20-12   Bulbs B and C are in parallel and the pair is in series with Bulb A.

A circuit breaker protects household circuits from drawing too much current and starting a fire.

Figure 20-13   Appliances are connected to household circuits in parallel. The circuit is protected from drawing too much current by a fuse or circuit breaker wired in series.

Some modern Christmas tree lights are wired in parallel. If one bulb burns out, it is the only bulb that goes out. The older style was wired in series; when one bulb went out, they all went out. That was rather inconvenient, because when one bulb burned out, all of them had to be tested until the defective one was found. The latest style of Christmas tree lights has the bulbs wired in series, but when a bulb burns out, a short circuit is formed across the burned-out bulb that allows the rest of the bulbs to stay lit.

The two bulbs in parallel are equally bright and each is as bright as our standard. Because each bulb has its own path, twice as much current is being supplied by the battery. Also, since each bulb is connected to both ends of the battery, the voltage across each one is equal to that across the battery and the total resistance of the combination must be one-half that of either bulb. This can be verified experimentally; in this arrangement the battery runs down in one-half its normal lifetime. The total resistance for parallel arrangements is always less than the resistance of any single pathway. Adding another pathway to any circuit increases the flow of charge from the battery and thus lowers the total effective resistance.

Three or more bulbs can be connected in series or parallel or in combinations of series and parallel. The relative brightnesses of these bulbs can be predicted using the ideas that we have discussed. As an example, consider the combination of bulbs in Figure 20-12. Bulb A will be the brightest because all of the current must pass through it but it will not be as bright as our standard. Bulbs B and C will be dimmer because the current splits—part going through bulb B and part through bulb C. The current flowing into the junction J must be equal to that flowing out of the junction. This is just a consequence of the conservation of charge. Because the bulbs are identical, the resistances of the two paths are equal and the current will split equally; bulbs B and C will be equally bright. In general, more current exists in the path with the lesser resistance.

The circuits in houses are wired in parallel so that electric devices can be turned on and off without affecting one another. As each new appliance is turned on, the electric company supplies more current. Each parallel circuit, however, is wired in series with a fuse, as shown in Figure 20-13, to deliberately put a "weak link" in the circuit. If too many devices

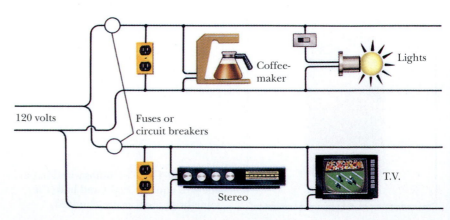

are plugged into one circuit, they will draw more current than the wires can safely carry. For example, the wires might heat up at a weak spot and start a fire. The fuse contains a short piece of wire that melts when the current exceeds the rated value, which interrupts the circuit, shutting everything off. Modern houses have "circuit breakers" rather than fuses. They serve the same purpose but only require resetting when tripped rather than replacing a burned-out fuse.

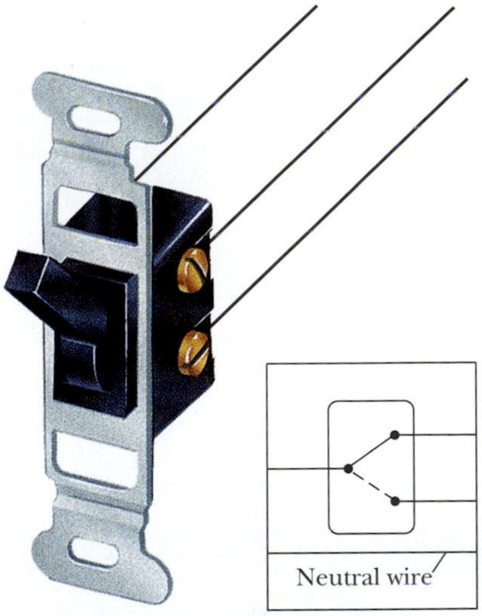

> **Physics on Your Own**   Obtain two "3-way" switches used in wiring hallway lights. (The switches act like the one in Fig. 20-14.) Wire them to a flashlight battery and bulb so that either switch can turn the light on or off independently of the other. Be sure that there are no short circuits when the light is off.

Over the years, a convention for drawing circuit elements has been developed. Like all symbols, the electric symbols capture the functional essence of the element and omit nonessential characteristics. A light-bulb, for example, has no directional characteristic, but a battery does. Their symbols reflect this difference. Figure 20-15 gives the common symbols paired with a diagram like those we have been using.

Circuits are also drawn with sharper corners than exist in the actual circuits. This is simply a technique that has helped communication among experimenters.

Figure 20-14   Schematic of a three-way switch. The wire on the left can be connected to either wire on the right by throwing the switch.

Electric Power

When you buy electricity from the power company, it is energy that you buy. This electric energy is converted to heat, light, or motion. Why is it, then, that we talk about power? As we discussed in Chapter 6, **power** is the amount of energy used per unit time. Formally, power is the energy transformed divided by the elapsed time. Most electric devices in our houses are rated by their power usage. Power is measured in **watts** (W). Many household light bulbs are rated 60, 75, or 100 watts. Electric heaters and hair dryers might use 1500 watts.

The electric meter connected between your house and the power company records the energy you use much like the odometer in your car records the miles you drive. Electric energy is usually billed at a few cents per kilowatt-hour. A typical house with an electric range and an electric clothes dryer (but no electric heat) uses about 900 kilowatt-hours per month.

$$\text{power} = \frac{\text{energy}}{\text{time}}$$

> **Physics on Your Own**   Monitor the use of electric energy in your home by taking periodic readings of the electric meter. How much electric energy does your household use in a typical day?
>
> You can also try to turn off all electric devices in your home. You can monitor your progress by looking at how fast the wheel in your electric meter turns.

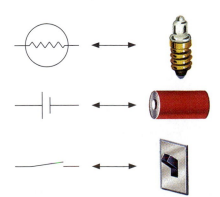

Figure 20-15   Common symbols used in diagrams of electrical circuits. (*Top to bottom:* resistor, battery, switch)

COMPUTING ELECTRIC POWER   Σ

We can obtain an expression for electric power by using the definition for electric energy we developed in Chapter 19 and the definition for current. Since energy is the charge multiplied by the voltage, we have

$$P = \frac{E}{t} = \frac{QV}{t} = \frac{Q}{t}V$$

This equation tells us that the power is equal to the current times the voltage.

$$P = IV$$

Therefore, an electric appliance that draws a current of 10 amperes at a voltage of 120 volts uses energy at a rate of

$$P = IV = (10 \text{ A})(120 \text{ V}) = 1200 \text{ W}$$

QUESTION   What power is required to operate a clock radio if it draws 0.05 ampere from the household circuit?

By rearranging the expression for power, you can determine the current used by a 100-watt bulb.

$$I = \frac{P}{V} = \frac{100 \text{ W}}{120 \text{ V}} = 0.83 \text{ A}$$

We can obtain an alternative expression for electric power by using the relationship that the voltage is equal to the current multiplied by the resistance.

$$P = IV = I(IR)$$
$$P = I^2R$$

In this form it is more obvious why a bulb in a circuit glows yet the connecting wires do not. The current through the circuit is the same everywhere but the section that has the highest resistance receives the most thermal energy per unit time. Because the filament is made of a very fine wire of high-resistance material, it has the highest resistance; its temperature goes up, and it glows. The connecting wires also heat up, but they just don't get hot enough to glow.

There are times when the energy losses in connecting wires become important. Sending electric energy long distances through wires from a power plant could result in significant energy losses. Two things can be

ANSWER $P = IV = (0.05 \text{ A})(120 \text{ V}) = 6 \text{ W}$.

power = current × voltage

This electric meter measures the energy used in a house or business. This meter reads 65,401 kilowatt-hours.

power = current squared × resistance

The Real Cost of Electricity

The real cost of electricity varies a great deal, depending on the source. We are most familiar with purchasing electric energy from the local power company for use in our homes. But what is the cost of this electric energy? Examining your electric bill shows that you are charged something between 5¢ and 20¢ per kilowatt-hour. Let's use a representative value of 10¢ per kilowatt-hour. Knowing that 1 kilowatt-hour is equal to 3.6 million joules, we can divide these two numbers to learn that we can use about 36 million joules for each dollar we spend.

We can also buy electric energy packaged in a variety of batteries. In this case, we pay for properties of the batteries that appeal to the manufacturer of the electric appliances. This might include the batteries' size, voltage, lifetime, or their ability to maintain current as they wear out. To discover what we are paying for the electric energy, we need to know the prices and the total electric energy that the batteries provide during their lifetimes. As shown in the text, the energy can be obtained by multiplying the total charge by the voltage rating of the battery. The total charge is obtained by multiplying the current rating by the battery's lifetime.

A standard flashlight battery—usually a carbon—zinc D cell—costs about 75¢ and has a rating of 1.5 volts. Its lifetime depends on the way it is used. Flashlight D cells are normally used for short intervals over a much longer time. Under these conditions the battery could supply 375 milliamperes for about 400 minutes, yielding a total charge of 9000 coulombs and a total energy of 13,500 joules. Therefore, you are getting 18,000 joules per dollar, 2000 times more expensive than household electricity. Alkaline batteries were a major improvement in performance over their carbon–zinc cousins, primarily in applications requiring fairly large, continuous currents, such as portable stereo radios. Alkaline batteries typically last about 50 to 75 percent longer than a standard battery, but under some condi-

tions can last much longer. However, their costs are also higher. Are alkaline batteries a good buy? Only if their increase in cost is less than their corresponding increase in lifetime.

The development of mercury, silver, and lithium batteries allowed manufacturers to pack more energy in smaller volumes. These devices have a much more stable current over their lifetimes, so they are especially useful in devices such as wristwatches and photocells in cameras, which require a particular value for the current. A watch battery delivers 1.3 volts and 10 millionths of an ampere for a year, giving a total energy of 410 joules. If these batteries sell for about $5, they have an energy–cost rating of 82 joules per dollar, or half a million times the cost of household electricity!

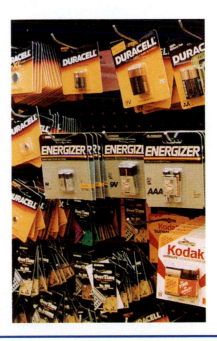

done to minimize these losses. First, the wires should have as little resistance as possible, which means using large-diameter wires and low-resistance materials. Second, transformers (discussed in Chapter 21) allow the utility company to send the same energy through the wire by raising the voltage and simultaneously lowering the current. Lower current means less thermal loss in the transmission lines. Economic and engineering considerations dictate the extent to which each of these methods is used to minimize energy loss. Electricity is carried between power companies and towns at very high voltages (up to 765,000 volts). The voltage is lowered by transformers for distribution around towns and then lowered once more to the safer 120 volts used in homes.

COMPUTING COST OF ELECTRIC ENERGY   Σ

A device running at constant power uses an amount of energy equal to the power multiplied by the time; the longer we use it, the more energy it consumes. For example, a 60-watt bulb burning for 10 hours uses

energy = power × time

$$E = Pt = (60 \text{ W})(10 \text{ h}) = 600 \text{ Wh} = 0.6 \text{ kWh}$$

of electric energy. It is important to recognize that a kilowatt-hour (kWh) is a unit of energy, not one of power. Because 1 kilowatt is equal to 1000 joules per second and 1 hour is the same as 3600 seconds, 1 kilowatt-hour is 3.6 million joules.

We can now calculate the monthly cost of electric energy for this typical household using 900 kilowatt-hours per month. Let's assume that electric energy costs 9¢ per kilowatt-hour.

$$\text{cost} = (900 \text{ kWh})(\$0.09/\text{kWh}) = \$81$$

QUESTION   At this price what would it cost to run a 1500-watt heater continuously during an 8-hour night?

SUMMARY

Volta used the electric potential difference between two different metals to build the first battery. Batteries make charges flow continuously as a current, producing heat and light and running motors and other appliances.

The voltage produced by an individual cell depends on the materials used but not on its size. The size determines the amount of chemicals and therefore the total amount of charge that can be transferred. The voltages of cells in series add. Cells placed in parallel do not increase the voltage, but increase the effective size of the battery.

ANSWER $E = Pt = (1500 \text{ W})(8 \text{ h}) = 12,000 \text{ Wh} = 12 \text{ kWh}$.
Cost = (12 kWh)(9¢/kWh) = $1.08.

The electricity in homes is usually supplied at 120 volts and 60-hertz alternating current. The electricity from a battery is direct current and usually at a much lower voltage.

Electric charges can only flow continuously when there is a complete pathway, or circuit, made of conducting materials. The current is determined by the voltage and the total resistance of the circuit, $V = IR$. The voltage between two points in a circuit is a measure of the change in electric potential between these two points, which is the work done in moving a unit of positive charge through the circuit. The most charge flows through the path with the least resistance. Conservation of charge requires that electricity flow out one end of the battery and back into the other. Charge does not get lost or used up along the way.

One or both types of charge may be free to move in a conductor. In liquids and gases, both charges can move, whereas only the negatively charged electrons are mobile in metals. For most macroscopic effects, positive charges moving in one direction are equivalent to negative charges moving in the opposite direction.

The resistance of wires increases with increasing length, decreases with increasing cross-sectional area, and depends on the type of material. Resistance R is the ratio of the voltage V across an object to the current I through the object, $R = V/I$. Resistance is measured in volts per ampere, or ohms (Ω).

A power company sells energy, not power. Power is the rate of using energy, and is used to rate most electrical devices in our houses. Power is measured in watts (W) or kilowatts (kW), whereas household electric energy usage is measured in kilowatt-hours (kWh).

CHAPTER 20 REVISITED

When electrons move through a conductor, they bump into the wire's atoms, raising the average kinetic energy of the atoms and, therefore, the temperature of the wire. The rise in temperature depends on the thermal properties of the material and the electrical resistance of the wire. The filaments in lightbulbs are made of very thin wires of high-resistance materials to enhance the heating. If the rise in temperature is great enough, the wire glows.

KEY TERMS

ampere: The SI unit of electric current, 1 coulomb per second.

complete circuit: A continuous conducting path from one end of a battery (or other source of electric potential) to the other end of the battery.

current: A flow of electric charge. Measured in amperes.

ohm: The SI unit of electric resistance. A current of 1 ampere will flow through a resistance of 1 ohm under 1 volt of potential difference.

Ohm's law: The resistance of an object is equal to the voltage across it divided by the current through it.

parallel circuit: An arrangement of resistances (or batteries) on side-by-side pathways between two points.

power: The rate at which energy is converted from one form to another. In electric circuits the power is equal to the current times the voltage. Measured in joules per second, or watts.

resistance: The impedance to the flow of electric current. The resistance is equal to the voltage across the object divided by the current through it. Measured in volts per ampere, or ohms.

series circuit: An arrangement of resistances (or batteries) on a single pathway so that the current flows through each element.

short circuit: A path in an electric circuit that has very little resistance.

watt: The SI unit of power, 1 joule per second.

CONCEPTUAL QUESTIONS

1. Galvani and Volta performed experiments with a frog's leg. Under what conditions did the frog's leg twitch?

2. Predict what you would feel if you placed pieces of silver and zinc on opposite sides of your tongue.

3. What effect does the size of a cell have?

4. When you stack three flashlight batteries in the same direction, you get a voltage of $3 \times 1\frac{1}{2}$ volts = $4\frac{1}{2}$ volts. What voltage do you get if one of the batteries is turned end for end?

5. What happens to the total voltage across two cells fastened in parallel?

6. What is the voltage of two cells connected in series?

7. What are the differences in the electricity provided by a car battery and a flashlight battery?

8. What are the differences in the electricity provided by a battery and by your local power company?

9. Which bulb(s) in the figure will be the brightest?

10. Which bulb(s) in the figure will not light?

11. Which arrangement(s) of batteries in the figure will light the bulb for the longest time?

12. Which arrangement(s) of batteries in the figure will burn out most quickly?

13. There are four different ways of connecting a battery, a bulb, and a wire so that the bulb lights. Two of these are shown in Figure 20-5. What are the other two?

14. Write a short statement describing how you would connect a battery, a bulb, and a single wire so that the bulb lights. Your statement should include all four ways without favoring any one way.

15. What happens to a circuit when you close the switch?

16. What happens when a lightbulb burns out?

17. You are given a box with two bare wires sticking out of it. Describe how you would use a battery and a bulb to determine if the wires are connected to each other inside the box.

18. Suppose you ran speaker wires from the stereo in the family room to your bedroom, but you forgot to mark them. What could you do to tell which wire you should connect to which speaker?

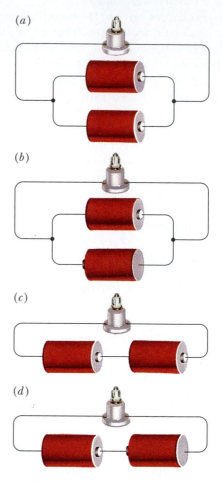

Questions 9–12.   Identical bulbs are connected to various combinations of identical batteries.

*19. You are given a mystery box with six bolt heads on top as shown in the figure. You connect a battery and a bulb across each pair of bolt heads and find that the bulb lights for the following pairs: AC, AD, CD, and EF. Draw diagrams showing all of the ways the box could be wired.

Questions 19–20.

*20. A friend examines a box similar to the one in the last question and finds that the bulb only lights for the following combinations: AC, CE, and DF. Are your friend's observations correct? Explain.

21. What is the difference between a volt and an ampere?

22. Car batteries are often rated in ampere-hours. What does this mean?

23. In the water model for electricity what are the analogs of charge, current, and voltage?

24. In the water model for electricity what are the analogs of a battery, a switch, a wire, and a lightbulb?

25. Which of the following affects the resistance of a wire: diameter, type of metal, length, or temperature?

26. Some power tools run poorly when connected to very long extension cords. Why? What could you do to improve the situation?

27. What happens to the current in a circuit when the voltage is doubled?

28. If the resistance connected to a battery is doubled, what happens to the current?

29. How does the atomic theory of matter that we studied in Chapter 7 account for the observation that the resistance of metallic wires increases with temperature?

30. If the electrons in a wire move slowly due to the wire's resistance, why do lights turn on "instantaneously" when you throw the switches?

31. Why is it dangerous to use a blow dryer while taking a bath?

32. Why is it possible for a bird to perch on a high-voltage transmission line without being electrocuted?

33. What is a short circuit?

34. Why does a short circuit cause a battery to run down quickly?

*35. If you connect a flashlight battery and bulb as shown in Figure 20-7 using water as part of the pathway, the bulb does not light. How do you reconcile this observation with the statement in the text that water is a conductor? (*Caution:* People are electrocuted each year when electric appliances fall into bathtubs.)

36. Why would seawater be a better conductor than freshwater?

37. In England, the voltage supplied to homes is 240 volts. What would happen if you used an English lightbulb in the United States?

38. If the only voltage you have is 120 volts, how could you light some 6-volt lightbulbs without burning them out?

39. How would you connect two batteries and two bulbs to get the most light?

40. How would you wire two batteries and two bulbs to produce light for the longest time?

41. The following arrangements of identical bulbs are connected to one battery: (a) one bulb, (b) two bulbs in series, (c) three bulbs in series, (d) two bulbs in parallel, and (e) three bulbs in parallel. In which arrangement(s) are the bulbs the brightest?

42. In which of the arrangement(s) in Question 41 will the battery last the longest time?

43. Three identical bulbs are connected in series to a battery. Which bulb is the brightest?

44. Three identical bulbs are wired in series. What happens to the brightness of the other two when one of the bulbs burns out?

45. Which bulbs are the brightest and dimmest in the figure?

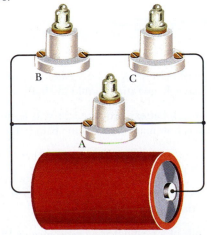

Questions 45 and 47–48.

46. Which bulbs are the brightest and dimmest in the figure?

47. What happens to the brightness of bulbs A and B in the figure when bulb C is removed from its socket?

48. What happens to the brightness of bulbs B and C in the figure when bulb A burns out?

49. What happens to the brightness of bulbs B, C, and D in the figure when bulb A burns out?

50. What happens to the brightness of bulbs A, B, and C in the figure when bulb D is removed from its socket?

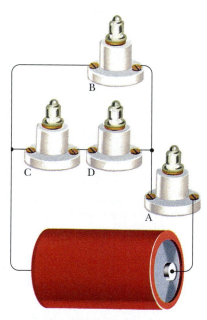

Questions 46 and 49–50.

*51. A box has three identical bulbs mounted on its top with the wires hidden inside the box. Initially bulb A is the brightest and bulbs B and C are equally bright. If you unscrew A, bulbs B and C go out. If you unscrew B, A gets dimmer and C gets brighter so that A and C are equally bright. If you unscrew C, A gets dimmer and B gets brighter so that A and B are equally bright. If you unscrew B and C, A goes out. How are the bulbs wired?

*52. A box has three identical bulbs mounted on its top with the wires hidden inside the box. Initially bulb A is the brightest and bulbs B and C are equally bright. If you unscrew A, bulbs B and C remain the same. If you unscrew B, A remains the same and C goes out. If you unscrew C, A remains the same and B goes out. If you unscrew B and C, A remains the same. How are the bulbs wired?

53. Are the headlights in cars and trucks wired in series or parallel?

54. Why should you not replace a 5-ampere fuse in your car with a 10-ampere fuse?

55. In what unit is electric power measured?

56. In what unit is a household's consumption of electricity measured?

57. What happens to the power if the resistance connected to a battery is cut in half?

58. What happens to the power if a resistor is connected to two batteries in series rather than a single battery?

59. Two bulbs have ratings of 60 watts and 120 watts. Which bulb carries the higher current?

*60. Two bulbs have ratings of 60 watts and 120 watts. Which bulb has the higher resistance?

61. Why do power companies use high voltages for their long-distance transmission lines?

EXERCISES

1. A long flashlight has four 1½-V batteries placed end to end. What voltage rating should the bulb have?

2. An old truck battery produced 24 V. If each cell produced 2 V, how many cells did the battery have?

3. What is the resistance of a light bulb that draws 0.8 A when it is plugged into a 120-V outlet?

4. If an iron draws 8 A on its high setting, what is its resistance at this setting?

5. If the coils of a heater have a resistance of 12 Ω when hot, what current does the heater require?

6. If a curling iron has a resistance of 750 Ω when hot, what current does it require?

7. If a lightbulb has a resistance of 6 Ω and a current of 0.6 A, at what voltage is it operating?

8. A lightbulb has a resistance of 200 Ω. What voltage is required for the bulb to draw a current of 0.6 A?

9. A 6-Ω resistor is connected in parallel with a 12-Ω resistor and the combination is connected to a 12-V battery. How much current does the battery supply?

10. Two 1½-V batteries are connected in parallel to a 3-Ω resistor. How much current does each battery supply?

11. A heating element is rated at 1500 W. How much current does it draw when it is connected to a 120-V line?

12. A VCR has a maximum power rating of 24 W. What is the maximum current that it requires?

13. What is the power used by a toaster that draws a current of 8 A?

14. What is the power rating of a heating coil with a resistance of 12 Ω that draws a current of 20 A?

15. A coffeemaker has a resistance of 12 Ω and draws a current of 10 A. What power does it use?

16. If a clock draws a maximum current of 4 mA, what is its maximum power consumption?

*17. What is the resistance of a 60-W light bulb?

*18. What is the resistance of the coil in a 1200-W heater?

19. If a hair dryer is rated at 1200 W, how much energy is used to operate the hair dryer for 6 min?

20. If a 60-W bulb is left on continuously in a secluded hallway, how much energy is used each month?

21. If electricity costs 15¢/kWh, how much does it cost to burn a 100-W bulb for one day?

22. A 1500-W heater for a sauna requires 40 min to heat the sauna to 190°F. What does this cost if electricity sells for 12¢/kWh?

During the 19th century scientists discovered that electricity and magnetism are not separate phenomena, but two different aspects of something called electromagnetism. It is electromagnetism that enables us to pop corn in our microwave ovens and brings television to us while we enjoy our popcorn. What is this connection between electricity and magnetism? (See p. 520 for the answer to this question.)

Electromagnetism

Satellite dishes receive TV signals from synchronous satellites.

Most of us have played with magnets. Through play we learned that these little pieces of material attract and repel each other and that they attract some objects but have no effect on others. For instance, magnets stick to refrigerator doors but not to the walls of a room, to paper clips but not to paper, and so on.

Although magnetic properties are interesting, they, like electricity, appear to be rather isolated phenomena. Unlike electricity, however, magnetism does seem more permanent. And although there are similarities between these two phenomena, they seem to be separate properties of matter: A charged rod has no effect on a nearby magnet and a magnet that has no net electric charge does not deflect the foils of an electroscope. Magnetism appears to be a different phenomenon from electricity. Yet some experimenters in the 17th century were more fascinated by the similarities than the differences between magnetism and electricity, so they searched for a connection.

A collection of permanent magnets.

Magnets

It has been known since at least the 6th century B.C. that lodestone, a naturally occurring mineral, attracts iron. It was also known that iron could be made magnetic by rubbing it with lodestone. The magnetized piece of iron then attracts other pieces of iron. The only elements that occur naturally in the magnetized state are iron, nickel, and cobalt. Many permanent magnets are made from alloys of these metals.

> **Physics on Your Own**   Examine a variety of materials and compile a list of the kinds that are attracted by magnets.

Imagine having three unmarked, magnetized rods. If we arbitrarily mark one end of one of the rods with an *X*, we find that it attracts one end of each of the other two magnets (Fig. 21-1), and it repels their other ends. Let's mark these ends *A* (for attract) and *R* (for repel), respectively. We then find that the two *A* ends repel each other, the two *R* ends repel each other, and the *A* and *R* ends attract each other. Because the two *A* ends were determined in the same way using the first magnet, they should be *alike*. Therefore, the ends of the magnets (called **magnetic poles**) behave like the electric charges we studied in Chapter 19. We can summarize the behavior of the magnetic poles with the simple statement that

Figure 21-1   Magnetic poles can be identified by their attraction and repulsion by known poles.

Like poles repel; unlike poles attract.

magnets

Both electricity and magnetism seem to have two kinds of "charge." In electricity, however, it is possible to separate the two charges, or at least to put more of one kind of charge on one object than on the other. But in magnetism the two poles always come in pairs that have the same strength.

Suppose you try to separate the poles by breaking a magnet in half. You get two magnets and each of these new magnets has two poles (Fig. 21-2). If you break each of these in half, you get four magnets, and so on. Even if you continue dividing down to the atomic level, you always obtain magnets with two poles. It does not seem to be possible to isolate a single magnetic pole, a **magnetic monopole.** Many extensive searches for magnetic monopoles have been conducted. Although the existence of a magnetic monopole is not ruled out by the present, well-established theories, all searches have ended in failure. The discovery of a magnetic monopole would most likely result in a Nobel Prize.

In naming the ends, or poles, of the magnets, we make use of the observation that a freely swinging magnet aligns itself along the north–south direction. The end that points north is called *north* and the other *south.* The compasses used for navigation are simply tiny magnets that are free to rotate. Magnets have been used as compasses since the 11th century, but it wasn't until 1600 that William Gilbert, an English physician, hypothesized that they work because the Earth itself is a giant lodestone. Gilbert even made a spherically shaped piece of lodestone to show that a compass placed near it behaved as it did on the Earth.

> **QUESTION**   How might you search for a third kind of magnetic pole?

> **Physics on Your Own**   You can make a simple compass by magnetizing a sewing needle and placing it on a cork floating in a glass of water. The needle can be magnetized by stroking it with a magnet.

Magnets are surrounded by **magnetic fields** like electric charges are surrounded by electric fields. These magnetic fields can be detected using a small compass. The direction of the field was arbitrarily chosen to be in the direction that the north pole of the compass points, and its size is a measure of the torque that aligns the compass. The photograph in Figure 21-3 was obtained by sprinkling iron filings on a piece of glass placed over a bar magnet. The iron filings line up along the direction of the magnetic field.

> **QUESTION**   How would the photograph in Figure 21-3 change if the poles of the magnet were reversed?

> **ANSWER**   You could search for a magnetized material that attracts or repels both north poles *and* south poles.

> **ANSWER**   It would not change, because the iron filings do not tell us which way the field points.

magnetic monopoles have not been found

Figure 21-2   Dividing a magnet results in two smaller magnets each with both poles.

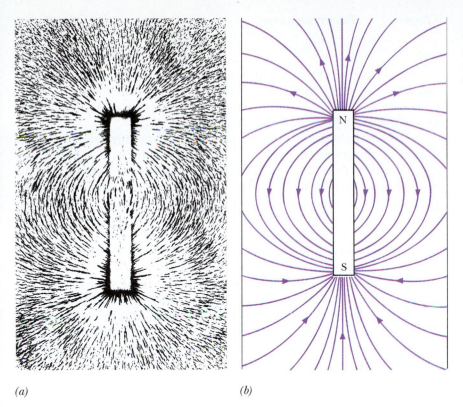

(a)   *(b)*

Electric Magnetism

Although electricity and magnetism were well known for centuries, any connection between the two phenomena eluded experimenters until the 19th century. In 1820, Hans Christian Oersted, a Danish scientist, discovered a connection while performing a lecture demonstration; a compass needle experiences a force when it is brought near a current-carrying wire. This means that the current-carrying wire produces a magnetic field in the surrounding space. This discovery was fascinating and yet confusing. It was known that stationary charges did not produce magnetic fields; so the field must be produced by the motion of the charges. The photograph in Figure 21-4 shows that the magnetic field lines form circles about the wire.

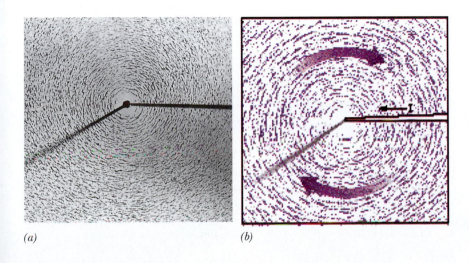

(a)   *(b)*

Figure 21-4 (a) Iron filings show that the magnetic field lines surrounding a long, straight, current-carrying wire are circles around the wire. The dark line at the left is the shadow of the wire. (b) The diagram shows the direction of the magnetic field for a current into the page.

501

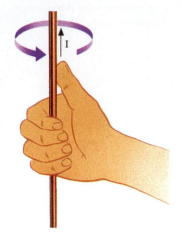

Figure 21-5 The direction of the magnetic field is given by the right-hand rule. If your thumb points in the direction of the current, your fingers encircle the wire in the direction of the field.

The direction of the magnetic field is given by the *right-hand rule:* If you grasp the wire with the thumb of your right hand pointing in the direction of the positive current, your fingers encircle the wire in the direction of the magnetic field, as shown in Figure 21-5.

Different field patterns can be produced by bending the wire. For instance, the magnetic field of a circular loop of wire is shown in Figure 21-6. The contributions to the magnetic field from each segment of the wire add together inside the loop to produce a field pointing into the page inside the loop. Adding more loops to form the cylindrical structure, called a *solenoid,* shown in Figure 21-7 produces a magnetic field like that of the bar magnet in Figure 21-3.

Observations like these led the French physicist André Ampère to suggest that *all* magnetic fields originate from current loops. We now believe that magnetism originates in current loops at the atomic level. In a simplified model of the atom, we visualize these currents arising from electrons orbiting atomic nuclei or spinning about their axes. The macroscopic magnetic properties of an object are determined by the superposition of these atomic magnetic fields. If their orientations are random, the object has no net magnetization; if they are aligned, the object is magnetized.

Making Magnets

Like electric charge, our Universe is filled with magnetism. We do not normally detect the magnetism because of the random orientation of the atomic current loops. The atoms can be aligned, however, by the presence of a magnetic field. In most materials this alignment disappears when the magnetic field is removed. However, in some materials the atoms remain aligned and therefore retain their macroscopic magnetism.

A piece of iron can be magnetized by placing it in a strong magnetic field, by stroking it with a magnet, or by hitting it while it is in a magnetic field. Tapping an iron rod on the floor while holding it vertically will magnetize it because the Earth's magnetic field has a vertical component. You

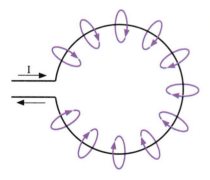

Figure 21-6 The magnetic field of this single loop of wire points into the page inside the loop and out of the page outside the loop.

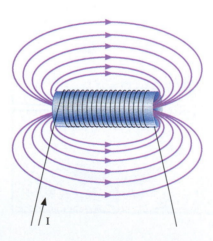

Figure 21-7 The magnetic field of a solenoid is like that of a bar magnet.

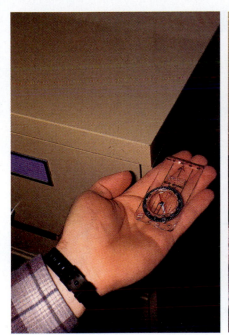

Figure 21-8   The magnetic fields near the top and bottom of a metal file cabinet point in opposite directions.

can check for the resultant magnetization by holding a compass near the top and bottom of a metal object, such as a filing cabinet. The opening and closing of the file drawers provides the tapping; the Earth's magnetic field does the rest. Figure 21-8 shows the orientation of compasses near the top and bottom of a file cabinet. On the other hand, an iron object that has been magnetized can lose its magnetism if it is dropped or struck repeatedly because the impact tends to randomize the alignment of the atoms. Heating the object also randomizes the atomic magnetic fields.

Physics on Your Own   Magnetize an iron rod by aligning it with the Earth's magnetic field and hitting it on the end with a hammer. In the United States, the Earth's magnetic field points north and down. Since the Earth's field has a vertical component, the rod could also be magnetized by bouncing its end on a hard surface.

A versatile magnet can be constructed by wrapping wire around an iron core and connecting the ends of the wire to a battery, as illustrated in Figure 21-9. When current passes through the coil of wire, it generates a magnetic field along the axis of the coil. The magnetic field that is induced in the iron adds to that of the solenoid, increasing the strength of the magnetic field. These **electromagnets** are very useful because they can be turned on and off, the strength of the field can be varied by varying the current, and they can produce very large magnetic fields. Electromagnets can be used for such tasks as moving cars or sorting some metals from other landfill materials.

Figure 21-9   A very simple electromagnet can be constructed from a nail and some wire.

A large electromagnet is used to load cargo containers onto a ship.

Physics Update

A magnetic resonance imaging (MRI) system developed by General Electric permits real-time MRI scans during surgery. The new system employs several innovations; the first is the use of a pair of solenoid magnets rather than a single magnet, enabling the part of the body under operation to lie outside the magnet (but still subject to the high magnetic fields necessary for MRI) in a place accessible to a surgeon. Also, the system's niobium–tin magnet coils are kept superconducting at a temperature of 10 K with a small cryocooler system rather than with liquid helium and its attendant large pressure vessel. High-temperature ceramic superconductors are used to connect the coils to an external power supply. The new system is undergoing tests at the Brigham and Women's Hospital in Boston.

The Ampere

A current-carrying wire exerts a force on compass needles (Figure 21-10). Therefore, by Newton's third law, a magnet should exert a force on the wire. This effect was very quickly verified by experiment. A wire between the jaws of a large horseshoe magnet jumps out of the gap when the current is turned on. This force is biggest when the wire is perpendicular to the magnetic field. The force on the wire is always perpendicular to the wire and to the magnetic field.

These experiences indicate that two current-carrying wires should exert forces on each other. They do. The magnetic field produced by each current exerts a force on the other. If the currents are in the same direction, they attract each other (Fig. 21-11); if the currents are in opposite directions, they repel.

This effect is used to define the unit of current—the basic electrical unit in the metric system. Consider two long parallel wires separated by 1 meter and carrying the same current. If the force between these wires is 2×10^{-7} newton on each meter of wire, the current is defined as 1 **ampere** (A). The **coulomb** is then defined as the amount of charge passing a given point in one of these wires during 1 second. This is the same as the charge on 6.24×10^{18} protons.

The source of the force on each wire is the magnetic field produced by the other wire. We can also use the interacting wires to define a field strength for magnetism. The strength of the magnetic field at a distance of 1 meter from a long straight wire carrying a current of 1 ampere is 2×10^{-7} **tesla** (T). Another unit for the magnetic field strength, the **gauss** (G), is often used instead of teslas, where 1 tesla = 10,000 gauss.

The magnets that are used to hold notes or pictures on refrigerator doors produce fields on the order of 0.01 tesla, whereas large laboratory magnets produce fields of 2.5 tesla. The theoretical limit for a permanent magnetic field is 5 tesla. Electromagnets made with ordinary wires have produced steady fields of 34 tesla, while those made with superconducting wires have not yet exceeded 22 tesla. Combinations of these two types of magnet have reached 35 tesla, and a combination magnet designed to

Figure 21-10 A magnetic field causes a force on a current-carrying wire. Notice that the compasses above and below the wire point in opposite directions.

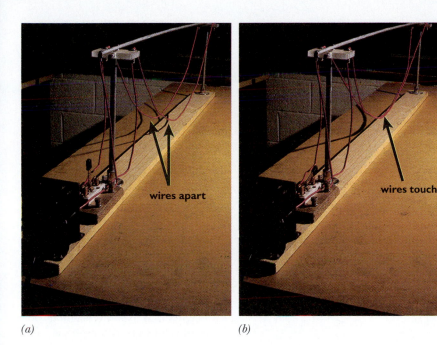

(a)   (b)

Figure 21-11 (a) When the switch is open, there is no attractive force between the wires. (b) When the switch is closed, the currents are in the same direction and the wires are attracted to each other.

produce 45 tesla is under construction. At the upper extreme, pulsed magnets have produced fields above 200 tesla for extremely short times, but the forces are so large that the magnets destroy themselves.

The Magnetic Earth

As mentioned earlier, a simple compass shows that the Earth acts as if it had a huge magnet in its core as illustrated by Figure 21-12. The strength of the Earth's magnetic field at the surface of the Earth is typically 5×10^{-5} tesla (0.5 gauss). In the United States its horizontal component points generally northward and an even larger component points downward.

> **QUESTION** Is the magnetic pole located in northern Canada a north or a south magnetic pole?

Measurements of the magnetic field show that one of the Earth's magnetic poles is located in northeastern Canada just north of Hudson Bay, about 1300 kilometers from the geographic North Pole. The Earth's other magnetic pole is almost directly on the other side of the Earth. A line through the Earth connecting these two poles is tilted about 12° with respect to the rotational axis of the Earth that passes through the Earth's geographic North and South Poles.

> **ANSWER** Because this pole attracts the north pole of a compass and since opposite poles attract, this magnetic pole must actually be a south pole. However, it is still known as the magnetic North Pole.

Figure 21-12   The Earth is a giant magnet with its south magnetic pole in the Northern Hemisphere.

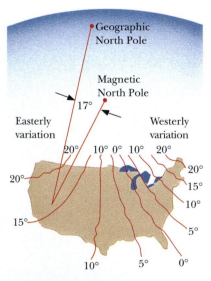

Figure 21-13   The magnetic variation between the directions to the geographic and magnetic North Poles changes across the United States.

The tilt of the Earth's magnetic axis with respect to its rotational axis makes finding true north with a compass a little bit complicated. The only time your compass will point to true north is when the magnetic pole is between you and the geographic pole. Because of local variations in the magnetic field this situation occurs along an irregular line running from Florida to the Great Lakes, as shown in Figure 21-13. If you are east of this line, your compass will point to the west of true north. If you are west of this line, your compass points to the east. The difference in the directions to the magnetic and geographic poles is known as the *magnetic variation.* As an example, the magnetic variation in Bozeman, Montana, is currently 15.5°E.

Airplane pilots use magnetic headings and refer to their aeronautical maps to find the true headings. The large numbers painted on the ends of runways correspond to the magnetic directions of the runways divided by 10.

At one time scientists imagined that the Earth's magnetic field was caused by magnetized solid iron in the Earth's interior, but they now believe this cannot be true. The interior of the Earth is known to be hot enough that the iron and nickel are in a liquid state. In a liquid state the atomic magnetic fields do not remain aligned, but take on random orientations, thus eliminating a macroscopic magnetic effect.

The best theory is that the Earth's magnetic field is caused by large electric currents circulating in the molten interior. Such currents could easily produce the field that we observe on Earth as well as the general features of the magnetic fields of the other planets. However, there are some difficulties with this theory. No one understands the details of the mechanisms for producing and maintaining the currents.

The most puzzling aspect of the Earth's magnetic field is its reversals; that is, the Earth's North and South Poles switching locations. There is strong evidence that the Earth's magnetic field has reversed directions 171 times in the last 17 million years. This evidence comes from the rocks on both sides of the mid-Atlantic rift. As the molten rock emerges from the rift, it cools and solidifies, and at that moment the direction of the Earth's magnetization is locked into the rocks and preserved. Samples from the ocean's floor show that the directions of the magnetic fields in the rocks alternate as we approach the rift. Although we know the reversals have occurred, no one has been able to propose a satisfactory mechanism for these reversals or to account for the source of the vast amount of energy required to reverse the field.

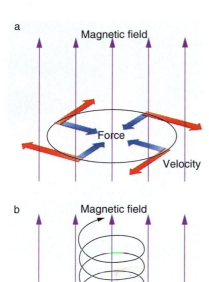

Figure 21-14   (a) The charged particle follows a circular path if its velocity is perpendicular to the magnetic field. (b) The path is helical if its velocity has components both perpendicular and parallel to the field.

Charged Particles in Magnetic Fields   Σ

Recall that a magnet has no effect on a charged object. On the other hand, a current-carrying wire in a magnetic field does experience a force. Because an electric current is a stream of charged particles, the motion of the charges must be important. As bizarre as it seems, the magnetic force on a charged particle is zero unless the charge is moving!

In addition, the direction of the magnetic force is probably not the direction you would predict. Recall that the current, and thus each charged particle, moves along the wire in Figure 21-10 but that the wire experiences a force in a direction perpendicular to the wire and perpendicular to the magnetic field. Thus, we expect a charged particle moving in a magnetic field to experience a force that is at right angles to its velocity and to the magnetic field.

The strength of the force depends on the charge q, the strength of the magnetic field B, the velocity v, and the angle between the field and the velocity. The magnetic force is maximum when the field and velocity are perpendicular,

$$F_{\max} = qvB$$

maximum magnetic force = charge × speed × magnetic field

When the velocity and the magnetic field are parallel, the force is zero.

The magnetic force produces some interesting phenomena. First, because the force is always perpendicular to the particle's velocity, it never changes the particle's speed, only its direction. If the particle's velocity is parallel to the magnetic field, the particle does not experience a force and moves in a straight line. On the other hand, if the velocity is perpendicular to the field, the particle experiences a centripetal force that causes it to move in a circular path [Fig. 21-14(a)]. If the velocity has components parallel and perpendicular to the magnetic field, these two motions will superimpose and the particle will follow a helical path [Fig. 21-14(b)] along the magnetic field.

The magnetic force on charged particles is the cause of dramatic effects known as the *aurora borealis* (northern lights) and the *aurora australis* (southern lights). As charged particles from outer space approach Earth, they interact with the Earth's magnetic field. This interaction causes them

The aurora borealis is caused by cosmic rays following magnetic field lines toward the magnetic North Pole.

to follow helical paths along the Earth's magnetic field so that they strike the Earth's atmosphere in the region of the magnetic poles. The collisions of these cosmic rays with the oxygen and nitrogen molecules in the atmosphere produce the spectacular evening light shows high over the north and south magnetic poles.

The magnetic forces on charged particles are important in many scientific and technologic devices ranging from containment vessels used in developing the future technology of fusion reactors, to particle accelerators used in research, to television sets in our homes. We will discuss many of these in the remaining chapters.

Magnetic Electricity

In the evolution of the physics world view, a small number of basic themes have emerged that reflect our biases but also fuel our searches. One such theme is symmetry. A number of times, progress has been made while searching for symmetrical effects. Early in the development of our understanding of electricity and magnetism, the questions of symmetry haunted some experimenters. In particular, they asked: Given that an electric current produces a magnetic field, does a magnetic field produce an electric current? To investigate this, you might try wrapping some wire around a magnet and connecting the wire to an instrument for measuring current. Although similar experiments were performed with large magnets and sensitive instruments, no effect was found.

British scientist Michael Faraday discovered the connection in 1831. He found that motion is the key to producing an electric current with a magnet. If a wire is moved through a magnetic field (but not parallel to the field), a current is produced in the wire. This current is due to the motion of the wire in the magnetic field; there are no batteries. The current is largest if the motion is perpendicular to the field and increases with the speed of the wire. Having the benefit of hindsight, we see that this is understandable because of the magnetic force on the charges in the wire.

Because the principle of relativity (Chapter 11) must be valid for electricity and magnetism, we know that it is the *relative* motion of the wire and the magnet that is important. It doesn't matter which is moving relative to the laboratory and which is stationary. Therefore, we should be able to generate a current in a stationary wire by moving the magnet. This principle can be demonstrated with a simple experiment. If a bar magnet is placed in a coil of wire and quickly withdrawn, a current is generated that is easily detected. Quickly inserting the magnet into the coil generates a current in the opposite direction. Turning the bar magnet end for end also reverses the direction of the current. The four possibilities are shown in Figure 21-15.

Faraday also discovered that motion is not the only way of producing a current with a magnetic field. A current is produced if the strength of the magnetic field varies with time, even when there is no relative motion of the wire and the magnet. An increasing field produces a current in one direction; a decreasing field produces a current in the opposite direction.

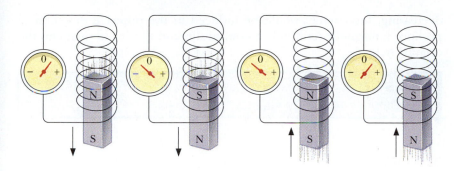

Figure 21-15   A magnet inserted into or removed from a coil of wire produces a current. When the meter is centered, there is no current.

After a long series of experiments conducted over several years, Faraday was able to generalize his results in terms of magnetic field lines. These lines are visualized as pointing along the direction of the magnetic field at all points in space. The number of lines in a given region of space represents the strength of the magnetic field. The lines are closer together where the magnetic field is stronger.

Faraday showed that if the number of magnetic field lines passing through a loop of wire changed *for any reason,* a current was produced in the loop (Fig. 21-16). The voltage (and hence the current) generated in the loop depends on how fast the number of field lines passing through the loop changes—the faster the change, the larger the voltage. These phenomena are now known as Faraday's law.

Faraday's law

The direction of the current induced in the coil by the changing number of magnetic field lines is given by Lenz's law, which states that the current always produces a magnetic field to oppose the change. For example, in Figure 21-16 the number of lines in the upward direction is increasing. Therefore, the induced current will produce a magnetic field that points downward to try to cancel out the increase. Notice that the current shown in Figure 21-16 does this. If the number of lines had been

Lenz's law

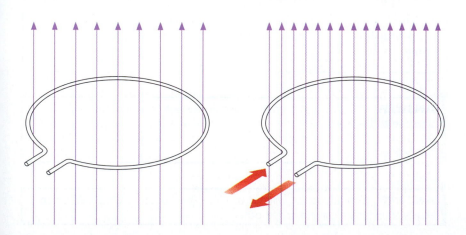

Figure 21-16   Changing the number of magnetic field lines passing through a loop of wire produces a current.

decreasing, the current would be induced in the opposite direction to produce a magnetic field in the upward direction to try to maintain the original field.

> **QUESTION**   What would you see on the current meter if you drop a magnet through the coil in Figure 21-15?

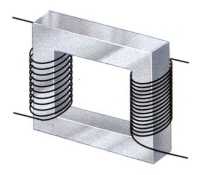

Figure 21-17   The essential features of a transformer.

Transformers   Σ

These discoveries about magnetism—especially its connections with electricity—have many practical uses. For example, they are fundamental to the operation of transformers used to change the voltage of alternating-current electricity. The same amount of electric power can be delivered through wires at low voltage and high current as at high voltage and low current (Chapter 20). For instance, 10 amperes at 12 volts supplies the same power as 1 ampere at 120 volts. The particular choice of what amperage and voltage are used depends on the circumstances. The large cylindrical containers on power poles are transformers for reducing the voltage to 120 volts before it enters our homes and businesses.

The schematic diagram in Figure 21-17 shows the essential features of a transformer. One coil, called the *primary coil,* is connected to a source of alternating current. The alternating current in the primary coil produces an alternating magnetic field which is transmitted to the *secondary coil* by the iron core. This alternating magnetic field produces an alternating voltage in the secondary coil.

The relative size of the voltage produced in the secondary coil depends on the ratio of the number of loops in the two coils. A transformer designed to reduce the voltage by a factor of 2 would have one-half as many loops in the secondary as in the primary coil.

> **QUESTION**   Why wouldn't a transformer work with direct-current electricity?

> **Physics on Your Own**   Call the power company and find out how many transformers there are between the generating plant and your home.

> **ANSWER**   As the magnet enters the coil, the needle would swing to one side due to the increase in the number of field lines passing through the coil. As the magnet exits, the needle will swing to the other side because the number of lines decreases.

> **ANSWER**   Direct current will not produce the varying magnetic field necessary to induce a current in the other coil.

Neighborhood transformers reduce the voltage for household use.

Generators and Motors

The electric generator is another application of Faraday's discovery. If we rotate the loop of wire in Figure 21-16 by 90°, the number of magnetic field lines passing through the loop drops to zero. The change in the number of field lines passing through the loop produces a voltage around the loop which causes a current to flow. The loop is usually rotated by a steam turbine or falling water in a hydroelectric facility. The steam can be produced by many means; burning wood, coal, petroleum, or natural gas, or by using heat from the Sun and nuclear reactors.

Figure 21-18 contains a series of drawings of the magnetic field and a wire loop in an electric generator that illustrate how the rotating loop produces electric voltages (and currents). Figure 21-18(a) shows the plane of the loop parallel to the magnetic field lines. The number of lines passing through the loop in this orientation is zero. As the loop rotates at a constant speed the number of lines initially increases rapidly, producing a large voltage. As it continues to rotate (b) the number of lines passing through the loop continues to increase but at a slower rate. The voltage drops to zero when the plane of the loop is perpendicular to the field lines (c) and the number of lines through the loop is a maximum. The number of lines through the loop now decreases (d) and the voltage increases in the opposite direction. It increases to a maximum (e) and then returns to zero when the plane of the loop is perpendicular to the field lines once again. The voltage increases to a maximum in the opposite direction and the entire cycle repeats. This generator produces the alternating voltage shown in Figure 21-19(a).

A simple change in the way the voltage is carried to the external circuit converts the generator to produce pulsating direct current [Fig. 21-19(b)]. A connector (called a *commutator*), shown in Figure 21-20, reverses the connections from the loop to the outside circuit each half turn. This pulsating current can then be electronically smoothed to produce a constant direct current like that from a battery.

A direct-current motor is basically a direct-current generator run backward. In fact, the discovery of the first such motor occurred during an exhibition in 1873 when a technician setting up a demonstration of generators hooked up one the wrong way and "discovered" a motor! In a generator we rotate the loop in a magnetic field, which produces a voltage that moves electric charges. In a motor the sequence is reversed; we apply a voltage, causing the charges to move. This current in a magnetic field produces a force that rotates the loop.

Figure 21-18 can be used to illustrate the operation of a motor. When a voltage is applied to the loop when it is in the position shown in Figure 21-18(a), the magnetic field exerts a torque on the current-carrying loop. The forces on the long sides of the loop are in opposite directions because the currents are in opposite directions. The torque decreases as the loop rotates and becomes zero when the plane of the loop is vertical. If nothing changes as the loop coasts through this position, it will oscillate and eventually come to a stop. However, the commutator reverses the direction of the current so that the torque continues to act in the same direction.

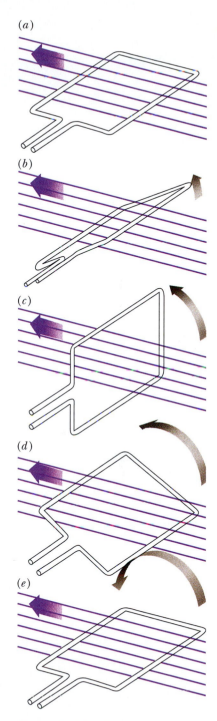

(a)

(b)

(c)

(d)

(e)

Figure 21-18   The number of magnetic field lines passing through a loop of wire changes as it rotates. This produces the alternating voltage shown in Figure 21-19.

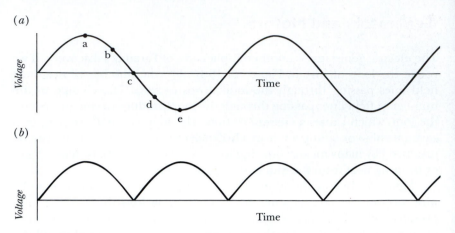

(a)

(b)

Figure 21-19 (a) A loop rotating in a magnetic field produces an alternating voltage. The letters show the voltage produced when the loop has the positions shown in Figure 21-18. (b) The pulsating direct current produced by using a commutator.

The similarity of a motor and a generator is important in some electric-powered vehicles. While the vehicle is speeding up or cruising, the engine acts like an electric motor. However, when the brakes are applied, the connections to the engine are changed so that it acts like a generator to recharge its batteries. The current exerts a torque on the engine, which in turn exerts a torque on the wheels to stop the vehicle. This is not a perpetual motion machine, as only part of the energy is recovered. However, regenerative braking can extend the range of the vehicle.

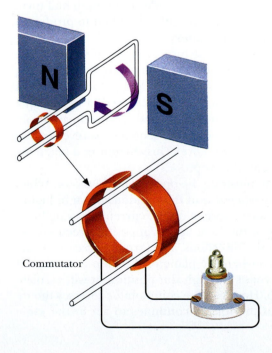

Commutator

Figure 21-20 Schematic of a direct-current generator.

A Question of Symmetry

The connection between electricity and magnetism appears to be complete—a *changing* magnetic field produces an electric current and an electric current produces a magnetic field. Notice, however, that the situation is not quite symmetric. It requires a changing magnetic field to produce the electric current, but the electric current produces a steady magnetic field.

Actually, there is more to this connection that only becomes apparent when everything is expressed in terms of the fields. Consider two parallel plates connected to a battery as shown in Figure 21-21. As current flows in the wires building up charges on the plates, an electric field is generated in the region between the plates. Even though no charges are flowing between the plates, a magnetic field is produced in the region surrounding the plates that matches the field surrounding the wires. Therefore, the origin of the new field cannot be the charges. Furthermore, the magnetic

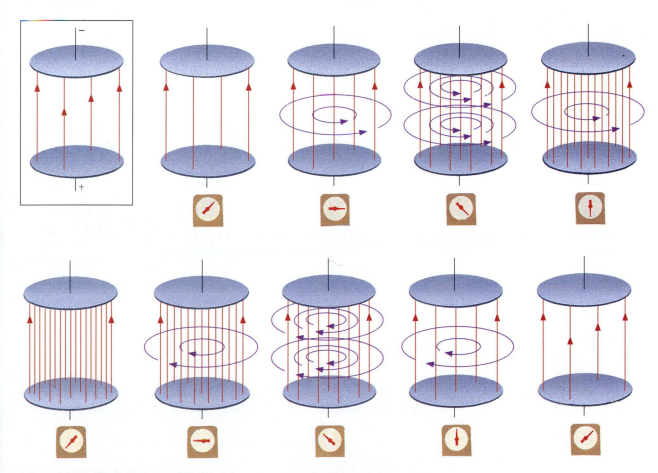

Figure 21-21   A steady electric field between the parallel plates shown in the boxed diagram at the left does not produce a magnetic field. An increasing electric field produces a magnetic field in one direction, while a decreasing field produces a magnetic field in the opposite direction.

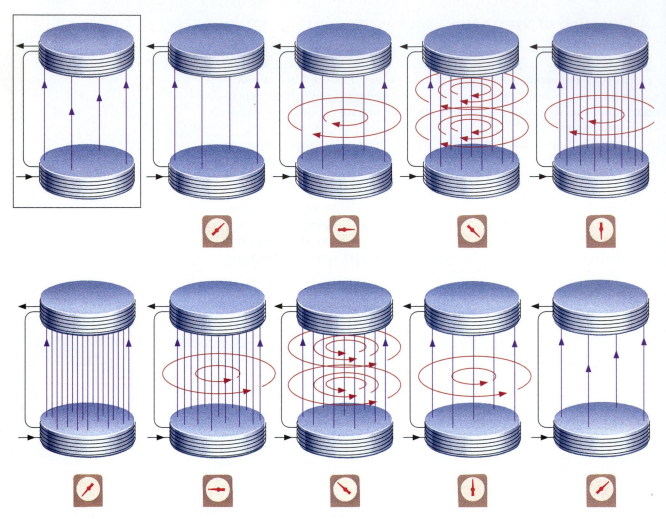

Figure 21-22 A steady magnetic field between the poles of the electromagnet shown in the boxed diagram at the left does not produce an electric field. An increasing magnetic field produces an electric field in one direction, while a decreasing field produces an electric field in the opposite direction.

field disappears when the current stops. The electric field between the plates continually increases as long as charge is flowing to the plates and remains constant when the charge stops. Therefore, the magnetic field is produced by a *changing* electric field. As the charges leave the plates, the electric field decreases and a magnetic field in the opposite direction is produced.

We can make the analogy closer by looking at the region between the poles of an electromagnet. As the magnetic field between the poles increases, an electric field is produced in the surrounding region (Fig. 21-22). When the magnetic field reaches its maximum strength and is no longer changing, the electric field disappears. As the magnetic field decreases, an electric field in the opposite direction is produced.

It is important to realize that a changing magnetic field produces an electric field in *empty* space. There is no need for a wire. If a wire is present, the electric field exerts forces on the charges within the wire and produces a current. But the important point is that even in the absence of the wire, the electric field is present.

If we use these results and focus on the fields rather than the currents, the situation is completely symmetrical: A *changing* magnetic field generates an electric field and a *changing* electric field generates a magnetic field. There is an intimate relationship between electricity and magnetism.

> A changing magnetic field produces an electric field and vice versa.

Electromagnetic Waves

If the magnetic field changes at a constant rate (that is, if it changes by the same amount each second), the electric field that is produced is constant. A rapidly changing magnetic field produces a large electric field, and a slowly changing magnetic field produces a smaller one. If the magnetic field starts out changing slowly and then increases its rate of change, the electric field starts out small and grows larger. Thus, it is possible for a changing magnetic field to produce a changing electric field.

The parallel plates in Figure 21-21 show the symmetric effect; a changing electric field produces a magnetic field. The rate of change of the electric field determines the size of the magnetic field. Therefore, a changing electric field can produce a changing magnetic field.

> **QUESTION**   How would you produce a magnetic field that is increasing in size?

We have discussed a lot of "rates of change" and this can be quite confusing, but the payoff for understanding the process is worth the effort. We have discovered a sequential chain of field productions; one changing field produces another changing field and then this new changing field produces the first kind of field again, and so on. The two fields generate each other in empty space.

We have only argued that this is possible, but the process was rigorously deduced in the 1860s by James Clerk Maxwell. Maxwell was able to show that this was a consequence of a set of four equations that he and others had developed to describe electricity and magnetism and the many connections between them. The equations, called Maxwell's equations in honor of his contributions, summarize all of electricity and magnetism.

Maxwell combined these equations into a single equation that had the same form as the equations that describe periodic waves, whether they be waves on a rope, water waves, or sound waves. Oscillating electric and magnetic fields can combine to produce waves that travel through

> **ANSWER**   The electric field must change slowly at the beginning and continually increase its rate of change.

Figure 21-23 An electromagnetic wave propagating through space. The electric and magnetic fields are perpendicular to each other and to the direction the wave is moving.

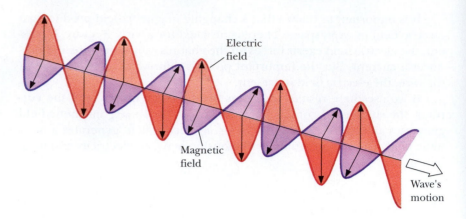

space (Fig. 21-23). As calculated by Maxwell, these **electromagnetic waves** take the form of oscillating electric and magnetic fields that travel with a speed equal to that of light. In 1887, the German physicist Heinrich Hertz was able to produce electromagnetic waves on one side of a room and detect them on the other. The existence of electromagnetic waves affirmed Faraday's belief that these fields had their own identities.

Making Waves $\boxed{\Sigma}$

Electromagnetic waves are produced whenever electric charges are accelerated. If the charges have a periodic oscillatory motion, the wave will have a fixed frequency and therefore a fixed wavelength according to the relationship we developed in Chapter 14 between speed, wavelength, and frequency,

speed = wavelength × frequency

$$v = \lambda f$$

Maxwell's equations require that the speed be that of light, but place no restrictions on the frequency, as shown in the diagram of the electromagnetic spectrum in Figure 21-24. Although these waves all have the same

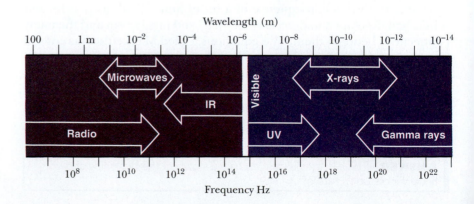

Figure 21-24 The electromagnetic spectrum.

basic nature, the different ranges in frequency are produced (and detected) by different means. The boundaries between the various named regions are not distinct and, in fact, they overlap quite a bit.

The lowest frequencies and the longest wavelengths belong to the radio waves. These are produced by large devices like broadcast radio antennas. Microwaves are also produced electronically, but the devices are smaller, ranging in size from a few millimeters to a few meters. These devices are used in microwave ovens, radar, and the long-distance transmission of telephone calls.

The frequencies of visible light extend from 4.0 to 7.5×10^{14} hertz. Although visible light occupies only a very small region of the complete spectrum, it is obviously very important to us. This region is bounded on the low side by the infrared radiation (IR), waves whose wavelengths are too long (beyond the red) to be seen by the human eye. Infrared radiation is most noticeable when given off by hot objects, especially those that are red hot. This is the radiation you feel across the room from a fire or a heating element. The radiation that lies beyond violet is known as ultraviolet (UV) waves. This is the component of sunlight that causes suntans (and sunburns if it is excessive). Visible light and its neighbors are produced at the atomic level, and their properties are valuable clues about the structure of matter at the atomic level.

We are all familiar with X rays from visits to the doctor or dentist. X rays have high frequencies and are very penetrating. They are produced by the rapid acceleration of electrons in X-ray machines and are emitted by atoms. Gamma rays are even higher frequency radiation that originate in the nuclei of atoms. We will study these in more detail in later sections of this book.

Infrared photo of the Washington, D.C., area looking north, showing the Potomac River in blue and vegetation in red. This view from the Space Shuttle covers 68 square miles.

Radio and TV

Radio is a means of coding electromagnetic waves with the information in sound waves so that they can be transmitted through space, intercepted, and converted back into sound. Television is the same sort of process with the addition of the video information. Sound waves are changed to an electric signal by a microphone. In one version the sound waves cause a coil to vibrate in a magnetic field. This produces a current in the coil which can then be amplified. Audio frequencies are in the range 20 hertz–20 kilohertz.

If audio frequencies were broadcast directly, there could only be one radio station in any geographic region. Instead, the audio signal is combined with a broadcast signal. A station broadcasting at "1450 on your dial" sends out waves with a frequency of 1450 kilohertz (often called kilocycles). This is the *carrier frequency*. The audio signal is used to vary, or modulate, the carrier signal. Two modulation methods are *amplitude modulation* (AM) and *frequency modulation* (FM). The frequency of the sound determines the frequency of the modulation, and the loudness of the sound determines the amplitude of the modulation. In AM radio the audio signal causes the amplitude of the carrier signal to vary; in FM radio it causes the frequency of the carrier signal to vary (Fig. 21-25).

Figure 21-25   (a) AM and (b) FM radio waves.

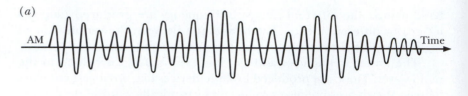

(*a*)

AM   Time

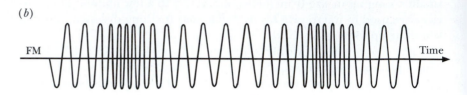

(*b*)

FM   Time

Restless Genius

James Clerk Maxwell

Curiosity may well kill the cat, but it can also be the impetus that turns a seemingly ordinary child into a brilliant physicist. When James Clerk Maxwell was nearly three years old, his mother wrote that he was forever demanding "Show me how it doos." Luckily for the world of physics, his insatiable curiosities were indulged by his parents. The young Maxwell could be found running around the house tracing the wires that connected bells in servant's quarters to more remote locations of the house. Examination of locks, keys, and reflections in tin plates were just a few of the everyday articles that did not fail to evoke a probing, "What's the go of that?"

On the other hand, Maxwell did poorly in school, probably due to the distractions of one research project or another he had designed for himself. Some of this research led to his publishing his first paper at the age of 15. Maxwell's academic career took off when he attended the University of Edinburgh. From there he went to Cambridge, where his interests in higher math crystallized. After holding several university positions, Maxwell resigned to pursue his own independent projects in electricity at his estate. He came out of retire-

ment to become director of the now renowned Cavendish Laboratory, but died two years later at the age of 48.

Maxwell's physical intuition and mathematical prowess led him to monumental discoveries in three major areas. Early in his career he worked with color perception and was the first to estimate the sensitivity of the human eye as an optical detector. In molecular physics, he was the first to use statistical methods to describe the molecular movement in a gas. He calculated the distribution of molecular speeds (Chapter 7) and developed the theory of diffusion.

Maxwell's greatest achievement came in the field of electricity and magnetism. He was able to combine his vast knowledge of experimental evidence, his fertile imagination, and his virtuoso mathematical ability to pull together electricity and magnetism into one coherent theory. His genius lay in his ability to see connections. Despite the fact that there was little evidence to support his assertions in his own time, there is much material today that confirms James Clerk Maxwell as one of physics' great minds.

Adapted from an essay written by Steven Janke for Pasco Scientific.

References: Campbell, L. and Garnett, W. *The Life of James Clerk Maxwell.* Sources of Science: House Series, Johnson Reproductions, 1970. Glazebrook, R.T. *James Clerk Maxwell and Modern Physics.* New York: Macmillan, 1900.

Stereo Broadcasts

The carrier frequencies of AM radio stations are separated by only 10 kilohertz. This means that the full 20-kilohertz range in audio frequencies cannot be broadcast without the stations overlapping. In practice, each station is limited to audio frequencies up to 5 kilohertz. The development of FM radio allowed the broadcasting of more of the audio frequency range (up to 15 kilohertz), as well as stereophonic signals.

But how do FM radio stations broadcast signals in stereo without affecting the performance of the older radios that were designed to receive monophonic signals? This is done by transmitting the two audio signals on the same carrier but separated by a fixed frequency. A 38-kilohertz frequency is added to one signal before broadcasting and then subtracted in the receiver. Similar techniques are used with the audio and video signals for TV broadcasts.

In order to create signals that can be played on either stereophonic or monophonic receivers, the signals that are broadcast are not the signals for the left and right channels, but the sum and difference of the two channels. A monophonic receiver plays the sum of the two signals and ignores the difference signal, whereas a stereophonic receiver electronically adds and subtracts these signals to recover the left and right channels.

Physics Update

The experimental ultraviolet index to be issued daily by the National Weather Service is a forecast of the noontime ultraviolet level in 58 U.S. cities. So in addition to smog, pollen, and the weather itself, people can now worry in a quantitative way about the cancer-causing or cataract-causing effects of UV. An index of 6, for example, is considered "moderate." At that level a person with skin very susceptible to sunburn would burn after 10 minutes of direct exposure, while a much less susceptible person would burn after 50 minutes. The general advice of the Environmental Protection Agency is that if you can see your shadow, you may want to use some protection against UV radiation.

In either case, the final signal is then amplified and sent to the antenna, where it causes electrons to move up and down the antenna wire. The accelerations of these electrons produce electromagnetic waves that are broadcast. As these waves hit the antenna in your radio, they cause the electrons in the antenna to move back and forth, producing oscillating currents. Although all nearby stations are received simultaneously, the radio is tuned so that it only resonates with one of them at a time. Your radio amplifies this station's signal.

The radio contains electric circuits to filter out the carrier frequency and retain only the electrical version of the sound information. These signals are then sent to a speaker for reconversion to sound waves. In one version of a speaker the electric signal generates a magnetic field which interacts with a magnet to move a diaphragm. This diaphragm then moves the air to generate the sound.

> **QUESTION**   Although modern police and fire sirens produce sound waves rather than radio waves, they are modulated. Is this modulation AM or FM?

The allocation of the possible ranges of carrier frequencies to various types of broadcast is a governmental responsibility that is complicated by the history of the development of such advances in broadcasting techniques as FM stereo and TV. AM radio stations broadcast between 550 and 1500 kilohertz, FM radio between 88 and 108 megahertz, and TV in three regions between 54 and 890 megahertz. Other regions are assigned to citizens'-band receivers, ships, airplanes, police, amateur radio operators, and satellite communication.

CHAPTER

21

REVISITED

The connection between your favorite TV show and your microwave oven lies in the electromagnetic waves that are created by combining electric and magnetic effects. These waves can resonate with water molecules in a microwave oven to raise the temperature of the popcorn and can be beamed through space to bring us television (and radio) programming.

SUMMARY

Magnetic poles behave similarly to electric charges; like poles repel and unlike poles attract. However, magnetic poles always occur in pairs. Magnets attract some objects but have no effect on others.

Magnets have no effect on stationary charges and do not deflect the foils of an electroscope. On the other hand, a current-carrying wire produces a magnetic field and is attracted or repelled by other magnets or current-carrying wires. The force on the wire is always perpendicular to the wire and to the magnetic field. Two current-carrying wires attract each other if the currents are in the same direction; they repel if the currents are in opposite directions. The field strength for magnetism is defined as 2×10^{-7} tesla at a distance of 1 meter from a wire carrying a current of 1 ampere.

All magnetic fields originate from current loops. Naturally occurring magnetism originates in current loops at the atomic level. The Earth's magnetism has a strength at the surface of about 5×10^{-5} tesla and is

> **ANSWER**   Because these sirens have oscillating frequencies, they must be FM.

caused by large electric currents circulating in the Earth's molten interior.

A charged particle moving in a magnetic field experiences a force at right angles to its velocity and to the magnetic field. The strength of the force depends on the angle between the field and the particle's motion. It is maximum when they are perpendicular and zero when they are parallel.

If a wire and magnetic field move relative to each other, a current is produced in the wire provided the motion is not parallel to either the wire or the field. This current is largest if the motion is perpendicular to the field and increases with the relative speed. A current also occurs in a loop of wire if the magnetic field inside the loop varies with time. An increasing field produces a current in one direction; a decreasing field produces a current in the opposite direction.

Field lines can be used to represent a magnetic field at all points in space, with the number of lines representing the strength of the magnetic field. If the number of magnetic field lines passing through a loop of wire changes *for any reason,* a current is produced in the loop. The voltage (and hence the current) generated in the loop depends on the rate of change—the faster the change, the larger the voltage.

The connections between electricity and magnetism are best expressed in terms of the fields. A changing magnetic field can generate a changing electric field, and a changing electric field can generate a changing magnetic field, creating electromagnetic waves that travel through empty space. These waves are produced whenever electric charges are accelerated. When the charges have a periodic oscillatory motion, the wave has a fixed frequency and wavelength. The spectrum of these waves ranges from very low-frequency radio waves, to visible light, to high-frequency X rays and gamma rays.

KEY TERMS

ampere: The SI unit of electric current, 1 coulomb per second.

coulomb: The SI unit of electric charge, the charge of 6.24×10^{18} protons.

electric field: The space surrounding a charged object, where each location is assigned a value equal to the force experienced by one unit of positive charge placed at that location.

electromagnet: A magnet constructed by wrapping wire around an iron core. An electromagnet can be turned on and off by turning the current in the wire on and off.

electromagnetic wave: A wave consisting of oscillating electric and magnetic fields. In a vacuum, electromagnetic waves travel at the speed of light.

gauss: A unit of magnetic field strength, 10^{-4} tesla.

magnetic field: The space surrounding a magnetic object, where each location is assigned a value determined by the torque on a compass placed at that location. The direction of the field is in the direction of the north pole of the compass.

magnetic monopole: A hypothesized, isolated magnetic pole.

magnetic pole: One end of a magnet; analogous to an electric charge.

tesla: The SI unit of magnetic field.

CONCEPTUAL QUESTIONS

1. To what branch of physics is magnetism most closely related?

2. What evidence do you have that the force of magnetism decreases with distance?

3. If you are given three iron rods, how could you use them to find the one that is not magnetized?

4. If the labels on a magnet are missing, how would you determine which pole is the north pole?

5. If a bar magnet is broken into two pieces, how many magnetic poles are there?

6. How would you isolate a single magnetic pole?

7. How is the direction of the magnetic field at each point in space defined?

*8. Why is it not possible for two magnetic field lines to cross?

9. In electric circuits and telephone lines two wires carrying currents in opposite directions are twisted together. How does this reduce the magnetic fields surrounding the wires?

10. Can you use a compass to locate the electric wires located inside the walls of your house?

11. How would the photograph in Figure 21-4 change if the current were in the opposite direction?

12. When you look down the axis of a solenoid so that the current circulates in the clockwise direction, are you looking in the direction of the magnetic field or opposite to it?

13. A horizontal wire carries a current westward. Which way does the magnetic field point directly above the wire?

14. A horizontal wire is aligned along the direction to magnetic north and carries a current northward. Which way will a compass needle point if the compass is placed just above the wire?

15. In the previous chapter we described how a charged rod attracted a neutral insulator. Assuming that the magnetic situation is analogous, describe a mechanism whereby a magnet can attract a piece of unmagnetized iron.

16. In our study of electricity we discovered that both positive and negative objects attracted the same uncharged object. Does an analogous behavior occur in magnetism?

17. Lodestone (magnetite) is an igneous rock, one that forms from molten material. How do you suppose it became magnetized?

18. What happens to the magnetic strength of a bar magnet when it is heated?

19. How would you go about producing a permanent magnet from an iron rod?

20. In Melville's *Moby-Dick* Captain Ahab regained the confidence of his crew when he fixed the compass that had been damaged by an electrical storm so that it pointed south. How might he have done this?

21. Would you expect the head of a steel hammer to be magnetized? Explain.

22. What do you expect would happen to the strength of the magnetism of a bar magnet that is repeatedly dropped on a hard floor?

23. How is the unit of current defined?

24. How is the unit of charge defined?

25. What effect do two parallel wires carrying currents in opposite directions have on each other?

26. Why don't the wires in high-voltage transmission lines attract each other?

27. One of the Earth's magnetic poles is located in northern Canada. Is it a magnetic north pole or a magnetic south pole?

28. Why might it be more proper to call a north magnetic pole a "north-seeking" magnetic pole?

29. How far does a compass in New York City point away from the geographic North Pole?

30. If you want to walk toward the geographic North Pole while in Portland, Oregon, what compass heading would you follow?

31. Why are there more cosmic rays in Antarctica than in Hawaii?

*32. In what direction must the electric current in the Earth's core circulate to produce the Earth's magnetic field?

33. Can you accelerate a stationary charged particle with a magnetic field? An electric field?

34. If a charged particle travels in a straight line, can you say that there is no magnetic field in that region of space?

35. A magnet produces a magnetic field that points vertically upward. Along what line does the force on a proton act if it enters this region with a horizontal velocity toward the south?

36. A proton and an electron have the same velocity. What happens if they enter a bending magnet with a magnetic field perpendicular to their velocity?

37. Quickly inserting the north end of a bar magnet into a coil of wire causes the needle of a meter to deflect to the right. Describe two actions that will cause the needle to deflect to the left.

38. Would you expect to produce an electric current if a wire is looped back on itself as shown in the figure and then passed through a magnetic field?

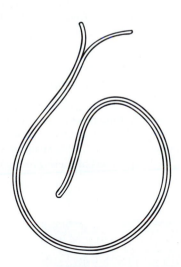

39. How could you produce a current in a solenoid by rotating a small magnet inside the coil?

40. Would you expect to produce an electric current when two bar magnets placed side by side but pointing in opposite directions are inserted into a coil of wire?

41. The plane of a rectangular loop of wire is perpendicular to a magnetic field that oscillates in strength. Would you expect to find a current in the loop?

*42. If a horizontal loop of wire is raised vertically upward from the Earth's magnetic North Pole, will there be a current in the loop?

43. When a transformer is plugged into a wall socket, it produces 9-volt alternating-current electricity for a portable tape player. Is the coil with the larger or smaller number of turns connected to the wall socket?

44. Would a transformer work with pulsating direct-current electricity?

45. If a loop of wire is rotated in a steady magnetic field, when will the voltage produced in the loop be at maximum?

*46. When the plane of a rotating loop of wire is parallel to the magnetic field, the number of magnetic field lines passing through the loops is zero. Why is it then that the current is at maximum?

47. What is the purpose of using a commutator in a motor?

48. What effect does a commutator have on the electrical output of a generator?

49. What determines whether a changing electric field produces a steady or a changing magnetic field?

50. What two effects take place when the coil in a generator is rotated more rapidly?

51. What two electric effects can produce magnetic fields?

52. Each of the following statements is correct as written. Which one is *not* true if you interchange the underlined words in each sentence?

    a. A changing <u>magnetic</u> field can produce a changing <u>electric</u> field.

    b. A steady <u>current</u> produces a steady <u>magnetic field</u>.

    c. A changing <u>current</u> produces a steady <u>magnetic field</u>.

    d. A changing <u>electric</u> field can produce a steady <u>magnetic</u> field.

53. Describe an electromagnetic wave propagating through empty space.

54. How are electromagnetic waves generated?

55. Which of the following are *not* electromagnetic waves: radio, television, blue light, infrared light, or sound?

56. Which of the following electromagnetic waves has the highest frequency: radio, microwaves, visible light, ultraviolet light, or X rays?

57. How fast do X rays travel through a vacuum?

58. What is the difference between X rays and gamma rays?

59. How is sound encoded by an AM radio station?

60. How is sound transmitted by an FM radio station?

61. What does it mean when your local disc jockey says that you are "listening to radio 970"?

62. At what frequency does radio station "FM 96.1" broadcast?

EXERCISES

1. The record steady magnetic field is 35.3 T. What is this field expressed in gauss?

2. What is a magnetic field of 500 G expressed in tesla?

3. A metal ball with a mass of 2 g, a charge of 1 μC, and a speed of 20 m/s enters a magnetic field of 20 T. What are the maximum force and acceleration of the ball?

4. What charge would the ball in the previous exercise need to experience a maximum acceleration equal to the acceleration due to gravity?

5. An electron has a velocity of 3×10^6 m perpendicular to a magnetic field of 2.5 T. What force and acceleration does the electron experience?

6. What force and acceleration would a proton experience if it has a velocity of 3×10^6 m perpendicular to a magnetic field of 2.5 T?

7. A transformer is used to convert 120-V household electricity to 9 V for use in a portable CD player. If the primary coil connected to the outlet has 800 turns, how many turns does the secondary coil have?

8. The voltage in the lines that carry electric power to homes is typically 20,000 V. What is the ratio of the turns in the primary and secondary coils of the transformer required to drop the voltage to 120 V?

9. How long does it take a radio signal from Earth to reach the Moon when it is 384,000 km away?

10. The communications satellites that carry telephone messages between New York City and London have orbits above Earth's Equator at an altitude of 13,500 km. Approximately how long does it take for each message to travel between these two cities via the satellite?

11. An X-ray machine used for radiation therapy produces X rays with a maximum frequency of 2.4×10^{20} Hz. What is the wavelength of these X rays?

12. Many microwave ovens use microwaves with a frequency of 2.45×10^9 Hz. What is the wavelength of this radiation and how does it compare with the size of a typical oven?

13. The radioactive isotope most commonly used in radiation therapy is cobalt-60. It gives off two gamma rays with wavelengths of 1.06×10^{-12} m and 9.33×10^{-13} m. What are the frequencies of these gamma rays?

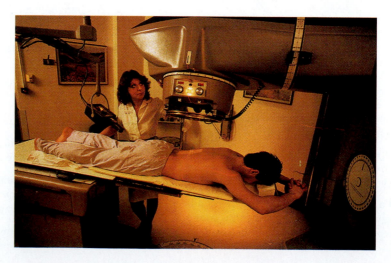

Exercise 13.

14. The ultraviolet light that causes suntans (and sunburns!) has a typical wavelength of 300 nm. What is the frequency of these rays?

15. What is the wavelength of the carrier wave for an AM radio station located at 1090 on the dial?

16. What is the longest wavelength used for TV broadcasts?

17. What is the range of wavelengths for AM radio?

18. What is the range of wavelengths for FM radio?

INTERLUDE

The shape of a snowflake is determined by the quantum-mechanical interactions of water molecules with each other.

The Story of the Quantum

In 1887, the German physicist Heinrich Hertz generated sparks in an electrostatic machine and successfully caused a spark to jump across the gap of an isolated wire loop on the other side of his laboratory. Hertz had transmitted the first radio signal a distance of a few meters. Hertz was not interested in the societal implications of sending signals; he was testing James Clerk Maxwell's prediction of the existence of electromagnetic waves. Hertz also mentioned the curious fact that when ultraviolet light illuminated his loop, more sparks were produced. A short time later, another German physicist, Max Planck, working in an entirely different area of physics, made an unusual calculation attempting to explain the radiation from hot objects and came up with an interesting result. Meanwhile, a group of British physicists were exploring the structure of atoms using the new technologies of high vacuum and high voltage. Although seemingly unrelated, in hindsight these events signaled the beginning of a major breakthrough in physics.

Physics was in turmoil during the first three decades of the 20th century as experimenters probed deeper and deeper into the fundamental nature of matter and radiation. The deeper they probed, the more confusing the situation became. They continually received information that was unbelievable. It wasn't the experiments, it wasn't the apparatus; it was nature returning information that defied common sense. There were serious contradictions in the physics world view. Experiments gave results that invited (in fact, demanded) bizarre explanations.

The only possible resolution was a revolution in the physics world view. Ideas that had been taken for granted were discarded, and the unbelievable became believable. The core of the revolution seemed innocent enough: Energy exists in "chunks," or quantum units. At first glance quantization may not seem that radical. The paintings of the French Impressionists look normal when viewed from afar, and

> **Physics was in turmoil during the first three decades of the 20th century as experimenters probed deeper and deeper into the fundamental nature of matter and radiation.**

our weight seems continuous even though it is really the sum of the weights of discrete atoms. But with this new discreteness it is not just a matter of adding many tiny components to get the whole; it is the tiny components that control the character of the whole.

The quantum nature of energy was the first of three new changes leading to this new physics. New insights into the nature of light and, finally, new insights into the nature of matter followed. Together these changes revealed the character of the Universe from the shape of snowflakes to the existence of neutron stars.

In the remaining chapters we will follow the development of these ideas. Although it might seem easier to simply give "the answers," they are unbelievable without knowing the experiments, the conflicts, the debates, and the eventual resolutions. As you track these developments, remember Sir Hermann Bondi's comment in the Prologue: "We should be surprised that the gas molecules behave so much like billiard balls and not surprised that electrons don't."

Heinrich Hertz.

We get much of our information about atoms from the light they emit. In fact, the pretty arrays of color emitted by atoms—known as atomic fingerprints—distinguish one type of atom from the others and hold many clues about atomic structure. What does the light tell us about atoms? (See p. 552 for the answer to this question.)

The Early Atom

Light is separated into its component colors by a diffraction grating, revealing properties of the atomic or molecular system that produced the light.

New world views don't emerge in full bloom—they start as seeds. Although a change may begin because of a single individual, it usually evolves from the collective efforts of many. Ideas are proposed. As they undergo the scrutiny of the scientific community, some are discarded and others survive, although almost always in a modified form. These survivors then become the beginnings of new ways of looking at the world.

During the early years of this century a number of phenomena were known or discovered that eventually led to several models for the atom. After exploring the phenomena, we will examine several atomic models that emerged in response to the experiments.

Periodic Properties

Once the values of the atomic weights of the chemical elements were determined (Chapter 7), it was tempting to arrange them in numerical order. When this was done, a pattern emerged; the properties of one element corresponded very closely to those of an element further down the list, and to another one even further down the list. A periodicity of characteristics became apparent.

Figure 22-1   Mendeleev's arrangement of the elements.

The Russian chemist Dmitri Mendeleev wrote the symbols for the elements on cards and spent a great deal of time arranging and rearranging them. For some reason he omitted hydrogen, the lightest element, and began with the next lightest one known at that time, lithium (Li). He laid the cards in a row. When he reached the eighth element, sodium (Na), he noted that it had properties similar to lithium and therefore started a new row. As he progressed along this new row, each element had properties similar to the one directly above it. The 15th element, potassium (K), is similar to lithium and sodium, and so Mendeleev began a third row. His arrangement of the first 16 elements looked like Figure 22-1.

The next known element was titanium, but it did not have the same properties as boron (B) and aluminum (A1); it was similar to carbon (C) and silicon (Si). Mendeleev boldly claimed that there was an element that had not yet been discovered. He went on to predict two other elements, even predicting their properties from those of the elements in the same column. Later, these elements were discovered and had the properties that Mendeleev had predicted. Mendeleev's work was the beginning of the modern periodic table of the elements. At that time there was no theory that explained why the elements showed these periodic patterns. However, the existence of the periodicity was a clue to scientists that there might be an underlying structure to atoms.

Atomic Spectra

Another thing fascinated and confused scientists around the beginning of the 20th century. When an electric current passes through a gas, the gas gives off a characteristic color. A modern application of this phenomenon is the neon sign. Although each gas emits light of a particular color, it isn't necessarily unique. However, if you pass the light through a prism or a grating to spread out the colors, the light splits up into a pattern of

The various colors of light emitted by neon-type signs are due to different gases.

Figure 22-2   The emission spectrum of a gas consists of a small number of distinctly colored lines.

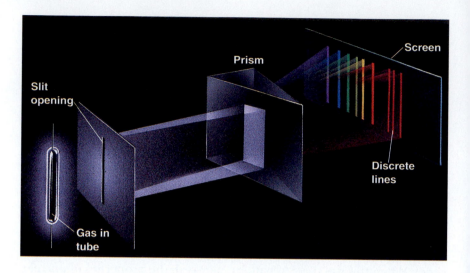

discrete colored lines (Fig. 22-2) which is always the same for a given gas but different from the patterns produced by other gases.

QUESTION   If a particular gas glows with a yellow color when a current is passed through it, will its spectrum necessarily contain yellow light?

Each gas has its own particular set of spectral lines, collectively known as its **emission spectrum.** The emission spectra for a few gases are shown in the schematic drawings in Figure 22-3. Since these lines are unique—sort of an atomic fingerprint—they identify the gases even if they are in a mixture. Vaporized elements that are not normally gases also emit spectral lines and these can be used to identify them.

Spectral lines also appear when white light passes through a cool gas (Fig. 22-4). In this case, however, they do not appear as bright lines but as dark lines in the continuous rainbow spectrum of the white light; that is, light is missing in certain narrow lines. These particular wavelengths are absorbed or scattered by the gas to form this **absorption spectrum.**

All of the lines in the absorption spectrum correspond to the lines in the emission spectrum for the same element. However, the emission spectrum has many lines that do not appear in the absorption spectrum, as illustrated in Figure 22-5.

Since the lines that appear in both spectra are the same, the absorption spectrum can also be used to identify elements. In fact, this is how helium was discovered. Spectral lines were seen in the absorption

ANSWER   The spectrum will not necessarily contain yellow light since the yellow sensation could result from a combination of red and green spectral lines. In fact, red and green light from two different helium–neon lasers produces a beautiful yellow even though each laser has a very narrow range of frequencies.

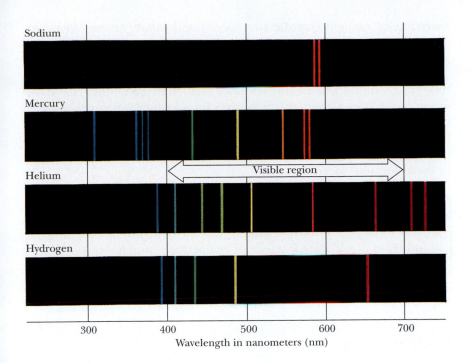

Figure 22-3 Schematic drawings of the emission spectra for sodium, mercury, helium, and hydrogen near the visible region.

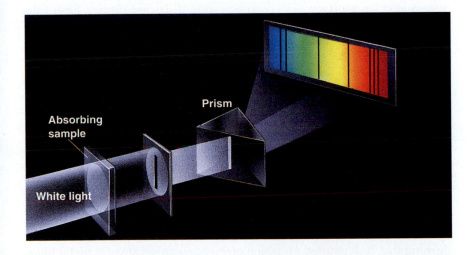

Figure 22-4 The absorption spectrum consists of a rainbow spectrum with a number of dark lines.

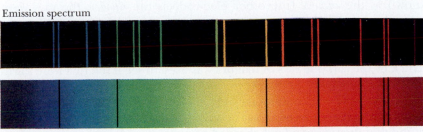

Figure 22-5 The absorption spectrum for an element has fewer lines than its emission spectrum.

spectrum of sunlight that did not correspond to any known elements. These lines were attributed to a new element that was named helium after the Greek name for the Sun, *helios*. Helium was later discovered on Earth.

Spectral lines are a valuable tool for identifying elements. Although this is of great practical use, the scientific challenge was to understand the origin of the lines. Clearly, the lines carry fundamental information about the structure of atoms. The lines pose a number of questions. The most important question concerns their origin: What do spectral lines have to do with atoms? Also, why are certain lines in the emission spectra missing in the absorption spectra?

Physics on Your Own   Observe the spectra of a variety of light sources through a diffraction grating. Inexpensive diffraction gratings mounted in $2'' \times 2''$ slide frames can be obtained from hobby stores. Don't look directly at the Sun!

Cathode Rays

During the latter half of the 19th century, two new technologies contributed to the next advance in our understanding of atoms: vacuum pumps and high-voltage devices. It was soon discovered that a high voltage applied across two electrodes in a partially evacuated glass tube produces an electric current and light in the tube. Furthermore, the character of the phenomenon changes radically as more and more of the air is pumped from the tube. Initially, after a certain amount of the air is removed, sparks jump between the electrodes, followed by a glow throughout the tubes like that seen in modern neon signs. As more air is pumped out, the glow begins to break up and finally disappears. When almost all of the air has been removed, only a yellowish green glow comes from one end of the glass tube.

These discoveries led to a flurry of activity. Experimenters made tubes with a variety of shapes and electrodes from a variety of materials. The glow always appeared on the end of the glass tube opposite the electrode connected to the negative terminal of the high-voltage supply. When the wires were exchanged, the glow appeared at the other end, showing that it depended on which electrode was negative. A metal plate placed in a tube cast a shadow (Fig. 22-6). These effects indicated that rays were coming from the negative electrode. Since this electrode was known as the *cathode* (the other electrode is the *anode*), the rays were given the name **cathode rays.**

The path of the cathode rays could be traced by placing a fluorescent screen along the length of the tube. The screen glowed wherever it was struck by the cathode rays. A magnet was used to see if the rays were particles or waves. We saw in the previous chapter that a magnetic field exerts a force on a moving charged particle, but this does not occur for a beam of light. A magnet placed near the tube deflected the cathode rays, indicating that they were indeed charged particles.

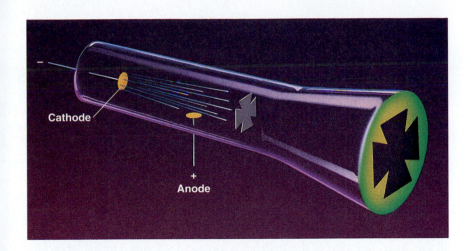

Figure 22-6 Cathode rays cast a shadow of the cross onto the end of the tube.

> **QUESTION**   Based on the information given, would you expect the cathode rays to be positively or negatively charged?

This being the case, cathode rays should also be deflected by an electric field. Early experiments showed no observable effect until the British scientist J. J. Thomson used the tube sketched in Figure 22-7. By placing the electric field plates inside the tube and by using a much better vacuum, he was able to observe electric deflection of the cathode rays. The direction of the deflection was that expected for negatively charged particles.

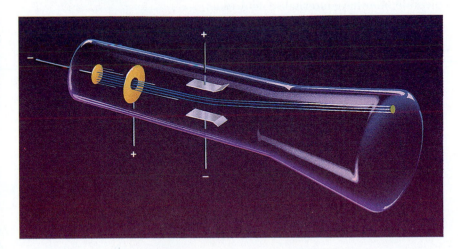

Figure 22-7 The cathode rays are deflected by an electric field.

> **ANSWER**   Because they *come from* the cathode and the cathode is negatively charged, we expect the cathode rays to also have a negative charge. The directions of the magnetic deflections were also consistent with this expectation.

J.J. Thomson.

The Discovery of the Electron   Σ

Thomson went beyond these qualitative observations. He compared the amount of deflection using a known electric field with that of a known magnetic field. Using these values he obtained a value for the ratio of the charge to the mass for the cathode particles. Furthermore, Thomson showed that the *charge-to-mass ratio* was always the same regardless of the gas in the tube or the cathode material. This was the first strong clue that these particles were universal to all matter and therefore an important part of the structure of matter.

This ratio could be compared with known elements. The charge-to-mass ratio for positively charged hydrogen atoms (called hydrogen **ions**) had previously been determined; Thomson's number was about 1800 times larger. If Thomson assumed that the charges on the particles were all the same, the mass of the hydrogen ion was 1800 times that of the cathode particle. Conversely, if he assumed the masses to be the same, the charge on the cathode ray was 1800 times larger. (Of course, any combination in between was also possible.)

Thomson felt that the first option—assuming all charges are equal—was the correct one. (There were other data that supported this assumption.) This assumption meant that the cathode particles had a much smaller mass than a hydrogen atom, the lightest atom. At this point, he realized that the atom had probably been split.

This was a special moment in the building of the physics world view. The discussion that had originated at least as far back as the early Greeks was centered on the question of the existence of atoms, never on their structure. It had always been assumed that these particles were indivisible. Thomson's work showed that the indivisible building blocks of matter are divisible; they have internal structures.

Around 1910, the American experimentalist Robert Millikan devised a way of measuring the size of the electric charge, thus testing Thomson's assumption. Millikan produced varying charges on droplets of oil between a pair of electrically charged plates (Fig. 22-8). By measuring the motions of individual droplets in the electric field, he determined the

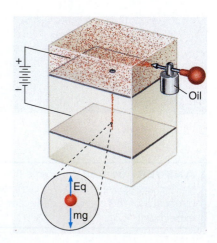

Figure 22-8   The Millikan oil-drop experiment showed that electric charges are quantized.

charge on each droplet and confirmed the widely held belief that electric charge comes in identical chunks, or **quanta.** A particle might have one of these elemental units of charge, or two, or three, but never two and a half. This smallest unit of charge is 1.6×10^{-19} coulomb, an extremely small charge (Chapter 19); a charged rod would have in excess of billions upon billions of these charges.

elementary charge

The mass of the cathode particles could now be calculated from the value of the charge and the charge-to-mass ratio. The modern value for the mass of the cathode particle is 9.11×10^{-31} kilogram. This particle—with a mass 1800 times smaller than the hydrogen atom, carrying the smallest unit of electric charge and a component part of all atoms—is known as an **electron.** The electron is so small that it would take a trillion trillion of them in the palm of your hand before you would notice their mass.

mass of the electron

> **Physics on Your Own**   Hold a small magnet near the face of a black-and-white television set while it's on. (*Caution:* Do not do this with color television sets, as you may cause permanent damage.) Is the picture generated by light or by cathode rays traveling inside the television tube?

Thomson's Model

Thomson's work showing that atoms have structure was an exciting breakthrough. A natural question emerged immediately: What does an atom look like? One of its parts is the electron, but what about the rest? Thomson proposed a model for this structure, picturesquely known as the "plum-pudding" model, with the atom consisting of some sort of positively charged material with electrons stuck in it much like plums in a pudding (Fig. 22-9). Because atoms were normally electrically neutral, he hypothesized that the amount of positive charge was equal to the negative charge and that atoms become charged ions by gaining or losing electrons.

Recall that gases were known to emit distinct spectral lines. Thomson suggested that forces disturbing the atom caused the electrons to jiggle in the positive material. These oscillating charges would then generate the electromagnetic waves observed in the emission spectrum. However, the model could not explain why the light was only emitted as definite spectral lines, nor why each element had its own unique set of lines.

Thomson also attempted to explain the repetitive characteristics in the periodic table. He suggested that each element in the periodic table had a different number of electrons. He hoped that with different numbers of electrons different arrangements would occur that might show a periodicity and thus a clue to the periodicity shown by the elements. For example, three electrons might arrange themselves in a triangle, four in a pyramid, and so on. Perhaps there was a periodicity to the shapes. Patterns did occur in his model, but the periodicity did not match that of the elements.

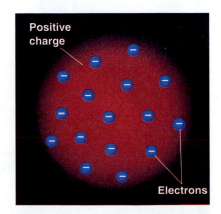

Figure 22-9   Thomson's model of the atom.

Rutherford's Model

A student of Thomson, Ernest Rutherford, did an experiment that radically changed his teacher's model of the atom. A key tool in Rutherford's work was the newly discovered radioactivity, the spontaneous emission of fast-moving atomic particles from certain elements. Rutherford had previously shown that one of these radioactive products is a helium atom without its electrons. These particles have two units of positive charge and are known as **alpha particles.**

helium atoms without electrons are alpha particles

Rutherford bombarded materials with alpha particles and observed how the alpha particles recoiled from collisions with the atoms in the material. This procedure was much like determining the shape of an object hidden under a table by rolling rubber balls at it (Fig. 22-10). Rutherford used the alpha particles as his probe because they were known to be about the same size as atoms. (One doesn't probe the structure of snowflakes with a sledgehammer.) The target needed to be as thin as possible to keep the number of collisions by a single alpha particle as small as

Powerfully Simple

Lord Ernest Rutherford.

Ernest Rutherford, the father of atomic physics, grew up on a small farm in New Zealand toward the end of the 19th century. He graduated from New Zealand University in Christchurch and immediately undertook various research projects of his own design. He was sufficiently brilliant and fortunate to receive New Zealand's only research-based scholarship. His mother told him of his award as he was digging potatoes on the family farm. He dropped his shovel and remarked, "That's the last potato I'll dig." The scholarship allowed Rutherford to travel to England and begin work at Cambridge University.

Rutherford had a very productive research career. Early on he studied radioactive decay (Chapter 24) and coined the term "half-life" to describe the decay process. He and Hans Geiger studied alpha particles and proved that they were helium atoms with their electrons removed. He is also credited with naming the

"proton." Rutherford's conclusions about the structure of atoms were so well supported with experimental findings that a contemporary was quoted as having said of the possibility of further subatomic research, "What Rutherford has left us of the large stone is scarcely worth kicking." For this work Rutherford was awarded the 1908 Nobel Prize in chemistry.

In 1919 Rutherford was appointed Cavendish Professor at Cambridge and remained there until he died suddenly in 1937. He was knighted in 1914 and made a Baron in 1931, which gave him a seat in the House of Lords. Rutherford directed the famous Cavendish Laboratory and became quite a popular figure. He was always accessible to his students, and heartily encouraged them. Rutherford's genius lay in his ability to simplify even the most complex problem. Beginning with his inevitable inquiry, "What are the facts?" Rutherford was able to get to the bottom of any problem quickly through a logical series of questions. His method of solving problems is an idealization of the experimental methodology of every physicist; from simple answers to direct fundamental questions, he formed powerful conclusions.

Adapted from an essay by Steven Janke for Pasco Scientific.

References: Eve, A.S. *Rutherford.* New York: Macmillan, 1939. Feather, N. *Lord Rutherford.* London: Blackie & Sons, 1940.

possible, ideally only one. Rutherford chose gold because it could be made into a very thin foil.

The alpha particles are about 4 times the mass of hydrogen atoms and 7300 times the mass of electrons. Thomson's model predicted that the deflection of alpha particles by each atom (and hence by the foil) would be extremely small. The positive material of the atom was too spread out to be a very effective scatterer. Consider the following analogy. Use 100 kilograms of newspapers to make a wall one sheet thick. Glue ping-pong balls on the wall to represent electrons. If you were to throw bowling balls at the wall, you would clearly expect them to go straight through with no noticeable deflection.

Initially, Rutherford's results matched Thomson's predictions. However, when Rutherford suggested that one of his students look for alpha particles in the backward direction, he was shocked with the results. A small number of the alpha particles were actually scattered in the backward direction! Rutherford later described his feelings by saying, "It was almost as incredible as if you fired a 19-inch shell at a piece of tissue paper and it came back and hit you!"

On the basis of his experiment, Rutherford proposed a new model in which the positive charge was not spread out but was concentrated in a very, very tiny spot at the center of the atom, which he called the **nucleus.** The scattering of alpha particles is shown in Figure 22-11. Returning to our analogy, imagine what would happen if all the newspapers were crushed into a small region of space: Most of the bowling balls would miss

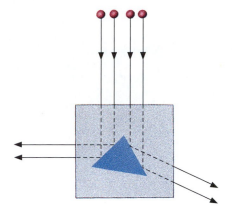

Figure 22-10   The shape of a hidden object can be determined by rolling rubber balls and seeing how they are scattered.

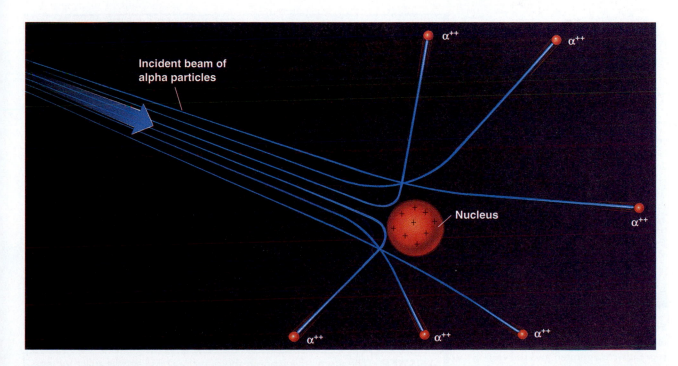

Figure 22-11   The positively charged alpha particles are repelled by the positive charges in the nucleus.

the paper and pass straight through, but occasionally a ball would score a direct hit on the 100-kilogram ball of newspaper and recoil.

Rutherford determined the speed of the alpha particles by measuring their deflections in a known magnetic field. Knowing their speeds, he could calculate their kinetic energies. Since he knew the charges on the alpha particle and the nucleus of the gold atom, he could also calculate the electric potential energy between the two. These calculations allowed him to calculate the closest distance an alpha particle could approach the nucleus of the gold atom before its kinetic energy had been entirely converted to potential energy. These experiments gave him an upper limit on the size of the gold nucleus. The nucleus has a diameter approximately 100,000 times smaller than its atom. This means that atoms are almost entirely empty space.

> **QUESTION**   If the nucleus were the size of a basketball, what would the diameter of the atom be?

Our work in electricity tells us that the negative electrons are attracted to the positive nucleus. Rutherford proposed that they don't crash together at the center of the atom because the electrons orbit the nucleus much like the planets in our Solar System orbit the Sun (Fig. 22-12). The force that holds the orbiting electrons to the nucleus is the electrical force. (The gravitational force is negligible for most considerations at the atomic level.)

> **QUESTION**   If atoms are mostly empty space, why can't we simply pass through each other?

Rutherford's work led to a new explanation for the emission of light from atoms. The orbiting electrons are accelerating (Chapter 3), and accelerating charges radiate electromagnetic waves (Chapter 21). A calculation showed that the orbital frequencies of the electron correspond to the region of visible light. But there is a problem. If light is radiated, the energy of the electron must be reduced. This loss would cause the electron to spiral into the nucleus. In other words, this model predicts that the atom is not stable and should collapse. Calculations showed that this should occur in about a billionth of a second! Since we know that most atoms are stable for billions of years, there is an obvious defect in the model.

> **ANSWER**   Assuming that a basketball has a diameter of about 24 centimeters, the diameter of the atom would be (100,000)(0.24 meters) = 24 kilometers (15 miles) across.

> **ANSWER**   The atom's electrons in the outer regions keep us apart. When two atoms come near each other, the electrons repel each other strongly.

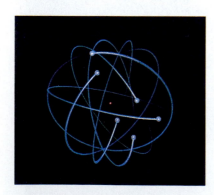

Figure 22-12 Rutherford's model of the atom. (The size of the nucleus is greatly exaggerated and the electrons are known to be even smaller than the nucleus.)

Rutherford's "solar system" model of the atom also failed to predict spectral lines and to account for the periodic properties of the elements. The resolution of these problems was accomplished by atomic models proposed during the two decades following Rutherford's work.

Radiating Objects

The Thomson and Rutherford models of the atom were able to explain some of the experimental observations, but they had serious deficiencies. Clearly, a new model was needed. Interestingly, the germination of the next model began in yet another area of physics, one having to do with the properties of the radiation emitted by a hot object.

All objects glow. At normal temperatures the glow is invisible; our eyes are not sensitive to it. As an object gets warmer, we feel this glow—we call it heat. At even higher temperatures, we can see the glow. The element of an electric stove or a space heater, for example, has a red-orange glow. In all these cases the object is emitting electromagnetic radiation. The frequencies, or colors, of the electromagnetic radiation are not equally bright as can be seen in the photographs of Figure 22-13. Furthermore, the color spectrum shifts toward the blue as the temperature of the object increases.

All solid objects give off continuous spectra that are similar, but not identical. However, the radiation is the same for all materials at the same temperature if the radiation comes from a small hole in the side of a hollow block. Typical distributions of the brightness of the electromagnetic radiation versus wavelength are shown in the graph of Figure 22-14 for several temperatures. When the temperature changes, the position of the hump in the curve changes; the wavelength with the maximum intensity varies inversely with the temperature. That is, as the temperature goes up, the highest point of the curve moves to smaller wavelengths.

> **QUESTION**   What happens to the frequency corresponding to the maximum intensity as the temperature increases?

The fact that the same distribution is obtained for cavities made of any material—gold, copper, salt, or whatever—puzzled and excited the experimenters. However, all attempts to develop a theory that would explain the shape of this curve failed. Most attempts assumed that the material contained some sort of charged atomic oscillators that generated light of the same frequency as their vibrations. Although the theories were able to account for the shape of the large-wavelength end of the curve, none could account for the small contribution at the shorter wavelengths (Fig. 22-15). Nobody knew why the atomic oscillators did not vibrate as well at the higher frequencies.

> **ANSWER**   Because the product of the frequency and the wavelength must equal the speed of light, the frequency must increase as the wavelength decreases. Therefore, the frequency increases as the temperature increases.

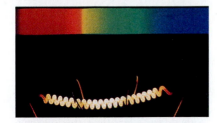
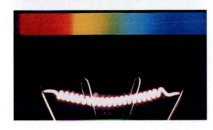
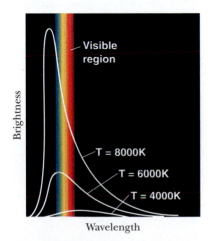

Figure 22-13   The color of the hot filament and its continuous spectrum change as the temperature is increased.

Figure 22-14   This graph shows the intensity of the light emitted by a hot object at various wavelengths for three different temperatures.

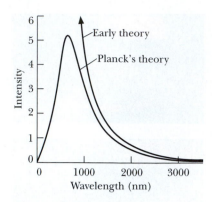

Figure 22-15 The theoretical ideas and the experimental data for the intensity of the light emitted by a hot body differed greatly in the low wavelength (high-frequency) region.

energy quanta = Planck's constant × frequency

Planck's constant

A solution was discovered in 1900 by Max Planck, a German physicist. He succeeded by borrowing a new technique from another area of physics. The technique called for temporarily pretending that a continuous property was discrete in order to make particular calculations. After finishing the calculation, the technique called for returning to the real situation—the continuous case—by letting the size of the discrete packets become infinitesimally small.

Planck was attempting to add up the contributions from the various atomic oscillators. Unable to get the right answer, he temporarily assumed that energy existed in tiny chunks. Planck discovered that if he didn't take the last step of letting the chunk size become very small, he got the right results! Planck announced his success in very cautious tones —almost as if he were apologizing for using a trick on nature.

Planck's "incomplete calculation" was, in effect, an assumption that the atomic oscillators could only vibrate with certain discrete energies. Because the energy could only have certain discrete values, it would be radiated in little bundles, or quanta. The energy E of each bundle was given by

$$E = hf$$

where h is a constant now known as *Planck's constant* and f is the frequency of the atomic oscillator. An oscillator could have an energy equal to 1, 2, 3, or more of these bundles, but not 2½. The value of Planck's constant, 6.63×10^{-34} joule-second, was obtained by matching the theory to the radiation spectrum.

COMPUTING   ENERGY OF A QUANTUM   Σ

We can get a feeling for the size of the energies of the atomic oscillators from our knowledge that red light has a frequency around 4.6×10^{14} hertz. Substituting these numbers into our relationship for the energy yields

$$E = hf = (6.63 \times 10^{-34}\,\text{J} \cdot \text{s})\,(4.6 \times 10^{14}\,\text{Hz}) = 3.05 \times 10^{-19}\,\text{J}$$

> **QUESTION**   What is the size of the energy quantum for blue light with a frequency of 7×10^{14} hertz?

Planck's idea explained why the contribution of the shorter wavelengths was smaller than predicted by earlier theories. Shorter wavelengths mean higher frequencies and therefore more energetic quanta. Because there is only a certain amount of energy in the macroscopic object and this energy must be shared among the atomic oscillators, the high-energy quanta are less likely to occur.

ANSWER 4.64×10^{-19} J.

Quantum of Energy

Max Planck.

Max Planck was born in Kiel, Germany, and received his college education in Munich and Berlin. After serving on the faculties of Munich and Kiel, Planck moved to the University of Berlin, where he remained until he retired. In 1918 Max Planck was awarded the Nobel Prize for the discovery of the quantized nature of energy, which laid the foundation for quantum theory (Chapter 23). The work leading to the "lucky" formula for the radiation from hot materials was described by Planck in his acceptance speech: "But even if the radiation formula proved to be perfectly correct, it would after all have been only an interpolation formula found by lucky guesswork and thus, would have left us rather unsatis-

fied. I therefore strived from the day of its discovery to give it a real physical interpretation."

Planck's life was filled with personal tragedies. One of his sons was killed in action in World War I and two daughters died in childbirth. His house was destroyed by bombs in World War II and another son was executed by the Nazis after being accused of plotting to assassinate Hitler.

In 1930 Planck became president of the Kaiser Wilhelm Institute of Berlin, which was renamed the Max Planck Institute in his honor after World War II. Although Planck remained in Germany during the Hitler regime, he openly protested the Nazi treatment of his Jewish colleagues, and consequently was forced to resign his presidency. Following the war, he was reinstated as president of the Max Planck Institute, and spent the last two years of his life as an honored and respected scientist and humanitarian.

Adapted from an essay by Steven Janke for Pasco Scientific.

This discreteness of energy was in stark disagreement with the universally held view that energy was a continuous quantity and any value was possible. At first thought you may want to challenge Planck's result. If this were true, it would seem that our everyday experiences would have demonstrated it. For example, the kinetic energy of a ball rolling down a hill appears to be continuous. As the ball gains kinetic energy, its speed appears to increase continuously, not jerking from one allowed energy state to the next. There doesn't seem to be any restriction on the values of the kinetic energy.

We are, however, talking about very, very small packets of energy. A baseball falling off a table has energies in the neighborhood of 10^{19} times larger than those of the quanta. These energy packets are so small we don't notice their size in our everyday experiences. On our normal scale of events energy seems continuous.

The Photoelectric Effect

Shortly before 1900 another phenomenon was discovered that connected electricity, light, and the structure of atoms. When light is shined on certain clean metallic objects, electrons are ejected from their surfaces (Fig. 22-16). This effect is known as the **photoelectric effect.** The electrons come off with a range of energies up to a maximum energy depending on the color of the light. If the light's intensity is changed, the rate of ejected electrons changes but the maximum energy stays fixed.

Figure 22-16   The photoelectric effect. Light shining on a clean metal surface causes electrons to be given off the surface.

> **QUESTION** Does the name *photoelectric effect* make sense for this phenomenon?

The ideas about atoms and light prevalent at the end of the 19th century were able to account for the possibility of such an effect. Because Planck's oscillators produced light, presumably light could jiggle the atoms. Light is energy, and when it is shined on a surface, some of this energy can be transferred to the electrons within the metal. If the electrons accumulated enough kinetic energy, some should have been able to escape from the metal. But no theory could account for the details of the photoelectric process. For instance, if the light source was very weak, the old theories predicted that it should take an average of several hours before the first electron gets enough energy to escape. On the other hand, experiments showed that the average time was much less than 1 second.

A range of kinetic energies made sense in the old theories—electrons bounced about within the metal accumulating energy until they got near the surface going in the right direction and escaped. There were bound to be differences in their energies. But a maximum value for the kinetic energy of the electrons was puzzling. The wave theory of light said that increasing the intensity of the light shining on the metal would increase the kinetic energy of the electrons leaving the metal. However, experiments showed that the intensity had no effect on the maximum kinetic energy. When the intensity was increased, the number of electrons leaving the metal increased but their distribution of energy was unchanged.

One thing did change the maximum kinetic energy. If the frequency of the incident light was increased, the maximum kinetic energy of the emitted electrons increased. This result was totally unexpected. If light were a wave, the amplitude of the wave would have been expected to govern the amount of energy that could be transferred to the electron. Frequency (color) should have had no effect.

The established ideas about light were inadequate to explain the photoelectric effect. The conflict was resolved by Albert Einstein. In yet another outstanding paper written in 1905, he claimed that the photoelectric effect could be explained by extending the work of Planck. Einstein said that Planck didn't go far enough. Not only are the atomic oscillators quantized, but the experimental results could only be explained if light itself is also quantized! Light exists as bundles of energy called **photons.** The energy of a photon is given by Planck's relationship,

energy of a photon

$$E = hf$$

Although it was universally agreed that light was a wave phenomenon, Einstein said that these photons acted like particles, little bundles of energy.

Einstein was able to explain all of the observations from the photoelectric effect. The ejection of an electron occurs when a photon hits an

> **ANSWER** Yes. The root *photo* is often associated with light; the most obvious example is photography. The *electric* root refers to the emitted electrons.

electron. A photon gives up its *entire* energy to a single electron. The electrons that escape have different amounts of energy depending on how much they lose in the material before reaching the surface. Also, the fact that the first photon to strike the surface can eject an electron easily accounts for the short time between turning on the light and the appearance of the first electron. A maximum kinetic energy for the ejected electrons is now easy to explain. A photon cannot give an electron any more energy than it has, and the chances of an electron getting hit twice before it leaves the metal are extremely small.

> **QUESTION**   How does this model explain the change in the maximum electron energy when the color of the light changes?

Einstein's model suggests a different interpretation of intensity. Because all photons of a given frequency have the same energy, an increase in the intensity of the light is due to an increase in the number of photons, not the energy of each one. Therefore, the model correctly predicts that more electrons will be ejected by the increased number of photons but their energies will be the same.

Einstein's success introduced a severe conflict into the physics world view. Diffraction and interference effects (Chapter 18) required that light behave as a wave; the photoelectric effect required that light behave as particles. How can light be both? These two ideas are mutually contradictory. This is not like saying that someone is fat and tall, but rather like saying that someone is short and tall. The two descriptions are incompatible. This conflict took some time to resolve. We will examine this important development in the next chapter.

Meanwhile, Einstein's work inspired a new, improved model of the atom.

Bohr's Model

The model of the atom proposed by Rutherford had several severe problems. It didn't account for the periodicity of the elements, it was unable to give any clue about the origins of the spectral lines, and it implied that atoms were extremely unstable. In 1913, the Danish theoretician Niels Bohr proposed a new model incorporating Planck's discrete energies and Einstein's photon into Rutherford's model.

Scientists are always using models when attempting to understand nature. The trick is to understand the limitations of the models. Bohr challenged the Rutherford model, stating that it was a mistake to assume that the atom is just a scaled-down solar system with the same rules; electrons do not behave like miniature planets. Bohr proposed a new model based on three assumptions, or postulates.

> **ANSWER**   This change occurs because the photon's color and energy depend on the frequency; the higher the frequency, the more energy the photon can give to the electron.

First, he proposed that the angular momentum of the electron is quantized. Only momenta L equal to whole-number multiples of a smallest angular momentum are allowed. This smallest angular momentum is equal to Planck's constant h divided by 2π. Bohr's first postulate can be written

$$L_n = n\frac{h}{2\pi}$$

where n is a positive integer (whole number) known as a **quantum number.**

In classical physics, the angular momentum L of a particle of mass m traveling at a speed v in a circular orbit of radius r is given by $L = mvr$ (Chapter 5). Therefore, the restriction on the possible values of the angular momentum puts restrictions on the possible radii and speeds. Bohr showed that this restriction on the electron's angular momentum means that the only possible orbits are those for which the radii obey the relationship

$$r_n = n^2 r_1$$

where r_1 is the smallest radius and n is the same integer that appears in the first postulate. This means that the electrons cannot occupy orbits of arbitrary size, but only a certain discrete set of allowable orbits, as illustrated in Figure 22-17. The numerical value of r_1 is 5.3×10^{-11} meter.

There is also a definite speed associated with each possible orbit. This means that the kinetic energies are also quantized. Because the value of the electric potential energy depends on distance, the quantized radii means that the potential energy also has discrete values. Therefore, there is a discrete set of allowable energies for the electron.

> **QUESTION**   Would you expect the energy of the electron to increase or decrease for larger orbits?

Bohr's second postulate states that an electron does not radiate when it is in one of the allowed orbits. This statement is contrary to the observation that accelerated charges radiate energy at a frequency equal to their frequency of vibration or revolution (Chapter 21). Bohr challenged the assumption that this property is also true in the atomic domain. Radically breaking away from what was accepted, Bohr said that an electron has a constant energy when it is in an allowed orbit.

But light is produced by atoms. If light isn't radiated by orbiting electrons, how does an atom emit light? Bohr's third postulate answered this question: A single photon is emitted whenever an electron jumps down from one orbit to another. Energy conservation demands that the photon

> **ANSWER**   Because the electric force is analogous to the gravitational force, we can think about putting satellites into orbit around the Earth. Higher orbits require bigger rockets and therefore more energy. Therefore, large orbits should have more energy.

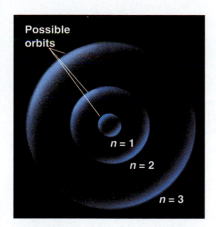

Figure 22-17 Electrons can only occupy certain orbits. These are designated by the quantum number n.

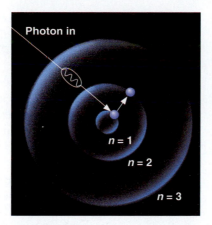

Figure 22-18 An electron can absorb a photon and jump to a higher energy orbit.

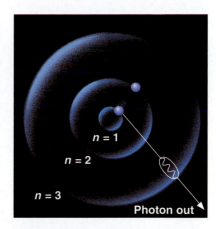

Figure 22-19 When an electron jumps to a lower energy orbit, it emits a photon.

has an energy equal to the difference in the energies of the two levels. It further demands that jumps up to higher levels can only occur when photons are absorbed. The electron is normally in the smallest orbit, the one with the lowest energy. If the atom absorbs a photon, the energy of the photon goes into raising the electron into a higher orbit (Fig. 22-18). Furthermore, it is not possible for the atom to absorb part of the energy of the photon. It is all or nothing.

The electron in the higher orbit is unstable and eventually returns to the innermost orbit, or ground state. If the electron returns to the ground state in a single jump, it emits a new photon with an energy equal to that of the original photon (Fig. 22-19). Again, this is different from our common experience. When a ball loses its mechanical energy, its temperature rises—the energy is transferred to its internal structure. As far as we know, an electron has no internal structure. Therefore, it loses its energy by creating a photon.

Occasionally, the phrase "a quantum leap" is used to imply a big jump. But the jump needn't be big. There is nothing big about the quantum jumps we are discussing. A quantum leap simply refers to a change from one discrete value to another.

Ordinarily, an amount of energy is stated in joules, the standard energy unit. However, this unit is extremely large for work on the atomic level, and a smaller unit is customarily used. The **electron volt** (eV) is equal to the kinetic energy acquired by an electron falling through a potential difference of 1 volt. One electron volt is equal to 1.6×10^{-19} joule. It requires 13.6 electron volts to remove an electron from the ground state of the hydrogen atom.

Atomic Spectra Explained

Bohr's explanation accounted for the existence of spectral lines, and his numbers even agreed with the wavelengths observed in the hydrogen spectrum. To illustrate Bohr's model, we draw energy-level diagrams like

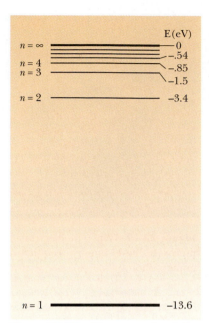

Figure 22-20 The energy-level diagram for the hydrogen atom.

the one in Figure 22-20. The ground state has the lowest energy and appears in the lowest position. Higher energy states appear above the ground state and are spaced to indicate the relative energy differences between the states.

To see how Bohr's model gives the spectral lines, consider the energy-level diagram for the hypothetical atom shown in Figure 22-21(a). Suppose that the electron has been excited into the $n = 4$ energy level. There are several ways that it can return to the ground state. If it jumps directly to the ground state, it emits the largest energy photon. We will assume it appears in the blue part of the emission spectrum in Figure 22-21(b). This line is marked $4 \rightarrow 1$ on the spectrum, indicating that the jump was from the $n = 4$ level to the $n = 1$ level.

The electron could also return to the ground state by making a set of smaller jumps. It could go from the $n = 4$ to the $n = 2$ level and then from the $n = 2$ to the ground level. These spectral lines are marked $4 \rightarrow 2$ and $2 \rightarrow 1$. The energies of these photons are smaller, so their lines occur at the red end of the spectrum. The lines corresponding to the other possible jumps are also shown in Figure 22-21(b).

QUESTION How do the energies of the photons that produce the $4 \rightarrow 2$ and $2 \rightarrow 1$ lines compare with the photon that produces the $4 \rightarrow 1$ line?

A single photon is not energetic enough for us to see. In an actual situation the gas contains huge numbers of excited atoms. Each electron takes only one of the paths we have described in returning to the ground state. The total effect of all the individual jumps is the atomic emission spectrum we observe.

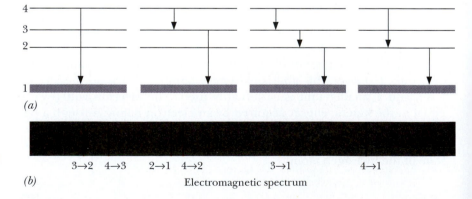

Figure 22-21 (a) Four possible ways for the electron in a hypothetical atom to drop from the $n = 4$ level to the $n = 1$ level. (b) The emission spectrum produced by these jumps.

ANSWER Conservation of energy requires that the two photons have the same total energy as the single photon.

Atomic Imagination

Niels Bohr.

Niels Bohr was the first son of a well-respected Danish professor. Despite the fact that he was mechanically adept, he soon established a reputation at the University of Copenhagen for breaking the most glassware in the chemistry laboratory. One professor even discounted a small series of explosions coming from the lab by casually noting, "It's just Bohr." His interests clearly lay in physics, although he was preoccupied by an occasional philosophical question.

Bohr received his doctorate in 1911. The next year he journeyed to England to work with Thomson at Cambridge, and then went on to Manchester to work with Rutherford. The two became friends and are responsible for the model of the atom discussed in this chapter. He was awarded a Nobel Prize in 1922.

In 1912 Bohr returned to Denmark as a professor at the University of Copenhagen. During the 1920s and 1930s he headed the Institute for Advanced Studies, which was sponsored by the Carlsburg brewery. The Institute was a magnet for many of the world's best physicists, providing a forum for the exchange of ideas, and fostering great progress in theoretical subatomic physics. Bohr was very much the lifeblood of the Institute, conducting physics on a "person-to-person" basis.

In 1939 Bohr visited the United States to attend a scientific conference, bringing news of the discovery of the fissioning of uranium, which led to the development of the atomic bomb during World War II. He returned to his homeland and remained there during the German occupation of Denmark in 1940. In 1943 he was smuggled out of Denmark in the bomb-bay of a British bomber.

Bohr returned to the United States to work on the Manhattan Project. Due partially to his role in developing "the bomb," Bohr felt strongly that openness between nations concerning nuclear weapons should be the first step in establishing their control. After the war Bohr committed himself to many humanitarian issues, including the development of peaceful uses of atomic energy. He was awarded the first Atoms for Peace Award in 1957. A colleague has perhaps summarized Bohr best by saying, "Bohr was a great scientist. He was a great citizen of Denmark, of the World. He was a great human being."

Adapted from essays by Steven Janke for Pasco Scientific and Serway, R.A. *Physics for Scientists and Engineers,* 3rd ed. updated. Philadelphia: Saunders College Publishing, 1993.

References: Moore, R. *Niels Bohr.* New York: Alfred Knopf, 1966. American Institute of Physics Niels Bohr Library. Special centennial issue of *Physics Today,* October 1985.

Bohr's scheme also explains the absorption spectrum. As we discussed earlier, the absorption spectrum is obtained when white light passes through a cool gas before it is dispersed by a prism or a diffraction grating. In this case, the photons from the beam of white light are absorbed and later reemitted. Because the reemitted photons are distributed in all directions, the intensities of these photons are reduced in the original direction of the beam. This removal of the photons from the beam leaves dark lines in the continuous spectrum.

The solution to the mystery of the missing lines in the absorption spectrum is easy within Bohr's model. For an absorption line to occur, there must be electrons in the lower level. Then they can be kicked up one or more levels by absorbing photons of the correct energy. Because almost all the electrons in the atoms of the gas occupy the ground state, only the lines that correspond to jumps from this state show up in the

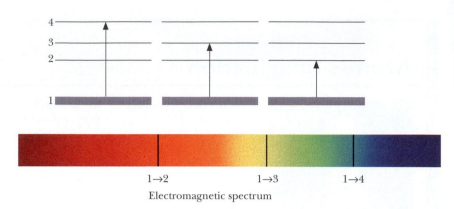

Electromagnetic spectrum

Figure 22-22   The absorption spectrum for the hypothetical atom in Figure 22-21.

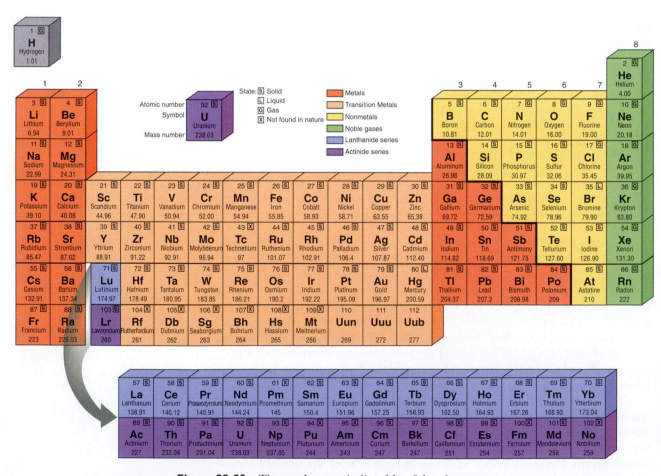

Figure 22-23   The modern periodic table of the elements.

spectrum. Jumps up from higher levels are extremely unlikely because electrons excited to a higher level typically remain there for less than a millionth of a second. Thus, there are fewer lines in the absorption spectrum (Fig. 22-22) than in the emission spectrum, in agreement with the observations.

The Periodic Table

Bohr's model helped unravel the mystery of the periodic properties of the chemical elements, a major unresolved problem for several decades. The table displaying the repeating chemical properties had grown considerably since Mendeleev's discovery and looked much like the modern one shown in Figure 22-23. About the time Bohr proposed his model, it became widely accepted that if the elements were numbered in the order of increasing atomic masses, their numbers would correspond to the number of electrons in the neutral atoms.

Bohr's theory gave a fairly complete picture of hydrogen (H): one electron orbiting a nucleus with one unit of positive charge (Fig. 22-24). However, it was not able to explain the details of the spectrum of the next element, helium (He), because of the effects of the mutual repulsion of the two electrons. But the theory was successful for the helium atom with one electron removed (the He$^+$ ion). Bohr correctly assumed that the helium atom has two electrons in the lowest energy level. Because the electrons presumably have separate orbits, the collection of orbits of the same size was called a **shell.**

The next element, lithium (Li), has spectral lines similar to those of hydrogen. This similarity could be explained if lithium had one electron orbiting a nucleus with three units of positive charge and an inner shell of two electrons. The two inner electrons partially shield the nucleus from the outer electron so it "sees" a net charge of approximately +1.

How many electrons can be in the second shell? The clue is found with the next element that has properties like lithium—sodium (Na), element 11. Sodium must have one electron outside two shells. Because the first shell holds two, the second shell holds eight. The next lithiumlike element is potassium (K), with 19 electrons. This time we have 1 electron orbiting 18 inner electrons: one shell of 2, and two shells of 8. And so on for succeeding elements.

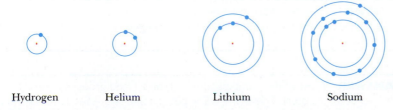

Hydrogen   Helium   Lithium   Sodium

Figure 22-24   The electron shell structure for hydrogen, helium, lithium, and sodium.

"The Periodic Table"

Physics Update

Element 112 has been discovered at the GSI Lab in Darmstadt, Germany, by the same researchers who first created elements 107–111. In the present experiment, led by Peter Armbruster, physicists smashed zinc atoms into a lead target. This resulted in the production of a single atom of the new element, the heaviest yet detected in a lab, with an atomic mass of 277.

In this scheme the chemical properties of atoms are determined by the electrons in their outermost shells. This idea can be extended to other vertical columns in the periodic table. The elements in the right-hand column (group 8) have their outer shells completely filled. These elements do not react chemically with other elements and rarely form molecules with themselves. They are inert. They are also called the *noble elements* because they don't mix with the common elements. The properties of group 8 elements led to the idea that the most stable atoms are those with filled outer shells.

The elements in group 7 are missing one electron in the outer shell and, like the elements in group 1, are chemically very active. They become much more stable by gaining an electron to fill this shell. Therefore, they often appear as negative ions. On the other hand, the elements in group 1 are more stable if they give up the single electron in the outer shell to become positive ions. Therefore, group 1 atoms readily combine with group 7 atoms. The atom from group 1 gives up an electron to that from group 7, and these two charged ions attract each other. Common table salt—sodium chloride—is such a combination. As soon as the two atoms combine, their individual properties—reactive metal and poisonous gas—disappear because the outer structure has changed.

QUESTION What other salts besides sodium chloride might you expect to find in nature?

Atoms can also share electrons. Oxygen (O) is missing two electrons in the outer shell, and hydrogen (H) is either missing one or has an extra. They can both have "filled" outer shells if they share electrons. Two hydrogens combine with one oxygen to give the very stable molecule H_2O.

ANSWER The next lithiumlike element is rubidium (Rb), with 37 electrons. Therefore, the 36 inner electrons must be arranged in shells with 2, 8, 8, and 18 electrons.

ANSWER Most of the combinations of group 1 and group 7 elements should occur in nature. Therefore, we expect to find such salts as sodium fluoride, sodium bromide, potassium fluoride, potassium chloride, and rubidium chloride.

These ideas were a great accomplishment for the Bohr model of the atom. Finally, there was a simple scheme for explaining the large collection of properties of the elements. The model provided a *physical* theory for understanding the *chemical* properties of the elements.

X Rays

The cathode-ray experiments that we discussed earlier in this chapter led to another important discovery that had an impact on the new atomic theory. In 1895, the German scientist Wilhelm Roentgen saw a glow coming from a piece of paper on a bench near a working cathode-ray tube. The paper had been coated with a fluorescent substance. Fluorescence was known to result from exposure to ultraviolet light, but there was no ultraviolet light present. The glow persisted even when the paper was several meters from the tube and the tube was covered with black paper!

Roentgen conducted numerous experiments demonstrating that while these new rays were associated with cathode rays, they were different. Cathode rays could penetrate no more than a few centimeters of air, but the new rays could penetrate several meters of air or even thin samples of wood, rubber, or aluminum. Roentgen was not able to deflect these rays using electric and magnetic fields, but did find that if he deflected the cathode rays, the rays came from the new spot where the cathode rays struck the glass walls. The new rays were also shown to be emitted by many different materials. Roentgen named them **X rays** because of their unknown nature.

The discovery of X rays quickly created public excitement because people could see their bones and internal organs. This discovery was quickly put to use by the medical profession. In 1901, Roentgen received the first Noble Prize in physics. We now know that X rays are high-frequency electromagnetic radiation created when the electrons in the cathode-ray tube are slowed down as they crash into the metal target inside the tube.

A typical X-ray spectrum (Fig. 22-25) consists of two parts: a continuous spectrum produced by the deceleration of the bombarding electrons, and a set of discrete spikes. These spikes are from X-ray photons emitted by the atoms and, therefore, carry information about the element. They occur when a bombarding electron knocks an atomic electron from a lower energy level completely out of the atom. The vacancy is filled by an electron in a higher energy level. When this electron falls into the vacated level, an X-ray photon is emitted. The energies of these X rays can be used to calculate the energy differences of the atom's inner electron orbits, which are characteristic of the element.

Shortly after Bohr proposed his model of the atom, Henry Mosely, a British physicist, was able to use Bohr's model and X rays to resolve some discrepancies in the periodic table. In several cases the order of the elements determined by their atomic masses disagreed with their order determined by their chemical properties. For instance, according to atomic masses, cobalt should come after nickel in the periodic table, yet its properties suggested that it should precede nickel.

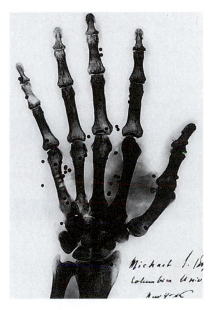

X ray of a hand hit by pellets.

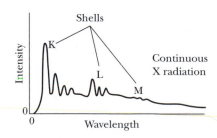

Figure 22-25 A typical X-ray spectrum contains a continuous background and characteristic lines from the inner atomic levels.

Bohr's model predicted that the energies of these characteristic X rays would increase as the number of electrons in the atom increased. Moseley bombarded various elements with electrons and studied the X rays that were given off. His experimental results agreed with Bohr's theoretical prediction. In almost all cases the results matched the order obtained from the atomic masses; however, they showed that cobalt should come before nickel. As a result of this work we now know that the correct ordering of the elements is by **atomic number** (the number of electrons) and not by atomic mass.

CHAPTER

22

REVISITED

The light emitted by an atom is created when an atomic electron moves from one energy level to a lower one. The size of the jump determines the energy of the photon and—via Planck's relationship— the color of the emitted light. Therefore, the colors provide us with information about the energy levels of the atom's electronic structure.

SUMMARY

When an electric current passes through a gas, it gives off a characteristic emission spectrum. The absorption spectrum is obtained by passing white light through a cool gas, producing dark lines in the continuous, rainbow spectrum. The emission spectrum has all the lines in the absorption spectrum plus many additional ones.

Cathode rays are electrons with a charge-to-mass ratio about 1800 times that for hydrogen ions. Electrons have an electric charge of 1.6×10^{-19} coulomb and a mass of 9.11×10^{-31} kilogram. The cathode-ray experiments also led to the discovery of X rays, a form of high-frequency electromagnetic radiation.

Rutherford used alpha particles to show that the positive charge was concentrated in a nucleus with a diameter 100,000 times smaller than the atom. Thus, Rutherford's "solar system" model of the atom is almost entirely empty space. It failed to explain the chemical periodicity and details of the spectral lines.

All objects emit electromagnetic radiation. The spectrum emitted by a heated object through a small hole in a hollow block is the same for all materials at the same temperature. The frequencies are not equally bright, and the maximum of the spectrum shifts toward the blue as the temperature of the cavity increases. The spectrum drops to zero on both the high-frequency (small-wavelength) and low-frequency (long-wavelength) sides.

Planck's analysis of the electromagnetic spectrum emitted by hot cavities provided the first clues that phenomena on the atomic level are quantized. The energy E of the quanta depends on Planck's constant h and the frequency f, $E = hf$.

Planck's idea was expanded to include the electromagnetic radiation by Einstein's efforts to explain the photoelectric effect. When light is shined on certain clean metallic surfaces, electrons are ejected with a range of energies up to a maximum energy depending on the color of the light. If the light's intensity is changed, the rate of ejecting electrons changes but the maximum energy stays fixed. This is explained by assuming that light is also quantized, existing as bundles of energy called photons. The energy of a photon is given by Planck's relationship, $E = hf$. A photon gives its entire energy to one electron in the metal.

Bohr's model for the hydrogen atom incorporated Planck's discrete energies and Einstein's photons into Rutherford's model. It was based on three postulates: (1) Allowed electron orbits are those in which the angular momentum is an integral multiple of a smallest one. (2) An electron in an allowed orbit does not radiate. (3) When an electron jumps from one orbit to another, it emits or absorbs a photon whose energy is equal to the difference in the energies of the two orbits.

The Bohr model was successful in accounting for many atomic observations, especially the emission and absorption spectra for hydrogen and the ordering of the chemical elements.

KEY TERMS

absorption spectrum: The collection of wavelengths missing from a continuous distribution of wavelengths. Caused by absorption of certain wavelengths by the atoms or molecules in a gas.

alpha particle: The nucleus of the helium atom.

atomic number: The number of electrons in the neutral atom of an element. This number also gives the order of the elements in the periodic table.

cathode ray: An electron emitted from the negative electrode in an evacuated tube.

electron: The basic constituent of atoms that has a negative charge.

electron volt: A unit of energy equal to the kinetic energy acquired by an electron or proton falling through an electric potential difference of 1 volt. Equal to 1.6×10^{-19} joule.

emission spectrum: The collection of discrete wavelengths emitted by atoms that have been excited by heating or electric currents.

ion: An atom with missing or extra electrons.

nucleus: The central part of an atom containing the positive charges.

photoelectric effect: The ejection of electrons from metallic surfaces by illuminating light.

photon: A particle of light. The energy of a photon is given by the relationship $E = hf$, where f is the frequency of the light and h is Planck's constant.

quantum: (pl., *quanta*) The smallest unit of a discrete property. For instance, the quantum of charge is the charge on the proton.

quantum number: A number giving the value of a quantized quantity. For instance, a quantum number specifies the angular momentum of an electron in an atom.

shell: A collection of electrons in an atom that have approximately the same energy.

X ray: A high-energy photon with a range of frequencies in the electromagnetic spectrum lying between the ultraviolet and the gamma rays.

CONCEPTUAL QUESTIONS

1. What determined the order of the elements in Mendeleev's periodic table?

2. What allowed Mendeleev to predict the existence of new elements?

3. What elements would you expect to have chemical properties similar to calcium (Ca)?

4. What element in Mendeleev's periodic table is most similar to chlorine (Cl)?

5. How could you determine if there is oxygen in the Sun?

6. Why are the spectral lines for elements sometimes called "atomic fingerprints"?

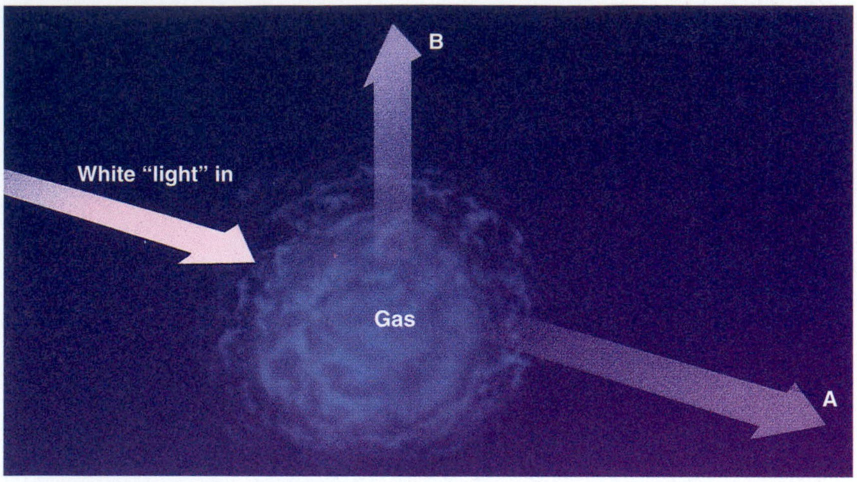
B
White "light" in
Gas
A

Questions 7–8.

7. What kind of spectrum is obtained at location A in the setup shown in the figure?

8. What kind of spectrum is obtained at location B in the setup shown in the figure?

9. The emission spectra shown in the figure were all obtained with the same apparatus. What elements are present in sample a?

10. What elements are present in sample b of the figure?

11. How does the number of lines in the absorption spectrum for an element compare with the number in the emission spectrum?

12. What are the differences between an emission spectrum and an absorption spectrum?

13. What observation(s) showed that cathode rays are not part of the electromagnetic spectrum?

14. How do we know that cathode rays have negative charges?

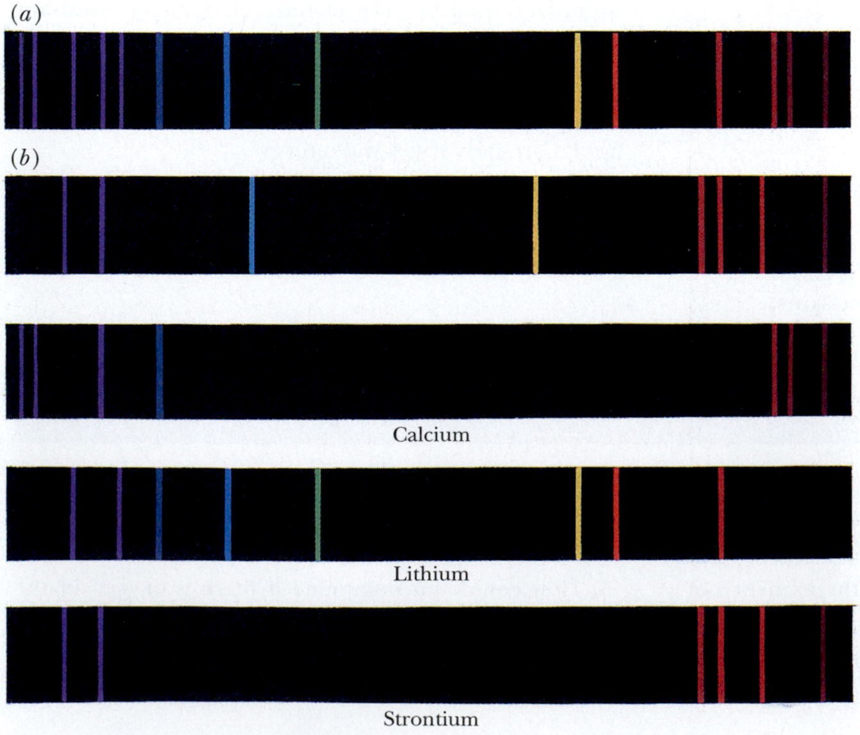
(*a*)
(*b*)
Calcium
Lithium
Strontium

Questions 9–10.

15. What do we mean when we say that a certain physical quantity is quantized?

16. What did Robert Millikan discover in his famous oil-drop experiment?

17. How does the amount of positive charge compare with the amount of negative charge in the Thomson atom?

18. Why did Thomson feel that electrons were pieces of atoms rather than atoms of a new element?

19. In Rutherford's model of the atom, what prevents the electrons from leaving the atoms?

20. List the following with the most massive first and the least massive last: atom, electron, and nucleus.

21. Did Rutherford's model explain the stability of atoms?

22. How does Rutherford's model of an atom explain why most alpha particles pass right through a thin gold foil?

23. How does the Rutherford model account for the backward scattering of alpha particles?

24. Could Rutherford's model of an atom as a very tiny, massive nucleus with the electrons orbiting at great distances explain why atoms emit a special set of discrete wavelengths?

25. Why can the curves for the intensities of the colors emitted by hot solid objects not serve as "atomic fingerprints" of the materials?

26. If all objects emit radiation, why don't we see most of them in the dark?

27. For approximately what temperature range does the radiation in Figure 22-14 have greater intensity in the ultraviolet than in the infrared?

28. For approximately what temperature range does the radiation in Figure 22-14 have greater intensity in the infrared than in the ultraviolet?

29. The curves in Figure 22-14 show the intensities of the various wavelengths emitted by an object at three different temperatures. The region corresponding to visible light is indicated. Describe the overall color that you see in each case.

30. What is the approximate temperature of something that is "white hot"?

31. What can be said about the relative temperatures of red, blue, and white stars?

32. Why are blue stars thought to be hotter than red stars?

33. What assumption(s) did Planck make that enabled him to obtain the correct curve for the spectrum of light emitted by a hot object?

34. What assumption(s) did Einstein make that enabled him to account for the experimental observations of the photoelectric effect?

35. Which of the following properties must be exceeded by the incident photons before electrons are emitted from a metallic surface: speed, charge, wavelength, frequency, or intensity?

36. What color light will produce photoelectrons with the highest average kinetic energy: red, blue, yellow, green, or orange?

37. What property of the emitted photoelectrons depends on the intensity of the incident light?

38. How does the maximum kinetic energy of the emitted photoelectrons depend on the frequency of the incident light?

39. How is it possible that UV light can cause sunburn but no amount of visible light will?

*40. A photographer can use a red safety light in a darkroom for hours without worrying about exposing the film. Explain why this is possible. What would happen if a regular bulb were used?

41. List the following photons in order of *increasing* energy: radio, infrared, X ray, visible, and ultraviolet.

*42. What happens if you shine ultraviolet light on the ball of a charged electroscope? Does it matter which charge the electroscope has?

43. What are the three assumptions of Bohr's model of the atom?

44. Why did Bohr assume that the electrons do not radiate when they are in the allowed orbits?

45. An electron in the $n = 3$ energy level can drop to the ground state by emitting a single photon or a pair of photons. How does the total energy of the pair compare with the energy of the single photon?

46. If electrons in hydrogen atoms are excited to the fourth Bohr orbit, how many different frequencies of light may be emitted?

47. What determines the frequency of a photon emitted by an atom?

48. How can the spectrum of hydrogen contain so many spectral lines when the hydrogen atom only has one electron?

49. How does Bohr's model explain that there are more lines in the emission spectrum than in the absorption spectrum?

50. Why does the spectrum of lithium (element 3) resemble that of hydrogen?

51. Why do certain families of elements in the periodic table have very similar properties?

*52. The chemical properties of elements 57 through 71 are similar. How might this be explained by Bohr's model?

53. Radon (element 86) is a gas. Would you expect the molecules of radon to consist of a single atom or a pair of atoms?

54. How many electrons would you expect to find in each shell for sulfur (S)?

55. Sodium does not naturally occur as a free element. Why?

56. What effective charge do the outer electrons in magnesium "see"? (Magnesium is element 12.)

57. What type of electromagnetic wave has a wavelength about the size of an atom? The electromagnetic spectrum is given in Figure 21-24.

58. Are X rays deflected by electric or magnetic fields or both? Explain.

59. How does an X ray differ from a photon of visible light?

60. What are the two different ways X rays can be produced?

61. Why would you not expect an X-ray photon to be emitted every time an inner electron is removed from an atom?

62. How does Bohr's model account for the observation that the energy of the most energetic X rays for each atom increases with atomic number?

EXERCISES

1. What is the charge-to-mass ratio for a cathode ray?

2. What is the charge-to-mass ratio for a hydrogen ion?

3. Given that the radius of a hydrogen atom is 5.3×10^{-11} m and that its mass is 1.682×10^{-27} kg, what is the average density of a hydrogen atom? How does it compare with the density of water?

4. What is the average density of the hydrogen ion given that its radius is 1.2×10^{-15} m and that its mass is 1.673×10^{-27} kg? It is interesting to note that such densities also occur in neutron stars.

5. A student decides to build a physical model of an atom. If the nucleus is a rubber ball with a diameter of 1 cm, how far away would the outer electrons be?

6. How does the volume of an atom compare with that of its nucleus? What does this say about the density of nuclear matter compared with ordinary matter?

7. What is the energy of a photon of orange light with a frequency of 5×10^{14} Hz?

8. What is the energy of the least energetic photon of visible light?

9. What is the angular momentum of an electron in the ground state of hydrogen?

10. What is the angular momentum of an electron in the $n = 4$ level of hydrogen?

11. What is the radius of the $n = 4$ level of hydrogen?

12. What is the quantum number of the orbit in the hydrogen atom that has 100 times the radius of the smallest orbit?

13. When a hydrogen ion captures an electron, a photon with an energy of 13.6 eV is emitted. What is the frequency of this photon? Does it lie in, above, or below the visible range?

14. According to Figure 22-20 it requires a photon with an energy of 10.2 eV to excite an electron from the $n = 1$ energy level to the $n = 2$ energy level. What is the frequency of this photon? Does it lie in, above, or below the visible range?

15. What is the ratio of the volumes of the hydrogen atom in the $n = 1$ state compared with the $n = 2$ state?

*16. The diameter of the hydrogen atom is 10^{-10} m. In Bohr's model this means that the electron travels a distance of about 3×10^{-10} m in orbiting the atom once. If the orbital frequency is 7×10^{15} Hz, what is the speed of the electron? How does this speed compare with that of light?

17. What difference in energy between two atomic levels is required to produce an X ray with a frequency of 5×10^{18} Hz?

18. What is the frequency of the X ray that is emitted when an electron drops down to the ground state from an excited state with 5000 eV more energy?

The Modern Atom

Modern lasers have allowed us to take truly three-dimensional pictures, perform surgery inside our bodies, weld retinas in our eyes, cut steel plates and fabric for blue jeans, and perform a myriad of other amazing tasks. What makes laser light so special? (See p. 580 for the answer to this question.)

These lasers are used to study antimatter at CERN, the European center for research in elementary particles.

Bohr's model of the atom lacked "beauty"; the way the theory blended classical and quantum ideas together in a seemingly contrived way to account for the experimental results was troublesome. Bohr used the ideas of Newtonian mechanics and Maxwell's equations until he ran into trouble; then he abruptly switched to the quantum ideas of Planck and Einstein. For example, the electron orbits were like Newton's orbits for the planets. The force between the electron and the nucleus was the electrostatic force. According to Maxwell's equations these electrons should have continuously radiated electromagnetic waves. Since this obviously didn't happen, Bohr postulated discrete orbits with photons being emitted only when an electron jumped from a higher orbit to a lower one. There was no satisfactory reason for the existence of the discrete orbits other than they gave the correct spectral lines.

Bohr's model of the atom was remarkably successful in some areas, but it failed in others. Even though it has been replaced by other models, it was important in advancing our understanding of matter at the atomic level.

Successes and Failures

Accounting for the stability of atoms by postulating orbits in which the electron did not radiate was a mild success of the Bohr theory. The problem was that nobody could give a fundamental reason why this was so. Why *don't* the electrons radiate? They are accelerating, and according to the classical theory of electromagnetism, they should radiate.

The Bohr model successfully provided the numerical values of the wavelengths in the spectra of hydrogen and hydrogenlike ions. It gave qualitative agreement for the elements with a single electron orbiting a filled shell, but could not predict the spectral lines when there was more than one electron in the outer shell.

But all along new data were being collected. Close examination of the spectral lines, especially when the atoms were in a magnetic field, revealed that the lines were split into two or more closely spaced lines. The Bohr theory could not account for this. Also, the different spectral lines for a particular element are not of the same brightness. Apparently some jumps were more likely than others. The Bohr theory provided no clue why this was so. Since the theory failed to show why, there was no way to calculate the likelihood of an electron jumping from one energy level to another.

Although Bohr's model successfully gave the general features of the periodic table, it was not able to provide the details of the subshells nor explain why the shells had a certain capacity for electrons. Why not just one or maybe three? And what's special about the next level that makes it accept eight electrons?

There were other loose ends. Einstein's theory of relativity was established and it was generally accepted that the speeds of the electrons were fast enough and the measurements precise enough that relativistic effects should be included. But Bohr's model was nonrealistic. Clearly, this was a transitional model.

De Broglie's Waves

In 1923, a French graduate student proposed a revolutionary idea to explain the discrete orbits. Louis de Broglie's idea is easy to state, but hard to accept. *Electrons behave like waves.* De Broglie reasoned that if light could behave like particles, then particles should exhibit wave properties. He viewed all atomic objects as having this dualism; each exhibits wave properties *and* particle properties.

De Broglie's view of the hydrogen atom had electrons forming standing-wave patterns about the nucleus like the standing-wave patterns on a guitar string (Chapter 15). Not every wavelength will form a standing wave on the string, only those that have certain relationships with the length of the string.

Imagine forming a wire into a circle. The traveling waves travel around the wire in both directions and overlap themselves. When a whole number of wavelengths fits along the circumference of the circle, a crest that travels around the circle arrives back at the starting place at the same time another crest is being generated. In this case the waves reinforce each other, creating a standing wave. This process is illustrated in Figure 23-1 for a wire and in Figure 23-2 for an atom.

De Broglie postulated that the wavelength λ of the electron is given by the expression

$$\lambda = \frac{h}{mv}$$

where h is Planck's constant, m is the electron's mass, and v is its speed. With this expression de Broglie was able to reproduce the numerical results for the orbits of hydrogen obtained by Bohr. The requirement that the waves form standing-wave patterns automatically restricted the possible wavelengths and therefore the possible energy levels.

electrons behave like waves

Louis de Broglie.

wavelength of an electron

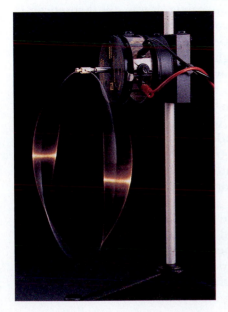
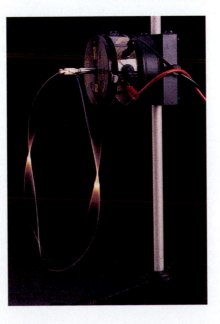

Figure 23-1 The shaft holding the wire executes simple harmonic motion that creates waves traveling around the circular wire in both directions. Standing waves form when the circumference is equal to a whole number of wavelengths.

Figure 23-2   The electrons form standing waves about the nucleus.

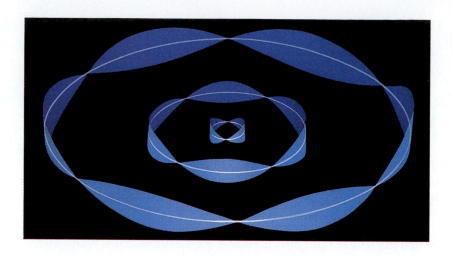

De Broglie's idea put the faculty at his university in a very uncomfortable position. If his idea proved to have no merit, they would look foolish awarding a Ph.D. for a crazy idea. The work was very speculative. Although he was able to account for the energy levels in hydrogen, there was no other experimental evidence to support his hypothesis. To make matters worse, de Broglie's educational background was in the humanities. He had only recently converted to the study of physics and was working in an area of physics that was not very well understood by the faculty. To make matters even worse, de Broglie came from a noble family that had played a leading role in French history. He carried the title of prince. His thesis supervisor resolved the problem by showing the work to Einstein. When Einstein indicated that the idea was basically sound, the dilemma was resolved.

Although de Broglie's idea was appealing, it was still theoretical. It was important to find experimental confirmation by observing some definitive wave behavior, such as diffraction or interference effects, for electrons. An electron accelerated through a potential difference of 100 volts has a speed that gives a de Broglie wavelength comparable to the size of an atom. Therefore, in order to see diffraction or interference effects produced by electrons, we need atom-sized slits. Since X rays have atom-sized wavelengths, they can be used to search for possible setups. Photographs of X rays scattered from randomly oriented crystals, such as the one shown in Figure 23-3, show that the regular alignment of the atoms in crystals can produce diffraction patterns like those produced by multiple slits. Therefore, we might look for the diffraction of electrons from crystals.

Experimental confirmation of de Broglie's idea came quickly. Two American scientists, C. J. Davisson and L. H. Germer, obtained confusing data from a seemingly unrelated experiment involving the scattering of low-energy electrons from nickel crystals. The scattered electrons had strange valleys and peaks in their distribution. While showing these data at a conference at Oxford, it was suggested to Davisson that this anomaly

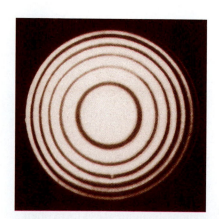

Figure 23-3   An interference pattern produced by light.

might be a diffraction effect. Further analysis of the data showed that electrons do exhibit wave behavior. The photograph in Figure 23-4 shows an interference pattern produced by electrons. De Broglie was the first of only two physicists to receive a Nobel Prize for his thesis work.

There is nothing in de Broglie's idea to suggest that it should only apply to electrons. This relationship for the wavelength should also apply to all material particles. Although it may not apply to macroscopic objects, assume for the moment that it does. A baseball (mass = 0.14 kilogram) thrown at 45 meters per second (100 miles per hour) would have a wavelength of 10^{-34} meter! This is an incredibly small distance. It is one trillion trillionth the size of a single atom. This wavelength is smaller compared with an atom than an atom compared with the Solar System. Even a mosquito flying at 1 meter per second would have a wavelength that is only 10^{-27} meter.

You shouldn't spend much time worrying that your automobile will diffract off the road the next time you drive through a tunnel. Ordinary-sized objects traveling at ordinary speeds have negligible wavelengths.

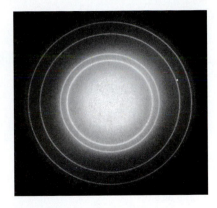

Figure 23-4   An interference pattern produced by electrons diffracting from randomly oriented aluminum crystallites.

> **QUESTION**   Why would you not expect baseballs thrown through a porthole to produce a measurable diffraction pattern?

Waves and Particles

Imagine the controversy de Broglie caused. Interference of light is believable because parts of the wave pass through each slit and interfere in the overlap region (Chapter 18). But electrons don't split in half. Every experiment designed to detect electrons has found complete electrons, not half an electron. So how can electrons produce interference patterns?

The following series of thought experiments was proposed by physicist Richard Feynman to summarize the many, many experiments that have been conducted to resolve the wave–particle dilemma. Although they are idealized, this sequence of experiments gets at crucial factors in the issue.

We will imagine passing various things through two slits and discuss the patterns that are produced on a screen behind the slits. In each case, there are three pieces of equipment: a source, two slits, and a detecting screen arranged as in Figure 23-5.

In the first situation we shoot indestructible bullets at two narrow slits in a steel plate. Assume that the gun wobbles, so that the bullets are fired randomly at the slits. Our detecting screen is a sand box. We simply count the bullets in certain regions in the sand box to determine the pattern.

Imagine that the experiment is conducted with the right-hand slit closed. After 1 hour we sift through the sand and make a graph of the

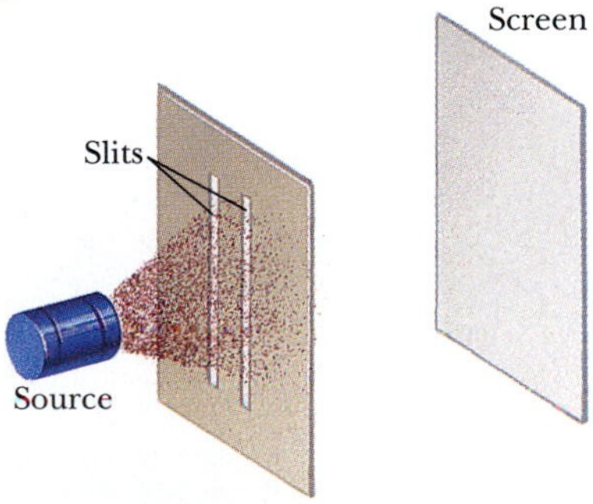

Figure 23-5   The experimental setup for the thought experiment concerning the wave-particle dilemma.

> **ANSWER**   For diffraction to be appreciable, the wavelength must be about the same size as the opening.

distribution of bullets. This graph is shown in Figure 23-6(a). The curve is labeled N_L to indicate that the left-hand slit was open. Repeating the experiment with only the right-hand slit opens yields the similar curve N_R shown in Figure 23-6(b).

QUESTION What curve would you expect with both slits open?

When both slits are opened, the number of bullets hitting each region during 1 hour is just the sum of the numbers in the previous experiments, that is, $N_{LR} = N_L + N_R$. This is just what we expect for bullets, or any other particles. This graph is shown in Figure 23-6(c).

Now imagine repeating the experiment using water waves. The source is now an oscillating bar that generates straight waves. The detecting screen is a collection of devices that measure the energy (that is, the intensity) of the wave arriving at each region. Since these are waves, the detector does not detect the energy arriving in chunks, but rather in a smooth, continuous manner. The graphs in Figure 23-7 show the average intensity of the waves at each position across the screen for the same three trials.

bullets add

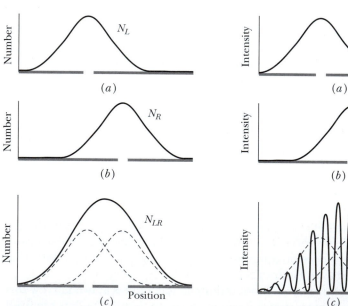

Figure 23-6 Distribution of bullets with (a) the left-hand slit open N_L, (b) the right-hand slit open N_R, and (c) both slits open N_{LR}.

Figure 23-7 Intensity of water waves with (a) the left-hand slit open I_L, (b) the right-hand slit open I_R, and (c) both slits open I_{LR}.

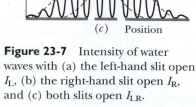

ANSWER The curve with both slits open should be the sum of the first two curves.

Once again, the three trials yield no surprises. We expect an interference pattern when both slits are open. We see that the intensity with two slits open is not equal to the sum of the two cases with only one slit open, $I_{LR} \neq I_L + I_R$, but this is what we expect for water waves, or any other waves.

Both sets of experiments make sense, in part, because we used materials from the macroscopic world—bullets and water waves. However, a new reality emerges when we repeat these experiments with particles from the atomic world.

This time we use electrons as our "bullets." An electron gun shoots electrons randomly toward two very narrow slits. Our screen consists of an array of devices that can detect electrons. Initially the results are similar to those obtained in the bullet experiment: Electrons arrive as whole particles. The patterns produced with either slit open are the expected ones and are the same as those in Figure 23-6(a and b). The surprise comes when we look at the pattern produced with both slits open: We get an interference pattern like the one for waves (Fig. 23-8)!

Note what this means: If we look at a spot on the screen that has a minimum number of electrons and close one slit, we get an *increase* in the number of counts at that spot. Closing one slit yields more electrons! This is not the behavior expected of particles.

One possibility that was suggested to explain these results is that the electrons are somehow affecting each other. We can test this by lowering the rate at which the source emits electrons so that only one electron passes through the setup at a time. In this case, we might expect the interference pattern to disappear. How can an electron possibly interfere with itself? Each individual electron should pass through one slit or the other. How can it even know that the other slit is open? But the interference pattern doesn't disappear. Even though there is only one electron in the apparatus at a time, interference effects are observed after a large number of electrons are measured.

The set of experiments can be repeated with photons. The results are the same. The detectors at the screen see complete photons, not half photons. But the two-slit pattern is an interference pattern. Photons behave like electrons. Photons and electrons exhibit a duality of particle and wave behavior. Table 23-1 summarizes the results of these experiments.

water waves interfere

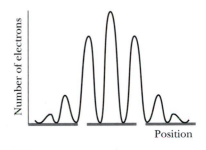

Figure 23-8 The distribution of electrons with two slits open.

electrons interfere!

photons interfere

Physics Update

A Feynman thought experiment has been demonstrated in an atom interferometer. In his lectures, Richard Feynman imagined an experiment that explored what would happen when you shine light at an object passing through an interferometer, a device that can split the object into a pair of wavelets, which are later recombined to produce an interference pattern. An MIT team has shone single photons on atoms inside such an interferometer. As Feynman correctly described, scattering a photon from an object inside the interferometer can destroy its wave properties. If one can, in principle, determine a pathway for the atom by detecting the position of the scattered photon, then the object acts as a particle. On the other hand, if the scattered photon cannot provide information on which path the atom traversed, the wave properties are not destroyed.

Table 23-1 Summary of experiments with two slits		
Source	**Detection**	**Pattern**
Bullets	Chunks	No interference
Water waves	Continuous	Interference
Electrons or photons	Chunks	Interference

Probability Waves

Nature has an underlying rule governing this strange behavior. We need to unravel the mystery of why electrons (or photons for that matter) are always detected as single particles and yet collectively they produce wave-like distributions. Something is adding together to give the interference patterns. To see this we once again look at water waves.

There is something about water waves that does add when the two slits are open—the displacements (or heights) of the individual waves at any instant of time. Their sum gives the displacement at all points of the interference pattern; that is, $h_{12} = h_1 + h_2$. The maximum displacement at each point is the amplitude of the resultant wave at that point.

With light and sound waves we observe the intensities of the waves, not their amplitudes. Intensity is proportional to the square of the amplitude. It is important to note that the resultant, or total, intensity is *not* the sum of the individual intensities. The total intensity is calculated by adding the individual displacements to obtain the resulting amplitude of the combined waves and then squaring this amplitude. Thus, the intensity depends on the amplitudes of each wave *and* their relative phase. As you can see in Figure 23-9, the intensity of the combined waves is larger than the sum of the individual intensities when the displacements are in the same direction and smaller when they are in opposite directions.

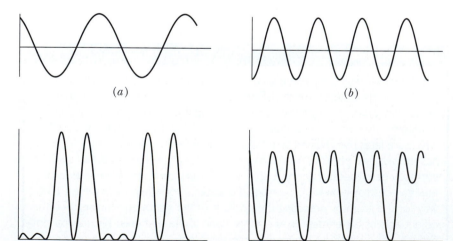

Figure 23-9 The two individual waves (a and b) are combined (c) by adding the displacements before squaring. Note the difference when (d) the individual displacements are squared before adding.

There is an analogous situation with electrons. The definition for a **matter-wave amplitude** (called a *wave function*) is analogous to that for water waves (this is usually represented by the symbol ψ and pronounced "sigh"). The square of the matter-wave amplitude is analogous to the intensity of a wave. In this case, however, the "intensity" represents the likelihood, or probability, of finding an electron at that location and time. There is one very important difference between the two cases. We can physically measure the amplitude of an ordinary mechanical wave, but there is *no* way to measure the amplitude of a matter wave. We can only measure the value of this amplitude squared.

These ideas led to the development of a new view of physics known as **quantum mechanics,** which are the rules for the behavior of particles at the atomic and subatomic levels. These rules replace Newton's and Maxwell's rules. A quantum-mechanical equation, called Schrödinger's equation after the Austrian physicist Erwin Schrödinger, is a wave equation that provides all possible information about atomic particles.

A Particle in a Box

It is instructive to look at another simple, though somewhat artificial, situation. Imagine you have an atomic particle that is confined to a box. Further imagine that the box has perfectly hard walls so no energy is lost in collisions with the walls and that the particle only moves in one dimension, as shown in Figure 23-10.

From a Newtonian point of view, there are no restrictions on the motion of the particle; it could be at rest or bouncing back and forth with any speed. Since we assume that the walls are perfectly hard, the sizes of the particle's momentum and kinetic energy remain constant. In this classical situation things happen much as we would expect from our commonsense world view. The particle is like an ideal superball in zero gravity. It follows a definite path; we can predict when and where it will be at any time in the future.

Until the early 1900s it was assumed that an electron would behave the same way. We now know that atomic particles have a matter-wave character and their properties—position, momentum, and kinetic energy—are governed by Schrödinger's equation. When these particles are confined, their wave nature guarantees that their properties are quantized. For example, the solutions of Schrödinger's equation for the particle in a box are a set of standing waves. In fact, the solutions are identical to those on the guitar string that we examined in Chapter 15. There is a discrete set of wavelengths that fit the conditions of confinement.

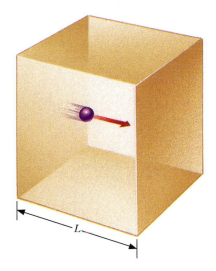

Figure 23-10   A particle in a box has quantized values for its de Broglie wavelength.

> **QUESTION**   How would you expect the wavelength of the fundamental standing wave to compare with the length of the box?

> **ANSWER**   In analogy with the string, we would expect the wavelength to be twice the length of the box.

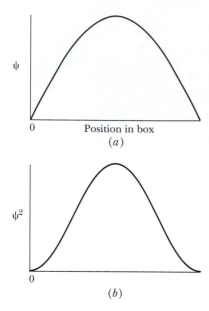

Figure 23-11 The (a) wave amplitude and (b) probability distribution for the lowest energy state for a particle in a box.

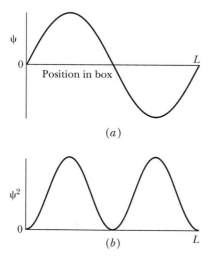

Figure 23-12 The (a) wave amplitude and (b) probability distribution for the first excited state for a particle in a box.

The fundamental standing wave corresponds to the lowest energy state (the ground level), and one-half of a wavelength fits into the box as shown in Figure 23-11(a). The probability of finding the particle at some location depends on the square of this wave amplitude. These probability values [Fig. 23-11(b)] help locate the particle. You can see that the most likely place for finding the atomic particle is near the middle of the box because ψ^2 is large there. The probability of finding it near either end is quite small.

A bizarre situation arises with the higher energy levels. The wave amplitude and its square for the next higher energy level are shown in Figure 23-12. Now the most likely places to find the particle are midway between the center and either end of the box. The least likely places are near an end or near the center. It is tempting to ask how the atomic particle gets from one region of high probability to the other. How does it cross the center where the probability of locating it is zero? This type of question, however, does not make sense in quantum mechanics. Particles do not have well-defined paths; they have only probabilities of existing throughout the space in question.

Suppose you try to find the particle. You would find it in one place, not spread out over the length of the box. However, you cannot predict ahead of time where it will be. All you can do is predict the probability of finding it in a particular region. In addition, even if you know where it is at one time, you still cannot predict where it will be at a later time. Atomic particles do not follow well-defined paths as classical objects do.

> **QUESTION** Where are the most likely places to find the particle in the third energy level?

Quantized wavelengths mean that other quantities are also quantized. Because de Broglie's relationship tells us the momentum is inversely proportional to the wavelength, momentum is quantized. The kinetic energy is proportional to the square of the momentum. Therefore, kinetic energy must also be quantized. Thus, the particle in a box has quantized energy levels that can be numbered with a quantum number to distinguish one level from another as we did for the Bohr atom.

The Quantum-Mechanical Atom

The properties of the atom are calculated from the Schrödinger equation just as they are for the particle in a box. Although it seems bizarre, quantum mechanics works; it not only accounts for all of the properties the Bohr model did, but it also corrects most of the deficiencies of that earlier attempt.

But this success has come at a price. We no longer have an atom that is easily visualized; we can no longer imagine the electron as being a little

> **ANSWER** The square of the third standing wave will have maxima at the center and between the center and each side of the box (see Fig. 15-6).

Psychedelic Colors

In Chapter 16 we discussed why objects have colors. The object's color depends on the colors reflected from the illuminating light, but there is an exception to this. When certain materials are illuminated with ultraviolet (UV) light, they glow brightly in a variety of colors. This process is called *fluorescence* and is responsible for the colors seen on many contemporary, psychedelic posters.

The "black lights" used to illuminate these posters radiate mostly in the UV region, which is invisible to our eyes. While the light gives off a little purple, there are certainly no reds, yellows, and greens present. And yet these bright colors appear in the posters. The atomic model explains their presence. UV photons are very energetic and can kick electrons to higher energy levels. If the electron returns by several jumps as shown in Figure A, it gives off several photons, each of which has a lower energy. When one or more of these photons lie in the visible region of the electromagnetic spectrum, the material fluoresces.

A laundry detergent manufacturer once claimed that its soap could make clothes "whiter than white" by pointing out that the soap contained fluorescent material. This makes sense scientifically because some of the ultraviolet light is converted to visible light, making the shirt give off more light in the visible region. You may wish to observe various types of clothing and laundry soap under ultraviolet light to see if they fluoresce. Do not look directly at the black light, as UV light is dangerous to your eyes.

This process is also used in fluorescent lights. The gas atoms inside the tube are excited by an electrical discharge and emit UV photons. These photons cause atoms in the coating on the inside of the tube to fluoresce, giving off visible photons.

A related phenomenon, *phosphorescence,* occurs with some materials. These materials continue to glow after the ultraviolet light is turned off. Some luminous watch dials are phosphorescent. In these materials the excited electrons remain much longer in upper energy states. In effect, phosphorescence is really a delayed fluorescence.

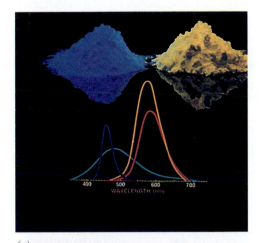

(a)

(b)

Figure B In a fluorescent lamp, a gas discharge produces ultraviolet light that is converted into visible light by a phosphor coating on the inside of the tube. (a) The two-component phosphor of praseodymium and yttrium fluorides produces the visible soft violet-pink light flooding the face of the researcher (b).

Figure A Fluorescence may occur when an electron makes several jumps in returning to its original state.

567

billiard ball moving in a well-defined orbit. It is meaningless to ask "particle" questions such as how the electron gets from one place to another or how fast it will be going after 2 minutes. The electron orbits are replaced by standing waves that represent probability distributions. The best we can do is visualize the atom as an electron cloud surrounding the nucleus. This is no ordinary cloud; the density of the cloud gives the probability of locating an electron at a given point in space. The probability is highest where the cloud is the densest and lowest where it is the least dense. An artist's version of these three-dimensional clouds is shown in Figure 23-13. The loss of orbiting electrons also means, however, that the issue of an accelerating charge continuously radiating energy disappears. The theory agrees with nature: Atoms are stable.

As with the particle in a box, the simple act of confining the electron to the volume of the atom results in the quantization of its properties. If we had allowed the particle in the box to move in all three dimensions, we would have standing waves in three dimensions and therefore three independent quantum numbers, one for each dimension. Similarly, the three-dimensionality of the atom yields three quantum numbers.

The particular form that these numbers take depends on the symmetry of the forces involved. In the case of atoms the force is spherically symmetric and the three quantum numbers are associated with the energy n, the size ℓ of the angular momentum, and its direction m_ℓ.

However, these three quantum numbers were not adequate to explain all of the features of the atomic spectra. The additional features could be explained by assuming that the electron spins on its axis. A fourth quantum number m_s was added that gave the orientation of the electron spin. There are only two possible spin orientations, usually called "spin up" or "spin down." Although the classical idea of the electron spinning like a toy top does not carry over into quantum mechanics, the effects analogous to those of a spinning electron were accounted for with this additional quantum number. The quantum number is retained and, for convenience, we use the Newtonian language of electron spin.

The most recent model of the atom combines relativity and the quantum mechanics of electrons and photons in a theory known as *quantum electrodynamics* (QED), which is even more abstract than the quantum-mechanical model. In this theory, however, the concept of electron spin is no longer just an add-on, but is a natural result of the combination of quantum mechanics and relativity.

The Exclusion Principle and the Periodic Table

Let us return to the periodicity of the chemical elements that we discussed in Chapter 22. The periodicity required that we think of electrons existing in shells, but the theory did not tell us how many electrons could occupy each shell. The introduction of the quantum numbers said that electrons could only exist in certain discrete states and that these states formed shells, but it did not say how many electrons could exist in each state.

Physics with Style

Richard Feynman.

Richard Phillips Feynman was a brilliant theoretical physicist who, together with Julian S. Schwinger and Schinichiro Tomonaga, shared the 1965 Nobel Prize in physics for fundamental work in the principles of quantum electrodynamics. His many important contributions to physics include the invention of simple diagrams to represent particle interactions graphically, the theory of the weak interaction of subatomic particles, a reformulation of quantum mechanics, the theory of superfluid helium, and his contribution to physics education through the magnificent three-volume text *The Feynman Lectures on Physics*.

Feynman did his undergraduate work at MIT and received his Ph.D. in 1942 from Princeton University. During World War II, he worked on the Manhattan Project at Princeton and then at Los Alamos, New Mexico. He then joined the faculty at Cornell University in 1945 and was appointed professor of physics at California Institute of Technology in 1950, where he remained for the rest of his career.

It is well known that Feynman had a passion for finding new and better ways to formulate each problem, or, as he would say, "turning it around." In the early part of his career, he was fascinated with electrodynamics, and developed an intuitive view of quantum electrodynamics. Convinced that the electron could not interact with its own field, he said, "That was the beginning, and the idea seemed so obvious to me that I fell deeply in love with it." Often called the outstanding intuitionist of our age, he said in his Nobel acceptance speech, "Often, even in a physicist's sense, I did not have a demonstration of how to get all of these rules and equations, from conventional electrodynamics. . . . I never really sat down, like Euclid did for the geometers of Greece, and made sure that you could get it all from a single set of axioms."

In 1986, Feynman was a member of the presidential commission to investigate the explosion of the Space Shuttle *Challenger*. In this capacity, he performed a simple experiment for the commission members which showed that one of the shuttle's O-ring seals was the likely cause of the disaster. After placing a seal in a pitcher of ice water and squeezing it with a clamp, he demonstrated that the seal failed to spring back into shape once the clamp was removed.[1]

Feynman worked in physics with a style commensurate with his personality, that is, with energy, vitality, and humor. The following quotes from some of his colleagues are characteristic of the great impact he made on the scientific community.[2]

"A brilliant, vital, and amusing neighbor, Feynman was a stimulating (if sometimes exasperating) partner in discussions of profound issues—we would exchange ideas and silly jokes in between bouts of mathematical calculation—we struck sparks off each other, and it was exhilarating." *Murray Gell-Mann*

"Reading Feynman is a joy and a delight, for in his papers, as in his talks, Feynman communicated very directly, as though the reader were watching him derive the results at the blackboard." *David Pines*

"He loved puzzles and games. In fact, he saw all the world as a sort of game, whose progress of 'behavior' follows certain rules, some known, some unknown. . . . Find places or circumstances where the rules don't work, and invent new rules that do." *David L. Goodstein*

"Feynman was not a theorist's theorist, but a physicist's physicist and a teacher's teacher." *Valentine L. Telegdi*

Laurie M. Brown, one of his graduate students at Cornell, noted that Feynman, a playful showman, was "undervalued at first because of his rough manners, who in the end triumphs through native cleverness, psychological insight, common sense and the famous Feynman humor. . . . Whatever else Dick Feynman may have joked about, his love for physics approached reverence."

Adapted from Serway, R. *Physics for Scientists and Engineers*. Philadelphia: Saunders, 1990.

[1]Feynman's own account of this inquiry can be found in *Physics Today*, 4:26, February 1988.

[2]For more on Feynman's life and contributions, see the numerous articles in a special memorial issue of *Physics Today*, 42, February 1989. For a personal account of Feynman, see his popular autobiographical books, *Surely You're Joking Mr. Feynman*, New York: Bantam Books, 1985, and *What Do You Care What Other People Think*, New York: W. W. Norton & Co., 1987.

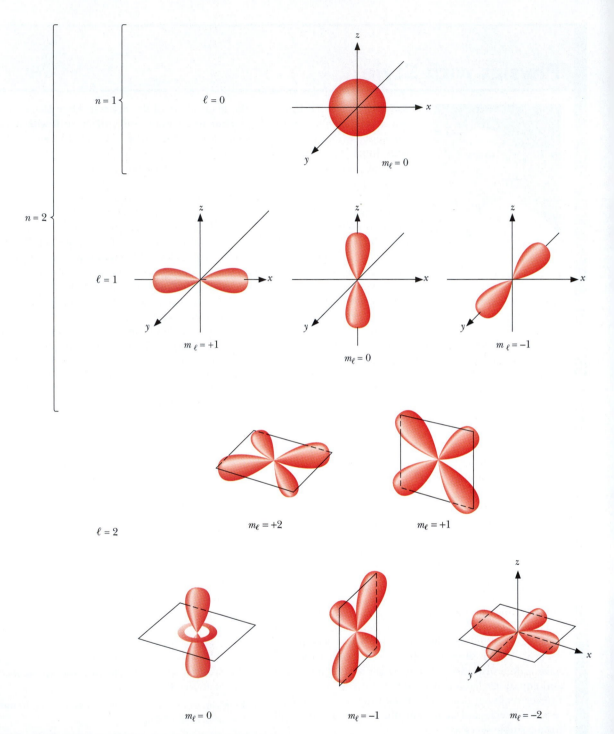

Figure 23-13   Probability clouds for a variety of electron states.

In 1924, Wolfgang Pauli suggested that no two electrons can be in the same state; that is, no two electrons can have the same set of quantum numbers. This statement is now known as the Pauli **exclusion principle.** When this principle is applied to the quantum numbers obtained from the Schrödinger equation and the electron spin, the periodicity of the elements is explained, as we will explain.

Table 23-2 gives the first two quantum numbers for the first 30 elements. The values of the angular momentum quantum number ℓ are restricted by Schrödinger's equation to be integers in the range from 0 up to $n-1$, and the values for the direction of the angular momentum m_ℓ are all integers from $-\ell$ to $+\ell$. Unlike the Bohr model, the angular momentum of the lowest energy state is zero, further supporting the notion that we cannot expect these atomic particles to act classically. For each value of n, ℓ, and m_ℓ two spin states are available; spin up and spin down.

The good news is that the relationships between the quantum numbers explain why the orbital shells have different capacities and this, in

Pauli exclusion principle

Wolfgang Pauli.

Table 23-2	Ground state quantum numbers for the first 30 elements							
Atomic Number	**Element**	$n=1$ $\ell=0$	2 0	2 1	3 0	3 1	3 2	4 0
1	hydrogen (H)	1						
2	helium (He)	2						
3	lithium (Li)	2	1					
4	beryllium (Be)	2	2					
5	boron (B)	2	2	1				
6	carbon (C)	2	2	2				
7	nitrogen (N)	2	2	3				
8	oxygen (O)	2	2	4				
9	fluorine (Fl)	2	2	5				
10	neon (Ne)	2	2	6				
11	sodium (Na)	2	2	6	1			
12	magnesium (Mg)	2	2	6	2			
13	aluminum (Al)	2	2	6	2	1		
14	silicon (Si)	2	2	6	2	2		
15	phosphorus (P)	2	2	6	2	3		
16	sulfur (S)	2	2	6	2	4		
17	chlorine (Cl)	2	2	6	2	5		
18	argon (Ar)	2	2	6	2	6		
19	potassium (K)	2	2	6	2	6	0	1
20	calcium (Ca)	2	2	6	2	6	0	2
21	scandium (Sc)	2	2	6	2	6	1	2
22	titanium (Ti)	2	2	6	2	6	2	2
23	vanadium (V)	2	2	6	2	6	3	2
24	chromium (Cr)	2	2	6	2	6	5	1
25	manganese (Mn)	2	2	6	2	6	5	2
26	iron (Fe)	2	2	6	2	6	6	2
27	cobalt (Co)	2	2	6	2	6	7	2
28	nickel (Ni)	2	2	6	2	6	8	2
29	copper (Cu)	2	2	6	2	6	10	1
30	zinc (Zn)	2	2	6	2	6	10	2

turn, explains the properties of the elements in the periodic table shown in Figure 22-23. The $n = 1$ state has only one angular momentum state and two spin states so its maximum capacity is two electrons, both with $n = 1$ and $\ell = 0$, but with different spins. The first element, hydrogen, has one electron in the lowest energy state, whereas helium, the second element, has both of these states occupied. Since there are no more $n = 1$ states available and the Pauli exclusion principle does not allow two electrons to fill any one state, this completes the first shell.

The next two electrons go into the states with $n = 2$ and $\ell = 0$ because these states are slightly lower in energy than the $n = 2$ and $\ell = 1$ states. This takes care of lithium (element 3) and beryllium (4). There are six states with $n = 2$ and $\ell = 1$ because m_ℓ can take on three values— −1, 0, 1— and each of these can be occupied by two electrons, one with spin up and the other with spin down. These six states correspond to the next six elements in the periodic table, ending with neon (element 10). This completes the second shell, and the completed shell accounts for neon being a noble gas.

The electrons in the next eight elements occupy the states with $n = 3$ and $\ell = 0$ or $\ell = 1$. This completes the third shell, ending with the noble gas argon (18). The two states with $n = 4$ and $\ell = 0$ are lower in energy than the rest of the $n = 3$ states and are filled next. These correspond to potassium (19) and calcium (20). Then the remaining ten states corresponding to $n = 3$ and $\ell = 2$ are filled, yielding the transition elements scandium (21) through zinc (30). Thus, the quantum mechanical picture of the atom accounts for the observed periodicity of the elements.

> **QUESTION**   There are 18 states with $n = 3$. How many are there with $n = 4$?

Werner Heisenberg.

The Uncertainty Principle Σ

The interpretation that atomic particles are governed by probability left many scientists dissatisfied and hoping for some ingenious thinker to rescue them from this foolish predicament. The German physicist Werner Heisenberg showed that there was no rescue. He argued that there is a fundamental limit to our knowledge of the atomic world.

Heisenberg's idea—that there is an indeterminacy of knowledge—is often misinterpreted. The uncertainty is not due to a lack of familiarity with the topic, nor is it due to an inability to collect the required data such as the data needed to predict the outcome of a throw of dice, or the Kentucky Derby. Heisenberg was proposing a more fundamental uncertainty—one that results from the wave–particle duality.

> **ANSWER**   With $n = 4$ we can now have $\ell = 3$ in addition to the other values possible for $n = 3$. This gives us seven values of m_ℓ ranging from −3 to +3. Because each of these has two spin states, there are 14 additional states for a total of 32.

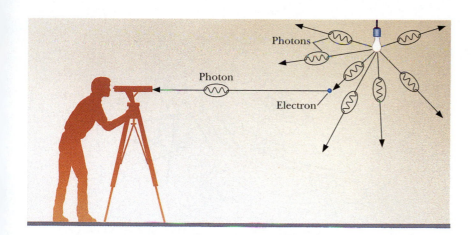

Figure 23-14 Heisenberg's thought experiment to determine the location of an electron.

Imagine the following thought experiment. Suppose you try to watch an electron moving through a room. The room is a complete vacuum, so there's no worry about the electron hitting gas molecules. To locate the electron you need something to carry information from the electron to your eyes. Suppose you use photons from a dim lightbulb. Because this is a thought experiment, we can also assume that you have a microscope so sensitive that you will see the electron as soon as one photon bounces off the electron and enters the microscope (Fig. 23-14).

You begin by using a bulb that emits low-energy photons. These have low frequencies and long wavelengths. The low energy means that the photon will not disturb the electron very much when it bounces off it. However, the long wavelength means that there will be lots of diffraction when the photon scatters from the electron (Chapter 18). Therefore, you won't be able to determine the location of the electron very precisely.

To improve your ability to locate the electron, you now choose a bulb that emits more energetic photons. The shorter wavelength allows you to determine the electron's position relatively well. But the photon kicks the electron so hard that you don't know where the electron is going next. The smaller the wavelength, the better you can locate the electron, but the more the photon alters the electron's path.

Heisenberg argued that we cannot make any measurements on a system of atomic entities without affecting the system in this way. The more precise our measurements, the more we disturb the system. Furthermore, he argued, the measured and disturbed quantities come in pairs. The more precisely we determine one half of the pair, the more we disturb the other. In other words, the more *certain* we are about the value of one, the more *uncertain* we are about the value of the other. This is the essence of the uncertainty principle.

Two of these paired quantities are the position and momentum along a given direction. (Recall from Chapter 5 that the momentum for a particle is equal to its mass multiplied by its velocity.) As we saw in the thought experiment described above, the more certain your knowledge of the position, the more uncertain was your knowledge of the momentum. The converse is also true.

Heisenberg *might* have been here!

"BUT YOU CAN'T GO THROUGH LIFE APPLYING HEISENBERG'S UNCERTAINTY PRINCIPLE TO EVERYTHING."

This idea is now known as Heisenberg's **uncertainty principle.** Mathematically, it says that the product of the uncertainties of these pairs has a lower limit equal to Planck's constant. For example, the uncertainty of the position along the vertical direction Δy multiplied by the uncertainty of the component of the momentum along the vertical direction Δp_y must always be greater than Planck's constant h.

Heisenberg's uncertainty principle

$$\Delta p_y \Delta y > h$$

This principle holds for the position and component of momentum along the same direction. It does not place any restrictions on simultaneous knowledge of the vertical position and a horizontal component of momentum.

Another pair of variables that is connected by the uncertainty principle is energy and time, $\Delta E \Delta t > h$. This mathematical statement tells us that the longer the time we take to determine the energy of a given state, the better we can know its value. If we must make a quick measurement, we cannot determine the energy with arbitrarily small uncertainty. Stated in another way, the energy of a stable state that lasts for a very long time is very well determined. However, if the state is unstable and exists for only a very short time, its energy must have some range of possible values given by the uncertainty principle.

QUESTION   What does the uncertainty principle say about the energy of the photons emitted when electrons in the $n = 2$ state of hydrogen atoms drop down to the ground state?

ANSWER   Because the electrons spend a finite time in the $n = 2$ state, the energy of that state must have a spread in energy. Therefore, the photons have a spread in energy that shows up as a nonzero width of the spectral line.

The Complementarity Principle

Frustrating questions emerged as physicists built a world view of nature on the atomic scale. Is there *any* underlying order? How can one contemplate things that have mutually contradictory attributes? Can we understand the dual nature of electrons and light: sometimes behaving as particles, other times as waves?

There was no known way out. Physicists had to learn to live with this wave–particle duality. A complete description of an electron or a photon requires both aspects. This idea was first stated by Bohr and is known as the **complementarity principle.**

The complementarity principle is closely related to the uncertainty principle. As a consequence of the uncertainty principle, we discover that being completely certain about particle aspects means that we have no knowledge about the wave aspects. For example, if we are completely certain about the position and time for a particle, the wave aspects (wavelength and frequency) have infinite uncertainties. Therefore, wave and particle aspects do not occur at the same time.

complementarity principle

> **QUESTION** If you could determine which slit the electron goes through, would this have any effect on the two-slit interference pattern?

The idea that opposites are components of a whole is not new. The ancient Eastern cultures incorporated this notion as part of their world view. The most common example is the yin–yang symbol of T'ai Chi Tu. Later in life, Bohr was so attracted to this idea that he wrote many essays on the existence of complementarity in many modes of life. In 1947, when he was knighted for his work in physics, he chose the yin–yang symbol for his coat of arms.

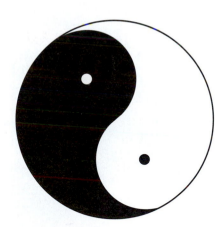

The yin-yang symbol reflects the complementarity of many things.

Determinism

Classical Newtonian mechanics and the newer quantum mechanics have been sources of much debate about the role of cause and effect in the natural world. With Newton's laws of motion came the idea that specifying the position and momentum of a particle and the forces acting on it allowed the calculation of its future motion. Everything was determined. It was like the Universe was an enormous machine. This idea was known as the *mechanistic view*. In the 17th century, René Descartes stated, "I do not recognize any difference between the machines that artisans make and the different bodies that nature alone composes."

These ideas were so successful in explaining the motions in nature that they were extended into other areas. Because the Universe is made of

the mechanistic Universe

> **ANSWER** It would destroy the interference pattern. Knowledge of the particle properties (that is, the path of the electron) precludes the wave properties.

William Blake's *The Creation of Error* was intended to ridicule the scientists' notion of God as the artisan.

Einstein felt that the probabilistic nature of modern physics did not reflect the true reality of nature.

Is the flight of the bumblebee predetermined or a result of free will?

particles whose futures are predetermined, it was suggested that the motion of the entire Universe must be predetermined. This notion was even extended to living organisms. Although the flight of a bumblebee seems random, its choices of which flowers to visit are determined by the motion of the particles that make up the bee. These generalizations caused severe problems with the idea of free will—that humans had something to say about the future course of events.

There were, however, some practical problems with actually predicting the future in classical physics. Because measurements could not be made with absolute precision, the position and momentum of an object could not be known exactly. The uncertainties in these measurements would lead to uncertainties about the calculations of future motions. However, at least in principle, certainty was possible. It was also impossible to measure the positions and momenta of all the atomic particles in a small sample of gas, let alone all those in the Universe. However, the motion of each atom was predetermined and therefore the properties of gases were predetermined. Even though humans could not determine the paths, nature knew them. The future was predetermined.

With the advent of quantum mechanics the future became a statistical issue. The uncertainty principle stated that it was impossible *even in principle* to measure simultaneously the position and momentum of a particle. The mechanistic laws of motion were replaced by an equation for calculating the matter waves of a system that only gave *probabilities* about future events. Even if we know the state of a system at some time, the laws of quantum mechanics do not permit the calculation of a future, only the probabilities for each of many possible futures. The future is no longer considered to be predetermined but is left to chance.

One of the main opponents of this probabilistic interpretation was Albert Einstein. His objections did not arise out of a lack of understanding. He understood quantum mechanics very well and even contributed to its interpretation. He believed that the path of an electron was governed by some (hidden) deterministic set of rules (sort of like atomic Newton's laws) not by some unmeasurable probability wave. The new physics didn't fit into his philosophy of the natural world. Einstein's famous rebellious quote is "I, at any rate, am convinced that [God] is not playing at dice."

The problem is that nobody has ever found those hidden rules, and quantum mechanics continues to work better than anything else that has been proposed. The point is that hoping doesn't change the physics world view. Einstein spent a lot of time trying to disprove the very theory that he helped begin. He did not succeed. New work has shown that quantum mechanics is a complete theory, proving that there are no hidden variables.

Lasers

The understanding of the quantized energy levels in atoms and the realization that transitions between these levels involved the absorption and emission of photons led to the development of a new device that produced a special beam of light. Assume that we have a gas of excited atoms. Further assume that an electron drops from the $n = 3$ to the $n = 2$ level in the energy diagram in Figure 23-15(a). A photon is emitted in some random direction. It could escape the gas, or it could interact with another atom with an electron in one of two ways. It could be absorbed by an atom with an electron in the $n = 2$ level, exciting the electron to the $n = 3$ level. This excited atom would then emit another photon at a random time in a random direction. On the other hand, the original photon could stimulate an electron in the $n = 3$ level to drop to the $n = 2$ level, causing the emission of another photon [Fig. 23-15(c)]. This latter process is known as **stimulated emission.** Moreover, this new photon does not come out randomly. It has the same energy, the same direction, and the same phase as the incident photon; that is, the two photons are coherent. These photons can then stimulate the emission of further photons, producing a coherent beam of light.

The device that actually produces coherent beams of light is called a **laser,** which is an acronym for *l*ight *a*mplification by *s*timulated *e*mission of *r*adiation. A laser produces a beam of light that is very different from that emitted by an ordinary light source such as a flashlight. The laser beam has a very, very narrow range of wavelengths, is highly directional, and can be quite powerful. The laser beam is a single color because all of the photons have the same energy. For instance, the helium–neon laser usually has a wavelength of 632.8 nanometers. The beam is highly directional because the stimulated photons move in the same direction as those doing the stimulating. The fact that all of the photons have the same phase means that the amplitude of the resulting electromagnetic wave is very large. The intensity of a collection of coherent photons is obtained by adding the amplitudes and then squaring the sum. The intensity of a collection of incoherent photons is obtained by squaring the amplitudes and then adding these squares. The difference can be illustrated by considering a collection of five photons. In the laser beam we have $(2 + 2 + 2 + 2 + 2)^2 = 100$, while in the ordinary light beam we have $(2^2 + 2^2 + 2^2 + 2^2 + 2^2) = 20$. The effect is even more drastic for the very large number of photons in a laser beam. These factors combine to allow us to clearly see a 1-milliwatt laser beam shining on the surface of a 100-watt lightbulb.

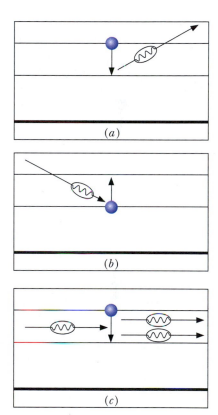

Figure 23-15   (a) Spontaneous emission. (b) Absorption. (c) Stimulated emission.

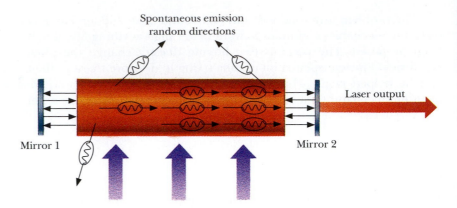

Figure 23-16   A schematic of a simple laser.

Making a working laser was more complicated than the above description implies. A method had to be found for "building" a beam of many, many photons. This is done by putting mirrors at each end of the laser tube to amplify the beam by passing it back and forth through the gas of excited atoms. One of the mirrors is only partially silvered, so that a small part of the beam is allowed to escape (Fig. 23-16).

The construction of a laser was difficult because a photon is just as likely to be absorbed by an atom with an electron in the lower level as it is to cause stimulated emission of an electron in an excited level. Usually most of the excited atoms are in the lower energy state, so most of the photons are absorbed and only a few cause stimulated emission. Therefore, building a laser depends on developing a population inversion, a situation in which there are many more electrons in the excited state than in the lower energy state. This is usually done by exciting the atom's electrons into a *metastable state* that decays into an unpopulated energy state. A metastable state is a state in which the electrons remain for a long time. The electrons can be excited by using a flash of light, an electric discharge, or collisions with other atoms.

Physics Update

Faster techniques for sequencing DNA may contribute to the Human Genome Project. With the conventional technique, known as gel electrophoresis, in which DNA fragments are separated by electric fields, it would take 20,000 human-years to determine the complete sequence of 3 billion "base pairs" that make up the human genetic code. Brian Chait of Rockefeller University has devised a sequencing method that completely bypasses the use of a gel. In his method, a laser pulse would zap DNA fragments, converting them into gaseous ions, which would then fly toward a detector to be analyzed. Once his technique is refined, Chait estimates that an amount of DNA code that normally takes several hours to sequence could be analyzed in less than a minute.

Lasers have a wide range of uses, from surveying to surgery. In surveying, the light beam defines a straight line, and by pulsing the beam the time for a round trip can be used to measure distances. This same method is used on a gigantic scale to determine the distance to the Moon to an accuracy of a few centimeters. A very short pulse (less than 1 nanosecond long) of laser light is sent through a telescope toward the Moon's surface. The beam is so well collimated that it only spreads out over an area a few kilometers in diameter. The Apollo astronauts left a panel of retroreflectors (Chapter 16) on the surface that reflects the light back to the telescope, allowing the round-trip time to be measured. These measurements serve as a test of the validity of the general theory of relativity because the theory predicts the detailed orbit of the Moon.

Laser "knives" are used in optometry and surgery. Two leading causes of blindness are glaucoma and diabetes. In treating glaucoma, a laser beam "drills" a small hole to relieve the high pressure that builds up in the eye. One of the complications of diabetes is the weakening of the walls of blood vessels to the point where they leak. The laser can be shined into the eye to cause coagulation to stop the bleeding. The energy in laser beams can also be used to weld detached retinas to the back of the eye. In laser surgery, the beam coagulates the blood as it slices through the tissue, greatly reducing bleeding. Laser beams can also be directed by fiber optics through tiny incisions to locations that would otherwise require major incisions and long healing times.

Lasers have widespread use in the marketplace. The audio and video information stored in the pits on CDs and videodiscs is read by laser beams. Likewise, the lasers at the checkout counters read the bar codes on the product identification labels, allowing the computer to print out a short description of the object and its current price. This has greatly reduced billing errors as well as the time required to check out.

Lasers have made practical holography possible (Chapter 18) and opened up a whole new research tool in holographic measurements.

A laser surveying instrument.

Light from an argon laser is carried by a fiber-optic cable to perform surgery deep within an ear.

SUMMARY

The Bohr model was successful in accounting for many atomic observations, especially the emission and absorption spectra for hydrogen and the ordering of the chemical elements. However, it failed to explain the details of some processes and could not give quantitative results for multi-electron atoms. It gave the general features of the periodic table but was not able to provide the details of the shells nor explain how many electrons could occupy each shell. Primary among the Bohr model's failures was the lack of intuitive reasons for its postulates. Finally, it was nonrelativistic.

The replacement of Bohr's model of the atom began with de Broglie's revolutionary idea that electrons behave like waves. The de Broglie wavelength of the electron is inversely proportional to its momentum. The primary consequence of this wave behavior is that

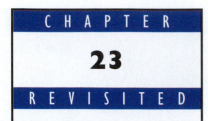

Laser light is special for a number of reasons. In holography, the most important attribute is its coherence, allowing a beam of light to be split and recombined while preserving the phase information. In many of the applications mentioned in the opening question, the key attribute is that the laser light is (nearly) unidirectional, allowing for a very concentrated beam. With proper optical focusing, this beam can be concentrated to a very small spot, drastically increasing the intensity—the amount of energy per unit area—striking a surface.

confined atomic particles form standing-wave patterns that quantize their properties.

Although successful, this new understanding led to a wave–particle dilemma for electrons and other atomic particles: They are always detected as single particles and yet collectively they produce wavelike distributions. The behavior of these particles is governed by a quantum-mechanical wave equation that provides all possible information about the particles. For example, the probability of finding the particle at a given location is determined by the square of its matter-wave amplitude. As a consequence, atomic particles do not follow well-defined paths as classical particles do. This quantum-mechanical view of physics replaced the classical world view containing Newton's laws and Maxwell's equations.

There are four quantum numbers associated with an electron bound in an atom: the energy n, the size ℓ, and the direction m_ℓ of the angular momentum, and the orientation m_s of the spin. The periodicity of the chemical elements is explained by the existence of shells of varying capacity determined by the constraints of quantum mechanics and the Pauli exclusion principle. This principle states that no two electrons can have the same set of quantum numbers.

According to the complementarity principle, all atomic entities require both wave and particle aspects for a complete description, but these cannot appear at the same time. These ideas also led to Heisenberg's uncertainty principle, a statement that there is an indeterminacy of knowledge that results from the wave–particle duality. Mathematically, the product of the uncertainties of paired quantities has a lower limit equal to Planck's constant.

Lasers produce coherent beams of light through the stimulated emission of radiation by electrons in excited, metastable states.

KEY TERMS

complementarity principle: The idea that a complete description of an atomic entity such as an electron or a photon requires both a particle description and a wave description, but not at the same time.

exclusion principle: No two electrons can have the same set of quantum numbers.

laser: An acronym for *l*ight *a*mplification by *s*timulated *e*mission of *r*adiation. A device that uses stimulated emission to produce a coherent beam of electromagnetic radiation.

matter-wave amplitude: The wave solution to Schrödinger's equation for atomic and subatomic particles. The square of the matter-wave amplitude gives the probability of finding the particle at a particular location.

quantum: (pl., *quanta*) The smallest unit of a discrete property. For instance, the quantum of charge is the charge on the proton.

quantum mechanics: The rules for the behavior of particles at the atomic and subatomic levels.

stimulated emission: The emission of a photon from an atom due to the presence of an incident photon. The emitted photon has the same energy, direction, and phase as the incident photon.

uncertainty principle: The product of the uncertainty in the position of a particle along a certain direction and the uncertainty in the momentum along this same direction must be greater than Planck's constant $\Delta p_x \Delta x > h$. A similar relationship applies to the uncertainties in energy and time.

CONCEPTUAL QUESTIONS

1. Make a list summarizing the successes and failures of the Bohr theory.

2. Why did it bother scientists that Bohr's theory was not relativistic?

3. What does it mean to call something a particle or a wave?

4. Which of the following technical terms can be used to describe both an electron and a photon: wavelength, velocity, mass, energy, or momentum?

5. Why do you think that the particle nature of the electron was discovered before its wave nature?

6. Why do we say that electrons have wave properties? Particle properties?

7. Bohr could never really explain why an electron was limited to certain orbits. How did de Broglie explain this?

8. What do standing waves have to do with atoms?

9. The wavelength of red light is 600 nanometers. An electron with a speed of 1.2 kilometers per second has the same wavelength. Will the electron look red?

10. An electron and a proton have the same speeds. Which has the longer wavelength?

11. Why is the wave behavior of bowling balls not observed?

12. Why doesn't a sports car diffract off the road when it is driven through a tunnel?

13. When we perform the two-slit experiment with electrons, do the electrons behave like particles, waves, or both?

14. When the photons in the two-slit experiment are detected at the screen, do they behave like particles, waves, or both?

15. What are the differences between an electron and a photon?

16. What are the similarities between an electron and a photon?

17. In the two-slit experiment with photons, what type of pattern do you expect to obtain if you turn the light source down so low that only one photon is in the apparatus at a time?

18. If the two-slit experiment is performed with a beam of electrons so weak that only one electron passes through the apparatus at a time, what kind of pattern would you expect to obtain on the detecting screen?

19. Two coherent sources of light shine on a distant screen. If we want to calculate the intensity of the light at positions on the screen, do we add the displacements of the two waves and then square the result or do we square the displacements and then add them?

20. Two lightbulbs shine on a distant wall. To obtain the brightness of the light, do we add the displacements or the intensities?

21. What kind of waves are de Broglie waves?

22. What meaning do we give to the square of the matter-wave amplitude?

23. If an atomic particle is confined to a box and if it has the lowest possible energy, where would you most likely find the particle?

24. Where would you most likely find an electron in the first excited state for a one-dimensional box?

25. A particle in a one-dimensional box is in the first excited state as shown in Figure 23-12. Why does it not make sense to ask how it gets from one region of high probability to the other?

*26. Quantum mechanics says that the kinetic energy of the lowest energy state is not zero. What does this say about the total kinetic energy of a system of particles as the temperature approaches absolute zero? Does all motion cease at absolute zero?

27. What are the main features of the quantum-mechanical model of the atom?

28. Why doesn't the quantum-mechanical model of the atom have the problem of accelerating charges emitting electromagnetic radiation?

29. Why does an atomic electron have four quantum numbers instead of the three that exist for a particle in a three-dimensional box?

30. Where would you most likely find the electron if it is in a quantum state with $n = 2$, $\ell = 1$, and $m_\ell = 0$ as shown in Figure 23-13?

31. What does the Pauli exclusion principle say about atoms?

32. Give an example that clearly illustrates the meaning of the Pauli exclusion principle.

33. How many electrons can have the quantum numbers $n = 4$ and $\ell = 2$?

34. How many electrons can have the quantum numbers $n = 4$ and $\ell = 3$?

35. Make a list showing the quantum numbers for each of the four electrons in the beryllium atom when it is in its lowest energy state.

36. Make a list showing the quantum numbers for each of the ten electrons in the neon atom when it is in its lowest energy state.

37. How is Heisenberg's uncertainty different from the other kinds of uncertainties mentioned in this chapter?

38. Give an example that clearly illustrates the meaning of the Heisenberg uncertainty principle.

39. According to the uncertainty principle, why does a tennis ball appear to have a definite position and velocity while an electron does not?

40. Explain why the Heisenberg uncertainty principle does not put any restrictions on our simultaneous knowledge of a particle's momentum and its energy.

41. Explain why we cannot get around the uncertainty principle by using electrons rather than photons to observe an electron travel through a room.

42. Explain why Bohr's model of the atom is not compatible with the uncertainty principle.

*43. What does the uncertainty principle say about the energy of an excited state of hydrogen that only exists for a short time? How will this affect the emission spectrum?

44. If the Bohr model of the atom has been replaced by the newer quantum-mechanical models, why do we still teach the Bohr model?

45. Quantum mechanics has had great success. What is its main drawback for people trained on physical models?

46. Einstein once complained, "The quantum mechanics is very imposing. But an inner voice tells me that it is still not the final truth. The theory yields much, but it hardly brings us nearer to the secret of the Old One. In any case, I am convinced that He does not throw dice." What about quantum mechanics was troubling him?

47. To what situations in our everyday world might the idea of complementarity apply?

48. Why are scientists willing to accept the complementarity principle?

49. Why did Bohr invent his principle of complementarity?

50. Give an example that illustrates the meaning of the complementarity principle.

*51. If we are certain about the position of a particle, what does the Heisenberg uncertainty principle say about the uncertainty in its wavelength?

*52. If we are certain about the energy of a particle, what does the Heisenberg uncertainty principle say about the uncertainty in its frequency?

53. What does quantum mechanics have to do with free will?

54. What does classical physics say about free will?

55. Why do some minerals glow when they are illuminated with ultraviolet light?

56. When ultraviolet light from a "black light" shines on certain paints, visible light is emitted. How can you account for this?

57. Phosphorescent materials continue to glow after the lights are turned off. How can you use the model of the atom to explain this?

58. Can infrared rays cause fluorescence? Why?

59. How does light from a laser differ from light emitted by an ordinary lightbulb?

60. What is stimulated emission?

EXERCISES

1. What is the de Broglie wavelength of a Ford (mass = 1500 kg) traveling at 30 m/s (67 mph)?

2. A bullet for a 30–60 rifle has a mass of 10 g and a muzzle velocity of 900 m/s. What is its wavelength?

3. Nitrogen molecules (mass = 4.6×10^{-26} kg) in room temperature air have an average speed of about 500 m/s. What is a typical wavelength for these nitrogen molecules?

4. What is the de Broglie wavelength for an electron traveling at 500 m/s?

5. What speed would an electron need to have a wavelength equal to the diameter of a hydrogen atom (10^{-10} m)?

6. What is the speed of a proton with a wavelength of 0.5 nm?

7. What is the size of the momentum for an electron that is in the lowest energy state for a one-dimensional box whose length is 1 nm?

8. What is the size of the momentum for an electron that is in the first excited energy state for a one-dimensional box whose length is 1 nm?

*9. What is the kinetic energy of an electron that is in the lowest energy state for a one-dimensional box whose length is 1 nm?

*10. What is the kinetic energy of an electron that is in the first excited energy state for a one-dimensional box whose length is 1 nm?

11. A Ford Escort passes through a narrow tunnel with a width of 4 m. What uncertainty is introduced in the Escort's momentum perpendicular to the highway?

12. A child runs straight through a door with a width of 0.8 m. What is the uncertainty in the momentum of the child perpendicular to the child's path?

13. An electron passes through a slit that has a width of 10^{-10} m. What uncertainty does this introduce in the momentum of the electron at right angles to the slit?

14. A proton passes through a slit that has a width of 10^{-10} m. What uncertainty does this introduce in the momentum of the proton at right angles to the slit?

15. What is the minimum uncertainty in the position along the highway of a Ford Escort (mass = 1000 kg) traveling at 30 m/s (67 mph)? Assume that the uncertainty in the momentum is equal to 1% of the momentum.

16. What is the uncertainty in the momentum of a Ford Escort with a mass of 1000 kg parked by the curb? Assume that you know the location of the car with an uncertainty of 0.1 mm.

17. What is the uncertainty in the location of a proton along its path when it has a speed equal to 0.1% the speed of light? Assume that the uncertainty in the momentum is 1% of the momentum.

18. What is the uncertainty in each component of the momentum of an electron confined to a box approximately the size of a hydrogen atom, say 0.1 nm on a side?

INTERLUDE

**Electrical discharges
visible in the pulse-
forming section of the
Particle Beam Fusion
Accelerator II. The
108-foot-diameter
accelerator is located at
Sandia National
Laboratories.**

The Subatomic World

The quantum-mechanical model of the atom has been very successful. From a large collection of seemingly unrelated phenomena has come a rather complete picture of the atom. This development has created a very profound change in our physics world view.

Our study of the physical world now goes one "layer" deeper, to that of the atomic nucleus. The search for the structure of the nucleus led to the discovery of new forces in nature, forces much more complicated than those of gravity and electromagnetism. It is not possible to write simple expressions for the nuclear forces like the one we wrote for gravity; the nuclear forces depend on much more than distance and mass. This discovery radically changed our basic concepts about forces. The models for the interactions between particles that began with Newton's action at a distance and progressed to the field concept in Maxwell's time now involve the exchange of fundamental particles.

Because the nuclear force is so strong, the energy associated with it is very large. The power of nuclear energy became evident with the explosion of nuclear bombs over Japan in World War II. The release of this energy in the form of bombs or nuclear power plants poses serious questions not only for scientists and engineers but for our entire society.

In our search for the ultimate structure of all matter we have discovered that atoms are not the most basic building blocks in nature. They are composed of electrons and very small, very dense nuclei. The nuclei are, in turn, composed of particles. These were originally believed to be electrons and protons. This notion was appealing because it required only three elemental building blocks: the electron, the proton, and the photon. Later the neutron replaced the electron in the nucleus, but the overall picture for the fundamental structure was still appealing.

But as scientists delved deeper and deeper into the subatomic world, new particles emerged. The number of "elementary" particles grew to more than 300! In current theories most of these are composed of a small number of more basic particles called quarks.

The search for the elusive elementary particles seems like a child playing with a set of nested eggs. Each egg is opened only to reveal another egg. However, the child eventually reaches the end of the eggs. In one version the smallest egg contains a bunny. Is there an end to the search for the elementary particles? Do quarks represent the bunny?

> The search for the structure of the nucleus led to the discovery of new forces in nature, forces much more complicated than those of gravity and electromagnetism.

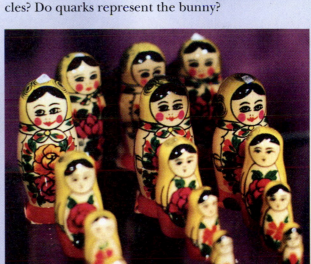

Are the layers of matter like the nested dolls or is there a final set of elementary particles?

The Nucleus

As our knowledge of nuclei grew, whole new technologies were developed to make use of this new knowledge. The new technologies range from obtaining medical images of our internal organs to determining the age of mummies or artifacts. How can nuclear techniques be used to determine the age of an artifact? (See p. 611 for the answer to this question.)

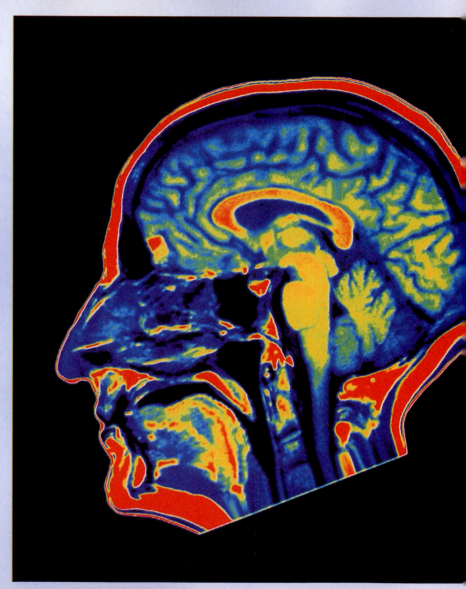

Magnetic resonance image of a normal human brain.

The nucleus of an atom is unbelievably small. If you could line up a trillion (10^{12}) of them, they would only stretch a distance equal to the size of the period at the end of this sentence. Another way of visualizing these sizes is to imagine expanding things to more human sizes and then comparing relative sizes. If we imagine a baseball as big as the Earth, the baseball's atoms would be approximately the size of grapes. Even at this scale, the nucleus would be invisible! To "see" the nucleus we need to expand one of these grape-sized atoms until it is as big as the Houston Astrodome. The nucleus would be in the middle and be about the size of a grape.

These analogies are a little risky since the quantum-mechanical view does not consider atomic entities to be classical particles. The images do, however, demonstrate the enormous amount of ingenuity required to study this submicroscopic realm of the Universe. You don't just pick up a nucleus and take it apart.

We learn about the nucleus by examining what pieces come out—either through naturally occurring radioactivity or artificially induced nuclear reactions. Studying the nucleus is more difficult than studying the atom because we already understood the electric force that held the atom together. There, the task was to find a set of rules that govern the behavior of the atom. These rules were provided by quantum mechanics. In nuclear physics the situation was turned around; quantum mechanics provided the rules, but at first there wasn't a good understanding of the forces.

The Discovery of Radioactivity

Discoveries in science are often not anticipated. Radioactivity is such a case. Radioactivity is a nuclear effect that was discovered 15 years before the discovery of the nucleus itself. Although the production of X rays is not a nuclear effect, their study led to the discovery of radioactivity. Four months after Roentgen's discovery of X rays (Chapter 22), a French scientist, Henri Becquerel, was looking for a possible symmetry in the X-ray phenomenon. Roentgen's experiment had shown that X rays striking certain salts produced visible light. Becquerel wondered if the phenomenon might be symmetrical—if visible light shining on these fluorescing salts might produce X rays.

His experiment was simple. He completely covered a photographic plate so that no visible light could expose it. Then he placed a fluorescing mineral on this package and took the arrangement outside into the sunlight (Fig. 24-1). He reasoned that the sunlight might activate the atoms and cause them to emit X rays. The X rays would easily penetrate the paper and expose the photographic plate.

The initial results were encouraging; his plates were exposed as expected. However, during one trial, the sky was clouded over for several days. Convinced that his photographic plate would be slightly exposed, he decided to start over. But being meticulous he developed the plate anyway. Much to his surprise he found that it was completely exposed. The fluorescing material—a uranium salt—apparently exposed the photographic plate without the aid of sunlight. Becquerel soon discovered

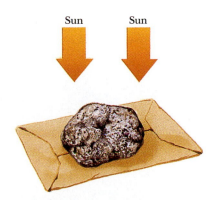

Sun   Sun

Figure 24-1 Becquerel's experimental setup to test whether X rays would penetrate the paper and expose the photographic plate.

that all uranium salts exposed the plates—even those that did not fluoresce with X rays.

This new phenomenon appeared to be a special case of Roentgen's X rays. They penetrated materials, exposed photographic plates, and ionized air molecules, but unlike X rays, the new rays occurred naturally. There was no need for a cathode-ray tube, a high voltage, or even the Sun!

Becquerel did a series of experiments that showed that the strength of the radiation depended only on the amount of uranium present—either in pure form or combined with other elements to form uranium salts. Later it was realized that this was the first clue that radioactivity was a nuclear effect rather than an atomic one. Atomic properties change when elements undergo chemical reactions; this new phenomenon did not. The nuclear origin of this radiation was further supported by observations that the exposure of the photographic plates did not depend on outside physical conditions such as strong electric and magnetic fields or extreme pressures and temperatures.

Pierre Curie, a colleague of Becquerel, and Marie Sklodowska Curie developed a way of quantitatively measuring the amount of radioactivity in a sample of material. Marie Curie then discovered that the element thorium was also radioactive. As with uranium, the amount of radiation depended on the amount of thoruim and not upon the particular thorium compound. It was also not affected by external physical conditions.

The Curies found that the amount of radioactivity present in pitchblende, an ore containing a large percentage of uranium oxide, was much larger than expected from the amount of uranium present in the ore. They then attempted to isolate the source of this intense radiation, a task that turned out to be difficult and time-consuming. Although they were initially unable to isolate the new substance, they did obtain a sample that was highly concentrated. The new element was named polonium after Marie Curie's native Poland. Further work led to the discovery of another highly radioactive element, radium. A gram of radium emits more than a million times the radiation of a gram of uranium. A sample of radium generates energy at a rate that keeps it hotter than its surroundings. This mysterious energy source momentarily created doubt about the validity of the law of conservation of energy.

Types of Radiation

These radioactive elements were emitting two types of radiation. One type could not even penetrate a piece of paper; it was called **alpha (α) radiation** after the first letter in the Greek alphabet. The second type could

Eight Tons of Ore

Marie Curie.

Her real name was Marya, her family called her Manya, in France she was Marie, and today she's often called Madame. She had little interest in her name and even her own well-being. She had one intense passion—to learn the laws of nature. In later years, she rebuffed a reporter's personal questions saying, "In science we must be interested in things, not in persons."

Marya Sklodowska was the youngest child born to two Polish teachers. She grew up thinking that the world was a giant school where there were only teachers and students. She learned to read at an early age and buried herself in book after book. One story reports that she would become so engrossed in reading that her sisters once built a pyramid of chairs around her. When she finished the book and stood up, the tower came crashing down. Although her sisters were ready to defend themselves, she simply muttered quietly, "That's stupid," and walked away.

Upon graduating from high school with honors, Sklodowska resigned herself to becoming a governess. The University of Warsaw did not accept women and her family could not afford to send her to Paris to study. However, her desire to learn was bigger than this obstacle and she soon designed a plan. She supported her sister Bronya while she attended medical school in Paris. Then Bronya worked to support her. In just three years Marya earned her degree at the Sorbonne. During this time she lived frugally, spending nothing on luxuries and too little on food. Several times she fainted in the cold laboratories, having eaten almost nothing. When the examinations were over, Sklodowska had placed first and earned a master's degree in mathematics and physics.

After her wedding to Pierre Curie, a young scientist, she began her doctor's dissertation in the new area of radioactivity. It wasn't long until she realized that the radiation coming from uranium ore was too large to be produced by uranium alone. The Curies discovered two new elements, polonium and radium, the latter in trace amounts. But the Curies wanted to isolate an amount that they could see and weigh. Chemists were skeptical when the Curies invested their life savings in 8 tons of waste ore and worked for four years in a lab with a dirt floor to produce just 10 milligrams of radium. This great effort won the recognition of the scientific community with the awarding of a Nobel Prize in physics. Eight years later Marie Curie was awarded a Nobel Prize in chemistry for the discovery of two elements. She is one of only three scientists to receive two Nobel Prizes. Element 96, curium, is named in honor of Marie and Pierre Curie.

Fame, however, irritated both Curies, who would have much preferred new scientific equipment in lieu of medals and honors. Marie Curie remained an incessant scientist until leukemia, most likely caused by long exposure to radiation, claimed her life. The Curies left an impressive legacy—their daughter and son-in-law also won Nobel Prizes.

Adapted from an essay written by Steven Janke for Pasco Scientific.

Reference: Vincent S., trans. *Madame Curie, a Biography by Eve Curie.* Garden City, NY: Doubleday, 1937.

travel through a meter of air or even thin metal foils; it was named **beta (β) radiation** after the second letter in the Greek alphabet.

By 1900, several experimentalists had shown that beta radiation could be deflected by a magnetic field, demonstrating that it was a charged particle. These particles had negative charges and the same charge-to-mass ratio as the newly discovered electrons. Later it was concluded that these **beta particles** were, in fact, electrons that were emitted by nuclei.

In 1903, Rutherford showed that alpha radiation could also be deflected by magnetic fields and that these particles had a charge of +2.

Figure 24-2 Alpha, beta, and gamma radiation behave differently in a magnetic field.

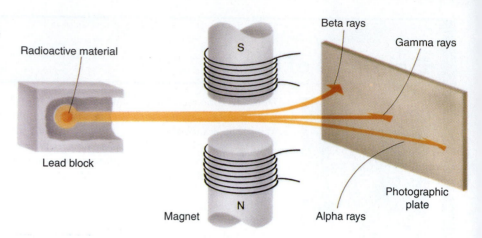

Because they had a larger charge than electrons and yet were much more difficult to deflect, it was concluded that they were much more massive than electrons. Six years later Rutherford was able to show that **alpha particles** are the nuclei of helium atoms.

The third type of radiation was discovered in 1900. Naturally, it was named after the third letter of the Greek alphabet and is known as **gamma (γ) radiation.** This radiation has the highest penetrating power; it can travel through many meters of air or even through thick walls. Unlike the other two types of radiation, gamma radiation was unaffected by electric and magnetic fields (Fig. 24-2). Gamma rays, like X rays, are now known to be very high-energy photons. Although the energy ranges associated with these labels overlap, gamma-ray photons usually have more energy than X-ray photons, which, in turn, have more energy than photons of visible light.

The Nucleus

The recognition that radioactivity was a nuclear phenomenon indicated that nuclei have an internal structure. Because nuclei emitted particles, it was natural to assume that nuclei were composed of particles. It was hoped there would only be a small number of different kinds of particles and all nuclei would be combinations of these.

Rutherford's early work with alpha particles showed that he could change certain elements into others. While bombarding nitrogen with alpha particles, Rutherford discovered that his sample contained oxygen. Repeated experimentation convinced him that the oxygen was being *created* during the experiment. In addition to heralding the beginning of artificially induced transmutations of elements, this experiment led to the discovery of the proton.

Rutherford noticed that occasionally a fast-moving particle emerged from the nitrogen. These particles were much lighter than the bombarding alpha particles and were identified as **protons.** Rutherford and James Chadwick bombarded many of the light elements with alpha particles and found ten cases in which protons were emitted. By 1919 they realized that the proton was a fundamental particle of nuclei.

The hydrogen atom has the simplest and lightest nucleus, consisting of a single proton. The next lightest element is helium. Its nucleus is approximately four times more massive than hydrogen. Since helium has two electrons, it must have two units of positive charge in its nucleus. If we assume that the helium nucleus contains two protons, we obtain the correct charge, but the wrong mass; it would only be twice that of hydrogen. What accounts for the other two atomic mass units? One possibility is that the nucleus also contains two electron–proton pairs. Each pair would be neutral in charge and contribute one atomic mass unit. Thus, the helium nucleus might consist of four protons and two electrons.

This scheme had some appeal. It accounted for the electrons emitted from nuclei and it could be used to "build" nuclei. However, it had several serious defects that caused it eventually to be discarded. For example, the spins of the electron and proton cause each of them to act like miniature magnets, and therefore the magnetic properties of a nucleus should be a combination of those of its protons and electrons. These magnetic values did not agree with the experimental results. The existence of electrons in the nucleus was also incompatible with the uncertainty principle (Chapter 23). If the location of the electron were known well enough to say that it was definitely in the nucleus, its momentum would be large enough to easily escape the nucleus. This left the "extra" mass unexplained.

The Discovery of Neutrons

By 1924, Rutherford and Chadwick began to suspect there was another nuclear particle. The **neutron** would have a mass about the same as the proton's but no electric charge. Chadwick received a Nobel Prize (1935) for the experimental verification of the neutron's existence in 1932. The "extra" mass in nuclei was now explained. Nuclei are combinations of protons and neutrons, as illustrated in Figure 24-3. Helium nuclei are made of 2 protons and 2 neutrons, oxygen nuclei have 8 protons and 8 neutrons, gold nuclei have 79 protons and 118 neutrons, and so on.

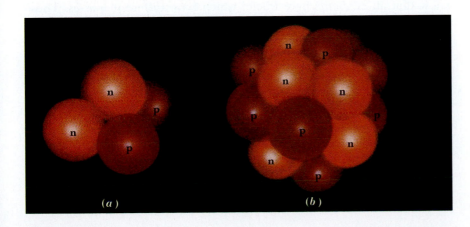

Figure 24-3 The helium nucleus (a) consists of 2 protons and 2 neutrons, and the oxygen nucleus (b) has 8 protons and 8 neutrons.

Table 24-1	Masses of electrons, protons, and neutrons in kilograms, atomic mass units, and units in which the mass of the electron is set equal to 1		
Particle	**kg**	**amu**	**$m_e = 1$**
Electron	9.109×10^{-31}	0.000549	1
Proton	1.6726×10^{-27}	1.007276	1836
Neutron	1.6750×10^{-27}	1.008665	1839

When it is not necessary to distinguish between neutrons and protons, they are often called **nucleons.** The number of nucleons in the nucleus essentially determines the atomic mass of the atom since the electrons' contribution to the mass is negligible. The scale of relative atomic masses has been chosen so that the atomic mass of a single atom in atomic mass units is nearly equal to the number of nucleons in the nucleus. This equivalence is attained by setting the mass of the neutral carbon atom with 12 nucleons equal to 12.0000 atomic mass units (Chapter 7). The masses of the electron, proton, and neutron are given in Table 24-1 for several different mass units. The masses of nuclei vary from 1 to about 260 atomic mass units and the corresponding radii vary from 1.2 to 7.7×10^{-15} meter.

Isotopes

While studying the electric and magnetic deflection of ionized atoms, J. J. Thomson discovered that the nuclei of neon atoms are not all the same. The neon atoms did not all bend by the same amount. Because each ion had the same charge, some neon atoms must be more massive than others. It was shown that the lighter ones had masses of about 20 atomic mass units, whereas the heavier ones had masses of about 22 atomic mass units. This discovery destroyed the concept that all atoms of a given element were identical, an idea that had been accepted for centuries. We now know that all elements have nuclei which have different masses. These different nuclei are known as **isotopes** of the given element. Isotopes have the same number of protons but differ in the number of neutrons.

QUESTION   Given that neon is the tenth element, how many neutrons and protons would each of its isotopes have?

ANSWER   Because each nucleon contributes about 1 atomic mass unit and neon must have 10 protons, the lighter isotope with 20 atomic mass units must have 10 neutrons and the heavier one with 22 atomic mass units must have 12 neutrons.

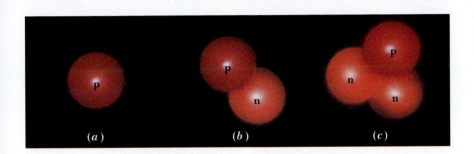

Figure 24-4   Hydrogen has three isotopes: (a) hydrogen, (b) deuterium, and (c) tritium.

Although each isotope of an element has its own nuclear properties, the chemical characteristics of isotopes do not differ. Adding a neutron to a nucleus does not change the electric charge; therefore the atom's electronic structure is essentially unchanged. Because it is the electrons that govern the chemical behavior of elements, the chemistry of all isotopes of a particular element is virtually the same. Some differences, however, can be detected. The slight changes in the electronic energy levels show up in the atomic spectra. The extra mass of the heavier isotopes also slows down the rates at which chemical and physical reactions occur.

Hydrogen has three isotopes (Fig. 24-4). The most common one contains a single proton. The next most common has one proton and one neutron; it has about twice the mass of ordinary hydrogen. This "heavy" hydrogen, known as *deuterium,* is stable and occurs naturally as 1 atom out of approximately every 6000 hydrogen atoms. The third isotope has one proton and two neutrons. This "heavy, heavy" hydrogen, known as *tritium,* is radioactive.

Isotopes of other elements do not have separate names. The symbolic way of distinguishing between isotopes is to write the chemical symbol for the element with two added numbers to the left of the symbol. A subscript gives the number of protons, and a superscript gives the number of nucleons, that is, the total number of neutrons and protons. The most common isotope of carbon has six protons and six neutrons and is written $^{12}_{6}C$. As there is some redundancy here (the number of protons is already specified by the chemical symbol), the isotope is sometimes written ^{12}C and read "carbon-12."

The Alchemists' Dream

The discovery of radioactivity changed another long-held belief about atoms. If the particle emitted during radioactive decay is charged, the resulting nucleus (commonly called the **daughter**) docs not have the same charge as the original (the **parent**). The parent and daughter are not the same element because the charge on the nucleus determines the number of electrons in the neutral atom and this number determines the atom's chemical properties. The belief that atoms are permanent was wrong; they can change into other elements. We see that nature has succeeded where the alchemists failed.

conservation of nucleons

Although we can't control the process of nuclear transformation, we can bombard materials like Rutherford did and change one element into another. But what kinds of reactions are possible? Can we, perhaps, change lead into gold? As scientists examined various processes that might change nuclei, they observed that the conservation laws are obeyed. These laws include the conservation of mass–energy, linear momentum, angular momentum, and charge, which we studied earlier. In addition, a new conservation law was discovered, the conservation of nucleons. Although the nucleons may be rearranged or a neutron changed into a proton, the total number of nucleons after the process is the same as before.

The reaction in which Rutherford produced oxygen by bombarding nitrogen with alpha particles can be written in the form

$$^4_2\alpha + ^{14}_7N \rightarrow ^{17}_8O + ^1_1p$$

Notice that the conservation of charge is obeyed; there are $2 + 7 = 9$ protons on the left-hand side of the arrow and $8 + 1 = 9$ protons on the right-hand side. The number of nucleons must also be conserved. There are $4 + 14 = 18$ nucleons on the left and $17 + 1 = 18$ nucleons on the right.

Neutrons are very effective at producing nuclear transformations because they do not have a positive charge and can penetrate to the nuclei, even with low energies. An example that occurs naturally in the atmosphere is the conversion of nitrogen to carbon via

$$^1_0n + ^{14}_7N \rightarrow ^{14}_7C + ^1_1p$$

Let's now look at changes in the nucleus that occur naturally. For example, when a nucleus emits an alpha particle, it loses two neutrons and two protons. Therefore, the parent nucleus changes into one that is two elements lower in the periodic chart. This daughter nucleus has a nucleon number that is four less. For instance, when $^{226}_{88}Ra$ decays by alpha decay, the daughter nucleus has $88 - 2 = 86$ protons and $226 - 4 = 222$ nucleons. The periodic table shown in Figure 22-23 tells us that the element with 86 protons is radon, which has the symbol Rn. Therefore, the daughter nucleus is $^{222}_{86}Rn$. This process can be written in symbolic form as

alpha decay

$$^{226}_{88}Ra \rightarrow ^{222}_{86}Rn + ^4_2\alpha$$

QUESTION   What daughter results from the alpha decay of $^{232}_{90}Th$?

The electron that is emitted in beta decay is not one of those around the nucleus, nor is it one that already existed in the nucleus. The electron is ejected from the nucleus when a neutron decays into a proton. (We will study the details of this process in Chapter 26.) A new element is pro-

ANSWER   The daughter will have $232 - 4 = 228$ nucleons and $90 - 2 = 88$ protons, so it must be $^{228}_{88}Ra$.

duced since the number of protons increases by one. The number of neutrons decreases by one, but it is the change in the proton number that causes the change in element. The daughter nucleus is one element higher in the periodic chart and has the same number of nucleons. For example,

$$^{24}_{11}Na \rightarrow {}^{24}_{12}Mg + {}^{0}_{-1}\beta^-$$

beta minus decay

QUESTION What is the daughter of $^{228}_{89}Ac$ if it undergoes beta decay?

An inverse beta decay process has also been observed. In this process an atomic electron in an inner shell is captured by one of the protons in the nucleus to become a neutron. This **electron capture** decreases the number of protons by one and increases the number of neutrons by one. Therefore, the daughter nucleus belongs to the element that is one lower in the periodic table and has the same number of nucleons as the parent. For instance,

$$^{7}_{4}Be + {}^{0}_{-1}e \rightarrow {}^{7}_{3}Li$$

electron capture

This process is detected by observing the atomic X rays given off when the outer electrons drop down to fill the energy level vacated by the captured electron. The evidence for this transformation is very convincing; the spectral lines are characteristic of the daughter nucleus and not the parent nucleus.

QUESTION What daughter is produced when $^{234}_{93}Np$ undergoes electron capture?

A third type of beta decay was observed in 1932. In this case the emitted particle has one unit of positive charge and a mass equal to that of an electron. This positive electron (**positron**) is the antiparticle of the electron, a concept that will be described further in Chapter 26. The emission of a positron results from the decay of a proton into a neutron. This process is called *beta plus decay* to distinguish it from *beta minus decay*, the process that produces a negative electron. Like electron capture, this decreases the number of protons by one and increases the number of neutrons by one. For instance,

$$^{15}_{8}O \rightarrow {}^{15}_{7}N + {}^{0}_{1}\beta^+$$

beta plus decay

ANSWER The daughter must have the same number of nucleons and one more proton. Therefore, it is $^{228}_{90}Th$.

ANSWER Once again the number of nucleons does not change, only this time the number of protons decreases by one to yield $^{234}_{92}U$.

Table 24-2 Changes in the number of protons, neutrons, and nucleons for each type of radioactive decay

Decay	Changes in the Number of		
	Protons	**Neutrons**	**Nucleons**
α	−2	−2	−4
β⁻	+1	−1	0
ec*	−1	+1	0
β⁺	−1	+1	0
γ	0	0	0

*ec stands for electron capture.

QUESTION   What is the result of $^{11}_{6}C$ decaying via beta plus decay?

Nuclei that emit gamma rays do not change their identities since gamma rays are high-energy photons and do not carry charge. The nucleus has discrete energy levels analogous to those in the atom. If the nucleus is not in the ground state, it may change to a lower energy state with the emission of the gamma ray. This process often happens after a nucleus has undergone one of the other types of decay to become an excited state of the daughter nucleus. Table 24-2 summarizes the changes produced by each type of decay.

Radioactive Decay   Σ

Equal amounts of different radioactive materials do not give off radiation at the same rate. For example, 1 gram of radium emits 20 million times as much radiation per unit time as 1 gram of uranium. The **activity** of a radioactive sample is a measure of the number of decays that take place in a certain time. The unit of activity is named the **curie** (Ci) after Marie Curie and has a value of 3.7×10^{10} decays per second, which is the approximate activity of 1 gram of radium.

1 curie = 3.7×10^{10} decays per second

There are two factors that determine the activity of a sample of material. First, the activity is directly proportional to the number of radioactive atoms in the sample. Two grams of radium will have twice the activity of 1 gram. Second, the activity varies with the type of nucleus. Some nuclei are quite stable, whereas others decay in a matter of seconds, and still others in millionths of a second. This enormous range of activity can be made plausible by realizing that the nucleus is not a static pile of nucleons, but

ANSWER   Because a proton turns into a neutron, we have $^{11}_{5}B$.

rather a dynamic, energy-packed center of activity. Apparently some nuclei are more unstable than others.

Both factors can be illustrated with an analogy that emphasizes the random nature of radioactive decay. Imagine that you have 36 dice. The radioactive decay process can be simulated by throwing the dice. Each throw represents a certain elapsed time. Assume that any die that has its number 1 face up represents an atom that has decayed during the last period.

Let's look at the "activity" of the sample of dice. How many dice would you expect to have 1's up after the first throw? Because each die has six faces that are equally likely to appear as the top face, there is a one-in-six chance of having 1's up. Since there are 36 dice in the sample, we expect that, on the average, 6 (that is, one-sixth of 36) will "decay" on the first throw. This number represents the activity of the sample.

If you double the number of dice in your sample, how many of the 72 dice should show 1's up after the first throw? On average, 12 (one-sixth of 72) will have 1's up. From this analogy we see that the level of radioactivity is directly proportional to the number of radioactive atoms in a sample.

In a sample of radioactive material the number of nuclei of the original isotope continually decreases, which decreases its activity. To illustrate this, imagine that we once again have 36 dice in our sample. After each throw of the dice, we remove the dice with 1's up. They are no longer part of the sample because they have changed to a new element. After the first throw there will be, on the average, six dice that decay. Removing them reduces these sample size to 30.

> **QUESTION**   On average, how many of these 30 dice do you expect to have 1's up on the next throw?

Removing the five dice that are expected to have 1's up on the second throw leaves a sample size of 25. Each successive throw of the dice yields a smaller sample size and consequently a lower level of activity on the next throw. The same is true of the radioactive sample. As with the dice, when a nucleus decays, it is no longer part of the sample. The activity of the radioactive sample decreases with time.

This change in the activity with time has a simple behavior, due to the probabilistic nature of radioactivity. The graph in Figure 24-5 shows this behavior. Notice that the time it takes the activity to drop to one-half its value is constant throughout the decay process. This time is called the **half-life** of the sample. In one half-life the activity of the sample decreases by a factor of 2. During the next half-life, it decreases by another factor of 2, to one-fourth of its initial value. After each additional half-life the activity would be ⅛, ⅟₁₆, ⅟₃₂, . . . that of the original value. It does not matter when you begin counting. If you wait 1 half-life, your counting rates drops by one-half.

> ANSWER   The number of 1's up will probably be ⅙ × 30 = 5.

On the average, 6 of the 36 dice will have 1's up.

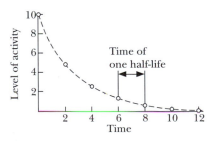

Figure 24-5 A graph of the activity of a radioactive material versus time. Notice that the time needed for the activity to decrease by a factor of 2 is always the same.

Twelve-sided Arabian dice have a smaller probability for a given number to be on top.

> **QUESTION**   If a sample of material has an activity of 20 millicuries, what activity do you expect after waiting 2 half-lives?

This decay law also applies to the number of radioactive nuclei remaining in the sample. The time it takes to reduce the population of radioactive nuclei by a factor of 2 is also equal to the half-life. The important point is that you do not get rid of the radioactive material in 2 half-lives. Let's say that you initially have 1 kilogram of a radioactive element. One half-life later you will have ½ kilogram. After 2 half lives, you will have ¼ kilogram, and so on. In other words, one-half of the remaining radioactive nuclei decays during the next half-life.

The second factor that determines the activity of a sample depends on the character of the nuclei. As we saw in the previous chapter, quantum mechanics determines the probability that something will occur. Therefore, the dice analogy is especially appropriate for nuclear decays. The probability in our dice analogy depends on the number of faces on each die. Six faces gives a probability of one in six for a given face being up on each throw of the dice. Imagine that instead of using six-sided dice, we use Arabian dice, which have twelve sides. In this case, there are more alternatives and the probability of 1's up is lower. Each die has a one-in-twelve chance of decaying. On the first throw of 36 dice we would expect an average of three dice with 1's up. The half-life of the Arabian dice is twice as long. This variation also occurs with radioactive nuclei, but with a much larger range of probabilities. Nuclear half-lives range from microseconds to trillions of years.

This process is all based on the probability of random events. With dice it is the randomness of the throwing process; with radioactive nuclei it is the quantum-mechanical randomness of the behavior of the nucleons within the nucleus. As with all probabilities any prediction is based on the mathematics of statistics. Each time we throw 36 dice, we should not expect exactly 6 dice to have 1's up. Sometimes there will be only 5; other times there might be 7 or 8. In fact, the number can vary from 0 to 36! However, statistics tells us that as the total number of dice increases, our ability to predict also increases. Because there are more than 10^{15} nuclei

Physics on Your Own   Experiment with the radioactivity analogy. To get good results you will need to make many throws of many dice. About 100 dice thrown six to eight times gives acceptable results. Because it is difficult to obtain 100 dice, use 100 sugar cubes with one side marked with a dot. If you get 16 "dots up" on the first throw of the 100 cubes, you would only use $100 - 16 = 84$ cubes for the second "time interval." Determine the half-life of your sample by graphing the number of cubes remaining after successive intervals.

> **ANSWER**   After waiting 1 half-life, the activity will be half as much, namely 10 millicuries. After waiting the other half-life, the activity will drop to half of this value. Therefore, the activity will be 5 millicuries.

in even a small sample of radioactive material, our predictions of the number of nuclei that will decay in a half-life are very reliable.

Although our predictions get better with large numbers, we must remember the probabilistic nature of dice and radioactive nuclei; we cannot predict the behavior of an individual nucleus. One particular nucleus might last a million years, whereas an "identical" one might only last a millionth of a second.

Radioactive Clocks

The decay rate of a radioactive sample is unaffected by physical and chemical conditions ordinarily found on Earth (excluding the extreme conditions found in nuclear reactors or bombs). Normal conditions involve energies that can rearrange electrons around atoms but cannot cause changes in nuclei. The nucleus is protected by its electron cloud and the large energies required to change the nuclear structure, which means that radioactive samples are good time probes into our history. Knowing the half-life of a particular isotope and the products into which it decays, we can determine the relative amount of parent and daughter atoms and calculate how long it has been decaying. In effect, we have a radioactive "clock" for dating events in the past.

One of the first such clocks gave us the age of the Earth. Prior to this, many different estimates were made by different groups. A 17th-century Irish bishop traced the family histories in the Old Testament and obtained a figure of 6000 years. Another estimate was obtained by calculating the length of time required for all the rivers of the world to bring initially freshwater oceans up to their present salinity. Another value was obtained by assuming the Sun's energy output is due to its slow collapse under the influence of the gravitational force. These latter two methods gave estimates on the order of 100,000 years. Charles Darwin suggested that the Earth was much older, as his theory of evolution required more time to produce the observed biodiversity.

The best value was obtained through the radioactivity of uranium. Uranium and the other heavy elements are only formed during supernova explosions that occur near the end of some stars' lives. Therefore, any uranium found on Earth was present in the gas and dust from which the Solar System formed. Uranium-238 decays with a half-life of 4.5 billion years into a stable isotope of lead ($^{206}_{82}$Pb) through a long chain of decays. Suppose we find a piece of igneous rock and determine that ½ of the ^{238}U has decayed into ^{206}Pb. We then know that the rock was formed at a time in the past equal to 1 half-life. The oldest rocks are found in Greenland and have a ratio of ^{206}Pb to ^{238}U a little less than 1, meaning that a little less than ½ of the original ^{238}U has decayed. This indicates that the rocks are about 4 billion years old. Incidentally, the rocks brought back from the Moon have similar ratios, establishing that the Earth and Moon were formed about the same time.

Another radioactive clock is used to date organic materials. Living organisms contain carbon, which has two isotopes of interest—the stable isotope ^{12}C and the radioactive isotope ^{14}C, which has a half-life of 5700 years. Cosmic rays bombarding our atmosphere continually produce ^{14}C

This is one of the rocks brought back from the Moon by the Apollo astronauts.

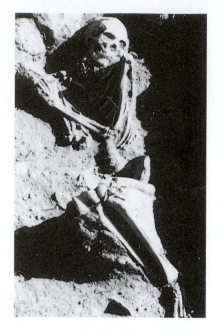

Anthropologists used radiocarbon dating to determine when this 11th-century South African died.

to replace those that decay. Because of this replacement the ratio of ^{12}C to ^{14}C in the atmosphere is relatively constant. While the plant or animal is living, it continually exchanges carbon with its environment and therefore maintains the same ratio of the two isotopes in the tissues that exists in the atmosphere. As soon as it dies, however, the exchange ceases and the amount of ^{14}C decreases due to the radioactive decay. By examining the ratio of the amount of ^{14}C to that of ^{12}C in the plant or animal, we can learn how long ago death occurred.

COMPUTING   RADIOACTIVE DATING

As an example, suppose a piece of charred wood is found in a primitive campsite. To find out how long ago the campsite was occupied, one examines the ratio of ^{14}C to ^{12}C in the wood. Suppose that the ratio is only one-eighth of the atmospheric value. Because ½ × ½ × ½ = ⅛, the tree died 3 half-lives ago. This gives an age of 3 × 5700 years = 17,100 years.

The age of the Dead Sea Scrolls was determined by this technique, because they were written on parchment. Similar studies indicate that the first human beings may have appeared on the North American continent about 27,000 years ago.

Radioactive dating with ^{14}C is not useful beyond 40,000 years (7 half-lives), as the amount of ^{14}C becomes very small. Other radioactive clocks can be used to go beyond this limit. These clocks indicate that humans, or at least prehumans, may have been around for more than 3.5 million years, mammals about 200 million years, and life for 3–4 billion years.

Radiation and Matter

How would you know if a chunk of material was radioactive? Radiation is usually invisible. However, if the material is extremely radioactive, such as radium, the enormous amount of energy being deposited in the material by the radiation can make it hotter than the surroundings. In extreme cases it may even glow. A less radioactive sample might leave evidence of its presence over a longer time. Pierre Curie developed a radioactive burn on his side because of his habit of carrying a piece of radium in his vest pocket.

We learn about the various types of radiation through their interactions with matter. In the majority of cases these interactions are outside the range of direct human observation. One exception is light. Although we don't see photons flying across the room, we do see the interaction of these photons with our retinas. Our eyes are especially tuned to a small range of photon energy. Other radiations are detected by extending our senses with instruments. For example, the electromagnetic radiation from radio and television stations is not detected by our bodies but interacts with the electrons in antennas.

A TV antenna is a radiation detector.

Physics Update

In the late heavy bombardment (LHB) epoch, a span of about 200 million years some 4 billion years ago, the Moon sustained many large impacts. Some astronomers believe that the projectiles responsible may have struck Mercury, Venus, Earth, and Mars as well. Others assert that the LHB phenomenon was unique to the Earth–Moon system or that it did not happen at all, at least not so suddenly. Now, a group of scientists at the University of Manchester (UK) has dated a rock found here on Earth but which is believed to have been a meteorite originating at Mars. The 4-billion-year age of the object, determined by isotope dating, is much older than previously studied Martian meteorites. The antiquity of the rock, say the researchers, provides evidence for a widespread LHB effect.

We also observe nuclear radiations through their interactions with matter. Alpha and beta particles interact with matter through their electric charge. For example, in Becquerel's discovery the charged particles interacted with the photographic chemicals to expose the film. For the most part these charged particles interact with the electrons in the material. As we saw in Rutherford's scattering experiment, they rarely collide with nuclei. As the alpha and beta particles pass through matter, they interact with atomic electrons, causing them to jump to higher atomic levels or to leave the atom entirely. This latter process is called **ionization.**

The relatively large mass of alpha particles compared with electrons means that the alpha particles travel through the material in essentially straight lines. This is like a battleship passing through a flotilla of canoes. Because beta particles are electrons, they are more like canoes hitting canoes and have paths that are more ragged. In both cases, the colliding particles deposit energy in the material in a more or less continuous fashion. Therefore, the distance they travel in a material is a measure of their initial energy. We can determine this energy experimentally by measuring the distance the particle travels; that is, its range. Some sample ranges are given in Table 24-3.

The interaction of gamma rays with matter is quite different. Gamma rays do not lose their energy in bits and pieces. (There is no such thing as

Energy*	**Range in Air (cm)**			**Range in Aluminum (cm)**		
(MeV)	α	**p**	e^-	α	**p**	e^-
1	0.5	2.3	314	0.0003	0.0014	0.15
5	3.5	34	2000	0.0025	0.019	0.96
10	10.7	117	4100	0.0064	0.063	1.96

Table 24-3 Ranges of alpha particles, protons, and electrons in air and aluminum

*MeV (million electron volts) is an energy unit equal to 1.6×10^{-13} joules.

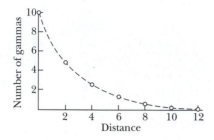

Figure 24-6 Graph of the number of gamma rays surviving at various distances through a material. Notice the similarity with the graph of the radioactivity in Figure 24-5.

Table 24-4
Half-distances for the absorption of gamma rays by aluminum

Energy (MeV)	Half-Distance (cm)
1	4.2
5	9.1
10	11.1

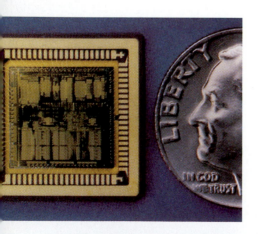

The 200,000 individual electronic components in this very small crystalline chip are especially sensitive to radiation damage.

half a photon.) Whenever a photon interacts with matter, it is completely annihilated in one of three possible ways. The photon's energy can ionize an atom, be transferred to a free electron with the creation of a new photon, or be converted into a pair of particles according to Einstein's famous equation $E = mc^2$. If we count the number of gamma rays surviving at various distances in a material and make a graph, we would obtain a curve like the one in Figure 24-6. This curve has the same shape as the one for radioactive decay in Figure 24-5, which means that gamma rays do not have a definite range but are removed from the beam with a characteristic half-distance. That is, one-half of the original gamma rays are removed in a certain length, half of the remaining are removed in the next region of this length, and so on. From Table 24-4 you can see that gamma rays are much more penetrating than alpha or beta particles of the same energy.

> **QUESTION**   How far must a beam of 10-million-electron-volt gamma rays travel in aluminum before it is all gone?

Any radiation passing through matter deposits energy along its path, resulting in temperature increases, atomic excitations, and ionization. The ionization caused by radiation can damage a substance if its properties depend strongly on the detailed molecular or atomic structure. For example, radiation can damage the complex molecules in living tissues and cause cancer or genetic defects. Damage can also occur in nonliving objects, such as transistors and integrated circuits. These modern electronic devices are made of crystals in which the atoms have very definite geometrical arrangements. In integrated circuits, such as those found in pocket calculators and computers, there are literally thousands of individual electronic components in very small crystalline chips. A disruption of these atomic structures can change the electronic properties of these components. If the radiation damage is severe, the device could fail. This hazard is obviously serious if the device is a control component for a nuclear reactor or part of the guidance system for a missile. There is, however, not enough radiation around the university computer for you to expect all of your grades to change into A's.

Biological Effects of Radiation

The ionization caused by radiation passing through living tissue can destroy organic molecules if the electrons are involved in molecular binding. If too many molecules are destroyed in this fashion or if DNA molecules are destroyed, cells may die or become cancerous.

The effects of radiation on our health depend on the amount of radiation absorbed by living tissue and the biological effects associated with this absorption. A **rad** (an acronym for *radiation absorbed dose*) is the

> **ANSWER**   Theoretically, this never happens; some of the gamma rays will travel very large distances. However, by the time the beam has traveled 7 half-distances, less than 1% of the beam remains.

unit used to designate the amount of energy deposited in a material. A rad of radiation deposits 0.01 joule per kilogram of material. Another unit, the **rem** (*r*adiation *e*quivalent in *m*ammals), was developed to reflect the biological effects caused by the radiation. Different radiations have different effects on our bodies. Alpha particles, for example, deposit more energy per centimeter of path than electrons with the same energy; therefore the alpha particles cause more biological damage. One rad of beta particles can result in 1–1.7 rem of exposure, 1 rad of fast neutrons or protons may result in 10 rem, and 1 rad of alpha particles produces between 10 and 20 rem. The rad and the rem are very nearly equal for photons. Because most human exposure involves photons and electrons, the two units are roughly interchangeable when talking about typical doses.

Knowing the amount of energy deposited or even the potential damage to cells from exposure to radiation is still not a prediction of the future health of an individual. The biological effects vary considerably among individuals depending on their age, health, the length of time over which the exposure occurs, and the parts of the body exposed. Large doses, such as those resulting from the nuclear bombs exploded over Japan or the accident at the nuclear reactor at Chernobyl in the former Soviet Union (Fig. 24-7), cause radiation sickness, which can result in vomiting, diarrhea, internal bleeding, loss of hair, and even death. It is virtually impossible to survive a dose of more than 600 rem to the whole body over a period of days. An exposure of 100 rem in the same period causes radiation sickness, but most people recover. For comparison, the maximum occupational dose is 5 rem per year, the maximum rate recommended for the general public by the U.S. Environmental Protection Agency is 0.5 rem per year, and the natural background radiation is about 0.3 rem per year.

It is very difficult to obtain data on the health effects of exposures to low levels of radiation, primarily because the effects are masked by effects not related to radiation and they do not show up for a long time. Although cancers such as leukemia may show up in 2–4 years, the more prevalent tumorous cancers take much longer. Imagine conducting a free-fall experiment with an apple in which the apple doesn't begin to move noticeably for 3 years. Even after you spot its motion you are compelled to ask if it occurred because of your actions or something else. When cause and effect are so widely separated in time, it is difficult to make connections. For example, because there is already a high incidence of cancer, it is hard to determine if any increase is due to an increase in radiation or to some other factor. On the other hand, it is known that large doses of radiation do cause significant increases in the incidence of cancer. Marie Curie died of leukemia, most likely the result of her long-term exposure to radiation.

Genetic effects due to low-level radiation are also difficult to determine. These effects can be passed on to future generations through mutation of the reproductive cells. Although mutations are important in the evolution of species, most of them are recessive. It is generally agreed that an increase in the rate of mutations is detrimental.

Any additional nuclear radiation hitting our bodies is detrimental. However, we live in a virtual sea of radiation during our entire lives. Exposure to some of this radiation is beyond our control, but other exposure is

Figure 24-7 One of the world's major nuclear reactor accidents happened in April 1986 at Chernobyl in the former Soviet Union. This photograph shows the structure built to contain the leaking radiation.

Air travelers implicitly accept the risk of the increased radiation for the benefits of quicker travel.

within our control. Indeed, many of the "nuclear age" debates center on this issue of assessing the effects of *additional* radiation on the health of the population and evaluating its risks and benefits. As an example, the increased radiation exposure during air travel due to the reduced shielding of the atmosphere is about 0.5 millirem per hour. This increase is not much for an occasional traveler, but may result in an additional 0.5 rem per year for an airplane pilot, which equals the maximum recommended dose.

The average dosage rates for a variety of sources are given in Table 24-5. A word of caution is appropriate here. Although the concept of av-

Table 24-5 Average annual radiation doses*		
Source	**Dose**	**(mrem/yr)**
Cosmic Rays	30	
Surroundings	30	
Internal	40	
Radon	200	
Naturally Occurring		300
Medical	50	
Consumer Products	5	
Nuclear Power	5	
Human-Made		60
Total		360

*Frank E. Gallagher III, at the California Campus Radiation Safety Officers Conference, September 1993.

eraging helps to make some comparisons, there are situations where the value of the average is quite meaningless. Consider, for example, the amount of radiation our population receives from the nuclear power industry. The bulk of this radiation exposure occurs for the workers in the industry. Most of us get very little radiation from this source. This situation is similar to putting one foot in ice water, the other foot in boiling water, and claiming that on the average you are doing fine.

Radiation Around Us

It is impossible to get away from all radiation. The radiation that is beyond our control comes from the cosmic rays arriving from outer space, from reactions in our atmosphere initiated by these cosmic rays, from naturally occurring radioactive decays in the material around us, and from the decay of radioactive isotopes (mostly ^{40}K) within our bodies. In fact, even table salts that substitute potassium for sodium are radioactive. Radiation is part of the world in which we live. The radiations within our control range from testing of nuclear weapons to nuclear power stations to medical radiations used for diagnosing and treating diseases.

Medical sources of radiation *on average* contribute less to our yearly dose than the background radiation. The risks of diagnostic X rays have decreased over the years with new improvements in technology. Modern machines and more sensitive photographic films have greatly reduced the exposure necessary to get the needed information. The exposure from a typical chest X ray is 0.01 rem.

We saw in Chapter 22 that X rays can be used to examine internal organs and other structures for abnormalities. During this process, however, some cells are damaged or killed. We need to ask: Is the possible damage worth the benefits of a better treatment based on the information gained? In most cases, the answer is yes. In other cases the answer ranges from a vague response to a definite no. If your physician suspects that your ankle is broken, the benefits of having an X ray are definitely worth the risks. On the other hand, having an X ray of your lungs as part of an annual physical exam is not felt to be worth the risks. If, on the other hand, there is good reason to suspect lesions or a tumor, the odds are in favor of having the X ray.

The largest doses of medically oriented radiation occur during radiation treatment of cancer. In these cases radiation is used to deliberately kill some cells to benefit the patient. Because a beam of radiation causes damage along its entire path, the beam is rotated around the body to minimize damage to normal cells. This procedure is shown in Figure 24-8. Only the cells in the overlap region (the cancerous tumor) get the maximum dosage. Alternative techniques use beams of charged particles. Because charged particles deposit energy at the highest rate near the end of their range, the damage to normal cells along the entrance path is minimized.

Sometimes radioactive materials are ingested for diagnostic or treatment purposes. Our bodies do not have the ability to differentiate between various isotopes of a particular chemical element. If a tumor or organ is known to concentrate a certain chemical element, a radioactive

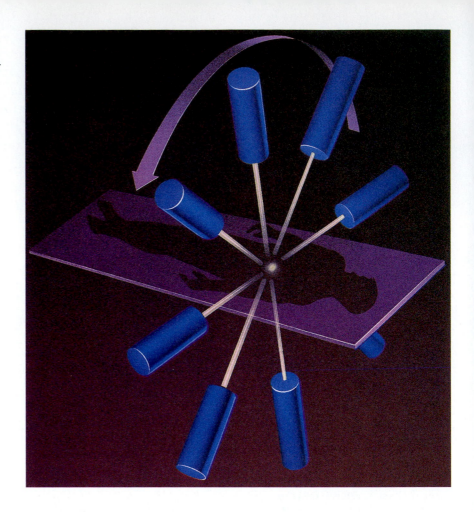

isotope of that element can be put into the body to treat the tumor or to produce a picture of the tumor or organ. For instance, because iodine concentrates in the thyroid gland, iodine-131, a beta emitter, can be used to kill cells in the thyroid with little effect on more distant cells.

Physics Update

A positron-emission tomography (PET) system has been developed that can resolve structures as small as 0.006 cubic centimeters (about the volume of a sesame seed), a nearly tenfold improvement in resolution over conventional scanners, at an estimated price of $300,000, less than one-fifth the cost of conventional PET systems. Used to provide images of such things as brain activity, PET detects pairs of gamma rays released when positrons from a radioactive tracer in the bloodstream annihilate electrons in the patient's body. MicroPET brings about increased resolution by employing smaller crystals for detecting the gammas. The crystals are small enough that their signals can be read out using optical fibers, which convey light to a multichannel phototube consisting of 64 individual elements, greatly reducing the need for numerous expensive tubes. (Much of the technology was adapted from high-energy particle physics.) This system is suitable for imaging the smaller-scale anatomies of laboratory animals used in drug trials.

Radon

A source of radiation that is partly under our control is the radioactive gas radon. Much attention has been focused by the media on the presence of indoor radon, especially in the air of some household basements. This awareness came about in part because of energy conservation efforts. In attempts to reduce heating bills, many homeowners improved their insulation and sealed many air leaks. This resulted in less air circulation within homes and allowed the concentration of radon in the air to increase.

The source of the radon is uranium ore that exists in the ground; the ore's quantities vary geographically. Through a long chain of decays shown in the table, the uranium eventually becomes a stable isotope of lead. All the radioactive daughters are solids except one, the noble gas radon (Rn). This gas can travel through the ground and enter basements of homes through small (or large) cracks in the basement floors or walls. Because radon is an inert gas, it doesn't deposit on walls and leave the air. Thus, it can be inhaled into a person's lungs and trapped there. In this case, each of the 8 subsequent decays causes the lungs to experience additional radiation and possible damage. The maximum "safe" level of radon has been established by the Environmental Protection Agency at 4×10^{-12} curies per liter. Kits to test for excessive concentrations are available through retail outlets. If high levels are found, the homeowner can take measures to ventilate the radon, and to block its entry into the house by sealing cracks around the foundation and in the basement.

Radioactive Decay

$^{238}_{92}\text{U} \rightarrow$	$^{234}_{90}\text{Th}$	$+ \alpha$	$(9.5 \times 10^9 \text{ years})$
	$\downarrow$		
	$^{234}_{91}\text{Pa}$	$+ \beta^-$	(24 days)
	$\downarrow$		
	$^{234}_{92}\text{U}$	$+ \beta^-$	(6.7 hours)
	$\downarrow$		
	$^{230}_{90}\text{Th}$	$+ \alpha$	$(2.5 \times 10^5 \text{ years})$
	$\downarrow$		
	$^{226}_{88}\text{Ra}$	$+ \alpha$	$(7.5 \times 10^4 \text{ years})$
	$\downarrow$		
	$^{222}_{86}\text{Rn}$	$+ \alpha$	(1622 years)
	$\downarrow$		
	$^{218}_{84}\text{Po}$	$+ \alpha$	(3.85 days)
	$\downarrow$		
	$^{214}_{82}\text{Pb}$	$+ \alpha$	(3 minutes)
	$\downarrow$		
	$^{214}_{83}\text{Bi}$	$+ \beta^-$	(27 minutes)
	$\downarrow$		
	$^{214}_{84}\text{Po}$	$+ \beta^-$	(19.7 minutes)
	$\downarrow$		
	$^{210}_{82}\text{Pb}$	$+ \alpha$	$(10^{-4} \text{ seconds})$
	$\downarrow$		
	$^{210}_{83}\text{Bi}$	$+ \beta^-$	(19 years)
	$\downarrow$		
	$^{210}_{84}\text{Po}$	$+ \beta^-$	(5 days)
	$\downarrow$		
	$^{206}_{82}\text{Pb}$	$+ \alpha$	(138 days)

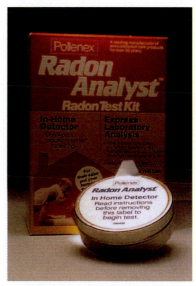

A commercially available kit for testing radon gas levels in homes.

With each source of radiation the debate returns to the question of risk versus benefit; does the potential benefit of the exposure exceed the potential risk?

Radiation Detectors

Many devices have been developed to detect nuclear radiation. The earliest was the fluorescent screen Rutherford used in his alpha-particle experiments. Alpha particles striking the screen ionized some of the atoms on the screen. When these atomic electrons returned to their ground states, visible photons were emitted. Rutherford and his students used a microscope to count the individual flashes when alpha particles arrived.

The fluorescent screen is just one example of a *scintillation detector,* a detector that emits visible light upon being struck by radiation. The most familiar device of this type is the television screen. Scintillation detectors can be combined with electronic circuitry to automatically count the radiation. When a photon enters the device, it strikes a photosensitive surface and ejects an electron via the photoelectric effect (Chapter 22). This electron is then accelerated by a voltage so that it strikes a surface with sufficient energy to free two or more electrons. These electrons are accelerated and release additional electrons when they strike the next electrode. A typical photomultiplier, like the one in Figure 24-9, may have ten stages, with a million electrons released at the last stage. This signal can then be counted electronically. Furthermore, the signal is proportional to the energy deposited in the scintillation material and hence it gives the energy of the incident radiation.

The Geiger counter, popularized in old-time movies about prospecting for uranium, utilizes the ionization of a gas to detect radiation. The essential parts of a Geiger counter are shown in Figure 24-10. The cylinder contains a gas and has a wire running along its length. As the radiation interacts with the gas, it ionizes gas atoms. The electrons gain kinetic energy from the electric field in the cylinder, which they then lose in collisions with other gas atoms. These collisions release more electrons and the process continues. This quickly results in an avalanche of electrons which produces a short electric current pulse that is detected by the electronics in the Geiger counter. In the movies the counter produces audible clicks. Many clicks per second means the count rate is high and our hero is close to fame and fortune.

Another type of detector produces continuous tracks showing the paths of charged particles. These tracks are like those left by wild animals in snow. The big difference, however, is that although we may eventually see a deer, we cannot ever hope to see the charged particles. One track chamber is a bubble chamber in which a liquid is pressurized so that its temperature is just below the boiling point. When the pressure is suddenly released, boiling does not begin unless there is some disturbance in the liquid. When the disturbance is caused by the ionization of a charged particle, tiny bubbles form along its path. These bubbles are allowed to grow so that they can be photographed, and then the liquid is compressed to remove them and get ready for the next picture.

(a)

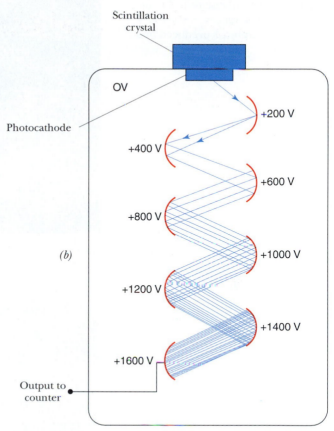

(b)

Figure 24-9   (a) Photograph and (b) schematic drawing
of a photomultiplier tube used to detect radiation striking
a scintillation counter.

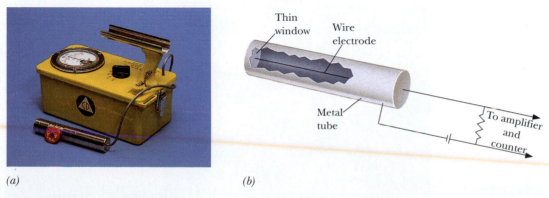

(a)   *(b)*

Figure 24-10   (a) A civil defense Geiger counter. (b) Schematic drawing of a
Geiger counter.

Figure 24-11 Photograph of tracks in a hydrogen bubble chamber show the production of some strange particles that we will discuss in Chapter 26. The spiral track near the top of the photograph was produced by an electron.

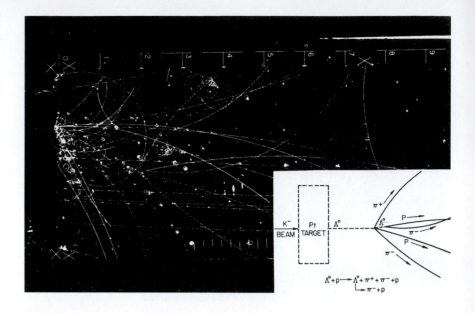

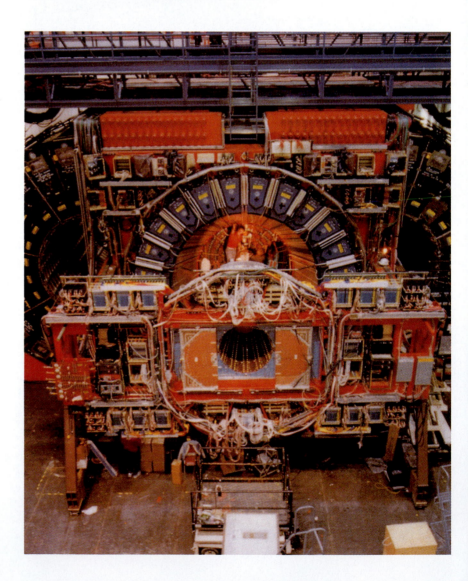

Figure 24-12 This very large particle detector is used at the Fermi National Accelerator Laboratory to study collisions of particles at very high energies.

The paths of the charged particles are bent by placing the bubble chamber in a magnetic field; positively charged particles bend one way and negatively charged particles the other way. Examination and measurement of the tracks in bubble chamber photographs, such as the one in Figure 24-11, yield information about the charged particles and their interactions.

Modern particle detectors come in a variety of forms and sizes. In many of these, the particles cannot be seen; their locations and identities are all determined by large arrays of electronics connected to large computers like the one shown in Figure 24-12.

SUMMARY

Radioactivity is a nuclear effect, rather than an atomic one. It is virtually unaffected by outside physical conditions, such as strong electric and magnetic fields or extreme pressures and temperatures. Three types of radiation can occur: alpha (α) particles, or helium nuclei; beta (β) particles, which are electrons; and high-energy photons, or gamma (γ) rays. These different radiations have different penetrating properties, ranging from the gamma ray that can travel through many meters of air or even through thick walls to the alpha ray that won't penetrate a piece of paper.

Nuclei are composed of protons and neutrons. Nuclei of the same element have the same number of protons, but isotopes of this element have different numbers of neutrons. Isotopes of a particular element have their own nuclear properties but nearly identical chemical characteristics, because adding a neutron to a nucleus does not change the atom's electronic structure appreciably.

Radioactive decays are spontaneous, obey the known conservation laws (mass–energy, linear momentum, angular momentum, electric charge, and nucleon number), and most often change the chemical nature of their atoms. There are five main naturally occurring nuclear processes. A nucleus (1) emits an alpha particle, producing a nucleus two elements lower in the periodic chart, (2) beta decays, emitting an electron and producing the next higher element, (3) inverse beta decays via electron capture, producing a daughter nucleus one lower in the periodic table, (4) beta decays with a positron, an antiparticle of the electron, decreasing the number of protons by one and increasing the number of neutrons by one, or (5) emits gamma rays, leaving its identity unchanged.

Equal amounts of different radioactive materials do not give off radiation at the same rate. The unit of activity, a curie, has a value of 3.7×10^{10} decays per second. Two factors determine the activity of a sample of material—the number of radioactive atoms in the sample, and the characteristic half-life of the isotope. The quantum-mechanical nature of nucleons results in a wide range of decay rates and prohibits us from predicting the behavior of individual nuclei. Radioactive samples are good clocks for dating events in the past.

Radiation interacts with matter, often outside the range of direct human observation. Alpha and beta particles interact with matter

CHAPTER

24

REVISITED

If the artifact is made from material that was once living, the organic material had the same ratio of carbon-14 to carbon-12 as the atmosphere at the time the artifact was made. Since that time, the ratio of these carbon isotopes has been decreasing because the carbon-14 is radioactive, while the carbon-12 is stable. This isotopic ratio serves as a clock, dating the artifact.

through their electric charge, whereas gamma rays ionize atoms or convert into a pair of particles. All radiation interactions with matter deposit energy, resulting in temperature increases, atomic excitations, and ionization. The ionization caused by radiation passing through living tissue can kill or damage cells. It is impossible to get away completely from radiation.

KEY TERMS

activity: The number of radioactive decays that take place in a unit of time. Measured in curies.

alpha particle: The nucleus of helium consisting of two protons and two neutrons.

alpha (α) radiation: The type of radioactive decay in which nuclei emit alpha particles (helium nuclei).

beta particle: An electron emitted by a radioactive nucleus.

beta (β) radiation: The type of radioactive decay in which nuclei emit electrons or positrons (antielectrons).

curie: A unit of radioactivity, 3.7×10^{10} decays per second.

daughter nucleus: The nucleus resulting from the radioactive decay of a parent nucleus.

electron capture: A decay process in which an inner atomic electron is captured by the nucleus. The daughter nucleus has the same number of nucleons but one fewer proton.

gamma (γ) radiation: The type of radioactive decay in which nuclei emit high-energy photons. The daughter nucleus is the same as the parent.

half-life: The time during which one-half of a sample of a radioactive substance decays.

ionization: The removal of one or more electrons from an atom.

isotope: An element containing a specific number of neutrons in its nuclei. Examples are $^{12}_{6}C$ and $^{14}_{6}C$, carbon atoms with six and eight neutrons, respectively.

neutron: The neutral nucleon. One of the constituents of nuclei.

nucleon: Either a proton or a neutron.

nucleus: The central part of an atom that contains the protons and neutrons.

parent nucleus: A nucleus that decays into a daughter nucleus.

positron: The antiparticle of the electron.

proton: The positively charged nucleon. One of the constituents of nuclei.

rad: An acronym for *r*adiation *a*bsorbed *d*ose; a rad of radiation deposits 1/100 joule per kilogram of material.

rem: An acronym for *r*adiation *e*quivalent in *m*ammals, a measure of the biological effects caused by radiation.

CONCEPTUAL QUESTIONS

(The periodic table of the elements in Figure 22-23 is useful for answering some of these questions.)

1. When Becquerel was trying to produce X rays by shining light on fluorescent materials, he discovered nuclear radiation. What observation convinced him that he was not simply observing X rays?

2. Why do we believe that radioactivity is a nuclear phenomenon?

3. Why can't we use visible light to look at nuclei?

4. Briefly summarize the differences in the three types of radiation.

5. How do gamma rays differ from X rays?

6. Which of the following names do *not* refer to the same thing: beta particles, cathode rays, alpha particles, and electrons?

7. If radium (which is radioactive) and chlorine (which is not radioactive) combine to form radium chloride, is the compound radioactive?

8. How would the activity of a gram of uranium oxide compare with that of a gram of pure uranium?

9. What are the constituents of nuclei?

10. Is the chemical identity of an atom determined by the number of electrons, neutrons, protons, or nucleons in its nucleus?

11. What do isotopes have in common?

12. Why are the atomic masses of many elements not close to whole numbers?

13. What is the name of the element represented by the X in each of the following?

    a. $^{204}_{82}X$ b. $^{26}_{12}X$ c. $^{107}_{47}X$

14. What is the name of the element represented by the X in each of the following?

    a. $^{27}_{13}X$ b. $^{245}_{96}X$ c. $^{197}_{79}X$

15. How many neutrons, protons, and electrons are there in each of the following atoms?

    a. $^{15}_{7}N$ b. $^{114}_{50}Sn$ c. $^{226}_{88}Ra$

16. How many neutrons, protons, and electrons are there in each of the following atoms?

    a. $^{11}_{5}B$ b. $^{64}_{29}Cu$ c. $^{239}_{93}Np$

17. What is the approximate mass of $^{40}_{19}K$ in atomic mass units?

18. What is the approximate mass of $^{64}_{30}Zn$ in atomic mass units?

19. What happens to the charge of the nucleus when it decays via beta plus decay?

20. What happens to the charge of the nucleus when it decays via electron capture?

21. How does the mass of a nucleus change when it decays via alpha decay?

22. What happens to the mass of a nucleus that undergoes beta decay?

23. Each of the following isotopes decays by alpha decay. What is the daughter isotope in each case?

    a. $^{236}_{92}U$ b. $^{220}_{86}Rn$

24. What daughter is formed when each of the following undergoes alpha decay?

    a. $^{254}_{99}Es$ b. $^{210}_{84}Po$

25. What daughter is formed when each of the following undergoes beta minus decay?

    a. $^{19}_{8}O$ b. $^{188}_{74}W$

26. Each of the following isotopes decays by beta minus decay. What is the daughter isotope in each case?

    a. $^{14}_{6}C$ b. $^{64}_{29}Cu$

27. Each of the following isotopes decays by electron capture. What is the daughter isotope in each case?

    a. $^{190}_{79}Au$ b. $^{204}_{84}Po$

28. If the following undergo electron capture, what daughter is produced?

    a. $^{22}_{11}Na$ b. $^{250}_{100}Fm$

29. Each of the following isotopes decays by beta plus decay. What is the daughter isotope in each case?

    a. $^{87}_{40}Zr$ b. $^{19}_{10}Ne$

30. What daughter is formed when each of the following undergoes beta plus decay?

    a. $^{22}_{11}Na$ b. $^{52}_{26}Fe$

31. This isotope $^{215}_{84}Po$ decays by alpha decay or by beta minus decay. What are the resulting nuclei?

32. The isotope $^{64}_{29}Cu$ can decay via beta plus and via beta minus decay. What are the daughters in each case?

33. What type of decay process is involved in each of the following:

    a. $^{228}_{92}U \rightarrow {}^{228}_{91}Pa + (?)$

    b. $^{254}_{100}Fm \rightarrow {}^{250}_{98}Cf + (?)$

34. What does the (?) stand for in each of the following decays?

    a. $^{233}_{90}Th \rightarrow {}^{233}_{91}Pa + (?)$

    b. $^{18}_{9}F \rightarrow {}^{17}_{8}O + (?)$

*35. Uranium-238 decays to a stable isotope of lead through a series of alpha and beta minus decays. Which of the following is a possible daughter of ^{238}U: ^{206}Pb, ^{207}Pb, ^{208}Pb, or ^{209}Pb?

*36. What isotopes of lead (element 82) could be the decay products of $^{234}_{90}Th$ if the decays are all alpha and beta minus decays?

37. What stable nucleus is formed when a nucleus of $^{13}_{7}N$ absorbs a neutron?

38. What nucleus results when $^{16}_{8}O$ captures a proton?

39. What isotope is produced when a neutron strikes a nucleus of $^{10}_{5}B$ and an alpha particle is emitted?

40. A proton strikes a nucleus of $^{20}_{10}Ne$. Assuming an alpha particle comes out, what isotope is produced?

41. A nucleus of $^{238}_{92}U$ captures a neutron to form an unstable nucleus that undergoes two successive beta minus decays. What is the resulting nucleus?

42. What changes in the numbers of neutrons and protons occur when a nucleus is bombarded with a deuteron (the nucleus of deuterium containing one neutron and one proton) and an alpha particle is emitted?

43. A radioactive material has a half-life of 50 days. How long would you have to watch a particular nucleus before you would see it decay?

44. How long does it take for a radioactive sample to completely decay?

45. How do radioactive clocks determine the age of things that were once living?

46. Can ^{14}C dating be used to date the stone pillars at Stonehenge?

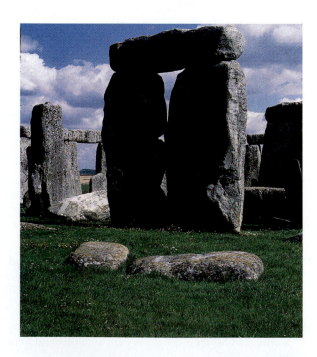

*47. It is known that the atmospheric ratio of ^{14}C to ^{12}C has varied over time. How might one correct for this variation in radioactive dating determinations?

*48. Assume that the atmospheric ratio of ^{14}C to ^{12}C increases with time. Would determinations of age by radiocarbon dating be too short or too long?

49. Is it possible to totally shield something against gamma rays?

50. Which has the largest range, a 10-MeV alpha particle, a 10-MeV electron, or a 10-MeV gamma ray?

*51. A beam containing electrons, protons, and neutrons is aimed at a concrete wall. If all particles have the same kinetic energy, which particle will penetrate the farthest into the wall?

52. A beam containing electrons, protons, and neutrons is aimed at a concrete wall. If all particles have the same kinetic energy, which particle will penetrate the wall the least?

53. Would you expect more radiation damage to occur with a rad of X rays or a rad of alpha particles?

54. How would you expect the range of a neutron to compare with that of a proton with the same energy?

55. Why are radioactive substances that emit alpha particles more dangerous inside the body than outside?

56. Which of the following would cause the most biological damage if they all have the same energy: photons, electrons, neutrons, protons, or alpha particles?

57. Which of the following sources of radiation contributes the least to the average yearly dose received by humans: cosmic rays, internal, nuclear power, surroundings, or medical?

58. Why is it difficult to determine the biological effects of low levels of exposure to radiation?

59. What are the methods that radiation can be used to treat cancerous tumors?

60. What factors should we consider when deciding whether to have an X ray?

EXERCISES

1. Given that a typical atom is 100,000 times bigger than its nucleus, how big would the atom be if the nucleus was scaled up so that it was 1 mm across?

*2. The radius of the Sun is 7×10^8 m and the average distance to Pluto is 6×10^{12} m. Are these proportions approximately correct for a scaled-up model of an atom?

3. What is the mass of the neutral ^{12}C atom in kilograms?

4. The atomic mass of neutral gold-197 is 196.97 amu. What is its mass in kilograms?

5. The activity of a radioactive sample is initially 64 µCi (microcuries). What will the activity be after 3 half-lives have elapsed?

6. If the activity of a particular radioactive sample is 40 Ci and its half-life is 200 years, what would you expect for its activity after 400 years?

7. A radioactive material has a half-life of 10 min. If you begin with 512 trillion radioactive atoms, approximately how many would you expect to have after 40 min?

8. If you start with 2 kg of a radioactive substance, how much will you have after 4 half-lives?

9. If the ratio of ^{14}C to ^{12}C in a piece of bone is only ⅛ of the atmospheric ratio, how old is the bone?

10. If the ratio of ^{14}C to ^{12}C in a piece of parchment is only 1/16 of the atmospheric ratio, how old is the parchment?

11. How thick an aluminum plate would be required to reduce the intensity of a 1-MeV beam of gamma rays by a factor of 4?

12. A beam of 5-MeV gamma rays is incident on a block of aluminum that is 18.2 cm thick. What fraction of the gamma rays penetrates the block?

CHAPTER 25

Nuclear Energy

Our world and its economy run on energy. Without an abundant supply of energy, people in third world countries will not be able to obtain our standard of living. With the depletion of fossil fuels—coal, oil, and natural gas—the need arises to explore alternatives. What options are available for using nuclear energy and what are the primary issues in extracting this nuclear energy? (See p. 637 for the answer to this question.)

In a nuclear power plant heavy nuclei are split, releasing energy which is used to generate electricity.

Although radioactivity indicates that nuclei are unstable and raises questions about why they are unstable, the more fundamental question is just the opposite: Why do nuclei stay together? The electromagnetic and gravitational forces do not provide the answer. The gravitational force is attractive and tries to hold the nucleus together, but it is extremely weak compared with the repulsive electric force pushing the protons apart. Nuclei should fly apart. The stability of the majority of known nuclei, however, is a fact of nature. There must be another force—a force never imagined before the turn of the century—that is strong enough to hold the nucleons together.

The existence of this force also means that there is a new source of energy to be harnessed. We have seen in previous chapters that there is an energy associated with each force. For example, an object has gravitational potential energy because of the gravitational force. Hanging weights can run a clock and falling water can turn a grinding wheel. The electromagnetic force shows up in household electric energy and in chemical reactions such as burning. When wood burns, the carbon atoms combine with oxygen atoms from the air to form carbon dioxide molecules. These molecules have less energy than the atoms from which they are made. The excess energy is given off as heat and light when the wood burns.

Similarly, the rearrangement of nucleons results in different energy states. If the energy of the final arrangement is less than that in the initial one, energy is given off, usually in the form of fast-moving particles and electromagnetic radiation. To understand the nuclear force, the possible types of nuclear reactions, and the potential new energy source, the early experimenters needed to take a closer look at the subatomic world. This was accomplished with nuclear probes.

The chemical energy released by the burning of the forests in Yellowstone National Park was a result of the electromagnetic forces between atoms.

Nuclear Probes

Information about nuclei initially came from the particles ejected during radioactive decays. This information is limited by the available radioactive nuclei. How can we study stable nuclei? Clearly we can't just pick one up and examine it. Nuclei are 100,000 times smaller than their host atoms.

The situation is much more difficult than the study of atomic structure. Although the size of atoms is also beyond our sensory range, we can gain clues about atoms from the macroscopic world because it is governed by atomic characteristics. That is, the chemical and physical properties of matter depend on the atomic properties, and some of the atomic properties can be deduced from these everyday properties. On the other hand, there is no everyday phenomenon that reflects the character of stable nuclei.

The initial information about the structure of stable nuclei came from experiments using particles from radioactive decays as probes. These probes—initially limited to alpha and beta particles—failed to penetrate the nucleus. The relatively light beta particle is quickly deflected by the atom's electron cloud. As the Rutherford experiments showed, the alpha particles are able to penetrate the electron cloud, but are repelled

The nuclear energy released in the explosion of a nuclear bomb is a result of the strong force between nucleons.

Figure 25-1 This Van de Graaff generator is used to accelerate protons to high velocities to probe the structure of nuclei.

by the positive charge on the nucleus. Higher energy alpha particles had enough energy to penetrate the smaller nuclei, but they could not penetrate the larger ones. Even when penetration was possible, the matter-wave aspect of the alpha particles hindered the exploration. In Chapter 23 we learned that the wavelength of a particle gets larger as its momentum gets smaller. Typical alpha particles from radioactive processes have low momenta, which places a limit on the details that can be observed since diffraction effects hide nuclear structures that are smaller than the alpha particle's wavelength.

Accelerators

This limit was lowered by constructing machines to produce beams of charged particles with much higher momenta and, consequently, much smaller wavelengths. **Particle accelerators,** such as the Van de Graaff generator shown in Figure 25-1, use electric fields to accelerate charged particles to very high velocities. In the process the charged particle acquires a kinetic energy equal to the product of its charge and the potential difference through which it moves (Chapter 19). The paths of the charged particles are controlled by magnetic fields.

There are two main types of particle accelerator: those in which the particles travel in straight lines, and those in which they travel in circles. The simplest linear accelerator consists of a region in which there is a strong electric field and a way of injecting charged particles into this region. The maximum energy that can be given to a particle is limited by the maximum voltage that can be maintained in the accelerating region without electric discharges. The largest of these linear machines can produce beams of particles with energies up to 10 million electron volts (10 MeV), where 1 electron volt is equal to 1.6×10^{-19} joule (Chapter 24). This energy is somewhat larger than the fastest alpha particles obtainable from radioactive decay.

The maximum energy can be increased by passing the charged particles through several consecutive accelerating regions. The largest linear accelerator is the Stanford Linear Accelerator shown in Figure 25-2. This machine is about 3 kilometers long (2 miles) and produces a beam of electrons with energies of 50 billion electron volts (that is 50 GeV, where the G stands for giga, a prefix that means 10^9). Electrons with this energy have wavelengths approximately one-hundredth the size of the carbon nucleus, greatly reducing the diffraction effects.

An alternative technology, the circular accelerator, allows the particles to pass through a single accelerating region many times. The maximum energy of these machines is limited by the overall size (and cost), the strengths of the magnets required to bend the particles along the circular path, and the resulting radiation losses. Because particles traveling in circles are continually being accelerated, they give off some of their energy as electromagnetic radiation. These radiation losses are more severe for particles with smaller masses so circular accelerators are not as suitable for electrons. The most energetic accelerator of this type is at the Fermi National Accelerator Laboratory outside of Chicago shown in Fig-

Figure 25-2 An aerial view of the Stanford Linear Accelerator Center.

(a)

(b)

Figure 25-3   (a) An aerial view of the Fermi National Accelerator Laboratory. (b) This interior view shows magnets surrounding the two vacuum tubes in the main tunnel. The conventional magnets around the upper tube are red and blue, whereas the newer superconducting magnets are yellow.

ure 25-3. It has a diameter of 2 kilometers and can accelerate a beam of protons to energies of nearly 1 trillion electron volts (1 TeV) with a corresponding wavelength thousands of times smaller than nuclei.

Physics Update

Major new accelerators in the United States are too costly, so some particle physicists hope to shift their operations into space. Scientists at the Stanford Linear Accelerator have proposed building the $100-million Gamma Large Array Space Telescope (GLAST), which would view gamma rays with stacks of silicon microstrip detectors, some 50 to 100 times more sensitive than the detectors used in the present Gamma Ray Observatory. Meanwhile, Sam Ting of MIT plans to build the Alpha Magnetic Spectrometer (AMS), a $20-million device to be mounted on the proposed Space Station Alpha. AMS would use a powerful permanent magnet to sort antiparticles from particles, the goal being to search for antimatter in the Universe and for the decay of dark matter particles.

The Nuclear Glue

Experiments with particle accelerators provide a great deal of information about the structure of nuclei and the forces that hold them together. The nuclear force between protons can be studied by shooting protons at protons and analyzing the angles at which they emerge after the collisions. Because it is not possible to make a target of bare protons, experimenters use the next closest thing, a hydrogen target. They also use energies that are high enough so that the deflections due to the atomic electrons can be neglected.

At low energies the incident protons are repelled in the fashion expected by the electric charge on the proton in the nucleus. As the incident protons are given more and more kinetic energy, some of them pass closer and closer to the nuclear protons. When an incident proton gets within a certain distance of the target proton, it feels the nuclear force in addition to the electric force. This nuclear force changes the scattering. Some of the details of this new force can be deduced by comparing the scattering data with the known effects of the electric interaction.

We know that the nuclear force has a very short range; it has no effect beyond a distance of about 3×10^{-15} meter (3 fermis). Inside this distance it is strongly attractive, approximately 100 times stronger than the electrical repulsion. At even closer distances, the force becomes repulsive, which means that the protons cannot overlap, but act like billiard balls when they get this close. The nuclear force is not a simple force that can be described by giving its strength as a function of the distance between the two protons. The force depends on the orientation of the two protons and on some purely quantum-mechanical effects.

> **QUESTION**   Because neutrons are neutral, they cannot be accelerated in the same manner as protons. How might one produce a beam of fast neutrons?

The nuclear force also acts between a neutron and a proton (the n–p force) and between two neutrons (n–n force). The n–p force can be studied by scattering neutrons from hydrogen. The force between two neutrons, however, cannot be studied directly since there is no nucleus composed entirely of neutrons. Instead, this force is studied by indirect means; for example, by scattering neutrons from deuterons (the hydrogen nucleus that has one proton and one neutron) and subtracting the effects of the n–p scattering. These experiments indicate that when the effects of the proton's charge are ignored, the n–n, n–p, and p–p forces are very nearly the same. Therefore, the nuclear force is independent of charge.

> **ANSWER**   A beam of fast neutrons can be produced by accelerating protons and letting them collide with a foil. Head-on collisions of these protons with neutrons produce fast neutrons. The extra protons can be bent out of the way by a magnet.

There is another force in the nucleus. Investigations into the beta decay process led to the discovery of a fourth force. This nuclear force is also short ranged but very weak. Although it is not nearly as weak as the gravitational force at the nuclear level, it is only about one-billionth the strength of the other nuclear force. The force involved in beta decay is thus called the **weak force,** and the force between nucleons is called the **strong force.** We will concentrate on the strong force in this chapter because it is this force that is involved in the release of nuclear energy.

Nuclear Binding Energy

Because nucleons are attracted to each other, a force is needed to pull them apart. This force acts through a distance and does work on the nucleons. Because it takes work to take a stable nucleus apart, the nucleus must have a lower energy than its separated nucleons. Therefore, we would expect energy to be released when we allow nucleons to combine to form one of these stable nuclei. This phenomenon is observed: A 2.2-million-electron-volt gamma ray is emitted when a neutron and a proton combine to form a deuteron (Fig. 25-4).

> **QUESTION**   How much energy would it require to separate the neutron and proton in a deuteron?

As an analogy, imagine baseballs falling into a hole. The balls lose gravitational potential energy while falling. This energy is given off in various forms such as heat, light, and sound. To remove a ball from the hole,

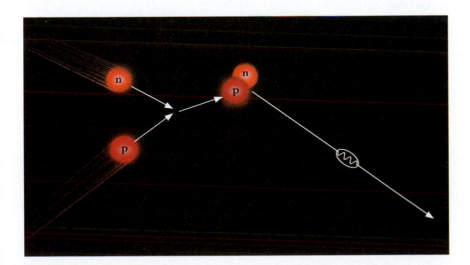

Figure 25-4   A 2.2-million-electron-volt gamma ray is emitted when a neutron and a proton combine to form a deuteron, indicating that the deuteron has a lower mass than the sum of the individual proton and neutron masses.

> **ANSWER**   Conservation of energy requires that the same amount of energy be supplied to separate them as was released when they combined—in the case above, it would be 2.2 million electron volts.

we must supply energy equal to the change in gravitational potential energy that occurred when the ball fell in. The same occurs with nucleons. In fact, it is common for nuclear scientists to talk of nucleons falling into a "hole," a nuclear potential well.

If we add the energies necessary to remove all the baseballs, we obtain the **binding energy** of the collection. This binding energy is a simple addition of the energy required to remove the individual balls since the removal of one ball has no measurable effect on the removal of the others. In the nuclear case, however, each removal is different since the removal of previous nucleons changes the attracting force. In any event, we can define a meaningful quantity called the *average binding energy per nucleon,* which is equal to the total amount of energy required to completely disassemble a nucleus divided by the number of nucleons.

We don't have to actually disassemble the nucleus to obtain this number. The energy difference between two nuclear combinations is large enough to be detected as a mass difference. The assembled nucleus has less mass than the sum of the masses of its individual nucleons. The difference in mass is related to the energy difference by Einstein's famous mass–energy equation, $E = mc^2$. Therefore, we can determine the binding energy of a particular nucleus by comparing the mass of the nucleus to the mass of its parts.

COMPUTING   Nuclear Binding Energies $\boxed{\Sigma}$

As an example, we calculate the total binding energy and the average binding energy per nucleon for the helium nucleus. We add the masses of the component parts and subtract the mass of the helium nucleus.

2 protons	$= 2 \times 1.00728$ amu	$=$	2.01456 amu
2 neutrons	$= 2 \times 1.00867$ amu	$=$	$+\,2.01734$ amu
mass of parts			4.03190 amu
1 helium nucleus		$=$	$-\,4.00150$ amu
mass difference			0.03040 amu

Therefore, the helium nucleus has 0.0304 atomic mass unit less mass than its parts. Using Einstein's relationship, we can calculate the energy equivalent of this mass. A useful conversion factor is that 1 atomic mass unit is equivalent to 930 million electron volts. Therefore, the total binding energy is 28.3 million electron volts. Dividing by 4, the number of nucleons, yields 7.1 million electron volts per nucleon.

Figure 25-5 shows a graph of the average binding energy per nucleon for the stable nuclei. The graph is a result of experimental work; it is simply a reflection of a pattern in nature. The graph shows that some nuclei are more tightly bound together than others.

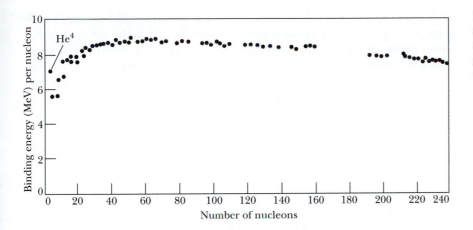

Figure 25-5 A graph of the average binding energy per nucleon versus nucleon number. Notice that the graph peaks in the range of 50 to 80 nucleons.

QUESTION Which nuclei are the most tightly bound?

These differences mean that energy can be released if we can find a way of rearranging nucleons. For example, combining light nuclei or splitting heavier nuclei would release energy. But all of these nuclei are stable. To either split or join them we need to know more about their stability and the likelihood of various reactions.

Stability

Not all nuclei are stable. As we saw in the previous chapter, some are radioactive, whereas others seemingly last forever. A look at the stable nuclei shows a definite pattern concerning the relative numbers of protons and neutrons. Figure 25-6 is a graph of the stable isotopes plotted according to their numbers of neutrons and protons. The curve, or **line of stability** as it is sometimes called, shows that the light nuclei have equal, or nearly equal, numbers of protons and neutrons. As we follow this line into the region of heavier nuclei, it bends upward, meaning that these nuclei have more neutrons than protons.

To explain this pattern we try to "build" the nuclear collection that exists naturally in nature. That is, assuming we have a particular light nucleus, can we decide which particle—a proton or neutron—is the best choice for the next nucleon? Of course, the best choice is the one that matches nature's choice.

Looking at the line of stability tells us that in the heavier nuclei adding a proton is not usually as stable an option as adding a neutron. Why is this so? Consider the characteristics of the two strongest forces involved. (We assume that gravity and the weak force play no role.) If you add a proton to a nucleus, it feels two forces—the electrical repulsion of

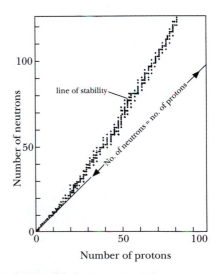

Figure 25-6 A graph of the number of neutrons versus the number of protons for the stable nuclei.

ANSWER The most tightly bound nuclei have about 60 nucleons.

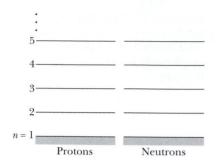

Figure 25-7 The proton and neutron energy levels in a hypothetical nucleus without electrical charge occur at the same values.

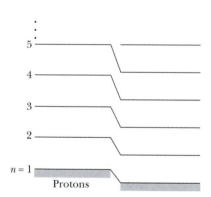

Figure 25-8 The proton energy levels in the hypothetical nucleus are raised due to the mutual repulsion of their positive charges.

the other protons and the strong nuclear attraction of the nearby nucleons. Although the electrical repulsion is weaker than the nuclear attraction, it has a longer range. The new proton feels a repulsion from *each* of the other protons in the nucleus but feels a nuclear attraction only from its nearest neighbors. With even heavier nuclei the situation gets worse; the nuclear attraction stays roughly constant—there are only so many nearest neighbors it can have—but the total repulsion grows with the number of protons.

An equally valid way of explaining the upward curve in the line of stability is to recall that quantum mechanics is valid in the nuclear realm. Imagine the nuclear potential-energy well described in the previous section. Quantum mechanics tells us that the well contains a number of *discrete* energy levels for the nucleons. The Pauli exclusion principle that governed the maximum number of electrons in an atomic shell (Chapter 23) has the same effect here, but there is a slight difference. Because we have two different types of particle, there are two discrete sets of levels. Each level can only contain two neutrons or two protons.

In the absence of electric charge the neutron and proton levels would be side-by-side (Fig. 25-7) because the strong force is nearly independent of the type of nucleon. If this were true—that is, level 4 for neutrons was the same height above the ground state as level 4 for protons—we would predict that there would be no preferential treatment when adding a nucleon. We would simply fill the proton level with two protons and the neutron level with two neutrons and then move to the next pair of levels. It would be unstable to fill proton level 5 without filling neutron level 4 because the nucleus could reach a lower energy state by turning one of its protons in level 5 into a neutron in level 4.

However, the electrical interaction between protons means that the separations of the energy levels are not identical for protons and neutrons. The additional force means that proton levels will be higher (the protons are less bound) than the neutron levels, as shown schematically in Figure 25-8. This structure indicates that at some point it becomes energetically more favorable to add neutrons rather than protons. Thus, the number of neutrons would exceed the number of protons for the heavier nuclei. Of course, too many neutrons will also result in an unstable situation.

Understanding the line of stability adds to our understanding of instability as well. We can now make predictions about the kinds of radioactive decay that should occur for various nuclei. If the nucleus in question is above the line of stability, it has extra neutrons. It would be more stable with fewer neutrons. In rare cases, the neutron is expelled from the nucleus, but usually a neutron decays into a proton and an electron via beta minus decay. If the isotope is below the line of stability, we expect alpha or beta plus decay or electron capture to take place to increase the number of neutrons relative to the number of protons. In practice alpha decay is rare for nucleon numbers less than 140, and beta plus decay is rare for nucleon numbers greater than 200.

When the nuclei get too large, there is no stable arrangement. None of the elements beyond uranium (element 92) occur naturally on Earth. Some of them existed on the Earth much earlier but have long since

Juggernaut Physicist

Enrico Fermi.

Enrico Fermi was nearly invincible in solving problems in theoretical physics. But in 1920, being a physicist was synonymous with being an experimentalist. Thus Fermi undertook a study of X-ray diffraction for his thesis work. Fermi eventually became quite adept at mechanical tasks. One story contends that on a lecture trip to America he experienced car trouble and repaired the car with such facility that the gas station owner immediately offered him a job—and this was during the Great Depression.

After the discovery of the neutron in 1932, Fermi headed a group at the University of Rome that investigated radioactivity induced with a neutron beam. The group was puzzled by differing results they obtained when the apparatus was placed on marble and wooden tables. Fermi found the answer and discovered that slower neutrons produced more radioactivity. He was rewarded for these efforts with the Nobel Prize in 1938.

Although Fermi was aware of the predictions concerning the possibility of fission, he found no evidence of fission in his work with neutron beams. When fission was actually discovered, Fermi chastised himself for his oversight. One day he and several colleagues were speculating about what a particular artist's drawing portrayed when Fermi suggested that it was "probably a scientist *not* discovering fission."

When World War II broke out in Europe, the United States declared Italian nationals to be "enemy aliens." In spite of this, Fermi was allowed to work on the American war effort. He designed the first atomic reactor, which went critical in 1942 under the football bleachers at the University of Chicago. Although the reactor produced only one-half watt of power for one-half minute, its success gave hope for the development of the atomic bomb.

Fermi's tremendous memory and powerful ability to simplify make him somewhat of a oracle of science. It was rumored that even if Fermi didn't know the correct answer, you could recite several numbers to him and his eyes would twinkle at the correct answer. Element number 100 was named fermium in his honor when it was discovered a year after his death.

Adapted from an essay written by Steven Janke for Pasco Scientific.

References: Segrè, E. *Enrico Fermi—Physicist.* Chicago: University of Chicago Press, 1970.

American Institute of Physics Niels Bohr Library.

The world's first nuclear reactor was assembled in Chicago in 1942. There were no photographs taken of this reactor due to wartime secrecy.

decayed. They can be artificially produced by bombarding lighter elements with a variety of nuclei. The list of elements (and isotopes) continues to grow through the use of this technique.

QUESTION What kind of decay would you expect to occur for $^{214}_{82}$Pb?

ANSWER Because this isotope is above the line of stability, it should undergo beta minus decay.

Nuclear Fission Σ

The discovery of the neutron added a new probe for studying nuclei and initiating new nuclear reactions. Enrico Fermi, an Italian physicist, quickly realized that neutrons made excellent nuclear probes. Neutrons, being uncharged, have a better chance of probing deep into the nucleus. Working in Rome in the mid-1930s Fermi was able to produce new isotopes by bombarding uranium with neutrons. These new isotopes were later shown to be heavier than uranium. Although this discovery is exciting, it perhaps isn't surprising; it seems reasonable that adding a nucleon to a nucleus results in a heavier nucleus.

The real surprise came later when scientists found that their samples contained nuclei that were much less massive than uranium. Were these nuclei products of a nuclear reaction with uranium, or were they contaminants in the sample? It seemed unbelievable that they could be products. Fermi, for example, didn't even think about the possibility that a slow-moving neutron could split a big uranium nucleus. But that was what was happening.

The details of this process are now known. The capture of a low-energy neutron by ^{235}U results in another uranium isotope ^{236}U. This nucleus is unstable and can decay in several possible ways. It might give up the excitation energy by emitting one or more gamma rays, it might beta decay, or it might split into two smaller nuclei. This last alternative is called **fission.** The fissioning of ^{235}U is shown in Figure 25-9. A typical fission reaction is

$$n + {}^{235}_{92}U \rightarrow {}^{142}_{56}Ba + {}^{91}_{36}Kr + 3n$$

The fission process releases a large amount of energy. This energy comes from the fact that the product nuclei have larger binding energies than the uranium nucleus. As the nucleons fall deeper into the nuclear potential wells, energy is released. The approximate amount of energy released can be seen from the graph in Figure 25-5. A nucleus with 236 nucleons has an average binding energy of 7.6 million electron volts per nucleon. Assume for simplicity that the nucleus splits into two nuclei of equal masses. A nucleus with 118 nucleons has an average binding energy

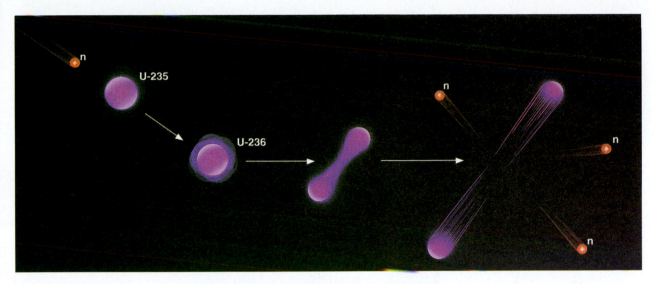

Figure 25-9   The fissioning of a ^{235}U nucleus caused by the capture of a neutron results in two intermediate mass nuclei, two or more neutrons, and some energy as gamma rays and kinetic energy.

of 8.5 million electron volts per nucleon. Therefore, each nucleon is more tightly bound by about 0.9 million electron volts and the 236 nucleons must release about 210 million electron volts. This is a *tremendously* large energy for a single reaction. Typical energies from chemical reactions are only a few electron volts per atom, 100 million times less.

> **QUESTION**   How many joules are there in 210 million electron volts?

Even though these energies are much larger than chemical energies, they are small on an absolute scale. The energy released by a single nuclear reaction as heat or light would not be large enough for us to detect without instruments. Knowledge of this led Rutherford to announce in the late 1930s that it was idle foolishness to even contemplate that the newly found process could be put to any practical use.

Chain Reactions

Rutherford's cynicism about the practicality of nuclear power seems silly in hindsight. But he was right about the amount of energy released from a single reaction. If you were holding a piece of uranium ore in your

> **ANSWER** $(2.1 \times 10^8 \text{ eV})(1.9 \times 10^{-19} \text{ J/eV}) = 4.0 \times 10^{-11}$ J. While this is indeed a very small amount of energy, it is huge relative to other single-reaction energies.

hand, you would not notice anything unusual; it would look and feel like ordinary rock. And yet, nuclei are continuously fissioning.

What is the difference between the ore in your hand and a nuclear power plant? The key to releasing nuclear energy on a large scale is the realization that a single reaction has the potential to start additional reactions. The accumulation of energy from many, many such reactions does get to levels that are not only detectable but can become tremendous. This occurs because the fission fragments have too many neutrons to be stable. The line of stability in Figure 25-6 shows that the ratio of protons to neutrons for nuclei in the 100-nucleon range is different from that of ^{235}U. These fragments must get rid of the extra neutrons. In practice they usually emit two or three neutrons within 10^{-14} second. Even then the resulting nuclei are neutron-rich and need to become more stable by emitting beta particles.

> **QUESTION**   How many neutrons must be emitted when $^{236}_{92}$U fissions to become $^{141}_{55}$Cs and $^{92}_{37}$Rb?

The release of two or three neutrons in the fissioning of ^{235}U means that the fissioning of one nucleus could trigger the fissioning of others, these could trigger the fissioning of still others, and so on. Because a single reaction can cause a chain of events as illustrated in Figure 25-10, this sequence is called a **chain reaction.**

Imagine the floor of your room completely covered with mousetraps. Each mousetrap is loaded with three marbles as shown in Figure 25-11, so that when a trap is tripped, the three marbles fly into the air. What happens if you throw a marble into the room? It will strike a trap, releasing three marbles. Each of these will trigger another trap, releasing its marbles. In the beginning the number of marbles released will grow geometrically. The number of marbles will be 1, 3, 9, 27, 81, 243, ... In a very short time the air will be swarming with marbles. Because the number of mousetraps is limited, the process dies out. However, the number of atoms in a small sample of ^{235}U is very large (on the order of 10^{20}), and the number of fissionings taking place can grow to be extremely large, releasing a lot of energy. If the chain reaction were to continue in the fashion we have described, a sample of ^{235}U would blow up.

But our piece of uranium ore doesn't blow up; it doesn't even get warm because most of the neutrons do not go on to initiate further fission reactions. Several factors affect this dampening of the chain reaction. One is size. Most of the neutrons leave a small sample of uranium before they encounter another uranium nucleus—a situation analogous to putting only a dozen mousetraps on the floor. If you trigger one trap, it is unlikely that one of its marbles will trigger another trap, but occasionally it happens. As the sample of material gets bigger, fewer and fewer of the neutrons escape the material.

> **ANSWER**   Because the total number of nucleons remains the same, we expect to have 236 − 141 − 92 = 3 neutrons emitted.

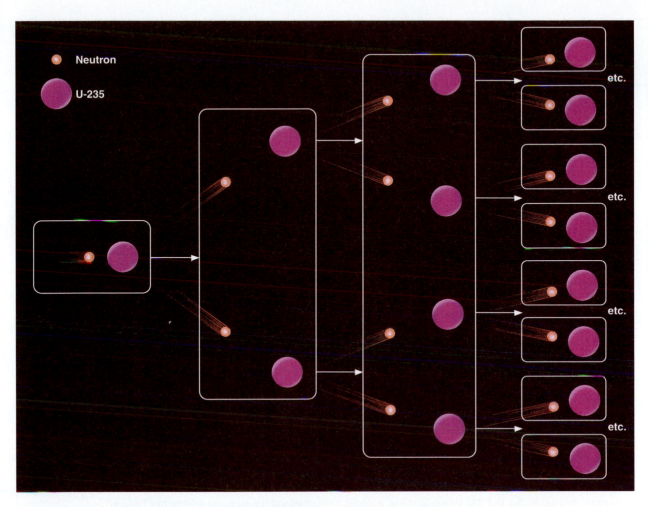

Figure 25-10 A chain reaction occurs because each fission reaction releases two or more neutrons that can initiate further fission reactions.

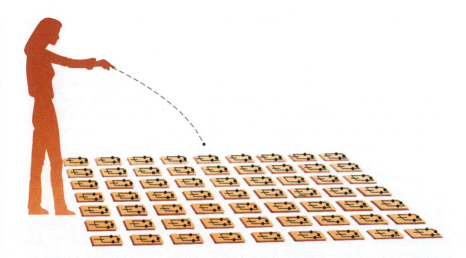

Figure 25-11 A chain reaction of mousetraps is initiated when the first marble is thrown into the array. Each mousetrap releases three marbles, which can trip additional mousetraps.

Fission in the Snow

Lise Meitner.

Lise Meitner once insisted, "I am not important; why is everybody making a fuss over me?" It is clear that we should make a fuss over her; she was a brilliant physicist responsible for many important contributions to modern physics. It wasn't easy for a woman to make such strides in a traditionally male field. She wrestled with prejudice against women and against Jews and endured two world wars. In retrospect, she knew that it was physics which had brought "light and fullness" into her life.

After studying with a private tutor, Meitner easily passed the entrance examination and entered the University of Vienna, where she had to get used to being the only woman in a class of 100 students. The men tolerated her, but silently. A particularly difficult calculus problem and an insensitive professor quickly turned Meitner from mathematics to physics. She studied long and hard until she mastered the subjects demanded of a physics student. But she hesitated to take the doctoral examinations. The university had awarded only 14 doctorates to women in the last 541 years, and not a single one was in physics. After continual prodding, she sat for the exams and received her doctorate in 1906.

Meitner traveled to Berlin to attend a series of lectures by Max Planck and ended up staying over 30 years. She found a position working as an assistant to Otto Hahn, which proved to be one of the most outstanding collaborations in science of all time. They worked in the area of radioactivity and had just hit on the idea of isotopes when World War I broke out. After the war, Meitner and Hahn announced the discovery of the 91st element, protactinium. Nuclear physics was buzzing with activity, when the Nazis took power and Meitner was forced to escape to Stockholm, where an appointment was arranged for her by Niels Bohr.

In 1938 Meitner hit upon her most famous discovery. Hahn sent a letter reporting that the bombardment of uranium with neutrons had produced barium, a much lighter element. Meitner's ideas gradually fell into place as she held the letter in her hands while trudging through the fresh snow to her nephew's house. Her nephew, Otto Frisch, listened intently as she carefully explained that the nucleus must be capturing the neutron, causing enough instability to pinch the nucleus in two. She made the appropriate calculations employing Bohr's model and Einstein's equation, and predicted that Hahn's experiment should yield barium, krypton, and energy. With the verification of the theory in early 1939, the concept of nuclear fission was introduced to the world. Meitner was the first woman to receive the Fermi Award from the U.S. Atomic Energy Commission. The name meitnerium (Mt) has been proposed for element 109 in her honor.

Adapted from an essay written by Steven Janke for Pasco Scientific.

But even a large piece of uranium ore does not blow up. Naturally occurring uranium consists of two isotopes. Only 0.7% of the nuclei are ^{235}U; the remaining 99.3% are ^{238}U. These ^{238}U will occasionally fission, but usually they capture the neutrons and decay by beta minus or alpha emission. Captured neutrons cannot go on to initiate other fission reactions. The chain reaction is **subcritical** and the process dies out. To make our analogy correspond to a piece of naturally occurring uranium, we would need 140 unloaded mousetraps for each loaded one.

Extracting useful energy from fission is much easier if the percentage of ^{235}U is increased through what is known as enrichment. Any enrichment scheme is very difficult because the atoms of ^{235}U and ^{238}U are chemically the same, making it difficult to devise a process that preferentially interacts with ^{235}U. Their masses differ by only a little over 1%. The various enrichment processes take advantage of the slight differences in the charge-to-mass ratios, the rates of diffusion of their gases through membranes, resonance characteristics, or their densities. The enrichment processes must be repeated many, many times, as only a small gain is made each time.

If enough enriched uranium is quickly assembled, the chain reaction can become **supercritical.** On average more than one neutron from each fission reaction initiates another reaction and the number of reactions taking place grows rapidly. This process was utilized in the nuclear bombs exploded near the end of World War II, showing that Rutherford vastly underestimated the power of the fission reaction.

> **Physics on Your Own**   Experience a chain reaction by gathering a large group of friends in the center of a room. Give each friend three styrofoam balls and instructions to throw the balls into the air whenever they are struck by a ball. Watch what happens as you throw a ball to initiate the process. How would you create a subcritical situation?

Nuclear Reactors

Harnessing the tremendous energy locked up in nuclei requires controlling the chain reaction. The process must not become either subcritical or supercritical, because the energy must be released steadily at manageable rates. In a nuclear reactor the conditions are adjusted so that an average of one neutron per fission initiates further fission reactions. Under these conditions the chain reaction is **critical;** it is self-sustaining. Energy is released at a steady rate and extracted from the reactor to generate electricity.

Several factors can be adjusted to ensure the criticality of the reactor. We have seen that the amount of uranium fuel (the core) must be large enough so that the fraction of neutrons escaping from it is small. It is also important that the nonfissionable ^{238}U nuclei not capture too large a fraction of the neutrons, which can be accomplished by enriching the fuel or by reducing the speed of the neutrons or both.

The likelihood that a neutron will cause a ^{235}U nucleus to fission varies with the speed of the incoming neutron. Initially, you might think that faster neutrons would be more likely to split the ^{235}U nucleus because they would impart more energy to the nucleus. This is not the case, as the splitting of the nucleus is a quantum-mechanical effect. Slow neutrons are much more likely to initiate the fission process. An added benefit is that the probability of a ^{238}U nucleus capturing a neutron decreases as the neutron speed decreases.

The neutrons are primarily slowed by elastic collisions with nuclei. A material (called the **moderator**) is added to the core of the reactor to slow down the neutrons without capturing too many of them. Neutrons can transfer the most energy to another particle when their masses are the same. (In fact, a head-on collision will leave the neutron with little or no kinetic energy.) Hydrogen would seem to be the ideal moderator, but its probability of capturing the neutron is large. The material with the next lightest nucleus, deuterium, is fine but costly, and since it is a gas, it is hard to get enough mass into the core to do the job. Canadian reactors use deuterium as the moderator but in the form of water, called heavy water because of the presence of the heavy hydrogen. The world's first reactor used graphite (a form of carbon) as the moderator. Most current U.S. reactors use ordinary water as the moderator, requiring the use of enriched fuel.

Reactors require control mechanisms for fine-tuning and for adjusting to varying conditions as the fuel is used. The chain reaction is controlled by inserting rods of a material that is highly absorbant of neutrons. Boron is quite often used. The control rods are pushed into the core to decrease the number of neutrons available to initiate further fission reactions. The mechanical insertion and withdrawal of control rods could not control a reactor if all of the neutrons were given off promptly in the usual 10^{-14} second. A small percentage of the neutrons are given off by the fission fragments and may take seconds to appear. These delayed neutrons allow time to adjust the reactor to fluctuations in the fission rate.

The last thing needed to make a reactor a practical energy source is a way of removing the heat from the core so that it can be used to run electric generators, which is accomplished in a variety of ways. In boiling-water reactors, water flows through the core and is turned to steam. Pressurized water reactors use water under high pressure so that it doesn't boil in the reactor. Still other reactors use gases. A diagram of a nuclear reactor is shown in Figure 25-12.

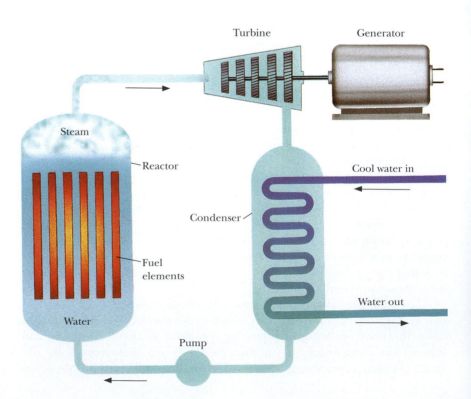

Figure 25-12   A schematic drawing of a nuclear reactor used to generate electricity.

A boiling water reactor and surrounding facilities.

Breeding Fuel

We are exhausting many of our energy sources. This is as true for fission reactors as it is for coal- and oil-powered plants. It is estimated that there is only enough uranium to run those reactors that have already been built or are under construction for the 30–40 years of expected operation of each reactor. If ^{235}U was the only fuel, the nuclear-power age would turn out to be short-lived.

Imagine, however, that somebody suggests that it is possible to make new fuel for these reactors. Does this claim sound like a sham? Does it imply that we can get something for nothing? The claim is not a sham, but it is also not getting something for nothing. The laws of conservation of mass and energy still hold. The new process takes an isotope that does not fission and through a series of nuclear reactions transmutes that isotope into one that does.

When a ^{238}U nucleus captures a neutron, it usually undergoes two beta minus decays to become plutonium, as diagrammed in Figure 25-13. The important point is that $^{239}_{94}$Pu is a fissionable nucleus that can be used as fuel in fission reactors. (A similar process transmutes $^{232}_{90}$Th into ^{232}U, another fissionable nucleus.)

Some plutonium is produced in a normal reactor because there is ^{238}U in the core. However, not much is produced because the neutrons are slowed to optimize the fission reaction. Special reactors have been designed so that an average of more than one neutron from each fission reaction is captured by ^{238}U. Such a reactor generates more fuel than it uses and is therefore known as a *breeder reactor*. Because most of the uranium is

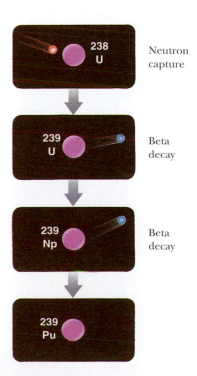

Figure 25-13   A scheme for converting $^{238}_{92}$U to $^{239}_{94}$Pu.

A breeder reactor at the National Reactor Laboratory in Idaho.

fusion of hydrogen

^{238}U and because there is about the same amount of $^{232}_{90}$Th, these reactors greatly extend the amount of available fuel.

Like most other energy options, there are serious concerns about breeder reactors. Briefly, these concerns center on the technology of running these reactors, the assessment of the risks involved, and the security of the plutonium that is produced. Plutonium-239 is bomb-grade material that can be separated relatively inexpensively from uranium because it has different chemical properties.

Fusion Reactors Σ

There are other nuclear energy options. Returning to the binding energy curve in Figure 25-5 reveals another way to get energy from nuclear reactions. The increase in the curve for the average binding energy per nucleon for the light nuclei indicates that some light nuclei can be combined to form heavier ones with a release of energy. For instance, a deuteron ($^2_1$H) and a triton ($^3_1$H) can be fused to form helium with a release of a neutron and 17.8 million electron volts.

$$^2_1\text{H} + ^3_1\text{H} \rightarrow ^4_2\text{He} + \text{n} + 17.8 \text{ MeV}$$

Although this process releases a lot of energy per gram of fuel, it is much more difficult to initiate than fission. The interacting nuclei have to be close enough together so that the nuclear force dominates. (This wasn't a problem with fusion because the incoming particle was an uncharged neutron.) The particles involved in the **fusion** reaction won't overcome the electrostatic repulsion of their charges to get close enough unless they have sufficiently high kinetic energies.

High kinetic energies mean very high temperatures. The required temperatures are on the order of millions of degrees, matching those found inside the Sun. In fact, the source of the Sun's energy is fusion. The first occurrence of fusion on Earth was in the explosion of hydrogen bombs in the 1950s. The extremely high temperatures needed for these bombs was obtained by exploding the older fission bombs (commonly called atomic bombs).

Making a successful fusion power plant involves harnessing the reactions of the hydrogen bomb and the Sun. Fusion requires not only high temperature but also the confinement of a sufficient density of material for long enough that the reactions can take place and return more energy than was necessary to initiate the process. At first this task might seem impossible and dangerous. Whether or not it is possible is still being determined. Most people in this field feel that it is a technological problem that can be solved. The characterization of fusion as dangerous is false and arises from confusion about the concepts of heat and temperature (Chapter 7). Heat is a flow of energy; temperature is a measure of the average molecular kinetic energy. Something can have a very high temperature and be quite harmless. Imagine, for example, the vast differences in potential danger between a thimble and a swimming pool full of boiling water. The thimble of water has very little heat energy and thus is relatively harmless. The same is true of fusion reactors. Although the fuel is very hot, it is a rarified gas at these temperatures. So the problem is not with it melting the container but rather the reverse—the container will cool the fuel.

Two schemes are being investigated for creating the conditions necessary for fusion. One method is to confine the plasma fuel in a magnetic "bottle" (Fig. 25-14). The *magnetic field* interacts with the charged particles and keeps them in the bottle. The other, *inertial confinement,* uses tiny pellets of solid fuel. As these pellets fall through the reactor, they are bombarded from many directions by laser beams. This produces rapid heating of the pellet's outer surface, causing a compression of the pellet and even higher temperatures in the center.

To date, no one has been able to simultaneously produce all of the conditions required to get more energy out of the process than was needed to initiate it. Successes have been limited to achieving some of

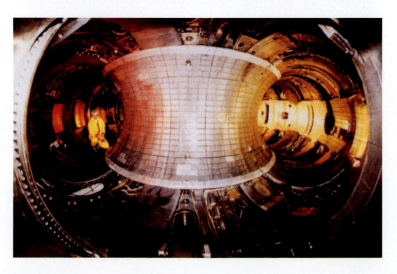

Figure 25-14 The Tokamak at Princeton University uses a magnetic bottle to confine the plasma so that fusion reactions will occur.

these conditions but not all at the same time. Fusion reactors on a commercial scale seem many years away.

Some of the features of fusion, however, make it an attractive option. The risks are believed to be much lower than for fission reactors and there is a lot of fuel. One of the possible reactions uses deuterium. Deuterium occurs naturally as one atom out of every 6000 hydrogen nuclei and thus is a constituent of water. This heavy water is relatively rare compared with ordinary water, but there is enough of it in a pail of water to provide the equivalent energy of 700 gallons of gasoline!

Solar Power   Σ

Throughout history people have puzzled about the source of the Sun's energy. What is the fuel? How long has it been burning? And how long will it continue to emit its life-supporting heat and light?

Many schemes have been suggested—some reasonable, some absurd. Early people thought of the Sun as an enormous "campfire" because fires were the only known source of heat and light. But calculations showed that if wood or coal were the Sun's fuel, it could not have been around for very long. Its lifetime would be much shorter than estimates of how long the Sun had already existed. Another idea involved a Sun heated by the constant bombardment of meteorites, which could account for the long lifetime of the Sun (the collisions continued indefinitely) but which also predicted that the mass of the Sun would increase. The Earth would then spiral into the Sun due to the ever increasing gravitational force. Another scheme had the Sun slowly collapsing under its own gravitational attraction. The loss in gravitational potential energy would be radiated into space. The flaw in this last scenario became apparent from geologic data that suggested that the Earth was much older than the Sun. Also, if we mentally uncollapse the Sun, going back in time, we find that the Sun would extend beyond the Earth's orbit at a time less than the assumed age of the Earth.

The source of the Sun's power was a major conflict at the turn of the last century. Astronomers were suggesting that the Sun was approximately 100,000 years old, but geologists and biologists were saying that the Earth was very much older. Both sides couldn't be right. The discovery of radioactivity and other nuclear reactions showed that the geologists and biologists were right. Scientists now believe that our entire Solar System formed from interstellar debris left over from earlier stars. As the matter collapsed, it heated up due to the loss in gravitational potential energy. At some point the temperature of the matter in the interior of the Sun got high enough to initiate nuclear fusion. Now we have a Sun in which hydrogen is being converted into helium via the fusion reaction. The amount of energy released by the Sun is such that the mass of the Sun is decreasing at a rate of 4.3 billion kilograms per second! Yet this is such a small fraction of the Sun's mass that the change is hardly noticeable.

Knowing the mechanism and the mass of the Sun, we are able to calculate its lifetime. Our Sun is believed to be about 4.5 billion years old and will probably continue its present activity for another 4.5 billion years.

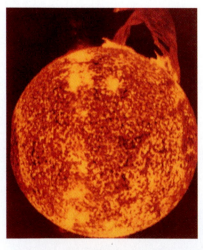

The source of the Sun's energy is the fusion of hydrogen into helium in its core.

SUMMARY

Nuclei stay together in spite of the electromagnetic repulsions between protons because of a nuclear force. This strong force between two nucleons has a very short range, is about 100 times stronger than the electric force, has a repulsive core, and is independent of charge. A second force in the nucleus, the weak force involved in the beta decay process, is also short-ranged but very weak, only about one-billionth the strength of the strong force.

Information about nuclei initially came from particles ejected during radioactive decays. If a nucleus is above the line of stability, it has extra neutrons, which usually results in a neutron decaying into a proton and an electron via beta minus decay. If the isotope is below the line of stability, alpha or beta plus decay or electron capture increases the number of neutrons relative to the number of protons.

Later, the particles from radioactive decays were used as probes to study the structure of stable nuclei. Finally, particle accelerators produced beams of charged particles with much higher momenta and, consequently, much smaller wavelengths. The largest of these accelerators produce beams with energies in excess of 1 trillion electron volts.

The average binding energy per nucleon varies for the stable nuclei; some nuclei are more tightly bound than others, reaching a maximum near iron. Combining light nuclei or splitting heavier nuclei releases energy. The energy differences between nuclei are large enough to be detected as mass differences.

Bombarding uranium with neutrons splits the uranium nuclei, releasing large amounts of energy. Typical reaction energies are 100 million times larger than those of chemical reactions. The fission reaction is a practical energy source because a single reaction emits two or three neutrons which can trigger additional fission reactions.

Another way of releasing energy, fusion, combines light nuclei to form heavier ones. Fusion requires high temperatures and the confinement of a sufficient density of material for long enough so the reactions can take place and return more energy than was needed to initiate the process. This technology is still being developed. The Sun, however, is a working fusion reactor.

CHAPTER 25 REVISITED

Nuclear energy can be used to heat water and make steam to turn turbines like the other options. The differences are at the front end—releasing the energy to make the steam. With coal, oil, and natural gas we burn the fuel. In conventional nuclear power plants we create a controlled chain reaction of nuclear disintegrations. Future fusion reactors will combine hydrogen nuclei to form heavier nuclei. With all nuclear reactors, the questions revolve around the use, production, and disposal of radioactive materials. Because radioactivity is unalterable by normal means, we need to isolate these materials from the biosphere. This is a formidable technological challenge.

KEY TERMS

binding energy: The amount of energy required to take a nucleus apart.

chain reaction: A process in which the fissioning of one nucleus initiates the fissioning of others.

critical chain reaction: A chain reaction in which an average of one neuron from each fission reaction initiates another reaction.

fission: The splitting of a heavy nucleus into two or more lighter nuclei.

fusion: The combining of light nuclei to form a heavier nucleus.

line of stability: The locations of the stable nuclei on a graph of the number of neutrons versus the number of protons.

moderator: A material used to slow down the neutrons in a nuclear reactor.

particle accelerator: A device for accelerating charged particles to high velocities.

strong force: The force responsible for holding the nucleons together to form nuclei.

subcritical: A chain reaction that dies out because an average of less than one neutron from each fission reaction causes another fission reaction.

supercritical: A chain reaction that grows rapidly because an average of more than one neutron from each fission reaction causes another fission reaction, an extreme example of which is the explosion of a nuclear bomb.

weak force: The force responsible for beta decay.

CONCEPTUAL QUESTIONS

1. What electric potential difference is required to accelerate an electron to an energy of 10 million electron volts?

2. What electric potential difference is required to accelerate an alpha particle to an energy of 20 million electron volts?

3. An electron, a positron, and a proton are each accelerated through a potential difference of 1 million volts. Which of these, if any, acquires the largest kinetic energy?

4. Does a proton or an alpha particle acquire a larger kinetic energy when accelerated by the same potential difference?

*5. An electron, a positron, and a proton are each accelerated through a potential difference of 1 million volts. Which of these acquires the largest momentum?

*6. Does a proton or an alpha particle acquire a larger momentum when accelerated by the same potential difference?

7. What is the main advantage of a linear accelerator over a circular one?

8. Why is a circular accelerator more suitable for protons than electrons?

9. What are the four fundamental forces in nature?

10. List the four fundamental forces in the order of decreasing strength.

11. How do we know that there must be a strong force?

12. Which of the forces in nature is responsible for the beta decay processes?

13. What evidence do we have to support the idea that the strong force is stronger than the electromagnetic force?

14. What are the general characteristics of the strong force?

15. What are the differences between the strong force and the electromagnetic force?

16. In what ways are the strong force and the electromagnetic force the same?

17. What is released when a neutron and a proton combine to form a deuteron?

18. Which elements have the largest average binding energies per nucleon?

19. How would the mass of a neutral hydrogen atom compare with the sum of the masses of a proton and an electron?

20. Which of the following has the largest mass: 90 protons and 150 neutrons, $^{240}_{90}$Th, or $^{110}_{40}$Zr plus $^{130}_{50}$Sn?

21. $^{14}_{6}$C decays to $^{14}_{7}$N via beta minus decay. Which nucleus has the smaller mass?

*22. Both $^{12}_{7}$N and $^{12}_{5}$B decay to the stable nucleus $^{12}_{6}$C. Which of these three nuclei has the smallest mass? How do you know this?

*23. Suppose that the curves for the average binding energy were those shown in the figure. Would fission and fusion be possible in each case? If so, for approximately what range of nucleon number would they occur?

*24. In a large star, when most of the hydrogen in the core of a star has been combined to form helium, the helium can undergo fusion to produce even heavier nuclei. What is the heaviest nucleus you would expect to be formed by the fusion process in the largest stars?

25. How do the numbers of neutrons and protons compare for most stable nuclei with small atomic numbers?

26. What general statement about the relative numbers of neutrons and protons can you make about nuclei with large atomic numbers?

27. How does our model of the nucleus account for the observation that the number of neutrons is nearly equal to the number of protons for light nuclei?

28. For heavy nuclei the number of neutrons exceeds the number of protons. How does our model of the nucleus account for this?

29. Why can't a stable nucleus contain only protons?

30. You may have learned the law "Matter can neither be created nor destroyed." Is this a true statement in modern-day physics?

31. Would you expect the nucleus $^{100}_{50}\text{Sn}$ to be stable? If not, how would you expect it to decay?

32. Would you expect the nucleus $^{130}_{50}\text{Sn}$ to be stable? If not, how would you expect it to decay?

33. How would you expect an unstable nucleus of $^{80}_{30}\text{Zn}$ to decay?

34. What is the most likely decay mode for $^{20}_{11}\text{Na}$?

35. What is nuclear fission?

36. Would a uranium nucleus release more or less energy by splitting into three equal mass nuclei rather than two?

37. Assume a $^{235}_{92}\text{U}$ nucleus absorbs a neutron and fissions with the release of three neutrons. If one of the fission fragments is $^{144}_{56}\text{Ba}$, what is the other?

38. How many neutrons are released in the following fission reaction?

$$^{1}_{0}\text{n} + {}^{235}_{92}\text{U} \rightarrow {}^{140}_{54}\text{Xe} + {}^{94}_{38}\text{Sr} + (?)\,{}^{1}_{0}\text{n}$$

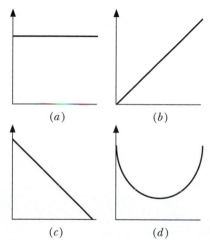

(a)   (b)

(c)   (d)   Question 23.

39. Why can the fissioning of $^{235}_{92}\text{U}$ produce a chain reaction?

40. What factors determine whether a piece of uranium undergoes a subcritical or supercritical reaction?

41. Why is it important for fission reactions to emit neutrons?

42. Why are fission fragments typically beta minus emitters?

43. Why don't chain reactions occur in naturally occurring deposits of uranium?

44. What are the differences between a nuclear reactor and a nuclear bomb?

45. Why does a nuclear reactor have control rods?

46. What role does a moderator play in a nuclear reactor?

47. How would the development of nuclear reactors have changed if the average number of neutrons released per fission of ^{235}U were 2.0 instead of 2.5?

48. The fissioning of ^{239}Pu yields an average of 2.7 neutrons per reaction compared to 2.5 for ^{239}U. Which substance would have the smaller critical mass?

49. What is bred in a breeder reactor?

50. Why do some people favor the development of breeder reactors?

51. What is the sequence of decays that turns $^{239}_{92}\text{U}$ into $^{239}_{94}\text{Pu}$?

52. Write down a sequence of decays by which $^{233}_{90}\text{Th}$ becomes $^{233}_{92}\text{U}$.

53. What is nuclear fusion?

54. Why are high temperatures required for nuclear fusion?

55. What are the two basic approaches to developing fusion reactors?

56. Why do some people favor fusion energy over fission energy?

57. What are the conditions in the interiors of stars that make fusion possible?

58. Why do scientists believe that the Sun is 4.5 billion years old?

59. Does $E = mc^2$ apply to both fusion and fission?

60. What is the basic difference between fusion and fission?

61. What are some of the public policy issues surrounding nuclear power?

EXERCISES

*1. A proton with an energy of 500 GeV has a momentum of 500 GeV/c, where c is the speed of light. What wavelength does this proton have?

*2. What is the wavelength of an electron with an energy of 50 GeV? Its momentum is 50 GeV/c, where c is the speed of light.

3. Given that 1 amu equals 1.66×10^{-27} kg, show that 1 amu has an energy equivalent of 930 MeV.

4. On average how many fission reactions of ^{235}U would it take to release 1 J of energy?

5. Calculate the total binding energy of the $^{12}_{6}$C nucleus.

6. Given that the oxygen nucleus with eight neutrons has a mass of 15.99052 amu, what is its total binding energy?

7. How much energy is released when $^{14}_{6}$C decays to form $^{14}_{7}$N? The masses of the neutral carbon and nitrogen atoms are 14.003242 and 14.003074 amu, respectively.

8. How much energy is released when $^{3}_{1}$H decays to form $^{3}_{2}$He? The masses of the neutral hydrogen and helium atoms are 3.016049 and 3.016029 amu, respectively.

9. How much energy is released in the alpha decay of $^{214}_{84}$Po? The masses of the neutral polonium, lead, and helium atoms are 213.995190, 209.990069, and 4.002603 amu, respectively.

10. How much energy is released in the alpha decay of $^{239}_{94}$Pu? The masses of the neutral plutonium, uranium, and helium atoms are 239.052158, 235.043925, and 4.002603 amu, respectively.

11. If $^{236}_{92}$U fissions to become $^{142}_{56}$Ba and $^{91}_{36}$Kr with the release of three neutrons, approximately how much energy is released?

12. Approximately how much energy is released when $^{239}_{94}$Pu fissions to become $^{96}_{40}$Zr and $^{141}_{54}$Xe with the release of two neutrons?

13. Show that the fusion reaction given below releases 17.6 MeV of energy. The masses of the deuteron and triton are 2.01355 and 3.01550 amu, respectively.

$$^{2}_{1}H + ^{3}_{1}H \rightarrow ^{4}_{2}He + ^{1}_{0}n$$

14. How much energy is released in the following fusion reaction? The masses of deuteron and triton are given in Exercise 13.

$$^{2}_{1}H + ^{2}_{1}H \rightarrow ^{3}_{1}H + ^{1}_{0}n$$

15. Given that the mass of the Sun is decreasing at the rate of 4.5×10^{9} kg/s, what is the present energy radiated by the Sun each second?

16. Approximately how much mass has the Sun lost during its lifetime? How does this compare with its current mass of 2×10^{30} kg?

Elementary Particles

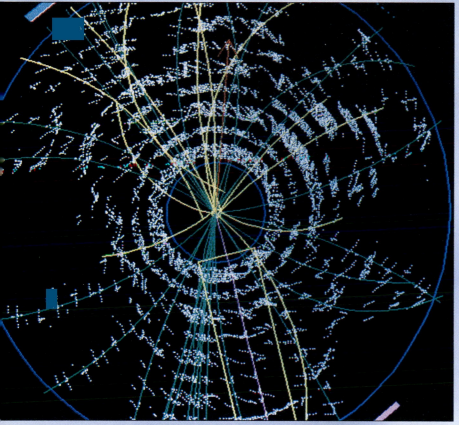

Computer-generated image of a collision in the Collider Detector at Fermilab. It is possible that the collision of a proton and an antiproton produced a top quark and antiquark.

Throughout recorded history we have searched for the primary building blocks of nature. Aristotle's four "elements" became the chemical elements, which gave way to electrons, protons, and neutrons. As we have delved deeper, we have found new, fascinating layers. Have we found the ultimate building blocks? If not, where will our search stop? (See p. 662 for the answer to this question.)

The idea that all of the diverse materials in the world around us are composed of a few simple building blocks is very appealing, and because of its appeal the idea has existed for more than 20 centuries. The search for these elementary components of matter is fueled by the desire to simplify our understanding of nature.

The search began with the Aristotelian world view, which assumed everything was made of four basic elements: earth, fire, air, and water. During the 18th and 19th centuries these four were eventually replaced by the modern chemical elements. Although the initial list numbered only a few dozen elements, it has grown to more than 100 entries. A hundred different building blocks are not as appealing as four. Things improved, however, when it was discovered that atoms were not indivisible but rather made up of three even more basic building blocks. The elementary particles described in 1932 consisted of the three constituents of atoms—the electron, proton, and neutron—and the quantum of light—the photon. These four building blocks restored an elegant simplicity to the physics world view.

This beautiful picture did not survive for long. As experimenters probed deeper and deeper into the subatomic realm, new particles were discovered. The experimenters were constantly confronted with new, seemingly bizarre observations. The first, and perhaps most bizarre, observation occurred during the same year the neutron was discovered and resulted in the discovery of an entirely new kind of matter.

Antimatter

A new type of particle was discovered in 1932. When high-energy photons (gamma rays) collided with nuclei, two particles were produced: one was an electron, the other an unknown. Double spirals in bubble chamber photographs showed tracks of the pairs of particles that were created (Fig. 26-1). This **pair production** is a dramatic example of Einstein's equation $E = mc^2$; energy (the massless gamma ray) is converted into mass (the pair of particles). The curvature of the new particle's path in the bubble chamber's magnetic field revealed that it has the same charge-to-mass ratio as an electron, but since it curved in the opposite direction, it must have a positive charge. All the properties of this new particle are of the same magnitude as the electron's although some have the opposite sign. For example, the masses are identical and the electric charges are the same size but opposite in sign. This positively charged electron was named the **positron** and is the **antiparticle** of the electron.

The prediction of the existence of the positron was contained in a quantum-mechanical theory for the electron developed by the English physicist P. A. M. Dirac in 1928. Until its discovery, however, Dirac's "other electron" seemed more like a mathematical oddity in the theory than a physical possibility. After its discovery, the full significance of Dirac's ideas was recognized.

Dirac's theory contained similar predictions for the antiparticles of the proton and neutron, and after the discovery of the positron they were assumed to exist. Finding these heavier antiparticles took two decades. A primary reason for the delay was the large energy needed. In order to cre-

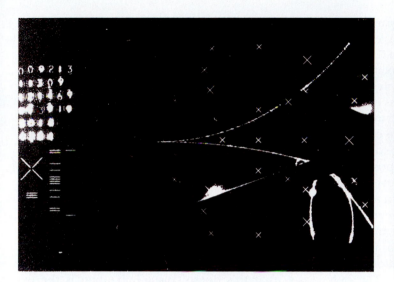 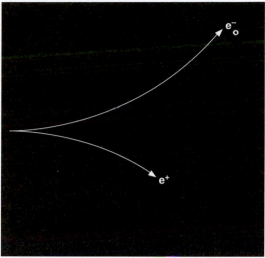

Figure 26-1   A bubble chamber photograph and drawing of the tracks of a positron-electron pair.

ate a pair of particles, the available energy has to be larger than the energy needed to create the rest masses of the particles. Protons and neutrons are so much more massive than electrons that the production of their antiparticles had to await the development of huge accelerators to achieve these great energies. The antiproton was observed in 1955 and the antineutron in 1956.

The antinucleons were discovered at the Bevatron, a particle accelerator in Berkeley, California.

Could any of these suns be composed of antimatter?

Antiparticles are usually designated by putting a bar over the symbol of the corresponding particle. Thus, an antiproton becomes $\bar{p}$ and an antineutron $\bar{n}$. However, the positron is usually written as e^+.

Antiparticles don't survive long around matter. When particles and antiparticles come into contact, they annihilate each other, converting their mass into energy. This process is the reverse of pair production. Because the world we know is assumed to be "regular" matter, antiparticles have little chance of surviving. Once they are created, their lifetimes are determined by how long it takes them to meet their corresponding particle. The positron is slowed down by collisions with particles and is eventually captured by an electron. They orbit each other briefly to form an "atom." In a time typically much less than a millionth of a second, the two annihilate, converting their combined mass back into photons.

If an antiparticle did not meet its counterpart, there would be no annihilation and the antiparticle would exist for a long time. This means that antiatoms could be formed from antielectrons, antiprotons, and antineutrons. The only reason that this doesn't usually happen is the extremely low probability of these antiparticles finding each other in a world so predominantly composed of ordinary particles. However, under special circumstances, this can be achieved. Antideuterium—an antinucleus consisting of one antiproton and one antineutron—has been observed and, in 1995, antihydrogen atoms were produced for very brief periods of time.

The existence of antiatoms leads to the fascinating question of whether antiworlds might exist somewhere in the Universe. There doesn't seem to be any reason to believe they don't. Antiatoms should behave the same as atoms. In particular, they would display the same spectral lines. And, since photons and antiphotons are identical, looking at a distant star won't reveal whether it is composed of matter or antimatter. However, there is evidence that each cluster of galaxies must be either matter or antimatter. If there were matter and antimatter galaxies in a single cluster, the intergalactic dust particles would annihilate, giving off characteristic photons. These photons have not been observed.

Because radio waves are composed of photons, we would have no trouble communicating via radio with antihumans on an antiworld. Although distances make the possibility extremely unlikely, any attempt to communicate by visiting would result in an explosion larger than any bomb that we could build. The entire mass of the spaceship plus an equal mass of the antiworld would annihilate each other.

The discovery of antiparticles provided a reassuring demonstration that the conservation laws of momentum and energy hold in the subatomic world. Every annihilation yields at least two photons. Suppose, for example, a positron–electron pair were orbiting each other as shown in Figure 26-2(a). If we are at rest relative to the pair, they have equal energies and equal but oppositely directed momenta. Before the annihilation they have a total energy which is twice that of one of them but their total momentum is zero due to their opposite directions. If only one photon were produced by the annihilation, the total momentum would not be zero but would be equal to that of the photon, which would be an obvious

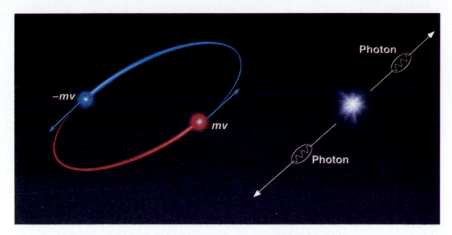

Figure 26-2   The two photons produced in an electron-positron annihilation have equal but oppositely directed momenta to match the zero momentum of the electron-positron pair.

violation of the conservation of momentum. This situation never occurs; there are always at least two photons produced in the annihilation. Their total momentum is zero, as illustrated in Figure 26-2(b). The conservation rules have been extremely helpful in understanding the details of the interactions of the elementary particles.

QUESTION   Would a decay into three photons be forbidden by the laws of conservation of linear momentum and energy?

Physics Update

Antihydrogen atoms have been created at CERN and the discovery confirmed at Fermilab. Physicists passed a beam of antiprotons through a jet of xenon gas. Occasionally, some of the antiproton's own energy can be converted into electron–positron pairs. In the case of nine events, the newly created positron's motion was well matched to that of the antiproton and they formed an atom, in effect an atom of antihydrogen. Antimatter has been produced in the lab artificially for decades, but not until now have antiatoms been made and detected. In the present experiment, the antihydrogen atoms were not trapped, and very quickly annihilated with ordinary matter in the vicinity. Scientists at CERN hope soon to actually capture and study the new exotic atoms. First of all, one wants to be sure that all the physical laws that pertain to atoms—such as quantum mechanics—govern the behavior of antimatter as well.

ANSWER   No. The momenta of three photons can be arranged in many ways to get a total momentum of zero and a given total energy. Therefore, these two laws do not forbid this process from occurring.

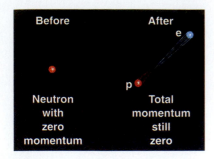

Figure 26-3   If beta decay of a neutron produced only a proton and an electron, they would have to emerge with equal but oppositely directed momenta.

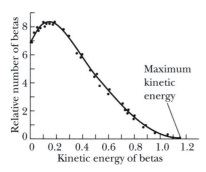

Figure 26-4   The spectrum of kinetic energies of the electrons emitted in the beta decay of the neutron.

discovery of the neutrino

The Puzzle of Beta Decay

In Chapter 24 we saw that one element can spontaneously change into another element by undergoing beta decay. On the nucleon level this means that a neutron turns into a proton, or vice versa. This process led to an interesting puzzle because beta decay did not appear to satisfy the very basic conservation laws of energy and linear momentum.

Imagine that you are sitting in a reference system in which a neutron is at rest. The linear momentum of the neutron is zero and its energy is that associated with its rest mass. If we assume that beta decay changes the neutron into a proton by emitting an electron, momentum can only be conserved if the electron goes off in one direction and the newly created proton recoils in the opposite direction with the same size momentum (Fig. 26-3). The value of these momenta is determined by the requirement that energy also be conserved. Because this can only happen in one way, the electron must always emerge with a fixed value of kinetic energy.

This was the expected result, but experiments showed that it is not what happens. The ejected electrons do not have a single kinetic energy. The graph of experimental data in Figure 26-4 shows a continuous range of kinetic energies from zero up to the value predicted above.

Scientists were in a dilemma. There seemed to be two choices: They could abandon the conservation laws or assume that one or more additional particles were emitted along with the electron. In 1930, Pauli proposed that a third particle, the **neutrino,** was involved. Using the conservation laws, he even predicted its properties. The neutrino has to be neutral because charge is already conserved. (*Neutrino* means "little neutral one.") The fact that the electron sometimes emerges with all the kinetic energy predicted for the decay without the neutrino means that the neutrino sometimes carries away little or no energy. For this to be possible the neutrino's rest mass must be very, very small because it would require some energy to produce its rest mass.

Even though the neutrino was not observed, faith in its existence continued to grow. It was such a nice solution to the beta decay puzzle that experimental verification of the neutrino's existence seemed like it would only be a matter of time. "Only a matter of time" eventually became 26 years. The neutrino was finally detected in 1956 by Clyde Cowan and Frederick Reines using an intense beam of radiation from a nuclear reactor (Fig. 26-5). The observed reaction had an antineutrino $\bar{\nu}$ strike a proton, yielding a neutron and a positron.

$$\bar{\nu} + p \rightarrow n + e^{+9}$$

The properties of the neutrino were confirmed by studying the dynamics of this interaction. Its rest mass has been shown to be very small; it is often assumed to be zero. However, in 1980 some experimental results indicated that the mass of the neutrino may not be exactly zero. If further experiments confirm this, our understanding of the evolution of stars and the Universe will be affected. Neutrinos may even contribute enough mass to the Universe that the Universe may eventually quit expanding and collapse under its own gravitational attraction.

Figure 26-5 This apparatus was used by Cowan and Reines at the nuclear reactor in Savannah River, South Carolina, to detect the neutrino.

The long delay in detecting neutrinos was due to the extremely weak interaction of neutrinos with other particles; neutrinos do not participate in the electromagnetic or the strong interactions, only in the weak interactions. In fact, the neutrinos' interaction with other particles is so weak that a trillion neutrinos pass through the Earth with only one of them being stopped.

Exchange Forces

During the development of Newton's law of universal gravitation there arose a nagging question about forces: What is it that reaches through empty space and pulls on the objects? Newton's idea of an action at a distance seemed unsatisfactory. A couple of hundred years later the concept of a field provided an alternative (Chapter 19). One object creates a change in space (the field) and a second object responds to this field. At least empty space was filled with something, but it was still somewhat unsatisfying.

With the discovery of the quantum of energy a new problem arose. An object moving through a field would gain energy continuously. However, the fact that the energy was quantized (Chapter 23) meant that the object should only receive energy in discrete lumps. This conclusion led to a picture of elementary particles interacting with each other through the exchange of still other elementary particles. Electrons, for example, exchange photons.

Instead of visualizing an electron being repelled by another electron due to an action at a distance or by a field, each electron continuously emits and absorbs photons. As we are talking about effects that are entirely quantum-mechanical, it is risky (if not foolhardy) to reply too heavily on classical analogies. Uncomfortable as it seems, we should be content to say that the interaction properties can be explained by

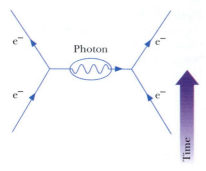

Figure 26-6 A Feynman diagram of the interaction between two electrons through the exchange of a photon.

assuming that photons are exchanged and, depending on the photon properties, the particles will attract or repel each other. Richard Feynman created a way of showing these interactions graphically. Imagine the Feynman diagram in Figure 26-6 as a graph where the vertical dimension is time. Two electrons approaching each other exchange a photon and repel each other.

We should rely on the quantum-mechanical explanation, and yet making analogies sometimes makes things plausible. Consider the following: Imagine two people standing on skateboards, as shown in Figure 26-7. One throws a basketball to the other. The person throwing the basketball gives it some momentum and therefore must acquire some momentum in the backward direction. The person catching the basketball must absorb this momentum and therefore acquires momentum in the direction away from the thrower. The total interaction behaves like a repulsive force between the two people. Although this analogy illustrates that exchanging particles can affect particles at a distance and is easy to visualize, it fails for an attractive force. We must return to the reality that these are quantum-mechanical effects and that our commonsense world view does not serve us very well in the subatomic world of elementary particles.

We can ask if the idea of exchanging photons accounts for the observations. It does; however, it also poses a new problem. Imagine an electron at rest in space. If we explain its effect on other particles by saying it emits photons, it must recoil after the emission due to the conservation of momentum. But if it recoils, it has energy. In fact the emitted photon also has some energy. These energies are over and above the rest mass of the electron and thus in violation of the law of conservation of energy (Fig. 26-8). The same argument can be made for the electron that feels the interaction by absorbing the photon.

Once again we are confronted with a situation in which energy and momentum are not conserved. Only in this case the solution is different; we are not bailed out by the discovery of a new particle. This violation exists. However, it does not mean the abandonment of the conservation

Figure 26-7 A classical analog of a repulsive exchange force is tossing a basketball back and forth.

rules. Overall, momentum and energy are conserved. The violations created by the emission of the photon are canceled by the absorption of the photon by the other particle. It is only during the time that the photon travels from one particle to the other that the violation exists.

We invoke the uncertainty principle (Chapter 23) to understand this situation. One form of the uncertainty principle says that we can only determine the energy of a system to within an uncertainty ΔE that is determined by the time Δt taken to measure the energy. In fact, the product of these two uncertainties is always greater than Planck's constant. This is our escape from the dilemma. If a violation of energy conservation by an amount ΔE takes place for less than a time Δt, no physical measurement can verify the violation. We now believe that unmeasurable violations of the conservation of energy (and momentum) can and do take place. (We have to be a bit careful with this resolution of the problem because we are assigning a classical type of trajectory to the photon like we tried to do with electrons in atoms. Once again it is really a quantum-mechanical effect; energy and momentum are conserved for the interaction.)

Because photons have no rest mass, their energies can be very small. Therefore, the violation of energy conservation due to the emission of a photon can be very small. Very small energy violations can last for very long times and these photons can travel large distances before they are reabsorbed, which explains the infinite range of the electromagnetic force. The fact that the electromagnetic force decreases with distance is explained by the observation that exchange photons that have long ranges have low energies (and momenta) and therefore produce smaller effects.

Figure 26-8 The emission of a real photon by an isolated particle violates the laws of conservation of energy and momentum.

Exchange Particles

Although perhaps sounding bizarre, the idea of exchanging particles explains more than previous models. This new concept of a force satisfies the requirements of quantum mechanics and provides a way of understanding all forces. As an example, because the different forces have different characteristics, presumably they have different exchange particles.

In 1935, the Japanese physicist Hideki Yukawa used this radical idea to show why the nuclear force abruptly "shuts off" after a very short distance. Yukawa reasoned that the short range of the strong force required the exchange particle to have a nonzero mass. The mere creation of its mass requires an energy violation. This minimum energy violation must be at least as large as the rest-mass energy, which means that there is a limit to the time that the violation can occur and therefore a maximum distance the particle can travel before it must be absorbed by another particle. In other words, the exchange of nonzero rest-mass particles means that the force has a limited range.

Finding the Yukawa exchange particle was difficult. It couldn't be detected in flight between two nucleons because of the consequences of the uncertainty principle. The hope was that it might show up in other interactions. We observe photons, for example, when they are created in jumps between atomic levels, not from being "caught" between two electrically charged particles.

During the 1930s cosmic rays were the only source of particles with energy high enough to create the new exchange particles. Cosmic rays are continually bombarding the Earth's atmosphere, creating many other particles that rain down on the Earth in extensive showers. In 1938, a new particle was discovered in a cosmic-ray shower that had a mass 207 times that of the electron and a charge equal to the electron's charge. For a while it was thought that this was the particle predicted by Yukawa, but it did not interact strongly with protons and neutrons. Therefore, the new particle, now known as the **muon,** could not be the exchange particle for the strong force.

The search for Yukawa's particle continued and it was finally discovered 10 years later in yet another cosmic-ray experiment. The **pion** (short for pi meson) has a mass between that of the electron and the proton and comes with three possible charges: $+1$, 0, and -1 times that on the electron. Although the pion is no longer considered to be an exchange particle, it played a pivotal role in the acceptance of the idea of exchange forces. (We return to the question of the exchange particle for the strong force in a later section.)

The success of this model for the interactions between particles led to the hypothesis that all forces are due to the exchange of particles. The gravitational force is presumably due to the exchange of **gravitons.** Because of the similarities between the gravitational force and the electromagnetic force, the graviton should have properties similar to the photon. It should have no rest mass and travel with the speed of light. Although the graviton has not been observed, most physicists believe it exists.

The exchange particles for the weak force, however, have been detected. The weak force occurs through the exchange of particles known as **intermediate vector bosons.** These three particles were discovered in 1983: the W comes in two charge states, $+1$ and -1, and the Z^0 is neutral. One of the reasons it was so difficult to discover these particles is that they are very massive, each one having more than 100 times the mass of the proton. Their discovery had to wait for the construction of new accelerators.

The Elementary Particle Zoo

By 1948, the discovery of antiparticles and exchange particles had nearly tripled the number of known elementary particles. And the situation got worse. During the next seven years four other particles were discovered. Because the behavior of these particles did not match that of the known particles, they became known as the **strange particles.** The existence of the neutrino was confirmed in 1956. A second type of neutrino was discovered in 1962, and even a third type exists.

The 1960s also witnessed the discovery of another new phenomenon. Particles were discovered that live for such a short time that they decay into other particles before they travel distances that are visible even under a microscope. Typical lifetimes for these particles are 10^{-23} second (the time it takes light to travel across a nucleus!).

QUESTION   What does the uncertainty principle say about the mass of a particle that has such a short lifetime?

Before long the number and variety of particles had become so large that physicists began calling the collection a zoo. The proliferation of new particles had once again destroyed the hope that the complex structures in nature could be built from a relatively small number of simple building blocks. The number of "elementary" particles exceeded a few hundred. And the number continued to grow. Particle physicists began to feel organizational problems similar to those of zookeepers.

Much as zookeepers build order into their zoos by grouping the animals into families, particle physicists began grouping the elementary particles into families. Making families helps organize information and may result in new discoveries. When confronted with many seemingly unrelated facts, scientists often begin by looking for patterns. Mendeleev developed the periodic table of chemical elements using this technique. A Swiss mathematics teacher, Johann Balmer, decoded the data on the spectral lines of the hydrogen atom by arranging and rearranging the wavelengths. In the process he discovered a formula that gave the correct results. Both of these discoveries were empirical relationships, not results derived from fundamental understandings of nature. They served, however, to classify the data and provide some guidance for further experimental work.

Elementary particle physics was in a similar condition. Large amounts of data had been accumulated, Scientists were looking for patterns that might provide clues for the development of a comprehensive theory. Just as any collection of buttons can be classified in many ways—size, color, shape, and so forth—the elementary particles can be classified in different ways. (Of course, macroscopic attributes like color and size don't apply here.) One fruitful way is to group them according to the types of interaction in which they participate.

All particles participate in the gravitational interaction—even the massless photon because of the equivalence of mass and energy. Therefore, this interaction doesn't yield any natural divisions for the particles. Furthermore, gravitation is so small at the nuclear level that it usually isn't included in discussions of particle behavior.

The particles that participate in the strong interaction are called **hadrons.** This family includes most of the elementary particles. The hadrons are further divided into two subgroups according to their spin quantum numbers. The **baryons** have spins equal to ½, ³⁄₂, ⁵⁄₂, . . . of the quantum unit of spin, whereas the **mesons** have the whole-number units of spin. The best known baryons are the neutron and proton. Table 26-1 lists some common hadrons and their properties. There are others but

ANSWER   The uncertainty in the energies (and consequently in their masses due to the equivalence of mass and energy) must be large since the product of the uncertainties in the energy and the lifetime must exceed Planck's constant. This has been confirmed by many experiments.

Table 26-1   Properties of some of the hadrons

Name	Symbol	Spin $(h/2\pi)$	Rest Mass (MeV/c^2)	Half-Life (s)	Strangeness
Baryons					
Proton	p	½	938.3	Stable	0
Neutron	n	½	939.6	616	0
Lambda	Λ^0	½	1116	1.82×10^{-10}	−1
Sigma	Σ^-	½	1189	0.55×10^{-10}	−1
	Σ^0	½	1193	5.1×10^{-20}	−1
	Σ^+	½	1197	1.03×10^{-10}	−1
Xi	Ξ^0	½	1315	2.01×10^{-10}	−2
	Ξ^-	½	1321	1.14×10^{-10}	−2
Omega	Ω^-	³⁄₂	1672	0.57×10^{-10}	−3
Mesons					
Pion	π^-	0	139.6	1.80×10^{-8}	0
	π^0	0	135.0	5.8×10^{-17}	0
	π^+	0	139.6	1.80×10^{-8}	0
Eta	η^0	0	548.8	5×10^{-19}	0
Kaon	K^-	0	493.6	0.86×10^{-8}	+1
	K^{0*}	0	497.7	6.18×10^{-11}	+1
				3.59×10^{-8}	

*The K^0 has two lifetimes; 50 percent decay via each mode.

their lifetimes are extremely short, more than a billion times shorter than most of those listed.

The **lepton** family includes the electron, the muon, the tau (discovered in 1977), and their associated neutrinos. The tau lepton is even more massive than the muon. The word *lepton* means "light particle" and refers to the observation that (with the exception of the tau) leptons are less massive than the hadrons. In fact, they appear to be pointlike, having no observable size and no evidence of any internal structure. Table 26-2 lists the known leptons.

All leptons and hadrons participate in the weak interaction. The only particles that fail to get listed in the hadron or lepton families are exchange particles.

Table 26-2   Properties of the leptons

Name	Symbol	Spin $(h/2\pi)$	Rest Mass (MeV/c^2)	Half-Life (s)
Electron	e^-	½	0.511	Stable
e-Neutrino	ν_e	½	<7.3 eV/c^2	Stable?
Muon	μ^-	½	105.7	1.52×10^{-6}
μ-Neutrino	ν_μ	½	<0.27	Stable?
Tau	τ^-	½	1784	2.11×10^{-13}
τ-Neutrino	ν_τ	½	<35	Stable?

Conservation Laws

The conservation laws provide insight into the puzzles of the elementary particles. We have already seen how the conservation laws were used to unravel beta decay. In some situations new conservation laws have been invented as a result of the experiences of viewing many, many particle collisions. The success of the conservation laws has been responsible for a guiding philosophy: *If it can happen, it will.* That is, any process not forbidden by the conservation laws will occur.

if a process can happen, it will

The classical laws of conserving energy (mass–energy), linear momentum, angular momentum (including spin), and electric charge are valid in the elementary particle realm. Any reaction or decay that occurs satisfies these laws. For instance, the neutron decays into a proton, an electron, and an antielectron neutrino via beta decay

$$n \rightarrow p + e^- + \overline{\nu}_e$$

allowed

but has never been observed to decay via

$$n \rightarrow p + e^+ + \overline{\nu}_e$$

forbidden

> **QUESTION** Why is this second alternative forbidden?

A little less obvious are situations that are forbidden because they violate energy conservation. A particle cannot decay in a vacuum unless the total rest mass of the products is less than the decaying particle's rest mass. To see this, view the decay from a reference system at rest with respect to the original particle. The principle of relativity (Chapter 11) states that conclusions made in one reference system must hold in another. In the rest system the total energy is the rest-mass energy of the particle. After the decay, however, the energy consists of the rest-mass energies of the products *plus* their kinetic energies. Even if they have no kinetic energy, there has to be enough energy to create the decay particles. Some of these decays that are forbidden by the conservation of energy can, however, occur in nuclei because the decaying particle can acquire kinetic energy from other nucleons to produce the extra mass. In reactions involving collisions of particles, kinetic energies must also be included in calculating the conservation of energy.

Additional conservation laws were created as more information about the elementary particles became known. The total number of baryons is constant in all processes. So the concept of baryon number and its conservation was invented to reflect this discovery. Baryons are assigned a value of +1, antibaryons a value of −1, and all other particles a value of 0. In any reaction the sum of the baryon numbers before the

> **ANSWER** The second alternative is forbidden by charge conservation since the initial state (the neutron) has zero charge and the final state (proton, positron, and neutrino) has a charge of +2.

reaction must equal the sum afterward. For example, suppose a positive pion collides with a proton. One result that could not happen is

forbidden

$$\pi^- + p \rightarrow \pi^+ + \overline{p}$$

$$0 + \ 1 \ \neq \ 0 \ - 1 \qquad \text{baryon numbers}$$

because the baryon number is +1 before and −1 after. On the other hand, we could expect to observe

allowed

$$\pi^- + p \rightarrow p + \overline{p} + p + \pi^-$$

$$0 + \ 1 \ = \ 1 - 1 + 1 + 0 \qquad \text{baryon numbers}$$

if the kinetic energy is sufficiently high.

Similarly, we don't expect the proton to decay by a process such as

forbidden

$$p \rightarrow \pi^0 + e^+$$

$$1 \ \neq \ 0 \ + 0 \qquad \text{baryon numbers}$$

In fact, the proton cannot decay at all because it is the baryon with the smallest mass. Any decay that would be allowed by baryon conservation would require more energy than the rest mass of the proton. Some recent results have indicated that the conservation of baryon number might not be strictly obeyed. The proton might decay, but its half-life is known to be at least a billion trillion times the age of the Universe!

There is no comparable conservation of meson number. Mesons can be created and destroyed provided the other conservation rules are not violated. There are no conservation laws for the number of any of the exchange particles.

A conservation law for leptons has been discovered but it is more complicated than that for baryons. There are separate lepton conservation laws for electrons and muons and presumably taus. Furthermore, the neutrino associated with the electron is not the same as that associated with the muon or tau. The bookkeeping procedure is slightly more detailed but the procedure is essentially the same. The electron, muon, and tau lepton numbers must be separately conserved.

Some quantities are conserved by one type of interaction but not another. Each new particle is studied and grouped according to its properties: how fast it decays, its mass, its spin, and so on. One group of particles became known as the strange particles because their half-lives didn't seem to fit into the known interactions. If they decayed via the strong interaction, their half-lives should be about 10^{-23} second. If they decayed via the electromagnetic interaction, the predicted half-lives should be about 10^{-16} second. However, these particles are observed to live about 10^{-10} second, at least a million times longer than they "should" live. Something must be prohibiting these decays. A property called **strangeness** and an associated conservation law were invented. The various strange particles were given strangeness values and the conservation law stated that any process that proceeds via the strong or electromagnetic interaction conserves strangeness, whereas those that proceed via the weak interaction can change the strangeness by a maximum of one unit.

The idea of a strangeness quantum number that is conserved seems quite foreign to our experiences. And it should. This attribute is clearly in the nuclear realm; we don't see its manifestation in our everyday world. In fact, we should probably be cautious about the feeling of comfort we have with other quantities. Consider electric charge. Most of us feel quite comfortable talking about the conservation of electric charge, perhaps because electricity is familiar to us. Imagine a world in which we had no experience with electricity. We would feel uneasy if someone suggested that if we assign a +1 to protons, a −1 to electrons, and 0 to neutrons, we might have conservation of something called "cirtcele" (electric spelled backward). It isn't unreasonable that unfamiliar quantities emerge as scientists explore the subatomic realm.

> **QUESTION**   Given the values of strangeness in Table 26-1, would you expect the decay of the lambda particle to a proton and a pion to proceed via the strong or weak interaction? Does this agree with its lifetime?

Quarks

The continual rise in the number of elementary particles once again raised the question of whether the known particles were the simplest building blocks. At the moment the leptons appear to be elementary; there is no evidence that they have any size or internal structure. The exchange particles also appear to be truly elementary for the same reasons. On the other hand, there is experimental evidence that the hadrons have some internal structure.

Particle physicists asked themselves: Is it possible to imagine a smaller set of particles with properties that could be combined to generate all the known hadrons? The most successful of the many attempts to build the hadrons is the **quark** model proposed by two American physicists, Murray Gell-Mann and George Zweig, in 1964. Their original model hypothesized the existence of three **flavors** of quark (and their corresponding antiquarks), now called the "up" (u), "down" (d), and "strange" (s) quarks. The strange quark has a strangeness number of −1, whereas the other quarks have no strangeness. Each quark has ½ unit of spin and a baryon number of ⅓.

Perhaps the boldest claim made in this model is the assignment of fractional electric charge to the quarks. There has never been any evidence for the existence of anything other than whole-number multiples of the charge on the electron. These fractional charges should (but

> **ANSWER**   Because the strangeness numbers assigned to these two particles differ by one, it should be a weak decay. The mean lifetime agrees with this conclusion.

Table 26-3 Properties of the quarks

Flavor	Symbol	Charge	Spin	Baryon No.	Strangeness	Charm	Bottomness	Topness
Down	d	$-\frac{1}{3}$	$\frac{1}{2}$	$\frac{1}{3}$	0	0	0	0
Up	u	$+\frac{2}{3}$	$\frac{1}{2}$	$\frac{1}{3}$	0	0	0	0
Strange	s	$-\frac{1}{3}$	$\frac{1}{2}$	$\frac{1}{3}$	−1	0	0	0
Charm	c	$+\frac{2}{3}$	$\frac{1}{2}$	$\frac{1}{3}$	0	+1	0	0
Bottom	b	$-\frac{1}{3}$	$\frac{1}{2}$	$\frac{1}{3}$	0	0	+1	0
Top	t	$+\frac{2}{3}$	$\frac{1}{2}$	$\frac{1}{3}$	0	0	0	+1

apparently don't) make the quarks easy to find. Yet the scheme works. The quark model has had remarkable successes describing the overall characteristics of the hadrons.

The properties assigned to the various flavors of quark are given in Table 26-3. The antiquarks have signs opposite to their related quarks for the baryon number, charge, and other properties such as strangeness.

To see how this concept works, let's "build" a proton. We must first list the proton's properties: A proton has baryon number +1, strangeness 0, and electric charge +1. Examination of the quarks' properties confirms that the proton can be made from two up quarks and one down quark (uud) as shown in Figure 26-9. Other baryons can be created with different combinations of three quarks.

QUESTION What three quarks form a neutron (baryon number +1, strangeness 0, and electric charge 0)?

The mesons have 0 baryon number so they must be composed of equal numbers of quarks and antiquarks. The simplest assumption is one of each. For instance, the positive pion is composed of an up quark and a down antiquark (ud̄), giving it a charge of +1 and a strangeness of 0 as shown in Figure 26-10.

QUESTION What two different combinations of up and down quarks and antiquarks would yield a neutral pion?

All of this was fine until 1974. In that year a neutral meson called J/ψ with a mass more than three times the mass of the proton was discovered. What made the particle unusual was its "long" lifetime. It was expected to decay in a typical time of 10^{-23} second, but lived 1000 times longer than this. The quark model was only able to account for this anomaly with the

ANSWER   udd

ANSWER   uū and dd̄.

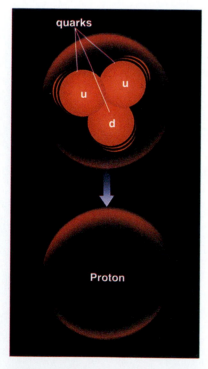

Figure 26-9 The proton is made of two up quarks and a down quark.

Figure 26-10 All three pions are composed of a quark and an antiquark.

addition of a fourth quark that possessed a property called **charm** (c). The J/ψ particle represents a bound state of a charmed quark and its antiquark (c$\bar{c}$). The next year a charmed baryon (udc) was observed, adding further support for the existence of this quark.

It is interesting to note that the existence of a fourth quark had been proposed several years earlier based on symmetry arguments. At that time, four leptons were known—the electron, the muon, and their two neutrinos. Why should there be four leptons and only three quarks? This discomfort with asymmetry led to the idea that there should be four quarks. Of course, the nice symmetry was quickly destroyed with the discovery of the tau lepton!

With the discovery of the tau and the presumed existence of its corresponding tau neutrino, two additional quarks were predicted so that there would be six leptons and six quarks. The discovery of the upsilon in 1977 was the first evidence for the fifth quark. The upsilon is a bound state of the **bottom** flavor of quark and its antiquark ($\bar{b}$b). There is also evidence for "bare bottom," baryons and mesons that contain a bottom quark without a bottom antiquark. The sixth quark has the flavor called **top** and its existence was confirmed in 1966.

Gluons and Color

Earlier we attributed the force between hadrons to the exchange of other hadrons such as pions. If the hadrons are actually composite particles made of quarks, we need to take our earlier idea one level further and ask what holds the quarks together in hadrons. The particles proposed to be exchanged by quarks are known as **gluons.** (They "glue" the quarks together.)

Another problem of the simple quark theory is illustrated by the omega minus (Ω^-) particle, which has a strangeness of −3. It should be composed of three "strange" quarks with all three spins pointing in the same direction to account for its 3⁄2 units of spin. However, the exclusion principle (Chapter 23) forbids identical particles with 1⁄2 unit of spin from having the same set of quantum numbers. The exclusion principle can be satisfied if quarks have a new quantum number. This new quantum number has been named *color* and has three values: red, green, and blue. All observable particles must be "white" in color; that is, if the colors are considered to be lights, they must combine to form white light (Chapter 16). Therefore, baryons consist of three quarks, one of each color, and mesons consist of a colored quark and an antiquark that has the complementary color (Fig. 26-11).

Although the idea of the color quantum number began as an ad hoc way of accommodating the exclusion principle, it soon became a central feature of the quark model. Each quark is assumed to carry a *color charge,* similar to electric charge, and the force between quarks is called the *color force.* This theory requires that there be eight varieties of gluon, which differ only in their color properties. The quarks interact with each other through the exchange of gluons.

The strong force that holds the nucleons together to form nuclei is due to the color force between the quarks making up the nucleons, but the strong force is much weaker than the color force between the quarks within a single nucleon. Therefore, gluons have replaced the mesons as the exchange particles of the strong force.

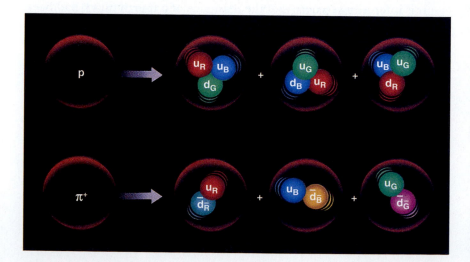

Figure 26-11 The addition of the color quantum number increases the possible combinations of quarks (and antiquarks) that make up particles such as the proton and the positive pion.

Physics Update

After years of speaking guardedly about tentative, preliminary evidence for the top quark, two rival collaborations at Fermilab jointly announced in 1995 that they had indeed discovered the top quark, the sixth and supposedly last of the quarks. Both experiments sampled about 6 trillion collision events, for which data were collected for about 40 million events. This vast mountain of information was sifted through a myriad of computer programs seeking just the right configurations of daughter particles, rare events in which top quarks are created out of the annihilation of high-energy protons and antiprotons. The experimenters found 54 top candidate events as against an expected background of 16 events. The mass of the top quark is about 180 GeV/c^2.

Unified Theories

The view that the elementary building blocks in nature are the 6 leptons, the 6 quarks, and the 12 exchange particles is a great simplification of the hundreds of subnuclear particles that have been discovered since 1932. The list of elementary particles totals 24, a reasonably small number for building the many, many diverse materials in the world. But can we reduce the complexity even more?

The present scheme contains two very distinct classes of particles—the leptons and the quarks. The leptons have whole-number units of charge, whereas the quarks have charges that are multiples of ⅓ of the basic unit. The quarks participate in the strong interaction through their color, whereas the leptons are colorless and do not participate. Leptons are observed as free particles, but quarks have yet to be isolated. There have been no observations that indicate that leptons can turn into quarks, and vice versa. Why should there be two classes? Why not only one?

Similarly, we have four different forces—gravitational, electromagnetic, weak, and strong (or color). Each of these appears to have its own strength, and there are three different dependencies on distance and 12 different exchange particles. Why should there be four forces? Why not only one?

Physicists have asked similar questions for a long time and have sought to reduce the number of particles and the number of forces as far as possible, ideally to one class of particles and one force. The theorists don't aim to eliminate (or overlook) the differences that are so apparent, but rather to show that these are all manifestations of something much more basic and therefore more elementary.

Much progress has been made. Before the turn of the century, Maxwell was able to show that electricity and magnetism were really two different aspects of a common force, the electromagnetic force. During the 1960s, Sheldon Glashow, Steven Weinberg, and Abdus Salam developed the electroweak theory that unified the weak and electromagnetic interactions.

The electroweak theory and the color theory of the strong interaction are the main components of the *standard model* of elementary

Cosmology

There is a cosmic connection between all of the forces and particles that we have studied. The connection is made in the Big Bang theory for the creation of the Universe. Experimental evidence indicates that the Universe had a beginning, and that this beginning involved such incredibly high densities and temperatures that everything was a primordial "soup" beyond which it is impossible to look. According to this theory, the Universe erupted in a Big Bang about 10–20 billion years ago.

Theorists divide the development of the Universe into seven stages. Immediately after the Big Bang, up until something like 10^{-43} second, the conditions were so extreme that the laws of physics as we now know them did not exist. As the Universe expanded, it cooled and the four forces developed their individual characteristics and the particles that we currently observe were formed. The particular details of what happened early in the process are very speculative, but rest on firmer ground as time unfolds.

During the second stage—between 10^{-43} and 10^{-35} second—the gravitational force emerges as a separate force, leaving the electromagnetic and the two nuclear forces unified, as described by the grand unification theories. During this time, the energies of particles and photons are so great that there is continual creation and annihilation of massive particle–antiparticle pairs.

During the third stage—between 10^{-35} and 10^{-10} second—temperatures drop from approximately 10^{29} K to approximately 10^{15} K, and the strong force splits off from the electroweak force, resulting in three forces.

Near the beginning of the fourth stage—which ends only a microsecond after the Big Bang—further cooling results in the electroweak force splitting into the weak force and the electromagnetic force, giving us the four forces we can now detect. The Universe was filled with radiation (photons), quarks, leptons, and their antiparticles. The temperatures were still too hot for quarks to stick together to form protons and neutrons.

The fifth stage takes us up to three minutes. By this time the temperature has dropped to 1 billion K, which is still very much hotter than the interior of our Sun. Typical photons can no longer form an electron–positron pair and quarks have formed into mesons and baryons.

During the sixth stage, neutrons and protons are able to form helium nuclei with about 25% of the Universe's mass as helium and the rest as hydrogen. Theory agrees that there should be 1 neutron for every 7 protons. (Due to the extra mass of the neutron, it is harder to convert protons into neutrons than it is to convert neutrons into protons.) This stage ends when the temperature drops to 3600 K at about 700,000 years after the Big Bang. Suddenly, matter becomes transparent to photons and the photons can travel throughout the Universe. This moment defines the boundaries of the observable Universe: Looking out into space is just like looking backward in time to this beginning. The radiation we see from objects at a distance of 1 billion light-years left there 1 billion years ago. Therefore, we are seeing the object as it was in the past, not as it is at the present.

The final stage begins with the formation of atoms and includes the formation of the galaxies, stars, and planets.

The most compelling piece of experimental evidence supporting the Big Bang theory was accidentally discovered in 1965 by Bell Laboratory scientists Arno

particles that evolved in the early 1970s. This model seems to be able to explain every experiment that can be done. However, even though the model seems to be complete, its mathematical complexity is such that many calculations cannot be done with presently available techniques. For instance, theorists are not able to calculate the mass of the proton.

The unification efforts continue; some success has been achieved in combining the electroweak interaction and the color interaction. However, a problem exists. These grand unification theories predict that

Penzias and Robert Wilson. While testing sensitive microwave communications equipment, Penzias and Wilson discovered a background "hiss" that would not go away. In addition, the intensity of this signal was the same, regardless of the direction they pointed their microwave detector. Further measurements showed that this radiation corresponded to blackbody radiation with a temperature of 2.7 K, consistent with a temperature of a gas that has been cooling from a temperature of 3600 K since 700,000 years after the Big Bang! Penzias and Wilson received the Nobel Prize in 1978 for discovering this cosmic background radiation.

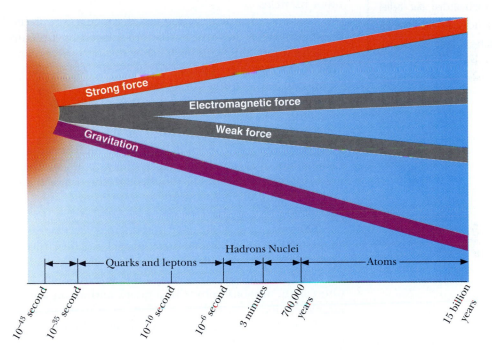

As the Universe expands and cools, the single original force splits into new forces that eventually become the four forces that we currently observe in nature.

baryon number is not strictly conserved and, as a consequence, the proton should decay. Present experiments have not seen the decay of a single proton and have determined that the lifetime for its decay is much larger than predicted by the simplest theory.

If efforts to produce a grand unified theory succeed, there is still the separate existence of the gravitational force. To reach the ultimate goal, gravity must also be included in an ultimate "theory of everything." Maybe, in the future, we will have one scheme that includes them all.

CHAPTER

26

REVISITED

Nobody knows where the search for the fundamental building blocks will end. The current theory of quarks and leptons is appealing in its completeness and no experiments have contradicted our belief that these particles do not have internal structures. Maybe the search has ended.

SUMMARY

The search for elementary components of matter began with Aristotle's basic elements—earth, fire, air, and water—and has led us through the chemical elements to the constituents of atoms—electrons, protons, and neutrons—to quarks.

Antimatter, an entirely new kind of matter, is produced by pair production or by the collisions of very energetic particles. Antiparticles of all the fundamental particles exist, but don't survive long because particles and antiparticles annihilate each other when they come into contact.

The apparent failure of the conservation laws in beta decay led Pauli to propose the existence of the neutrino. The neutrino is electrically neutral, has a very small or zero rest mass, and interacts extremely weakly with other particles.

Newton's idea of an action at a distance was replaced by the concept of a field, which, in turn, has been replaced by a picture of elementary particles interacting with each other through the exchange of still other elementary particles. The electromagnetic force occurs via the exchange of photons, the strong (or color) force via the eight gluons, the weak force via the intermediate vector bosons (W and Z^0), and the gravitational force via gravitons.

The discovery of the antiparticles, the muon, the tau, their neutrinos, and the very large collection of mesons and baryons quickly destroyed the concept of the elementary particles as being elementary. Some sense was brought to the particle zoo through the use of the conservation laws and classifying them according to their interactions. Further simplification came with the quark model.

The quark model has had remarkable success describing the overall characteristics of the elementary particles. The Universe appears to be made of six leptons and six quarks (each in three colors). The baryons, for example, consist of three quarks, one in each of the three colors, whereas the mesons consist of a quark and an antiquark of the complementary color.

Theorists continue to work on possible theories to unify all the particles and all the forces.

KEY TERMS

antiparticle: A subatomic particle with the same size properties as those of the particle although some may have the opposite sign. The positron is the antiparticle of the electron.

baryon: A type of hadron having a spin of ½, ³⁄₂, ⁵⁄₂, . . . times the smallest unit. The most common baryons are the proton and the neutron.

bottom: The flavor of the fifth quark.

charm: The flavor of the fourth quark.

flavor: The types of quark: up, down, strange, charm, bottom, or top.

gluon: An exchange particle responsible for the force between quarks. There are eight gluons that differ only in their color quantum number.

graviton: The exchange particle responsible for the gravitational force.

hadrons: The family of particles that participate in the strong interaction. Baryons and mesons are the two subfamilies.

intermediate vector bosons: The exchange particles of the weak nuclear interaction: the W^+, W^-, and Z^0 particles.

lepton: A family of elementary particles that includes the electron, muon, tau, and their associated neutrinos.

meson: A type of hadron with whole-number units of spin. This family includes the pion, kaon, and eta.

muon: A type of lepton; often called a heavy electron.

neutrino: A neutral lepton; one exists for each of the charged leptons—the electron, the muon, and the tau.

pair production: The conversion of energy into matter in which a particle and its antiparticle are produced. This usually refers to the production of an electron and a positron (antielectron).

photon: The exchange particle for the electromagnetic interaction.

pion: The least massive meson. The pion has three charge states: +1, 0, and −1.

positron: The antiparticle of the electron.

quark: A constituent of hadrons. Quarks come in six flavors and three colors each. Three quarks make up the baryons, whereas a quark and an antiquark make up the mesons.

strangeness: The flavor of the third quark.

strange particle: A particle with a nonzero value of strangeness. In the quark model it is made up of one or more quarks carrying the quantum property of strangeness.

strong force: The force responsible for holding nucleons together to form nuclei.

top: The flavor of the sixth quark.

weak force: The force responsible for beta decay. This force occurs through the exchange of the W and Z particles. All leptons and hadrons interact via this force.

CONCEPTUAL QUESTIONS

1. What particles were on the list of "elementary particles" in 1932?

2. Which particles on the list of "elementary particles" in 1932 are not on the current list?

3. What is the antiparticle of an electron?'

4. What is the antiparticle of the photon?

5. Would the emission spectrum of hydrogen differ from that of antihydrogen? If so, how?

6. How much energy would be given off if an antiproton and an antineutron combined to form an antideuteron?

7. If we ignore signs, which (if any) properties of particles and their corresponding antiparticles do *not* have the same size?

8. Do antiparticles have negative mass?

9. What is the ultimate fate of an antiparticle here on Earth?

10. How much energy is released when a positron is annihilated?

*11. You are in communication with a civilization on what you suspect is an antiworld. Are there any simple questions you might ask to find out if your suspicions are correct?

12. Which elementary particles could be used to communicate with an antiworld?

13. Explain why momentum conservation requires the emission of at least two photons when a positron and an electron annihilate. (The photon has a momentum equal to E/c.) For simplicity assume that the electron and positron are at rest in your reference system.

*14. Use the idea that the physics of elementary particles is the same whether time runs forward or backward and the answer to the previous question to explain why a single photon cannot produce an electron–positron pair without the presence of a nearby particle.

15. The initial observations of beta decay indicated that some of the classical conservation laws might be violated. Which (if any) were obeyed without inventing the neutrino?

16. Why don't we feel the hundreds of billions of neutrinos that pass through each square centimeter of our bodies each second?

17. Why did it take so long for experimentalists to detect the neutrino?

18. Through which force does the neutrino interact with the rest of the world?

19. A gravitational or electromagnetic signal travels at the speed of light. Therefore, if an object changes position or explodes, a detecting instrument cannot register this change in a time shorter than it takes light to travel the distance. Is the same thing true for the strong force?

*20. One attempt at creating an analogy of an attractive exchange force utilizes boomerangs. The person on the right throws a boomerang toward the right, gaining momentum toward the left. The boomerang travels along a semicircle and is caught coming in from the left. When he catches the boomerang, the person on the left gains some momentum toward the right. Why is this not a good analogy?

21. Under what conditions can the laws of conservation of energy and linear momentum be violated?

*22. What must happen to a photon emitted by an electron if there is nothing else in the neighborhood?

23. How does the uncertainty principle explain the infinite range of the electromagnetic interaction?

24. How does the uncertainty principle account for the decrease in the strength of the Coulomb (electrostatic) force with increasing distance?

25. What feature of the strong interaction requires the exchange particles to have nonzero masses?

26. At one time a small part of the strong interaction was hypothesized to be due to the exchange of mesons

called kaons. Use Table 26-1 to predict whether this component of the force would have a longer or a shorter range than that due to the exchange of pions.

27. What argument can be used to support the idea that the graviton has zero rest mass?

28. Given that the masses of the intermediate vector bosons are more than 100 times that of the proton, what can you say about the range of the weak interaction?

29. Which particles do not participate in the strong interaction?

30. What are the differences between baryons and mesons?

31. What particles belong to the lepton family?

32. What are the important differences between the hadrons and the leptons?

33. Do protons interact via the weak force?

34. Particles and antiparticles have the same properties except for the sign of some of them. Which particles in Table 26-1 could be particle–antiparticle pairs?

35. What quantity is conserved in strong and electromagnetic interactions but can change by one unit in weak interactions?

36. Roughly what would you expect for the lifetime of the decay $\Omega^- \rightarrow \Xi^0 + \pi^-$?

37. Why can't a proton beta decay outside the nucleus?

38. Why can't an electron decay?

39. Name at least one conservation law that prohibits each of the following:

    a. $\pi^- + p \rightarrow \Sigma^+ + \pi^0$

    b. $\mu^- \rightarrow \pi^- + \nu_\mu$

    c. $\Sigma^0 \rightarrow \Lambda^0 + \pi^0$

40. Name at least one conservation law that prohibits each of the following:

    a. $\mu^- \rightarrow e^- + \nu_e + \bar{\nu}_\mu$

    b. $p \rightarrow \pi^+ + \pi^+ + \pi^-$

    c. $\Omega^- \rightarrow \Lambda^0 + \pi^-$

41. A particle X is observed to decay by $X \rightarrow \pi^+ + \pi^-$ with a lifetime of 10^{-10} second. What are possible values for the (a) baryon number, (b) strangeness number, and (c) charge of X?

42. If you observe the decay $X \rightarrow \Lambda^0 + \gamma$ with a lifetime of approximately 10^{-20} second, what can you say about the (a) baryon number, (b) strangeness, and (c) charge of X?

43. That the lifetime of the π^0 is roughly 10^{-16} second indicates that it decays via the electromagnetic interaction. What would you guess would be the products of the decay?

44. The mean lifetime of the π^- is roughly 10^{-8} second. What decay products would you expect?

45. What combinations of quarks correspond to the antiproton and the antineutron?

46. What combination of quarks makes up the π^- meson?

47. Which hadron corresponds to the combination of a strange quark and an up antiquark?

48. Which hadron is composed of two down quarks and a strange quark?

49. What combination of quarks corresponds to the Σ^+?

50. What quarks make up a Ξ^0?

51. What quark and antiquark make up a K^0?

52. What combinations of quarks and antiquarks could make up an η^0?

53. In the original quark model, the Ω^- was believed to be composed of three strange quarks. This assumption causes problems since quarks are expected to obey the Pauli exclusion principle. How might you get around this problem?

54. Why was color proposed as an additional quantum number for quarks?

55. What charge would a baryon have if it was composed of a down quark, a strange quark, and a bottom quark?

56. What properties would a particle made of a top quark and its antiquark have?

EPILOGUE

The Orion Nebula as seen by the Hubble Space Telescope.

The Search Goes On

In the Prologue we set ourselves the task of expanding your world view. We began with a common-sense world view and carefully added bits and pieces of the physics world view. It is impossible for us to know which pieces have become part of your own world view. Experience has shown us that expansion of a world view enables a person to see new connections between events. Some "see" sound waves, others feel the pull of gravity in a new way, and still others report experiencing new beauties in a rainbow or a red sunset.

This expansion of the world view produces different kinds of wonderment in different people. Some of us look at individual phenomena and marvel at the connections that can be made between seemingly unrelated things. There are others of us who wonder about the possibility of making sense out of the Universe. The German philosopher Friedrich Nietzsche once remarked that "the most incomprehensible thing about this Universe is that it is comprehensible."

We hope that you have gained some insight into the ways the physics world view evolves. We hope, for example, that you know that there is no single, static physics world view. Rather, there is a central core of relatively stable components surrounded by a fuzzy, very fluid boundary. Metaphorically, it is an organism with spurts of growth, and regions of maturation, decay, death, and even rebirth.

Part of our purpose was to show you that this growth proceeds within very definite constraints; it is certainly not the case that "anything goes." Although intuitive feelings motivate new paths, they often result in dead ends. There were many more dead-end paths than we mentioned in this book. We picked up new ideas and discarded old ones. But within the space of this book we couldn't follow the many, many blind paths that occurred in building the current physics world view. For the most part we had to follow the main paths.

Even the main paths, however, demonstrate why scientists believe what they believe. As one journeys from Aristotle's motions to Newton's inventions of concepts like gravity and force, it is tempting to look back in amusement at the more primitive ideas. Perhaps you had these feelings. This superiority should have quickly vanished as we adopted the spacetime ideas of Einstein and the wave–particle duality of the submicroscopic world. All of these notions are creations of the human mind. Our task in building a world view is to create, but to create in a way different from the artist or poet.

> **"The most incomprehensible thing about this Universe is that it is comprehensible."**

The creations are different because we have a different answer to the question of why we believe what we believe. Our ideas must agree with nature's results. For a new idea to replace an old one, it must do all that the old idea did and more. Old ideas fail because experiments give results that don't agree with these ideas. The new ideas must meet the challenge: They must account for the observations that were in agreement with the old theory and encompass the anomalous data. Einstein's idea of a warped spacetime includes Newton's concepts of gravity, but does much more.

People might claim that a particular idea is a lot of malarkey. For example, they might claim that the speed of light is not the maximum speed limit in the Universe. For these critics to make a contribution, they will need to determine ways of testing their ideas. Their theories must be able to make predictions that can be tested by experiments.

So science has an interesting "split personality." During the birth of an idea, its personality is strongly dependent on the very personal, intuitive feeling of the scientist. But science also has a strong, cold, and very impersonal style of ruthlessly discarding ideas which have failed. There is no idea that escapes this threat of abandonment. No single idea is so appealing that it can circumvent the tests of nature.

It is very appealing to think that perpetual motion machines could exist and would solve our energy crisis, that the visual positions of the celestial objects give insight into the future, or that there is some magic potion that can free us from the entropy of old age. These ideas are appealing to most people. However, the appeal of an idea is not the only criterion for inclusion into the physics world view.

Now we arrive at our final point: The search goes on and the world view continually evolves. The locations of future growth spurts, however, are virtually impossible to predict. A survey of physicists would yield various possibilities. The differences depend on the responder's area of expertise and, perhaps, a few most cherished ideas.

And so, the search goes on.

Physics Update

An experiment at Fermilab has now gathered data which could be interpreted as evidence for the existence of a new level of matter. Monitoring high-energy proton–antiproton collisions, physicists are on the lookout for a variety of interesting phenomena. In 1995 their vigilance paid off handsomely with the discovery of the top quark. Now they are reporting on a class of violent interactions in which a quark inside a proton scatters directly with a quark inside one of the oncoming antiprotons, sending jets of particles sideways out of the interaction area. The data can be explained if one of the incoming quarks scatters from something hard inside the other quark. This can be compared with two other notable chapters in physics history. In 1911 Ernest Rutherford surmised that atoms had a substructure (consisting of a nucleus surrounded by electrons) by scattering alpha particles from a thin foil. Many years later at SLAC, physicists discovered that protons also had a substructure; that is, that they consist of constituent quarks. If this result is verified, it would constitute a departure from the venerable "standard model" of particle physics and would be an indication that quarks, which we currently regard as the most basic of building blocks, would themselves have constituents.

If someone offered to sell a bar of gold for $200, you would immediately ask, "How large is the bar?" The size of the bar obviously determines whether it is a good buy. A similar problem existed in the early days of commerce. Even when there were standard units of measure, they were not the same from time to time and region to region. Later, several standardized systems of measurement were developed.

The two dominant systems are the British system based on the foot, pound, and second and the metric system based on the meter, kilogram, and second. Thomas Jefferson advocated that the United States adopt the metric system, but his advice was not taken. As a result the metric system is used very little by most people in the United States. It is used primarily by the scientific community and those who work on such things as cars. Most of the countries, such as England and Canada, that originally adopted the British system have now officially changed over to the metric system. The United States is the only major country not to have made the change.

There are obvious advantages in having the entire world use a single system. It avoids the cumbersome task of converting from one system to another and aids in worldwide commerce. The disadvantages for the countries which must change are the expense of converting the machinery, signposts, and standards; the maintenance of dual systems for a time; and the abandonment of the familiar units for new ones for which they do not have intuitive feelings.

The metric system has advantages over the British system and was the system chosen in 1960 by the General Conference on Weights and Measures. The official version is known as **Le Système International d'Unités** and abbreviated SI. In this book we begin our studies using the more familiar British system and slowly phase in major portions of the SI system. In many places we give the approximate British equivalent in parentheses.

One problem of the British system of measurement is illustrated by Table A-1 in which we have listed a sample of units used to measure lengths. Some of these units are historic; others have been developed for use in specialized areas. Many of them may be unfamiliar. Few of us have any idea how long a fathom is, let alone how many inches there are in 1 fathom. Even among the more familiar units, it is often a difficult task to convert from one unit to another. For instance, to determine how many inches there are in a mile, we must make the following computation.

$$1 \text{ mile} = 5280 \text{ feet} \times \frac{12 \text{ inches}}{1 \text{ foot}} = 63,360 \text{ inches}$$

The metric system eliminates the confusion of having many different unfamiliar units and the difficulty of converting from one size unit to

Table A-1 Partial list of length measures

angstrom	micron
astronomical unit	mil
barley corn	mile
cubit	nautical mile
fathom	pace
fermi	palm
foot	parsec
furlong	pica
hand	point
inch	rod
league	stadium
light-year	thumbnail
meter	yard

another by adopting a single standard unit for each basic measurement and a series of prefixes that make the unit larger or smaller by factors of 10. For instance, the basic unit for measuring length is the **meter.** It is a little longer than a yard. This unit is inconveniently small for measuring distances between cities, so road signs everywhere else in the world display kilometers as a unit of length. The prefix **kilo** means one thousand and indicates that the kilometer is equal to 1000 meters. The kilometer is about ⅝ mile, so a distance of 50 miles will appear as 80 kilometers. Speed limits appear as 100 kilometers per hour rather than as 65 miles per hour.

Smaller distances are measured in such units as centimeters. The prefix **centi** means one-hundredth. It takes 100 centimeters to equal 1 meter. The other prefixes are given in Table A-2 along with their abbreviations and various forms of their numerical values. These units are pronounced with the accent on the prefix. Note also that the terms *billion* and *trillion* do not have the same meanings in all countries. In this textbook 1 billion is 1,000,000,000 and 1 trillion is 1000 billion.

Because all the prefixes are multiples of ten, conversions between units are done by moving the decimal point, that is, multiplying and dividing by 10s. For instance, since **milli** means one-thousandth, we can convert from meters to millimeters by multiplying by 1000. There are 5670 millimeters in 5.67 meters.

In its purest form, the SI system allows no other units for length. However, other units that have grown up historically will remain in use for some time (and maybe forever). For instance, the terms *micron* and *fermi* continue to be used for micrometer and femtometer. A common unit of length on the astronomical scale is the light-year, the distance light travels in 1 year. Although it is not an SI unit, it is a naturally occurring unit of length on this scale and will continue to be used. On the other hand, the angstrom (10^{-10} m) has been very popular, but is now being replaced by the nanometer (10^{-9} m).

The metric system also differs from the British system in that mass is considered the primary unit and weight (force) the secondary unit. In the British system the situation is reversed. (The distinction between mass

Table A-2 The metric prefixes

Prefix	Symbol			Value
tera	T	trillion	10^{12}	1,000,000,000,000
giga	G	billion	10^{9}	1,000,000,000
mega	M	million	10^{6}	1,000,000
kilo	k	thousand	10^{3}	1,000
		one	10^{0}	1
centi	c	hundredth	10^{-2}	0.01
milli	m	thousandth	10^{-3}	0.001
micro	μ	millionth	10^{-6}	0.000001
nano	n	billionth	10^{-9}	0.000000001
pico	p	trillionth	10^{-12}	0.000000000001
femto	f	quadrillionth	10^{-15}	0.000000000000001

and weight is explained in Chapter 2.) The basic unit in the British system is the pound, but the basic unit in the SI system is the **kilogram.** The weight of 1 kilogram is 2.2 pounds. A U.S. nickel has a mass that is very close to 5 grams. The term *megagram* (1000 kilograms) is not often used in the metric system. This is known as a metric ton and has a weight equal to a British long ton (2200 pounds).

Since the invention of the metric system in 1791, many unsuccessful attempts have been made to change the time system over to a decimal basis so the units would also be multiples of 10. They have all failed. The SI system has the same units of time as the British system.

QUESTION: What is 10^{-12} boo?
ANSWER: A picoboo (peek-a-boo).

QUESTION: What is 10^{-3} pede?
ANSWER: A millipede.

QUESTION: What is 10^{12} dactyl?
ANSWER: A teradactyl (pterodactyl).

In this book we study objects and events that go far beyond the normal human scale of things. When we look at phenomena on very large and very small scales, the sizes of the numbers quickly get out of hand. For instance, the masses of the Sun and an electron are

$$\text{mass of Sun} = 1,989,000,000,000,000,000,000,000,000,000,000 \text{ grams}$$

$$\text{mass of electron} = 0.000\ 000\ 000\ 000\ 000\ 000\ 000\ 000\ 000\ 910\ 94 \text{ gram}$$

These numbers are very difficult to read, write, and manipulate mathematically. It is even easy to make errors in counting the zeros unless they are grouped in threes as we have done.

In this text we use the **powers-of-ten notation** that displays the count of the number of zeros in these numbers. In mathematics the notation 10^2 means 10×10, which is equal to 100. Similarly, 10^3 means $10 \times 10 \times 10$ = 1000. The superscript is called an *exponent* and is equal to the number of 10s that are multiplied together. The exponent is also equal to the numbers of zeros in these numbers.

Using this notation the mass of the Sun is written as 1.989×10^{33} grams. This indicates that the number in front is to be multiplied by ten 33 times to get the actual number. Because multiplication by 10 just moves the decimal point one position to the *right,* the superscript 33 indicates the total number of places the decimal point must be moved. By convention the number out in front is usually written so that it has a value between 1 and 10. You should check that moving the decimal point in 1.989 thirty-three places to the right gives the value in the first paragraph.

Positive values for exponents indicate that the numbers are large. Small numbers are indicated by negative exponents. For instance, 10^{-1} = $\frac{1}{10}$ and $10^{-2} = \frac{1}{10} \times \frac{1}{10} = \frac{1}{100}$. The minus sign indicates that the power of 10 is in the denominator; that is, it is to be divided into 1. Using this convention the mass of the electron is written as 9.1094×10^{-28} grams. The number in front must be divided by ten 28 times. Or equivalently, the number can be obtained by moving the decimal point 28 positions to the *left*.

Sometimes you may see a number written with only the power of 10. If the usual number preceding a power of 10 is missing, it is assumed to be 1; that is, $10^5 = 1 \times 10^5$. Similarly, if the exponent is missing, it is assumed to be zero; that is, $4 = 4 \times 10^0$.

The greatest power of using this notation comes when you have to multiply or divide these very large or small numbers. To *multiply* you multiply the two numbers in front and *add* the exponents. To *divide* you divide the two numbers in front and *subtract* the exponent in the denominator from that in the numerator. These procedures are shown below.

$$M_{sun} \times M_{electron} = 1.989 \times 10^{33} \text{ g} \times 9.1094 \times 10^{-28} \text{ g}$$
$$= (1.989 \times 9.1094) \times 10^{33 + (-28)} \text{ g} \times \text{g}$$
$$= 18.12 \times 10^{5} \text{ g}^2$$
$$= 1.812 \times 10^{6} \text{ g}^2$$

$$\frac{M_{sun}}{M_{electron}} = \frac{1.989 \times 10^{33} \text{ g}}{9.1094 \times 10^{-28} \text{ g}} = \frac{1.989}{9.1094} \times 10^{33-(-28)} \frac{\text{g}}{\text{g}}$$
$$= 0.2183 \times 10^{61}$$
$$= 2.183 \times 10^{60}$$

EXERCISES

1. Write each of the following numbers in powers-of-ten notation:
   a. 2,378,000 000 m
   b. 0.003 24 ft

2. Write each of the following numbers in powers-of-ten notation:
   a. 89,760 in.
   b. 0.000 000 000 000 707 g

3. Write each of the following numbers as ordinary numbers:
   a. 5.782×10^{6} s
   b. 6.9×10^{-3} ft

4. Write each of the following numbers as ordinary numbers:
   a. 4.3×10^{3} g
   b. 8.12×10^{-5} m

5. Complete the following computations:
   a. $4.2 \times 10^{8} \times 2.2 \times 10^{4} =$
   b. $\dfrac{4.4 \times 10^{4}}{1.1 \times 10^{2}} =$

6. Complete the following computations:
   a. $6.8 \times 10^{-6} \times 2.3 \times 10^{4} =$
   b. $\dfrac{6.8 \times 10^{5}}{2.0 \times 10^{-3}} =$

The Nobel Prizes are awarded under the will of Alfred Nobel (1833–1896), the Swedish chemist and engineer who invented dynamite and other explosives. The annual distribution of prizes began on December 10, 1901, the anniversary of Nobel's death. No Nobel Prizes were awarded for 1916, 1931, 1934, 1940, 1941, or 1942.

1901 **Wilhelm Roentgen*** (Germany), for the discovery of X rays.

1902 **Hendrik Lorentz** and **Pieter Zeeman** (both of Netherlands), for investigation of the influence of magnetism on radiation.

1903 **Henri Becquerel*** (France), for the discovery of radioactivity; and **Pierre*** and **Marie Curie***† (France), for the study of nuclear radiation.

1904 **Lord Rayleigh** (Great Britain), for the discovery of argon.

1905 **Philipp Lenard** (Germany), for research on cathode rays.

1906 **Sir Joseph Thomson*** (Great Britain), for research on the electrical conductivity of gases.

1907 **Albert A. Michelson*** (U.S.), for spectroscopic and metrologic investigations.

1908 **Gabriel Lippmann** (France), for photographic reproduction of colors.

1909 **Guglielmo Marconi** (Italy) and **Karl Brun** (Germany), for the development of wireless telegraphy.

1910 **Johannes van der Waals*** (Netherlands), for research concerning the equation of the state of gases and liquids.

1911 **Wilhelm Wien** (Germany), for laws governing the radiation of heat.

1912 **Gustaf Dalén** (Sweden), for the invention of automatic regulators used in lighting lighthouses and light buoys.

1913 **H. Kamerlingh Onnes*** (Netherlands), for investigations into the properties of matter at low temperatures and the production of liquid helium.

1914 **Max von Laue** (Germany), for the discovery of the diffraction of X rays by crystals.

1915 **Sir William Bragg** and **Sir Lawrence Bragg** (both Great Britain), for analysis of crystal structure by X rays.

1916 No award.

1917 **Charles Barkla** (Great Britain), for the discovery of the characteristic X rays of elements.

1918 **Max Planck*** (Germany), for discoveries in connection with quantum theory.

1919 **Johannes Stark** (Germany), for the discovery of the Doppler effect in canal rays and the division of spectral lines by electric fields.

1920 **Charles Guillaume** (Switzerland), for the discovery of anomalies in nickel–steel alloys.

1921 **Albert Einstein*** (Germany), for explanation of the photoelectric effect.

1922 **Niels Bohr*** (Denmark), for investigation of atomic structure and radiation.

1923 **Robert A. Millikan*** (U.S.), for work on the elementary electric charge and the photoelectric effect.

1924 **Karl Siegbahn** (Sweden), for investigations in X-ray spectroscopy.

1925 **James Franck** and **Gustav Hertz** (both Germany), for the discovery of laws governing the impact of electrons on atoms.

1926 **Jean B. Perrin** (France), for work on the discontinuous structure of matter and the discovery of equilibrium in sedimentation.

1927 **Arthur H. Compton** (U.S.), for the discovery of the Compton effect; and **Charles Wilson** (Great Britain), for a method of making the paths of electrically charged particles visible by vapor condensation (cloud chamber).

1928 **Sir Owen Richardson** (Great Britain), for discovery of the Richardson law of thermionic emission.

*These winners are mentioned in the text.

†Marie Curie also received a Nobel Prize in chemistry in 1911.

1929 **Prince Louis deBroglie*** (France), for the discovery of the wave nature of electrons.

1930 **Sir Chandrasekhara Raman** (India), for work on light diffusion and discovery of the Raman effect.

1931 No award.

1932 **Werner Heisenberg*** (Germany), for the development of quantum mechanics.

1933 **Erwin Schrödinger*** (Austria) and **Paul A. M. Dirac*** (Great Britain), for the discovery of new forms of atomic theory.

1934 No award.

1935 **James Chadwick*** (Great Britain), for the discovery of the neutron.

1936 **Victor Hess** (Austria), for the discovery of cosmic radiation; and **Carl D. Anderson** (U.S.), for the discovery of the positron.

1937 **Clinton J. Davisson*** (U.S.) and **George P. Thomson** (Great Britain), for the discovery of the diffraction of electrons by crystals.

1938 **Enrico Fermi*** (Italy), for identification of new radioactive elements and the discovery of nuclear reactions effected by neutrons.

1939 **Ernest Lawrence** (U.S.), for development of the cyclotron.

1940–1942 No awards.

1943 **Otto Stern** (U.S.), for the discovery of the magnetic moment of the proton.

1944 **Isador I. Rabi** (U.S.), for work on nuclear magnetic resonance.

1945 **Wolfgang Pauli*** (Austria), for discovery of the Pauli exclusion principle.

1946 **Percy Bridgman** (U.S.), for studies and inventions in high-pressure physics.

1947 **Sir Edward Appleton** (Great Britain), for discovery of the Appleton layer in the ionosphere.

1948 **Patrick Blackett** (Great Britain), for discoveries in nuclear physics and cosmic radiation using an improved Wilson cloud chamber.

1949 **Hideki Yukawa*** (Japan), for prediction of the existence of mesons.

1950 **Cecil Powell** (Great Britain), for the photographic method of studying nuclear process and discoveries about mesons.

1951 **Sir John Cockroft** (England) and **Ernest Walton** (Ireland), for work on the transmutation of atomic nuclei.

1952 **Edward Purcell** and **Felix Bloch** (both U.S.), for the discovery of nuclear magnetic resonance in solids.

1953 **Frits Zernike** (Netherlands), for the development of the phase contrast microscope.

1954 **Max Born** (Great Britain), for work in quantum mechanics; and **Walter Bothe** (Germany), for work in cosmic radiation.

1955 **Polykarp Kusch** (U.S.), for measurement of the magnetic moment of the electron; and **Willis E. Lamb, Jr.** (U.S.), for discoveries concerning the hydrogen spectrum.

1956 **William Shockley, Walter Brattain,** and **John Bardeen** (all U.S.), for development of the transistor.

1957 **Tsung-Dao Lee** and **Chen Ning Yang** (both China), for discovering violations of the principle of parity.

1958 **Pavel Cerenkov, Ilya Frank,** and **Igor Tamm** (all U.S.S.R.), for discovery and interpretation of the Cerenkov effect (emission of light waves by electrically charged particles moving faster than light in a medium).

1959 **Emilio Segrè** and **Owen Chamberlain** (both U.S.), for confirmation of the existence of the antiproton.

1960 **Donald Glaser** (U.S.), for development of the bubble chamber for the study of subatomic particles.

1961 **Robert Hofstadter** (U.S.), for determination of the size and shape of the atomic nucleus; and **Rudolf Mössbauer** (Germany), for the discovery of the Mössbauer effect of gamma ray absorption.

1962 **Lev D. Landau** (U.S.S.R.), for theories about condensed matter (superfluidity in liquid helium).

1963 **Eugene Wigner, Maria Goeppert Mayer** (both U.S.), and **J. Hans D. Jensen** (Germany), for research on the structure of the atomic nucleus.

1964 **Charles Townes** (U.S.), **Nikolai Basov,** and **Alexandr Prokhorov** (both U.S.S.R.), for work in quantum electronics leading to the construction of instruments based on maser-laser principles.

1965 **Richard Feynman,*** **Julian Schwinger*** (both U.S.), and **Shinichiro Tomonaga*** (Japan), for research in quantum electrodynamics.

1966 **Alfred Kastler** (France), for work on atomic energy levels.

1967 **Hans Bethe** (U.S.), for work on the energy production of stars.

1968 **Luis Alvarez** (U.S.), for the study of subatomic particles.

1969 **Murray Gell-Mann*** (U.S.), for the study of subatomic particles.

1970 **Hannes Alfvén** (Sweden), for theories in plasma physics; and **Louis Néel** (France), for discoveries in antiferromagnetism and ferrimagnetism.

1971 **Dennis Gabor*** (Great Britain), for the invention of the hologram.

1972 **John Bardeen,*** **Leon Cooper,*** and **John Schrieffer*** (all U.S.), for the theory of superconductivity.

1973 **Ivar Giaever** (U.S.), **Leo Esaki** (Japan), and **Brian Josephson** (Great Britain), for theories and advances in the field of electronics.

1974 **Anthony Hewish** (Great Britain), for the discovery of pulsars; **Martin Ryle** (Great Britain), for radio-telescope probes of outer space.

1975 **James Rainwater** (U.S.), **Ben Mottelson,** and **Aage Bohr** (both Denmark), for showing that the atomic nucleus is asymmetrical.

1976 **Burton Richter** and **Samuel Ting** (both U.S.), for discovery of the subatomic J/ψ particles.

1977 **Philip Anderson, John Van Vleck** (both U.S.), and **Nevill Mott** (Great Britain), for work underlying computer memories and electronic devices.

1978 **Arno Penzias*** and **Robert Wilson*** (both U.S.), for work in cosmic microwave radiation; **Pyotr Kapitsa** (U.S.S.R.), for research in low temperature physics.

1979 **Steven Weinberg,* Sheldon Glashow*** (both U.S.), and **Abdus Salam*** (Pakistan), for developing the theory that the electromagnetic force and the weak nuclear force are facets of the same phenomenon.

1980 **James Cronin** and **Val Fitch** (both U.S.), for work concerning the asymmetry of subatomic particles.

1981 **Nicolaas Bloembergen, Arthur Schawlow** (both U.S.), and **Kai Siegbahn** (Sweden), for developing laser technology to study the form of complex forms of matter.

1982 **Kenneth Wilson** (U.S.), for the study of phase changes in matter.

1983 **Subrahmanyan Chandrasekhar** and **William Fowler** (both U.S.), for research on the processes involved in the evolution of stars.

1984 **Carlo Rubbia** (Italy) and **Simon van der Meer** (Netherlands), for work in the discovery of three subatomic particles in the development of a unified force theory.

1985 **Klaus von Klitzing** (Germany), for developing an exact way to measure electrical conductivity.

1986 **Ernst Ruska, Gerd Binnig** (both Germany), and **Heinrich Rohrer** (Switzerland), for work on microscopes.

1987 **K. Alex Muller** (Switzerland) and **J. Georg Bednorz** (Germany), for development of a "high temperature" superconducting material.

1988 **Leon Lederman, Melvin Schwartz,** and **Jack Steinberger** (all U.S.), for the development of a new tool for studying the weak nuclear force.

1989 **Norman Ramsay** (U.S.), for various techniques in atomic physics; and **Hans Dehmelt** (U.S.) and **Wolfgang Paul** (Germany), for the development of techniques for trapping single charge particles.

1990 **Jerome Friedman, Henry Kendall** (both U.S.), and **Richard Taylor** (Canada), for experiments important to the development of the quark model.

1991 **Pierre de Gennes** (France), for discovering methods for studying order phenomena in complex forms of matter.

1992 **Georges Charpak** (France), for his invention and development of particle detectors.

1993 **Russell Hulse*** and **Joseph Taylor*** (both U.S.), for discovering evidence of gravity waves.

1994 **Bertram N. Brockhouse** (Canada) and **Clifford G. Shull** (U.S.), for pioneering contributions to the development of neutron scattering techniques for studies of condensed matter.

1995 **Martin L. Perl** and **Frederick Reines*** (both U.S.), for pioneering experimental contributions to lepton physics.

1996 **David M. Lee, Douglas D. Osheroff,** and **Robert C. Richardson** (all U.S.), for their discovery of superfluidity in helium-3.

ANSWERS TO ODD-NUMBERED QUESTIONS AND EXERCISES

CHAPTER 1

Questions

1. The upper puck speeds up and then slows down. The lower puck travels at a constant speed.
3. At the right end
5. ●●● ● ● ● ● ● ● ● ● ● ● ● ● ● ● ● ●●●
7. **a.** ● ● ● ● ● ● ● ● ●
   b. ● ● ● ● ● ● ●
   c. ● ● ● ● ● ● ● ● ●
9. 4 meters
11. The average speeds are the same.
13. A world-class sprinter can run 100 meters in 10 seconds or 10 meters per second; world-class long-distance runners about half as fast.
15. Determine the time it takes to travel between mile markers.
17. It requires both to obtain a speed.
19. No, the average speed doesn't tell us any instantaneous speeds.
21. Time; length; length/time
23. Direction
25. It is opposite the velocity.
27. Anything that changes the speed is an accelerator.
29. The lower one, because the separations increase by a constant amount.
31. The average accelerations are the same.
33. Runner
35. They are falling in a vacuum.
37. It stays the same.
39. Galileo concluded that the object falls with a constant acceleration when air resistance is ignored. Aristotle hypothesized that the object quickly reaches a constant speed.
41. The heavier ball hits first.
43. They hit at the same time.
45. Constant acceleration
47. Motion along the ramp is a slower version of vertical motion and air resistance can be ignored.
49. The accelerations are the same.
51. The increasing upward force due to the air resistance causes the downward acceleration to continually decrease.

Exercises

1. 109 yards
3. 8.73 mph
5. 5.96 mph
7. 840 miles
9. 5,040,000 miles
11. 4 s, compared with 10 seconds for humans
13. 3.33 hours
15. 2 m/s^2
17. $6 \text{ km/h} \cdot \text{s}$
19. 17 m/s; 27 m/s
21. 45 m
23.

time (s)	height (m)	velocity (m/s)
0	80	0
1	75	10
2	60	20
3	35	30
4	0	40

25. 8 s; 80 m

CHAPTER 2

Questions

1. No, Galileo gave a logical argument using a thought experiment.

3. The surface exerts a frictional force on the book.

5. All the forces add to zero.

7. Your explanation should include the two-ramp thought experiment.

9. Because of its inertia

11. Because of its inertia

13. The inertia of the anvil keeps it from moving.

17. The larger the inertia, the larger the force required to produce a given *change* in velocity.

19. 13 newtons; 3 newtons

21. Motion with a constant acceleration

23. West; still west

25. The ball *accelerates* in the direction of the larger force.

27. It triples.

29. 2 (meters per second) per second

31. Weight is a force whereas mass is a measure of the quantity of matter. They are proportional to each other.

33. Newtons; newtons; kilograms

35. The weight doubles.

37. Aristotle would say that heavier objects naturally fall faster; Galileo would attribute the difference to air resistance.

39. Zero

41. Before the parachute is deployed because of the smaller air resistance

43. Newton's first law is always valid. With friction there must be other forces opposing the friction so that the net force is zero.

45. The force of friction cancels the push. Because there is no acceleration, the net force must be zero.

47. According to Newton's first law the net force must be zero. The 300-newton force is canceled by some other force, such as friction.

51. 500,000 newtons

53. Zero

55. According to Newton's third law there is a reaction force on the cannon.

57. The frictional force of the floor on your feet

59. Ceiling on string & string on ceiling; string on ball & ball on string; ball on Earth & Earth on ball. Each pair is equal by Newton's third law. Ceiling on string & ball on string are the same by Newton's second law. Likewise for string on ball & Earth on ball. Therefore, all are equal.

61. The forces act on different objects. The frictional force of the ground on the horse's hooves allows the horse to move the cart.

Exercises

1. 14 N; 4 N; 10.3 N

3. 150 N backward

5. 3 m/s^2

7. 0.08 m/s^2

9. 100 N

11. 3600 N

13. 80 kg

15. The force of gravity acting 20 N downward

17. 12 m/s^2

19. 300 N

21. 20 N

CHAPTER 3

Questions

1. See Figure A3-1. In the direction of its velocity because there would be no net force

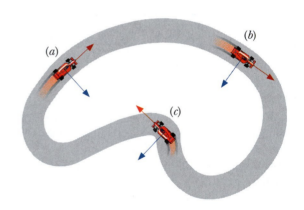

3. Gravity provides the required centripetal force.

5. The velocity is tangent to the path; all the others point radially inward.

7. They point in opposite directions; they have the same length.

9. The speed changes.

11. The vine must exert extra force to make Tarzan move in a circle.

13. The frictional force of the ground on the tires acting toward the inside of the curve

15. Because the Earth's velocity is nearly perpendicular to the direction between the Earth and the Sun, the Earth's speed is nearly constant.

17. See Figure A3-2.

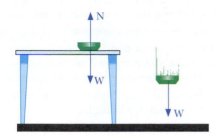

19. The ball's velocity is horizontal toward home plate; the net force and acceleration are vertically downward.

21. The same time

23. They hit at the same time.

25. If the two sides of the canyon are at the same height, he fails because he falls during the time required to cross the canyon.

27. Kicks the ball at a steeper angle

29. The center of mass

31. In rotational acceleration the angle is divided by the time interval twice; in rotational speed only once.

33. F_2, because it acts through a larger moment arm

35. The force on the nail is larger because it acts through a smaller radius than that of the force applied to the handle.

37. The force of the chain produces a larger torque when the gear has a larger radius.

39. A net torque

41. Increases; decreases

43. To smooth the motion

45. Hoop

49. Yes

51. Below the foot of the figure

53. One diagram should show that the force acting at the center of mass lies inside the base for the upside-down cone even when it is tipped.

55. Moving the center of mass toward one face makes the die more stable when this face is down.

57. The wall prevents you from rocking forward to put your center of mass over your toes.

59. Near the top edge because it would be in stable equilibrium

Exercises

1. a. 3 m/s west; **b.** 3 m/s east; **c.** 9 m/s east

3. 5 m/s at 37° north of west

5. a. 10 m/s²; **b.** 900 N

7. 2.74×10^{-3} m/s² 2.03×10^{20} N

9. 14 m/s horizontal and 4 m/s vertical

11. 0.24 m

13. 1 s; 40 m

15. 300 rev/min

17. -10 rev/min²

19. 210 N · m

21. 4 m from the pivot point

23. 100 g

CHAPTER 4

Questions

1. None

3. Ordinary projectile motion

5. 50,000 newtons

7. The same acceleration as the Moon

9. The surface area quadruples

11. d

13. Increase; decreases

15. To infinity

17. By measuring the attraction between two masses in a laboratory

19. Because it is also moving sideways fast enough to match the Earth's curvature

21. No, it is still *mg*.

23. No

25. The force is too small.

27. To determine its mass

29. Yes

31. The form *mg* gives a very good approximation to the force and is easier to use.

33. Send a satellite to orbit Venus.

35. New York City because it is closer to the center of the Earth

37. The radius of the Moon's orbit would slowly increase.

39. Because they are orbiting the Earth

41. The star appears to move in the night sky.

43. Because the mass of the Earth is so much larger than that of the Moon, it has a much smaller orbit about their common center of mass.

45. Yes

47. 4 hours 55 minutes due to the Sun

49. b and d

51. b and d

53. The extra tidal force is minuscule.

Exercises

1. 9.8 m/s^2
3. $g/4 = 2.5 \text{ m/s}^2$
5. 400 N
7. 2000 N
9. It is one-fourth as large in the spacecraft.
11. 40 N
13. 1.68×10^{-7} N compared with 50 N
15. 200 N
17. 25

CHAPTER 5

Questions

1. Its numerical value does not change.
3. $m\mathbf{v}$
5. They have large inertia.
7. The net force is equal to the change in momentum divided by the time required to make the change. Or the change in momentum is equal to the impulse.
9. They lengthen the time for the body to stop.
11. Same
13. The impulses are equal.
15. More mass means the fist has more momentum. The fist is more rigid, decreasing the interaction time.
17. At each interaction there are equal and opposite momentum changes. The total momentum is conserved at all times.
19. 4 newtons acting for 4 seconds.
21. Conservation of linear momentum requires the bullet to go one way and the rifle the other.
23. If the air molecules leave the board moving to the side, the momentum transferred by the fan to the air molecules is canceled by that due to the air molecules colliding with the board.
25. When there is no net outside force
27. 200 kg · m/s
29. Zero
31. They give the rowboat momentum away from the dock.
33. It is conserved.
35. Not without outside help unless the astronaut has something to throw
37. To counteract the torque of the main rotor acting on the helicopter. This torque is a reaction to the torque applied to the main rotor.
39. Conservation of angular momentum requires the sprinkler arm to rotate.
41. Tuck
43. The lower rotational inertia requires a larger rotational speed to conserve angular momentum.
45. From the somersaulting motion
47. No, one part of its body has angular momentum in one direction while the rest of its body has an equal angular momentum in the opposite direction.
49. The grooves give the bullet angular momentum so that it doesn't tumble in flight.
51. The speed increases.
53. Horizontally
55. South Pole
57. The torque caused by its weight acting at its center of mass
59. There must be torques acting on the rattleback.

Exercises

1. 27,000 kg · m/s
3. Defensive tackle has 900 kg · m/s; quarterback has 800 kg · m/s.
5. 6000 N
7. 36,000 kg · m/s (= 36,000 N · s)
9. Impulse = 42,000 N · s; $\overline{F} = 6000$ N
11. 203 N
13. 2 m/s
15. 2 kg · m/s to the left; it is the same.
17. 16 kg · m/s (right)
19. 20.8 m/s (north)
21. 180 kg · m^2/s
23. 57.8 km/s

CHAPTER 6

Questions

1. When the system is isolated
3. It is conserved.
5. $\frac{1}{2} mv^2$
7. Motorboat pulling a water skier
9. White car
11. **a.** Object A has four times the kinetic energy; **b.** They have the same kinetic energies.
13. They will rebound with the same speeds.
15. Work is the force in the direction of motion times the distance moved. In everyday language we use the word much more loosely, sometimes using *work* to describe efforts that have no force or no movement.
17. Decrease
19. Both will stop in the same distance.

21. The speed will change.
23. No, since we don't know that the two forces act through the same distances.
27. 3 newtons acting through 3 meters
29. All of them
31. Most of it changes to thermal energy due to the friction with the pole.
33. The gain in the kinetic energy equals the loss in the gravitational potential energy, keeping the mechanical energy the same.
35. None
37. At either end
39. It will decrease.
41. Frictional forces did 10 joules of work on the block.
43. Changes in its gravitational potential energy are compensated by changes in its kinetic energy so that its mechanical energy is conserved.
49. The work done in plowing through the dirt or sand reduces the kinetic energy of the truck.
51. The kinetic energy of the car is converted to thermal energy in the brakes.
53. We cannot recover the energy lost due to frictional effects.
55. Power is the energy transformed divided by the time required.
57. Kilowatt-hour

Exercises

1. 300,000 J
3. Momentum is conserved, but kinetic energy is not.
5. 10 J
7. 0.36 J
9. 6 J
11. Zero
13. 11.6 J
15. 1800 J
17. 80 J
19. 96,000 J
21. 1.82 horsepower
23. 120 Wh = 0.12 kWh

CHAPTER 7

Questions

1. It agrees with existing data and makes predictions that can be tested.
3. They believed it was foolish to expect that something could be divided an infinite number of times.

5. Water
7. It is a compound, because the mercury combines with something in the air.
9. For a given compound, the elements must combine in a fixed mass ratio.
11. Compounds are new substances with their own properties that result from substances combining according to the law of definite proportions.
13. 65 amu
15. The gold atom has about 197 times the mass.
17. They have the same number of molecules.
19. 16 grams
21. A large number of tiny particles separated by large distances, no internal structure, indestructible, interact elastically only when they collide.
23. The smoke particle moves when more atoms happen to hit it on one side.
25. mass/(length × time × time)
27. The pressure results from the force of the molecules rebounding from the walls.
29. The pressure times the area of the tire in contact with the ground must equal the weight of the car supported by the tire.
31. The perfume molecules travel very crooked paths.
33. To amplify the rise in the narrow tube
35. Body temperature varies among people and with time for a given person.
37. 39.0°C
39. 21°C
41. 273 K
43. Average kinetic energy of the molecules
45. The average kinetic energies are equal because they are at the same temperature.
47. The particles have more kinetic energy. Therefore, they are moving faster and strike the walls more frequently with more momentum.
49. Volume
51. Temperature drops to one-third on the Kelvin scale.
53. The pressure increases as the temperature increases.
55. The more energetic molecules leave via evaporation, lowering the average kinetic energy and, therefore, the temperature of the remaining water.
57. From the cooling due to evaporation

Exercises

1. $(10 \times 1 \text{ g})/(10 \times 2 \text{ g}) = 1/2 = (1 \text{ g})/(2 \text{ g})$
3. 3 g
5. 18 g
7. 5.02×10^{22} atoms

9. 3.35×10^{25} molecules

11. 1 nitrogen and 3 hydrogen

13. 14 amu

15. $\approx 10^{24}$

17. 2500 kPa $\approx$ 24 atm

19. 1/3 L

21. 160 balloons

23. 1.6 atm

CHAPTER 8

Questions

1. Solid, liquid, gas, and plasma

3. They have the same densities.

5. Silver

7. The water expands upon freezing.

9. Crystals have regular geometric structure.

11. They are not the same since the crystals have different shapes.

13. Diamond has strong bonds in all directions; graphite has stronger bonds between atoms in two-dimensional layers.

15. The interatomic forces are stronger.

17. Measure its temperature as it cools and notice the lack of a plateau as it hardens.

19. Alloys are a mixture of metals.

21. Solid oxygen has a lower melting temperature.

23. The forces between water molecules are stronger than the forces between water molecules and wax molecules.

25. The surface tension allows the surface to rise without overflowing.

27. Gas particles are neutral, whereas the particles in a plasma are electrons and charged ions.

29. The larger pressures exerted by the heels can dent floors.

31. Because the pressure of the gas inside the balloon balances the atmospheric pressure

33. The lower atmospheric pressure means that it is easier for molecules to escape.

35. Barometric pressure in weather reports is corrected to sea level.

37. Even though the pressure is large, the area of the fingertip is small, and the required force is not large.

39. It is the atmospheric pressure that determines the maximum height.

41. Yes, because the air pressure decreases with height

43. The loaded freighter must displace more water before the buoyant force is equal to its weight.

45. The density of gasoline is less than water.

47. Your average density increases with less air in your lungs.

49. Higher in saltwater

51. It stays the same, because the volume below the surface is the same as the volume of the ice when melted.

53. To take advantage of the Bernoulli effect

55. So the balls will drop over the net

57. The water flowing by the curtain causes the air to move, reducing the air pressure.

59. The wind blowing by the base reduces the outside pressure.

Exercises

1. 3.75 g/cm^3

3. 600 kg

5. 7.62 cm^3

7. 1 cm^3

9. $340'' = 28'\ 4''$

11. 19 cm = 7.5 in.

13. 15 lb/in.^{2}

15. $0.9 = 90\%$

17. 2 cm

19. 68,600 N

CHAPTER 9

Questions

1. The kinetic energy is converted to thermal energy (and energy of distortion).

3. To show that heat was not a material substance

5. To determine the equivalence between mechanical energy and heat

7. If the kinetic energy was not taken into account, the number of joules per calorie would be larger.

9. Two objects in thermal equilibrium have the same temperature.

11. Yes, if they have the same temperature

13. The unit of heat is the joule and the unit of temperature is the kelvin. The units do not represent the same physical quantity.

15. Heat is a flow of thermal energy, whereas temperature is a measure of the average kinetic energy of the atoms and molecules. Notice that temperature is not an energy.

17. The gallon of water has much more internal energy.

19. Under all conditions

23. The heat added divided by the product of the mass and the resulting temperature change

25. The water because it has a larger specific heat. However, if the temperature of the ice is initially higher than −6°C, it would take more energy to melt the ice before the temperature of the resulting water could be raised.

27. It takes a tremendous amount of thermal energy to change the temperature of the water in the oceans.

29. It will freeze if heat is removed, melt if heat is added.

31. It requires a lot of thermal energy to melt the ice without changing its temperature. Also, ice is not a very good conductor of thermal energy.

33. The water would vaporize instantly when it reached the boiling temperature.

35. The ice in the steak sublimes.

37. A lot of thermal energy is lost from the head.

39. The average temperature of the slab

41. Stationary air

43. Wood is a poor thermal conductor.

45. Convection currents distribute the thermal energy throughout the water.

47. The ground beneath the road keeps the road warm for a while.

49. They cool by emitting radiation.

51. The vacuum reduces conduction and convection, the silvering reduces radiation.

53. The nonuniform contraction of the glass can cause large stresses.

55. Cross-sectional area

57. The spacing would be smaller on the wide portion.

Exercises

1. 4000 cal

3. 1200 J; 286 cal

5. 42 min

7. 9 J increase

9. 70 J

11. **a.** 1600 cal; **b.** 1600 cal; **c.** It stays the same.

13. $0.25 \text{ cal/g} \cdot \text{°C}$

15. $460 \text{ J/kg} \cdot \text{°C}$

17. 124 cal

19. 52.3 kJ

21. 334 kJ = 80 kcal

23. 0.8 mm

CHAPTER 10

Questions

1. It converts thermal energy to mechanical energy.

3. This would violate the second law of thermodynamics

because heat will not naturally flow between reservoirs at the same temperature.

7. Yes

9. Different regions of the water must have different temperatures.

11. You cannot get more energy out than you put in.

13. First: You cannot get more energy out than you put in. Second: You cannot convert all of the thermal energy to mechanical work.

15. About 20% of the thermal energy in the food shows up as mechanical work.

17. The efficiency increases.

19. The efficiency increases with increased operating temperature and decreases with increased temperature of the surroundings.

21. Because an efficiency of one means that all of the input heat is converted to mechanical work

23. No, because the thermal energy exhausted into the room is more than that removed from the room

27. Yes

29. No

31. The sums 3 and 18 have the highest order since they can each only occur one way.

33. There are more ways of getting this sum.

37. The entropy increases.

39. Increased

41. The entropy decreases. It is consistent because the entropy of the surroundings increases by a larger amount.

43. **a.** The mechanical energy is converted to thermal energy. **b.** The temperature rises. **c.** The entropy increases.

45. If the coin is at room temperature, it will not absorb any net energy from its surroundings. Even if it did, the chances of converting the internal energy into macroscopic kinetic energy are minuscule.

47. No; yes

49. b

51. Because it does not involve a heat engine (at the power plant)

53. Using electricity because it is less efficient

Exercises

1. 600 kJ

3. 3000 cal

5. 0.2 = 20%

7. 3000 cal; 0.375 = 37.5%

9. 0.0274 = 2.74%

11. 500 K = 227°C

13. 100 kW

15. 500 J

17. 200 J

19. HHHH, HHHT, HHTH, HTHH, THHH, HHTT, HTTH, TTHH, HTHT, THTH, THHT, HTTT, THTT, TTHT, TTTH, TTTT

21. $^6/_{36} = ^1/_6 = 17\%$

23. $^8/_{216} = ^1/_{36} = 2.8\%$

CHAPTER 11

Questions

1. Alice would not see any motion of the jar relative to her. Someone sitting on a shelf would see the jar fall freely.

3. A reference system in which Newton's first law is obeyed

5. It falls freely with only a vertical motion.

7. Both observers agree that the ball lands on the white spot.

9. There are no horizontal forces.

11. The normal acceleration due to gravity.

13. They are valid.

15. Forward because of its inertia

17. Both observers agree that the ball will land behind the white spot.

19. 50 km/h

21. There is a constant horizontal force toward the rear of the train.

23. Greater since the inertial force acts in the same direction as the gravitational force

25. Less since the inertial force acts in the opposite direction as the gravitational force

27. Same

29. An inertial force acts in the backward direction.

31. The same because you are comparing masses. The weights increase by the same amount during the acceleration.

33. In the car your date slides over to join you. From the roadway you (and the car) move over to join your date, who is traveling in a straight line.

35. **a.** up; **b.** up; **c.** down

37. Up is not defined.

39. **a.** center; **b.** behind center; **c.** in front of center; **d.** to the outside

41. It banks so that the effective gravity remains normal to the floor.

43. Your figure should have the plants leaning toward the center and then curving upward.

45. The centrifugal force pulls the mud off.

47. The centrifugal force is an inertial force, whereas the centripetal force is due to one of the fundamental forces in nature.

49. Toward the axis of rotation

51. In a straight line perpendicular to the line from the ball to the axis of the cylinder

53. In the geocentric model, the stars revolve around the Earth. In the heliocentric model, the Earth rotates on its axis.

55. The Foucault pendulum, effects due to the Coriolis force, the bulging of Earth at the Equator, etc.

57. No, there is no Coriolis force.

59. Smaller at the Equator due to the rotation of the Earth

Exercises

1. 45 m/s; 15 m/s

3. **a.** 70 mph; **b.** 30 mph backward

5. **a.** 16 m/s^2 downward; **b.** 10 m/s^2 downward

7. 1800 N

9. 450 N

11. 225 N

13. 270 N

15. 10 m/s^2

CHAPTER 12

Questions

1. Because the physical laws are the same in all inertial systems, it is impossible to determine your speed.

3. Only in inertial reference frames

5. On the white spot

7. All observers measure the speed to be c in a vacuum.

9. All observers measure the speed to be c in a vacuum.

11. Yes, this does not pose a problem.

13. No, observers agree on events at a single location.

15. The observers would agree because the clocks move toward or away from the light by the same amount.

17. The firecracker in the caboose exploded first.

19. The one at the back of the skateboard. The third person must have a speed to the right that is larger than that of the skateboard.

21. No, unless the events occur at the same place.

23. It will take 9 months according to clocks in the ship, but greater than 9 months according to clocks on Earth.

25. As a person approaches the speed of light, external observers will see the person's clocks run slower and slower.

27. Yes, if the person is traveling fast enough to dilate the time enough

29. Longer

31. Greater

33. Peter will be the younger.

35. It is not possible to travel backward in time.

37. It is not possible to travel backward in time. This would wreak havoc in our beliefs about cause and effect.

39. Shorter

41. Longer

43. Same

45. The effects are extremely small.

47. Lengths and time intervals change by the same factor. This factor cancels when the speed is computed.

49. **a.** The caboose entered first; **b.** The tunnel is longer; **c.** yes

51. If the force remains the same, the acceleration decreases.

53. Objects that have mass cannot be accelerated to the speed of light. It would take an exceedingly large amount of energy to accelerate it *close* to the speed of light.

55. No, just include mass-energy.

57. Yes

59. When the speeds involved are a sizeable fraction of the speed of light, we need to use the relativistic expressions.

61. No, they can be arbitrarily large.

Exercises

1. 149 million km

3. 373 s = 6.22 min

5. 1.091

7. 2.53 millionths of a second; 15.6 millionths of a second

9. 7 min

11. 14.2 years

13. 89.9 m

15. 13.1 trillion miles

17. 67.4 N

19. $0.866c$

CHAPTER 13

Questions

1. Both theories include inertial reference systems, but the general theory also includes noninertial reference systems.

3. Yes

5. The inertial mass determines the acceleration of the object under a given force, whereas the gravitational mass determines the strength of its gravitational attraction for other masses.

7. The team on Earth thinks it has the larger acceleration. Neither can tell anything about their velocities.

9. By rotating the space station

13. You cannot distinguish the two.

15. The bending is less than the diameter of an atom.

17. The positions of those originally seen would not change since changing the size of the Sun does not affect the strength of the gravitational field outside the original radius. However, stars more in line with the Sun could be seen.

19. On the ground floor where the gravitational field is stronger (although you would not be able to detect the difference)

21. Clocks run slower in stronger gravitational fields.

23. They would lose time.

25. The light would experience a gravitational redshift relative to a stationary clock.

27. Greater

29. There is an attractive force between them. Parallel lines converge in this curved space.

31. Yes

33. Yes

35. Supernovas that form black holes, rapidly rotating pulsars, close binary stars, and colliding galaxies

39. The gravitational field of the galaxy bends the light of a distant object to form images.

Exercises

1. 8 m/s^2 backward

3. 5.36×10^{11} s = 170,000 y; the same

5. 5 m

7. $3 \times 180° = 540°$. It is almost a great circle, but it has slight bends in three places; for instance, at one pole and both crossings of the Equator.

CHAPTER 14

Questions

1. Its inertia carries it through the equilibrium point.

3. Downward

5. A larger mass or a smaller spring constant causes the period to increase.

7. To a very good approximation the period stays the same.

9. $1/(60 \text{ s}) = 0.017$ hertz

11. 60 s

13. Move the mass down to lengthen the period because longer pendulums have longer periods.

15. To avoid matching a resonant frequency of the bridge

17. 2

19. Energy

21. It must be a longitudinal wave because the medium is a fluid.

23. It is not possible because the two pulses have (essentially) the same speed.

25. Tension and mass per unit length

27. It returns as a trough from a free end, but as a crest from a fixed end.

29. Crest

31. See Figure A14-1.

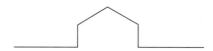

33. Amplitude

35. The wavelength increases if the speed stays the same.

37. The wavelength increases.

39. The wavelength is cut in half.

41. The wavelength doubles.

43. One wavelength

45. Five

47. Three times

49. One-half the length

51. Four times the length of the rope

53. Twice the length of the rod

55. Antinode

57. Decreases

59. Increase

61. Decreases

Exercises

1. 3 s

3. 2 Hz

5. 0.2 s

7. 0.628 s

9. 6.28 s

11. See Figure A14-2.

13. See Figure A14-3.

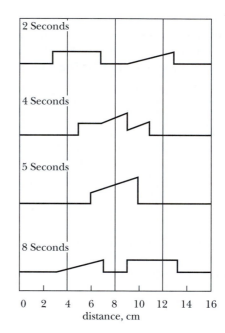

15. 100 cm/s

17. 10 m

19. 256 Hz

21. 4 m; $(4 \text{ m})/2 = 2$ m; $4/3$ m; $(4 \text{ m})/4 = 1$ m; $(4 \text{ m})/5 \ldots$

23. 2.5 Hz

CHAPTER 15

Questions

1. Sound waves travel through fluids such as air and water.

3. At the same speed

5. Water

7. The speed of light is much faster than the speed of sound.

9. Sound reflecting from surfaces

11. 10 dB

13. Loudness

15. Mainly frequency

17. They have different resonate frequencies.

19. Second

21. Increase tension, finger lower down on the fret board, or replace with a string with less mass per unit length.

23. The shorter wavelength will have the higher frequency.

25. Three

27. No

29. To produce lower notes

31. Nodes

33. Four times the length of the pipe

35. The wavelength doubles.

37. Antinode

39. The wavelength decreases.

41. Nothing

43. One

45. They produce harmonics with different intensities.

47. The interference of two waves with slightly different frequencies

49. Decrease

51. The frequency of the whistle decreases as the train passes by.

53. You hear a higher frequency.

55. The frequency is higher.

57. The speed of the boat is higher than the speed of the water waves.

59. Sound absorption and reflection

Exercises

1. 0.00382 s

3. 0.655 m

5. 343 Hz

7. 17.2 cm

9. 2.74 km or 1.6 miles

11. 300 m

13. 1000 times as intense

15. 784 Hz

17. 400 m/s

19. $f_1 = 71.5$ Hz, $f_3 = 214$ Hz, $f_5 = 357$ Hz, . . .

21. 17.4 cm

23. 438 Hz or 442 Hz

CHAPTER 16

Questions

1. Unless there is dust in the air, none of the light gets scattered into the eyes until the light hits the wall.

3. Because light is scattered into the shadow from the surroundings

5. Point source

7. The shadow increases in size.

9. There will be no umbra when the shadow is not long enough to reach the screen. This occurs when the Moon's shadow does not reach the Earth.

11. The image gets bigger and dimmer.

13. c

15. The image is the same size as you and also walks toward the mirror.

17. Midway between the heights of the top of your head and your eyes

19. 2 m

21. One; infinity

23. A mirror doesn't reverse right and left, it reverses front and back.

25. Seven

27. Concave

29. No

31. Very close to the mirror on the back side and slightly magnified; moves away and gets larger

33. Concave

35. The light actually converges to form a real image; there is no light at the location of a virtual image.

37. Only real images

39. The only change is that the image becomes dimmer.

41. The new image is formed at the original location of the object. This is a consequence of the reversibility of light rays.

43. We can say nothing for sure.

45. The speed of light is much faster than the speed of sound.

47. Cyan; the shade depends on how much of each color is reflected.

49. Red; black

51. Black

53. Red

55. A dark blue

57. With lights, two colors that add to white

59. Black because there is no scattered light

Exercises

1. 15-cm diameter
5. 4 m
7. Your image should be virtual, erect, and magnified.
9. The image is located 37 cm in front of the mirror and is half as large as the object.
11. The image is located 10 cm behind the surface of the mirror and is ⅔ as big.
13. 0.1 ms; it is ½₀₀₀ as big.
15. 500 s = 8.33 min
17. 9.46×10^{15} m

CHAPTER 17

Questions

1. c
3. Looking through goggles is like looking into an aquarium.
5. The light from objects behind the drop is changed.
7. Greater than 49°
9. The light rays hit the sides at angles greater than the critical angle and are totally internally reflected.
11. The astronauts do not look through the atmosphere.
13. Smaller (remember that the Moon is shortened in the vertical direction)
15. The Sun appears to rise before it actually does. Therefore, sunrise is earlier.
17. Assuming that there is no deflection of the spear as it enters the water, you would aim low, since the fish appears to be higher than it actually is.
19. The diamond has a higher index of refraction and therefore more total internal reflection.
21. The drawing should be similar to Figure (b) in the feature on "Mirages."
23. The colors are reversed.
25. The angle formed by the line from the Sun to the raindrop and the line from your eye to the raindrop must have a fixed value for a given color.
27. Midafternoon
29. Yes, but it would be very dim.
31. The refraction of light in ice crystals
33. 22° to either side of the Sun
35. Converging to create a real image on the screen
37. Through the principal focal point
39. The only change is that the image becomes dimmer.
41. The lens made from glass with the larger index of refraction will have the shorter focal length because the light is bent more at the surfaces.

43. You could argue that light is required to expose the film.
45. Image of background is formed in front of the film.
47. To regulate the amount of light entering the eye
49. Dispersion occurs when light is refracted, not reflected; the law of reflection is independent of wavelength.
51. Converging
53. Converging
55. A little less than the focal length
57. The length of the telescope is equal to the sum of the two focal lengths.
59. Consider the symmetry of the letters.

Exercises

1. 13°; 15°
3. From Figure 17-2, we get 42°.
5. 10°–11°
7. 48°
9. Your diagram should show rays diverging even more when they exit the glass. When the rays are extended backward, they form a virtual image that is closer than the object.
11. The image is virtual, located at the other focal point, and the magnification is 2.
13. 60 cm on the other side of the lens
15. Between f and $2f$ from the lens
17. The image is virtual, erect, and reduced in size.
19. +5

CHAPTER 18

Questions

1. Yes, because particles moving at the speed of light will not have noticeable deflections due to gravity.
3. If the surface had friction, the component of the momentum parallel to the surface would change due to the impulse of the frictional force. If the collision were inelastic, the normal component of the momentum would change. Both changes would affect the ratio of the components and hence the angle.
5. Newton hypothesized that light particles experience a force normal to the surface.
7. The direction of the waves would be reversed.
9. Frequency
11. Amplitude
13. The particle model predicts a faster speed in materials, the wave model predicts a slower speed.
15. Red light has a longer wavelength.

17. All colors emerge at the same angle.

19. No, it depends on the wavelength or frequency.

21. Decrease

23. Orange light

25. The lights do not maintain a fixed phase relationship.

27. Diffraction effects are more pronounced in sound because it has a much longer wavelength.

29. On the order of the wavelength of visible light

31. Decrease

33. Orange light because angles of diffraction are proportional to wavelength

35. Thin film interference

37. Either

39. Yes

41. Polarization

43. Rotate the sunglasses and look for variations in brightness.

45. The amount of transmitted light increases.

47. Blue light has the least diffraction.

49. The pupils are smaller during the day.

51. Objects at different depths in the hologram exhibit parallax; that is, they move relative to each other when the hologram is viewed from different angles.

53. Monochromatic light with a constant phase relationship

55. White lights do not have a constant phase relationship.

57. We are free to use the description that works best for a given situation.

Exercises

1. 1.76×10^8 m/s

3. 2.42

5. 2.1

7. 4.74×10^{14} Hz

9. 600 nm

11. 372 nm

13. 244 nm

15. 150 nm

17. 100 nm

CHAPTER 19

Questions

1. Insulator

3. Because the charge will flow throughout the rod

5. The moisture allows some of the charge to leave the balloon.

7. Because some clothes have acquired electric charges

9. Negative because it must be opposite the charge on the glass rod.

13. It has equal amounts of positive and negative charge.

15. Either an excess of positive charge or a deficiency of negative charge

17. It could be negative or neutral.

19. By inducing a charge in the wall

21. **a.** By attracting positive charges to the near side and/or repelling negative charges to the far side of the object

   b. By inducing a charge separation in atoms and molecules or by rotating polar molecules

23. The charges flow through the hand to ground.

25. Touch the rod to the electroscope.

27. The charges are the same and one-half of the original.

29. Touching a charged sphere to a neutral one yields two spheres with ½ charge on each one. Using two neutral spheres yields charges of ⅓, and so on.

31. The direction of the force is reversed.

33. The force increases by a factor of 4.

35. The force would double.

37. Directly toward the negative charge

39. The gravitational force is always attractive, the elementary electric charge is not proportional to inertial mass, electrical charge comes in one size, and gravity is much weaker.

41. The Sun and the planets have almost no net electric charge.

43. Electric

45. Because they have different masses

47. The metal body shields you.

49. It is equal to the force experienced by one unit of positive charge placed at that point.

51. It points toward the ball and decreases in strength as the square of the distance from the ball's center.

53. The forces are equal in size, but opposite in direction with the force on the proton being upward.

55. It is equal to the work required to bring that object from the location chosen as zero to the point.

57. It is equal to the work required to bring 1 coulomb of positive charge from the location chosen as zero to the point.

59. Lower electric potential energy and lower electric potential

Exercises

1. 1.47×10^{-17} C

3. 1.8×10^{10} N

5. 2.13×10^{-6} N

7. 1.23×10^{36}

9. 4.5×10^9 N/C

11. 5.13×10^{11} N/C away from the proton

13. 9×10^9 N/C toward the 2-C charge

15. 8×10^{-16} N; 4.79×10^{11} m/s^2

17. 72 J

19. 27 mJ

21. 39,000 V

CHAPTER 20

Questions

1. When it was touched by two different metals

3. The size determines the total amount of charge available.

5. The voltage remains the same.

7. Voltage, maximum current available, and total charge available

9. C

11. A

13. Turn the battery end for end.

15. Closing the switch allows the charges to flow.

17. Connect the battery and bulb to the wires so that you would have a complete circuit if the wires were connected. If the bulb lights, they are connected.

19. A   B   A   B   A   B   A   B
    | |
  C D   C—D   C—D   C—D
  E—F   E—F   E—F   E—F

21. A volt is a measure of potential difference, while an ampere is a measure of current.

23. Water, volume per unit time, and pressure

25. All of them

27. The current doubles.

29. The increased movement of the atoms with increased temperature causes more collisions with the electrons.

31. You could be electrocuted as the current flows through you to ground.

33. A path with very little resistance

35. Pure water is a poor conductor. There will be little current unless the voltage is high or the water contains impurities such as salt or soap.

37. It would not draw very much current and the filament would be very dim.

39. Batteries in series with the bulbs in parallel

41. Arrangements a, d, and e are equally bright for an ideal battery.

43. The bulbs are equally bright.

45. A is the brightest; B and C are equally dim.

47. A stays the same and B goes out.

49. They all go out.

51. B and C are wired in parallel and the combination is wired in series with A.

53. Parallel

55. Watt

57. The power doubles.

59. 120-W bulb

61. To reduce energy losses

Exercises

1. 6 V

3. 150 Ω

5. 10 A if the voltage is 120 V

7. 3.6 V

9. 3 A

11. 12.5 A

13. 960 W

15. 1200 W

17. 240 Ω

19. 120 Wh = 0.12 kWh

21. 36¢

CHAPTER 21

Questions

1. Electricity

3. Both ends of the unmagnetized rod will attract both ends of the other two rods.

5. 4

7. The value is determined by the torque on a compass placed at the point; the north pole of the compass points in the direction of the field.

9. The magnetic field from one wire cancels that from the other wire.

11. The photograph would not change.

13. Northward

15. The presence of the magnet lines up (at least partially) the atomic current loops in the iron, inducing magnets in the iron.

17. As the magnetite solidified it locked in the magnetism induced by the Earth's magnetic field.

19. Place it in a strong magnetic field, stroke it with a magnet, or hit it in a magnetic field.

21. Yes, because of the pounding it receives while in the Earth's magnetic field.

23. If two parallel wires each carry a current of 1 ampere, the force on each wire will be 2×10^{-7} newton per meter.

25. They attract each other.

27. Magnetic south pole since the north magnetic pole of the compass is attracted to it.

29. 15° W

31. The cosmic rays spiral along the magnetic field lines toward the South Pole.

33. No; yes

35. Along an east–west line

37. Inserting the south end into the coil or removing the north end

39. Rotate the magnet end for end.

41. Yes

43. Larger

45. When the plane of the loop is parallel to the field lines

47. To reverse the direction of the current in the loop, ensuring that the torque is in the same direction

49. If the rate of change of the electric field is constant, the magnetic field will be constant. Otherwise, the magnetic field will vary.

51. A current and a varying electric field

53. The electric and magnetic fields are perpendicular to the direction of travel and to each other. They oscillate in phase and travel at the speed of light.

55. Sound

57. At the speed of light

59. By modulating the amplitude of the carrier wave

61. The carrier wave has a frequency of 970 kilohertz.

Exercises

1. 3.53×10^5 G

3. 4×10^{-4} N; 0.2 g

5. 1.2×10^{-12} N

7. 60 turns

9. 1.28 s

11. 1.22×10^{-12} m

13. 2.83×10^{20} Hz; 3.22×10^{20} Hz

15. 275 m

17. 200–545 m

CHAPTER 22

Questions

1. The order was determined by their masses and chemical properties.

3. Beryllium, magnesium, strontium, barium, and radium

5. Look for the oxygen absorption lines in the Sun's spectrum.

7. Absorption

9. Calcium and lithium

11. There are many fewer lines in the absorption spectrum.

13. Their deflections by electric and magnetic fields

15. It only exists in multiples of some smallest amount.

17. They are equal.

19. The electric force

21. No

23. The alpha particles encounter a massive, compact, positively charged nucleus.

25. The intensity curves are the same for all materials.

27. From the graph we estimate 6000 K and above.

29. Red-orange for 4000 K; white for 6000 K; blue for 8000 K

31. Red, white, and blue have increasing temperatures.

33. That the atomic oscillators have quantized energies that are whole-number multiples of a lowest energy, $E = hf$

35. Frequency

37. The maximum kinetic energy does not depend on the intensity of the light.

39. Because each photon gives its energy to a single electron, the photon's energy must exceed that required for electrons to escape the atoms.

41. Radio, infrared, visible, ultraviolet, X ray

43. (1) Only angular momenta equal to whole-number multiples of a smallest angular momentum are allowed. (2) Electrons do not radiate when they are in allowed orbits. (3) A single photon is emitted or absorbed when an electron changes orbits.

45. They are equal.

47. The energy of the photons given off as electrons jump between the discrete orbits

49. Absorption lines only occur between the ground-state orbit and higher orbits.

51. They have the same basic structure for their outer electrons.

53. Because it has a complete outer shell, it should not form molecules.

55. It easily gives up its single outer electron and forms a bond with other atoms.

57. X ray

59. The X-ray photon has higher energy, shorter wavelength, and higher frequency.

61. Sometimes the inner orbit is not very deep and sometimes an electron from a nearby orbit fills the vacancy, thereby emitting a less energetic photon.

Exercises

1. 1.76×10^{11} C/kg
3. 2.7×10^3 kg/m^3; 2.7 times as much
5. about 500 m
7. 2.07×10^{-19} J
9. $h/2\pi = 1.06 \times 10^{-34}$ J $\cdot$ s
11. 8.48×10^{-10} m
13. 3.28×10^{15} Hz; above
15. 1:64
17. 20.7 keV

CHAPTER 23

Questions

1. *Successes:* Accounting for the stability of atoms, the numerical values for wavelengths of spectral lines in hydrogen and hydrogenlike atoms, and the general features of the periodic table. *Failures:* Could not account for why accelerating electrons didn't radiate, the spectral lines in non-hydrogenlike atoms, the splitting of spectral lines into two or more lines, the relative intensities of the spectral lines, details of the periodic table including the capacity of each shell, and relativity.

3. That the observed property can best be described in terms of wave or particle language

5. The wave nature only becomes important in interactions about the same size as the wavelengths, which are very small.

7. The electrons must form standing waves around the nucleus.

9. No, the wavelength has nothing to do with color.

11. The wavelength is too small.

13. Both; waves when producing the pattern and particles when detected

15. Electrons would be deflected by electric and magnetic fields, electrons have charge and mass, photons travel at the speed of light.

17. You would still get an interference pattern.

19. Add the displacements and then square the result.

21. Matter waves whose squares give the probability of finding electrons

23. In the middle

25. Because this is a particle-type question

27. The nucleus is surrounded by an electron cloud in which the probability of finding an electron at a given

place is determined by the square of a matter-wave amplitude. The electron has quantized values for its energy, its angular momentum, and the directions of its angular momentum and spin.

29. The fourth quantum number is for its spin.

31. No two electrons within the atom can have the same set of quantum numbers.

33. 10

35. $n = 1, \ell = 0, m_\ell = 0, m_s = +\frac{1}{2}$
    $n = 1, \ell = 0, m_\ell = 0, m_s = -\frac{1}{2}$
    $n = 2, \ell = 0, m_\ell = 0, m_s = +\frac{1}{2}$
    $n = 2, \ell = 0, m_\ell = 0, m_s = -\frac{1}{2}$

37. The Heisenberg uncertainty is an uncertainty in making measurements due to the fundamental character of nature.

39. Planck's constant is a very small number.

41. Electrons have wave properties like those of photons.

43. The energy of the excited state is not precisely determined. This will cause the spectral line to broaden.

45. It is very difficult to visualize.

49. To explain the behavior of subatomic particles that exhibited both wave and particle properties

51. The uncertainty in its momentum must be infinite. Therefore, its wavelength is completely uncertain.

53. The random nature of quantum mechanics makes free will possible from a scientific viewpoint.

55. The electrons return to their lower energy levels through a series of jumps, emitting visible light.

57. The electrons must remain in the excited energy levels for a relatively long time.

59. It is monochromatic, has a fixed phase relationship, and is well collimated.

Exercises

1. 1.47×10^{-38} m
3. 2.88×10^{-11} m
5. 7.28×10^6 m/s
7. 3.32×10^{-25} kg $\cdot$ m/s
9. 6.03×10^{-20} J $= 0.377$ eV
11. 3.32×10^{-34} kg $\cdot$ m/s
13. 1.33×10^{-23} kg $\cdot$ m/s
15. 2.21×10^{-36} m
17. 0.132 nm

CHAPTER 24

Questions

1. The radiation did not depend on external physical conditions.

3. The wavelengths of visible light are very much bigger than the sizes of nuclei.

5. Gamma rays are emitted by nuclei and typically have more energy, higher frequencies, and shorter wavelengths.

7. Yes

9. Neutrons and protons

11. Numbers of protons and electrons

13. **a.** lead; **b.** magnesium; **c.** silver

15. **a.** 8 neutrons, 7 protons, and 7 electrons; **b.** 64 neutrons, 50 protons, and 50 electrons; **c.** 138 neutrons, 88 protons, and 88 electrons

17. 40

19. Decreases by 1

21. Decreases by 4 amu

23. **a.** $^{232}_{90}$Th; **b.** $^{216}_{84}$Po

25. **a.** $^{19}_{9}$F; **b.** $^{188}_{75}$Re

27. **a.** $^{190}_{78}$Pt; **b.** $^{204}_{83}$Bi

29. **a.** $^{87}_{39}$Y; **b.** $^{19}_{9}$F

31. $^{211}_{82}$Pb or $^{215}_{85}$At

33. **a.** beta plus; **b.** alpha

35. ^{206}Pb

37. ^{14}N

39. 17

41. $^{239}_{94}$Pu

43. It could happen at any time.

45. By determining the fraction of the radioactive ^{14}C that has decayed to ^{12}C

47. You need an independent method for determining the age of a sample of material. For instance, you could look at the wood in each ring of a tree to determine the ratio of ^{14}C to ^{12}C.

49. No

51. Neutrons, because they are electrically neutral and do not lose as much energy to ionization

53. Alpha particles

55. Substance cannot be washed off and the skin provides some protection.

57. Nuclear power

59. External beams and ingestion

Exercises

1. 100 m

3. 1.99×10^{-26} kg

5. 8 μCi

7. 32 trillion

9. 17,100 years

11. 8.4 cm

CHAPTER 25

Questions

1. 10 million volts

3. They all acquire the same kinetic energy.

5. Proton

7. Reduction of energy losses due to the radiation of the accelerating charges

9. Strong, electromagnetic, weak, and gravitational

11. There must be a strong force to counteract the electric repulsive force and hold the nucleons together.

13. Nuclei are bound together in spite of the electric repulsion of the protons.

15. The strong nuclear force is stronger, has a finite range, and changes from attractive to repulsive at very short distances, whereas the electromagnetic force has an infinite range and remains either attractive or repulsive.

17. A 2.2-million electron-volt gamma ray

19. The mass of the neutral hydrogen atom would be slightly smaller due to the binding energy.

21. ^{14}N

23. **a.** None possible; **b.** only fusion; **c.** only fission; **d.** fission for light nuclei and fusion for heavy nuclei

25. They are about equal.

27. The energy spacing between proton states and neutron states is about the same.

29. It is energetically more favorable to add neutrons than additional protons.

31. It is unstable and would decay via beta plus decay.

33. Beta minus decay

35. The splitting of a heavy nucleus into two or more lighter ones

37. $^{89}_{36}$Kr

39. Because it releases more than 1 neutron on the average

41. To initiate additional fission reactions

43. On average less than 1 neutron from each fission process initiates another.

45. To absorb enough neutrons to assure that an average of 1 neutron from each fission process initiates another

47. Criticality would have been more difficult to obtain because fewer neutrons could be lost.

49. New fuel

51. $^{239}_{92}$U $\rightarrow$ $^{239}_{93}$Np $+ ^{0}_{-1}\beta$, then $^{239}_{93}$Np $\rightarrow$ $^{239}_{94}$Pu $+ ^{0}_{-1}\beta$

53. The combining of two or more light nuclei to form a heavier one

55. Magnetic confinement and inertial confinement

57. High temperature, high density, and long confinement

59. Yes

61. Safety issues, including management of reactors and storage and disposal of radioactive wastes

Exercises

1. 2.49×10^{-18} m

5. 92 MeV

7. 156 keV

9. 2.35 MeV

11. approx. 190 MeV

15. 4.05×10^{26} W

CHAPTER 26

Questions

1. Electron, proton, neutron, and photon

3. Positron

5. No

7. None

9. It will mutually annihilate with its corresponding particle.

11. There are no simple questions, but there are some complex ones involving special properties of the elementary particle interactions.

13. In the reference system at rest relative to the electron and positron, the total linear momentum is zero. If a single photon were emitted, it would carry away momentum E/c along its direction of travel, yielding a nonzero total momentum.

15. Conservation of charge and nucleon number (or baryon number)

17. Because it does not participate in the electromagnetic or the strong nuclear interactions

19. Yes, but the time is even longer than it takes light to travel the distance, because exchange particles with mass travel slower than this.

21. The violations must not last long enough to be detectable.

23. Exchange photons with very little energy can exist for very long times and can, therefore, travel infinite distances.

25. Its finite range

27. The gravitational force has an infinite range.

29. The leptons and some of the exchange particles (the graviton, the photon, and the intermediate vector bosons)

31. The electron, muon, tau, and their associated neutrinos

33. Yes, as evidenced by its beta decay

35. Strangeness

37. It doesn't have enough rest-mass energy.

39. **a.** charge or strangeness; **b.** energy; **c.** energy

41. **a.** 0; **b.** 0 or ±1; **c.** 0

43. Two photons

45. $\overline{u}\,\overline{u}\,\overline{d}$ for the antiproton and $\overline{u}\,\overline{d}\,\overline{d}$ for the antineutron

47. K^-

49. uus

51. $d\overline{s}$

53. By assigning the color quantum number to the quarks

55. −1

APPENDIX B

Exercises

1. **a.** 2.378×10^9 m; **b.** 3.24×10^{-3} ft

3. **a.** 5,782,000 s; **b.** 0.0069 ft

5. **a.** 9.2×10^{12}; **b.** 4.0×10^2

FIGURE CREDITS

Cover

Photographic Images by Alex Pietersen. This beautiful and intriguing drawing of the cyclists clearly conveys that the cyclists are traveling with a fairly high speed. We've superimposed arrows on the drawing to indicate how a physicist might quantify the motions and make predictions of how the motions might evolve.

Prologue

Prologue Opener: Jerry Yulsman/The Image Bank; unnum. fig.(a) p. 2: Albert Einstein Archives, The Hebrew University of Jerusalem, Israel; unnum. fig. p. 3 telescope: Scott Goldsmith/Tony Stone Images; unnum. fig. p. 3 cartoon: © 1992 by Nick Downes from BIG SCIENCE.

Chapter 1

Chapter Opener: Gerald F. Wheeler; unnum. fig. p. 5 football player: © Renee Lynn/Allstock/Tony Stone Images; unnum. fig. p. 5 waterfall: Gerald F. Wheeler; unnum. fig. p. 6 mile marker: Gerald F. Wheeler; unnum. fig. p. 7 metric elevation sign: Larry D. Kirkpatrick; Fig. 1-1: Tony Stone Images; Fig. 1-2: David Rogers; unnum. fig. p. 9 cartoon: © Sidney Harris; unnum. fig. p. 10 Olympic runners: © J.O. Atlanta 96/GAMMA; Fig. 1-3: Gerald F. Wheeler; unnum. fig. p. 13 amusement park ride: Superstock; unnum. fig. p. 15 cartoon: © 1992 by Nick Downes from BIG SCIENCE; unnum. fig. p. 16 Galileo: North Wind Picture Archives; Fig. 1-7: E/D/C Distribution Center; unnum. fig. p. 20 skydivers: U.S. Air Force Academy, Photo by Sgt. West C. Jacobs, 94ATS; Figs. 1Q-1, 1Q-29: David Rogers.

Chapter 2

Chapter Opener: Courtesy of U.S. Army Parachute Team, Golden Knights; Fig. 2-1: Charles D. Winters; Figs. 2-3, 2-4, 2-11: David Rogers; unnum. fig. p. 31 Newton: National Portrait Gallery, London; unnum. fig. p. 39 floating astronaut: NASA; unnum. fig. p. 39 supermarket scales: Charles D. Winters; unnum. fig. p. 41 downhill skier: Jean Y. Ruszniewski/Tony Stone Images; Fig. 2-14: Ben Rose/The Image Bank; Fig. 2Q-13: A. Copley/Visuals Unlimited.

Chapter 3

Chapter Opener: © J.O. Atlanta 96/GAMMA; Fig. 3-3: David Rogers; unnum. fig. p. 56 Olympic hammer thrower: © David Madison/DUOMO; unnum. fig. p. 57 bullet and gun: David Rogers; Fig. 3-7: PSSC Physics, 2nd ed., 1965, D.C. Heath & Co. and Educational Development Center, Inc., Newton, Mass.; unnum. fig. p. 58 cars on race track: Al Satterwhite/The Image Bank; unnum. fig. p. 58 Michael Jordan: UPI/Corbis-Bettmann; Fig. 3-13: PSSC Physics, 2nd ed., 1965, D.C. Heath & Co. and Educational Development Center, Inc., Newton, Mass.; unnum. fig. p. 65 ballet dancers: © John Terence Turner/FPG International; Figs. 3-16, 3-17,

3-18, 3-20, 3-23, unnum. fig. p. 67 children on tracks: David Rogers; Fig. 3Q-25: Simon McComb/Tony Stone Images; Fig. 3Q-28: George Semple; Figs. 3Q-51, 3Q-52: David Rogers; Fig. 3Q-58: Kathy Ferguson/PhotoEdit.

Chapter 4

Chapter Opener: NASA; unnum. fig. p. 79 Kepler: Corbis-Bettmann; unnum. fig. p. 80 Newton's farm: courtesy of J. Pasachoff, Collection of the Royal Society; Fig. 4-3: University of California Press; unnum. fig. p. 88 lunar rover: NASA; unnum. fig. p. 89 airplane: United Airlines; unnum. fig. p. 90 satellite dish in desert: Jeff Smith/FOTOSMITH; unnum. fig. p. 90 weather satellite: Courtesy of NOAA; unnum. fig. p. 91 digital satellite: Courtesy of Thomson Consumer Electronics; unnum. fig. p. 94 tides: Courtesy of Nova Scotia Tourism; unnum. fig. p. 95 star clusters: NASA; unnum. fig. p. 95 Andromeda Galaxy: California Institute of Technology/Palomar Observatory; unnum. fig. p. 96 cartoon: © 1992 by Sidney Harris; Figs. 4Q-6, 4Q-20, 4E-18: NASA.

Interlude 1

Interlude Opener: Charles D. Winters; unnum. fig. p. 103 optical illusion: Paul Doherty, Exploratorium, San Francisco.

Chapter 5

Chapter Opener: David Rogers; unnum. fig. p. 106 canoe/ship: Joel W. Rogers/AllStock/Tony Stone Images; unnum. fig. p. 107 Darren Daulton: © AllSport USA/Jeff Hixon, 1992; unnum. fig. p. 108 air bags: Courtesy of American Plastics Council; unnum. fig. p. 108 pole vaulter: Superstock, Inc.; Figs. 5-2, 5-3, unnum. fig. p. 115 billiards parlor, unnum. fig. p. 115 Cessna, 5-4, 5-8: David Rogers; unnum. fig. p. 113 car crash: Tony Freeman/PhotoEdit; unnum. fig. p. 121 falling cat: © Gerard Lacz/NHPA; Fig. 5Q-29: David Rogers; Fig. 5Q-38: Arnulf Husmo/Tony Stone Images; Fig. 5Q-43: Bob Martin/Tony Stone Images; Fig. 5Q-38: Arnulf Husmo/Tony Stone Images; Fig. 5Q-43: Bob Martin/Tony Stone Images; Fig. 5Q-46: NASA.

Chapter 6

Chapter Opener: A&L Sinibaldi/Tony Stone Images; Fig. 6-1: Courtesy of Arbor Scientific, Inc.; unnum. fig. p. 136 human pyramid: David Young-Wolff/PhotoEdit; unnum. fig. p. 141 roller coaster: Bob Torrey & Tony Stone Images; unnum. fig. p. 143 Grand Coulee Dam: UPI/Corbis-Bettmann; unnum. fig. p. 144 sun: NASA; unnum. figs. p. 146 wheat grains: Larry D. Kirkpatrick; unnum. fig. p. 148 weightlifter: © J.O. Atlanta 96/GAMMA; unnum. fig. p. 148 Gossamer Albatross: Jack Lambie; Fig. 6Q-16: David Young-Wolff/PhotoEdit; Fig. 6Q-31: Richard Hutchings/PhotoEdit; Fig. 6Q-45: David Young-Wolff/Tony Stone Images; 6Q-49: © David R. Frazier.

Interlude 2

Interlude Opener: BMRL/Science VU/Visuals Unlimited; unnum. figs. p. 157 Pompeii mosaic: Marc Sherman.

Chapter 7

Chapter Opener: Nicholas DeVore/Tony Stone Images; unnum. fig. p. 160 fly: Gerald F. Wheeler; unnum. fig. p. 161 copper ore woodcut: Burndy Library; unnum. fig. p. 163 electrolysis: Charles D. Winters; unnum. fig. p. 168 iodine atoms: Bruce Schardt, image taken using Nanoscope II, Digital Instruments, Inc.; Fig. 7-4: Charles D. Winters; unnum. fig. p. 177 canvas bag: Jeff Smith/FOTO-

SMITH; Fig. 7Q-28 © Vanessa Vick/Photo Researchers, Inc.; Fig. 7Q-41: Jeff Greenberg/Visuals Unlimited; Fig. 7Q-55: Gerald F. Wheeler.

Chapter 8

Chapter Opener: Tony Stone Images; unnum. fig. p. 186 silica aerogel: University of California, Lawrence Livermore National Laboratory and the U.S. Dept. of Energy; unnum. fig. p. 187 wallpaper: David Rogers; unnum. fig. p. 187 snowflakes: Richard C. Walters/Visuals Unlimited; Fig. 8-1; Courtesy of Geoffrey Sutton; Fig. 8-2: Leonard Fine; Fig. 8-3: David Rogers; unnum. fig. p. 189 King Tut: Superstock, Inc.; Fig. 8-4: Charles D. Winters; unnum. fig. p. 190 Notre Dame window: Gerald F. Wheeler; unnum. fig. p. 190 digital watch: Charles D. Winters; Figs. 8-5, 8-6: George Semple; Fig. 8-7: Aaron Haupt/Photo Researchers, Inc.; unnum. fig. p. 192 honey: Charles D. Winters; unnum. fig. p. 193 aurora borealis: Mark Kelly/AllStock, Inc./Tony Stone Images; unnum. fig. p. 194 scuba diver: Neville Coleman/Visuals Unlimited; unnum. fig. p. 196 bathysphere: Rod Catanach, Woods Hole Oceanographic Institute; unnum. fig. p. 197 ocean liner: T. Nakamura/Superstock, Inc.; unnum. fig. p. 198 exercisers: Bally's Health & Tennis Corporation; unnum. fig. p. 199 tornado damage: David J. Sams/Tony Stone Images; Figs. 8Q-11, 8Q-23, 8Q-25, 8Q-46: George Semple.

Chapter 9

Chapter Opener: Courtesy of Dr. Terrie Williams/University of California, Santa Cruz; unnum. fig. p. 207 Joule: Oesper Collection in the History of Chemistry, University of Cincinnati; unnum. fig. p. 217 Glacier National Park: Gerald F. Wheeler; Unnum. fig. p. 218 branding iron: David Rogers; Fig. 9-5: Courtesy of Corning Glass Works; Fig. 9-6: A.A. Bartlett, University of Colorado, Boulder; unnum. fig. p. 219 snow cave: Phil Schofield, Tony Stone Images; Fig. 9-7 Courtesy of Soaring Adventures of America, Wilton, Conn.; Fig. 9-9: Gene Schultz; Fig. 9-10; Courtesy of Honeywell, Inc.; unnum. fig. p. 223 expansion slots: Edward M. Wheeler; unnum. fig. p. 224 otter: Gijsbert van Frankenhuyzen/Dembinsky Photo Associates; Fig. 9Q-8: Marc Sherman; Fig. 9Q-30: George Semple; Fig. 9Q-31: Ira Rubin/Dembinsky Photo Associates.

Chapter 10

Chapter Opener: Gerald F. Wheeler; Fig. 10-1: Courtesy of Central Scientific Company; unnum. fig. p. 234 steam locomotive: Rufus Cone, Montana State University, Physics Department; unnum. fig. p. 234 car engine: David Rogers; unnum. fig. p. 238 coal plant: Bob Webster, Montana Power Company; unnum. fig. p. 238 Three Mile Island: Larry LeFever/Grant Heilman Photography, Inc.; unnum. fig. p. 240 heat pump: Marc Sherman; unnum. fig. p. 241 cards: David Rogers; unnum. fig. p. 244 ghost town: Gerald F. Wheeler; unnum. fig. p. 246 water heater: © Comstock, Inc.; Fig. 10Q-41(a): Pearl Levi; Fig. 10Q-41(b): Courtesy of Anne Sherman; Fig. 10Q-48: Charles D. Winters.

Interlude 3

Interlude Opener: Courtesy of Hersheypark.

Chapter 11

Chapter Opener: National Optical Astronomy Observatories; unnum. fig. p. 256 moving cars: David Rogers; unnum. fig. p. 263 inertial forces; unnum. figs. p. 264 astronauts floating, female astronaut: NASA; Fig. 11-9: Cedar Point photos by Dan Feicht; unnum. fig. p. 266 space station: NASA; unnum. fig. p. 270 Foucault pendulum: Courtesy of the Smithsonian Institution; unnum. fig. p. 270 merry-go-round: Toby Rankin/The Image Bank; Fig. 11-14: NASA; unnum. figs. p. 272

Neptune, Jupiter: NASA; Fig. 11Q-23: Adam Jones/Dembinsky Photo Associates; Fig. 11Q-37: NASA.

Chapter 12

Chapter Opener: David Rogers; unnum. figs. pp. 281, 296: Einstein: AIP Niels Bohr Library; unnum. fig. p. 299 cartoon; © Sidney Harris; Fig. 12E-3: Dr. Seth Shostak/Science Photo Library/Photo Researchers, Inc.

Chapter 13

Chapter Opener: Painting by George Kelvin; unnum. fig. p. 309 Cygnus X-1: Gerald F. Wheeler; unnum. fig. p. 315 gravity wave detector: Joe Weber, University of California at Irvine; Fig. 13E-3: NASA.

Interlude 4

Interlude Opener: © Raymond Gendreau/AllStock/Tony Stone Images; unnum. fig. p. 321 water waves: © Steve Gottlieb/FPG International.

Chapter 14

Chapter Opener: Herman Eisenbeiss/Photo Researchers, Inc.; unnum. fig. p. 327 pendulum: Richard Megna/Fundamental Photographs, NYC; unnum. fig. p. 328 clock: David Rogers; unnum. fig. p. 328 cesium clock: Courtesy of National Institute of Standards and Technology, U.S. Dept. of Commerce; unnum. fig. p. 329 swings: David Rogers; unnum. fig. p. 331 bridge: Special Collections Div., Univ. of Washington Libraries, photo by Farquharson; Fig. 14-7: Herman Eisenbeiss/Photo Researchers, Inc.; Figs. 14-20(b), 14-22(a), 14-23: PSSC Physics, 2nd ed., 1965, D.C. Heath & Co. and Educational Development Center, Inc., Newton, Mass.; Fig. 14Q-14: John D. Cunningham/Visuals Unlimited; Fig. 14Q-59: Courtesy of Central Scientific Company.

Chapter 15

Chapter Opener: © 1996 West Chester University of Pennsylvania; unnum. fig. p. 358 jackhammer: © Jim Cummins/AllStock/Tony Stone Images; Fig. 15-3 Courtesy of Brook/Cole Publishing Co.; Fig. 15-4: Catgut Acoustical Society, Inc.; unnum. fig. p. 361 Eric Clapton: Sam Mircovich Stringer/Corbis-Bettmann; Fig. 15-7: David Rogers; unnum. fig. p. 366 cartoon: © 1992 by Nick Downes from BIG SCIENCE; Fig. 15-14: Gerald F. Wheeler; Fig. 15Q-21: Courtesy of C.F. Martin & Co., Nazareth, Penn.; Fig. 15Q-32: Courtesy of Henry Leap.

Interlude 5

Interlude Opener: Steven Gray; unnum. fig. p. 375 stained glass: Bill Kamin/Visuals Unlimited.

Chapter 16

Chapter Opener: Mark G. Wheeler; unnum. fig. p. 380 total eclipse of sun: Jim Anderson, Montana State University, Physics Department; unnum. fig. p. 380 lunar eclipse: Courtesy of Mike Murray; Figs. 16-4, 16-7(a), 16-13, 16-14, unnum. fig. p. 386 fun mirrors, unnum. fig. p. 388 convex mirrors, 16-15, 16-16, 16-17, 16-18, 16-19, 16-20, 16-21, 16-22, 16-23, 16-24, 16-25, 16-26, 16-28: David Rogers; unnum. fig. p. 388 cartoon: © Sidney Harris; unnum. fig. p. 387 solar farm mirrors: Randy Montoya, Sandia National Laboratories; Figs. 16-27, 16Q-54: George Semple.

Chapter 17

Chapter Opener: © 1996 Pekka Parviainen/Dembinsky Photo Associates; Figs. 17-1, 17-4, 17-5: David Rogers; Fig. 17-9: Paul Swenson; Fig. 17-10: David Parker/Science Photo Library/Photo Researchers, Inc.; Fig. 17-11: Paul Markovits; unnum. fig.p. 411 sprinkler: David Cavagnaro/Visuals Unlimited; Fig. 17-15: Courtesy of Robert Greenler; unnum. fig. p. 418 retina: Gerald F. Wheeler; unnum. figs.p. 423 Hubble telescope: NASA; Figs. 17-16, 17-18, 17-19, 17Q-59: David Rogers.

Chapter 18

Chapter Opener: Peter Aprahamian/Science Photo Library/Photo Researchers, Inc.; Figs. 18-1, 18-3: PSSC Physics, 2nd ed., 1965, D.C. Heath & Co. and Educational Development Center, Inc., Newton, Mass.; Figs. 18-4, 18-5, 18-7: Kodansha Publications, Japan; unnum. fig. p. 437 diffraction patterns: M. Cagnet, M. Francon, and J.C. Thierr, Atlas of Optical Phenomena, Berlin, Springer-Verlag 1962, plate 16; Fig. 18-8: Gerald F. Wheeler and Henry Cruz; Fig. 18-12: Charles D. Winters; unnum. fig. p. 440 colors on puddle: George Semple; unnum. fig. p. 441 visor: NASA; Fig. 18-14: Vincent Mallet, Georgia Institute of Technology; unnum. fig. p. 443 polarizing filter: George Semple; unnum. fig. p. 445 chessmen: Gerald F. Wheeler; Fig. 18-17: David Rogers; unnum. fig. p. 447 structure: Robert Mark, Princeton University.

Interlude 6

Interlude Opener: Leif Skoogfors/The Franklin Institute Science Museum/ Woodfin Camp & Associates.

Chapter 19

Chapter Opener, unnum. fig. p. 466 lightning bolts: Bob Webster, Montana Power Company; unnum. fig. p. 458 Franklin: biography and portrait by Steven and Craig Janke, Courtesy PASCO Scientific; unnum. fig. p. 455 amber, unnum. fig. p. 456 fuel truck, Figs. 19-5, 19-6, 19-7: David Rogers; unnum. fig. p. 465 Faraday cage: Courtesy of Prof. Clint Sprott, University of Wisconsin-Madison; unnum. fig. p. 468 ice crystals: William J. Wever/Visuals Unlimited; unnum. fig. p. 468 galaxy: NASA; Fig. 19-10: PSSC Physics, 2nd ed., 1965, D.C. Heath & Co. and Educational Development Center, Inc., Newton, Mass.

Chapter 20

Chapter Opener: Courtesy of the Philadelphia Convention and Visitors Bureau; unnum. fig. p. 477 grapefruit battery: Charles D. Winters; unnum. fig. p. 478 car battery, Fig. 20-2, unnum. fig. p. 483 resistors: David Rogers; unnum. fig. p. 486 permanent magnet: Argonne National Laboratory and the U.S. Dept. of Energy; unnum. fig. p. 488 circuit breaker, unnum. fig. p. 491 battery packs: Gerald F. Wheeler; unnum. fig. p. 490 electric meter: Marc Sherman.

Chapter 21

Chapter Opener: Richard Megna, Fundamental Photos, NYC; Figs. 21-1, 21-8, 21-10, 21-11: David Rogers; Figs. 21-3, 21-4: PSSC Physics, 2nd ed., 1965, D.C. Heath & Co. and Educational Development Center, Inc., Newton, Mass.; unnum. fig. p. 504 electromagnet: Dembinsky Photo Associates; unnum. fig. p. 508 aurora borealis: © Mark Kelley/AllStock/Tony Stone Images; unnum. fig. p. 510 transformer: George Semple; unnum. fig. p. 518 Maxwell: AIP Niels Bohr Library; unnum. fig. p. 519 stereo: Courtesy of SONY Electronics, Inc.; unnum. fig. p. 517 aerial view of D.C.: NASA; Fig. 21E-13: Martin Dohrn/Science Photo Library/ Photo Researchers, Inc.

Interlude 7

Interlude Opener: Richard C. Walters/Visuals Unlimited; unnum. fig. p. 527 Hertz: Corbis-Bettmann.

Chapter 22

Chapter Opener: AT&T Bell Laboratories; unnum. fig. p. 529 neon: David Rogers; unnum. fig. p. 534 Thomson: AIP Emilio Segre Visual Archives, W.F. Meggers Collection; unnum. fig. p. 536 Rutherford: U.K. Atomic Energy Authority, Courtesy AIP Emilio Segre Visual Archives; Fig. 22-13: Kodansha Publications, Japan; unnum, fig. p. 541 Planck, unnum. fig. p. 547 Bohr: AIP Niels Bohr Library, W.F. Meggers Collection; unnum. fig. p. 550 cartoon: © Sidney Harris; unnum. fig. p. 551 X-ray hand: Courtesy of Burndy Library, Electra Square, Conn.

Chapter 23

Chapter Opener: Courtesy of CERN; unnum. fig. p. 559 de Broglie: AIP Niels Bohr Library, W.F. Meggers Collection; Fig. 23-1: Courtesy of PASCO Scientific; Fig. 23-3: ZYGO Corporation; Fig. 23-4: Courtesy of RCA/General Electric Corporate Research and Development; unnum. figs. p. 568 colors: General Electric Company; unnum. fig. p. 563 Feynmann: Linn Duncan/Univ. of Rochester, Courtesy AIP/Emilio Segre Visual Archives, unnum. fig. p. 571 Pauli: Oesper Collection in the History of Chemistry/Univ. of Cincinnati; unnum fig. Heisenberg p. 572: Courtesy of University of Hamburg; unnum. fig. p. 574 cartoon: © 1992 by Sidney Harris; unnum. fig. p. 576 Blake's Creation of Error: The Bettmann Archive; unnum. fig. p. 576 Einstein: AIP Niels Bohr Library; unnum. fig. p. 578 laser surveying: Edward M. Wheeler; unnum. fig. p. 579 laser microsurgery: Alexander Tsiaras/Science Source/Photo Researchers, Inc.; Fig. 23E-16: Courtesy of the Ford Motor Company.

Interlude 8

Interlude Opener: Courtesy of Sandia National Laboratories; unnum. fig. p. 585 nested dolls: Gerald F. Wheeler.

Chapter 24

Chapter Opener: Scott Camazine/Photo Researchers, Inc.; unnum. fig. p. 589 Curie: AIP Niels Bohr Library, W.F. Meggers Collection; unnum. figs. p. 597 dice, p. 598 Arabian dice, Figs. 24-9, 24-10: David Rogers; unnum. fig. p. 599 moon rocks: NASA; unnum. fig. p. 600 skeleton: Nikolas J. van der Merwe, American Scientist, 70: 596–606, 1982; unnum. fig. p. 600 antenna: George Semple; unnum. fig. p. 602 electronic components: AT&T Bell Laboratories; Fig. 24-7: Sovfoto/Eastfoto; unnum. fig. p. 604 airplane: Courtesy of The Boeing Company; unnum. fig.p. 607 radon test: Charles D. Winters; Figs. 24-11, 24-12: Fermi National Accelerator Laboratory; Fig. 24Q-46: Courtesy of Anne Gibby; Fig. 24Q-59: Art Stein/Photo Researchers, Inc.

Chapter 25

Chapter Opener: Courtesy of Public Service Electric & Gas Co.; unnum. fig. p. 617 forest fire: Don Collins, Montana State University, Biology Dept.; unnum. fig. p. 617 atomic bomb: U.S. Army White Sands Missile Range; Fig. 25-1: Courtesy of Gene Sprouse, SUNY at Stony Brook; Fig. 25-2: Stanford Linear Accelerator and the U.S. Dept. of Energy; Fig. 25-3: Fermi National Accelerator Laboratory; unnum. fig. p. 625 Fermi: Univ. of Chicago, Courtesy AIP Emilio Segre Visual Archives; unnum. fig. p. 630 Meitner: AIP Emilio Segre Visual Archives, Herzfeld Collection; unnum. fig. p. 633 boiling water reactor: Washington Public Power

Supply System; unnum. fig. p. 636: Idaho National Laboratory and the U.S. Dept. of Energy; Fig. 25-14: Princeton University Plasma Physics Laboratory; unnum. fig. p. 636 sun, Fig. 25E-15: NASA.

Chapter 26

Chapter Opener: Fermilab, Visual Media Services; Fig. 26-1: Stanford Linear Accelerator and the U.S. Dept. of Energy; unnum. fig. p. 643 antinucleons: Univ. of California, Lawrence Livermore National Laboratory and the U.S. Dept. of Energy; unnum. fig. p. 644 stars: NASA; Fig. 26-5: Savannah River Laboratory.

Epilogue

Epilogue Opener: NASA.

GLOSSARY

aberration   A defect in a mirror or lens causing light rays from a single point to fail to focus at a single point in space.

absolute temperature scale   The temperature scale with its zero point at absolute zero and degrees equal to those on the Celsius scale. This is the same as the Kelvin temperature scale.

absolute zero   The lowest possible temperature; 0 K, $-273°$C, or $-459°$F.

absorption spectrum   The collection of wavelengths missing from a continuous distribution of wavelengths. Caused by the absorption of certain wavelengths by the atoms or molecules in a gas.

acceleration   The change in velocity divided by the time it takes to make the change. An acceleration can result from a change in speed, a change in direction, or both.

activity   The rate at which a collection of radioactive nuclei decay. One curie corresponds to 3.7×10^{10} decays per second; also called radioactivity.

alloy   A metal produced by mixing other metals.

alpha particle   The nucleus of helium consisting of two protons and two neutrons.

alpha (α) radiation   The type of radioactive decay in which nuclei emit alpha particles (helium nuclei).

ampere   The SI unit of electric current, 1 coulomb per second.

amplitude   The maximum distance from the equilibrium position that occurs in periodic motion.

angular momentum   A vector quantity giving the rotational momentum. For an object orbiting a point, the angular momentum is the product of the linear momentum and the radius of the path. For a solid body, it is the product of the rotational inertia and the rotational velocity.

antinode   One of the positions in a standing wave or interference pattern where there is maximal movement; that is, the amplitude is a maximum.

antiparticle   A subatomic particle with the same-size properties as those of the particle although some may have the opposite sign. The positron is the antiparticle of the electron.

Archimedes' principle   The buoyant force is equal to the weight of the displaced fluid.

astigmatism   An aberration, or defect, in a mirror or lens that causes the image of a point to spread out into a line.

atom   The smallest unit of an element that has the chemical and physical properties of that element. An atom consists of a nucleus surrounded by an electron cloud.

atomic mass   The mass of an atom in atomic mass units. Sometimes this refers to the atomic mass number, the number of neutrons and protons in the nucleus.

atomic mass unit   One-twelfth the mass of a neutral carbon atom containing six protons and six neutrons.

atomic number   The number of protons in the nucleus or the number of electrons in the neutral atom of an element. This number also gives the order of the elements in the periodic table.

average speed   The distance traveled divided by the time taken.

Avogadro's number   The number of molecules in 1 mole of any substance. Equal to 6.02×10^{23} molecules.

baryon   A type of hadron having a spin of ½, ⅗, ⅚, . . . times the smallest unit. The most common baryons are the proton and neutron.

beats   A variation in the amplitude resulting from the superposition of two waves that have nearly the same frequencies. The frequency of the variation is equal to the difference in the two frequencies.

Bernoulli's principle   The pressure in a fluid decreases as its velocity increases.

beta particle   An electron emitted by a radioactive nucleus.

beta (β) radiation   The type of radioactive decay in which nuclei emit electrons or positrons (antielectrons).

binding energy   The amount of energy required to take a nucleus apart. The analogous amount of energy for other bound systems.

black hole   A massive star that has collapsed to such a small size that its gravitational force is so strong that not even light can escape from its "surface."

bottom   The flavor of the fifth quark.

British thermal unit   The amount of heat required to raise the temperature of 1 pound of water by $1°$ Fahrenheit.

buoyant force   The upward force exerted by a fluid on a submerged or floating object. *See* Archimedes' principle.

calorie   The amount of heat required to raise the temperature of 1 gram of water by 1° Celsius.

camera obscura   A room with a small hole in one wall used by artists to produce images.

cathode ray   An electron emitted from the negative electrode in an evacuated tube.

Celsius temperature   The temperature scale with the values of 0 and 100 for the temperatures of freezing and boiling water, respectively. Its degree is ⅝ that of the Fahrenheit degree.

center of mass   The balance point of an object. The location in an object that has the same translational motion as the object if it were shrunk to a point.

centi   A prefix meaning ¹⁄₁₀₀. A centimeter is ¹⁄₁₀₀ meter.

centrifugal force   A fictitious force arising in a rotating reference system. It points away from the center, in the direction opposite to the centripetal acceleration.

centripetal   An adjective meaning "center-fleeing."

centripetal acceleration   The acceleration of an object moving along a curved path. For uniform circular motion, the acceleration points toward the center of the circle and has a magnitude given by v^2/r.

centripetal force   The force causing an object to change directions. For uniform circular motion, the force is directed toward the center of the circle and has a magnitude given by mv^2/r.

chain reaction   A process in which the fissioning of one nucleus initiates the fissioning of others.

change of state   The change in a substance between solid and liquid or between liquid and gas.

charge   A property of elementary particles that determines the strength of its electric force with other particles possessing charge. Measured in coulombs, or in multiples of the charge on the proton.

charged   Possessing a net negative or positive charge.

charm   The flavor of the fourth quark.

chromatic aberration   A defect in lenses that causes different colors (wavelengths) of light to have different focal lengths.

coherent   A property of two or more sources of waves that have the same wavelength and maintain constant phase differences.

complementarity principle   The idea that a complete description of an atomic entity, such as an electron or a photon, requires both a particle description and a wave description.

complementary color   For lights, two colors that combine to form white.

complete circuit   A continuous conducting path from one end of a battery (or other source of electric potential) to the other end of the battery.

compound   A combination of chemical elements that forms a new substance with its own properties.

conduction, thermal   The transfer of thermal energy by collisions of the atoms or molecules within a substance.

conductor   A material that allows the passage of electric charge or the easy transfer of thermal energy. Metals are good conductors.

conservation of angular momentum   If the net external torque on a system is zero, the total angular momentum of the system does not change.

conservation of charge   In an isolated system the total charge is conserved.

conservation of energy   The total energy of an isolated system does not change.

conservation of mass   The total mass in a closed system does not change even when physical and chemical changes occur.

conservation of momentum   If the net external force on a system is zero, the total linear momentum of the system does not change.

conserved   This term is used in physics to mean that a number associated with a physical property does not change; it is invariant.

convection, thermal   The transfer of thermal energy in fluids by means of currents such as the rising of hot air and the sinking of cold air.

Coriolis force   A fictitious force that occurs in rotating reference frames. It is responsible for the direction of the winds in hurricanes.

coulomb   The SI unit of electric charge, the charge of 6.24×10^{18} protons.

covalent bonding   The binding together of atoms by the sharing of their electrons.

crest   The peak of a wave disturbance.

critical angle   The minimum angle of incidence for which total internal reflection occurs.

critical chain reaction   A chain reaction in which an average of one neutron from each fission reaction initiates another reaction.

critical mass   The minimum mass of a substance that will allow a chain reaction to continue without dying out.

crystal   A material in which the atoms are arranged in a definite geometric pattern.

curie   A unit of radioactivity, 3.7×10^{10} decays per second.

current   A flow of electric charge. Measured in amperes.

cycle   One complete repetition of a periodic motion. It may start anywhere in the motion.

daughter nucleus   The nucleus resulting from the radioactive decay of a parent nucleus.

definite proportions, law of   When two or more elements combine to form a compound, the ratios of the masses of the combining elements have fixed values.

density   A property of material equal to the mass of the material divided by its volume. Measured in kilograms per cubic meter.

diaphragm   An opening that is used to limit the amount of light passing through a lens.

diffraction   The spreading of waves passing through an opening or around a barrier.

diffuse reflection   The reflection of rays from a rough surface. The reflected rays do not leave at fixed angles.

diopter   A measure of the focal length of a mirror or lens, equal to the inverse of the focal length measured in meters.

disordered system   A system with an arrangement equivalent to many other possible arrangements.

dispersion   The spreading of light into a spectrum of color. The variation in the speed of a periodic wave due to its wavelength or frequency.

displacement   In wave (or oscillatory) motion, the distance of the disturbance (or object) from its equilibrium position.

Doppler effect   A change in the frequency of a periodic wave due to the motion of the observer, the source, or both.

efficiency   The ratio of the work produced to the energy input. For an ideal heat engine, the Carnot efficiency is given by $1 - T_c/T_h$.

elastic   A collision or interaction in which kinetic energy is conserved.

electric field   The space surrounding a charged object where each location is assigned a value equal to the force experienced by one unit of positive charge placed at that location.

electric potential   The electric potential energy divided by the object's charge. The work done in bringing a positive test charge of 1 coulomb from the zero reference location to a particular point in space.

electric potential energy   The work done in bringing a charged object from some zero reference location to a particular point in space.

electromagnet   A magnet constructed by wrapping wire around an iron core. The electromagnet can be turned on and off by turning the current in the wire on and off.

electromagnetic wave   A wave consisting of oscillating electric and magnetic fields. In a vacuum electromagnetic waves travel at the speed of light.

electron   A basic constituent of atoms, a lepton.

electron capture   A decay process in which an inner atomic electron is captured by the nucleus. The daughter nucleus has the same number of nucleons but one fewer proton.

electron volt   A unit of energy equal to the kinetic energy acquired by an electron or proton falling through an electric potential difference of 1 volt. Equal to 1.6×10^{-19} joule.

element   Any chemical species that cannot be broken up into other chemical species.

emission spectrum   The collection of discrete wavelengths emitted by atoms that have been excited by heating or by electric currents.

entropy   A measure of the order of a system. The second law of thermodynamics states that the entropy of an isolated system tends to increase.

equilibrium position   A position where the net restoring force is zero.

equivalence principle   Constant acceleration is completely equivalent to a uniform gravitational field.

ether   The hypothesized medium through which light traveled.

exclusion principle   No two electrons can have the same set of quantum numbers. This statement also applies to protons, neutrons, and other baryons.

Fahrenheit temperature   The temperature scale with the values of 32 and 212 for the temperatures of freezing and boiling water, respectively.

field   A region of space where each location is assigned a value. *See* electric, gravitational, and magnetic fields.

first postulate of special relativity   The laws of physics are the same for all inertial reference systems.

fission   The splitting of a heavy nucleus into two or more lighter nuclei.

flavor   The types of quark: up, down, strange, charm, bottom, or top.

fluorescence   The property of a material whereby it emits visible light when it is illuminated by ultraviolet light.

focal length   The distance from a mirror or the center of a lens to its focal point.

focal point   The location at which a mirror or a lens focuses rays parallel to the optic axis or from which such rays appear to diverge.

force   A push or a pull. Measured by the acceleration it produces on a standard, isolated object. Measured in newtons.

frequency   The number of times a periodic motion repeats in a unit of time. It is equal to the inverse of the period. Measured in hertz.

fundamental frequency   The lowest resonant frequency for the oscillating system.

fusion   The combining of light nuclei to form a heavier nucleus.

Galilean principle of relativity   The laws of motion are the same in all inertial reference systems.

gamma (γ) radiation   The type of radioactive decay in which nuclei emit high-energy photons. The daughter nucleus is the same as the parent. The range of frequencies of the electromagnetic spectrum that lies beyond the X rays.

gas   Matter with no definite shape or volume.

gauss   A unit of magnetic field strength, 10^{-4} tesla.

general theory of relativity   An extension of the special theory of relativity to include the concept of gravity.

geocentric model   A model of the Universe with the Earth at its center.

gluon An exchange particle responsible for the force between quarks. There are eight gluons that differ only in their color quantum numbers.

gravitational field The space surrounding an object where each location is assigned a value equal to the gravitational force experienced by one unit of mass placed at that location.

gravitational mass The property of a particle that determines the strength of its gravitational interaction with other particles.

gravitational potential energy The work done by the force of gravity when an object falls from a particular point in space to the location assigned the value of zero.

gravitational redshift The decrease in the frequency of electromagnetic waves due to a gravitational field.

graviton The exchange particle responsible for the gravitational force.

gravity wave A wave disturbance caused by the acceleration of masses.

grounding Establishing an electric connection to the Earth in order to neutralize an object.

ground state The lowest energy state of a system allowed by quantum mechanics.

hadrons The family of particles that participate in the strong interaction. Baryons and mesons are the two subfamilies.

half-life The time during which one-half of a sample of a radioactive substance decays.

halo A ring of light that appears around the Sun or Moon. It is produced by refraction in ice crystals.

harmonic A frequency that is a whole-number multiple of the fundamental frequency.

heat A flow of energy due to a difference in temperature.

heat engine A device for converting heat into mechanical work.

heat pump A reversible heat engine that acts as a furnace in winter and an air conditioner in summer.

heliocentric model A model of the Universe with the Sun at its center.

hologram A three-dimensional record of visual information.

holography The photographic process for producing three-dimensional images.

hyperopia Farsightedness. Images of distant objects are formed beyond the retina.

ideal gas An enormous number of very tiny particles separated by relatively large distances. The particles have no internal structure, are indestructible, do not interact with each other except when they collide, and all collisions are elastic.

ideal gas law $PV = cT$, where P is the pressure, V is the volume, T is the absolute temperature, and c is a constant that depends on the amount of gas.

impulse The product of the force and the time during which it acts. This vector quantity is equal to the change in momentum.

index of refraction An optical property of a substance that determines how much light bends upon entering or leaving it. The index is equal to the ratio of the speed of light in a vacuum to that in the substance.

inelastic A collision or interaction in which kinetic energy is not conserved.

inertia An object's resistance to a change in its velocity. *See* inertial mass.

inertia, law of *See* Newton's first law of motion.

inertial force A fictitious force that arises in accelerating (noninertial) reference systems. Examples are centrifugal and Coriolis forces.

inertial mass An object's resistance to a change in its velocity. Measured in kilograms.

inertial reference system Any reference system in which the law of inertia (Newton's first law of motion) is valid.

in phase Two or more waves with the same wavelength and frequency that have their crests lined up.

instantaneous speed The limiting value of the average speed as the time interval becomes infinitesimally small. The magnitude of the velocity.

insulator A material that does not allow the passage of electric charge or is a poor conductor of thermal energy. Ceramics are good insulators.

interference The superposition of waves.

intermediate vector bosons The exchange particles of the weak nuclear interaction: the W^+, W^-, and Z^0 particles.

internal energy The total microscopic energy of an object, which includes its atomic and molecular translational and rotational kinetic energies, vibrational energy, and the energy stored in the molecular bonds.

inverse proportionality A relationship in which a quantity is related to the reciprocal of a second quantity.

inverse-square A relationship in which a quantity is related to the reciprocal of the square of a second quantity. Examples include the force laws for gravity and electricity; the force is proportional to the inverse-square of the distance.

ion An atom with missing or extra electrons.

ionic bonding The binding together of atoms through the transfer of one or more electrons from one atom to another.

ionization The removal of one or more electrons from an atom.

isotope An element containing a specific number of neutrons in its nuclei. Examples are $^{12}_{6}C$ and $^{14}_{6}C$, carbon atoms with six and eight neutrons, respectively.

joule The SI unit of energy, equal to 1 newton acting through a distance of 1 meter.

Kelvin temperature The temperature scale with its zero point at absolute zero and a degree equal to that on the Celsius scale. Also called the absolute temperature scale.

kilo A prefix meaning 1000. A kilometer is 1000 meters.

kilogram The SI unit of mass, the approximate mass of 1 liter of water. A kilogram of material weighs about 2.2 pounds on Earth.

kilowatt-hour A unit of energy; 3,600,000 joules. One kilowatt-hour of energy is transformed to other forms when a machine runs at a power of 1000 watts for 1 hour.

kinetic energy The energy of motion, $\frac{1}{2}mv^2$, measured in joules.

laser An acronym for *l*ight *a*mplification by *s*timulated *e*mission of *r*adiation. A device that uses stimulated emissions to produce a coherent beam of electromagnetic radiation.

latent heat The amount of heat required to melt (or vaporize) 1 gram of a substance. The same amount of heat is released when 1 gram of the same substance freezes (or condenses).

lepton A family of elementary particles that includes the electron, muon, tau, and their associated neutrinos.

Le Système International d'Unités The French name for the metric, or System International (SI), system of units.

light ray A line that represents the path of light in a given direction.

linear momentum A vector quantity equal to the product of an object's mass and its velocity.

line of stability The locations of the stable nuclei on a graph of the number of neutrons versus the number of protons.

liquid Matter with a definite volume that takes the shape of its container.

liquid crystal A liquid that exhibits a rough geometrical ordering of its atoms.

longitudinal wave A wave in which the vibrations of the medium are parallel to the direction the wave is moving.

macroscopic The bulk properties of a substance such as mass, size, and temperature.

magnetic field The space surrounding a magnetic object, where each location is assigned a value determined by the torque on a compass placed at that location. The direction of the field is in the direction of the north pole of the compass.

magnetic monopole A hypothesized, isolated magnetic pole.

magnetic pole One end of a magnet; analogous to an electric charge.

magnitude The size of a vector quantity. For example, speed is the magnitude of a velocity.

mass *See* inertial mass, gravitational mass, critical mass, and center of mass.

matter-wave amplitude The wave solution to Schrödinger's equation for atomic and subatomic particles. The square of the matter-wave amplitude gives the probability of finding the particle at a particular location.

mechanical energy A sum of the kinetic energy and various potential energies, which may include the gravitational and the elastic potential energies.

meson A type of hadron with whole-number units of spin. This family includes the pion, kaon, and eta.

metallic bonding The binding together of atoms through the sharing of electrons throughout the material.

meter The SI unit of length equal to 39.37 inches, or 1.094 yards.

microscopic Properties not visible to the naked eye such as atomic speeds.

milli A prefix meaning $\frac{1}{1000}$. A millimeter is $\frac{1}{1000}$ meter.

mirage An optical effect that produces an image that looks like it has been reflected from the surface of a body of water.

moderator A material used to slow down the neutrons in a nuclear reactor.

molecule A combination of two or more atoms.

momentum Usually refers to linear momentum. *See* angular momentum, linear momentum, and conservation of momentum.

muon A type of lepton; often called a heavy electron.

myopia Nearsightedness. Images of distant objects are formed in front of the retina.

neutrino A neutral lepton; one exists for each of the charged leptons (electron, muon, and tau).

neutron The neutral nucleon in nuclei. A member of the baryon and hadron families of elementary particles.

newton The SI unit of force. A net force of 1 newton accelerates a mass of 1 kilogram at a rate of 1 (meter per second) per second.

Newton's first law Every object remains at rest or in motion in a straight line at constant speed unless acted upon by an unbalanced force.

Newton's second law $\mathbf{F}_{net} = m\mathbf{a}$; the net force on an object is equal to its mass times its acceleration and points in the direction of the acceleration.

Newton's third law If an object exerts a force on a second object, the second object exerts an equal force back on the first object.

node One of the positions in a standing wave or interference pattern where there is no movement; that is, the amplitude is zero.

noninertial reference system Any reference system in which the law of inertia (Newton's first law of motion) is not valid. An accelerating reference system is noninertial.

normal A line perpendicular to a surface or curve.

nucleon Either a proton or a neutron.

nucleus The central part of an atom that contains the protons and neutrons.

ohm The SI unit of electrical resistance. A current of 1 ampere flows through a resistance of 1 ohm under 1 volt of potential difference.

Ohm's law The resistance of an object is equal to the voltage across it divided by the current through it.

optic axis A line passing through the center of a curved mirror and the center of the sphere from which the mirror is made. A line passing through a lens and both focal points.

ordered system A system with an arrangement belonging to a group with the smallest number (possibly one) of equivalent arrangements.

oscillation A vibration about an equilibrium position or shape.

pair production The conversion of energy into matter in which a particle and its antiparticle are produced. This usually refers to the production of a electron and a positron (antielectron).

parallel circuit An arrangement of resistances (or batteries) on side-by-side pathways between two points.

parent nucleus A nucleus that decays into a daughter nucleus.

particle accelerator A device for accelerating charged particles to high velocities.

penumbra The transition region between the darkest shadow and full brightness. Only part of the light from the source reaches this region.

period The shortest length of time it takes a periodic motion to repeat. It is equal to the inverse of the frequency.

periodic wave A wave in which all the pulses have the same size and shape. The wave pattern repeats itself over a distance of 1 wavelength and over a time of 1 period.

phosphorescence The property of a material whereby it continues to emit visible light after it has been illuminated by ultraviolet light.

photoelectric effect The ejection of electrons from metallic surfaces by illuminating light.

photon A particle of light. The energy of a photon is given by the relationship $E = hf$, where f is the frequency of the light and h is Planck's constant. The exchange particle for the electromagnetic interaction.

pion The least massive meson. The pion has three charge states: $+1$, 0, and -1.

plasma The fourth state of matter in which one or more electrons have been stripped from the atoms forming an ion gas.

polarized A property of a transverse wave when its vibrations are all in a single plane.

polymer A material produced by linking carbon–hydrogen molecules to form very long macromolecules.

positron The antiparticle of the electron.

pound The unit of force in the British system. The weight of 0.454 kilogram on Earth.

power The rate at which energy is converted from one form to another. Measured in joules per second, or watts.

powers-of-ten notation A method of writing numbers in which a number between 1 and 10 is multiplied or divided by 10 raised to a power.

pressure The force per unit area of surface. Measured in newtons per square meter, or pascals.

projectile motion A type of motion that occurs near the surface of the Earth when the only force acting on the object is that of gravity.

proton The positively charged nucleon in nuclei. A member of the baryon and hadron families of elementary particles.

quantum (pl., **quanta**) The smallest unit of a discrete property. For instance, the quantum of charge is the charge on the proton.

quantum mechanics The rules for the behavior of particles at the atomic and subatomic levels.

quantum number A number giving the value of a quantized quantity. For instance, a quantum number specifies the angular momentum of an electron in an atom.

quark A constituent of hadrons. Quarks come in six flavors and three colors each. Three quarks make up the baryons, whereas a quark and an antiquark make up the mesons.

rad Any acronym for *r*adiation *a*bsorbed *d*ose. A rad of radiation deposits $\frac{1}{100}$ joule per kilogram of material.

radiation The transport of energy via electromagnetic waves. Particles emitted in radioactive decay.

real image An image formed by the convergence of light.

reference system A collection of objects not moving relative to each other that can be used to describe the motion of other objects. *See* inertial and noninertial reference systems.

reflecting telescope A type of telescope using a mirror as the objective.

reflection, law of The angle of reflection (measured relative to the normal to the surface) is equal to the angle of incidence. The incident ray, the reflected ray, and the normal all lie in the same plane.

refracting telescope A type of telescope using a lens as the objective.

refraction The bending of light that occurs at the interface between two transparent media. It occurs when the speed of light changes.

refrigerator A heat engine running backward.

rem An acronym for *r*adiation *e*quivalent in *m*ammals, a measure of the biological effects caused by radiation.

resistance The impedance to the flow of electric current. The resistance is equal to the voltage across the object divided by the current through it. Measured in volts per ampere, or ohms.

resonance   A large increase in the amplitude of a vibration when a force is applied at a natural frequency of the medium or object.

rest-mass energy   The energy associated with the mass of a particle. Given by $E_0 = mc^2$, where c is the speed of light.

retroreflectors   Three flat mirrors at right angles to each other that reflect light back to its source.

rotational acceleration   The change in rotational speed divided by the time it takes to make the change.

rotational inertia   The property of an object that measures its resistance to a change in its rotational speed.

rotational speed   The angle of rotation or revolution divided by the time taken.

second postulate of special relativity   The speed of light in a vacuum is a constant regardless of the speed of the source or the speed of the observer.

series circuit   An arrangement of resistances (or batteries) on a single pathway so that the current flows through each element.

shell   A collection of electrons in an atom that have approximately the same energy.

shock wave   The characteristic cone-shaped wave front that is produced whenever an object travels faster than the speed of the waves in the surrounding medium.

short circuit   A pathway in an electric circuit that has very little resistance.

sliding friction   The frictional force between two surfaces in relative motion.

solid   Matter with a definite size and shape.

sonar   Sound waves in water.

spacetime   A combination of time and three-dimensional space that forms a four-dimensional geometry expressing the connections between space and time.

special theory of relativity   A comprehensive theory of space and time that replaces Newtonian mechanics when velocities get very large.

specific heat   The amount of heat required to raise the temperature of 1 gram of a substance by 1° Celsius.

spherical aberration   A defect caused by grinding the surface of a lens or mirror to a spherical rather than a parabolic shape.

spring constant   The amount of force required to stretch a spring by 1 unit of length. Measured in newtons per meter.

stable equilibrium   An equilibrium position or orientation to which an object returns after being slightly displaced.

standing wave   The interference pattern produced by two waves of equal amplitude and frequency traveling in opposite directions. The pattern is characterized by alternating nodal and antinodal regions.

static friction   The frictional force between two surfaces at rest relative to each other.

stimulated emission   The emission of a photon from an atom due to the presence of an incident photon. The emitted photon has the same energy, direction, and phase as the incident photon.

strange particle   A particle with a nonzero value of strangeness. In the quark model it is made up of one or more quarks carrying the quantum property of strangeness.

strange   The flavor of the third quark.

strong force   The force responsible for holding the nucleons together to form nuclei.

subcritical   A chain reaction that dies out because an average of less than one neutron from each fission reaction causes another fission reaction.

supercritical   A chain reaction that grows rapidly because an average of more than one neutron from each fission reaction causes another fission reaction. An extreme example of this is the explosion of a nuclear bomb.

superposition   The combining of two or more waves at a location in space.

terminal speed   The speed obtained in free fall when the upward force of air resistance is equal to the downward force of gravity.

tesla   The SI unit of magnetic field.

thermal energy   Internal energy.

thermal equilibrium   A condition in which there is no net flow of thermal energy between two objects. This occurs when the two objects obtain the same temperature.

thermal expansion   The expansion of a material when heated.

thermodynamics   The area of physics that deals with the connections between heat and other forms of energy.

thermodynamics, first law of   The increase in internal energy of a system is equal to the heat added plus the work done on the system.

thermodynamics, second law of   There are three equivalent forms: (1) It is impossible to build a heat engine to perform mechanical work that does not exhaust heat to the surroundings. (2) It is impossible to build a refrigerator that can transfer heat from a lower temperature region to a higher temperature region without expending mechanical work. (3) The entropy of a system tends to increase.

thermodynamics, third law of   Absolute zero may be approached experimentally but can never be reached.

thermodynamics, zeroth law of   If objects A and B are each in thermodynamic equilibrium with object C, then A and B are in thermodynamic equilibrium with each other.

top   The flavor of the sixth quark.

torque   The rotational analog of force. It is equal to the radius multiplied by the force perpendicular to the radius. A net torque produces a change in an object's angular momentum.

total internal reflection   When light is traveling from a material with a higher index of refraction into one with a lower index of refraction, if the angle of incidence exceeds the critical angle, the light is totally reflected.

translational   Motion through space as opposed to rotation.

transverse wave   A wave in which the vibrations of the medium are perpendicular to the direction the wave is moving.

trough   A valley of a wave disturbance.

umbra   The darkest part of a shadow where no light from the source reaches.

uncertainty principle   The product of the uncertainty in the position of a particle along a certain direction and the uncertainty in the momentum along this same direction must be greater than Planck's constant, $\Delta p_x \Delta x > h$. A similar relationship applies to the uncertainties in energy and time.

universal gravitation, law of $F = GM_1 M_2/r^2$, where F is the force between any two objects, G is a universal constant, M_1 and M_2 are the masses of the two objects, and r is the distance between their centers.

unstable equilibrium   An equilibrium position or orientation from which an object leaves after being slightly displaced.

van der Waals bonding   A weak binding together of atoms or molecules due to their electrical attraction.

vector   A quantity with a magnitude and a direction.

velocity   A vector quantity that includes the speed and direction of an object.

vibration   An oscillation about an equilibrium position or shape.

virtual image   The image formed when light only appears to come from the location of the image.

viscosity   A measure of the internal friction within a fluid.

volt   The SI unit of electric potential. One volt produces a current of 1 ampere through a resistance of 1 ohm.

watt   The SI unit of power, 1 joule per second.

wave   The movement of energy from one place to another without any accompanying matter.

wavelength   The shortest repetition length for a periodic wave. For example, it is the distance from crest to crest or from trough to trough.

weak force   The force responsible for beta decay. This force occurs through the exchange of the W and Z^0 particles. All leptons and hadrons interact via this force.

weight   The support force needed to maintain an object at rest relative to a reference system. For inertial systems, the weight is sometimes taken to be the force of attraction of the Earth for an object, $W = mg$.

work   The product of the force along the direction of motion and the distance moved. Measured in energy units, joules.

X ray   A high-energy photon, usually produced by cathode rays or emitted by electrons falling to lower energy states in atoms. The range of frequencies in the electronic spectrum lying between the ultraviolet and the gamma rays.

INDEX

c = computing box
g = definition in glossary
t = table

X

Y

Z

Physical Constants

Atomic Mass Unit	$\text{amu} = 1.66 \times 10^{-27}$ kg
Avogadro's Number	$N_A = 6.02 \times 10^{23}$ particles/mol
Bohr Radius	$r_o = 5.29 \times 10^{-11}$ m
Electron Volt	$\text{eV} = 1.60 \times 10^{-19}$ J
Elementary Charge	$e = 1.60 \times 10^{-19}$ C
Coulomb's Constant	$k = 8.99 \times 10^9$ N·m²/C²
Gravitational Constant	$G = 6.67 \times 10^{-11}$ N·m²/kg²
Mass of Electron	$m_e = 9.11 \times 10^{-31}$ kg
Mass of Neutron	$m_n = 1.675 \times 10^{-27}$ kg
Mass of Proton	$m_p = 1.673 \times 10^{-27}$ kg
Planck's Constant	$h = 6.63 \times 10^{-34}$ J·s
Speed of Light	$c = 3.00 \times 10^8$ m/s

Physical Data

Acceleration due to Gravity	$g = 9.80$ m/s²
Density of Water	$D_w = 1.00 \times 10^3$ kg/m³
Earth–Moon Distance	$R_{em} = 3.84 \times 10^8$ m
Earth–Sun Distance	$R_{es} = 1.50 \times 10^{11}$ m
Mass of Earth	$M_e = 5.98 \times 10^{24}$ kg
Mass of Moon	$M_m = 7.36 \times 10^{22}$ kg
Mass of Sun	$M_s = 1.99 \times 10^{30}$ kg
Radius of Earth	$R_e = 6.37 \times 10^6$ m
Radius of Moon	$R_m = 1.74 \times 10^6$ m
Radius of Sun	$R_s = 6.96 \times 10^8$ m
Speed of Sound (20°C, 1 atm)	$v_s = 343$ m/s
Standard Atmospheric Pressure	$P_{atm} = 1.01 \times 10^5$ Pa

Standard Abbreviations

A	ampere
amu	atomic mass unit
atm	atmosphere
Btu	British thermal unit
C	coulomb
°C	degree Celsius
cal	calorie
eV	electron volt
°F	degree Fahrenheit
ft	foot
g	gram
h	hour
hp	horsepower
Hz	Hertz
in.	inch
J	joule
K	kelvin
kg	kilogram
lb	pound
m	meter
min	minute
mph	mile per hour
N	newton
Pa	pascal
psi	pound per square inch
rev	revolution
s	second
T	tesla
V	volt
W	watt
Ω	ohm